T0189212

Advances in Intelligent Systems and Computing

Volume 944

The series "Advances in Intelligent Systems and Computing" contains publications on theory, applications, and design methods of Intelligent Systems and Intelligent Computing. Virtually all disciplines such as engineering, natural sciences, computer and information science, ICT, economics, business, e-commerce, environment, healthcare, life science are covered. The list of topics spans all the areas of modern intelligent systems and computing such as: computational intelligence, soft computing including neural networks, fuzzy systems, evolutionary computing and the fusion of these paradigms, social intelligence, ambient intelligence, computational neuroscience, artificial life, virtual worlds and society, cognitive science and systems, Perception and Vision, DNA and immune based systems, self-organizing and adaptive systems, e-Learning and teaching, human-centered and human-centric computing, recommender systems, intelligent control, robotics and mechatronics including human-machine teaming, knowledge-based paradigms, learning paradigms, machine ethics, intelligent data analysis, knowledge management, intelligent agents, intelligent decision making and support, intelligent network security, trust management, interactive entertainment, Web intelligence and multimedia.

The publications within "Advances in Intelligent Systems and Computing" are primarily proceedings of important conferences, symposia and congresses. They cover significant recent developments in the field, both of a foundational and applicable character. An important characteristic feature of the series is the short publication time and world-wide distribution. This permits a rapid and broad dissemination of research results.

** Indexing: The books of this series are submitted to ISI Proceedings, EI-Compendex, DBLP, SCOPUS, Google Scholar and Springerlink **

More information about this series at http://www.springer.com/series/11156

Kohei Arai · Supriya Kapoor
Editors

Advances in Computer Vision

Proceedings of the 2019 Computer Vision
Conference (CVC), Volume 2

 Springer

Editors
Kohei Arai
Saga University
Saga, Saga, Japan

Supriya Kapoor
The Science and Information
(SAI) Organization
Bradford, West Yorkshire, UK

ISSN 2194-5357   ISSN 2194-5365   (electronic)
Advances in Intelligent Systems and Computing
ISBN 978-3-030-17797-3   ISBN 978-3-030-17798-0   (eBook)
https://doi.org/10.1007/978-3-030-17798-0

This Springer imprint is published by the registered company Springer Nature Switzerland AG
The registered company address is: Gewerbestrasse 11, 6330 Cham, Switzerland

Preface

It gives us the great pleasure to welcome all the participants of the Computer Vision Conference (CVC) 2019, organized by The Science and Information (SAI) Organization, based in the UK. CVC 2019 offers a place for participants to present and to discuss their innovative recent and ongoing research and their applications. The prestigious conference was held on 25–26 April 2019 in Las Vegas, Nevada, USA.

Computer vision is a field of computer science that works on enabling the computers to identify, see and process information in a similar way that humans do and provide an appropriate result. Nowadays, computer vision is developing at a fast pace and has gained enormous attention.

The volume and quality of the technical material submitted to the conference confirm the rapid expansion of computer vision and CVC's status as its flagship conference. We believe the research presented at CVC 2019 will contribute to strengthen the great success of computer vision technologies in industrial, entertainment, social and everyday applications. The participants of the conference were from different regions of the world, with the background of either academia or industry.

The published proceedings has been divided into two volumes, which covered a wide range of topics in Machine Vision and Learning, Computer Vision Applications, Image Processing, Data Science, Artificial Intelligence, Motion and Tracking, 3D Computer Vision, Deep Learning for Vision, etc. These papers are selected from 371 submitted papers and have received the instruction and help from many experts, scholars and participants in proceedings preparation. Here, we would like to give our sincere thanks to those who have paid great efforts and support during the publication of the proceeding. After rigorous peer review, 118 papers were published including 7 poster papers.

Many thanks go to the Keynote Speakers for sharing their knowledge and expertise with us and to all the authors who have spent the time and effort to contribute significantly to this conference. We are also indebted to the organizing committee for their great efforts in ensuring the successful implementation of the conference. In particular, we would like to thank the technical committee for their constructive and enlightening reviews on the manuscripts in the limited timescale.

We hope that all the participants and the interested readers benefit scientifically from this book and find it stimulating in the process. See you in next SAI Conference, with the same amplitude, focus and determination.

Regards,
Kohei Arai

Contents

Selection of Personnel Based on Multicriteria Decision Making and Fuzzy Logic

Zapata C. Santiago[⊠], Escobar R. Luis, and Lopez N. Ricardo

Department of Informatics and Computing, Faculty of Engineering,
Technological Metropolitan University of the State of Chile (UTEM),
Jose Pedro Alessandri 1242, Macul, 8330378 Santiago, Chile
{szapata,lescobar,rlopez}@utem.cl
http://www.utem.cl

Abstract. The recruitment of new workers in a company is a critical and crucial decision due to the fact that employees are the basis, fundament and main support of their working environment. Following the objective of minimizing subjectivity and the excessive use of assumptions as a valid skill for choosing the proper candidate to fulfill a work place in a company, an attractive method for choosing staff based on Multicriterio Decisions Making and Fuzzy Logic is proposed, in a mathematical model through the multicriterio technique known as TOPSIS.

Keywords: Fuzzy Logic · Personnel selection ·
Multicriteria Decision Making · TOPSIS

1 Introduction

At present, the strategic management of human resources has become a priority concern of organizations due to the change of thinking about employees, which are not considered an ordinary productive factor, but a strategic factor, since they implement organizational strategies and are a source Competitive advantage by generating a sustainable profit in the long term.

The knowledge and experience of the organization's staff are no longer sufficiently differentiating elements to create competitive advantage and add value to the organization, but must also take into account motivation, commitment, behavior, etc., of the people who are intended to obtain a perfect match between the worker and the job [1]. Thus, excellent and not merely satisfactory performance of the tasks and activities of the position can be achieved and gain an advantage over competitors because it is difficult to copy or imitate.

2 Research and Problem Theme

At present the problem of the selection of personnel is used to cover certain jobs optimally in an organization. It is a widely discussed and studied problem whose success determines the fulfillment of objectives and projects, especially in small

© Springer Nature Switzerland AG 2020
K. Arai and S. Kapoor (Eds.): CVC 2019, AISC 944, pp. 1–10, 2020.
https://doi.org/10.1007/978-3-030-17798-0_1

organizations and is easily extrapolated to other practical fields of application, such as the management of training in the career of employees.

Objective of the Research: To propose a method of Selection of Personnel based on Multicriteria Decision Making and Fuzzy Logic.

3 Theoretical Framework

The process of selection of personnel in some organizations is currently based on subjectivity, so this is conducive to the loss of human resources with talent and skills, which is eminent to have a method for the selection process of personnel with theoretical bases Solids that allow reducing the level of uncertainty and subjectivity of this process.

The means of personnel selection can be highlighted in two fundamental groups: traditional procedures and scientific procedures.

Traditional Procedures: Recommendations, Reference Letters, Curriculum, Interview, Test Period, among others.

Scientific Procedures: They were developed trying to find more objective means to measure the qualities of the aspirants to the job, and also to use them with employees who are candidates for transfer or promotion [3].

A. *Personnel Selection Process*
   The selection process is as follows (Fig. 1):

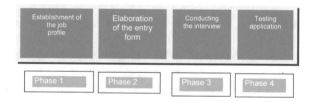

Fig. 1. Personnel selection process

B. *Multicriteria Decision Making*
   Multicriteria Decision Analysis is a very useful tool that helps the decision maker during the decision making process, since the methods that make it possible to approach the problem in an orderly manner, facilitating the consensus of the final decision and the treatment of the great Amount of information, which is usually found in different magnitudes of measure and meanings.
C. *TOPSIS Method*
   The method was originally developed by Hwang and Yoon in 1981. TOPSIS [3] is based on the concept that the chosen alternative must have the shortest geometric distance from the positive ideal solution (PIS) and the longest geometric distance of the solution negative ideal (NIS).

Figure 2 represents five alternatives (A, B, C, D and E) for a problem of two criteria, where the ideal and anti-ideal points are shown:

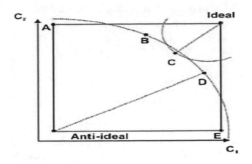

Fig. 2. Distances to the ideal and the anti-ideal

D. *Fuzzy Logic*

Fuzzy Logic was first investigated in the mid-1960s at the University of Berkeley (California) by engineer Lotf A. Zadeh when he realized what he called the principle of incompatibility: "According to the complexity of a System increases, our ability to be precise and construct instructions on its behavior decreases to the threshold beyond which precision and meaning are exclusive characteristics".

E. *Hamming Distance*

Given a reference set $X = \{x_1, x_2, \ldots, x_n\}$ and two sets $\Phi - fuzzy$, $\tilde{A}^{\Phi}, \tilde{B}^{\Phi}$ with membership functions

$$\mu_{\tilde{A}}^{\Phi}(x_j) = \left[a_{x_j}^1, a_{x_j}^2\right]$$

$$\mu_{\tilde{B}}^{\Phi}(x_j) = \left[b_{x_j}^1 - b_{x_j}^2\right] j = 1, 2, \ldots, n,$$

The standard Hamming distance is defined as:

$$d\left(\tilde{A}^{\Phi}, \tilde{B}^{\Phi}\right) = \frac{1}{2n} = \frac{1}{2n}\left(\sum_{i=1}^{n}\left(\left|a_{x_j}^1 - b_{x_j}^1\right| + \left|a_{x_j}^2 - b_{x_j}^2\right|\right)\right)$$

F. *Coefficient of Adequacy*

Given a reference set is $X = \{x_1, x_2, \ldots, x_n\}$ and two sets $\Phi - fuzzy$, $\tilde{A}^{\Phi}, \tilde{B}^{\Phi}$ with membership functions

$$\mu_{\tilde{A}}^{\Phi}(x_j) = \left[a_{x_j}^1, a_{x_j}^2\right],$$

$$\mu_{\tilde{B}}^{\Phi}(x_j) = \left[b_{x_j}^1 - b_{x_j}^2 \right], \, j = 1, 2, \ldots, n,$$

The coefficient of adequacy between them is defined as:

$$\mu_{\tilde{B}^{\Phi}}\left(\tilde{A}^{\Phi}\right) = \frac{1}{n}\sum_{j=1}^{n}\mu_{\tilde{B}^{\Phi}}^{x_j}\left(\tilde{A}^{\Phi}\right)$$

4   Methodology

The methodological design consists of the following procedure:

For the job of "Seller" and the full-time working hours, using the chips for application to the case of fuzzy method, both methods apply all workers who are carrying out the position of Seller.

In each method we obtained the best seller with the following structure:

a. Test No. 1 best candidate to Seller by classic TOPSIS algorithm.
b. Test No. 2, two best candidates for Sellers fuzzy method based on two metrics: the Hamming distance and the Adequacy Ratio.

5   Analysis and Results

They were used as input data information obtained from the process of recruitment in the Independent Network Library, a branch; these data were obtained through an interview.

A. *Results Test No. 1:*
   The most suitable candidate, according to the TOPSIS multicriterio method, is candidate N° 5 (V_5), it obtains the value closest to 1, which implies a higher priority of the alternative.
   Then, the order of preference of the candidates is as follows in Table 1:

Table 1. Relative proximity to test 1

$\bar{R}_i$	Value
$\bar{R}_5$	0,76752
$\bar{R}_3$	0,68265
$\bar{R}_1$	0,39637
$\bar{R}_2$	0,35318
$\bar{R}_4$	0,30548

B. Results Test No. 2:

According to the Hamming Distance (see Table 2), the best candidate for the post of "Full time Seller" is the applicant N° 5 with a value of 0.48, which is the more approaches to 0 and hence the best applicant. In the case of the coefficient of fitness, higher will be the best candidate, and the results points to the N° 4 applicant.

Table 2. Best seller distance hamming test 2

Seller	D. Hamming
Seller 5	0,48
Seller 3	0,72
Seller 2	0,96
Seller 1	3,38
Seller 4	4,6

Table 3. Best seller adequacy coefficient test 2

Seller	C. Adequacy
Seller 4	4,42
Seller 1	3,31
Seller 2	1,62
Seller 3	1,4
Seller 5	1,36

Having two best sellers with Hamming Distance (Table 2) and the Coefficient of Adequacy (Table 3), both metrics measure different things; Hamming Distance calculated the difference between the ends and the Coefficient of Adequacy implicitly includes a correction of excesses. It is for this reason that different results are obtained for the same position.

C. Results Proposed Method Fuzzy TOPSIS

The most suitable candidate, according to the proposed method TOPSIS, Fuzzy is the candidate N° 2 (V_2), given that gets the value closer to 1, which means a higher priority of the alternative. Then, the order of preference of candidates is the following Table 4, which is of greater to lesser.

Table 4. Order of relative proximity of proposed method fuzzy TOPSIS

$\bar{R}_i$	Value
$\bar{R}_2$	0,5053
$\bar{R}_5$	0,2722
$\bar{R}_4$	0,2719
$\bar{R}_1$	0,1894

It is concluded that the candidate N°2 is the optimum for the position of full time Seller, with a relative proximity of 0.5053.

6 Proposal of the Fuzzy TOPSIS Method

The proposed method of Personnel Selection called "Fuzzy TOPSIS" consists of eight steps to obtain the best candidate. These are:

1. Construction of the Fuzzy Matrix of Valuation
2. Introduction of Fictive Alternatives
3. Formalization of the Selection Criteria for Candidates, Positive and Negative
4. Standardization of the Fuzzy Decision Matrix
5. Construct the Weighted Normalized Fuzzy Decision Matrix
6. Calculation of Distance Measures
7. Construction of Distance Measurement Matrix
8. Calculation of the relative proximity to the ideal solution
9. Selection of the best candidate based on proximity.

For this case, a multi-criteria decision-making is considered, which includes both numerical values and linguistic tags, which can be expressed in a Fuzzy valuation matrix (Table 5).

The structure of the matrix can be expressed as follows:

Table 5. Fuzzy matrix of valuation

	$\underset{\sim}{w_1}\underset{\sim}{w_2}$	$\cdots$	$\underset{\sim}{w_j}$	$\cdots$	$\underset{\sim}{w_n}$
	$C_1 C_2$	$\cdots$	C_j	$\cdots$	C_n
A_1	$X_{11}X_1\underset{\sim}{X_{1j}}\underset{\sim}{X_{1n}}$				
A_2	$X_{21}X_{22}\underset{\sim}{X_{2j}}\underset{\sim}{X_{2n}}$				
$\cdots$	$\cdots\cdots\cdots\cdots\cdots\cdots$				
A_m	$X_{m1}X_{m2}\underset{\sim}{X_{mj}}\underset{\sim}{X_{mn}}$				

Where:

- $C_j, j = 1,\ldots, n$, they represent the criteria according to which it will evaluate alternatives.
- $A_i, i = 1,\ldots, m$, denotes the possible alternatives;
- $\underset{\sim}{W} = \left[\underset{\sim}{w_1}, \underset{\sim}{w_2},\ldots, \underset{\sim}{w_n}\right]$ it is the vector of weights associated with C_j;

- $Z_{ij} = \left\{ x_{ij}, \underset{\sim}{x_{ij}} \right\}]] >$ represents the valuations of, with respect to some attribute/criterion; Being $X_{ij} = \left[x_{ij}^a, x_{ij}^b, x_{ij}^c \right]$ the representation of a Fuzzy numerical value and $X_{ij} = \left[\underset{\sim}{x_{ij}^a}, \underset{\sim}{x_{ij}^b}, \underset{\sim}{x_{ij}^c} \right]$ the representation of a language tag.

The following procedure proposed is called Fuzzy TOPSIS to the selection of the best candidate.

i. Step 1: Establishing the Fuzzy valuation matrix
   The establishment of fuzzy assessment matrix is of importance, to optimize the management of the best alternatives. As a result, this matrix can be of three types:

   - Numeric Values Matrix (Classic TOPSIS Method Case)
   - Matrix of Linguistic Values
   - Matrix of Numeric and Linguistic Values (Mix of both)

   The classical multicriteria decision methods only require precise values for a finite set of alternatives. However, in the context of real decision making is composed of precise and imprecise values and the decision matrices contain different types of information to analyze.
   Therefore, in the case that all or some of the weights of the criteria are imprecise, the use of fuzzy set theory is a useful way to solve these problems.

ii. Step 2: Introduction of fictitious alternatives
   For the introduction of fictitious alternatives, you must add the alternative corresponding to the excellent value (this alternative would be the best valued labels on all criteria), $F_1 = (\text{Max } S_1, \text{Max } S_2, \ldots, \text{Max } S_n$ and the alternative that corresponds to the worst value (the one that would be valued by the worst of the labels, for all criteria) $F_1 = (\text{Min } S_1, \text{Min } S_2, \ldots, \text{Min } S_n)$ where S is the space of valuations. These variables are understood as fictitious.

iii. Step 3: Formalization of the Selection Criteria for Candidates, Positives and Negatives.
   As for the formalization of the positive and negative criteria, it is necessary to mention that, it will be defined based on the vertex method [16] proposed by Chen [17], which define:

$$\tilde{V}_j^+ = (1, 1, 1) \text{ y } \tilde{V}_j^- = (0, 0, 0)$$

$$A^+ = \left\{ \tilde{v}_1^+, \tilde{v}_j^+, \ldots, \tilde{v}_m^+ \right\}$$

$$A^- = \left\{ \tilde{v}_1^-, \tilde{v}_j^-, \ldots, \tilde{v}_m^- \right\}$$

This is internalized in the analysis of the personnel selection process, when defining these criteria, it is suggested that the alternatives of full-time sellers should have positive criteria, since the objective is to analyze the skills,

iv. Step 4: Normalization of the Fuzzy decision matrix

As reviewed in this paper, one of the modifications to the technique Multicriteria TOPSIS classic is the change of standardization using the procedure, which considers the role of membership of triangular type with three values, and following the procedure used in the classic TOPSIS method.

An element $\bar{n}_{ij}$ of the standard decision matrix $N = \left[\bar{n}_{ij}\right]_{mxn}$ is calculated as follows:

$$\bar{n}_{ij} = \frac{X_{ij}}{\sqrt{\sum_{j=1}^{m} \left(X_{ij}\right)^2}}, j = 1, \ldots, n; i = 1, \ldots, m,$$

In the case of the Fuzzy TOPSIS procedure, it is required: $\bar{n}_{ij} = \left(\bar{n}_{ij}^a, \bar{n}_{ij}^b, \bar{n}_{ij}^c\right)$ is a Fuzzy number.

v. Step 5: Construct the weighted normalized Fuzzy decision matrix

We proceed to weight the normalized Fuzzy matrix taking into account the mathematical operations for Fuzzy numbers.

vi. Step 6: Calculation of the measures of distance

The TOPSIS method uses the Euclidean distance to measure the distance of the alternatives with its PIS and NIS, so that each alternative has the shortest distance to the PIS and the longest distance to the NIS. In the Fuzzy version of the method the calculation of the distance to the FPIS Fuzzy positive ideal solution and the FNIS Fuzzy negative ideal solution is performed.

There are many possible distances in $\mathbb{R}^n$ but the most common ones are those of Minkowski's unifying approach [9]: The Minkowski metric M_p between two points $x = (x_1, x_2, \ldots, x_n)$ and $y = (y_1, y_2, \ldots, y_n)$ of $\mathbb{R}^n$ is defined by:

$$M_p = \left[\sum_j |x_j - y_j|^p\right]^{1/p}, para \ p \geq 1$$

Among the distances addressed by the Minkowski metric are: Distance from Manhattan, Euclidean Distance and Tchebycheff Distance.

You can choose any of these distances or coefficients in order to measure the proximity to the ideal or the remoteness of the anti-ideal. For our case we will work with a Euclidean distance between two Fuzzy triangular numbers. However, the distance between two triangular Fuzzy numbers "h" and "k" can also be calculated by eliminating the Fuzzy and considering a crisp number, by the procedure:

$$d(h, k) = \sqrt{\frac{1}{3}\left[(a_h - a_k)^2 + (b_h - b_k)^2 + (c_h - c_k)^2\right]}$$

vii. Step 7: Construction of Distance Measurement Matrix
With the distance measurements found, we proceed to construct a new matrix that contains the sums of each associated distance measure, that is calculated with the following expressions, both for positive and negative distances:

$$d_i^+ = \sum_{j=1}^n d_v\left(\tilde{v}_{ij}, \tilde{v}_j^+\right)$$

$$d_i^- = \sum_{j=1}^n d_v\left(\tilde{v}_{ij}, \tilde{v}_j^-\right)$$

viii. Step 8: Calculating the relative proximity to the ideal solution
Similar to the classical TOPSIS, the relative proximity is calculated with the consideration of operations corresponding to the use of Fuzzy numbers. The formula for the calculation is as follows, the greater the value close to 1, it is considered as the best candidate:

$$\bar{R}_i = \frac{\bar{d}_i^-}{\bar{d}_i^+ + \bar{d}_i^-} \quad i = 1, \ldots, m$$

ix. Step 9: Selection of the best candidate based on the vicinity.
The best alternatives are ordered according to $\bar{R}_i$ in descending order, so the method ends by presenting in order of the relative proximities and the ideal candidates to the position of Seller.

7 Conclusions

The Multicriteria Decision Making Theory has made the decision-making process a sufficiently structured methodology to offer the user tools to deal with conflicting problems for the usefulness of these, especially in the field of Industrial Organization, and for having worked throughout this paper.

Fuzzy logic is a powerful tool for representing data in a convenient way about a problem. In the theory of fuzzy sets, complex mathematics appears, in order to model the data and select the best candidate, turns out to be a useful tool in this work.

The management of the multi-criteria technique TOPSIS facilitated the first calculations for this method and, together with the support of various documents and the bibliography, simplified the treatment of the data to achieve the expected results. In the case of the method based on fuzzy logic is more complex in calculations of hamming distance and adequacy coefficient.

8 Future Work

- Conduct a study covering all the independent branches of Santiago of Chile, in order to have a list of all personnel selection processes carried out in each one, in this way a totally representative result will be obtained and no relevant information will be left out Time to propose some method of staff selection.

- Analyze this process of selection of personnel with different membership functions to review the behavior of the selection process of candidates at the branch level.
- Implement a computational tool at the coding level developed with the standards and processes of selection of Independent Library Networks of Santiago, with the aim of making a more efficient and secure selection of the candidates presented to these organizations.
- Implement a web platform for administrators and applicants focused on the selection of candidates: Sellers who wish to apply to the Independent Library Network.

References

1. Canós Darós, L., Casasús Estellés, T., Lara Mora, T., Liern Carrión, V., Pérez Cantó, J.C.: Flexible Personnel Selection Models based on competency assessment. Universidad Politécnica de Valencia, Valencia (España) (2008)
2. López, F.J.H., Trejo, C.O.: Method of Selection of Personnel by competences based on fuzzy logic. Universidad Autónoma de Querétaro, Querétaro (Mexico) (2013)
3. López, D.L., Segura, S.L.: Design and development of concept maps for Multicriteria Decision Making. Universidad de Sevilla, Sevilla (España) (2014)
4. Ramírez, M.L.: The method of analytical hierarchies of Saaty in the weighting of Variables in the level of mortality and morbidity in the province of Chaco. Universidad Nacional del Nordeste, Chaco (Argentina) (2004)
5. Contreras, C.D., Aguilera Rojas, A., Guillén Barrientos, N.: Fuzzy Logic v/s Multiple Regression Model for Personnel Selection. Universidad de Tarapacá, Arica (Chile) (2014)
6. del Socorro García Cascales, M., Jiménez, M.T.L., Merino, R.R.: Methods for comparing alternatives through a Decision Support System (S.A.D) and Soft Computing. Universidad Politécnica de Cartagena, Cartagena (España) (2009)
7. Ercole, R.A., Alberto, C.L., Carignano, C.E.: TOPSIS in actions with linguistic variables. In: XXXIII Argentine Congress of University Teachers of Costs, Mar del Plata (Argentina), Octubre (2010)
8. Chagoya, E.R.: Investigation methodology. Minatitlan (Mexico), 5 Noviembre (2014)
9. Contreras, G.P., Cáceres, S.Z.: Diagnosis of the emotional and relational Capital in the students of the school of computer science – UTEM. Universidad Tecnológica Metropolitana, Santiago (Chile), Diciembre (2015)
10. Cheesman de Rueda, S.: Research Basics. Universidad del Reyno de Guatemala, Guatemala (Guatemala), Abril (2011)
11. Buendía, L., Colás, P., Hernández, F.: Research Methods in Psychopedagogy. Universidad Autónoma de Madrid, Madrid (España) (1997)
12. Mortis Lozoya, M.S., Rosas Jiménez, R., Chairez Flores, E.: Research Designs. Instituto Tecnológico de Sonora, Obregón (Mexico) (2012)
13. Lima Junior, F.R., Carpinetti, C.R.: Comparison between TOPSIS Fuzzy and AHP Fuzzy methods to support the decision making process for Supplier Selection. Escuela de Ingenieria de San Carlos, Sao Paulo (Brasil) (2013)
14. Kahraman, C.: Fuzzy Multi-Criteria Decision Making: Theory and Applications with Recent Developments. Universidad Técnica de Estambul, Estambul (Turquia) (2016)

Data-Driven Multi-step Demand Prediction for Ride-Hailing Services Using Convolutional Neural Network

Chao Wang[1(✉)], Yi Hou[2], and Matthew Barth[1]

[1] University of California Riverside, Riverside, CA 92507, USA
cwang061@ucr.edu, barth@ece.ucr.edu
[2] National Renewable Energy Laboratory, Golden, CO 80401, USA
Yi.Hou@nrel.gov

Abstract. Ride-hailing services are growing rapidly and becoming one of the most disruptive technologies in the transportation realm. Accurate prediction of ride-hailing trip demand not only enables cities to better understand people's activity patterns, but also helps ride-hailing companies and drivers make informed decisions to reduce deadheading vehicle miles traveled, traffic congestion, and energy consumption. In this study, a convolutional neural network (CNN)-based deep learning model is proposed for multi-step ride-hailing demand prediction using the trip request data in Chengdu, China, offered by DiDi Chuxing. The CNN model is capable of accurately predicting the ride-hailing pick-up demand at each 1-km by 1-km zone in the city of Chengdu for every 10 min. Compared with another deep learning model based on long short-term memory, the CNN model is 30% faster for the training and predicting process. The proposed model can also be easily extended to make multi-step predictions, which would benefit the on-demand shared autonomous vehicles applications and fleet operators in terms of supply-demand rebalancing. The prediction error attenuation analysis shows that the accuracy stays acceptable as the model predicts more steps.

Keywords: Ride-hailing · Demand prediction · Convolutional neural network

1 Introduction

In recent years, ride-hailing companies such as Uber, Lyft, DiDi Chuxing (China), and RideAustin have emerged as new and disruptive on-demand mobility services. In major cities, 21% of adults use ride-hailing services [1]. Accurate prediction of ride-hailing trip demand not only enables cities to better understand people's activity patterns, but also helps ride-hailing companies and drivers make informed decisions to reduce deadheading vehicle miles traveled, traffic congestion, and energy consumption. In the next wave of transportation innovations, such as mobility-as-a-service where on-demand shared automated vehicles transport people and goods in cities, accurate trip demand prediction will provide decision-making tools for automated vehicle fleet operators to optimize shared automated vehicle assignment holistically for the entire city.

© Springer Nature Switzerland AG 2020
K. Arai and S. Kapoor (Eds.): CVC 2019, AISC 944, pp. 11–22, 2020.
https://doi.org/10.1007/978-3-030-17798-0_2

Considerable interest in predicting the trip demand for taxi and ride-hailing trips has grown in the research community for the last a few years. Chang et al. [2] mined historical data to predict taxi demand distributions using clustering algorithms. Moreira-Matias et al. [3, 4] applied time series techniques to forecast taxi passenger demand. Gong et al. [5] proposed a machine learning model, XGBoost, to predict New York City taxi demand. Most of these efforts have focused on taxi trip demand, whereas the studies on ride-hailing demand prediction have been relatively limited. Recently, Ke et al. [6] introduced the fusion convolutional long short-term memory (LSTM) network (FCL-Net) to forecast passenger demand for on-demand ride services in Hangzhou, China, using real-world data provided by DiDi Chuxing. Wang et al. [7] developed a LSTM to predict the number of Uber pickups in New York City. Xu et al. [8] also developed an LSTM to predict taxi passenger demand for each small area in New York City. Liao et al. [9] conducted a thorough comparison between two deep neural network structures, deep spatio-temporal residual network (ST-ResNet) and FLC-Net, for taxi demand prediction using New York City taxi data. They found that deep neural networks outperform most traditional machine learning models when predicting taxi trip demand.

Most of the previous studies adopted a time-series model or recurrent neural network architecture, which are good at capturing time dependencies from historical data. One drawback of these models is that spatial information is usually lost in the modeling process. This paper proposes a convolutional neural network (CNN)-based deep learning model to predict the ride-hailing demand considering both temporal and spatial features. CNN is a class of deep and feed-forward artificial neural networks that is most commonly applied to analyzing visual imagery and proven to be a very efficient and well-performed image recognition algorithm [10, 11]. CNN was inspired by biological processes [12]. The connectivity pattern between neurons resembles the organization of the animal visual cortex. Individual cortical neurons respond to stimuli only in a restricted region of the visual field known as the receptive field. The receptive fields of different neurons partially overlap such that they cover the entire visual field. CNNs use a variation of multilayer perception designed to require minimal preprocessing [13]. Similarly, the convolutional layers of a CNN not only enable CNN to capture the local feature of the image data, but also reduce model parameters that need to be estimated. In recent years, CNNs have been utilized to solve transportation problems [10, 14–16]. In this study, we uncover the possibility of learning the ride-hailing service demand patterns as images and make predictions by constructing the CNN input with proper spatial-temporal demand data.

Last year, DiDi Chuxing, the largest ride-hailing company in China, opened a new opportunity window for transportation researchers by sharing one month of trip data at city of Chengdu, China [17]. Thus, the proposed model was evaluated using DiDi Chuxing trip data. The remainder of the paper is organized as follows: the next section presents an overview of the data used for this research effort and formulates the problem to be solved in this paper; The third section discusses the methodology adopted for predictive analysis. The fourth section explains and presents the model results. The fifth and final section offers concluding thoughts and directions for future research.

2   Data and Problem Formulation

2.1   Data

The data used in this study are the ride-hailing trip request data in Chengdu, China offered by DiDi Chuxing [17]. DiDi Chuxing is the world's largest and most valuable ride-sharing behemoth, with a monopolistic investment and merger and acquisition portfolio in the ride and bike sharing industry across the globe [18]. As a leading mobile transportation platform, DiDi receives more than 25 million trip orders, collects more than 70 TB of new route data, processes more than 4,500 TB of data, and obtains more than 20 billion queries for route planning and 15 billion queries for geolocation. In 2017, DiDi announced its GAIA Initiative and decided to share a complete sample of the route and ride request data with the academic community. The dataset was collected from November 1 to November 30, 2016, at Chengdu, China. The request dataset contains 7 million ride request records with origin-destination (OD) points and the trip start and end times. The route dataset offers 1 billion data points of the trip global positioning system trajectories with a sampling rate of 2–4 s. The rich, dense records enable us to pursue high-resolution demand prediction at a local level. The dataset mainly used for this study is the ride request records. The information includes the pick-up and drop-off location and time for each ride. Table 1 lists the details of the data description. Figure 1 visualizes all the pick-up locations, which indicate the ride-hailing demand, over the observed month on a map of Chengdu.

The weather information is important for demand prediction considering the large impact of weather condition on the ride demand. We collected the open source hourly weather data from World Weather Online [19], including temperature, humidity, and weather conditions. Table 2 lists the parameters that we used in this study. For the convenience of model training, we assign each weather condition a numerical code.

Table 1. Ride request data from DiDi.

Parameter	Sample	Comment
Order id	mjiwdgkqmonDFvCk3ntBpron5mwfrqvI	Anonymized
Trip start time	1501581031	Unix timestamp, in seconds
Trip end time	1501582195	Unix timestamp, in seconds
Pick-up longitude	104.11225	GCJ-02 coordinate system
Pick-up latitude	30.66703	GCJ-02 coordinate system
Drop-off longitude	104.07403	GCJ-02 coordinate system
Drop-off latitude	30.6863	GCJ-02 coordinate system

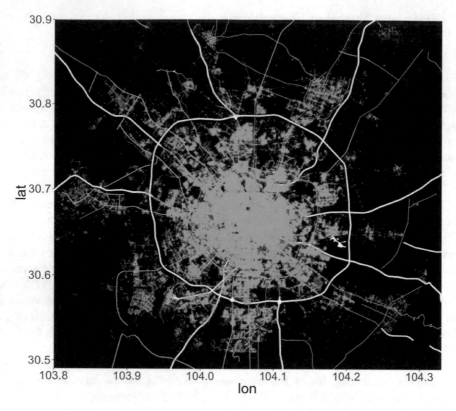

Fig. 1. Visualization of the pick-up locations in Chengdu, China.

Table 2. Weather data parameters and weather condition encoding.

Parameter	Unit/Code	
Temperature	F°	
Humidity	%	
Weather condition	Unknown	0
	Clear	1
	Fog	2
	Haze	3
	Light rain	4
	Light rain showers	5
	Mist	6
	Mostly cloudy	7
	Partly cloudy	8
	Patches of fog	9
	Scattered clouds	10

2.2   Problem Description

This study is a continuing work based on [7], in which the ride-hailing service demand prediction for New York City was proposed. We were able to predict the next 1-h total number of demands for a certain region in the city using an LSTM network trained by previous demand in that region. The goal of this study is to improve the capability of the prediction method in terms of accuracy, computational efficiency, temporal granularity, and spatial scalability. Since the DiDi dataset offers denser ride request data, we can divide the city region into smaller zones for demand prediction while still having enough data points at each region for model training. As shown in Fig. 2, we select the urban core of the city and split it into 10×10 square grid cells. The longitude and latitude boundaries of the study region are also labeled. The high density of the ride-hailing demand also allows us to segment the prediction time interval in a smaller size and make more in-time prediction (e.g., predict the ride-hailing demand in the next 10 min). Another improvement we want to make is to involve the context information to the learning model, which will potentially increase the accuracy of the prediction. For instance, weather conditions heavily impact the ride-hailing demands, e.g., there might be more rides when it's raining or snowing. Predicting pick-up demand further into the future is always desirable. Therefore, instead of only predicting the pick-up demand in the next time step, the proposed model can predict the demand for multiple time intervals in the future.

Let $D_t = [d_{1_t}, d_{2_t}, \ldots d_{100_t}]$ be the number of demands for all 100 cells in time slot t, let $C_t = [min_t, day_t, t, temp_t, humi_t, wc_t]$ be the context-aware information of time t, where $temp_t, humi_t, wc_t$ represent the temperature, humidity, and weather conditions of time t, and min_t, day_t, t are the minutes of the day, day of week, and global time step. Based on the above considerations, we defined the problem that needs to be solved as follows:

> **Given** $[D_{t-m}, D_{t-m+1}, \ldots, D_{t-1}]$ and $[C_{t-m}, C_{t-m+1}, \ldots, C_{t-1}]$ as input,
> **provide** a prediction model that outputs D_{t+k},

where $m \geq 1$ is the number of past time steps used for prediction, $k \geq 0$ indicates the time step of ride-hailing demand that is predicted. In this study, m is set to be 6 and $k \in [0, 5]$, and the duration for each time slot is 10 min. Therefore, the past 60 min ride-hailing demands and weather records are used to predict the next 10–60 min of ride-hailing demand.

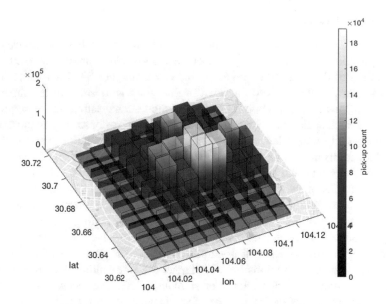

Fig. 2. Ride-hailing demand distribution at Chengdu.

3   Methodology

In this section, the methods to develop the context-aware multi-step ride-hailing demand prediction model are presented. Two deep learning methods are designed and tested for this study, the LSTM and the CNN. The LSTM methods are developed in author's previous work [7, 20] as a comparison. CNN is the major method that we focus on and explore in this paper.

Referring to the problem description section, the input of the model is $[D_{t-m}, D_{t-m+1}, \ldots, D_{t-1}]$ and $[C_{t-m}, C_{t-m+1}, \ldots, C_{t-1}]$, where D_t is a 10×10 matrix representing the ride-hailing demand at time t for all 100 zones. Inspired by computer vision and image processing techniques, D_t can be treated as the frame of the image with 10 by 10 pixels. In order to involve the context-aware parameters, we added one more row of pixels to the frame and filled the first six pixels with the context-aware parameters $C_t = [min_t, day_t, t, temp_t, humi_t, wc_t]$. Thus, an 11×10 image input for the CNN is defined as shown in Fig. 3. The "look-back" parameter is $m = 6$, meaning six frames are input to the network at one time. So the input we constructed is a tensor, and its size is $6 \times 11 \times 10$, as shown in Fig. 4.

The designed CNN structure for this study is shown in Fig. 4. The kernel size for the convolution and the max-pooling layer are labeled in the graph. The input is six frames of the matrices as we constructed above, representing that the model looks at the demand of the previous six time steps (60 min) and predicts the future time steps. There are two convolution layers, each followed by a max-pooling layer for down sampling. It is common to periodically insert a pooling layer between successive convolutional layers in a CNN architecture [21]. The pooling operation provides another form of translation invariance, operates independently on every depth slice of the input, and resizes it

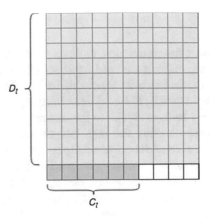

Fig. 3. CNN input construction.

spatially [22]. The most common form of the pooling layer contains filters of size 2×2 with a stride of 2. It downsamples at every depth slice in the input by 2 along both the width and height, discarding 75% of the activations. The depth dimension remains unchanged. To increase the nonlinear properties of the decision function and of the overall network without affecting the receptive fields of the convolution layer, rectifier is used as activation function for the convolutional layer, also known as rectified linear unit (ReLU). The rectifier function gives an output x if x is positive and 0 otherwise:

$$f(x) = \max(0, x) \tag{1}$$

Following the two convolutional-pooling layer pairs is a flattening layer, which flattens the output from the last layer to a vector. The last two layers in the network are fully connected layers (256 hidden units) and the output layer. Since there are 10×10 regions in total for demand prediction, the output of the whole CNN is a 1×100 vector. For the multi-step prediction, we do not change the main characteristic of the CNN but only change the size of the output layer to output a $1 \times 100t$ vector, where t is the number of steps predicted ahead. We applied zero padding to pad the input volume with zeros around the border while calculating the convolution to keep the output having same size with the input of the previous layer.

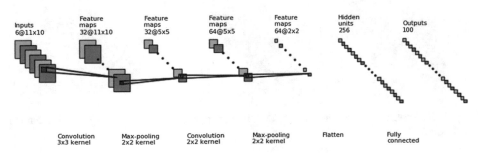

Fig. 4. CNN architecture design.

4   Results

In this section, we present the results and findings of the proposed CNN model. In addition to CNN, two other models are used for comparison. One is a trivial instanton model that uses the current time slot-observed ride-hailing demand as the prediction of the next time slot, another is a LSTM based deep learning model. The main design of the LSTM network for this study is similar to the author's previous work [7, 20] with a few modifications. Instead of using the one-dimensional inputs and outputs in the previous work, multi-dimensional inputs and outputs are used to enable including the context-aware information in the model. The modified model structure allows the LSTM network to predict the ride-hailing demand of all zones simultaneously. Figure 5 visualizes the observation vs. the prediction of the three models. It shows that the predictions from the LSTM and CNN models are far better than the instanton model. For both LSTM and CNN, the correlation of the predicted demand and the observed demand is close to the ideal condition represented by the red line. More details of the prediction error are presented in Table 3. The error metrics that we used are weighted mean absolute percentage error (WMAPE) and mean absolute error (MAE). They are defined as follows:

$$WMAPE = 100\% \cdot \frac{\sum_{i=1}^{n}|y_i - \hat{y}_i|}{\sum_{i=1}^{n}|y_i|} \tag{2}$$

$$MAE = \sum_{i=1}^{n}|y_i - \hat{y}_i| \tag{3}$$

where y_i is the observation, $\hat{y}_i$ is the prediction, and n is the number of samples. The WMAPE is a variant of mean absolute percentage error (MAPE). It is designed for measuring the percentage error but avoids problems where a series of small or zero denominators are present. Table 3 shows that CNN performs slightly better than LSTM in terms of error measures WMAPE and MAE.

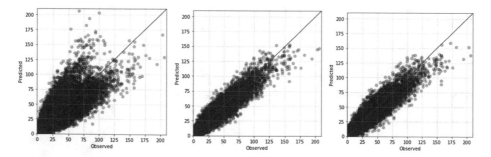

Fig. 5. Observation vs. prediction for next 10-min time step (from left to right: instantons, LSTM, CNN)

Table 3. Model prediction error comparison (for next 10-min time step).

Model	Prediction error	
	WMAPE	MAE
Instanton	46.84%	4.6783
LSTM	24.97%	3.1078
CNN	23.59%	3.0616

The computational efficiency of LSTM and CNN is further examined. Table 4 lists the computing time for training and prediction for the two models. The comparison shows that despite more trainable parameters in the CNN model, the efficiency of the CNN model is over 30% faster for both the training and predicting when compared with LSTM.

Table 4. Model efficiency comparison.

Model	Training time	Prediction time	# Net parameters
LSTM	17.59 s	0.2057 s	133,220
CNN	11.73 s	0.1407 s	172,452
Improvement	33.31%	31.60%	

For ride-hailing services, fleet operators need time to reassign vehicles to meet the future trip demand. Providing trip demand prediction further into the future is more useful than providing prediction of the immediate next time interval. Therefore, the accuracy of multi-step ride-hailing demand prediction was examined to test the long-term prediction capability of the proposed CNN model. The CNN architecture is slightly modified to accommodate 10, 20, 30, 40, 50, and 60 min ahead demand predictions. We only need to change the output size to fit the number of demands that are predicted for multiple steps, which is 100×6. Figure 6 indicates that as the prediction becomes further ahead in the future, the prediction error of the CNN model is increase very slowly within an acceptable level. This indicates that the proposed CNN model performs well in long-term prediction. Figure 7 shows the 60-min ahead predictions vs. observations along time. It can be observed that the predictions align closely with the observed values.

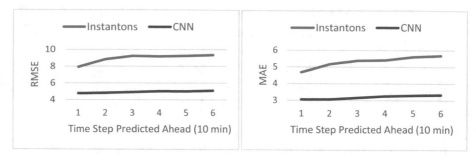

Fig. 6. Error attenuation with multi-step prediction.

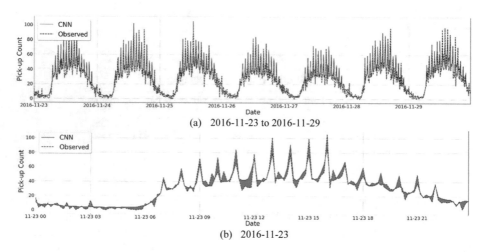

Fig. 7. 60-minute-ahead ride-hailing demand predictions vs. observation (Zone #56, which is represented by the pixel at row 6, column 6, refer to Fig. 3).

5   Conclusion and Future Work

In this study, a CNN-based deep learning model is proposed for context-aware multi-step ride-hailing demand prediction. We utilized the 7-million trip records collected in Chengdu, China, provided by DiDi Chuxing to train and test the model. The outcomes and findings are promising. We split Chengdu City into 10×10 square zones with 1-km side lengths. The CNN model can provide accurate demand predictions for all 100 zones every 10 min. The prediction accuracy significantly outperforms the baseline model and produces slightly lower error measures than the LSTM model, which the authors proved to effectively model ride demand in earlier published works. The computational efficiency of the CNN model is further examined. The result shows that although the number of trainable parameters in the CNN model are higher than in the LSTM model, the CNN model is 30% more computationally efficient for both training and predicting. The proposed model can also be easily extended for multi-step predictions that could be applied for on-demand shared automated vehicle operations.

We found that the CNN model prediction accuracy is still satisfied when predicting ride-hailing demand 60 min ahead.

Predicting ride-hailing demand can benefit ride-hailing vehicle operation efficiency. For future work, a ride-hailing fleet dispatching system will be developed based on the demand prediction. Prior knowledge of the demand distribution around the city would help operators dispatching vehicles to the passengers' nearby locations before they make a ride request to provide more in-time service. Proactive fleet management will save vacant time and travel distance for the vehicles between rides.

Acknowledgment. The authors want to thank DiDi Chuxing for providing the data for this study.

This work was authored in part by the National Renewable Energy Laboratory, operated by Alliance for Sustainable Energy, LLC, for the U.S. Department of Energy (DOE) under Contract No. DE-AC36-08GO28308. Funding provided by U.S. Department of Energy Office of Energy Efficiency and Renewable Energy Vehicle Technologies Office. The views expressed in the article do not necessarily represent the views of the DOE or the U.S. Government. The U.S. Government retains and the publisher, by accepting the article for publication, acknowledges that the U.S. Government retains a nonexclusive, paid-up, irrevocable, worldwide license to publish or reproduce the published form of this work, or allow others to do so, for U.S. Government purposes.

References

1. Clewlow, R.R., Mishra, G.S.: The adoption, utilization, and impacts of ride-hailing in the United States. University of California, Davis, Institute of Transportation Studies, Davis, Research Report UCD-ITS-RR-17-07 (2017)
2. Chang, H.W., Tai, Y.C., Hsu, J.Y.J.: Context-aware taxi demand hotspots prediction. Int. J. Bus. Intell. Data Min. **5**(1), 3 (2010)
3. Moreira-Matias, L., Gama, J., Ferreira, M., Damas, L.: A predictive model for the passenger demand on a taxi network. In: 15th International IEEE Conference on Intelligent Transportation Systems (ITSC), pp. 1014–1019. IEEE (2012)
4. Moreira-Matias, L., Gama, J., Ferreira, M., Mendes-Moreira, J., Damas, L.: Predicting taxi–passenger demand using streaming data. IEEE Trans. Intell. Transp. Syst. **14**(3), 1393–1402 (2013)
5. Predict New York city taxi demand | NYC Data Science Academy Blog, https://nycdata-science.com/blog/student-works/predict-new-york-city-taxi-demand/. Accessed 21 July 2018
6. Ke, J., Zheng, H., Yang, H., Chen, X.M.: Short-term forecasting of passenger demand under on-demand ride services: a spatio-temporal deep learning approach. Transp. Res. Part C Emerg. Technol. **85**, 591–608 (2017)
7. Wang, C., Hao, P., Wu, G., Qi, X., Barth, M.: Predicting the Number of Uber Pickups by Deep Learning (No. 18–06738) (2018)
8. Xu, J., Rahmatizadeh, R., Bölöni, L., Turgut, D.: Real-time prediction of taxi demand using recurrent neural networks. IEEE Trans. Intell. Transp. Syst. (2017)
9. Liao, S., Zhou, L., Di, X., Yuan, B., Xiong, J.: Large-scale short-term urban taxi demand forecasting using deep learning. In: Proceedings of the 23rd Asia and South Pacific Design Automation Conference, pp. 428–433. IEEE Press (2018)

10. Krizhevsky, A., Sutskever, I., Hinton, G.E.: Imagenet classification with deep convolutional neural networks. In: Advances in Neural Information Processing Systems, pp. 1097–1105 (2012)
11. Sun, S., Zhang, C., Yu, G.: A Bayesian network approach to traffic flow forecasting. IEEE Trans. Intell. Transp. Syst. **7**(1), 124–132 (2006)
12. Matsugu, M., Mori, K., Mitari, Y., Kaneda, Y.: Subject independent facial expression recognition with robust face detection using a convolutional neural network. Neural Netw. **16**(5–6), 555–559 (2003)
13. LeCun, Y.: LeNet-5, convolutional neural networks (2015). http://yann.lecun.com/exdb/lenet/. Accessed 1 June 2016
14. Ma, X., Dai, Z., He, Z., Ma, J., Wang, Y., Wang, Y.: Learning traffic as images: a deep convolutional neural network for large-scale transportation network speed prediction. Sensors **17**(4), 818 (2017)
15. Ma, X., Tao, Z., Wang, Y., Yu, H., Wang, Y.: Long short-term memory neural network for traffic speed prediction using remote microwave sensor data. Transp. Res. Part C Emerg. Technol. **54**, 187–197 (2015)
16. Duan, Y., Lv, Y., Kang, W., Zhao, Y.: A deep learning based approach for traffic data imputation. In: IEEE 17th International Conference on Intelligent Transportation Systems (ITSC), pp. 912–917. IEEE (2014)
17. GAIA Open Dataset. https://outreach.didichuxing.com/research/opendata/en/. Accessed 21 July 2018
18. DiDi – Wikipedia. https://en.wikipedia.org/wiki/DiDi. Accessed 01 Nov 2018
19. World Weather Online. https://www.worldweatheronline.com/lang/en-us/. Accessed 21 July 2018
20. Hou, Y., Garikapati, V., Sperling, J., Henao, A., Young, S.: A deep learning approach for TNC trip demand prediction considering spatial-temporal features. In: 98th Annual Meeting of Transportation Research Board (2019)
21. Performing Convolution Operations. https://developer.apple.com/library/archive/documentation/Performance/Conceptual/vImage/ConvolutionOperations/ConvolutionOperations.html. Accessed 01 Aug 2018
22. Pooling Layer - Artificial Intelligence. https://leonardoaraujosantos.gitbooks.io/artificial-inteligence/content/pooling_layer.html. Accessed 01 Aug 2018

Sentiment Classification of User Reviews Using Supervised Learning Techniques with Comparative Opinion Mining Perspective

Aurangzeb Khan[1(✉)], Umair Younis[2], Alam Sher Kundi[2],
Muhammad Zubair Asghar[2], Irfan Ullah[3], Nida Aslam[3],
and Imran Ahmed[4]

[1] Department of Computer Science, University of Science and Technology,
Bannu, Pakistan
aurangzeb.ustb@gmail.com
[2] Institute of Computing and Information Technology (ICIT), Gomal University,
Dera Ismail Khan, KP, Pakistan
zubair@gu.edu.pk
[3] Department of Computer Science, College of Computer Science
and Information Technology, Imam Abdulrahman Bin Faisal University,
Dammam, Kingdom of Saudi Arabia
[4] IMS, University of Peshawar, Peshawar, Pakistan

Abstract. Comparative opinion mining has received considerable attention from both individuals and business companies for analyzing public feedback about the competing products. The user reviews about the different products posted on social media sites, provide an opportunity to opinion mining researchers to develop applications capable of performing comparative opinion mining on different products. Therefore, it is an important task of investigating the applicability of different supervised machine learning algorithms with respect to classification of comparative reviews. In this work different machine learning algorithms are applied for performing multi-class classification of comparative user reviews into different classes. The results show that Random Forest outperforms amongst all other classifiers used in the research.

Keywords: Comparative opinion mining · Machine learning algorithms · Sentiment analysis · Sentiment classification · Supervised machine learning

1 Introduction

With the growing interest of the online community in comparing and analyzing products online, comparative reviews about products are becoming popular, providing an opportunity to opinion mining experts to develop applications based on comparative opinion mining paradigm [1]. The comparative opinion mining is an important task in sentiment-based applications. Due to the fact that there is a huge amount of opinions posted online every day, analyzing comparative opinions from a user perspective is an important application that needs to be explored further [2].

The comparative opinion mining is a branch of opinion mining with emphasis on acquiring and analyzing comparative reviews of the public about different products for

© Springer Nature Switzerland AG 2020
K. Arai and S. Kapoor (Eds.): CVC 2019, AISC 944, pp. 23–29, 2020.
https://doi.org/10.1007/978-3-030-17798-0_3

making their purchase decisions [3]. Furthermore, it assists business companies to analyze the strength and weaknesses of their manufactured items. For example, the input text: *"IOS is secure than Android"*, is a comparative review comparing two operating systems, namely IOS and Android, and showing the superiority of IOS in terms of security features. Acquisition and analysis of such opinions are beneficial for individuals for their purchase decision and companies, for improving the quality of the products [4]. Therefore, it is an important task to develop a comparative opinion mining application.

The previous studies [3, 5] on comparative opinion mining have used supervised machine learning techniques. The study conducted by Khan et al. [3] applied only one classifier namely Naïve Bayes on a dataset of 400 reviews for performing comparative opinion mining. Due to use of only one classifier and very small dataset comprising of few hundred reviews, non-satisfactory results were achieved. The proposed work is an extension of the work performed by Khan et al. [3] with respect to increasing the number of classifiers, applied on three (3) extended datasets. Additionally, an increased number of Joint Label Comments (ideally divided into 9 classes) are used.

In this work, we evaluate the overall performance and efficiency of the seven machine learning classifiers. It is focused on the implementation of different algorithms on different datasets containing comparative reviews and recommends the classifier with best results.

Rest of the article is organized as follows: Sect. 2 presents a review of relevant literature; Sect. 3 elaborates proposed methodology; in Sect. 4, we present result and discussion; and finally work is concluded with the future directions.

2   Related Work

This section presents a review of related studies performed on comparative opinion mining with respect to online user reviews.

Khan et al. [3] proposed a supervised machine learning classification technique based on Naïve Bayes classifier for classifying comparative comments express by viewers on YouTube. The system is based on multi-label classification technique for a dataset of two products, namely iOS and Android. The system can be made more efficient by incorporating data pre-processing steps along with different lexicons. Additionally, experimentation on other machine learning classifiers can give efficient results.

Bach et al. [5] proposed a comparative opinion mining system for analyzing comparative sentences in the Vietnamese language on product reviews domain. Different supervised learning algorithms are applied and satisfactory results are achieved. Other advanced methods for performing comparative opinion mining needs to be investigated.

Jin et al. [6] proposed a model for comparative customer reviews on online products review sites. The method is based on the two supervised approaches and three greedy algorithms for the optimal solution and satisfactory results are obtained. Work can be extended in the direction to present visual summary reports for improved customer satisfaction.

Varathan et al. [7] proposed a comparative opinion mining paradigm based on a combination of supervised machine learning techniques (SVM), and unsupervised techniques (clustering). Assorted results were obtained with respect to baseline methods. The work can be extended by experimenting on datasets in other domains.

Ganapathibhotla and Liu [8] presented a comparative opinion model with emphasis on comparative, superlatives sentences, and identifying preferred entities. Different benchmark datasets were acquired and also, a dataset is compiled from Howard forums, CNet and Amazon. Experimental results are satisfactory with respect to other methods. The system can be extended by adopting a new method called One-Side Association (OSA).

Ejaz et al. [9] proposed a comparative opinion mining system using lexicons and a set of n-gram features with machine learning algorithms. The product reviews dataset was acquired from Amazon and the experimental results were promising with respect to baseline methods. Further enhancement can be made by assigning weights to reviews and to perform analysis with respect to product popularity a long time dimensions.

Vargas et al. [10] proposed a supervised learning method for contrastive opinion summarization, based on blog threads on social media in the product domain. The method achieved improved results over the baseline methods in terms of accuracy, precision, recall, and F-measure. The system can be extended by performing experiments on real datasets with improved feature-set and M.L classifiers.

Liang et al. [11] proposed a model for the contrastive opinion summarization and to generate a summary, which can replicate the statistics associated with contrastive viewpoints. For this purpose, different feature rating schemes such as tf*idf and Lex Rank are used. Experimental results show the efficiency of the proposed system with respect to efficient contrastive opinion summary generation. The system can be enhanced by generating more refined contrastive summaries.

3 Proposed Methodology

The proposed system is comprised of the following modules: (i) Data Collection and Compilation, (ii) Pre-processing, (iii) Applying machine learning classifiers, (iv) comparing the efficiency of different classifiers, and (v) Suggesting classifier with best classification results for comparative opinion mining.

3.1 Dataset Collection

Due to the limited size of a publically available dataset of the baseline study [3] and to achieve more robust results, we manually acquired business-related comparative reviews from YouTube about different products: iOS vs Android, Facebook vs Twitter and Microsoft vs Google. The review sentences are annotated manually into nine different combinations of positive, negative and neutral classes. For this purpose, beautiful soup [12] is used to scrap the required reviews. The acquired reviews are stored dataset in machine-readable CSV files for conducting experiments. The resulting dataset is approximately 6,000 reviews.

3.2 Preprocessing

We performed different pre-processing steps on the acquired reviews, such as tokenization, part of speech tagging, spell correction, removal of hashtags and other special characters [13]. To accomplish the pre-processing tasks, we will apply python-based coding using Anaconda Framework [12].

3.3 Reviews Classification

The data set acquired is split into training (80%) and testing (20%). In next step, comparative reviews are classified to different classes (pos_neg, pos_pos, pos_ neu, neg_pos, neg_neg, neg_neu, neu_neu, neu_neg, neu_pos) using different supervised M.L Classifiers namely, Naïve Bayes, Logistic Regression, Support Vector Machine (SVM), KNN (K-Nearest Neighbor), Decision Tree, Random Forest, Gradient Boosting by applying Anaconda-based python framework. The labeled comparative reviews dataset are passed through different machine learning classifiers to evaluate the performance and efficacy of each of the classifier individually. Figure 1 shows overall working the proposed system.

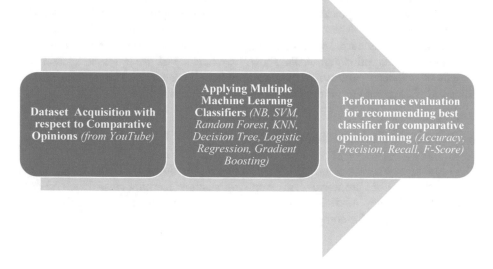

Fig. 1. Proposed system.

Tables 1 and 2 show sample outputs obtained from the training and testing phases, respectively.

Table 1. Set of reviews from the dataset (Training).

Review no.	Review	Polarity label
1	I own Android and iOS both are fantastic operating systems	Pos-Pos
2	Android is better than iOS	Pos-Neg
3	iOS is hit Android is just okay	Pos-Neu
4	Android and iOS both sucks	Neg-Neg
5	iOS has no ability to easy file sharing with Android	Neg-Neu
6	Android is ok iOS is rough	Neu_Neg

Table 2. Set of reviews from the dataset (Testing).

Review no.	Review	Polarity label
1	iOS holds the quality Android don't have quality	pos_neg
2	iOS has less number of apps than Android	neg_pos
3	I am sure Android and iOS will catch up with features	neu_neu

4   Discussions and Results

A. Performance Evaluation

For analyzing the performance of above-mentioned machine learning classifiers with respect to sentiment classification of comparative reviews, different evaluation measures, namely Accuracy, Precision, Recall, and F1-score, are applied.

Experiment#1(RQ1: What is the efficacy of different machine learning algorithms for performing comparative opinion mining for online user reviews).

This experiment assists in obtaining the efficiency of each of the supervised algorithm w.r.t to comparative opinion mining. To answer RQ1, we performed an experiment on the acquired dataset of comparative user reviews (6000) about *"iOS vs Android"*. The performance evaluation results are reported in Table 3, showing that the Random Forest achieves the best results as compared to other classifiers. After this, the same mechanism is followed for a dataset of size 3000 reviews about *"Microsoft vs Google"*. Results presented in Table 4 show that Decision Tree has outperformed other classifiers.

Table 3. Performance evaluation results on data set of 6000 reviews (*iOS vs Android*)

Classifier	Accuracy	Precision (%)	Recall (%)	F-measure (%)
Naïve Bayes	62%	0.67	0.63	0.55
Logistic Regression	69%	0.68	0.69	0.66
Support Vector Machine (SVM)	78%	0.79	0.79	0.78
K-Nearest Neighbor (KNN)	61%	0.59	0.60	0.61
Decision Tree	77%	0.77	0.77	0.77
Random Forest	**80%**	**0.80**	**0.80**	**0.80**
Gradient Boosting	77%	0.70	0.68	0.65

Table 4. Performance evaluation results on data set of 3000 reviews (*Microsoft vs Google*)

Classifier	Accuracy	Precision (%)	Recall (%)	F-measure (%)
Naïve Bayes	71%	0.74	0.71	0.66
Logistic Regression	76%	0.77	0.77	0.75
Support Vector Machine (SVM)	85%	0.86	0.85	0.85
K-Nearest Neighbor (KNN)	61%	0.60	0.61	0.59
Decision Tree	**93%**	**0.94**	**0.94**	**0.94**
Random Forest	89%	0.90	0.89	0.89
Gradient Boosting	84%	0.86	0.84	0.84

B. Comparison with the state-of-the-art method

The performance of the proposed system is evaluated with the state-of-the-art (baseline) methods and results are reported in Table 5. It is obvious that the proposed system (SVM classier) for performing comparative opinion mining outperforms the baseline method in terms of Accuracy, Precision, Recall, and F-measure.

Table 5. Comparison of results of the baseline method

Method	Accuracy	Precision (%)	Recall (%)	F-measure (%)
Khan et al. [3]	70%	0.74	0.71	0.65
Proposed (SVM)	**95%**	**0.96**	**0.95**	**0.95**

C. Recommending the Classifier for Comparative Opinion Mining with Best Classification Results

After applying different machine learning classifiers with respective sentiment classification of comparative opinions, we recommend the algorithm with best results as follows.

Experiment#2(RQ. 2: What is an efficient machine learning classifier for comparative opinion mining?)

For performing comparative opinion mining, different machine learning algorithms are applied on different datasets of varying sizes to inspect which classifier yields best results.

The performance evaluation results received from different classifiers (experiment#1) are reported in Table 3, showing that for the dataset of size 6000 reviews, the Random Forest classifier is best amongst the other classifiers. Similarly, Table 4 (data size = 3000) shows that Decision tree performs best.

5   Conclusion and Future Work

This study investigates the applicability of different machine learning algorithms for performing comparative opinion mining, i.e. classifying the comparative opinions into 9 polarity classes, i.e. pos_neg, pos_pos, pos_neu, neg_pos, neg_neg, neg_neu, neu_neu, neu_neg, neu_pos. We experimented different classifiers and the results show that Random Forest and Decision Tree outperformed other classifiers on two data sets of different sizes (6000 and 3000 reviews).

Additionally, further experimentation is required to investigate the viability of different machine learning classifiers in multiple domains.

References

1. Asghar, M.Z., Kundi, F.M., Ahmad, S., Khan, A., Khan, F.: T-SAF: Twitter sentiment analysis framework using a hybrid classification scheme. Expert Syst. **35**(1), e12233 (2018)
2. Khan, A., Asghar, M.Z., Ahmad, H., Kundi, F.M., Ismail, S.: A rule-based sentiment classification framework for health reviews on mobile social media. J. Med. Imaging Health Inform. **7**(6), 1445–1453 (2017)
3. Khan, A.U.R., Khan, M., Khan, M.B.: Naïve multi-label classification of YouTube comments using comparative opinion mining. Procedia Comput. Sci. **82**(16), 57–64 (2016)
4. Asghar, M.Z., Khan, A., Zahra, S.R., Ahmad, S., Kundi, F.M.: Aspect-based opinion mining framework using heuristic patterns. Cluster Comput. **20**(1), 1–19 (2017)
5. Bach, N.X., Van, P.D., Tai, N.D., Phuong, T.M.: Mining Vietnamese comparative sentences for sentiment analysis. In: Merialdo, B., Nguyen, L.M., Li, D.D., Duong, D.A., Tojo, S. (eds.) CONFERENCE 2015, KSE, vol. 7, pp. 162–167. IEEE, Ho Chi Minh (2015)
6. Jin, J., Ji, P., Gu, R.: Identifying comparative customer requirements from product online reviews for competitor analysis. Eng. Appl. Artif. Intell. **49**(C), 61–73 (2016)
7. Varathan, K.D., Giachanou, A., Crestani, F.: Comparative opinion mining: a review. J. Assoc. Inf. Sci. Technol. **68**(4), 811–829 (2017)
8. Ganapathibhotla, M., Liu, B.: Mining opinions in comparative sentences. In: Scott, D., Uszkoreit, H. (eds.) CONFERENCE 2008, COLING, vol. 1, pp. 241–248. Association for Computational Linguistics, Manchester (2008)
9. Ejaz, A., Turabee, Z., Rahim, M., Khoja, S.: Opinion mining approaches on Amazon product reviews: a comparative study. In: Mahmood, T., Rauf, I., Khoja, S., Ghani, S. (eds.) CONFERENCE 2017, ICICT, vol. 17, pp. 173–179. IEEE, Karachi (2017)
10. Vargas, D.S., Moreira, V.: Identifying sentiment-based contradictions. J. Inf. Data Manag. **8**(3), 242 (2017)
11. Liang, X., Qu, Y., Ma, G.: Research on contrastive viewpoint summarization for opinionated texts. J. Interconnect. Netw. **14**(03), 1360003 (2013)
12. Hetland, M.: Python and the Web. Beginning Python: From Novice to Professional, 2nd edn. Apress, New York (2005)
13. Asghar, M.Z., Khan, A., Khan, F., Kundi, F.M.: RIFT: a rule induction framework for Twitter sentiment analysis. Arab. J. Sci. Eng. **43**(2), 857–877 (2018)

Asymmetric Laplace Mixture Modelling of Incomplete Power-Law Distributions: Application to 'Seismicity Vision'

Arnaud Mignan[(✉)]

Swiss Federal Institute of Technology, Sonneggstr. 5, 8092 Zurich, Switzerland
arnaud.mignan@sed.ethz.ch

Abstract. Data used in statistical analyses are often limited to a narrow range over which the quantity of interest is observed to be reliable. The power-law behavior of many natural processes is only observed above a threshold below which information is discarded due to detection limitations. These incomplete data can however also be described by a power law, and the distribution over the full quantity range reformulated as an asymmetric Laplace (AL) distribution. With the detection process heterogeneous in space and time in realistic conditions, the data can be modelled by a mixture of AL components. Using seismicity as example, we describe an asymmetric Laplace mixture model (ALMM), which considers ambiguous overlapping components - as observed in Nature - based on a semi-supervised hard Expectation-Maximization algorithm. We show that the ALMM fits reasonably well incomplete data and that the number of AL components can be related to the seismic network density. We conclude that the full range of data can be used in statistical analyses, including in computer vision. In the case of seismicity, a ten-fold increase in sample size is possible which provides, for example, a better spatial pattern resolution to improve the correlation between fault features and earthquake labels.

Keywords: Expectation-Maximization · Incomplete data · Data optimization

1 Introduction

Power-law distributions $p(x)$ occur in many domains but are most often restricted to a narrow range of the quantity of interest x [1]. Whereas deviations at the tail are subject to under-sampling, use of a lower bound x_{min} avoids divergence at very small values of x (Eq. 1). In practice, x_{min} represents the minimum completeness of the data above which the power-law behavior holds. Due to the nature of the power-law, the majority of the data may be contained below that threshold, which represents the range of 'unreliable' data (i.e. incomplete, under-reported, error-prone).

$$p(x) = \frac{\alpha - 1}{x_{min}} \left(\frac{x}{x_{min}} \right)^{-\alpha} \tag{1}$$

K. Arai and S. Kapoor (Eds.): CVC 2019, AISC 944, pp. 30–43, 2020.
https://doi.org/10.1007/978-3-030-17798-0_4

Could we use those incomplete data in statistical analyses, instead of discarding them? This would be possible if the measures below x_{min} could be shown to follow a definable distribution.

The under-reported part of a Pareto Law (in the case of reported income or property data) was early on described by a second power-law [2] such that

$$p(x) \propto \begin{cases} x^{\gamma} & x < x_{min} \\ x^{-\alpha} & x \geq x_{min} \end{cases} \qquad (2)$$

and it was noted soon after that Eq. (2) could be reformulated in terms of an asymmetric Laplace (AL) distribution (Eq. 3) [3].

$$p(x) \propto \begin{cases} \exp(\gamma \ln x) & x < x_{min} \\ \exp(-\alpha \ln x) & x \geq x_{min} \end{cases} \qquad (3)$$

This function is a generalization of the first law of Laplace, the second law of Laplace being the well-known Normal distribution [4].

Independently of the findings made in Econometrics [2, 3], it has recently been observed in the field of Geophysics that seismicity also follows Eq. (3) when considering the full data range, which includes both reliable and unreliable observations [5]. This suggests that the Asymmetric Laplace distribution could be a reasonable choice to describe data that can be approximated by a power-law and are incomplete at low x values. In a more realistic setting where the detection or reporting process is heterogeneous, a mixture of asymmetric Laplace distributions should then be used. Whereas Gaussian Mixture Modelling (GMM) represents a standard machine learning method [6], Asymmetric Laplace Mixture Modelling remains seldom used [7].

The present paper presents an Asymmetric Laplace Mixture Model (ALMM) with its parameters estimated using a semi-supervised hard Expectation-Maximization (E-M) method. The case of ambiguous overlapping AL populations is also addressed. The method is illustrated using earthquake data as input. Earthquake catalogs are publicly available and there already exists an empirical relationship relating seismic network detection capabilities and x_{min} [8]. Earthquake data, due to their pattern complexity in space, time and magnitude, form an excellent testbed for machine learning methods. Recent examples include the use of Random Forest for tracking precursory seismic signals in the laboratory [9] or deep neural networks for aftershock spatial distribution prediction [10]. We shall show that 'seismicity vision', or the prediction of spatial seismicity patterns based on physics-based or geometric features, could be enhanced by the proposed ALMM. The proposed approach can be seen as a prerequisite step to improve the input data available for computer vision applications.

2 Asymmetric Laplace Mixture Model for Seismicity

2.1 Seismological Literature Review

The earthquake frequency-energy distribution is commonly described by a power-law, better known in the form of the exponential Gutenberg-Richter law, defined from the magnitude scale m with $E \propto \exp(m)$ the seismic energy [11]. It has recently been observed in local datasets with constant completeness magnitude m_c (equivalent to x_{min}) that the full magnitude range can be described by the following AL distribution

$$p_{AL}(m; m_c, \kappa, \beta) = \frac{1}{\frac{1}{\kappa-\beta} + \frac{1}{\beta}} \begin{cases} \exp[(\kappa - \beta)(m - m_c)], & m < m_c \\ \exp[-\beta(m - m_c)], & m \geq m_c \end{cases} \quad (4)$$

where κ is a detection parameter and β the size ratio [5]. Equation (4) has an angular shape in log-linear space although the frequency-magnitude distribution $p(m)$ is curved in regional datasets (see Fig. 1 and results of Sect. 3.3). The curvature of power-law distributions around x_{min} is observed in many data domains [1].

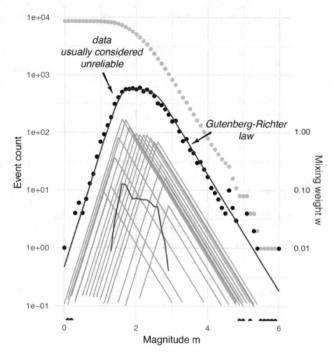

Fig. 1. Simulated earthquake frequency-magnitude distribution (in red), defined as the sum of AL components (in orange). The distribution of the mixing weights $w(m_c)$ controls the shape of the mixture distribution. Magnitudes are binned in $\Delta m = 0.1$ intervals. Black dots represent the number of events per bin while the grey dots the cumulative number of events per bin. Earthquakes with magnitude $m < \max(m_c)$ are usually discarded - here corresponding to the m-range (0.0, 2.9].

For seismicity, this curvature can be due to m_c being a function of the spatial configuration of the seismic network with

$$m_c(x, y) = c_1 d_i(x, y)^{c_2} + c_3 \pm \sigma \tag{5}$$

where d [km] is the distance to the ith nearest seismic station, c_1, c_2 and c_3 empirical parameters, σ the standard deviation, and (x, y) spatial coordinates [8]. Hence p_{AL} is rarely directly observed but any observed $p(m)$ can be described by a mixture of p_{AL} components. This is illustrated in Fig. 1 based on a synthetic earthquake catalog: A seismic network is simulated with station locations sampled from a normal distribution and $m_c(x, y)$ estimated from Eq. (5) with $i = 4$, $c_1 = 5.96$, $c_2 = 0.0803$, $c_3 = -5.80$ and $\sigma = 0$ [8]. For each cell (x, y), events are generated by applying the Inversion Method [12] to Eq. (4). The regional frequency-magnitude distribution finally corresponds to the set of events over all cells. It will be shown in Sect. 3.3 that seismicity follows such curved distribution - The distribution can also be curved with multiple maxima when several seismic networks with different spatial configurations are present in a same region. This will also be illustrated in Sect. 3.

Data below $\max(m_c)$ are usually considered unreliable and are thus discarded. This is due to the fact that different distributions can be observed for different catalogs and sub-catalogs. It is only since recently that the variations from an angular shape to a curved shape with potentially multiple maxima have been explained by an increase in m_c heterogeneities [5]. Other detection models proposed in the Seismological literature include the cumulative Normal distribution [13] and the generalized Gamma distribution [14]. None of those, however, can explain the angular shape observed locally, such as in Southern California and Nevada [5], Greece [15], or Taiwan [16]. This suggests that a mixture of AL distributions is so far the only model flexible enough to represent the variability observed in incomplete earthquake data. This will be tested in Sect. 3.

In the case of seismicity, incomplete data represent approximately 90% of a full catalog, which demonstrates the importance of being able to predict the behavior below m_c. The same may be said for other domains dominated by power-law statistics (e.g. other hazards, networks), although the curvature around x_{min} might be explained by a log-normal distribution in some other domains [17]. The selected model should depend on the underlying physical process. For seismicity, the ALMM is preferred due to the known relationship to seismic network detection capabilities [5, 8, 16].

2.2   E-M Algorithm

The ALMM is defined by the following probability density function

$$p_{ALMM}(m; m_{ck}, \kappa_k, \beta_k) = \sum_{k=1}^{K} w_k p_{AL}(m; m_{ck}, \kappa_k, \beta_k) \tag{6}$$

with K the number of AL components, m_{ck}, κ_k and β_k the parameters of the kth component, and w_k the mixing weight of the kth component such that $\sum_{k=1}^{K} w_k = 1$. The ALMM can be fitted by using the E-M algorithm [18]. The approach presented here is a simple case of hard E-M, as our goal is not to assign the probability of having m belonging to a given cluster, but to define a surrogate of the true (but unknown) frequency-magnitude distribution in order to estimate the value of K, w_k, m_{ck}, κ and β. By doing so, one can estimate whether incomplete data are reliable to be used in other statistical analyses (i.e. if the model can fit the incomplete data within realistic and stable parameter ranges). The proposed E-M algorithm, applied for $K = \{1, 2, \ldots, K_{\max}\}$ components, is defined as follows:

We set the initial parameter values m_{ck}, κ and β by applying k-means [19], with $k = \{1, 2, \ldots, K\}$, w_k the normalized number of events per cluster, and m_{ck} the cluster center. The magnitude vector is defined as $\mathbf{M} = \{\mathbf{M}_1, \mathbf{M}_2, \ldots, \mathbf{M}_K\}$, ordered by increasing m_{ck} and with each component defined as $\mathbf{M}_k = \{m_1, m_2, \ldots\}$, the vector of magnitude scalars m to be labelled to cluster k. We obtain parameters κ and β from the clusters of centers $m_{c1} = \min(m_{ck})$ and $m_{cK} = \max(m_{ck})$, or $\mathbf{M}_1$ and $\mathbf{M}_K$, respectively, by using the maximum likelihood estimation (MLE) method:

$$\begin{cases} \chi = 1/\left(\left(\min(m_{ck}) - \frac{\Delta m}{2}\right) - \overline{\mathbf{M}}_{left}\right) \\ \beta = 1/\left(\overline{\mathbf{M}}_{right} - \left(\max(m_{ck}) - \frac{\Delta m}{2}\right)\right) \end{cases} \tag{7}$$

where $\chi = \kappa - \beta$ is the slope of the incomplete part of the distribution in log-linear scale, $\mathbf{M}_{left} = \{m \in \mathbf{M}_1 : m \leq m_{c1} - \frac{\Delta m}{2}\}$ and $\mathbf{M}_{right} = \{m \in \mathbf{M}_K : m > m_{cK} - \frac{\Delta m}{2}\}$ [20]. Although k-means may provide biased estimates of m_{ck}, it nevertheless reliably finds the local maxima of the m-space, hence avoiding one of the common difficulties of mixture modelling [21]. We will however refine the m_{ck} values using the mode in an iterative E-M.

At each iteration i, a deterministic version of the expectation step (E-step) attributes a hard label k to each m-event from the parameter set $\theta_k^{[i-1]} = \{m_{ck}, \kappa, \beta\}$ defined in the previous iteration $i - 1$ ($i = 0$ corresponding to the k-means estimates). Hard labels are assigned as:

$$k = \text{argmax}_k p_{AL}\left(\theta_k^{[i-1]}, m\right) \tag{8}$$

The maximization step (M-step) updates the component parameters: w_k is the normalized number of m-events per component k, $m_{ck} = \text{mode}(\mathbf{M}_k)$, and κ and β are calculated from Eq. (7). It should be mentioned that when m_{ck} extrema lead to component under-sampling and therefore to errors in κ and β, $\mathbf{M}_{1+j}$ and $\mathbf{M}_{K-l}$ are used instead in Eq. (7) (with j and l increased incrementally until no error is found). When the AL components are relatively well separated, m_{ck} estimates rapidly fall into the local distribution maxima (see Sect. 3.1). However, when the AL components significantly overlap (see Fig. 1 and Sect. 3.3 - which is the most realistic case), the estimates tend to migrate towards the unique maximum. This problem is solved by shifting m_{ck} to the nearest free magnitude bin when a bin is already occupied, which is the semi-supervised part of the clustering method. To determine whether this semi-supervision should be done, we first apply a classifier 'curved/not-curved' at iteration $i = 1$ (see below).

The E- and M-steps are repeated until log-likelihood LL convergence (difference between two iterations lower than 10^{-6}) or until $i = i_{max} = 5$ (a higher i_{max} does not significantly improve the results). Once the procedure has been repeated K_{max} times, the best number of components is K_{BIC}, the number of components with the lowest Bayesian Information Criterion (BIC) estimate $\text{BIC}(K) = -LL + 1/2 \cdot n_{par} \ln N_{tot}$ with $n_{par} = 2 + K$ with 2 representing the free parameters κ and β [22]. Note that computing the log-likelihood from the function $\sum \ln(p_{ALMM}(m))$ is inconclusive due to higher weights on m_{ck} components with the largest $\mathbf{M}_k$ size. To avoid this bias towards the main mode of the distribution, we compute instead the log-likelihood of a Poisson process:

$$LL\left(\theta_k^{[i]}, X = \{n_j; j = 1, \dots, N_j\}\right) = \sum_{j=1}^{N_j} \left[n_j \ln\left(v_j\left(m_j \Big| \theta_k^{[i]}\right)\right) - v_j\left(m_j \Big| \theta_k^{[i]}\right) - \ln(n_j!)\right]$$
(9)

for the observed magnitude rate $n_j\left(m \in \left(m_j - \frac{\Delta m}{2}, m_j + \frac{\Delta m}{2}\right]; m_j = 0.0, 0.1, \dots, 8.0\right)$ and predicted rate

$$v\left(m \Big| \theta_k^{[i]}\right) = N_{tot} p_{ALMM}\left(m; \theta_k^{[i]}\right) \Delta m$$
(10)

Hence the present MLE method is an estimator of the shape of the frequency-magnitude distribution represented by the rate $v(m)$, instead of the population of magnitudes m. As for the 'curved/not-curved' classifier, we compare the BIC estimates of the mixture model and of the curved frequency-magnitude distribution model of Ogata and Katsura [13] (see their Eq. 6) using the LL definition of Eq. (9). If the distribution is not curved, the mixture model will lead to a lower BIC; however, if the distribution is curved, the mixture model will lead to a higher BIC since, even if it would fit reasonably well its shape at $i = 1$, it is penalized for the higher number of parameters compared to the simple 3-parameter Ogata-Katsura model.

Finally, it is noted that in the cases where the E-M algorithm fails for a given stochastic realization, this realization is not recorded (the k-means iteration would always provide a result, but likely biased). The E-M algorithm may fail in convoluted cases and more so in real cases. However, we did not come upon any shape in earthquake data in which the ALMM would systematically fail (see results below).

3   ALMM Results

3.1   Validation on Simulated AL Mixtures

Multi-angular Distributions (i.e. well separated components). Figure 2 shows four examples of simulated frequency-magnitude distributions where the true m_{ck} values are regularly spaced and with equal weights w_k, for $K_{true} = 1$ to 4. Our ALMM fitting procedure retrieves K and θ_k reasonably well, as demonstrated in the second and third columns. Note that using random weights only alters K_{BIC}, which decreases when the

E-M algorithm does not find components with very low weight w_k. This becomes systematic in more realistic m_c distributions, as shown below.

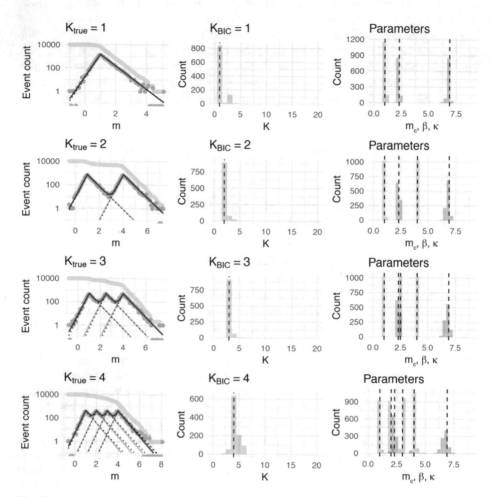

Fig. 2. ALMM fitting results for multi-angular distributions (in log-linear scale). The first column shows one distribution from 1,000 simulations, the true model (black dotted curves), and the simulation fit (components in orange and their mixture in red). The central column shows the distribution of K_{BIC} for the 1,000 simulations, and the third column the θ_k distribution. True values of K and θ_k are represented by vertical dashed lines.

Curved Distributions (i.e. significantly overlapping components). Examples of simulated m_c maps are shown in the top row of Fig. 3 based on Eq. (5) for two seismic network configurations leading to two different m_c distributions. One example of distribution out of 1,000 simulations is shown on the second row for those two classes. The color coding is the same as in Fig. 2. The ALMM retrieves the curved and curved-with-two-maxima shapes reasonably well. The K_{BIC} and θ_k distributions are shown on the third and fourth rows with the true values represented by vertical lines. The parameter set θ_k is again reasonably well recovered although κ is slightly underestimated for the most convoluted case where two networks are mixed; note also the bimodal m_{ck} distribution obtained in that case. In contrast to previous tests (Fig. 2), the number of components K is now systematically underestimated, which can be explained by the presence of low-weight components in realistic m_c distributions. This however does not seem to have a significant impact on the estimation of θ_k, although it might explain the slight bias now observed in κ.

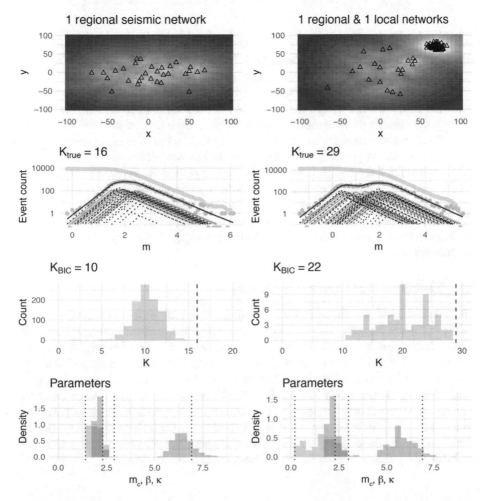

Fig. 3. ALMM fitting results for curved distributions with one or two maxima. The first row shows the m_c maps of the synthetic seismic networks represented by the open triangles. Low m_c values are represented in yellow and high ones in purple. The second row represents one distribution from 1,000 simulations, with the true model (black dotted curves), and the simulation fit (components in orange and their mixture in red). The third and fourth rows show the distributions of K_{BIC} and θ_k, respectively. True values of K and θ_k are represented by vertical lines.

3.2   Relationship Between Number of Components K and Seismic Network Density

We observed that the parameter K can be underestimated when component weights are low. K, which can be seen as a proxy to the degree of m_c heterogeneity, can however be estimated independently. Equation (5) suggests that m_c evolves faster in the dense part of the seismic network than in the sparser areas, from which we can derive

$$L(K, d_i) = \left(\frac{c_1 d_i^{c_2} + \left(K - \frac{3\sigma}{\Delta m} \right) \frac{\Delta m}{2}}{c_1} \right)^{\frac{1}{c_2}} - \left(\frac{c_1 d_i^{c_2} - \left(K - \frac{3\sigma}{\Delta m} \right) \frac{\Delta m}{2}}{c_1} \right)^{\frac{1}{c_2}} \tag{11}$$

where d_i is the distance to the ith nearest seismic station and $L = 2\sqrt{A/\pi}$ is the characteristic length of the area of interest A. Equation (11) means that, over distance L, m_c has varied by K times the bin Δm (subject to noise σ).

Figure 4 shows how K depends on d_5 (distance to the 5th nearest seismic station) and L for simulations of m_c in a regional network (as defined in Sect. 2.1). We estimate d_5 from the nucleus of Voronoi cells [23, 24] and L from the Voronoi cell area A (see example of Voronoi tessellation on the left column). The true K is estimated as the sum of unique m_c values per cell, constrained by the bin Δm. The resulting $K(d_5, L)$ distribution, obtained for 100 Voronoi tessellations, is plotted on the right column with $\sigma = 0$ in the top row (no noise) and $\sigma = 0.18$ in the bottom row (with noise) [8]. The curves defined from Eq. (11) for different K values are consistent with the simulation results, hence proving that K can be approximated independently of the ALMM. Due to the large scattering observed, we so far do not use Eq. (11) for K semi-supervision in the mixture modelling.

3.3   Testing on Real Earthquake Catalogs

The ALMM E-M algorithm is then tested on two real earthquake catalogs, from Taiwan [25] and Nevada [5]. These catalogs are selected for their different shapes: the former distribution is curved with one maximum, representative of the regional Taiwan Central Weather Bureau Seismic Network; the latter distribution is curved with two maxima, representative of the combination of the Nevada regional network and the local Southern Great Basin Digital Seismic Network around the Yucca Mountain nuclear waste repository. To estimate the accuracy of the ALMM as done previously with simulations, we bootstrap the real data 100 times [26]. Results of the fitting are shown in Fig. 5. Once again, the distribution shapes are retrieved. A final example of ALMM fitting will be shown for an aftershock sequence in Sect. 5.

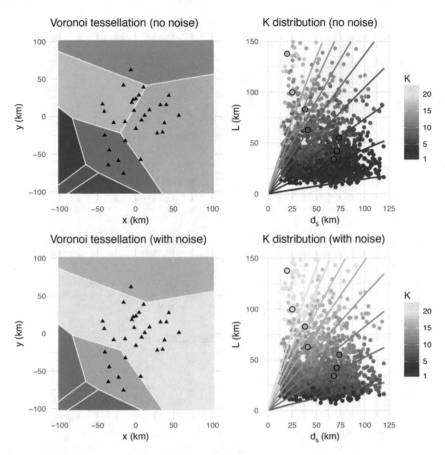

Fig. 4. Spatial scale of detected seismicity described by the number of mixture components K, as a function of the distance to the 5th nearest seismic station d_5 and of the Voronoi cell characteristic length L. Seismic stations are represented by triangles, in map view (left). For each cell, d_5 is calculated from the Voronoi nucleus coordinates and L from the cell area A. The $K(d_5, L)$ distribution of 100 Voronoi models (right) is represented by dots (the circled dots represent the results of the Voronoi model instance shown on the left column). The $K(d_5, L)$ distribution is reasonably well predicted by Eq. (11), represented by curves for different K values.

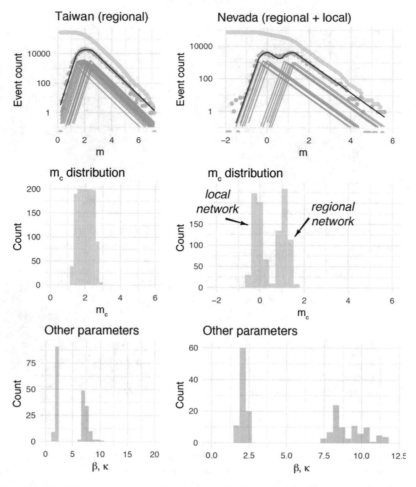

Fig. 5. ALMM fitting on real data (Taiwan and Nevada earthquake catalogs). The first row represents one distribution from 100 bootstraps and the matching fit (components in orange and their mixture in red). The second and third rows show the distributions of θ_k for the 100 bootstraps.

4   Conclusion

We presented an Asymmetric Laplace Mixture Model (ALMM; Eqs. 4 and 6) to describe power-law distributions which are incomplete, due to detection limitations of the quantity of interest x. The piecewise power-law distribution centered on completeness threshold x_{min} (Eq. 2) was first reformulated in terms of a Laplace distribution (Eq. 3). The rationale behind the ALMM was that x_{min} follows itself a distribution (e.g. Eq. 5), which can explain why incomplete data may show different behaviors and why it was so far discarded from statistical analyses. We used earthquake catalogs to test a semi-supervised hard Expectation-Maximization algorithm (Sect. 2.2) and to demonstrate that the ALMM can fit reasonably well different distribution shapes observed in both simulations and Nature, including the realistic case of overlapping components (Sect. 3).

5 Future Scope: Implications for 'Seismicity Vision'

With up to 90% of seismicity data potentially discarded in regional seismicity analyses (i.e. the ratio of events with $m < \max(m_c)$ observed in simulations and real catalogs), the ALMM has the potential to improve seismicity forecasts [27] by significantly increasing the sample size. Indeed, the ALMM can prove that incomplete data, otherwise claimed unreliable, can in fact be included in statistical analyses since they can be modelled. With the parameters of the incomplete m-range now trackable in space and time, the incomplete data can be used whenever the ALMM parameters are shown to be stable.

To illustrate the possible impact of incomplete data on seismicity vision, let us consider the case of aftershock spatial distribution forecasting, for which deep neural networks were recently applied [10]. We take as example the 2011 $m = 6.2$ New-Zealand Christchurch aftershock sequence [28]. Figure 6 (right panel) shows the corresponding ALMM fit with $\max(m_c) = 3.1$ obtained. The rest of Fig. 6 shows that considering the full magnitude range, instead of just the complete data, may provide a better match to the fault features (overlaid in orange on the aftershock binary grid).

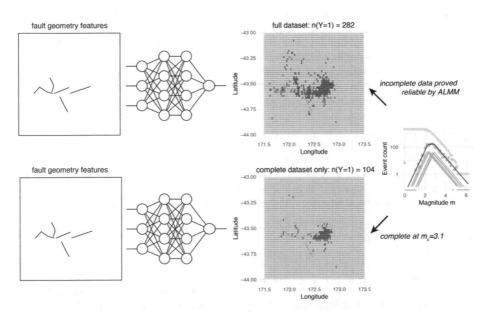

Fig. 6. Sketch illustrating how using both incomplete ($m < m_c$) and complete ($m \geq m_c$) seismicity data would improve the correlation between features (e.g. fault geometrical constraints) and labels ($Y = 1$ meaning aftershocks present). 'Full dataset' means the set of both incomplete and complete data (i.e. full m-range).

Although the method was illustrated for the geophysical domain, it could find applications in any data domain that is dominated by power-law statistics combined to heterogeneous data detection.

References

1. Clauset, A., Shalizi, C.R., Newman, M.E.J.: Power-Law distributions in empirical data. SIAM Rev. **51**(4), 661–703 (2009)
2. Hartley, M.J., Revankar, N.S.: On the estimation of the Pareto Law from under-reported data. J. Econometrics **2**, 327–341 (1974)
3. Hinkley, D.V., Revankar, N.S.: Estimation of the Pareto Law from under-reported data. J. Econometrics **5**, 1–11 (1977)
4. Kotz, S., Kozubowski, T.J., Podgorski, K.: The Laplace Distribution and Generalizations. Birkhäuser, Boston (2001)
5. Mignan, A.: Functional shape of the earthquake frequency-magnitude distribution and completeness magnitude. J. Geophys. Res. **117**, B08302 (2012)
6. Bishop, C.: Pattern Recognition and Machine Learning. Springer, New York (2006)
7. Franczak, B.C., Browne, R.P., McNicholas, P.D.: Mixtures of shifted asymmetric laplace distributions. IEEE Trans. Pattern Anal. Mach. Intell. **36**(6), 1149–1157 (2014)
8. Mignan, A., Werner, M.J., Wiemer, S., Chen, C.-C., Wu, Y.-M.: Bayesian estimation of the spatially varying completeness magnitude of earthquake catalogs. Bull. Seismol. Soc. Am. **101**(3), 1371–1385 (2011)
9. Rouet-Leduc, B., Hulbert, C., Lubbers, N., Barros, K., Humphreys, C.J., Johnson, P.A.: Machine learning predicts laboratory earthquakes. Geophys. Res. Lett. **44**, 9276–9282 (2017)
10. DeVries, P.M.R., Viegas, F., Wattenberg, M., Meade, B.J.: Deep learning of aftershock patterns following large earthquakes. Nature **560**, 632–634 (2018)
11. Utsu, T.: Representation and analysis of the earthquake size distribution: a historical review and some new approaches. Pure. Appl. Geophys. **155**, 509–535 (1999)
12. Devroye, L.: Non-uniform Random Variate Generation. Springer, New York (1986)
13. Ogata, Y., Katsura, K.: Immediate and updated forecasting of aftershock hazard. Geophys. Res. Lett. **33**, L10305 (2006)
14. Kijko, A., Smit, A.: Estimation of the frequency-magnitude Gutenberg-Richter b-value without making assumptions on levels of completeness. Seismol. Res. Lett. **88**, 311–318 (2017)
15. Mignan, A., Chouliaras, G.: Fifty Years of Seismic Network Performance in Greece (1964-2013): Spatiotemporal Evolution of the Completeness Magnitude. Seismol. Res. Lett. **85**, 657–667 (2014)
16. Mignan, A., Chen, C.-C.: The spatial scale of detected seismicity. Pure. Appl. Geophys. **173**, 117–124 (2016)
17. Broido, A.D., Clauset, A.: Scale-free networks are rare. https://arxiv.org/abs/1801.03400. Accessed 28 Oct 2018
18. Dempster, A.P., Laird, N.M., Rubin, D.B.: Maximum likelihood from incomplete data via the EM algorithm. J. R. Stat. Soc. Ser. B (Methodological) **39**(1), 1–38 (1977)
19. Jain, A.K.: Data clustering: 50 years beyond K-means. Pattern Recogn. Lett. **31**, 651–666 (2010)
20. Aki, K.: Maximum likelihood estimate of b in the formula log N=a-bM and its confidence limits. Bull. Earthq. Res. Inst. **43**, 237–239 (1965)
21. Celeux, G., Hurn, M., Robert, C.P.: Computational and inferential difficulties with mixture posterior distributions. J. Am. Stat. Assoc. **94**(451), 957–970 (2000)
22. Schwarz, G.: Estimating the dimension of a model. Ann. Stat. **6**(2), 461–464 (1978)

23. Voronoi, G.F.: Nouvelles applications des paramètres continus à la théorie des formes quadratiques, 2ème Mémoire: Recherches sur les parallélloèdres primitifs. J. fûr die. Reine und Angewandte Mathematik **134**, 198–287 (1908)

24. Lee, D.T., Schachter, B.J.: Two algorithms for constructing a Delaunay triangulation. Int. J. Comput. Inf. Sci. **9**, 219–242 (1980)

25. Wu, Y.M., Chang, C.-H., Zhao, L., Teng, T.L., Nakamura, M.: A comprehensive relocation of earthquakes in Taiwan from 1991 to 2005. Bull. Seismol. Soc. Am. **98**(3), 1471–1481 (2008)

26. Efron, B.: Second thoughts on the bootstrap. Stat. Sci. **18**(2), 135–140 (2003)

27. Mignan, A.: The debate on the prognostic value of earthquake foreshocks: a meta-analysis. Sci. Rep. **4**, 4099 (2014)

28. Bannister, S., Fry, B., Reyners, M., Ristau, J., Zhang, H.: Fine-scale relocation of aftershocks of the 22 February M_w 6.2 Christchurch earthquake using double-difference tomography. Seismol. Res. Lett. **82**(6), 839–845 (2011)

Iceberg Detection by CNN
Based on Incidence-Angle Confusion

Yongli Zhu[1（✉）] and Chengxi Liu[2]

[1] GEIRI North America, San Jose, CA 95127, USA
yzhul6@vols.utk.edu
[2] Aalborg University, Fredrik Bajers Vej 5, 9100 Aalborg, Denmark

Abstract. In this paper, an image object identification problem for the Kaggle Iceberg Classifier Challenge was tackled by deep neural network. Basic convolutional neural network (CNN) was implemented and tested firstly. Then, deeper networks including VGG16 and ResNet50 are adopted to improve the accuracy. The deep learning-based methods are also compared with the conventional machine learning method i.e. SVM (support Vector Machine). Three feature augmentation approaches are utilized and compared, i.e. incidence angle confusion of satellite radar signals, multi-band composition and data augmentation of the original image data. Tentative results by GAN (Generative Adversarial Network) and Capsule Network are also presented. Results demonstrate the applicability and superiority of CNN over the conventional method (SVM) on the given dataset.

Keywords: Object detection · Capsule · GAN · Iceberg · SVM

1 Introduction

Modern satellite imaging technology uses the motion of the airborne antenna over a target region to provide finer spatial resolution than conventional camera imaging technologies, which is widely used to create two or three-dimensional images of objects, such as landscapes [1]. Typically, the larger the aperture of antenna is, the higher the image resolution will be, regardless of whether the aperture is physical (e.g. a large antenna) or synthetic (e.g. a small but moving antenna). This can contribute to create higher-resolution images with physically-small antennas. Due to the scattering mechanism and the large amount of speckle noises in the satellite images, the interpretation of those images is much more challenging than that of the conventional images obtained by optical camera. Identification of objects in a massive image dataset by human-being eyes is not practical and cost-expensive. Therefore, object detection by modern computer vision techniques for those images has become more and more popular in both industry and academic communities, such as using (SVM) support vector machine [2], graphical models [3], sparse representation [4], compressive sensing [5], etc. With the rapid development of deep learning theory and techniques [6–9], its applications in computer vision area have been widely explored. In the Kaggle Iceberg Classifier Challenge [10] (in short, "Kaggle Iceberg Competition"), a typical remote sensing data set is provided, where the task is to identify whether a given image containing icebergs or not.

© Springer Nature Switzerland AG 2020
K. Arai and S. Kapoor (Eds.): CVC 2019, AISC 944, pp. 44–56, 2020.
https://doi.org/10.1007/978-3-030-17798-0_6

The organization of this paper is as follows: Sect. 2 introduces the background and problem description of the Kaggle Iceberg Competition. Section 3 presents the data preprocessing part. In Sect. 4, nine approaches based on different network structures are implemented and tested on the given data set, i.e.

(1) LeNet [11]
(2) VGG-16 [12]
(3) ResNet [13]
(4) SVM
(5) VGG-16 + Incidence angle features
(6) VGG-16 + Multi-band composition
(7) VGG-16 + Data augmentation
(8) DCGAN for Data augmentation
(9) Capsule network

Results and performance for each method are analyzed. Conclusions and future directions are provided in Sect. 5. Related source code can be found in https://github.com/yonglizhu/Iceberg-Detection-by-CNN-based-on-Incidence-Angle-Confusion.

2 Kaggle Iceberg Competition

2.1 Problem Description [10]

Drifting icebergs present threats to navigation and activities in areas such as offshore of the East Coast of Canada. Currently, many institutions and companies use aerial reconnaissance and shore-based support to monitor environmental conditions and assess risks from icebergs. However, in remote areas with particularly harsh weather, these methods are not feasible, and the only viable monitoring option is via satellite.

Statoil, an international energy company operating worldwide, has worked closely with companies like C-CORE. C-CORE has been using satellite data for over 30 years and has built a computer vision-based surveillance system. To keep operations safe and efficient, Statoil is interested in getting a fresh new perspective on how to use machine learning to more accurately detect and discriminate against threatening icebergs as early as possible.

In this competition, the goal is to build an algorithm that automatically identifies if a remotely sensed target is a ship or iceberg. Improvements made will help cut the costs down for maintaining safe working conditions.

2.2 Challenges

Several challenges are:

- The remote sensing systems used to detect icebergs are housed on satellites over 600 km above the Earth.
- Satellite antenna works as: It bounces a signal off an object and records the echo, then that data is translated into an image.

- However, when the antenna detects an object, it can't tell an iceberg from a ship or any other solid object. The object needs to be analyzed for certain characteristics - shape, size and brightness - to find that out. The area surrounding the object (in this case, ocean) can also be analyzed or modeled.
- The Sentinel-1 satellite is a side looking antenna, which means it photographs the image area at certain angle (incidence angle). Generally, the ocean background will be darker at a higher incidence angle.

Thus, the motivations of this problems are:

- Human labelling is cost-expensive and time-consuming.
- Even human work is allowed, the image is pretty vague, thus eye-sight identicality can easily yield wrong result.
- How to maximumly utilize the hidden information in the two channel images sensed by the Satellite antenna.
- How to fine-tune the hyper-parameters to improve accuracies.

Image examples is shown in Fig. 1. Two examples of the raw input images are shown in Fig. 2, respectively for the ships and the icebergs.

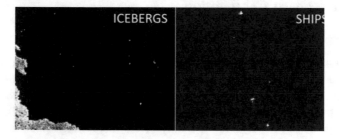

Fig. 1. Example of a raw satellite imaging figure.

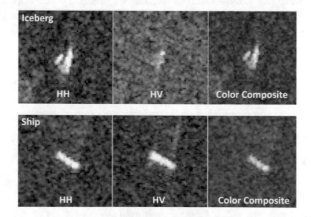

Fig. 2. Example two-channel images for the icebergs and ships, respectively.

3   Data Wrangling

3.1   Data Description

In this competition, the task is to predict whether an image contains a ship or an iceberg. The labels are provided by human experts and geographic knowledge on the target. All the images are 75×75 images with two bands.

The data (*train*.json, *test*.json) is presented in **json** format. The files consist of a list of images. For each image, there are following fields:

id - the id of the image

band_1, band_2 - the flattened image data. Each band has 75×75 pixel values in the list, so the list has 5625 elements. Note that, these values are not integers like that in the conventional image format but having physical meanings, i.e. they are float numbers with unit dB. **band_1** and **band_2** are signals characterized by antenna backscatter produced from different polarizations at a particular incidence angle. The polarizations correspond to HH (transmit/receive horizontally) and HV (transmit horizontally and receive vertically).

inc_angle - the incidence angle at which the image was taken. Note that, this field has missing data marked as "**na**", and those images with "**na**" incidence angles are all in the training data to prevent information leakage.

is_iceberg - the target variable to predict, set to 1 if the image contains iceberg (s), and 0 otherwise. This field only exists in *train*.json.

Please note that machine-generated images in the test set have been included to prevent hand labeling. They are excluded in the final ranking and scoring.

3.2   Data Preprocessing

The data is read by the json package in python. For "*train*.json", it is unpacked to two arrays, i.e. "**band_1**" and "**band_2**" respectively. Physically, **band_1** and **band_2** are measurements from different electromagnetic wave polarization direction (similar to the concept of multiple "channels" in a color image). As an example, input data of the first image of the training set is shown in Fig. 3.

Key	Type	Size	Value
band_1	list	5625	[-27.878361, -27.15416, -28.668615, -32.834259 ...
band_2	list	5625	[-27.154118, -29.537888, -31.0306, -31.030729, ...
id	str	1	dfd5f913
inc_angle	float	1	43.9239
is_iceberg	int	1	0

Fig. 3. Example raw data of one image in the training set (in *Spyder* Python IDE).

For label information, the original list is like [0, 1, 0, 0, …]; because the final output should be probability, and on the other hand, the one-hot encoding for the label vector is adopted to avoid unnecessary layer change, i.e. in a form like [0 1; 1 0; 0 1; …].

Histogram of the feature **inc_angle** is shown in Fig. 4 for the two classes.

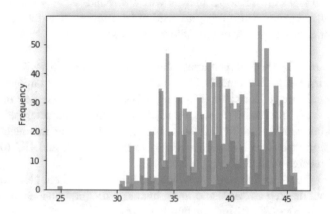

Fig. 4. Histogram of the feature "incidence angle" (orange: class-0, blue: class-1)

From the histogram, it is hard to conclude that the **inc_angle** can provide too much information in helping improve the accuracy. Thus, at the beginning step of the following study, only "band_1" is used for training. Since the true label information is not given for the testing data set, thus, 2/3 of the original training data is used for "training" and the remaining 1/3 for "validation".

4   Experiments and Results

4.1   LeNet

Baseline settings: Epoch = 100, Batch_size = 32, Momentum = 0, Optimizer = SGD, Padding = 'VALID', Activation = 'relu', 1st_Conv_outdim = 32. Accuracies on testing dataset with different hyper-parameter settings are shown in Table 1.

Table 1. LeNet results.

Experiment no.	Different settings	Validation accuracy
1	Baseline setting	77.01
2	Batch_size = 64	77.86
3	Padding = 'SAME'	78.50
4	Momentum = 0.7	78.32
5	1^{st}_Conv_outdim = 64	79.63
6	Activation = 'tanh'	75.14

From the results of LeNet, we can observe that:

(1) 1st_Conv_outdim has more effect than other hyper parameters;
(2) 'Tanh' activation layer is worse than 'Relu' activation layer
(3) Padding using 'SAME' scheme can help as well.

4.2   VGG-16

Baseline setting: Optimizer = 'Adam', Epoch = 100, Batch_size = 32, 1st_Conv_outdim = 64, Padding = 'VALID'. Accuracies on testing dataset with different hyperparameters are shown in Table 2.

Table 2. VGG-16 results.

Experiment no.	Different settings	Validation accuracy
1	Baseline setting	84.67
2	Padding = 'SAME'	84.12
3	Optimizer = 'SGD	38.00
4	Batch_size = 64	84.48
5	Optimizer = 'SGD, Momentum = 0.7	83.18

Example performance curves for the LeNet (six experiments marked as 1 to 6) and VGG-16 (showing its 2nd experiment) on the validation dataset are shown in Figs. 5 and 6 for comparison. As expected, VGG-16 performs better than LeNet on this data set.

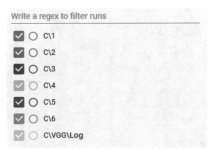

Fig. 5. Training curves id.

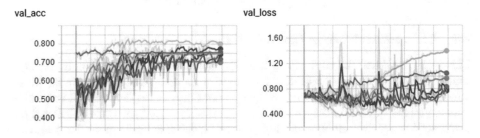

Fig. 6. Performance curves for LeNet and VGG-16.

4.3   ResNet

As currently the deepest network which has been successfully applied into industrial application, the ResNet is chosen and tested on the given data set. Due to the smaller data set size compared to ImageNet, only ResNet-18 and 50 are considered here. The performance in terms of the validation accuracy are not improved compared with the shallower networks (VGG-16). The result is shown in Table 3.

Table 3. ResNet results.

Experiment no.	Different settings	Validation accuracy
1	ResNet-50, Batch_size = 256	80.00
2	ResNet-18, Batch_size = 64	81.31

Performance curves for the ResNet on the validation dataset are shown in Fig. 7.

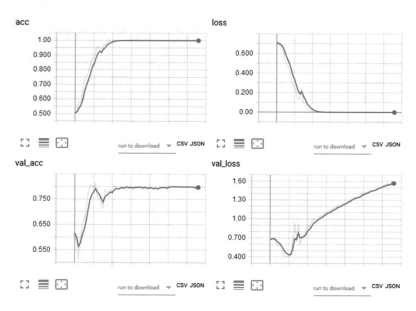

Fig. 7. Performance curves for ResNet.

4.4   Conventional Machine Learning Methods

The SVM is chosen for comparison here. Input feature is X = [**inc_angle**, **band_1**, **band_2**] for each image. 2/3 of data set is used for training and the remaining 1/3 for validation. Since **band_2** data are relatively noisy than **band_1** by visual inspection, thus experiment by using only **band_1** as input is also conducted. It is observed that using all data set for training considering **band_1** and **inc_angle** gives the best result for SVM classifier. The result is shown in Table 4.

Table 4. SVM results.

Experiment no.	Configuration	Validation accuracy
1	2/3 for training (use both bands)	70.66%
2	All for training, also all for validation (use both bands)	84.12%
3	All for training, also all for validation (use only band_1)	94.26%

4.5   Using Incidence Angle Confusion

Previously, no angle information is utilized in the deep learning network. To use the incidence angle, a novel confusion method is put forward in the following formula:

$$X_{new} = band\_1 * \cos\theta + band\_2 * \sin\theta \qquad (1)$$

Network parameters and results are listed in the following table. It can be concluded that adding inc_angle indeed improves the performance of CNN (VGG-16) on the given data set. The result is shown in Table 5.

Table 5. Using angle information in VGG-16.

Configuration	Validation accuracy
VGG16, 1st_Conv_outdim = 64, Batch_size = 32, Epoch = 100, Kernel_size = 3, Data split = 1/4	87.75

However, by adding the weighted two matrices, some hidden information might be lost. Thus, in the follow sections, both two bands will be considered together by merging them into a "pseudo-multi-channel" tensor as input.

4.6   Using Feature Composition and Data Augmentation

Besides using multiple bands as "pseudo-channel" image, the following feature augmentation approach is exploited shown in Eq. (2). To better inspect the effect of augmentation, here the incidence angle information is not included. The input feature is:

$$X_{new} = [band\_1, band\_2, band\_3] \qquad (2)$$

where, band_3 = (band_1 + band_2)/2

Also, the "vertical flip" and "horizontal flip" are adopted to augmenting the data set. The result is shown in Table 6.

Table 6. Using feature composition and data augmentation in VGG-16.

Configuration	Validation accuracy
VGG16, 1st_Conv_outdim = 64, Batch_size = 32, Epoch = 100, Kernel_size = 3, Data split = 1/4	93.21

4.7   Combining All Techniques

In this section, an approach combined all the above techniques, i.e. using Feature Composition, Data Augmentation and Incidence Angle Confusion is adopted. based on the above discussion, the input tensor for each image is as follows:

$$X_{new} = [\text{band\_1} * \cos(\theta), \text{band\_2} * \sin(\theta), \text{band\_3}] \tag{3}$$

where, band_3 = (band_1 + band_2)/2

Also, the "vertical flip" and "horizontal flip" are adopted similar to the previous section. The result is shown in Table 7.

Table 7. Combining all techniques in VGG-16.

Configuration	Validation accuracy
VGG16, 1st_Conv_outdim = 64, Batch_size = 32, Epoch = 100, Kernel = 3, Data split = 1/4	92.57

It can be observed that the result here is slightly inferior to that of the previous section without using angle information. One potential reason for this fact is that, about 100 samples in the training data set are **deliberately** set to "**na**" value for those incident angles by the Kaggle committee to prevent information leakage.

4.8   DCGAN

The DCGAN (Deep Convolutional Generative Adversarial Network) [14] is experimented here for data augmentation purpose if it can be successfully trained. As a tentative study in this paper, a simple two Conv-layer DCGAN is adopted for discriminator and Generator network respectively. Generated sample images by using **band_1** data as input are shown as a 6-by-6 grid (36 images) in Figs. 8 and 9, respectively by the vanilla-GAN and conditional-GAN techniques.

Intuitively, the qualities of sampled images are not good enough to help training. More research must be done to investigate the potential improvement by GAN techniques, e.g. try other variants of GAN structure like WGAN [15] for more stable training process and better forgery effect.

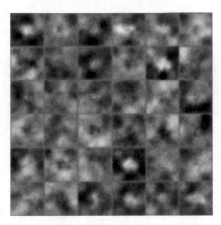

Fig. 8. Generated images from the vanilla-GAN.

4.9   Capsule Network

The Capsule network has been published by Dr. Hinton [16], which claims an obvious improvement on MNIST data set with several novel properties introduced, e.g. enhanced spatial/location feature recognition for the same object in different images.

The idea of a "Capsule" can be regarded as a natural dimensional augmentation of the conventional concept of "neuron", i.e. the output "neuron" is a vector (with dim > 1) as shown in Fig. 10. With those augmented dimensions, the expressing power of a "Capsule Network" is expected to outperform the conventional neuron-based deep network.

Fig. 9. Generated images from the conditional-GAN.

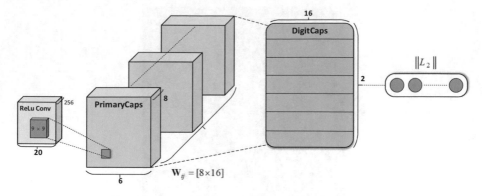

Fig. 10. Structure of a typical capsule network.

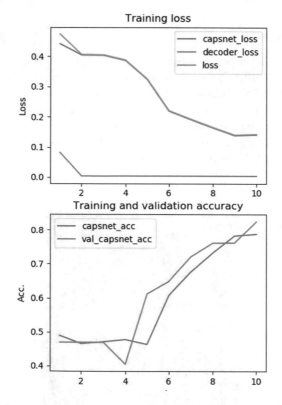

Fig. 11. Training loss and performance curves for a simple capsule network on this data set.

Unfortunately, due to the slow training speed on this Kaggle competition data set, a preliminary study using only **band_1** data for 10 epochs is presented here. The whole training data set are used during both the training and validating stages. The training curves are shown in Fig. 11. With only 10 epochs and no fine-tuning of parameters, the validation accuracy is 82.4%, which is not better than previous VGG-16 based methods. It can be expected that, after training more epochs with proper hyper-parameters, a promising result can be hopefully obtained by Capsule Network.

5 Conclusion

In this study, the following conclusions can be drawn:

(1) Deeper networks will necessarily bring improvements on this satellite data set;
(2) Data set preprocessing and data augmentation is critical in achieving better performance by reducing potential overfittings.
(3) Advanced network structure like GAN and Capsule might help but needs more elaborated fine-tuning work and more intricate feature engineering.

Therefore, the next step is to explore any potential improvements on this dataset by fine-tuning of network hyperparameters and experimenting different feature combinations.

References

1. Novak, L.M., Owirka, G.J., Brower, W.S., Weaver, A.L.: The automatic target-recognition system in SAIP. Lincoln Lab. J. **10**, 187–202 (1997)
2. Zhao, Q., Principe, J.C.: Support vector machines for SAR automatic target recognition. IEEE Trans. Aerosp. Electron. Syst. **37**, 643–654 (2001)
3. Srinivas, U., Monga, V., Raj, R.G.: SAR automatic target recognition using discriminative graphical models. IEEE Trans. Aerosp. Electron. Syst. **50**, 591–606 (2014)
4. Dong, G., Wang, N., Kuang, G.: Sparse representation of monogenic signal: with application to target recognition in SAR images. IEEE Signal Process. Lett. **21**(8), 952–956 (2014)
5. Zhang, X.Z., Qin, J.H., Li, G.J.: SAR target classification using bayesian compressive sensing with scattering centers features. Prog. Electromagnet. Res. Pier **136**, 385–407 (2013)
6. Hinton, G.E., Salakhutdinov, R.R.: Reducing the dimensionality of data with neural networks. Science **313**, 504–507 (2006)
7. Krizhevsky, A., Sutskever, I., Hinton, G.E.: Imagenet classification with deep convolutional neural networks. In: Advances in Neural Information Processing Systems, pp. 1097–1105. USA (2012)
8. Zeiler, M.D., Fergus, R.: Visualizing and understanding convolutional networks. In: Computer Vision–ECCV 2014, pp. 818–833. Springer, USA (2014)
9. Hinton, G.E., Srivastava, N., Krizhevsky, A., Sutskever, I., Salakhutdinov, R.R.: Improving neural networks by preventing co-adaptation of feature detectors. arXiv preprint arXiv:1207. 0580 (2012)
10. Statoil/C-CORE Iceberg Classifier Challenge. https://www.kaggle.com/c/statoil-iceberg-classifier-challenge

11. Lecun, Y., Bottou, L., Bengio, Y., Haffner, P.: Gradient-based learning applied to document recognition. Proc. IEEE **86**(11), 2278–2324 (1998)
12. Simonyan, K., Zisserman, A.: Very Deep Convolutional Networks for Large-Scale Image Recognition. eprint arXiv:1409.1556, September 2014
13. He, K., Zhang, X., Ren, S., Sun, J.: Deep Residual Learning for Image Recognition. eprint arXiv:1512.03385, December 2015
14. Mirza, M., Osindero, S.: Conditional generative adversarial nets. eprint arXiv:1411.1784, November 2014
15. Arjovsky, M., Chintala, S., Bottou, L.: Wasserstein GAN. arXiv:1701.07875 [stat.ML], January 2017
16. Sabour, S., Frosst, N., Hinton, G.E.: Dynamic Routing Between Capsules. arXiv:1710.09829 [cs.CV], November 2017

The Effectiveness of Distinctive Information for Cancer Cell Analysis Through Big Data

Babu Kaji Baniya[1($\boxtimes$)] and Etienne Z. Gnimpieba[2]

[1] Department of Computer Science, Grambling State University,
403 Main Street, Grambling, LA 71245, USA
baniyab@gram.edu
[2] Department of Biomedical Engineering, University of South Dakota,
GEAR Center, 4800 N. Career Avenue, Sioux Falls, SD 57107, USA
etienne.gnimpieba@usd.edu

Abstract. In the healthcare system, a huge volume of multi-structured patient data is generated from in-hospital clinical examinations, wearable body sensors, and doctor memos. These data play a deterministic role in finding the patient's cause of disease and corresponding cure. However, there are many challenges underlying healthcare system such as integration of different data format, appropriate selection of healthcare parameters and disease prediction. Besides these, concern whether this data is equally significant for decision making or not exists. To address these problems, we introduced the infinite latent feature selection (ILFS) to find highly informative gene from a gene pool and feed it to the classifier. We adopted a hybrid approach for gene classification. The hybrid scheme integrated two (sophisticated) algorithms: speed advantage of extreme learning machine (ELM) and accuracy advantage of sparse representation classifier (SRC). We validated our proposed model with 198 tumor samples of global cancer map (GCM) dataset and divided them into 14 common tumor categories having 11,370 expression level of genes. The performance of this system was measured using five different matrices (i.e., accuracy, sensitivity, specificity, precision, and F-score). This achieved a notable improvement of accuracies in both scenarios in selective (considering only highly expressive genes) and original genes (74.75% and 81.82% respectively).

Keywords: Precision · Recall · Sensitivity · Predication · Multi-structure

1 Introduction

Big data plays a significant role in healthcare system. Several sources contribute to the growth of healthcare data. Tremendous amount of data is collected in the healthcare through sources including in-hospital patient's questionnaires, doctors and other care providers' memos, radiological investigations,

© Springer Nature Switzerland AG 2020
K. Arai and S. Kapoor (Eds.): CVC 2019, AISC 944, pp. 57–68, 2020.
https://doi.org/10.1007/978-3-030-17798-0_7

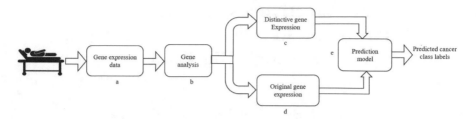

Fig. 1. Overview of cancer gene analysis system: a. acquiring the gene from cancer patients, b. removing low-level noise values and very high saturation effects, c. probabilistic latent graph-based algorithm for highly expressive gene selection, d. preparing original gene expressions to feed the classifier, e. gene classification through the EA-SRC using the leave-one-out cross-validation and predicated class labels of different cancers

and different biochemical and pathological tests. Similarly, data is also gathered from outdoor monitoring system through the Internet of Things (IoT) [1] in which patients are equipped with different smart devices including in-plant pacemaker, motion sensors, and mobile devices. Wearable devices collect health-related data such as body temperature, heart rate, and blood pressure from each instant/activity, which coordinate to track the personal fitness, raising the awareness of personal healthcare, and facilitate better medical treatment. These diverse data sets appear in the different format (structures) and size ranging from megabytes to terabytes [2].

Healthcare data are presented as a big data, which are specified by 5Vs: Variety, Volume, Velocity, Veracity, and Value. Variety depicts the diversified datasets with structured, semi-structured, and unstructured data sets such as clinical reports, electronic health records (EHRs), and radiological images. The space required to store patients data ranges from mega-to-terabytes, which describes the volume. The velocity is data arrival (streaming) rate in milliseconds to second from patients and veracity is the truthfulness of data sets with respect to data availability [7]. The collected data should comprehensively provide meaningful information. This highlights the value of 5Vs.

In our proposed method, we sought to address the problem of cancer cell discrimination through learning algorithm. An overview of our proposed method is shown in Fig. 1. Initially, we collected a large volume of gene expression data (gene pool) from patients. The gene expression data were bound into a specific box. Genes whose ratios and absolute variations across gene samples were under 5 and 500 respectively were excluded. In the next stage, we explored more influential genes from the gene pool in order to find the substantial cause of the disease. Infinite latent feature selection (ILFS) was introduced to find the highly expressed gene. This algorithm deals with the feature (gene) selection strategy in the different set of data and computes the rank by considering all the possible subsets of features (genes), as a path of a graph, and bypasses the combinatorial challenge analytically [4]. It provides highly expressed gene sequence list from gene pool. In the next stage, the original and selective (only highly expressive) genes were fed to the classifier to measure the discriminative capability (same as feature selection [5] and [6]). In doing so, we obtained a convincing and higher classification accuracies in both scenarios. The detailed experimental results and comparison are presented in the result and discussion section.

We adopted the integrated approach to a cascade of ELM and SRC for cancer classification. Since ELM is not effective in handling the noisy data, it is worthy to feed the noisy data and perform the classification with the more effective SRC classifier. To improve the performance of ELM and maximize the separation boundary, the regularized ELM adopting the leave-one-out cross-validation (LOO) scheme with a sub-dictionary selection strategy (i.e., ELM output is used to construct the sub-dictionary) for each query data was considered for sparse representation rather than using the whole dictionary [8]. The scheme is sum-up regularized ELM and adaptive SRC for cancer classification (in short it is called as EA-SRC), description is given in Algorithm 3.

Regularized ELM carries out classification by exploring the largest entry of the output vector for a query input where optimal regularized ELM can produce the maximal decision boundary for discrimination. In practice, the difference between the first and second largest entries for each output vector reflects the decision boundary whenever ELM employs the one-against-all labeling [9]. For the sake of convenience, the output difference is calculated as $o_{diff} = (o_f - o_s)$, where o_f and o_s represent the first and second largest entries in the ELM output vector. If $(o_{diff} > \sigma)$ the classification based on ELM otherwise feature vector will be into SRC (σ is threshold).

We validated our proposed model in two distinct stages by using GCM dataset. In the first stage, we only considered the most distinctive genes and fed them to the classifier. In the next stage, all genes were employed to the classifier and measured the overall performance of the system. The reported classification accuracies are 74.75% and 81.82% from 725 and 11,370 feature dimensions in 30 s and 6.2 min respectively. The rest of the paper is organized in different sections. Section 2 describes the related work of machine learning approaches, Sect. 3 includes the algorithm for cancer cell classification. Section 4 describes the dataset. Section 5 contents the result and discussion, and the final section includes the conclusion and future work with the proposed method.

2   Related Work

Different machine learning algorithms are already implemented for healthcare data analysis and label prediction [1]. We also introduced the two algorithms ELM and SRC and their corresponding algorithms 1 and 2 respectively. ELM is single layer feedforward neural networks (SLFNs) [3] and later extended to generalized feedforward networks. The output weights of ELM are determined based on randomly selected hidden node parameters (input weights and bias).

Algorithm 1. ELM Classifier

Input: A training set $(x_j, t_j)_{j=1}^{N}$, activation function $g(x)$,
hidden node L, and query object y
Output: Class label y
  a. Randomly initialize the hidden node parameters (w_i, b_i), $i = 1, 2, ..., L$
  b. Calculate $H(w_1, ..., w_L, x_1, ...x_N, b_1, ..., b_L)$
  c. Determine the output weight matrix $\hat{\beta} = H^{\dagger}T$
  d. Determine the actual output o of ELM
  e. $Label(y) = \underset{d \in [1,2,...,m]}{\arg\max}\ (o)$

For a given set of training samples $(x_j, t_j)_{j=1}^{N}$ with N samples and m classes, the SLFN with L hidden nodes and activation $g(x)$ is written as

$$\sum_{i=1}^{L} \beta_i h_i(x_j) = \sum_{i=1}^{L} \beta_i g(w_i.x_j + b_i) = o_j, j = 1, 2, ..., N \tag{1}$$

where $x_j = [x_{j1}, x_{j2}, ..., x_{jn}]^T$, $t_j = [t_{j1}, t_{j2}, ..., t_{jm}]^T$, $w_i = [w_{i1}, w_{i2}, ..., w_{in}]^T$, and b_i are the input, it corresponding output, the connecting weights of ith hidden node to input nodes. Similarly, $\beta_i = [\beta_{i1}, \beta_{i2}, ..., \beta_i m]$ is the connecting weights of the ith hidden node to the output nodes and o_j is actual network output. For minimizing the training error and improving the generalization performance of neural networks, the training error and output weight should be simultaneously minimized:

$$||H\beta - T||, ||\beta|| \tag{2}$$

where H is called the hidden layer output matrix of neural network. Equation (2) is a least squares problem whose solution can be addressed by $\hat{\beta} = H^{\dagger}T$, where $H^{\dagger}$ is the pseudo-inverse of H [3].

SRC algorithm estimates the spare representation coefficients of query sample through the over complete dictionary which objectives are extracted features along with available class labels. Then, the classification is performed by determining minimum residual error based on sparse coefficient [8,9]. Assuming that the dataset $\{A_d\}_{d=1}^{m}$ for m classes $A \in R^{n \times k_d}$ contains k_d training objects belonging to the dth class, cancer gene being concatenated as a column vector. The coding dictionary is presented as $A = [A_1, A_2, ..., A_m]$. The objective is to explore the optimal sparse representation coefficient via the following optimization

$$\hat{x} = \underset{x}{\arg\min} ||x||_0\ s.t.\ Ax = y \tag{3}$$

where $||.||_0$ is the l_0-norm, denoting the support of a vector. The convex l_1-norm minimization problem can be solved as

$$\hat{x} = \underset{x}{\arg\min} ||x||_1\ s.t. ||Ax - y||_2^2 < \varepsilon \tag{4}$$

Algorithm 2. SRC based Classifier

Input: A dictionary with m classes $A = [A_1, A_2, ..., A_m]$, a query object y
Output: Class label y
   a. Normalize the columns of A to have unit l_2-norm
   b. Solve the optimization problem (7)
   c. Calculate the residuals $r_d(y) = ||y - A\delta_d(\hat{x})||_2^2, d = 1, 2, ..., m$
   e. $Label(y) = \underset{d \in [1,2,...,m]}{\arg\min} \ (r_d(y))$

---

where ε is a small error tolerance. In SRC, the characteristic function $\delta_d(\hat{x})$ maps to a new vector whose nonzero elements are the entries in $\hat{x}$ associated to the dth class [14]. Thus the query object y belongs to the dth class with minimum residual $r_d(y) = ||y - A\delta_d(\hat{x})||_2^2$ Finally, the classification result can be determined by calculating the minimum residual as

$$Label(y) = \underset{d \in \{1,2,...,m\}}{\arg\min} \ r_d(y) \ with \ r_d(y) = ||y - A\delta_d(\hat{x})||_2^2 \qquad (5)$$

3 Algorithm Description

The leave-one-out cross validation (LOO) method is one of the most effective approaches for model selection and parameter optimization [10]. The basic principle of LOO is to partition the dataset into N different training datasets each of which has one samples left out. The left-out sample is then employed to evaluate the trained model and measured generalization performance [11]. The predicted residual sum of squares (PRESS) is employed to compute the mean square error (MSE) of the LOO. The value of PRESS determined based on MSE (MSE^{PRESS}) is explained in [10]. The description and calculation of HAT matrix and optimal regularization parameter (λ_{opt}) are available on [10]. The pseudo code of cancer cell classification is given in Algorithm 3.

4 Dataset Description

We performed an experiment by taking the well-known GCM dataset. It contains the different gene expression profiles of 198 tumor samples representing 14 common human cancer types [1]. It is publically available on the website: http://www.broadinstitute.org/cgi-bin/cancer/datasets.cgi and it can also be downloaded from another website https://zenodo.org/record/21712. The training and testing set of these gene expression are used for cancer classification [13]. Thus, the combined dataset contains 198 samples with 16,063 genes. The brief description of dataset is shown in Table 1.

The gene expression data were bound into a specific box constraint ranging from 20 to 16,000 units. Those genes whose ratios and absolute variations across the samples were under 5 and 500, respectively were excluded [12] resulting in the gene expression dataset containing 11,370 genes (total dimension).

Algorithm 3. Proposed EA-SRC Classifier for cancer cell classification

Input: A training database A with m classes, a query object y, a candidate set of ELM regularization parameter λ $[\lambda_{\min}, \lambda_{\max}]$, $\tau > 0, threshold\,\sigma$,

Output: Class label y

1. Randomly generated hidden node parameters
2. Calculate $H(w_1, ..., w_L, x_1, ..., x_N, b_1, ..., b_L)$
3. Find the optimal λ_{opt} and give the corresponding output matrix $\hat{\beta}$
4. Calculate the output o of ELM with respect to y through
$o = H(w_1, ..., w_L, y, b_1, ..., b_L)\hat{\beta}$

if $o_f - o_s > \sigma$ **then**

$\quad\quad$ Label(y) = $\underset{d\in[1,...,m]}{\arg\max}(o)$

$\quad$ **else**

$\quad\quad$ Find the indexes of k largest entries in o;

$\quad\quad$ Implement the adaptive sub-dictionary A_y^* ;

$\quad\quad$ $\hat{x} = \arg\min_x \left\| A_y^* x - y \right\|_2^2 + \tau \|x\|_l$;

$\quad\quad$ **for** $d \in m(1), ..., m(k)$ **do**

$\quad\quad\quad$ Find A_d and $\delta_d(\hat{x})$;

$\quad\quad\quad$ Calculate the residuals $r_d(y) = \|y - A_d\delta_d(\hat{x})\|_2^2$;

$\quad\quad$ **end**

$\quad\quad$ Label(y) = $\underset{d\in\{m(1),...,m(k)\}}{\arg\min} r_d(y)$

$\quad$ **end**

---

5   Experimental Results and Discussion

We performed the experiments in two stages: first, by selecting highly expressive genes from gene-pool using ILFS algorithm before feeding to a classifier and second, taking all of the original gene expression data. The ILFS algorithm calculated the rank while considering all the possible subsets of genes. Based on available ranks of gene expressions, we performed LOO cross-validcatione in a chosen interval (25 features in each validation at a time). We also plotted the result of gene expression versus classification accuracy as shown in Fig. 2. We determined that classification accuracy increases consistently with the increment of gene expressions in a certain level in Fig. 2. We achieved around 74.75% accuracy in 30 s by consuming 725 gene expressions out of 11,370. It is only 6.38% in total expressions. More importantly, ILFS listed out the order of highly expressed genes which has the highest discriminative capability. In the next stage, cross-validation was performed by considering gene expressions (11,370) which resulted in 81.82% classification accuracy in 6.2 min. An accuracy is increased by around 7% by consuming the remaining 93% gene expressions (remaining all expression).

During the experiment, we also calculated the confusion matrices of both scenarios: discriminative (in which the classification accuracy is 74.75%) and all (accuracy is 81.82%). The confusion matrix is a matrix in which each column represents the instances in a predicted class, while each row represents

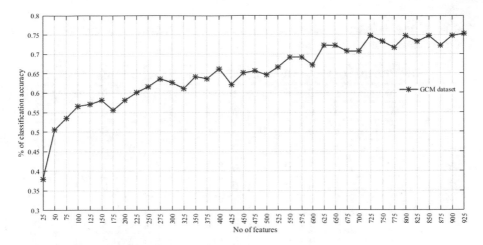

Fig. 2. Classification accuracy of distinctive gene expressions in different intervals (25 features in each validation) by using EA-SRC classifier

the instances in an actual class. The diagonal entries of the confusion matrix correspond to correctly classified rates, while the off-diagonal entries correspond to misclassification rates. We measured both overall classifications accuracies and confusion matrixes in both cases by using EA-SRC. The class-label distribution of discriminative gene expressions is presented in Table 2. The cancer

Table 1. GCM dataset contains 14 different cancer types and they have variable number of samples in each class

No. of class	Cancer types	Samples of each category
1	Breast adenocarcinoma (BR)	12
2	Prostate adenocarcinoma (PR)	14
3	Lung adenocarcinoma (LU)	12
4	Colorectal adenocarcinoma (CO)	12
5	Lymphoma (LY)	22
6	Bladder transitional cell carcinoma (BL)	11
7	Melanoma (ML)	10
8	Uterus adenocarcinoma (UT)	10
9	Leukemia,(LE)	30
10	Renal cell carcinoma (RE)	11
11	Pancreas adenocarcinoma (PA)	11
12	Ovarian adenocarcinoma (OV)	12
13	Pleural mesothelioma (MS)	11
14	Central nervous system (CNS)	20
Total		198

Table 2. The calculation of confusion matrix of GCM dataset by considering only highly expressive genes using EA-SRC classifier

	BR	PR	LU	CO	LY	BL	ML	UT	LE	RE	PA	OV	MS	CMS	Total
BR	8	0	0	1	0	0	1	1	0	0	0	1	0	0	12
PR	1	10	0	0	0	0	1	0	0	0	1	1	0	0	14
LU	4	1	5	1	0	0	0	0	0	0	0	1	0	0	12
CO	5	0	0	3	0	0	1	0	0	1	0	0	2	0	12
LY	0	0	0	0	20	0	0	0	1	0	0	0	0	1	22
BL	4	0	0	1	0	3	0	0	0	0	0	3	0	0	11
ML	1	0	0	0	1	1	5	0	0	1	0	0	1	0	10
UT	2	0	0	0	0	0	2	3	0	1	0	1	1	0	10
LE	0	0	0	0	0	0	0	0	30	0	0	0	0	0	30
RE	2	0	0	0	0	1	0	1	0	7	0	0	0	0	11
PA	3	0	0	0	0	1	0	0	0	0	7	0	0	0	11
OV	0	1	0	0	0	3	0	1	1	2	0	3	1	0	12
MS	0	0	0	0	0	0	0	0	0	0	0	0	11	0	11
CMS	0	0	0	0	0	0	0	0	0	0	0	0	0	20	20
Total	30	11	5	6	21	9	10	6	32	12	8	10	16	21	198

Table 3. Confusion matrix of GCM dataset by considering all gene expressions using EA-SRC

	BR	PR	LU	CO	LY	BL	ML	UT	LE	RE	PA	OV	MS	CMS	Total
BR	3	1	2	1	0	0	1	0	0	1	0	1	1	1	12
PR	0	11	2	0	0	0	0	0	0	0	0	1	0	0	14
LU	2	0	9	0	0	0	0	0	0	0	0	1	0	0	12
CO	0	0	1	11	0	0	0	0	0	0	0	0	0	0	12
LY	0	0	0	0	21	0	0	0	0	0	0	0	0	1	22
BL	1	1	0	0	0	7	0	0	1	0	0	1	0	0	11
ML	1	0	0	0	0	1	7	1	0	0	0	0	0	0	10
UT	1	0	0	0	0	0	0	8	0	0	0	1	0	0	10
LE	0	0	0	0	0	0	0	0	30	0	0	0	0	0	30
RE	0	0	1	0	0	1	0	0	0	9	0	0	0	0	11
PA	2	0	0	0	0	2	1	0	0	0	6	0	0	0	11
OV	0	2	0	0	1	1	0	1	1	0	0	6	0	0	12
MS	0	0	0	0	0	0	0	0	0	0	0	0	11	0	11
CMS	0	0	0	0	0	0	0	0	0	0	0	0	0	20	20
Total	10	15	15	12	22	12	9	10	32	10	6	11	12	22	198

Table 4. Sensitivity, specificity, precision, and F-score by considering the distinctive features of GCM datset using a EA-SRC

	Sensitivity	Specificity	Precision	F-score
BR	0.6667	0.8523	0.2667	0.3810
PR	0.7143	0.9843	0.8333	0.7692
LU	0.4167	1	1	0.7882
CO	0.2500	0.9778	0.5000	0.3333
LY	0.9091	0.9914	0.9524	0.9302
BL	0.2727	0.9565	0.3333	0.3000
ML	0.5000	0.9630	0.5000	0.5000
UT	0.3000	0.9778	0.5000	0.3750
LE	1	0.9813	0.9375	0.9677
RE	0.6364	0.9624	0.5833	0.6087
PA	0.6364	0.9922	0.8750	0.7368
OV	0.2500	0.9496	0.3000	0.2727
MS	1	0.9612	0.6875	0.8148
CNS	1	0.9914	0.9524	0.9756

sub-types namely colorectal adenocarcinoma (CO), bladder transitional cell carcinoma (BL), uterus adenocarcinoma (UT), and ovarian adenocarcinoma (OV) are highly overlap. In the same table, lung adenocarcinoma (LU) and melanoma (ML)have around 50% accuracies in each. Rest of cancer types have more than 70% accuracy with each other.

Similarly, Table 3 depicts the class-wise performance of cancer types by considering all gene expressions. Except for breast adenocarcinoma (BR), all other cancer types yielded a convincing classification accuracy (by taking all gene expression). Furthermore, we also calculated the sensitivity, specificity, precision, and F-score in both scenarios and presented in Tables 4 and 5.

We also compared the experimental results with semi-PNMF (semi-Supervised Projective Non-Negative Matrix Factorization) [12] in terms of sensitivity, specificity, precision, and F-score in both scenarios (selective and all features). The results of class-wise comparison are presented in Tables 6 and 7. It showed that the outcome of selective features is highly competitive with semi-PNMF. Later, we also measured the outcomes of all features than the semi-PNMF. Proposed scheme outperformed than the semi-PNMF the results of which are shown in Table 7. Finally, we compared accuracies with semi-PNMF which yielded only 70.71% accuracy [12]. The classification results of selective and better in all-features are 74.75% and 81.82% respectively. These results highlights that the proposed method is much efficient and accurate than semi-PNMF.

Table 5. Sensitivity, specificity, precision, and F-score by considering the all features of GCM dataset using a EA-SRC

	Sensitivity	Specificity	Precision	F-score
BR	0.25	0.9571	0.3	0.2727
PR	0.7857	0.9737	0.7333	0.7586
LU	0.75	0.9615	0.6	0.7586
CO	0.9167	0.9933	0.9167	0.9167
LY	0.9545	0.9928	0.9545	0.9545
BL	0.6364	0.9681	0.5833	0.6087
ML	0.7	0.9870	0.7778	0.7368
UT	0.8	0.9869	0.8	0.8
LE	1	0.9847	0.9375	0.9677
RE	0.8181	0.9934	0.9	0.8571
PA	0.5454	1	1	0.7159
OV	0.5	0.9683	0.5454	0.5217
MS	1	0.9933	0.9167	0.9565
CNS	1	0.9858	0.9091	0.9523

Table 6. Comparison of proposed method (only selected features) with semi-PNMF

	Class-wise performance of proposed method				Semi-pnmf result [12]			
	Sensitivity	Specificity	Precision	F-score	Sensitivity	Specificity	Precision	F-score
BR	**0.6667**	0.8523	0.2667	0.3810	0.416667	**0.944056**	**0.384615**	0.4
PR	**0.7143**	**0.9843**	**0.8333**	**0.7692**	0.5	0.956835	0.538462	0.518519
LU	**0.4167**	1	1	**0.7882**	0.416667	0.985401	0.714286	0.526316
CO	0.2500	0.9778	0.5000	0.3333	0.25	**0.992754**	**0.75**	**0.375**
LY	**0.9091**	0.9914	0.9524	0.9302	0.954545	**0.991667**	**0.954545**	**0.954545**
BL	0.2727	0.9565	0.3333	0.3000	**0.636364**	**0.963768**	**0.583333**	**0.608696**
ML	0.5000	0.9630	0.5000	0.5000	**0.9**	**0.992424**	**0.9**	**0.9**
UL	0.3000	**0.9778**	**0.5000**	0.3750	**0.9**	0.916084	0.428571	**0.580645**
LE	1	0.9813	0.9375	0.9677	0.966667	1	1	**0.983051**
RE	0.6364	**0.9624**	**0.5833**	**0.6087**	**0.727273**	0.891892	0.333333	0.457143
PA	**0.6364**	0.9922	**0.8750**	**0.7368**	0.545455	**0.992593**	0.857143	0.666667
OV	**0.2500**	0.9496	0.3000	**0.2727**	0.083333	**0.992857**	0.5	0.142857
MS	1	0.9612	0.6875	0.8148	0.909091	**0.992366**	**0.909091**	**0.909091**
CNS	1	**0.9914**	**0.9524**	**0.9756**	1	0.97561	0.869565	0.930233

Table 7. Comparison of the outcome by considering all features with semi-PNMF

	Class-wise performance of proposed method				Semi-pnmf result [12]			
	Sensitivity	Specificity	Precision	F-score	Sensitivity	Specificity	Precision	F-score
BR	0.25	**0.9571**	0.3	0.2727	**0.416667**	0.944056	**0.384615**	0.4
PR	**0.7857**	**0.9571**	**0.7333**	**0.7586**	0.5	0.956835	0.538462	0.518519
LU	**0.75**	0.9615	0.6	**0.7586**	0.416667	**0.985401**	**0.714286**	0.526316
CO	**0.9167**	**0.9933**	**0.9167**	**0.9167**	0.25	0.992754	0.75	0.375
LY	**0.9545**	**0.9928**	0.9545	0.9545	0.954545	0.991667	0.954545	0.954545
BL	**0.6364**	**0.9681**	0.5833	0.6087	0.636364	0.963768	0.583333	0.608696
ML	0.7	0.9870	0.7778	0.7368	**0.9**	**0.992424**	**0.9**	**0.9**
UL	0.8	**0.9869**	**0.8**	**0.8**	0.9	0.916084	0.428571	0.580645
LE	1	0.9869	0.9375	0.9677	0.966667	1	1	**0.983051**
RE	**0.8181**	**0.9847**	**0.9375**	**0.9677**	0.727273	0.891892	0.333333	0.457143
PA	0.5454	1	1	**0.7159**	0.545455	0.992593	0.857143	0.666667
OV	0.5	0.9683	**0.5454**	**0.5217**	0.083333	**0.992857**	0.5	0.142857
MS	1	**0.9933**	**0.9167**	**0.9565**	0.909091	0.992366	0.909091	0.909091
CNS	1	**0.9858**	**0.9091**	**0.9523**	1	0.97561	0.869565	0.930233

6 Conclusion

Our model proved the complement strength of two distinct classifiers for cancer classification. We complimented the speed advantage of ELM with discriminability advantage of SRC. Initially, cancer cell was classified using a regularized ELM where output difference was greater than a threshold. Otherwise, SRC was adopted for the classification. This strategy significantly reduced the testing time of SRC because it only tested limited data whose output difference was less than a threshold. Similarly, classification accuracies was also measured in two stages: considering only highly expressed and all genes to show the outcome difference by using an EA-SRC classifier. We validated the proposed method by using GCM dataset. The recorded classification accuracies of GCM dataset are 74.75% and 81.82% in 30 s and 6.2 min. Most importantly, 74.75% accuracy was achieved by considering 725 highly expressive genes out of 11370 in a significantly reduced validation time for the selective genes. Indeed, this proposed method minimizes the computational burden in big data and provides a list of the highly expressive genes making it feasible to predict about a cancer type with higher accuracy. In the meantime, this method also justifies that we have to consider all features for tumor classification because human health is not comparable to computational time and cost even though we achieved 74.75% accuracy by using 725 genes.

From the big data perspective, we covered velocity, volume, and value, however, we are still far from the variety and veracity. Our immediate future goal is to integrate these remaining factors for examining the impact of tumor/cancer classification.

References

1. Islam, S.R.M., Kwak, D., Kabir, H., Hossain, M., Kwak, K.-S.: The internet of things for health care: a comprehensive survey. IEEE Access **3**, 678–708 (2015)
2. Lin, K., Xia, F., Wang, W., Tian, D., Song, J.: System design for big data application in emotion-aware healthcare. IEEE Access **4**, 6901–6909 (2016)
3. Huang, G.-B., Zhu, Q.-Y., Siew, C.-K.: Extreme learning machine: theory and applications. Neurocomputing **70**, 489–501 (2006)
4. Roffo, G., Melzi, S., Castellani, U., Vinciarelli, A.: Infinite latent feature selection: a probabilistic latent graph-based ranking approach. In: IEEE International Conference on Computer Vision (ICCV), pp. 1407-1415 (2017)
5. Baniya, B. K., Lee, J., Li, Z.: Audio feature reduction and analysis for automatic music genre classification. In: 2014 IEEE International Conference on Systems, Man and Cybernetics (SMC), pp. 457–462 (2014)
6. Baniya, B.K., Lushbough, C., Gnimpieba, E.: Significance of reduced features for subcellular bioimage classification. In: International Symposium on Bioinformatics Research and Applications (ISBRA), Minsk, Belarus (2016)
7. Sahoo, P.K., Mohapatra, S.K., Wu, S.L.: Analyzing healthcare big data with prediction for future health condition. IEEE Access **4**, 9786–9799 (2016)
8. Liu, H., Guo, D., Sun, F.: Object recognition using tactile measurements: kernel sparse coding methods. IEEE Trans. Instrum. Measure. **65**(3), 656–665 (2016)
9. Cao, J., Lin, Z., Huang, G.B., Liu, N.: Voting based extreme learning machine. Inf. Sci. **185**(1), 66–77 (2012)
10. van Heeswijk, M., Miche, Y.: Binary/ternary extreme learning machines. Neurocomputing **149**, 187–197 (2015)
11. Cao, J., Zhang, K., Luo, M., Yin, C., Lai, X.: Extreme learning machine and adaptive sparse representation for image classification. Neural Netw. **81**, 91–102 (2014)
12. Zhang, X., Guan, N., Jia, Z., Qiu, X., Luo, Z.: Semi-supervised projective non-negative matrix factorization for cancer classification. PLOS One **10**, 1–20 (2015)
13. Ramaswamy, S., Tamayo, P., Rifkin, R., Mukherjee, S., Yeang, C.-H., Angelo, M., Ladd, C., Reich, M., Latulippe, E., Mesirov, J.P., Poggio, T., Gerald, W., Loda, M., Lander, E.S., Golub, T.R.: Multiclass cancer diagnosis using tumor gene expression signatures. Proc. Natl. Acad. Sci. **98**(26), 15149–15154 (2001)
14. Cao, J., Lin, Z.: Bayesian signal detection with compressed measurements. Inf. Sci. **289**, 241–253 (2014)

Forecasting Food Sales in a Multiplex Using Dynamic Artificial Neural Networks

V. Adithya Ganesan[1], Siddharth Divi[1(✉)], Nithish B. Moudhgalya[1],
Uppu Sriharsha[2], and Vineeth Vijayaraghavan[3]

[1] Sri Sivasubramaniya Nadar College of Engineering, Chennai, India
{v.adithyaganesan,siddharthdivi,nithish.moudhgalya}@ieee.org
[2] Rajalakshmi Engineering College, Chennai, India
uppusriharsha@ieee.org
[3] Solarillion Foundation, Chennai, India
vineethv@ieee.org

Abstract. In India, food sales are emerging to be a major revenue generator for multiplex operators currently amounting to over \$367 million a year. Efficient food sales forecasting techniques are the need of the hour as they help minimize the wastage of resources for the multiplex operators. In this paper, the authors propose a model to make a day-ahead prediction of food sales in one of the top multiplexes in India. Online learning and feature engineering by data correlative analysis in conjecture with a densely connected Neural Network, address the concept drifts and latent time correlations present in the data respectively. A scale independent metric, η_{comp} is also introduced to measure the success of the models across all food items from the business perspective. The proposed model performs better than the traditional time-series models, and also performs better than the corporate's currently existing model by a factor of 7.7%. This improved performance also leads to a saving of 170 units of food everyday.

Keywords: Time-series · Neural networks · Concept drift ·
Forecasting · Feature engineering · Online learning

1 Introduction

In India, there are more than 6,000 single movie-screens and around 2,500 multiplex screens, selling more than 2 billion tickets every year. The food sales in these multiplexes, amount to over \$1.1 million a day, contributing to almost a quarter of the theaters' revenue. In the last half decade alone, this contribution increased by over 4%. This increase in the food sale coupled with its high profit margins makes it a very important revenue stream to the multiplex operators.

An excess of food leads to wastage of both food and money, as most of the food sold at the theaters have a shelf life of less than a day, after which they

© Springer Nature Switzerland AG 2020
K. Arai and S. Kapoor (Eds.): CVC 2019, AISC 944, pp. 69–80, 2020.
https://doi.org/10.1007/978-3-030-17798-0_8

are disposed off. On the other hand, shortage of food leads to a potential loss of revenue for the operator. This necessitates an efficient system that can forecast the food sales at the multiplex to minimize wastage of food and money.

The food sale values in real world scenarios, depend on mainly three class of factors:

- Movie based factors - rating, genre, star cast, release date, show-time, etc.
- Crowd based factors - age groups, sex ratio, area of residence, personal preferences, etc.
- External/generic factors - weather, climate, government holidays, financial and tax based laws/bills, school or college exams, etc.

The external factors are elaborated further in Sect. 6.2. The movie based factors are either too subjective or too variant and keeps changing from time to time. To ensure non-disclosure of personally identifiable information (PII) of the customer, the crowd based details are abstracted while extracting data from the database. Thus forecasting the food sales using such limitations and constrained feature set is a challenge that needs to be tackled while modeling the corporate data. The multiplex operators placed orders for various ingredients from markets and confectioneries on a daily basis. These necessitate the requirement for a robust and effective model to make daily food sale predictions and to tackle the complexities and constraints present in the corporate data.

2   Related Works

Forecasting of food sales has traditionally been a tricky problem because factors like price, changing consumer preferences et al, affect the sales, and oftentimes these are hard to model [1]. Adebanjo et al. [2], have found that 48% of food companies are poor at forecasting. Sales forecasting has usually been approached using a variety of methods. It ranges from linear mathematical models like ARIMA, Linear Regression etc. to non-linear time series models like Artificial Feed Forward neural Networks. With the introduction of Long Short-Term Memory architecture in 1997, the Deep Neural Network models with Recurrent Layers have also been deployed to make sales forecasting better. However, a separate family of belief exists wherein the sales forecasting is done using features that could determine the values at logical levels. These include tree based models like Extreme Random Forest regressors and XGBoost models, and domain based models like Expert Systems.

Linear models, like ARIMA [3] are limited by their assumption of linear characteristics of the data and are not always satisfactory [4]. To overcome this problem, fuzzy logic systems and ANNs [5] are used, which can build generic model structures that can handle complex data. Moreover, a number of studies have been conducted to determine which method would be the best for forecasting, and the results were not particularly in favor of one single method [6]. Stock and Watson [7] in one of their studies had found that in terms of forecasting performance, a combination of non-linear methods were better than a combination of linear methods. In addition to this, ANNs were found to perform better than or equal to the traditional methods in more than half of the instances. Zhang [4] pointed out that no single method can work in every scenario and that a combination of models is more effective.

Usually in sales forecasting, tree based models are used, that factor independent relations of the features with sales during the tree construction. In the Kaggle competition for restaurant sales prediction [8], the best kernel was a Gradient Boosting model. Even here, the winning solution consisted of an ensemble of weak learners powered by gradient boosting. In another Kaggle competition to predict the grocery sales [9], the winning solution consisted of an ensemble of 2 Linear Gradient Boosting models and 2 Neural Network models with different training methodologies. An interesting point to note here is that it is mentioned that not all features were useful in building the model. In both these cases, we observe that the right set of features helped in improvising on any baseline mathematical models. However they fail when the features get constrained, i.e when there are very less features and they do not offer substantial information. Expert systems have also been used [10] where statistical reasoning and logics are used to help in the prediction of food sales. The above stated examples just reinforce the fact that there is no one single method which works in all scenarios of forecasting. Given the complexity in this real world dataset, the knowledge base would have to be altered very often to accommodate the frequent shifts in patterns and since the logic used to represent this knowledge is sophisticated, it is difficult to accommodate such changes.

In this paper, the authors propose an Artificial Neural Network (ANN) model to predict the food sales in the multiplex. To deal with frequent variations in patterns and recognize latent time correlations, we trained the model dynamically, with recent data segments. Data correlation analysis was done to engineer compound features that provide better insights on data patterns.

A model to make a day ahead prediction of food sales in a multiplex that is capable of handling concept drifts has been proposed in this paper. The authors first describe a scale independent metric to evaluate the model's performance. Following this, the dataset used and the outcomes of the analysis have been explained with relevant representations. The features engineered are listed in the forthcoming section. Finally, the strategy adopted to tackle the phenomenon of concept drift has been explained along with the model's performance.

3 Metric Introduction

Conventional metrics like Mean Absolute Error (MAE) and Mean Absolute Percent Error (MAPE) portray an incomplete picture of the model's efficiency. The results projected by MAE lack contextual meaning without the range of the value being predicted. The MAPE values are inflated, even for acceptable errors, when the target variable takes a small range of values. The new metric proposed, η_{comp}, takes into account the range of sales for each food, thus giving a better representation of the model efficiency across food items and also weights the values to avoid inflation of error. For any food item f, η_{comp} is computed as,

$$\eta_{comp_f} = MAPE_f * \frac{\sum_{i=1}^{n} SaleOfFood_f}{\sum_{i=1}^{n} \sum_{j=1}^{m} SaleOfFood_j} \qquad (1)$$

where, n is the total number of days and m is the total number of food items considered.

Table 1 explains the different scenarios that traditional metrics fail to depict and how η_{comp} captures it. Lower the metric value, better the model's performance.

Table 1. Comparing metrics on different case scenarios

Average food sale	MAE	MAPE	Weight	η_{comp}
100	5 (very good)	5	0.2	1
40	5 (good)	12.5	0.08	1
100	15 (bad)	15	0.2	3
40	15 (very bad)	37.5	0.08	3

The MAPE is multiplied by a weight factor, calculated using the sales contribution of that food item. This helps to compare various food items despite their varying sales values. Two similar numbers on the scale, could only mean either of the following. One, a high MAPE on a low yielding food item or two, a comparatively low MAPE on a high yielding food item.

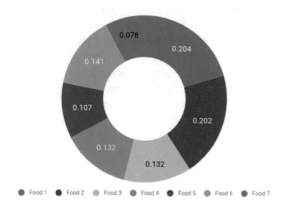

Fig. 1. Weight ratios of food items

The pie chart in Fig. 1 shows the weight ratios of the seven food items considered.

4   Dataset Description

The dataset is a collection of transactions that pertain to various food items sold at the counters of the multiplex, which on an average caters to about 4000 people everyday. The database maintained by the multiplex operator spreads over 3 years, from 2015–2017 comprising nearly 13 million transaction records. Each transactional record consists of the time-stamp of the sale, a unique transaction id, food item code, quantity and its price as shown in Fig. 2. The database also consists of the occupancy for each show at the multiplex and the language of the movies screened.

Time-stamp	Transaction ID	Food item Code	Quantity sold	Price per piece

Fig. 2. Sample transactional record

The transactional data was grouped by food items followed by day-wise sales aggregation. It was found that each food item had its own characteristic patterns that could best be modeled independently, and also over 93% of the food sales was constituted by just seven out of 60 items. Thus, only these items were considered for modeling crowd behavioral patterns of food sales.

5   Data Mining and Analysis

Conventional time series datasets exhibit certain traits or characteristics that are repetitive over short and long periods of time. These could be broadly classified as trends, stationarity and seasonality, wherein the trends showcase the short term characteristics of the data while the seasonality and stationarity showcase its long term characteristics. The violin plot in Fig. 3(a) shows the probability distribution of food sales of a particular food item across the days of a week. The denser region of each violin plot implies that the distribution of sales is maximum in those range of values. The tips of the violin plots shows the range of values without the removal of outliers and noise from the data. It can be observed that Fridays, Saturdays and Sundays have their denser regions higher than any other day implying that the sale values on these days could be higher thus showcasing a form of weekday-weekend trend. It is also observable that the sale distributions on a weekday - Monday, Tuesday, Wednesday and Thursday, are almost identical thus making the day of the week a poor discriminating feature. The bar chart in Fig. 3(b) shows that every weekday have somewhat similar values. The pie chart in Fig. 4 shows that on an average, weekend sales are $(2/3)^{rds}$ of weekday sale values. These show that there exists a faint time series characteristic in the data.

The multiplex provides the visiting customers a multi-storeyed entertainment center with air conditioned hallways, thus isolating them from the environment outside. The isolation attenuates the impact of weather and season based factors

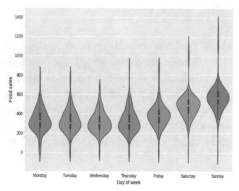

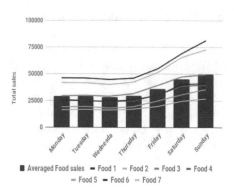

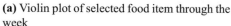

(a) Violin plot of selected food item through the week

(b) Similarity among weekdays

Fig. 3. Short-term characteristics

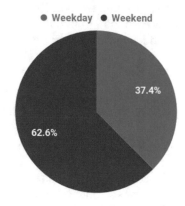

Fig. 4. Pie chart showing weekday v/s weekend average sale ratio

that could influence the food choices of the crowd, which is one of the reason why there is no long term periodicity. The food sale value to a great extent, depends on PII that describes the nature of the crowd visiting the complex - the ratio of male to female audience, age groups of the audience, their personal preferences and movie specific information like genre of the movie, language of the movie et al. Thus the food sales in a multiplex has a distorted and diminished long term time series characteristics. As observed in Fig. 5, there exists insignificant monthly, seasonal, and yearly pattern among the sale values of a particular food item under consideration.

A stationary time series is one whose statistical properties such as mean, variance etc. are all constant over time. Most statistical forecasting methods are based on the assumption that the time series can be rendered approximately stationary through the use of mathematical transformations. Figure 6(a) shows

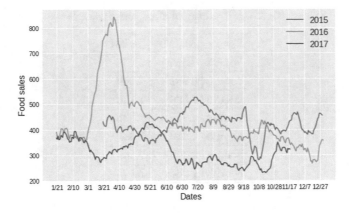

Fig. 5. Year wise sales patterns

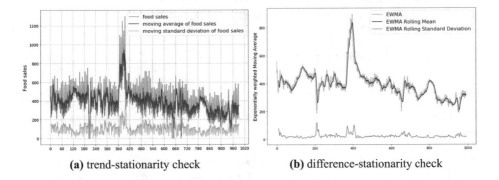

(a) trend-stationarity check   **(b)** difference-stationarity check

Fig. 6. Stationarity check

that statistical measures like standard deviation and mean are not even close
to constant for the given data. It is also observable from Fig. 6(b), differencing
and transforming the data into period-to-period differences using log-transform,
log-difference transform or exponentially weighted moving average, doesn't make
the aforementioned statistical measures constant. Hence it can be concluded that
the data is neither trend-stationary nor difference-stationary and cumulatively,
it not a stationary time series dataset.

The poor performance of S-ARIMA (Seasonal Autoregressive Integrated
Moving Average) model presented in Table 2 shows that there exists no constant
auto-correlative characteristics in the data, thus validating the lack of stationarity in the data. Moreover, the undeniably poor performances of LSTM (Long
Short-Term Memory) models as observed from Table 2, proves that there exists
no long short-term characteristics that the LSTMs could capture. Both these
are conventional approaches to time series forecasting and their failure proves

Table 2. Traditional time series models

Food	ARIMA	LSTM
Food 1	4.550	5.251
Food 2	3.951	4.049
Food 3	2.560	4.138
Food 4	2.449	2.531
Food 5	2.619	2.631
Food 6	2.459	2.655
Food 7	1.584	1.531

that the dataset at hand needs to be treated as both time series forecasting and regression modeling.

The above reasons make it necessary to approach such complex time sequenced datasets in a more non-conventional methodology.

6   Feature Engineering

6.1   Feature Set

The aim of building the feature set for the model is to make it movie agnostic because of two major reasons:

I The attributes relating to each movie could be distinct and might not to be categorical or quantitative but could be subjective. Movie specific features like cast, crew, review etc. were highly subjective and simply increase the complexity.

II There were several movies that were running at the same time in a multiplex and the food counters are common to them. Aggregation of movie specific information might lead to ambiguity due to its subjective nature.

Keeping this in mind, the feature set was aimed to have quantitative measures that directly impact the food sale. The occupancy of the movies had a direct implication on the food sale. Hence, the total occupancy of the day was extracted from the database.

Also, the language of the film played affects the crowd visiting the multiplex and their food preferences, thus affecting the food sale patterns greatly. These differences in patterns show that the food sales can be mapped as a function of language. Hence the day wise occupancy extracted was split into four main language wise occupancies namely English, Hindi, Regional and Others. Moreover, language-wise occupancy provides more specific characteristics of the crowd than what total occupancy could provide. Figure 7 shows comparatively lower error rates obtained using language wise occupancy as opposed to the total occupancy as the feature set for the ANN model.

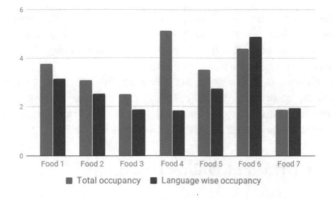

Fig. 7. η_{comp} comparing total and language wise occupancy as features

Section 5 highlights the existence of the weekday-weekend trend. Hence to incorporate this information, a boolean categorical variable to signify whether a day is a weekend or not was added to the feature set.

Aiming to capture the short-term trends in the data, a new set of features was incorporated into the existing set that accounted for the change in the amount of food sold per person on an average. Moving average of the ratio of food sale to the total occupancy was taken over a set of varying window sizes to account for the aforementioned behavior. The results of this feature set have been tabulated in Table 3. It is observed that, given the same feature set, ANN model outperforms XGBoost (Extreme Gradient Boosting) and ET (Extremely Randomized Trees) models across most food items.

Table 3. η_{comp} comparison of models after feature engineering

Food items	XGBoost	Extra trees	Static ANN
Food 1	4.719	5.995	3.135
Food 2	2.982	4.287	2.527
Food 3	3.250	3.706	1.875
Food 4	1.726	1.882	1.831
Food 5	1.660	1.971	2.751
Food 6	1.663	1.724	4.866
Food 7	1.275	1.683	1.960

6.2 External Factors

A lot of external factors like concurrent airing of a high TRP program along with a good movie, a public strike, university or school exams, any financial policies or bills passed by the State or Central Government affect the food sale patterns directly or indirectly. However it is infeasible to quantify the tangible impact of these factors on the food sales. These factors influence the sale values at discrete instances of time thus disrupting the continuity in the values and making it harder to quantify the actual sales.

Also the choices of the crowd are affected by age, sex, profession, socio-economic status, residential locality and other diverse factors (that can't be mined keeping in mind privacy concerns), which are rudimentary in modeling user behavior patterns are infeasible to mine and quantify.

7   Concept Drift and Online Learning

Data collected in real-world scenarios with non-stationary environments exhibit Concept Drift [11], a situation when the relation between the input and target variables change over time. In our data, an instance of this phenomenon can be observed from Fig. 6, where the food sales rise for a duration of 50 days marked by the x-axis range - (360, 410). To tackle concept drift, online learning is used every 15 days with an overlap of 7 days with the previous training period. This helps the model focus more on the recent trends in the data thus giving it relevant context. Table 4 shows the comparison of online learning v/s static training across all food items for the ANN model and it is evident that online learning method outperforms the other algorithms and techniques.

8   Results and Observations

1. From Table 4, it is observable that the ANN model outperforms LSTMs and tree based models given the same feature set. LSTMs try to capture the non-linear correlations between the features and target sales in time domain. However since the data possesses no distinct long term time series characteristics, the memory unit in LSTMs fail to adapt to the changes.
2. The regression trees constructed in Extra trees or XGBoost algorithms use statistical measures to find the split points in the data and use logical boolean conditions to split the nodes at each level to determine the sale values in the leaf nodes. To determine which column to choose as the split feature and assign it to a node, only independent correlations of features with target sales is considered in these models. However, the statistical measures as observed from Sect. 5 prove to be fluctuating frequently making it harder for these models that use such measures to improve their efficiency.
3. ANN combines the properties of both LSTMs and tree based models that use features' correlations with the target values in a non-linear and multi-dimensional level thus giving an upper hand over both the algorithms mentioned. ANNs use the given feature set and transform them to higher order combinations that could better distinguish the data points [12]. These higher order features prove to be useful in capturing latent time correlations of food sales prediction as compared to LSTMs and Extra Tree models.

Table 4. η_{comp} comparison results of all models

Food items	LSTM	Extra trees	ANN static	ANN dynamic
Food 1	5.251	5.995	3.135	2.641
Food 2	4.049	4.287	2.527	2.011
Food 3	4.138	3.706	1.875	1.497
Food 4	2.531	1.882	1.831	1.381
Food 5	2.631	1.971	2.751	1.597
Food 6	2.655	1.724	4.864	1.620
Food 7	1.531	1.683	1.960	1.170

9   Conclusion

From the results above, it is evident that in real-time datasets traditional models like ARIMA or S-ARIMA due to lack of features fail to capture the variance in the trends. Also, the deep sequence models like recurrent networks also fails in this case as no long term characteristics of the data exists or is visible, Thus an ANN model that combines the cross correlations of features and target values and represents them in higher dimensions disentangles the variations present in the data better than the former methods mentioned. With basic features that could aggregate the crowd characteristics and online training outperforms the current model of the corporate by 7.7% as shown in Fig. 8. The proposed model helped the multiplex save on an average, around 170 units of food everyday which translates to a saving of $450000 over a period of nine months.

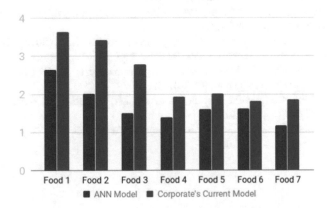

Fig. 8. η_{comp} comparing dynamic ANN and corporate's current model

10   Future Works

Every multiplex keeps track of their occupancy and food sales. As these form the basic feature set in this work, this model has the potential to be applied to datasets collected from other multiplex operators.

The efficiency of the current ANN model can be further improved by factoring in external variables mentioned in Sect. 6.2. Moreover, a ensemble of linear and non-linear models could be used for forecasting, that could better capture the intricacies present in the data.

References

1. Vorst, J., Beulens, A., Wit, W., Beek, P.: Int. Trans. Oper. Res. **5**(6), 487. https:// doi.org/10.1111/j.1475-3995.1998.tb00131.x, https://onlinelibrary.wiley.com/doi/ abs/10.1111/j.1475-3995.1998.tb00131.x
2. Adebanjo, D., Mann, R.: Benchmarking: Int. J. **7**(3), 223 (2000). https://doi.org/ 10.1108/14635770010331397
3. Box, G.E.P., Jenkins, G.: Time Series Analysis, Forecasting and Control. Holden-Day Inc, San Francisco (1990)
4. Zhang, G.P. (ed.): Neural Networks in Business Forecasting. IGI Global, Hershey (2003)
5. Haykin, S.: Neural Networks: A Comprehensive Foundation, 1st edn. Prentice Hall PTR, Upper Saddle River (1994)
6. Doganis, P., Alexandridis, A., Patrinos, P., Sarimveis, H.: J. Food Eng. **75**(2), 196 (2006). https://doi.org/10.1016/j.jfoodeng.2005.03.056, http://www.sciencedirect. com/science/article/pii/S0260877405002402
7. Stock, J.H., Watson, M.W.: J. Monetary Econ. **44**(2), 293 (1999). https://ideas. repec.org/a/eee/moneco/v44y1999i2p293-335.html
8. Arsenal: Restaurant Revenue Prediction (2015). https://www.kaggle.com/c/ restaurant-revenue-prediction/
9. Eureka: CorporaciÓn Favorita Grocery Sales Forecasting (2018). https://www. kaggle.com/c/favorita-grocery-sales-forecasting/
10. Zliobaite, I., Bakker, J., Pechenizkiy, M.: Expert Syst. Appl. **39**(1), 806 (2012). https://doi.org/10.1016/j.eswa.2011.07.078
11. Gama, J., Zliobait, I., Bifet, A., Pechenizkiy, M., Bouchachia, A.: ACM Comput. Surv. (CSUR) **46**(4), 44 (2014)
12. Pascanu, R., Gülçehre, Ç., Cho, K., Bengio, Y.: CoRR **abs/1312.6026** (2013). http://arxiv.org/abs/1312.6026

Automatic Detection of Vibration Patterns During Production Test of Aircraft Engines

Jérôme Lacaille[1]([envelope]), Julien Griffaton[1]([envelope]), and Mina Abdel-Sayed[2]([envelope])

[1] Safran Aircraft Engines, 77550 Moissy-Cramayel, France
{jerome.lacaille, julien.griffaton}@safrangroup.com
[2] Université Paris Diderot, 75013 Paris, France
mina.abdelsayed.7@gmail.com

Abstract. Every day, several aircraft engines exit Safran plant in the south of Paris. Each engine is assembled and sent for a last bench test before shipment to the aircraft. Among all operations implemented during this hour length phase and after the first run-in, we realize a slow acceleration and a slow deceleration. During those two steps the engine is almost stabilized and a Fourier spectrogram may be evaluated for each rotating speeds. The engine has two shafts and two accelerometers. Hence for each operation we draw four spectrograms with a rotating speed on the x-axis and frequency ordinate. These images are transferred to a team of specialized engineers whose work is to look at the spectrograms and annotates zones they found unusual or patterns they already saw on past observations. Then if something may be of concern, a decision is taken to send the engine back for inspection. This still is today's process, but several years ago we tried image analysis algorithms on the spectrograms and discover a way to help our team in the detection of abnormal patterns. We finally implement two algorithms which employed together give a pretty good detection of interesting zones. Subsequent sections in this paper will present the original data, show some abnormal patterns and describe our set of algorithms. Now, Safran develops a big-data environment with a datalake that capitalizes all measurements and our production test vibrations analysis should be operational in the coming years among lots of other algorithms.

Keywords: Spectrograms · NMF · Wavelets · Curvelets

1 The Original Data

Most damages due to rotating parts like bearings or gears appear in the very beginning of an aircraft engine life. In fact some IFSD (In Flight Shut Down) are the result of oil pollution by particles in a bearing during the assembly of the low pressure shaft or a scratch initiated by an inadequate tool manipulation. Sure enough, we implement embedded algorithms on the engine computer, the FADEC (Full Authority Digital Engine Control) and those algorithms most of the time detect the problem, but as anyone can understand it is a lot better to prevent a shop visit by a good inspection during the production test.

As one can see on Fig. 1, it is impossible to detect a big difference between a normal engine and an engine with a problem just looking at the raw data. The acceleration (or deceleration) tests are done with a very slow acceleration rate of about

© Springer Nature Switzerland AG 2020
K. Arai and S. Kapoor (Eds.): CVC 2019, AISC 944, pp. 81–93, 2020.
https://doi.org/10.1007/978-3-030-17798-0_9

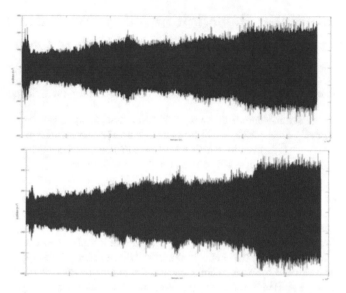

Fig. 1. Those two images represent acquisition from accelerometers during a slow acceleration phase from idle to max power. The top one (blue) is a normal signal and the bottom one (red) is abnormal.

40 rpm/s enabling the computation of spectrum with a fast Fourier transform (FFT) algorithm. This is done on temporal windows of 0.32 s which with a 51.2 kHz sensor gives already 16384 points and is sufficient to observe our frequencies of interest.

Each window should be associated to a shaft speed. The rotation speed is acquired from a phonic wheel which signal is converted in electric tension (see Fig. 2).

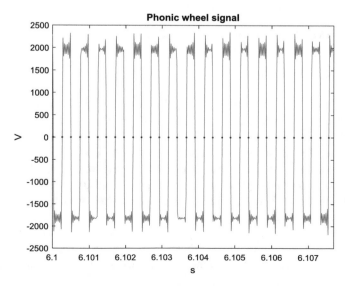

Fig. 2. On this figure one can observe the passage of each phonic wheel tooth.

The rotation speed is estimated as the average delay between two periods of the signal computed from zero-crossing detection.

Finally, the spectrums are computed at regularly spaced (10 rpm) samplings of the high pressure shaft rotating speed to form the spectrogram. Each spectrogram is an image with 3200 lines and 500 columns hence we are dealing with very high dimension (1.6 million of pixels) but those images show many specific regularities as will be explained below.

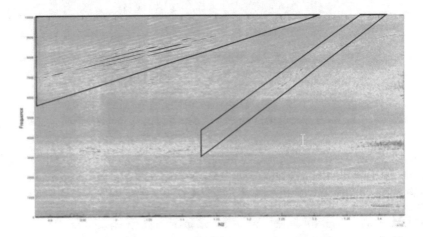

Fig. 3. A normal spectrogram (frequencies on y-axis) sampled according to high pressure (HP) shaft N2 (x-axis). The black triangle on top-left shows some HP shaft rays and the blue trapeze zone in the middle-right corresponds to low pressure (LP) shaft rays N1.

Figure 3 presents such a spectrogram where normal rays are identified. They correspond to the shaft rotation speed harmonics. As we said before, this engine has two shafts which rotation speeds are approximately linear during most of the operation regimes; the relationship is the result of aerodynamic balance between turbines and compressors. On this image, horizontal bands correspond to structural modes of some engine parts and combustion noise. Rays with a rational slope multiple of a shafts speed corresponds to blade passing frequencies and gear mesh frequencies harmonics, while rays that are almost synchronous with a multiple of a rotating speed may be the excitation of a damage in a bearing. Rays may also be a linear combination of the two rotations speeds for inter-shaft bearings.

Figure 4 represents the exact same acquisition as the one on Fig. 3 but from a defective engine. We add a red rectangle which zoom is shown on the next image (Fig. 5).

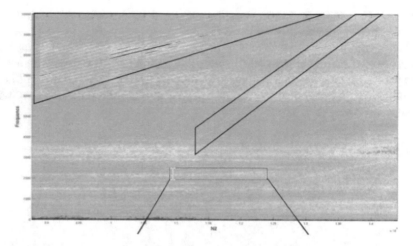

Fig. 4. The same spectrogram acquired on an engine with a damaged bearing.

In fact damages are approximately localized and the experts already know where to look at them in the spectrogram. The tools we defined use this a priori knowledge (rough location, continuity and parallelism of rays) by specifically looking at small parts of the spectrograms (for example 128 × 128 pixels) instead of the whole image. The next image (Fig. 6) shows example of abnormality signatures. The automatic detection will look specifically at rectangular parts positioned around those known locations.

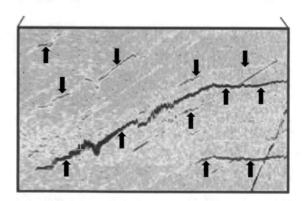

Fig. 5. A zoom of a pattern representative of some sort of damage. Each black arrow corresponds to an anomaly observed by at least one mechanical vibration expert.

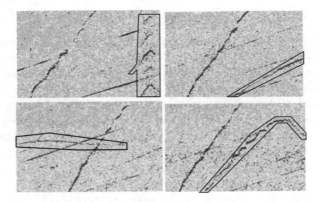

Fig. 6. Four example patterns of anomalies.

2 Our Methodology

Mechanical problems in fact are really rare on aircraft engines and instead of using a classification tool to detect the abnormal patterns we select to model the statistic distribution of normal images and hope to discover abnormalities by observing a lower likelihood value when the observation needs some review. The way we want to proceed is to automatically identify and select spectrograms needing review and helping specialized engineers to take a decision. As our production almost double the last couple of years we really need to ease the engineers work.

2.1 How to

Identifying normality in images is a complex challenge and if it can be realized with artificial intelligence algorithms it may be very difficult to transfer as an operational tool. To facilitate confidence by our experts we decide to use a reconstruction method. The idea is to use the model to compress the data, letting the mathematics find regularities in the observations, and then use a reconstruction method and compare the result to the original image. The "bigger" is the difference between input and reconstruction, more important should be the abnormality. Our score is based on this difference and the experts can see by themselves a proposition of "normal" reconstruction and compare with the original data.

It seems important to note the limited amount of data for this problem. We produce around 500 engines each year, so is the number of available spectrograms. Hence, a solution based only on data like deep neural networks was doubtful. Our experience shows that the use of knowledge is a real gain in the quality of the result, particularly to avoid just retrieving well-known facts, which may be a killer for any industrial application.

Our goal is to capture the regularities behind the observations. Two approaches were tested, each one with its own advantages and drawbacks. Finally we use both of them with a fusion method to suppress ambiguities and improve the global detection level [1].

2.2   Related Work

Different teams already tried some abnormality detections based on image processing. In [2], Clifton and Tarassenko first identify the noise in spectrograms as a probabilistic Gamma distribution, then after separation of noise and significant spectral rays, they classify the patterns as normal (usual) and abnormal.

Klein, Masad, Rudyk and Winkler [3] use a representation of the spectrum in order domain which can be computed from FFT or even wavelets. In this new representation, they estimate a baseline distribution to identify abnormal patterns as outliers.

Finally, in [4], Griffaton, Picheral and Tenenhaus combine the previous steps in an efficient tool. They first eliminate the background noise using a Gamma filter as in [2], then they suppress known observations link to normal physical behavior and apply a last image analysis using local baseline estimation like in [3] to identify abnormal patterns.

In those previous works, the image-processing step was either a filter, or a baseline distribution estimation. Our new proposal is to use underlying relations stored in the specific images we analyse by inserting a compression step to reduce de dimensionality of the problem. For this purpose, we propose two algorithms that complement each other: NMF and curvelets. The first one uses a set of specific observations to build an ad-hoc dictionary and the second exploits an a-priori knowledge given by the choice of the representation basis.

2.3   About Validation and Confidence

Before explaining each of those methods one needs to present our validation procedure. As explained just above, abnormalities leading to IFSD are really rare[1], less than one over 300,000 flight hours. Anyway with more than 35,000 engines, and almost ten flights per day, those are events we can observe at Safran even if the probability that a normal passenger ever experience this problem is really low. We even announced that assuming a B737 pilot logs 900 flight hours annually that rate equates to one event every 185 years [5]. With such a so low rate of events the validation may be difficult. To solve this problem and higher experts' confidence we asked them to label suspicious patterns, even the ones that finally do not lead to a mechanical damage. The analysts will accept the demonstration if the algorithm selects at least the same spectrograms.

2.4   The Mathematics

The first model we select was a non-negative matrix factorization (NMF). This sparse linear representation model has the advantage of classical linear approach like singular value decomposition (SVD) to project the spectrogram on positive samples. Thus the image of frequencies is represented as a combination of a very low number of images which may also be seen as specific spectrogram patterns and labelled by the experts. Hence a given observation is immediately interpreted as part of such and such kind of anomaly [6].

---

[1] I just have to remind here that an IFSD is not a catastrophic event according to EASA and FAA regulation CS 25.1309 [16].

The second one, is a more classic mathematic approach on images because it looks for regularities with a wavelet decomposition [7] and more precisely curvelets which are kind of oriented wavelets. This solution gives a less parsimonious result but is able to find sharpest patterns [8].

One of the main differences between the two methods is that NMF does not take into account local correlations between pixels. Correlations are important for the curvelet decomposition. It partly explains why the fusion of the two methods improves the final identification of abnormal patterns.

3   NMF Decomposition

NMF is a factorization method of one positive matrix V into two positive matrices W and $H(V \approx W \times H)$. The constraints imposed on the decomposition are that every elements are positive [9].

$$[W^*, H^*] = \arg \min_{W \geq 0, H \geq 0} \|V - WH\|_2^2 \tag{1}$$

In our case V is a matrix where each column represents a spectrogram for a given engine, then the columns of W define a dictionary of spectrogram examples and H is a list of coefficients that explains each spectrogram as a linear combinations (columns of H) of the dictionary elements. These two matrices are estimated alternatively with two update rules given in Eq. (2).

$$H_{ij}^{k+1} = H_{ij}^k \frac{\left(W^{k'} V\right)_{ij}}{\left(W^{k'} W^k H^k\right)_{ij}}$$
$$W_{ij}^{k+1} = W_{ij}^k \frac{\left(V H^{k'}\right)_{ij}}{\left(W^k H^k H^{k'}\right)_{ij}} \tag{2}$$

Initialization is done by choosing random initial positive values for W^0. Then as described in [6] we initialize H^1 and W^1 as given in Eq. (3) before applying the normal NMF iterations of Eq. (2).

$$H^0 = \left[\left(W^{0'} W^0\right)^{-1} W^{0'} V\right]_+$$
$$W^1 = \left[V H^{0'} \left(H^0 H^{0'}\right)^{-1}\right]_+$$
$$H^1 = \left[\left(W^{1'} W^1\right)^{-1} W^{1'} V\right]_+ \tag{3}$$

For each matrix V there exists an infinity of solutions. Once a solution (W, H) is found, it is easy to find another solution by multiplying W by a positive diagonal matrix and H by its inverse. The solution would be the same. To prevent from this identifiability issue, at each step, the columns of the matrix W are normalized by a matrix A to unitary vectors, matrix H is multiplied by the inverse of A.

Since there is no constraint on the size of the matrices, this method can be applied as a dimension reduction with a chosen rank. The rank corresponds to the number of columns of matrix W, the number of pattern examples. This number must be estimated by optimization of the square error on the right part of Eq. (1). A good idea for this number should be given as the number of independent vibration sources observed on the spectrogram part we try to decompose, as vibrations energies tends to be additives on the spectrogram. However on a two shafts engine, classical source decomposition is not easy due to the high dependencies between sources. Our approach was to test successive increasing rank values until the difference in relative error becomes negligible. This may lead to different rank values for each parts of the spectrogram.

Figure 7 shows an example of decomposition of a small pattern extracted from a spectrogram; the successive examples are given in the magnitude order of the coefficients of the linear combination used for reconstruction.

Figure 8 shows reconstruction and residuals for a pattern extracted from an abnormal engine. The small hooks on the right clearly appear as well as some extra energy on the N2 shaft.

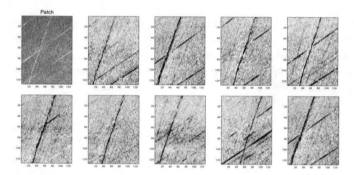

Fig. 7. A patch example (first image, top left) and a set of patterns examples (columns of W) obtained after training on a database of normal engines.

Fig. 8. From left to right: original pattern, reconstruction and residuals. This pattern was issued from an abnormal engine spectrogram.

4   Curvelets Decomposition

2D wavelet decomposition is also a sparse representation over a dictionary of patterns. However in this case the patterns are not learnt from examples but selected among a set of mathematic functions.

The types of wavelets we use in 2 dimensions are curvelets [10] which are a special case of ridgelets. The ridgelet coefficients are indexed by three parameters: the scale, the localization and the orientation in Eq. (4), so are the curvelet coefficients with a local characteristic.

$$\psi_{a,b,\theta}(x,y) = \frac{1}{\sqrt{a}}\psi\left(\frac{x\cos\theta + y\sin\theta - b}{a}\right) \tag{4}$$

In this equation we have a function ψ vanishing at infinity with regularity conditions defined in [11] so the decomposition computed by convolution may be reversible. The ridgelet is clearly constant on the line in the direction defined by the angle θ and behaves like a wave in the orthogonal direction (see ridgelets examples on Fig. 9).

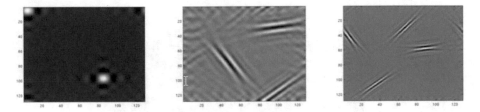

Fig. 9. Dictionary atoms at three different scales.

However, the vibrations are represented by curves in the spectrogram, hence the ridgelet transform (characterizing straight lines) is not sufficient to characterize the vibrations; but a curve can be approximated by a succession of small straight lines as shown on Fig. 10. Therefore by applying the ridgelet transform on dyadic squares at fine scale, it is possible to characterize a curve. This is the idea on which the curvelet transform is based. For the curvelet transform, the orthonormal ridgelets [12] are used. The construction of the curvelet transform is based on a dyadic decomposition [13].

Fig. 10. Curvelets characterize a curve with ridgelets of successive scales.

The dyadic curvelet decomposition is complete and reversible, hence keeping 100% of all coefficients give a perfect reconstruction. To build an interesting reconstruction score we select, for each spectrogram sub-image a sparse decomposition by minimization of the reconstruction error with a penalty coefficient in norm 1 as in Eq. (5). (Norm 1 is the sum of absolute values of coordinates.)

$$[W^*, H^*] = \arg \min_{W \geq 0, H \geq 0} \|V - WH\|_2^2 + \lambda \|H\|_1 \tag{5}$$

Where this time V is a spectrogram part, W is the set of curvelets (defined on each point) and H is a vector of coefficients. But this time we add a LASSO (least absolute shrinkage and selection operator) constraints [14]. This selects a subset of curvelets depending to the value of parameter λ. This parameter is usually selected from a cross-validation test. Finally we retain only the elements uses on a training set of normal engines, assuming that those functions should be enough to represent any normal pattern observed in the training set. The reconstruction of any other spectrogram part will be limited to this set of curvelets and the difference may represent unusual patterns.

Figure 11 shows reconstruction and residuals for 3 engines, one normal and two unusual.

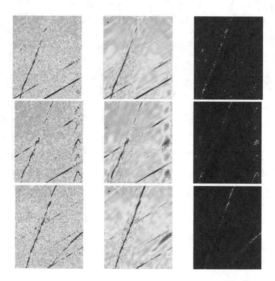

Fig. 11. From left to right: an original spectrogram part, the reconstruction and the residuals. The first line corresponds to a normal engine and the two last are abnormal.

5   Results

Each reconstruction error gives a score value we had to threshold if we want an automatic selection of suspicious patterns. For each algorithm we identify a threshold that corresponds to an optimization of the identification done by the experts. To obtain better statistical values for this threshold and as we have many pixels on each sub-image we identify the parameter of an extreme value distribution on the residuals. This distribution is of Gumbel type with cumulative distribution function given in Eq. (6). Only two parameters (μ and σ) should be estimated and provide very accurate threshold values for unilateral tests.

$$P\left(\max_{i \in S, \#S = n} r_i > r\right) = e^{-e^{-\left(\frac{r-\mu}{\sigma}\right)}} \tag{6}$$

In Eq. (6), n represent the number of pixels i we look at simultaneously in random subsets S of size n, r_i is a residual value and r is a threshold.

However there is still one issue in our process. Even if the engines are of similar conception there is always some minor differences, moreover the production tests are not exactly realized in the same meteorological conditions. This mainly leads to a small shift of the relation between the two shaft speeds N1 and N2 (see Fig. 12). Hence if we draw the corresponding spectrograms in N2 we expect to see a shift on the N1 rays.

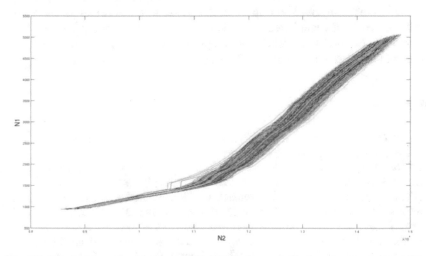

Fig. 12. N1 (LP shaft speed) according to N2 (HP shaft speed). Each curve represents a different engine (among 500), a different production test.

Table 1. Mean detection rates (%) and standard deviations for different types of algorithms and different types of abnormalities.

(%)	Normal	Unusual	Shift
NMF	35.8 ± 3.2	83.1 ± 1.2	85.5 ± 3.3
Curvelets	3.6 ± 3.4	67.0 ± 2.0	60.0 ± 3.4

This will naturally impact an automatic detection, and we are actually working on a new representation of spectrograms where this phenomenon will be removed by a normalization procedure. This will give us another challenge as the specialized engineers are clearly used to spectrograms and will not easily accept a new set of coordinate even less a new representation.

In the mean time we separate identification of shifts and real abnormalities from our validation. Table 1 gives detections rates obtained on a set of test spectrograms that where not used for calibration.

6 Conclusion

- NMF algorithm is adapted to the data and is more able to identify abnormal patterns; however its false alarm rate is high.
- Curvelets based algorithm is very good to model normal behavior, but its detection rate is lower than NMF's.

Together both options form a good alternative to help engineers analyzing spectrograms. The simultaneous analysis of both spectrogram types, in N2 for the x-axis and frequency, N1 or N2 orders for the y-axis, also is a solution to eliminate ambiguities due to the shifts discussed previously. To work with both representations we also build a morphologic graphic tool that presents a continuity between the two representations [15].

References

1. Abdel-Sayed, M.: Étude de représentations pour la détection d'anomalies - Application aux données vibratoires des moteurs d'avions (2018)
2. Clifton, D.A., Tarassenko, L.: Novelty detection in jet engine vibration spectra. Int. J. Cond. Monit. **5**, 2–7 (2015). https://doi.org/10.1784/204764215815848393
3. Klein, R., Masad, E., Rudyk, E., Winkler, I.: Bearing diagnostics using image processing methods. Mech. Syst. Signal Process. **45**, 105–113 (2014). https://doi.org/10.1016/j.ymssp.2013.10.009
4. Griffaton, J., Picheral, J., Tenenhaus, A.: Enhanced visual analysis of aircraft engines based on spectrograms. In: ISMA, pp. 2809–2822. Leuven (Belgium) (2014)
5. CFM International: Now That's a Reliable Engine. https://www.cfmaeroengines.com/press-articles/now-thats-a-reliable-engine/
6. Abdel-sayed, M., Duclos, D., Faÿ, G., Lacaille, J., Mougeot, M.: NMF-based decomposition for anomaly detection applied to vibration analysis. In: CM & MFPT, pp. 1–14 (2015)
7. Mallat, S.A.: A Wavelet Tour of Signal Processing (2008)
8. Abdel-Sayed, M., Duclos, D., Faÿ, G., Lacaille, J., Mougeot, M.: Dictionary comparison for anomaly detection on aircraft engine spectrograms. In: Machine Learning and Data Mining in Pattern Recognition, pp. 362–376 (2016)
9. Lee, D., Seung, H.: Algorithms for non-negative matrix factorization. Adv. Neural Inf. Process. Syst. 556–562 (2001). https://doi.org/10.1109/ijcnn.2008.4634046
10. Candes, E.J., Donoho, D.L.: Curvelets: a surprisingly effective nonadaptive representation of objects with edges. Curves Surf. Fitting. 105–120 (2000). https://doi.org/10.1016/j.biopsycho.2009.12.002

11. Candes, E.J., Donoho, D.L.: Ridgelets: a key to higher-dimensional intermittency? Philos. Trans. Math. Phys. Eng. Sci. **357**, 2495–2509 (1999). https://doi.org/10.1098/rsta.1999.0444
12. Donoho, D.L.: Orthonormal ridgelets and linear singularities. SIAM J. Math. Anal. **31**, 1062–1099 (2000). https://doi.org/10.1137/S0036141098344403
13. Abdel-Sayed, M., Duclos, D., Faÿ, G., Lacaille, J., Mougeot, M.: Anomaly detection on spectrograms using data- driven and fixed dictionary representations. In: ESANN, pp. 3–8. Springer (2016)
14. Tibshirani, R.: Regression shrinkage and selection via the lasso. In: Royal Statistical Society, pp. 267–288 (1996)
15. Griffaton, J.: Morphing of spectrograms for vibration sources identification in aircraft engines. In: CM & MFPT. BINDT, Paris (2016)
16. EASA: Certification Specifications for Large Aeroplanes (2007)

Cooperation of Virtual Reality and Real Objects with HoloLens

Jindřich Cýrus[✉], David Krčmařík, Michal Petrů, and Jan Kočí

Institute for Nanomaterials, Advanced Technologies and Innovation,
Technical University of Liberec, Studentská 2, 461 17 Liberec, Czech Republic
jindrich.cyrus@tul.cz

Abstract. We propose several approaches when programming applications for Microsoft HoloLens. These applications aim at close cooperation of real world with holographic objects (virtual reality). Often a precise placement that does not change with the position of a user can be challenging. One way of avoiding such problems is to use technology of anchors. Another way proposed in this paper is a prior scanning of object of interest with an application aimed for Vuforia plugin. Both concepts use close cooperation with user-friendly environment for developing of mixed reality applications – Unity. We present and discuss examples in this article. The area of precise location for mixed reality (augmented reality mixed with real reality in a way that one can interact witch each other) is of paramount importance. If we know precise position of virtual objects in the scene and precise position of the device in the scene, we can alleviate the burden of massive computing within HoloLens. As a result, we gain better experience from mixed reality.

Keywords: Microsoft HoloLens · Anchors · Vuforia · Unity

1 Introduction

Over two recent decades, virtual reality (VR) became widely spread mainly in gaming industry. The field can be roughly divided into mobile and tethered devices. Tethered devices are more expensive but offer more convincing experience. Among the most popular are HTC Vive, Sony PlayStation VR and Oculus Rift. These technologies require a powerful PC to be connected to it where all the massive computation is done.

On the other hand, mobile branch of VR is based on current cell phones: Samsung Gear VR or Google Daydream View. The device is simply a holder with special lenses which divide the image from cell phone into two images one for each eye. Nice table comparison of mentioned technologies is available in [1].

Virtual reality is becoming a standard tool for education, science, promotion, military etc. As the VR becomes more common, a few drawbacks are apparent. The user is in the case of tethered devices tightly connected to a processing computer what is uncomfortable due to potential risk of stumbling over the cable. In the case of mobile devices with special holders, the VR experience is far less impressing than with a tethered device with a lot of computational power. The user is closed within the virtual reality and does not have usually any clue where he is in the real world. The user can

© Springer Nature Switzerland AG 2020
K. Arai and S. Kapoor (Eds.): CVC 2019, AISC 944, pp. 94–106, 2020.
https://doi.org/10.1007/978-3-030-17798-0_10

hardly cooperate with other colleagues and perform other tasks while wearing VR glasses. Another aspect to mention is a health problem when using VR. Many people feel uncomfortable when wearing such glasses [2].

The possible remedy of mentioned issues with VR can be an augmented reality (AR) or mixed reality (MR). Augmented and mixed realities are technologies where the user still can have contact with real environment and on top of it he/she sees some additional information (pictures, objects). The virtual objects can interact with the real world in the mixed reality. One of the first examples was systems used for pilots in military. These systems enabled pilots to have current information from several systems in their field of view without the necessity of watching at the system's displays directly. Google glass [3] was several years ago a promising technology enabling the users to have additional information about the things which they saw. Unfortunately mainly due to privacy issues when using such glasses the project was halted.

In 2016 a technology called Microsoft HoloLens was delivered as a developer version to end customers. Huge positive response to the offered product followed since it enabled the virtual reality to be seamlessly mixed with the real world. Microsoft HoloLens are based on Windows 10, 32 bit version. It is a standalone tool looking like glasses weighting 579 g. Within the device a computer with many sensors and batteries is embedded. The sensors are abundant: inertial measurement unit (rotation and acceleration), 4 environment understanding cameras, depth camera, 2Mpx photo/HD video camera, mixed reality capture, 4 microphones, ambient light sensor [4]. The battery life is about 2, 5 h when actively used.

HoloLens can be programmed using a Direct 3D libraries or take the advantage of universal Unity environment. It requires Windows 10 operating system with Visual Studio 2015 or later.

Since HoloLens can interact with a real world, a very new dimension of simulations and modeling of various things can be performed. The technology enables to visualize requested holograms in specific places without the need of any additional computing from other devices. One can simulate for example a fall of a hologram in real environment where the environment is changing (hence if underneath the hologram there is a table the hologram drops on to the table otherwise it drops directly on to the floor).

We have to mention also drawbacks which have mixed reality compared to VR. The most visible thing is smaller field of view. It is due to the fact that the displays (one for each eye) cannot be bended due to special requirements for waveforms used [4].

To start with programming, tutorials from Microsoft are available [5]. The ways a user can interact with HoloLens are spatial sound, gaze tracking, gesture inputs – Fig. 1 – and a voice support. A user can take advantage of using a pack of libraries called Holotoolkit that is available for download and ready for deploy into the code. Such libraries simplify common tasks like the human understanding (gestures, gaze, ...), spatial understanding (recognition of walls, ceiling, floor) or even cooperation of several HoloLens together. Since the HoloLens has internet connectivity via WiFi one can take advantage of Cortana when controlling the device.

HoloLens can perform a spatial mapping in real time. This can be also used in the process of programming when the programmer wants to place the holograms into different places within a room. First the programmer using the HTML interface (Microsoft Edge) of HoloLens takes the spatial map of the room – Fig. 2 and after that

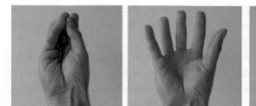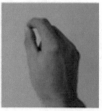

Fig. 1. Two leftmost pictures – system gesture to end the application and show menu, two rightmost pictures – click gestures (and drag gestures).

the exported mesh of polygons is imported into programming interface (e.g. Unity) as a standalone object within which he/she can position the holograms – Fig. 3.

Fig. 2. Spatial mapping of a room using HTML interface

2 Current State

Partly exceptional are HoloLens due to its ability of precise itself localization and tracking of other real object in the scene. The fundamental concept is described in [4]. Unfortunately the whole thing is more complicated for OST devices [6] since one have to take into account the movement of eye. AR device have to be capable of 6-degree-of-freedom (6 DoF) movement tracking (x, y, z, roll, pith, yaw). Every pixel in the virtual scene (with index B) has to be transformed using position and rotation into a real scene (with index A) according to an equation

$$x_B = R_B^A x_A + t_B^A \tag{1}$$

where R stand for rotation and t for translation. In the case of moving eye the coefficients for transformation (1) have to be heavily and periodically computed and updated. For that reason the field of view in HoloLens is limited in order to minimize eye movement.

Fig. 3. Result of spatial mapping with holograms in Unity (spatial mapping depicted as a mesh of white polygons – upper right big table, lower right real car, behind green car tables with flowers).

If we want to used mixed reality platform even for applications where extremely accurate positioning and stability is demanded [7] we have to improve overall location error – for example using calibration [8]. Online temporal tracking [9] or well-know Kalman filter based solutions [10] seem to give promising results. Descriptor utilization is a concept where objects are compared from different views, classified and the system is in this way trained in such a way that only 16-dimensional descriptors can help with object recognition and their 3D pose estimation [11].

It is necessary to mention that there are also attempts to use low-cost RFID based devices, which can to a certain extent track 3D objects in space [12]. A signal strength indicator (RSSI) was measured using commercial off-the-shelf device called RFTrack which uses discrete wavelet transform for detection target motion. Unfortunately precision of only 32 cm was reported. Tele-rehabilitation was tested using Microsoft Kinect (predecessor of HoloLens) based on optical system with the help of calibration [13]. The calibration is provided as complementary information concerning bone lengths, position of elbows etc. A Google Tango based project with additional visual markers [14] is another way of precise localization of moving devices in the scene which does not require any additional infrastructure.

As shown above there are many challenges in newly appearing platform of mixed reality. The thing of paramount importance remains precise localization which is not processor demanding and without needed ambient infrastructure.

3 Anchors and Unity

First attempt to solve the problem with stability of virtual content in the scene was made using anchor approach. It can alleviate the positioning inaccuracy due to the movement of the user around the scene. We have programmed a sample application where a cooperation of a real robot and holographic copy of such a robot is demonstrated – Fig. 4.

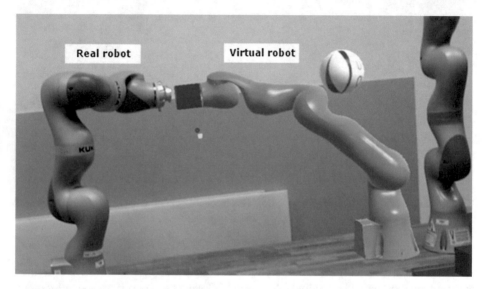

Fig. 4. Cooperation of virtual and real robot when handing over a virtual blue cube

The real Kuka robot has a web server enabled. The application communicates with the web server via requests in a way that when the cube has to be handed over from holographic robot to real robot a special request is made. Then the real Kuka robot moves according to some trajectory and the application fires other requests to which the robot responses with its current position and end tip rotation. The position and rotation information is processed and the cube is accordingly moved. The result is that the holographic cube moves with the end tip of the real Kuka robot. We have used a coroutine technology as shown in a piece of program below.

```
myCouroutine = StartCoroutine(GetData());
UnityWebRequest CreateRequest() {
    DownloadHandler download = new DownloadHandlerBuffer();
        var dummy = new byte[] { 0x20, 0x20 };
        UploadHandler upload = new UploadHandlerRaw(dummy);
        string url = "http://172.31.1.215:8096/api/values/1";
        UnityWebRequest www = new UnityWebRequest(url,
                                    "GET", download, upload);
        www.SetRequestHeader("Content-Type", "text/html");
        return www;
}
IEnumerator GetData() {
    using (UnityWebRequest www = CreateRequest()) {
        yield return www.Send();
        if (www.isError) {
                response = "error";
        } else {
                string myString = www.downloadHandler.text;
                response = myString;
        }
    }
}
```

The code above explains how we have established a communication link between the robot and the HoloLens. We have used TCP based HTTP protocol which periodically reports the end point position of the robot. In Unity environment this is done through concept know as coroutines. The coroutine has a parameter of the type IEnumerator which actually performs http requests (or in other words sending information about its position packed into an HTTP packet). In the method CreateRequest the user specifies the backend IP address (in our case HoloLens IP address).

However we have found out that when moving with HoloLens around the scene the cube and the holographic robot moves slightly with respect to real robot. This is a common problem when not using anchors. Anchors are special objects placed to the scene which contain a cluster of positioning information of neighboring objects. A sample program is provided below.

```
if (Placing) {
   WorldAnchor
   attachingAnchor = gameObject.AddComponent<WorldAnchor>();
   if (attachingAnchor.isLocated) {
       Debug.Log("Saving persisted position immediately");
       bool saved =
              anchorStore.Save(ObjectAnchorStoreName,
              attachingAnchor);
       Debug.Log("saved: " + saved);
   } else {
       attachingAnchor.OnTrackingChanged
              +=AttachingAnchor_OnTrackingChanged;
   }
} else {
   WorldAnchor anchor = gameObject.GetComponent<WorldAnchor>();
   if (anchor != null) {
       DestroyImmediate(anchor);
   }
   string[] ids = anchorStore.GetAllIds();
   for (int index = 0; index < ids.Length; index++) {
       Debug.Log(ids[index]);
       if (ids[index] == ObjectAnchorStoreName) {
              bool deleted = anchorStore.Delete(ids[index]);
              Debug.Log("deleted: " + deleted);
              break;
       }
   }
}
```

The code above waits using *if* block for the scene being stabilized (HoloLens user should not move quickly). If the condition is fulfilled (*Placing* is set to *true*) a new component WorldAnchor is added to the scene. This anchor is a special object that cares for exceptionalities of the scene/surrounding space to which the virtual content is fixed.

It is important to notice that in order to anchors working in an optimal way a more variable background of the scene is needed. The background as shown in Fig. 4 is very monotonic. It helped a lot for this case to put some colored posters on the wall and HoloLens immediately started to place the holographic objects more precisely and constant in time when moving with the device around the scene.

4   Vuforia

We have developed another application simulating a virtual car showroom where a user with HoloLens can go around several checkpoints in the room and interesting features are presented to him/her via spatial sound, videos started on demand and holographic models that change according to wish of the user – Fig. 5.

Fig. 5. Example of controlling the scene using the checkpoint arrangement (see bubbles 1, 2, 3, 5).

Within the car showroom was also a real car that was intended for presentation of technological features of the car, which are hard to see since they are covered by chassis (e.g. engine). This task was different from the previous task solved by anchors. This time we wanted to precisely localize a predefined real world object and then to place some features within this object. A way to accomplish this is via Vuforia software [15].

Vuforia is a mobile device suite that allows users to build AR-based applications (Enhanced Reality) when the actual environment enriches the virtual object. The main feature of Vuforia is to identify and analyze the captured image. This feature is available for SDK (Software Development Kit), which can be freely translated as a development toolkit.

Using the SDK developers can encode application mechanisms to detect and track individual image characteristics. The SDK allows us to track up to five objects in the field of view simultaneously. Target recognition not only focuses on objects in 2D space, such as images and tags, but also 3D objects can be detected. Vuforia supports 2D image targeting (ImageTarget function), marker (FrameMarker function), text recognition (TextRecognition function), 3D object (cube, square, pyramid) and the last target we discuss here is any object in the space.

Implementation requires the correct version of Unity that supports the Vuforia and installed the latest version of Vuforia, which was downloaded at official website [16]. Free registration has to be made on the official site so that one can make full use of the Vuforia.

For detection of any object in space, we need a special template, which is downloadable as a page in A4 format. We also need a cell phone (preferably Samsung S8 due to high quality camera and latest Android) with software called Vuforia Object Scanner.

Vuforia Object Scanner allows us to create a targeted object using an Android phone. The program is only supported by phone Samsung S6 and above. We have installed the application downloaded from the official Vuforia site, placed the object on a 3 × 4 m vector-sized magnified template (Fig. 6) and started scanning. The template is needed in order to correctly align the scanned object within the coordinate system.

Fig. 6. Magnified template for scanning real 3D objects with Vuforia.

After the scan is complete the program creates a *.od file with which we work later. During the scan, the program on the cell phone allows us to check if scanning has been successful from a particular direction. In Fig. 7 is visible a scanning process where the application just successfully scanned part of the car with front right headlight – the corresponding polygon has changed to green. In the picture are also slightly visible small green points which are features of the object to which some "anchors" are made. These points are then in the scene taken using HoloLens searched for in order to recognize the wanted object (in our case whole car).

Fig. 7. Scanning real 3D objects for use in Unity with Vuforia.

After the scan is completed, we can check if a sufficient number of anchor points were taken and the recognition of real object in the scene works well. We have parked the scanned car among others cars and started the application. If the application recognizes the object, a green block is shown at the position of origin of coordinate system as shown in Fig. 8. This block does not change position as one moves with the phone around the car.

At the Vuforia Developer Portal site in the Developer section, a target was created to display the relevant virtual object (in our case green block). The file extracted from the phone was then placed in a form to create a recognized target object called ObjectTarget. In this form, it is possible to use, for example, block, cylindrical, and 2D target in addition to the 3D object target – Fig. 9.

Fig. 8. Recognition of previously scanned car among other cars.

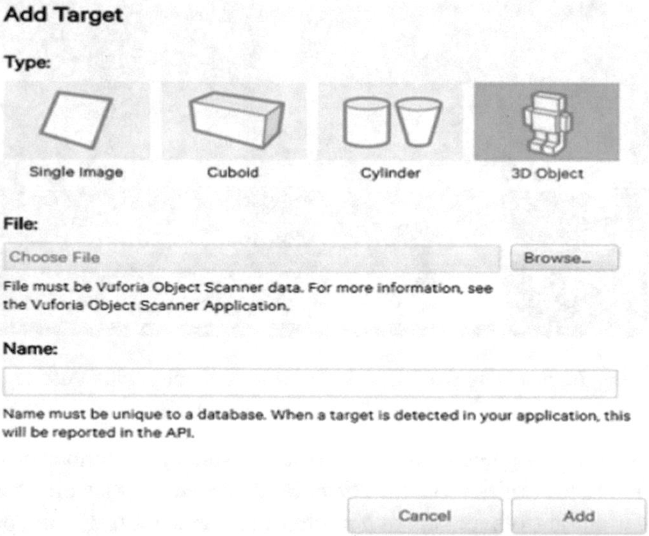

Add Target

Type:

Single Image   Cuboid   Cylinder   3D Object

File:

Choose File   Browse...

File must be Vuforia Object Scanner data. For more information, see
the Vuforia Object Scanner Application.

Name:

Name must be unique to a database. When a target is detected in your application, this
will be reported in the API.

Cancel   Add

Fig. 9. Form for creation of 3D object target for Vuforia.

The next step for creating ObjectTarget was its download offered by Vuforia for two development platforms: SDK and Unity Editor. The file with the name *.unity-package is then easily uploaded to Unity via the Import Package feature and we can create our own application. The final step was to create a license key in the Develop tab in Vuforia section. This license key is later placed in the corresponding section of the

Unity AR Camera. Without this license key the application that we have created does not work. An example when a working application with engine placement within the hood of a car is presented in Fig. 10.

Fig. 10. Engine placement within the hood of a car.

5   Conclusion

Mixed reality represented up to date with Microsoft HoloLens is a promising technology that enables the user to have additional information available when needed without user interface complexity. The user can control the virtual scene with voice, gestures, gaze or just simply moving himself around the scene. The technology gives him/her full 3D experience (visual and audio) and at the same time he/she has unoccupied hands and has full visual control of the real world surrounding. We have presented two methods when implementing interaction of virtual reality and real world. Such methods for fixing the holograms at predefined position are crucial for user's experience. The best practice is to use recognition of objects (via Vuforia) and anchors together. The anchors work best in not monotonic scenes. Using these techniques, we can reach a precision of 1 mm in placement of hologram into real scene. Viable solution for the real world "clever" devices can profit from usage of some kind of network server (UDP, TCP) that provides with changing position information and can be easily incorporated into the MR/AR applications. The field of MR/AR is developing very rapidly and future substantial improvements are be expected.

Acknowledgment. The result was obtained through the financial support of the Ministry of Education, Youth and Sports of the Czech Republic and the European Union (European Structural and Investment Funds - Operational Programme Research, Development and Education) in the frames of the project "Modular platform for autonomous chassis of specialized electric vehicles for freight and equipment transportation", Reg. No. CZ.02.1.01/0.0/0.0/16_025/0007293.

References

1. Singh, H., Singh, S.: Virtual reality: a brief survey. In: 2017 International Conference on Information Communication and Embedded Systems (ICICES), India (2017)
2. Sharples, S., Cobb, S., Moody, A., Wilson, J.R.: Virtual reality induced symptoms and effects (VRISE): comparison of head mounted display (HMD), desktop and projection display systems. Displays **29**(2), 58–69 (2008)
3. He, J.B., et al.: Does wearable device bring distraction closer to drivers? Comparing smartphones and Google Glass. Appl. Ergon. **70**, 156–166 (2018)
4. Hoffman, M.A.: The future of three-dimensional thinking. Science **353**(6302), 876 (2016)
5. Robinett, W., Holloway, R.: The visual display transformation for virtual reality. Presence Teleoperators Virtual Environ. **4**(1), 1–23 (1995)
6. Grubert, J., et al.: A survey of calibration methods for optical see-through head-mounted displays. IEEE Trans. Visual Comput. Graphics **24**(9), 2469–2662 (2017)
7. Vassallo, R., et al.: Hologram stability evaluation for Microsoft HoloLens. In: SPIE Medical Imaging, Florida, USA (2017)
8. Garon, M., et al.: Real-time high resolution 3D data on the HoloLens. In: IEEE International Symposium on Mixed Reality and Augmented Reality, Mexico (2016)
9. Tan, D.J., et al.: A versatile learning-based 3D temporal tracker: scalable, robust, online. In: IEEE International Conference on Computer Vision, Chile (2015)
10. Yang, Q., et al.: Kalman filter based localization and tracking estimation for HIMR RFID systems. In: IEEE International Conference on RFID, Florida, USA (2018)
11. Wohlart, P., Lepetit, V.: Learning descriptors for object recognition and 3D pose estimation. In: IEEE Conference on Computer Vision and Pattern Recognition, Boston, USA (2015)
12. Li, L., et al.: Accurate device-free tracking using inexpensive RFIDs. Sensors **18**(9), 2816 (2018)
13. Yu, K., et al.: On the accuracy of low-cost motion capture systems for range of motion measurements. In: Medical Imaging, Texas, USA (2018)
14. Marques, B., et al.: Evaluating and enhancing Google Tango localization in indoor environments using fiducial markers. In: IEEE International Conference on Autonomous Robot Systems and Competitions, Portugal (2018)
15. Yang, X., et al.: Robust and real-time tracking for augmented reality on mobile devices. Multimedia Tools Appl. **77**(6), 6607–6628 (2018)
16. Bajana, J., et al.: Mobile tracking system and optical tracking integration for mobile mixed reality. Int. J. Comput. Appl. Technol. **53**(1), 13–22 (2016)

Ontology of Ubiquitous Learning: WhatsApp Messenger Competes Successfully with Learning Management Systems (LMS)

William Kofi Koomson(✉)

Valley View University, Oyibi, Dodowa Road, Accra, Ghana
william.koomson@vvu.edu.gh

Abstract. The purpose for this research was to add to the body of research and further study how mobile learning can help to remedy the limitations e-learning poses for students who live in the sub-Saharan Africa Region with lack of access to electrical power and internet connectivity issues. Qualitative approach was employed with a total sample size of 807 students, composed of 58% male and 42% female. In this study, I made several assertions that, for WhatsApp to work properly in any classroom in sub-Saharan Africa, there must be intentional designs and step-by-step approach to teach both the faculty and the students how to use the application to achieve the utmost outcomes. I, therefore, concluded that using WhatsApp Messenger in a blended mobile learning context can help resolve many of the contextual difficulties that plague students in an e-learning situation.

Keywords: Ubiquitous Learning · Blended m-Learning · Connectivity

1 Introduction

The evolution of the Third Generation Web and smartphone applications which have been created to run on mobile devices of the 21st Century have transformed the entire universe in all areas, including the way we communicate, function, in our daily living, and even the way we study. Almost every corner of the universe, including developing countries such as, sub-Saharan Africa, South Asia, and South America, use these technologies for transactional purposes. Farmers use these applications (apps) to transact farm businesses; communities use them to preserve family traditions and memories; and business executives are able to bridge transactional gaps. Meanwhile, researchers acquire access to research artifacts through apps, and teachers engage their students in the classrooms using these innovative technologies.

Mobile technologies with cellular connectivity continue to dominate the information communication technology market in sub-Saharan Africa. Prior to the advent of mobile phones, many sub-Saharan African countries whose citizens lived in the most remote parts of the country were cut-off regarding the use of telephones, powered by landlines to connect to the global world. However, the abundance of the mobile network systems has changed the face of telecommunication and has transformed the way business is transacted in sub-Saharan Africa and the rest of the developing world.

K. Arai and S. Kapoor (Eds.): CVC 2019, AISC 944, pp. 107–117, 2020.
https://doi.org/10.1007/978-3-030-17798-0_11

Citizens are able to skip the landline developmental stage of telecommunication to digitalization.

According to the Pew Research Center [1], cell phone usage in Africa pales in comparison to that of developed countries like the United States of America. However, there has been a dramatic surge in the growth of smartphone usage in sub-Saharan Africa. As of 2014, the following countries recorded high percentages of cell phone usage; Uganda 65%, Tanzania 73%, Kenya 82%, Ghana 83%, and South Africa, 89%. In the same year, the United States' cell phone usage was 89%, the same as in South Africa and only in single-digits, higher than Ghana and Kenya. Among the many uses of cell phones in Africa for a twelve-month period, texting was the most [1].

Joy Online [2] reported that Ghana was ranked by the International Telecoms Union Report as number one in Africa with more people using or connected to mobile broadband. Laary [3] stated that for the period ending December 2015, Ghana's mobile phone voice penetration rate surged to 128%, far above earlier projections by telecommunication experts. The adoption of mobile technology with its diverse apps can serve as a conduit for mobile learning.

2   Related Work

The objective of this paper was to create an ontology to demonstrate new approaches to study and understand mobile learning through the use of WhatsApp Messenger as a learning tool in a distance learning program. The goal was to add to the body of research and further study how mobile learning can help to remedy the limitations online learning poses for students who live in sub-Saharan Africa with lack of access to electrical power and internet connectivity issues. Motlik [4] suggested that mobile learning will pave the way for online learning as the internet is not stable and is unavailable in many parts of rural areas in developing nations. Also mobile learning is more affordable to less developed nations and financially constrained groups [5].

Notwithstanding these positive developments, some, including academics in higher education in sub-Saharan Africa, refuse to accept the fact that online learning can be done through mobile devices. They still believe that because of the unstableness of Internet connectivity, few institutions of formal learning can successfully go online in sub-Saharan Africa, including Ghana [6]. However, with mobile learning technologies like "WhatsApp Messenger," developing countries have no excuses as to why they are not able to adopt online learning in the remotest parts of the country where connectivity is a major setback. Everywhere a mobile phone is used, whether for WhatsApp, Email, SMS, video or photo sharing, online learning is possible. In the academic environments, just as in the community, households and business places, WhatsApp Messenger has been used to create group chats for work teams, social networking, and learning.

In Ghana, the most common format adopted in Distance Learning is the tutorial format, where very few online interactions occur; in most instances, there are no online interactions. The universities that enroll their students through the distance learning mode, rely heavily on print materials in the form of course modules and students meet regularly during weekends in tutorial centers throughout the nation where they receive face-to-face instructions. Very few programs include videos and voice presentations in their distance learning pedagogy [6, 7].

2.1 Mobile Learning

UNESCO [8] defines mobile learning (m-Learning) as involving; "the use of mobile technology, either alone or in combination with other information and communication technology (ICT), to enable learning anytime and anywhere. Learning can unfold in a variety of ways: people can use mobile devices to access educational resources, connect with others, or create content, both inside and outside classrooms (p. 6). Quin [9] also defines m-Learning as learning that is done through mobile computational devices, such as "Palms, Windows CE machines, even your digital cell phone."

UNESCO has compiled many unique benefits of mobile learning. Among the list are; (1) Reach and equity of education – making learning accessible to people in the world who would otherwise be cut out from receiving education via online learning; (2) Personalized learning – learners can carry their mobile devices to and from places of employment, schools, bedrooms, boardrooms, and to recreational venues; (3) Provides immediate feedback and assessment – allowing learners to quickly pinpoint problems of understanding and review explanations of key concepts; (4) Productive use of time spent in classrooms – students can learn anywhere anytime; (5) Build new communities of learners. (6) Support situated learning; (7) Enhance seamless learning – students can access learning materials from wide varieties of devices; (8) Bridge formal and informal learning; (9) Minimize educational disruption in conflict and disaster areas; (10) Assist learners with disabilities; (11) Improve communication and administration; and (12) Maximize cost-efficiency by leveraging technology students already own, rather than providing new devices that are not tested [8, 10–26].

2.2 WhatsApp Messenger for m-Learning

Research on the application of WhatsApp Messenger in the classroom is new and developing, however, its usage as a social media tool on smartphones is widespread [6, 10–12]. WhatsApp is the most popular mobile messaging application widely used worldwide and is ranked as the number one in terms of monthly active users, based on a study of over 22,500 sources worldwide [13].

WhatsApp features include:

- Text – simple and reliable;
- Group Chat – keeping in touch with love ones, people in your network, business partners, and parishioners;
- On the Web and Desktop – keeping the conversation going anytime, anyplace, anywhere;
- Voice and Video Calls – free face-to-face conversation, when voice and text are not enough;
- End-to-End Encryption – provides security by default;
- Photos and Videos – opportunity to share moments that matter;

- Voice Messaging – using the voice messaging system to convey emotional moments; and
- Documents – attaching and sharing documents including PDFs, spreadsheets, slideshows, photos, and Word documents (http://www.whatsapp.com/features/).

WhatsApp Messenger features make it easy for teaching and learning. The app uses phone internet connections (4G/3G/2G/EDGE of Wi-Fi) of users to send and receive messages. That is, as long as there is data on users' phones, sending and receiving messages are free. (https://faq.whatsapp.com/en/android/20965922/). WhatsApp announced in May 2018 at its F8 developer conference in San Jose, California, that over 65 billion messages have been sent by users with more than 2 billion minutes of voice and video calls made everyday on the app platform, and about 1 billion people uses this messaging app each day [14].

Though few studies have researched into the educational benefits of the WhatsApp Messenger platform; students at the university level have used the texting feature to send and receive short messages through mobile devices; institutions of higher learning are gradually adopting WhatsApp for educational purposes; and discussion forums that are prominent in Learning Management Systems (LMS) are also available on some mobile learning platforms such as WhatsApp Messenger [15–17].

2.3 Key Players of the Ontology

Effective distance learning program requires inter-relationships among key players. This section identifies four important key players, which include: students, faculty, support staff, and administrators [18]. Kathman [19] posited that, higher education institutions are engaging students more and more through text messaging and fostering of one-to-one relationships. In the past, distance learning students were not as able to freely interact with each other to share their backgrounds and interests. However, new technologies are bringing students together and helping to build communities of learners through distance education [18, 30–32]. According to Bull [20], faculty acts as tour guides by directing and redirecting the attention of learners toward key concepts and ideas in teaching and learning. Barry [18] described the support staff as "silent heroes of a successful distance education program".

The support staff assist in promoting persistence and participation to avoid students' dropout. Their services include academic, administrative, and technological support. In most institutions, the support staffs' services are offered through extended hours [21]. Regarding administrators, Barry [18], indicated that their duties include, planning for technological resources, deploying manpower resources, financial and the necessary capital expenditures to enhance the institution's online learning mission. They also "lead and inspire faculty and staff in overcoming obstacles that arise. Most importantly, they maintain an academic focus, realizing that meeting the instructional needs of distant students is their ultimate responsibility" [18].

Factors that plague online learners in Ghana are many, including; computer illiteracy, access, long-term power outages, and connectivity. Relevant skills deficiency hinders successful facilitation of online learning through discussion boards, timely response to students, and promotion of active learning strategies. "Online instructors specifically need to be able to facilitate online discussions that are rich and meaningful, respond in a timely manner, and model active learning strategies" [22]. However, universities in Ghana lack the appropriate infrastructure to offer courses online, thereby resorting to face-to-face tutorial sessions in the form of tutorial center operations throughout the country by using untrained tutors who have little or no experience in distance learning. Students who are enrolled in the distance learning programs have to attend tutorials every weekend or bi-weekly.

3   The Context

3.1   Description of the Ontology

This paper used qualitative approach, framed under the paradigm which postulates that reality is relative and depends on multiple systems for meaning. *On-to-logy*, a Greek word, relates to the nature of reality as seen in the lens of a person in his experiences, this experience may lead the individual to seek meaning. There are two schools of thoughts: the objectivist and subjectivist. The objectivist approach correlates with a quantitative research paradigm, while the subjectivist approach sees the world as socially constructed – a qualitative paradigm [23–28].

In a typical university distance learning classroom in Ghana; there are students who come directly from the Senior High Schools (17 to 21 year olds); Top-up students (21–27 year olds) coming from the nation's Polytechnics, Colleges of Education, and diploma (or associate) degree programs who enrol into the universities for degree purposes; mature students who are permitted to enrol into university degree programs after attaining the age of 25 (25–40 year olds); and similar age groups who are workers and are seeking university degrees for professional advancements. A vast majority of these students live in the nation's hinterlands and are not able to take full advantage of online learning due to factors beyond their control.

This paper describes an ongoing research study, which began in January 2017, about how to create an effective distance learning program in a hybrid mode that integrates WhatsApp Messenger as the learning platform for students who live in Ghana's remote areas where connectivity and electrical power supply are limited. The purpose was to better understand the application of WhatsApp Messenger by using its features to construct meaning for learners and instructors in an online learning context. The study was based in a university in Ghana with three campuses and two learning centers with total student population of about 10,000. Demographic characteristics of the sample participants in the study consisted of 807, composed of 58% males and 42% females. Students above the age of 25 formed the dominant age group for the study, scoring a total of 83%. Sixty percent of the students were married with about 44% indicating that about 4 persons depend on them for their sustenance. About 51% of the students indicated that they entered the university with other qualifications apart from associate degree or high school diploma. Concerning commitment to study, about 89% of the students indicated that they work, while about 54% of them were engaged in full time employment. Forty-three percent of the students in the study were committed to study for about ten hours a week.

4   Blended Mobile Learning Structure

I present in Table 1, a mobile learning structure indicating a summary comparison between a typical Learning Management System (LMS) and the proper application of the use of WhatsApp as mobile learning platform in a Ghanaian context. The following assumptions were made to explain Table 1:

Assumption #1
Why it will not work

(a) WhatsApp Messenger as a social media tool is not fit for the classroom – for learning purposes.
(b) A typical LMS delivers courseware over the internet – lack of internet connectivity and prolonged power outages in Ghana, especially, in the countryside makes it impossible to sustain online learning. Therefore, LMS will not work for students in Ghana who live far away from the cities.

Assumption #2
How it will work

(a) For WhatsApp to function properly in online learning environment, the features must be properly integrated to fit the purpose of teaching and learning in a mobile learning context.
(b) WhatsApp Messenger uses phone internet connections of users to send and receive messages. That is, as long as there is data on users' phones, sending and receiving messages are free. Therefore, students in Ghana, who live far away from the cities can also access online learning benefits through their mobile devices.

Jurado, Pattersson, Regueiro-Gomez, and Scheja [29], classified learning management systems features into four different tool groups, namely: distribution, communication, interaction, and administration.

1. **Tools for distribution** allow lecturers to upload documents, available to students. Earlier it was mainly text documents and today it may also be different kinds of media files. Nevertheless, the process is still one-way, that is, teacher-to-learner distribution of information.
2. **Tools for communication** allow information to go either way as well as from student-to-student. The most common example is E-mail.
3. **Tools for interaction** call for reaction and feedback. Discussion boards are the most typical example. These tools are of great interest since they may promote student activity and cooperation, hence enhancing the learning experience.
4. **Tools for course administration** are used to monitor and document the educational process, rather than to facilitate teaching or learning [4].

Table 1. Mobile learning structure using WhatsApp Messenger – Ghanaian context

Application: Online learning tools by Jurado, et al.,	Key Players: Students/ Faculty/ Support Staff/ Administrators	Purpose: Teaching and learning using blended mode of online learning	Learning Management System (LMS): Why LMS will not work in Ghana due to lack of Internet Access	WhatsApp Messenger for Learning: How WhatsApp will work despite lack of Internet Access in Ghana
Distribution	1. Faculty 2. Student Interaction flows from teacher to student	One-way: from teacher to student – one way process	1. Teacher sends course information to students via the course management system 2. Students retrieve course information 3. LMS delivers courseware over the internet 4. Students lack access to retrieve and view course content via the internet.	1. Teacher sends course information via PDFs or Word document attachments to students 2. Students sign their name (forum signature) before each WhatsApp post; 3. Students retrieve course information 4. WhatsApp Messenger uses phone internet connections 5. As long as there is data on students' phones, viewing course content is possible.
Communication	1. Faculty 2. Student Interaction flows both ways	Information go either way. Teacher to student, student to teacher	1. Students respond to teacher via the course management system 2. Teacher grades students work and post comments on course management system	1. Students post completed assignments in more than one format via: a. PDF or Word attachments b. Direct text message 2. Teacher grades students work and post comments via WhatsApp Messenger 3. Teacher sends transcripts of WhatsApp communication to course administrators.
Interaction	1. Students Peer interactions. Student to student	Discussion boards, students reactions and feedbacks.	1. Student to student interaction through LMS discussion forums 2. Teacher as facilitator guides students	1. Students to student interaction through: a. WhatsApp 'group- chat' b. Possible video and voice calls 2. Teacher as facilitator guides students

(*continued*)

Table 1. (*continued*)

Application: Online learning tools by Jurado, et al.,	Key Players: Students/ Faculty/ Support Staff/ Administrators	Purpose: Teaching and learning using blended mode of online learning	Learning Management System (LMS): Why LMS will not work in Ghana due to lack of Internet Access	WhatsApp Messenger for Learning: How WhatsApp will work despite lack of Internet Access in Ghana
Course Administration	1. Support Staff 2. Administrators Back-end interaction	Course monitoring, management, documentation, and evaluation.	1. Teacher setup the courses via the LMS 2. Teacher post course syllabus and assignments for class discussions 3. Teacher grades students work and post grades online 4. Students perform teacher and course evaluations online	1. Support staffs create WhatsApp groups for students and faculty 2. Support staff monitors students and faculty interactions through WhatsApp transcripts 3. Support staff receives transcripts from teachers regularly and monitors for course content and interactions 4. Teacher sends WhatsApp transcripts to program office for archival purposes. 5. Support staff archives course materials for quality control purposes.

5   Discussion and Conclusion

5.1   Discussion

This paper depicted an ontology of an ongoing research study. The purpose of the research was to better understand the application of WhatsApp Messenger by using its features to construct meaning for learners and instructors in a blended mobile online learning context. The study was based in a university in Ghana with three campuses and two learning centers with total student population of about 10,000. A sampled total of 807 students from three campuses and two learning centres of the university adopted the use of WhatsApp Messenger in a blended online learning mode.

Total sample size for the study was 807, composed of 58% male and 42% female. Students above the age of 25 formed the dominant age group for the study, scoring a total of 83%. Sixty percent of the students were married with about 44% indicating that about 4 persons depend on them for their sustenance. About 51% of the students indicated that they entered the university with other qualifications apart from associate degree or high school diploma.

Concerning commitment to study, about 89% of the students indicated that they work, while about 54% of them were engaged in full time employment. Forty-three percent of the students in the study were committed to study for about ten hours a week. The results from the demographics report fit traditional adult learners as described in the literature. According to Ross-Gordon [30], adult students, referred to as – non-traditional students form sizeable presence on university campuses and also constitute a substantial share of the undergraduate student body. The National Center for Educational Statistics [31] survey reported that 38% of student enrolment for the 2007 academic year were 25 years of age or older. Choy [32] cited the 2002 NCES statistics that defined seven characteristics of non-traditional students as follows:

1. Entry to college delayed by at least one year following high school,
2. Having dependents,
3. Being a single parent,
4. Being employed full time,
5. Being financially independent,
6. Attending part time, and
7. Not having a high school diploma.

Ross-Gordon [30] described characteristics that separate re-entry adults from other traditional university students to be; "the high likelihood that they are juggling other life roles while attending school, including those of worker, spouse or partner, parent, caregiver, and community member" [27].

5.2 Conclusion

In designing the blended mobile learning structure, I applied agile methodologies using WhatsApp Messenger as a learning platform, that meets the current infrastructural conditions in Ghana.

Earley [33], stated that, there must be the need to interpret user signals accurately to "enable the system to present the right content for the user's context," this may "require not only that our customer data is clean, properly structured, and integrated across multiple systems and processes but also that the system understand the relationship between the user, his or her specific task, the product, and the content needed".

According to Yeboah and Ewur [6], the adoption of WhatsApp in the classroom is anathema. To them, the technology is nuisance to university students. They concluded that, "if students bring their mobile phones to class, they get bored of the lesson and find their way onto WhatsApp. These detracts their attention from the main lesson, and are not able to fully understand what is going on, hindering participation and drawing them even further into WhatsApp making it more difficult for them at the end of the day".

Contrary to Yeboah and Ewur's, assertions, the current paper has proven otherwise. In this study, I made several assertions that, for WhatsApp to work properly in any classroom in Ghana, there must be intentional designs and step-by-step approach to teach both the faculty and the students how to use the application to achieve the utmost outcomes (see Table 1, above). Because, I believe that, "seemingly intractable problems have been solved by advances in processing power and capabilities. Not long ago, autonomous vehicles were considered technologically infeasible due to the volume of data that needed to be processed in real time. Speech recognition was unreliable and required extensive speaker-dependent training sessions. Mobile phones were once "auto-mobile" phones, requiring a car trunk full of equipment" [33].

References

1. Pew Research Center Surveys.: Spring 2014 global attitudes survey: Q68 (2015). www.pewglobal.org/2015/04/15/cell-phones-in-Africa-communication-lifeline/
2. Joy Online: Increase use of smartphones made Ghana No-1 in mobile broadband penetration (2013). business.myjoyonline.com/pages/news/201301/99976.php
3. Laary, D.: Ghana: Mobile phone penetration soars to 128%. The Africa Report (2016). www.theafricareport.com/West-Africa/ghana-mobile-phone-penetration-soars-to-128.html
4. Motlik, S.: Mobile learning in developing nations. Technical Evaluation Report. The International Review of Research in Open and Distance Learning, 63, 9(2) (2008). http://www.irrodl.org/index.php/irrodl/article/viewArticle/564/1039
5. Gronlund, Å., Islam, Y.: A mobile e-learning environment for developing countries: The Bangladesh virtual interactive classroom. Inf. Technol. Dev. 16(4), 244–259 (2010). https://doi.org/10.1080/02681101003746490
6. Yeboah, J., Ewur, G.D.: The impact of WhatsApp Messenger usage on students' performance in tertiary institutions in Ghana. J. Educ. Pract. 5(6), 157–164 (2014)
7. Larkai, A.T., Ankomah-Asare, E.T., Nsowah-Nuamah, N.N.N.: Distance education in Ghana: an overview of enrolment and programme patterns. In: Proceedings of INCEDI 2016 Conference, Accra, Ghana, pp. 184–190 (2016)
8. UNESCO: UNESCO policy guidelines for mobile learning. Paris, France (2013). ISBN 978-92-3-001143-7
9. Quin, C.: mLearning: mobile, wireless, in-your-pocket learning. /L1 = LQH, Fall (2002)
10. Cetinkaya, L.: The impact of WhatsApp use on success in education process. Int. Rev. Res. Open Distrib. Learn. 18(7) (2017). https://doi.org/10.19173/irrodl.v18i7.3279
11. Bouhnik, D., Deshen, M.: Whatsapp goes to school: mobile instant messaging between teachers and students. J. Inf. Technol. Educ. Res. 13, 217–231 (2014)
12. Church, K., de Oliveira, R.: What's up with WhatsApp? Comparing mobile instant messaging behaviors with traditional SMS. In: Proceedings of the 15th International Conference on Human-Computer Interaction with Mobile Devices and Services (MobileHCI), Munich, Germany, pp. 352–361 (2013)
13. Statista: Most popular mobile messaging apps worldwide as of April 2018, based on number of monthly active users (in millions). The Statistics Portal (2018). https://www.statista.com/statistics/258749/most-popular-global-mobile-messenger-apps/
14. Al-Heeti, A.: WhatsApp: 65B messages sent each day, and more than 2B minutes of calls. C/net News. https://www.cnet.com/news/whatsapp-65-billion-messages-sent-each-day-and-more-than-2-billion-minutes-of-calls/. Accessed 14 July 2018
15. Chan, L.: WebCT revolutionized e-learning. UBC Reports, 51(7) (2005)
16. Johnson, G.M.: College student internet use: convenience and amusement. Can. J. Learn. Technol. 33(1) (2007)
17. Smith, S.D., Salaway, G., Borrenson, C.J.: The ECAR study of undergraduate students and information technology, EDUCAUSE Center (2009)

18. Barry, W.: Effective distance education: a primer for faculty and administrators. In: Monograph Series in Distance Education No. 2. Fairbanks, Alaska (1992)
19. Kathman, B.: 3 biggest trends impacting higher education communication this year. Higher Education Technology: SignalVine (2017). https://www.signalvine.com/higher-education-communication-trends/. Accessed 18 July 2018
20. Bull, B.: Eight roles of an effective online teacher. Higher Education Teaching Strategies from Magna Publications: Faculty Focus. https://www.facultyfocus.com/articles/online-education/eight-roles-of-an-effective-online-teacher/. Accessed 18 July 2013
21. Moisey, S., Hughes, J.: Supporting the online learner. In: Anderson, T. (ed.) The Theory and Practice of Online Learning, 2nd edn., pp. 419–432. AU Press, Athabasca (2008)
22. Burns, M.: Distance Education for Teacher Training: Modes, Models, and Methods. Education Development Center Inc., Washington, DC (2011)
23. Hudson, L., Ozanne, J.: Alternative ways of seeking knowledge in consumer research. J. Consum. Res. 14(4), 508–521 (1988)
24. Lincoln, Y., Guba, E.: Naturalistic Inquiry. In: Löfgren, K. (ed.) What is ontology? Introduction to the Word and the Concept. Examily. Sage, London (2013). https://examily.com/article/what-is-ontology-introduction-to-the-word-and-the-concept-10090. Accessed 14 July 2018
25. Neuman, L.W.: Social Research Methods: Qualitative and Quantitative Approaches, 4th edn. Allyn and Bacon, Boston (2000)
26. Bogdan, R., Biklen, S.K.: Qualitative Research for Education: An Introduction to Theories and Methods, 5th edn. Pearson, London (2006)
27. Corbin, J., Strauss, A.: Basics of Qualitative Research: Techniques and Procedures for Developing Grounded Theory, 3rd edn. SAGE Publications, Thousand Oaks (2008)
28. Creswell, J.W.: Research Design: Qualitative, Quantitative, and Mixed Methods Approaches, 4th edn. Sage, Los Angles (2014)
29. Jurado, R.G., Pattersson, T., Regueiro-Gomez, A. Scheja, M.: Classification of the feature in learning management systems. In: XVII Scientific Convention on Engineering and Architecture, Havana City, Cuba, 24–28 November28, Conference Paper (2014)
30. Ross-Gordon, J.M.: Research on adult learners: supporting the needs of a student population that is no longer nontraditional. Peer Rev. 13(1), 26–29 (2011)
31. National Center for Education Statistics.: Total fall enrollment in degree-granting institutions by control and type of institution, age, and attendance status of student: Digest of Educational Statistics 2009. Table 192.2007 (2009). http://nces.ed.gov/programs/digest/d09/tables/dt09_192.asp?referrer=list
32. Choy, S.: Findings from the Condition of Education 2002: Nontraditional Undergraduates. National Center for Education Statistics, Washington, DC (2002). http://nces.ed.gov/programs/coe/2002/analyses/nontraditional/index.asp
33. Earley, S.: There's no AI (artificial intelligence) without IA (information architecture). IT Professionals 18(3), 58–64 (2017). https://doi.org/10.1109/MITP.2016.43

Feature Map Transformation
for Multi-sensor Fusion
in Object Detection Networks
for Autonomous Driving

Enrico Schröder[1(✉)], Sascha Braun[1], Mirko Mählisch[1], Julien Vitay[2],
and Fred Hamker[2]

[1] Department of Development Sensor Data Fusion and Localisation,
AUDI AG, 85045 Ingolstadt, Germany
{enrico.schroeder,sascha-alexander.braun,mirko.maehlisch}@audi.de
[2] Department of Computer Sciences, Technical University Chemnitz,
Straße der Nationen, Chemnitz, Germany
{julien.vitay,fred.hamker}@informatik.tu-chemnitz.de

Abstract. We present a general framework for fusing pre-trained object
detection networks for multiple sensor modalities in autonomous cars
at an intermediate stage. The key innovation is an autoencoder-inspired
Transformer module which transforms perspective as well as feature activation characteristics from one sensor modality to another. Transformed
feature maps can be combined with those of a modality-native feature
extractor to enhance performance and reliability through a simple fusion
scheme. Our approach is not limited to specific object detection network
types. Compared to other methods, our framework allows fusion of pre-
trained object detection networks and fuses sensor modalities at a single
stage, resulting in a modular and traceable architecture. We show effectiveness of the proposed scheme by fusing camera and Lidar information
to detect objects using our own as well as the KITTI dataset.

Keywords: Autonomous driving · Perception · Sensor fusion ·
Object detection · Lidar

1 Introduction

One of the most important tasks in automotive perception is detecting objects
and obstacles like cars or pedestrians in the vehicle's surroundings. For reasons of
performance and robustness, autonomous cars combine multiple sensor modalities such as camera, Lidar- and Radar sensors to exploit individual strengths of
each sensor type.

Traditionally, discrete objects detected by each sensor are fused via a model-
based Bayesian Framework. Recently, methods based on convolutional neural
networks have been used to detect objects not only in raw images [9] but also

© Springer Nature Switzerland AG 2020
K. Arai and S. Kapoor (Eds.): CVC 2019, AISC 944, pp. 118–131, 2020.
https://doi.org/10.1007/978-3-030-17798-0_12

Lidar pointclouds [10,14], outperforming methods based on traditional hand-crafted features. This makes it viable to explore methods which fuse sensor modalities at a lower level of abstraction directly within these object detection networks to potentially make use of additional information that has not yet been abstracted away by an underlying model. Our method achieves this by fusing feature maps of pre-trained object detection networks, while still providing some of the benefits of a traditional late-fusion scheme such as modularity and traceability.

2   Related Work

In contrast to traditional image classification networks, object detection networks predict position and class of *multiple objects* in the frame. Most object detection networks first feed input images through a convolutional *feature extractor* network such as ResNet [5] or VGG [11]. The resulting feature maps are then passed through a combination of convolutional and fully-connected layers (the *detector*) to generate object bounding boxes and classes. In the context of autonomous driving we are especially interested in multi-modal variations of these object detection networks, detecting objects using both camera and Lidar sensors. Research in this area has made significant progress since introduction of the 2D bird view and 3D object detection challenges to the automotive KITTI dataset [3], providing labeled camera and Lidar data.

Most approaches implement involved network architectures with cross connections between the individual sensor paths at multiple locations. **Frustum PointNets** [8] do object detection on the ego-view camera image and then, for each object proposal, generate a 3D perspective frustum which defines a region-of-interest in the 3D point cloud. This region-of-interest is then fed to a segmentation network to generate 3D objects. While this approach performs well, it relies on camera alone to detect objects in the first place and then only enhances class and depth prediction by use of the Lidar pointcloud. **Multi-View 3D Object Detection** (MV3D) [1] projects Lidar pointclouds into front view and bird view images and generates 3D object proposals from bird view. These 3D proposals are then used to generate proposals for the projected views and to extract candidate regions from the individual feature maps via region-of-interest pooling (see [4]). Feature map region proposals from the three input views (front view Lidar, bird view Lidar and camera) are then fused via concatenation and fed to an object detector which outputs 3D bounding boxes. **Aggregate View Object Detection** (AVOD) [7] is similar, but additionally fuses input data before region proposal generation. Both MV3D and AVOD, while performing well, form quite complex architectures with multiple links between the two sensor modalities at different layers of abstraction. These architectures strictly require both sensors to be operational and hence would not be able to deal with sensor failure. Additionally, the complex nature of these architectures makes it difficult to analyze system behavior.

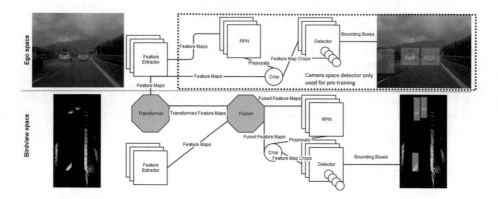

Fig. 1. Our proposed fusion architecture. White elements denote parts of the network belonging to an arbitrary state-of-the-art object detector (Faster-RCNN in our case). Blue elements denote our Transformer and Fusion modules mapping feature maps from ego to bird view. This allows using camera features in addition to Lidar for predicting objects' locations and classes in bird view. The camera-only detector is used only for training the feature extractor and can be removed afterwards.

A modular approach, fusing information at a single point, would be beneficial for real-world use cases. Our approach focuses on making feature map activations from different sensors compatible in locality and properties to allow for fusion. The author in [12] uses an approach that is not dissimilar to ours. The authors perspectively transform feature maps from the camera feature extractor network into the corresponding bird view space by using the camera's inverse perspective transformation matrix. Then they concatenate the two feature maps and perform region proposal generation and classification on this fused feature representation. However, the process of using a static inverse perspective transformation from camera space to bird view (and hence implicitly calculating depth information from a 2D image through inverse camera projection and ground plane assumption) is inherently lossy and produces features with strong distance-dependent distortion which are not suited well for processing via convolutional layers in a detection network. We thus improve upon this idea by learning the perspective transformation in order to generate non-distorted features.

The main component of our proposed architecture is influenced by the **Perspective Transformer** [13], originally used to learn perspective-invariant features from 2D projections of objects. They propose an encoder-decoder architecture that takes as input a 2D image of an object and generates 3D volumetric features. 2D input data is downsampled via convolutional operations, fed through a fully-connected bottleneck layer and then upsampled using transposed convolutions to a 3D volume. From this 3D volume, a 2D projection image is created. Similarity between projection of the generated 3D volume and the original 2D image is used as training objective. This forces the encoder to extract perspective invariant features of the object in order for the decoder to create meaningful 3D information from the fully-connected intermediate representation.

3 Fusion Architecture

In our implementation of our proposed architecture we use camera and Lidar sensors to detect objects in the autonomous vehicle's environment. Our employed architecture is shown in Fig. 1. The architecture respects that most relevant sensors can be processed efficiently in a 2D bird view representation, since all relevant sensors with exception of monocular cameras produce distance information. Thus, our framework mainly operates in 2D bird view and outputs 2D object bounding boxes. It consists of two standard object detection networks (one for each sensor modality) and the Transformer as well as a Fusion module. Fusion of the sensor modalities takes place at one single location within the object detection networks, after feature extraction and before region proposal generation.

3.1 Object Detectors

In our experiments, we use two Faster RCNN [9] object detection networks with Resnet feature extractors [5], one for camera ego view and one for Lidar bird view. Our proposed architecture does not have any special requirements for the object detection network (other then separation of feature extraction and detection), so it would work with any state-of-the-art object detection network. We modified the standard Faster RCNN network to output six scalars per bounding box, four for the object location and dimensions relative to the anchor with position (y_a, x_a) and dimension h_a, w_a and two representing object orientation angle ϕ as imaginary and real components t_{im}, t_{re} of a complex number $e^{i\phi}$ with unit length. Given ϕ, $t_{im} = sin(\phi)$ and $t_{re} = cos(\phi)$ or given t_{im}, t_{re}, $phi = arctan_2(t_{im}, t_{re})$. This representation of an oriented bounding box was suggested by [10] as a simple modification to add orientation to a bounding box without modifying the regression loss of the network, as the *euclidean distance* metric can be applied to this angle representation as well. Our predicted bounding box parameters are thus:

$$t_y = \frac{(y - y_a)}{h_a} \quad t_x = \frac{(x - x_a)}{w_a} \quad t_h = log(\frac{h}{h_a}) \quad t_w = log(\frac{w}{w_a}) \tag{1}$$

$$t_{im} = sin(\phi) \quad t_{re} = cos(\phi) \tag{2}$$

Note that this box encoding scheme is used not only on the bird view Lidar network, but also for the ego camera space network.

3.2 Transformer

Key component of our proposed fusion architecture is the Transformer module, shown in Fig. 2. It is used to transform feature maps from camera space into bird view, creating locality in the feature maps needed for further processing via convolutional layers in the detector network. We adopt a convolutional encoder-decoder scheme as employed in [13].

The general setup of the Transformer is comparable to that of a regular autoencoder, with the difference that the system is not tasked to recreate the original input after decoding. Instead, the task is to output features which are similar in *locality* and *properties* to that of the other sensor modality. The convolutional encoder-decoder represents a function $T : F_{ego} \rightarrow F_{birdview}$ from ego (camera) feature maps F_{ego} to bird view (Lidar) feature maps $F_{birdview}$. The Transformer has to perform two tasks: It has to transform *perspective* of the feature maps from ego to bird view while at the same time producing features in bird view *matching* those of the native Lidar feature extractor. We could model the perspective transformation via traditional methods (inverse perspective mapping), but this only works by ground plane assumption (every point in the image lies on the ground plane), resulting in strong distance-dependent distortions which are not suited for convolutional operations. Instead, we train the Transformer to perform both tasks in a joint manner, thereby being able to learn which features at what location in ego view are useful for detecting objects in bird view.

Input to the encoder are feature maps F_{ego} from the camera space feature detector. These feature maps contain feature activations with locality in regard to the input camera space. A cascade of convolutions downsamples these feature maps, which are then fed through two fully-connected network layers and are upsampled through transposed convolutions to generate the output feature map $F_{birdview}$. We use a second fully connected layer in the bottleneck in order to get matching dimensions of $F_{birdview}$ for fusion of the feature maps with those from the Lidar in the next step, so the exact dimensions are tailored to our particular experiment setup. A Rectified Linear Unit (ReLU) non-linearity follows after every layer.

The main idea behind the encoder-decoder architecture is that by squeezing information through a fully-connected bottleneck layer and then upsampling this information to the desired output space, the Transformer is forced to represent information in an perspective invariant manner. The characteristics of the transformation are entirely defined by the training procedure and not by the Transformer architecture: The output space is enforced by letting the system predict objects in the desired output space. Generated output features are implicitly specified by using them with a pre-trained detector network.

3.3   Fusion

Transformed feature maps $F_{ego \rightarrow birdview}$ from the prior step possess the same width, height and feature locality as those of the bird view Lidar detector $F_{birdview}$ (see Fig. 5 for an example of the transformed feature maps). This allows fusion with the native bird view Lidar feature map through various schemes, such as mean, weighted mean or concatenation. Following [1,7], we employ depth concatenation, i.e. concatenation along the depth axis of the feature map tensors. Since the resulting concatenated feature maps have incompatible depth for the bird view Lidar object detector, we feed them through two convolution operations (stride one, kernel 3×3) to bring the feature map depth down to the original depth of the bird view object detector.

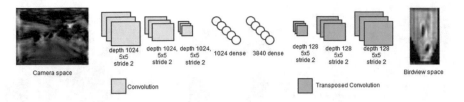

Fig. 2. Transformer: Encoder-decoder architecture for mapping feature maps from ego space to bird view space. Rectangular elements denote convolutional operations while circle-shaped elements are fully-connected layers. A *ReLU* non-linearity is applied after every layer.

4   Training

The training scheme to train the fusion network is outlined below. We use Momentum-based Stochastic Gradient Descent (SGD) with a learning rate of 10^{-4} for optimization. Additionally we use L2-regularization, Batch Normalization and Dropout with $p = 0.5$ on the convolutional layers of Transformer and Fusion modules.

4.1   Training Procedure

As outlined above, characteristics of the Transformer depend entirely on the Training procedure. Therefore we investigate two slightly different variants, denoted *variant A* and *variant B*. Both variants have the first step in common (Fig. 3a): Training of the **single-modality baseline** Faster-RCNN models for camera ego view and Lidar bird view individually with random weight initialization until convergence (i.e. no more significant improvement on validation error). The next training steps are described below.

Variant A. Trains Transformer and Fusion modules in separate steps. First, we train the **Transformer** (Fig. 3c): For this, we take the camera feature extractor trained in the previous step, connect it to a random-initialized Transformer module and feed its output through the bird view detection network that has also been trained in the previous step. The task is to predict objects in bird view. All weight parameters except those of the Transformer are frozen, meaning that only its weight parameters are affected by the optimizer. This forces the Transformer to learn a transformation from camera ego to Lidar bird view feature maps. After training until convergence, transformed feature maps now *mimic* those of the native Lidar bird view feature extractor. We could use this network in stand-alone mode to predict bird view objects (and hence implicitly predict depth information) from a monocular camera sensor. For fusion (Fig. 3d), we insert a random-initialized **Fusion** module between Transformer, Lidar bird view feature extractor and detector. We now freeze all weights except those of the Fusion module and again train the entire system until convergence with detecting bird view objects as task.

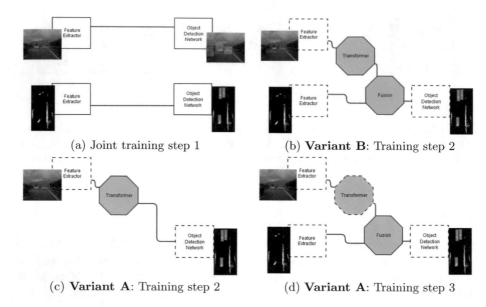

(a) Joint training step 1

(b) **Variant B**: Training step 2

(c) **Variant A**: Training step 2

(d) **Variant A**: Training step 3

Fig. 3. Top-level network layouts for the employed training steps. Solid boxes indicate the modules which are being trained in this step. Dashed boxes indicate pre-trained, weight-fixed modules from a previous training step. Training (a) Joint training of single modality baseline networks. (b) Variant B: Training of Transformer and Fusion Module. (c) Variant A: Training of Transformer. (d) Variant A: Training of Fusion module with pre-trained Transformer.

Variant B. Trains Transformer and Fusion modules in a joint manner (Fig. 3b). The setup is the same as that of the last step of Variant A, however the Transformer has not been trained on its own and its weights are thus also randomly initialized. All weights except those of the Transformer and Fusion modules are frozen and the whole system is again trained until convergence with the task of detecting bird view objects.

4.2   Objective Function

Objective (loss) function for all training phases is the default loss function as employed in the original Faster-RCNN paper [9], minimizing error over class and object location for each of the anchor boxes i that the network predicts:

$$L_{\text{det}} = \frac{1}{N_{\text{cls}}} \sum_i L_{\text{cls}}(p_i, p_i^*) + \lambda \frac{1}{N_{\text{reg}}} \sum_i L_{\text{reg}}(t_i, t_i^*) \tag{3}$$

L_{cls} denotes the *classification loss* while L_{reg} is the bounding box regression loss (which for us includes six regression parameters as we also include bounding box orientation as outline in Sect. 3.1). p_i, p_i^* are predicted and groundtruth classes and t_i, t_i^* are predicted and groundtruth bounding box parameters. For any of the training phases described above we use this prediction-centric loss function while freezing different parts of the network weights.

When training the Transformer, we additionally use a reconstruction loss L_r, which is the mean squared distance between transformed camera ego view feature map $\mathsf{F}_{ego\rightarrow birdview}$ and native Lidar bird view feature map $\mathsf{F}_{birdview}$, as this helps convergence by encouraging similarity to the native Lidar feature map:

$$L_{\mathrm{r}} = \frac{1}{N} \sum_i^N (\mathsf{F}^i_{birdview} - \mathsf{F}^i_{ego\rightarrow birdview})^2 \qquad (4)$$

where i is the index running over all entries of the feature maps. The full loss term L thus becomes

$$L = L_{\mathrm{det}} + \lambda L_{\mathrm{r}} \qquad (5)$$

with a hyperparameter λ for weighting the reconstruction loss (set to 0.5 for our experiments).

5   Data

We evaluate our method on two datasets: Our own automotive dataset as well as the public KITTI dataset [3]. Our own dataset is comprised of 16 mixed driving sequences (mostly European highway and rural) with an overall length of approx. 8 h sampled at 10 hz, yielding 284690 samples. The employed sensor set consists of a monochrome camera with resolution 1280×960 and a field of view of $45°$ as well as a front-facing four channel Lidar with a field of view of $140°$. Additionally a Velodyne HDL-64E 64 channel Lidar sensor is used as reference sensor to manually label objects in 3D space (x/y/z position, width/length/height, yaw orientation and class) in the ego vehicles own lane and the adjacent left and right lanes up to a distance of 100 m. From each of the 16 sequences, two subsequences amounting to 20% length are set aside as validation set. Furthermore we have a separate 40 min long mixed driving sequence that we only use as test set. See Table 1 for statistics on our dataset and Fig. 4 for distribution of groundtruth objects. For experiments on the KITTI set we use a custom training/validation split of approximately 7:1, where we made sure that all samples in the validation set are from different sequences as those in the training set.

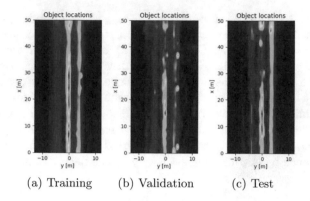

(a) Training   (b) Validation   (c) Test

Fig. 4. Distribution of groundtruth object locations (bird view) in our dataset, encoded as heatmap. Hotter locations denote areas of increased occurrence of groundtruth objects. Note that the labeled area includes only the own vehicles' lane as well as the ones right next to it.

5.1 Preprocessing

Camera images are preprocessed through VGG-style per-channel mean substraction as in [11]. We project Lidar pointclouds to two 2D bird view grid representations as suggested by [1], one containing the normalized height of the highest Lidar point which falls into a particular grid location and one containing the normalized mean intensity of all points falling into a grid location. The two grids are then concatenated by depth, resulting in two-channel 2D input images for the Lidar object detector. We use a [50 m, 25 m] area as viewport for Lidar bird view projection, with the ego vehicle located at the bottom of the frame and facing upwards. Figures 5a and c show an example of our input camera and lidar data. For KITTI experiments we employ the same data preprocessing, however due to the higher field of view of the camera of 90° we use a viewport of [50 m, 50 m].

Table 1. Statistics of our employed dataset. Note that for our experiments we only use the class *car* and thus only report numbers for this class.

	Frames	Duration	Number of instances (Car)
Training	225745	6:16:14	110662
Validation	58945	1:38:14	31059
Test	31090	0:38:20	25809

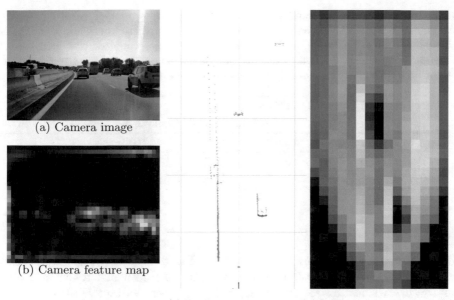

(a) Camera image

(b) Camera feature map

(c) Bird view projection of (d) Transformed camera
Lidar pointcloud   feature map

Fig. 5. (a) shows the front camera image of an example scene. (b) shows one of the 1024 feature maps output by the camera feature extractor. (c) shows our bird view Lidar projection as input to the Lidar feature extractor. (d) shows the feature map of (a) transformed from ego to bird view by the Transformer after being trained using *Variant A* of the training procedure (see Sect. 4.1). Note that activations are generated from the 2D camera input image only, so the Transformer implicitly has to learn to predict depth from the image. You can clearly identify areas in the feature maps belonging to the two cars in front of the ego vehicle.

6   Experiments

We use an extended version of the Object Detection API framework [6] for our implementation, which has been heavily modified to support multiple synchronous input sensor streams and predicting objects in multiple spaces. Reported metrics are Pascal VOC metrics [2] with mean average precision at 0.5 intersection over union (IOU). IOU is calculated on axis-aligned bounding boxes regardless of an object's orientation, so we also require the predicted orientation to be within $[+\frac{\Pi}{16}, -\frac{\Pi}{16}]$ of the groundtruth orientation to count as a true positive. We investigate both variant A and B of the training algorithm. For comparison, we also provide results for the Lidar-only detector which has been used as baseline for Transformer and Fusion module training. Our networks have been trained either entirely on our own dataset or entirely on KITTI (due to the different employed sensors in both datasets).

Table 2. Results on our validation set. We report mean average precision (mAP) at bounding box intersection-over-union (IOU) 0.5 for class *car*. Numbers labeled *Cam+Lidar* describe results with both sensor modalities enabled. Numbers labeled *Lidar* or *Cam* describe results with that particular sensor modality only (in case of Fusion, the corresponding input feature maps to the Fusion stage are set to zero).

	Baseline	Fusion Variant A			Fusion Variant B		
	Lidar	Cam+Lidar	Lidar	Cam	Cam+Lidar	Lidar	Cam
Own validation	0.85	0.90	-	-	**0.94**	-	-
Own test	0.78	0.78	0.56	0.16	**0.82**	0.70	0.0
KITTI validation	0.75	0.76	0.64	0.01	**0.77**	0.68	0.0

6.1   Results

Table 2 shows results of the experiments while Fig. 6 contains some sample scenes with predictions of the fusion system. We see that both fusion schemes show improved classification results on our validation set. Variant B also improves performance on the test set as well. Variant A however allows to "switch off" one of the sensor paths (simulated in our experiments by setting the tensor coming from this sensors' feature extractor to zero). In case of disabling the camera, the network is still able to show good detection performance, though worse than the Lidar-only network. This shows that the network relies on a combination of features from both Lidar and camera. When disabling the Lidar, detection performance drops significantly, although near objects directly in front of the car are still detected relatively well. These are very desirable characteristics for any real-world perception system which has to deal with sensor failures or blindness of individual sensors due to weather. Variant B (joint training of Transformer and Fusion from scratch) does not result in the Transformer learning any features that are useful on their own. However, running this network as intended with both sensor modalities enabled improves detection performance compared to the Lidar-only baseline network. Qualitatively, fusion significantly reduces false-positives compared to Lidar-only (which tends to erroneously detect objects on barriers or other static objects resembling the characteristic L-shapes in the projected Lidar data).

We can confirm these results on the KITTI set as well. Variant A camera-only however does not yield proper detections even though the feature maps are interpretable like those on our own dataset. Note that the KITTI dataset is several magnitudes smaller than our dataset, while the employed Lidar sensor has a much higher resolution resulting in much denser pointcloud projection. In order for the Transformer to produce feature maps that are similar enough to the dense ones from the Lidar feature extractor, it would need a higher capacity. However, due to the limited amount of data we could not significantly increase capacity of the Transformer. Nonetheless, we see improved detection performance when using both sensor modalities together.

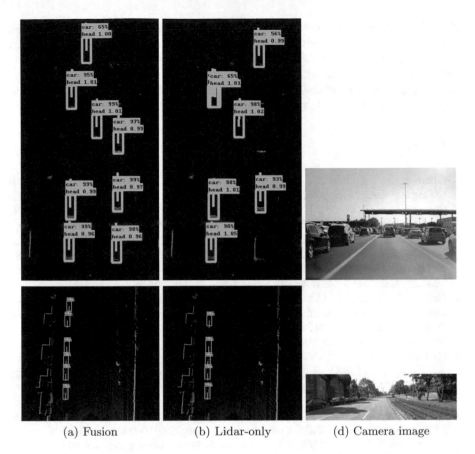

(a) Fusion   (b) Lidar-only   (d) Camera image

Fig. 6. Inference detection samples on our test set (top row) as well as on the KITTI validation set (bottom row). (a) and (b) show detected objects in bird view for the Fusion system and the Lidar-only system as reference (plotted over the projected Lidar pointcloud). We see that in accordance to the quantitative results the fusion system manages to detect more of the objects in the scene. Qualitatively, using both sensors in conjunction also improves false positives which the Lidar-only network tends to generate on isolated frames.

7   Conclusion

We propose a novel and general method for fusing pre-trained single-modality object detection networks. Key component is a Transformer module which learns transformation of feature map activations from one sensor space to another, using one pre-trained detection network for another sensor modality. Using this method, we show that we can fuse features from a camera sensor in ego space with those of a Lidar sensor in bird view to improve detection performance compared to using only the Lidar. Our fusion method differs from the state-of-the-art in that it fuses sensor modalities at a single location, leaving the underlaying single modality networks intact. This results in a simple network architecture, increasing traceability of the entire system and allowing to deal with real-world challenges such as sensor failures or blindness.

References

1. Chen, X., Ma, H., Wan, J., Li, B., Xia, T.: Multi-view 3D object detection network for autonomous driving. In: 2017 IEEE Conference on Computer Vision and Pattern Recognition (CVPR), pp. 6526–6534 (2017). https://doi.org/10.1109/CVPR.2017.691. http://doi.ieeecomputersociety.org/10.1109/CVPR.2017.691
2. Everingham, M., Van Gool, L., Williams, C.K.I., Winn, J., Zisserman, A.: The pascal visual object classes (VOC) challenge. Int. J. Comput. Vis. **88**(2), 303–338 (2010). https://doi.org/10.1007/s11263-009-0275-4, http://link.springer.com/10.1007/s11263-009-0275-4
3. Geiger, A., Lenz, P., Urtasun, R.: Are we ready for autonomous driving? The kitti vision benchmark suite. In: 2012 IEEE Conference on Computer Vision and Pattern Recognition (CVPR), pp. 3354–3361. IEEE (2012). http://ieeexplore.ieee.org/abstract/document/6248074/
4. Girshick, R.: Fast R-CNN. In: Proceedings of the IEEE International Conference on Computer Vision, pp. 1440–1448 (2015). http://www.cv-foundation.org/openaccess/content_iccv_2015/html/Girshick_Fast_R-CNN_ICCV_2015_paper.html
5. He, K., Zhang, X., Ren, S., Sun, J.: Deep residual learning for image recognition. In: Proceedings of the IEEE Conference on Computer Vision and Pattern Recognition, pp. 770–778 (2016). https://www.cv-foundation.org/openaccess/content_cvpr_2016/html/He_Deep_Residual_Learning_CVPR_2016_paper.html
6. Huang, J., Rathod, V., Sun, C., Zhu, M., Korattikara, A., Fathi, A., Fischer, I., Wojna, Z., Song, Y., Guadarrama, S., Murphy, K.: Speed/accuracy trade-offs for modern convolutional object detectors. arXiv:1611.10012 [cs] (2016)
7. Ku, J., Mozifian, M., Lee, J., Harakeh, A., Waslander, S.: Joint 3D Proposal Generation and Object Detection from View Aggregation. arXiv:1712.02294 [cs] (2017)
8. Qi, C.R., Liu, W., Wu, C., Su, H., Guibas, L.J.: Frustum PointNets for 3D object detection from RGB-D data. In: Proceedings of the IEEE Conference on Computer Vision and Pattern Recognition, pp. 918–927 (2018). http://openaccess.thecvf.com/content_cvpr_2018/html/Qi_Frustum_PointNets_for_CVPR_2018_paper.html
9. Ren, S., He, K., Girshick, R., Sun, J.: Faster R-CNN: Towards Real-Time Object Detection with Region Proposal Networks. arXiv:1506.01497 [cs] (2015)
10. Simon, M., Milz, S., Amende, K., Gross, H.M.: Complex-YOLO: Real-time 3D Object Detection on Point Clouds. arXiv:1803.06199 [cs] (2018)
11. Simonyan, K., Zisserman, A.: Very Deep Convolutional Networks for Large-Scale Image Recognition. arXiv:1409.1556 [cs] (2014)
12. Wang, Z., Zhan, W., Tomizuka, M.: Fusing Bird View LIDAR Point Cloud and Front View Camera Image for Deep Object Detection (2017). https://arxiv.org/abs/1711.06703

13. Yan, X., Yang, J., Yumer, E., Guo, Y., Lee, H.: Perspective transformer nets: learning single-view 3D object reconstruction without 3D supervision. In: Lee, D.D., Sugiyama, M., Luxburg, U.V., Guyon, I., Garnett, R. (eds.) Advances in Neural Information Processing Systems 29, pp. 1696–1704. Curran Associates, Inc. (2016)
14. Zhou, Y., Tuzel, O.: VoxelNet: end-to-end learning for point cloud based 3D object detection. In: Proceedings of the IEEE Conference on Computer Vision and Pattern Recognition, pp. 4490–4499 (2018). http://openaccess.thecvf.com/content_cvpr_2018/html/Zhou_VoxelNet_End-to-End_Learning_CVPR_2018_paper.html

Cross-Safe: A Computer Vision-Based Approach to Make All Intersection-Related Pedestrian Signals Accessible for the Visually Impaired

Xiang Li[1,2], Hanzhang Cui[1,2], John-Ross Rizzo[2,3], Edward Wong[1,2], and Yi Fang[1,2,4($\boxtimes$)]

[1] NYU Multimedia and Visual Computing Lab, New York, USA
yfang@nyu.edu
[2] NYU Tandon School of Engineering, Brooklyn, USA
[3] NYU Langone Medical Center, New York, USA
[4] NYU Abu Dhabi, Abu Dhabi, UAE

Abstract. Intersections pose great challenges to blind or visually impaired travelers who aim to cross roads safely and efficiently given unpredictable traffic control. Due to decreases in vision and increasingly difficult odds when planning and negotiating dynamic environments, visually impaired travelers require devices and/or assistance (i.e. cane, talking signals) to successfully execute intersection navigation. The proposed research project is to develop a novel computer vision-based approach, named Cross-Safe, that provides accurate and accessible guidance to the visually impaired as one crosses intersections, as part of a larger unified smart wearable device. As a first step, we focused on the red-light-green-light, go-no-go problem, as accessible pedestrian signals are drastically missing from urban infrastructure in New York City. Cross-Safe leverages state-of-the-art deep learning techniques for real-time pedestrian signal detection and recognition. A portable GPU unit, the Nvidia Jetson TX2, provides mobile visual computing and a cognitive assistant provides accurate voice-based guidance. More specifically, a lighter recognition algorithm was developed and equipped for Cross-Safe, enabling robust walking signal sign detection and signal recognition. Recognized signals are conveyed to visually impaired end user by vocal guidance, providing critical information for real-time intersection navigation. Cross-Safe is also able to balance portability, recognition accuracy, computing efficiency and power consumption. A custom image library was built and developed to train, validate, and test our methodology on real traffic intersections, demonstrating the feasibility of Cross-Safe in providing safe guidance to the visually impaired at urban intersections. Subsequently, experimental results show robust preliminary findings of our detection and recognition algorithm.

Keywords: Visual impairment · Assistive technology · Pedestrian safety · Portable device

© Springer Nature Switzerland AG 2020
K. Arai and S. Kapoor (Eds.): CVC 2019, AISC 944, pp. 132–146, 2020.
https://doi.org/10.1007/978-3-030-17798-0_13

1   Introduction

1.1   Backgrounds

The World Health Organization circa 2014 concluded that there are more than 39 million people suffering from blindness and more than 246 million people suffering from low vision. In the United States alone, billions of dollars are spent per year towards direct and indirect medical cost for vision-related illness and accidents. Visual impairment is well-known to impede both spatial perception and object detection. This creates a myriad of mobility and related functional difficulties, including but not limited to trips, falls, and traffic accidents. Even in situations where an object may be partially visualized or roughly localized, difficulties still abound.

To make traffic intersections a comfortable and functional environment for the blind or visually impaired traveler, Accessible Pedestrian Signals (APS), a special kind of devices affixed to normal pedestrian signal poles, were designed. APSs convey critical real-time traffic information (i.e. red light, walk signal) in an audible and vibrotactile multi-media format to assist blind or low vision pedestrians in street crossing [1]. Figure 1 gives an illustration of the APS at the road intersection. Wired to a pedestrian signal, an APS is able to synchronize with traffic light information, concurrently transcribe audible and/or vibrotactile messages, and consequently provide the visually impaired traveler with information about the traffic signal at road intersections. APS is a great advance for accessible environments for the visually impaired but unfortunately are not systematically installed in many metropolises across the United States. In NYC alone, there are 12,460 signalized intersections and presently (as of writing this paper) 288 intersections with APS technology deployed, representing approximately 1/50th of the necessary installations. Considering the high cost associated with installations ($45,000), it is currently planned to install 75 sets of equipment per year, which is expected to be completed in 162 years [1].

Administration conceived of and implement High-Intensity Activated Crosswalk (HAWK) signals, firstly installed at Georgia Ave. and Hemlock St, NW in 2009 [3]. The HAWK usually features APS devices to provide the blind or visually impaired pedestrians with real-time signal information at intersections through an audible message and/or vibrating arrow, plus additional motorist signals to indicate messages about pedestrian crossings; in certain cases, HAWK signals are paired with automated pedestrian detectors to assist detection and motorist signaling, but light conversion can give severe safety concerns [3].

Given these progressively increasing safety concerns due to a drastic dearth of APS technology and the massive increases in electric and more silent hybrid vehicles on the road, there is a dire need to innovate past conventional approaches to these identified accessibility predicaments. Object recognition is a technology in the field of computer vision, for finding and identifying objects of interest in images. Humans can recognize objects of different sizes, scales, and angles with little effort. However, while computer-based recognition was a challenging task and multiple approaches, such as edge or gray-scale matching, had been tested

Fig. 1. Illustration of APS at road intersection [2].

and implemented over the years, more recently, these bottlenecks have been alleviated. The development of Deep Learning (DL), especially convolutional neural networks (CNN), has enabled object detection and recognition algorithms to become more accurate and efficient, and the implementation in many real-world applications, such as autonomous driving, face recognition, military target detection, etc.

1.2   Related Work

Helping visually impaired people traveling in urban environments is an important consideration for civil engineers, urban planners, and for local, state and federal government. Within navigation and mobility challenges faced daily by those with visual impairment, intersection crossing ranks highly as one of the most dangerous. Several types of technologies have been developed to assist the visually impaired crossing traffic intersections. For example, governmental authorities install voice-assistance devices on traffic light posts for disabled pedestrians (see above on APS) to guide visually impaired individuals across intersections safely; HAWKs enable accessibility to pedestrians and also to motorists on unsignalized intersections (see above); Talking Signs allow visually impaired travelers to locate signs in intersections [4]. Despite effective results, these solutions all require substantial installation costs and hardware/componentry at the intersection of interest and create deployment bottlenecks, stifling accessibility.

Several smartphone applications including Intersection Explorer and Nearby Explorer for Android, and Intersections and Sendero GPS LookAround [5] for iPhone were developed to provide a lower-cost solution. Such applications are promising but are incapable of providing detailed information about the intersection nor real-time traffic control patterns or dynamics. Another attempt of combining the two strategies above is the Mobile Accessible Pedestrian Signal

(MAPS) system [6]. Again, however, it also requires hardware installations at intersections.

Some work on computer vision algorithms for detecting crosswalks, zebra 'stripes' or lines and traffic lights [7–12] have been implemented to help the visually impaired. One advantage afforded by computer vision-based approaches and algorithms is that they can detect and recognize the regions of interests in the scene (walking signals, zebra lines and traffic lights, etc.) without the installation of expensive intersection-dependent hardware and can often come 'on-line' following initial iterations with much higher 'success' rates (e.g., capturing the majority of pedestrian traffic signals, rather than the underwhelming minority given hardware implementation, i.e. 1/50th, as described above).

With recent improvements in powerful GPU's and algorithms, the processing time for object recognition has been reduced without jeopardizing accuracy. Convolutional Neural Networks (CNN) are particularly useful for object detection/recognition and scene segmentation because of the ability to extract the features of objects from various angles and distances in images. One such example of this pioneering work using a CNN for object detection is the R-CNN model [13]. However, the detection efficiency is limited because it adopts a multi-stage pipeline and region proposals are extracted before the CNN network, which also limits the overall detection accuracy. Following the success of using the CNN for object detection, many superior deep learning algorithms were proposed. Fast R-CNN [14] improved the detection speed using the Spatial Pyramid Pooling Network (SPP-Net) and ROI pooling layers; Faster R-CNN [15] proposed to use a Region Proposal Network (RPN) for object proposal extraction and improved the computational speed; YOLO [16] and SSD [17] significantly boosted detection speed by making a trade-off between speed and accuracy. Projects implemented on lighter embedded devices such as that used for Door Handle Detection [18] were developed based on YOLO.

1.3   Our Solution: Cross-Safe

Our solution is to develop a novel computer vision-based approach, named Cross-Safe, that provides accurate and accessible guidance to the visually impaired as one crosses intersections, as part of a larger unified smart wearable device. As a first step, we focused on the red-light-green-light, go-no-go problem, as accessible pedestrian signals are drastically missing from urban infrastructure in New York City. Cross-Safe leverages state-of-the-art deep learning techniques for real-time pedestrian signal detection and recognition. Our long-term goal is to reverse the increased morbidity and mortality resulting from low vision and blindness by providing wearable technology solutions with pan-applicability to visual deficits from all etiologies, leading to true functional independence. Our proposed solution, the Visually Impaired Smart Service System, provides real-time situational and spatial awareness in one's immediate environment, allowing the individual to travel more safely in a complicated city environment. Cross-Safe is a feature set that will be included as part of our automated platform. The Cross-Safe real-time pedestrian signal detection and recognition component will leverage the larger platform's portable GPU unit which provides mobile visual computing, along with a cognitive assistant component to provide real-time voice-based guidance. Figure 2 gives an overview of the proposed system.

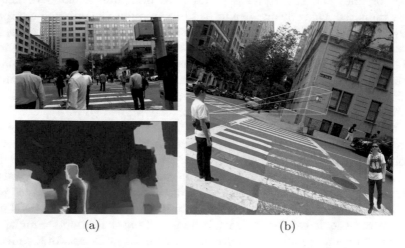

(a)   (b)

Fig. 2. Illustration of the proposed Cross-Safe system. (a) example of captured RGB image (top) and depth image (bottom). (b) illustration of traffic light detection at the road intersection.

2   Method

The core of this larger service system technology consists of four main elements: sensors, micro-controller, human-machine interfaces, and user (as illustrated in Fig. 2). The first (sensory) element set includes a collection of various sensors (infrared sensor, monocular/stereo camera sensor and ultrasonic sensor as shown in the Fig. 2). The sensors are ergonomically equipped on the scaffold to consistently and reliably acquire scene data/information from the immediate (surrounding) environment. The second (processing) element, a micro-controller unit, is responsible for coordinating the communication flows among different functional system units. Along with a portable GPU unit, the platform is able to complete a list of the scene and audio understanding tasks including scene parsing, object detection as well as audio recognition. In our project, Nvidia Jetson TX2, is chosen as the GPU unit for our portable platform for its high computing performance for deep learning and computer vision algorithms, power efficiency, and portability. The third (feedback) element set, the audio interfaces, displays processed and filtered environmental information to the end user in real-time via an audio feedback that is delivered through bone-conduction transducers in a paired headset. The end user, as the fourth (user) element, receives live scene information and dynamic alerts in an ongoing fashion upon full-system initiation (core) and responds accordingly. The detailed description of each component is given as follows.

2.1 Traffic Signal Detection

Model. To achieve real-time traffic signal detection and recognition, we first divide the large wild scene image into small gridded patches, and a light convolutional neural network is used to distinguish whether each patch contains green light, red light or no light. This provides efficient and accurate classification of the patches in a larger scene, and the boundary of each patch can be viewed as an object localization. The overview of our CNN architecture is shown in Fig. 3. Our CNN model contains 4 convolutional layers, 2 maxpooling layers, and 2 fully connected layers.

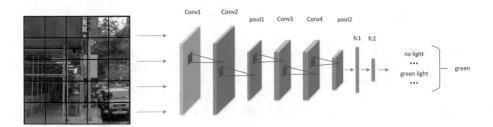

Fig. 3. Overview of our CNN model. 'Conv' indicates convolutional layer, 'pool' indicates pooling layer, and 'fc' indicates fully connected layer. In the inference stage, the predicted labels of each patch are fused to get the final prediction.

One should note that there exists a sample imbalance problem as each large wild scene is divided into many smaller patches while only a few of them contains pedestrian traffic lights (green or white/red). Directly training this CNN will lead the model to pay much more attention to the negative examples (no traffic light) and fail to recognize positive ones (green/red lights) accurately. To address this problem, the AdaBoost algorithm is integrated into our base CNN to enable the adaption of the weight of each sample on the fly. This strategy is easy to implement and introduces no additional parameters.

AdaBoost [19] was first introduced by Yoav Freund and Robert Schapire in 1995. It works by adapting the weights of samples based on the previous basic classifier; the weighted samples are used again to train the next basic classifier. At the same time, a new weak classifier is added in each round until the error rate is small enough or a predetermined maximum number of iterations is reached. Our final model with Adaboost is shown in Fig. 4.

Dataset Preparation. As we mentioned above, there is only a small portion of images/patches with a traffic light (red/green). In addition to dealing with this sample imbalance problem in the model design through sample weight adaption using the Adaboost algorithm, data reinforcement is another effective way of solving this issue. After the dataset is split into training and validation sets and

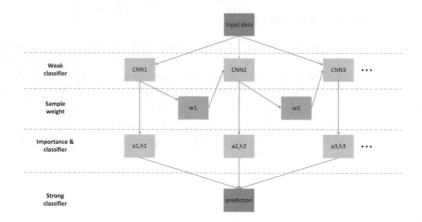

Fig. 4. Overview of our pedestrian traffic signal detection model. The overall pipeline contains a successive of CNN networks as weak classifier, each classifier CNN_i is trained independently, and takes as input both input data and the sample weight w_{i-1} calculated from its preceding classifier CNN_{i-1}. The importance of each classifier a_i and output probability h_i are summarized to get the final prediction.

divide each image into small patches, we augment all the positive images (here, positive means images with a pedestrian traffic light).

The dataset is augmented by randomly flip 30% of the positive patches horizontally and randomly flip 10% of positive patches vertically; randomly choose 70% of the positive patches and crop by −5% to 10% of their height and width; randomly choose 70% of the positive patches and make the following transformations: translate by −10% to 10% per axis; scale images to 90% to 110% of their original size, individually per axis; rotate by −10° to 10°. The procedure above is repeated 5 times. Together with the original set of patches, the total number of positive patches is augmented 6 times.

The negative patches (negative means patches without a traffic light) are generated by randomly cropping areas that have no overlap with traffic lights. We pick 6 negative patches per image to make a balanced dataset so that our model receive the same numbers of positive and negative patches.

2.2   Wearable Devices

Figure 5 shows the components used in the prototype system that contains the Cross-Safe feature. Detailed information about the hardware is as follows:

Zed Camera: In this project, we use the Zed Camera, which captures 1080p HD video at 30FPS or WVGA at 100FPS.

Processor: We use Nvidia Jetson TX2 as the processor because of its portability, high performance and power-efficiency. Nvidia Jetson TX2 is an embedded module with Nvidia Pascal GPU. It has 256 CUDA cores, ARM 128-bit CPU,

8 GB LPDDR4 memory and 32 GB eMMC SDIO SATA storage. When Cross-Safe system receives a command from the user, it will request an image from the camera and run the recognition model over OpenCV and Keras backend.

Vocal Device: After the image runs through the classifier algorithm, the result will be converted into voice feedback, telling the user whether it is safe to cross the intersection. In our prototype, we use a bone-conduction transducer in a paired headset (binaural). The detection results will be translated into meaningful phrases or sentences using a Text-To-Speech (TTS) toolkit.

Power Bank: We use a MAXOAK 50000 mAh 5/12/20v Portable Charger External Battery as the power bank in our prototype. It allows the Jetson TX2 to run for 24 h with a full-load operation, which allows a visually impaired individual to go outdoor for a full day.

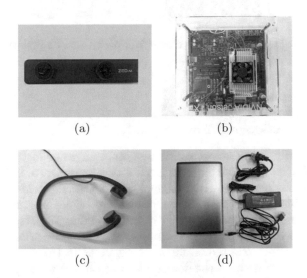

Fig. 5. Hardwares. (a) Zed camera used in our prototype, (b) Nvidia Jetson TX2, (c) bone-conduction headset for voice feedback, and (d) power bank and its accessories (power cables of power bank itself and TX2, and AC adapter).

2.3   Human Machine Interaction

A bi-directional communication route is designed in our platform. The communication route in blue represents the default information flow where the system obtains environmental information through the sensors, detects and recognizes traffic signals at street intersections and provide feedback to the user through audio interfaces. Specifically, in this communication route, the environmental information captured by the sensor will be processed in the GPU unit for both scene parsing and object detection. The results will be structured and organized for further post-processing and ultimately sent back to the user via audio feedback. The communication route in red is for user-selected information flow where the end user can explore nearby scenes and send their instructions to the system. Specifically, the system understands the user's instructions by recognizing his or her voice and a set of pre-defined speech commands. The overall human-machine interaction pipeline is shown in Fig. 6.

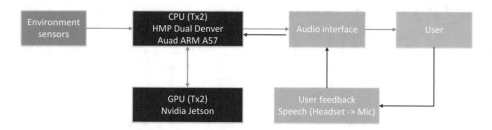

Fig. 6. Major components of the proposed Cross-Safe platform.

3 Experiments

3.1 Data Collection and Annotation

Data Collection. Our dataset contains 3,693 images collected by volunteers from New York City with Zed camera. Each image has a size of 900×1200 pixels, and most images contain one traffic light located near the center. We manually annotated the images by providing a bounding box and a class label for each traffic light. These images are captured from New York City streets and contain common street scenes like cars, trees, buildings, and pedestrians, etc. Figure 7 gives examples of the collected traffic scene dataset. In this paper, we focus on traffic light detection without the depth information and we only used the R, G, B channels of the collected images.

3.2 Experimental Details

We implement our model based on the Keras Deep Learning Framework [20]. Our segmentation network is trained from scratch using stochastic gradient descent (SGD) algorithm, with an initial learning rate set to 1e–3. We use a weight decay of 1e–6 and a dropout rate of 0.25. To facilitate network training, each training and validation image was divided into patches of size 96×96 pixels. Data augmentation is performed on the positive patches in the training set as described above. Details on the number of patches in the training, validation and test sets are shown in Table 1. Note that no cropping is performed for the test set.

Fig. 7. Examples from the traffic scene dataset (cropped at the center of traffic lights). 'g' stands for green light, 'r' stands for red light.

Table 1. Dataset details. The first two rows show the number of patches in training and validation sets. The third row indicates the number of patches after data augmentation. The last row indicates the number of images in the test set.

	Red	Green	No light
Train	1600	1020	15000
Valid	499	307	1348
Train (aug)	6606	4314	15000
Test	136	131	-

In the testing stage, each image is divided into patches using a sliding windowing of size 96×96 and with a stride of 50 pixels. After generating the predicted label for each patch, we count the number of 'red' and 'green' patches. If there are more 'red' patches than 'green' patches, the whole scene will be determined as 'red'; otherwise, it will be determined as 'green'.

3.3   Model Evaluation

First, we evaluate the classification performance on the patches in the validation set. Table 2 shows the confusion matrix. As shown in Table 2, our model gives a satisfactory accuracy of 98.5% for patch classification. Figure 8 gives some challenging examples of misclassification. One can see that our model fails to distinguish blurry patches, which can be caused by camera jitter, strong/weak sunlight or limited focal length. In addition, there are a few numbers of labeling errors which affects our classification accuracy.

Table 2. The confusion matrix for the patches in the validation set (2154 patches in total). The rows stand for the ground truth labels, and the columns indicate the predicted labels.

	Red	Green	No light
Red	479	4	16
Green	2	300	5
No light	3	3	1342

Then, we test the performance of our model on the test images using a sliding technique as described above. Numerical results are shown in Table 3. As it can be seen, our proposed model has an overall accuracy of 94.0%. Figure 9 gives examples of correct traffic signal detection. In Fig. 9(a), our model generated 9 predicted labels for the patches, 2 of which are predicted as green and none is red, so our model predicts green for this image. Figure 10 shows the precision-recall curve for red/green light detection. Our model got a mAP (mean average precision) of 0.96 for traffic light detection. It is worth mentioning that our model has a higher recall value for red light detection than green light detection when the precision is high enough. The numerical results in Table 3 verify this finding and show that our model didn't miss detecting the red light. Missing a red light will be more critical than the false alarms. Hence, this feature of our proposed model is more pragmatic.

Fig. 8. Selected examples of prediction errors. 'gt' indicates ground truth, 'pred' indicates predicted label. 'g' for green light, 'r' for red light, 'n' for no light. One should note that among all prediction errors, (1, 1), (1, 2), (1, 3), (2, 1), (2, 4), (2, 5), (3, 5), (4, 1) are caused by blurry images, and (1, 5), (2, 2), (2, 3), (3, 2), (4, 3), (4, 5) may be caused by labeling errors.

Table 3. The confusion matrix for the test set images (267 images in total, each image contains either a red light or a green light). The rows stand for the ground truth labels, and the columns indicate the predicted labels.

	Red	Green	No light
Red	131	0	0
Green	16	120	0
No light	0	0	0

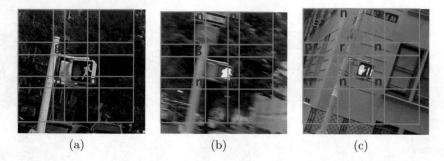

(a)   (b)   (c)

Fig. 9. Selected examples of correct predictions. Text in the upper-left corner of each patch indicates the predicted label, in which 'g' stands for green light, 'r' stands for red light, and 'n' stands for no light. (a) and (b) show images with green light, and (c) shows an image with red light.

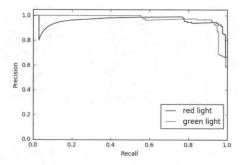

Fig. 10. Precision-recall curves for red/green light detection. The proposed model got a satisfying accuracy for both red and green light detection, with a mAP of 0.96. The drop to the left of the red line is most likely due to a mislabeling error in the test set.

4   Conclusion and Future Work

This paper presents a novel and effective technique for the blind or visual impaired to safely cross signalized intersections. State-of-the-art convolutional neural network-based algorithms are used to implement pedestrian traffic signal detection and recognition, along with portable CPU/GPU units to establish an effective computer vision-based assistive platform. A speech synthesizer is deployed in our module to support friendly human-machine interaction. We provide details for the overall structure of our feature set and, by extension, platform, as well as clarifying how the user can interact with the intelligent system. Experimental results on a large-scale wild scene dataset indicate that the proposed model can achieve an accuracy of 98.46% on a patch-wise classification test set and an overall whole-image accuracy of 94.0% on pedestrian traffic signal recognition given aforementioned testing. In the future work, we are going to collect more pedestrian traffic signal data and further refine our approach to improve the accuracy and robustness of our proposed model.

Traffic signal detection calls for an extremely high accuracy because any singular small mistake can bring a catastrophic result to the blind or visually disabled, especially the missed detection of a red light signal. Given this fact and that different types of traffic signals, misclassification will lead to very different conclusions for an end user engaged in navigation and wayfinding, we will design a cost-sensitive model to address this issue. More specifically, the missing detection of red lights will be given higher penalties than its false alarms, and the missing detection of green lights will be given lower penalties than its false alarms. Additionally, stereoscopic or depth information is also of great importance to the blind or visually impaired travelers, as 3D content can not only object label and assist in go-no-go but also map local environments. Scene information will help inform visually impaired travelers how far he or she is from the traffic lights or corresponding curb 'cut', and/or zebra 'stripe' or lines, and/or other permanent spatial cues in the relevant environments relative to dynamic hazards, such as pedestrians. We will leverage this information to provide more applicable feature and function sets for the blind and visually impaired in our platform, as it scales.

References

1. Resources, N.: NYC resources. http://www.nyc.gov/html/dot/html/infrastructure/accessiblepedsignals.shtml. Accessed 22 Apr 2018
2. APS: Accessible pedestrian signal features. http://accessforblind.org/aps/aps-features/. Accessed 22 Apr 2018
3. HAWK: Hawk pedestrian signal guide. https://ddot.dc.gov/sites/default/files/dc/sites/ddot/publication/attachments/dc_hawk_pedestrian_signal_guide.pdf. Accessed 22 June 2018
4. Crandall, W., Bentzen, B.L., Myers, L., Brabyn, J.: New orientation and accessibility option for persons with visual impairment: transportation applications for remote infrared audible signage. Clin. Exp. OPTOMETRY **84**(3), 120–131 (2001)
5. Lookaround: Lookaround turorial. https://getlookaround.com/. Accessed 22 Apr 2018
6. Liao, C.F.: Using a smartphone app to assist the visually impaired at signalized intersections (2012)
7. Utcke, S.: Grouping based on projective geometry constraints and uncertainty. In: Sixth International Conference on Computer Vision, 1998, pp. 739–746. IEEE (1998)
8. Uddin, M.S., Shioyama, T.: Bipolarity and projective invariant-based zebra-crossing detection for the visually impaired. In: IEEE Computer Society Conference on Computer Vision and Pattern Recognition-Workshops, 2005. CVPR Workshops, pp. 22–22. IEEE (2005)
9. Se, S.: Zebra-crossing detection for the partially sighted. In: IEEE Conference on Computer Vision and Pattern Recognition, 2000, Proceedings, vol. 2, pp. 211–217. IEEE (2000)
10. Chung, Y.C., Wang, J.M., Chen, S.W.: A vision-based traffic light detection system at intersections. J. Taiwan Normal Univ. Math. Sci. Technol. **47**(1), 67–86 (2002)
11. Aranda, J., Mares, P.: Visual system to help blind people to cross the street. In: International Conference on Computers for Handicapped Persons, pp. 454–461. Springer (2004)

12. Se, S., Brady, M.: Road feature detection and estimation. Mach. Vis. Appl. **14**(3), 157–165 (2003)
13. Girshick, R., Donahue, J., Darrell, T., Malik, J.: Rich feature hierarchies for accurate object detection and semantic segmentation. In: Proceedings of the IEEE Conference on Computer Vision and Pattern Recognition, pp. 580–587 (2014)
14. Girshick, R.: Fast R-CNN. arXiv preprint arXiv:1504.08083 (2015)
15. Ren, S., He, K., Girshick, R., Sun, J.: Faster R-CNN: towards real-time object detection with region proposal networks. In: Advances in Neural Information Processing Systems, pp. 91–99 (2015)
16. Redmon, J., Divvala, S., Girshick, R., Farhadi, A.: You only look once: unified, real-time object detection. In: Proceedings of the IEEE Conference on Computer Vision and Pattern Recognition, pp. 779–788 (2016)
17. Liu, W., Anguelov, D., Erhan, D., Szegedy, C., Reed, S., Fu, C.Y., Berg, A.C.: SSD: single shot multibox detector. In: European Conference on Computer Vision, pp. 21–37. Springer (2016)
18. Niu, L., Qian, C., Rizzo, J.R., Hudson, T., Li, Z., Enright, S., Sperling, E., Conti, K., Wong, E., Fang, Y.: A wearable assistive technology for the visually impaired with door knob detection and real-time feedback for hand-to-handle manipulation. In: Proceedings of the IEEE Conference on Computer Vision and Pattern Recognition, pp. 1500–1508 (2017)
19. Freund, Y., Schapire, R.E., et al.: Experiments with a new boosting algorithm. In: ICML, vol. 96, pp. 148–156, Bari, Italy (1996)
20. Chollet, F., et al.: Keras (2015)

AutoViDev: A Computer-Vision Framework to Enhance and Accelerate Research in Human Development

Ori Ossmy[1(⊠)], Rick O. Gilmore[2(⊠)], and Karen E. Adolph[1(⊠)]

[1] Department of Psychology, New York University, New York, NY, USA
{oo8, karen.adolph}@nyu.edu
[2] Department of Psychology, The Pennsylvania State University,
State College, PA, USA
rogilmore@psu.edu

Abstract. Interdisciplinary exchange of ideas and tools can accelerate scientific progress. For example, findings from developmental and vision science have spurred recent advances in artificial intelligence and computer vision. However, relatively little attention has been paid to how artificial intelligence and computer vision can facilitate research in developmental science. The current study presents AutoViDev—an automatic video-analysis tool that uses machine learning and computer vision to support video-based developmental research. AutoViDev identifies full body position estimations in real-time video streams using convolutional pose machine-learning algorithms. AutoViDev provides valuable information about a variety of behaviors, including gaze direction, facial expressions, posture, locomotion, manual actions, and interactions with objects. We present a high-level architecture of the framework and describe two projects that demonstrate its usability. We discuss the benefits of applying AutoViDev to large-scale, shared video datasets and highlight how machine learning and computer vision can enhance and accelerate research in developmental science.

Keywords: Computer vision · Human development · Behavioral science · Body recognition · Convolutional pose machines

1 Introduction

Many artificial intelligence (AI) and computer vision researchers draw important insights from developmental science [1–3]. Conversely, relatively few developmental researchers use AI and computer-vision tools to facilitate the study of behavioral development [4–10]. Most developmental researchers lack expertise in AI and computer vision, so the enormous potential of AI and computer vision to speed progress in video-based developmental research remains untapped. To address core questions in behavioral development, useful tools must be adapted for use by non-experts and provide information more efficiently than—and just as accurately as—human observers. Here, a novel framework, **AutoViDev**, is proposed to exploit the power of video to capture behavior and to harness the power of computer vision to reveal patterns in video data.

© Springer Nature Switzerland AG 2020
K. Arai and S. Kapoor (Eds.): CVC 2019, AISC 944, pp. 147–156, 2020.
https://doi.org/10.1007/978-3-030-17798-0_14

We focus on video-based data for several reasons. First, video captures the subtleties and complexities of behavior and the surrounding context [11, 12]. Because video data are so rich, video is uniquely suited for reuse—researchers can ask new questions beyond the scope of the original study. Second, video also documents research procedures with high fidelity—impossible to achieve by the text-based descriptions and static images commonly used in the method sections of empirical articles [13].

Third, video provides the backbone of most developmental research programs. Developmental scientists routinely collect video recordings of infants, children, caregivers, and family members in lab, home, classroom, and museum settings. Trained human observers then annotate the videos, often on a frame-by-frame basis, with time-locked codes that mark the onset, offset, category, and magnitude of specific behaviors of interest or of inferred mental or emotional states. Datavyu (datavyu.org), for example, is a free, open source, video coding tool used by large numbers of developmental researchers. The resulting coding files provide the raw materials for the quantitative analysis of the frequency, duration, and co-occurrence of behavioral patterns within individuals and across groups.

Finally, the videos and accompanying coding files constitute a valuable source of ground-truth-labeled training data for machine-learning models. But AI and computer-vision researchers need access to the data. Until recently, developmental researchers rarely shared their video and coding files with other researchers because of privacy restrictions and technical constraints on file sizes and formats. The web-based Databrary video library (databrary.org [14, 15]) has reduced most of these barriers [16]. Databrary has standardized the process of securing participants' permission for data sharing and the approval process required for researchers to access recordings. Databrary has also standardized stored video data formats and the API calls to securely access data.

However, to exploit the power of video to advance developmental science, researchers must overcome a number of remaining challenges. Human video annotation is expensive, labor intensive, and time consuming, and thereby slows progress. Variations in video quality (due to differences in camera type and position, ambient lighting, operator expertise, and the fact that children are notoriously unpredictable, difficult research participants) increase the burden of human annotation and limit sharing and reuse.

Furthermore, although video typically captures behaviors across multiple domains—locomotion, emotion, gaze direction, object interaction, gesture, language, and so on—most researchers have specialized expertise in only one domain. Thus, to exploit the information buried in video, teams of researchers with diverse expertise must collaboratively annotate the videos with codes specific to their own domain of expertise. Accordingly, 65 researchers with diverse expertise recently embarked on the Play & Learning Across a Year (PLAY) project to collaboratively collect and annotate 900+ video hours of parent/child natural activity in the home [17]. The annotated dataset will be shared with the larger research community on Databrary. Video quality assurance is required to ensure high common standards for recording and reuse.

Given the promise of video and the challenges facing developmental researchers, the time is ripe to apply computer vision tools to developmental science. The AutoViDev framework can accelerate video annotation and provide real-time feedback to experimenters about whether the ongoing video stream meets quality assurance criteria needed for capturing specific desired behaviors.

2 AutoViDev High-Level Architecture

AutoViDev has two modes: 'offline' and 'online'. In the offline mode, the system connects to existing datasets and analyzes video files, providing researchers with automatic temporal and spatial descriptions of body movements and postures. In the online mode, the system analyzes real-time video streams and provides researchers with automatic feedback about how well the recording fits specific criteria for capturing target behaviors, and how to alter the camera position, angle, zoom, and so on, to better capture behaviors of interest.

To handle the two modes of operation, AutoViDev was designed as a modular program based on a 'separation of concerns.' AutoViDev consists of four major components—the Extractor, Core, UI, and Database (Fig. 1). Conceptually, each component does the following:

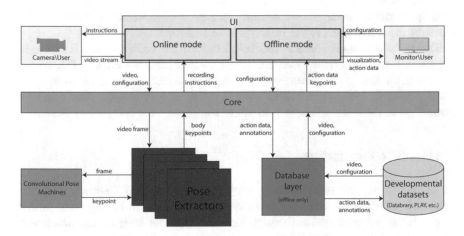

Fig. 1. AutoViDev high-level architecture. The architecture reflects the conceptual model of the system including its 4 main components and their communication with external entities. The design is based on a separation of concerns and it accounts for both online and offline modes.

The Core component is responsible for interaction between the Extractor, UI, and Database components. Each frame in a given video (either a real-time stream in the online mode or a video file in the offline mode) is first sent to the Extractor component. Then, the body key points are communicated back to the Core, where data are dispatched based on a configuration defined by the researcher (see UI component) and the Database guidelines (see Database component). The interface between components is managed by researcher-defined configurations that include information on the desired behaviors and body parts. That is, the configurations contain directives on *how* to analyze the body key points from the video. For example, if a researcher is interested in facial expressions, the Core component compares the key points (provided from the Extractor component) with pre-defined configurations (e.g., what facial features must appear in the frame, what angle to record) to provide the researcher with data for analyses in the offline mode or instructions on how to record in the online mode. This information will be sent to the UI component to be presented on the screen. In the offline mode, the configurations also include guidelines of the specific dataset regarding standardization and credentials.

The Extractor component is responsible for detecting key points on the human torso, arms, hands, palms, legs, feet, or face from each video frame. The input is a frame given by the Core component to which the extractor applies convolutional pose machine algorithms [18] that are based on effective receptive fields, using the same procedures described in previous research [19–21]. Extracted key points are sent back to the core component for presentation and further analyses. Multiple Extractor components (red; Fig. 1) serve the single Core component to speed up performance and prevent delays in the online mode.

The UI component manages the system's graphical interface with the researcher. In the offline mode, the component receives instructions from the researcher about how to analyze the data and visualizes the key points from the Extractor component on the video. In the online mode, the component receives input from the Core component and visualizes the instructions to the researchers in real-time.

The Database component is functional only in the offline mode. It manages all communications with existing datasets (e.g., PLAY or other volumes on Databary) and contains guidelines about videos in the dataset including access credentials and any ethical considerations applicable to specific files. The database communicates this information to the Core component to guide analyses of the body's key points.

3   Usability Evaluation

The current version of AutoViDev is implemented in a .NET framework (C#). The framework is still being actively developed and the code will be improved with new features including advanced action detection, body-object interactions, and multiple-camera support. Our major guiding principle is that the framework must be inexpensive, easy to use, and readily available to developmental researchers who often do not have access to high performance computing (HPC) resources or lack the required funds to use these resources. The system was evaluated in two use cases. We deliberately avoided the use of an expensive HPC during our evaluations, although system performance could have been significantly improved. Specifically, the system was evaluated using a standard laptop (CPU: Intel Core i7-7700HQ, Quad-Core, 6 MB Cache, up to 3.8 GHz w/Turbo Boost; RAM: 16 GB DDR4 at 2400 MHz (2 × 8 GB); and a graphics processor: NVIDIA GeForce GTX 1060 with 6 GB GDDR5).

3.1   Use Case 1: Using AutoViDev Offline Mode for Automatic Detection of Infant Locomotion

To track the location of infants and children during natural activity, researchers must obtain information about moment-to-moment body locations. Currently, researchers use video coding and manual digitizing procedures [10, 22–24] to record spatial locations and the temporal onset of key events. Trained human observers first code each time infants took a step and then manually digitize the location of each step from a large overhead camera view using custom-built software (e.g., DLTDataViewer; https://www.unc.edu/~thedrick/software1.html). The coordinates from each video frame are then used to generate infants' paths (Fig. 2a). Video coding 20 min of activity in this fashion takes ~5 person hours. Assuming one developmental study includes 30 infants, the dataset includes 10 h of video data. The cost of video coding and digitizing the dataset is ~150 person-hours.

In the first use case to evaluate AutoViDev, we analyzed thirty 20-min videos of infants' natural locomotor activity from an existing shared dataset (Databrary.org [25]). The offline mode of AutoViDev was used, with infants' feet as the body part to track in the configuration (Fig. 2b). AutoViDev completed the analysis in 90 min (100 times faster than manual coding and digitizing). The paths (x, y coordinates) produced by AutoViDev significantly matched (within an error range of 5 pixels) the manually digitized paths produced by human coders (M = 97.1% of frames across infants; p < .01; unequal variance t-test between coordinate differences and zero; see one example in Fig. 2).

Use case 1 demonstrates the power of automatic detection of infants' locomotor behavior, and highlights the exciting potential to accelerate the pace of developmental research by embedding computer vision tools within standard laboratory workflows.

3.2   Use Case 2: Using AutoViDev Online Mode to Standardize Video Recordings of Infants' Faces

Given that video is a uniquely informative source of behavioral data and is widely used in developmental research, an important concern is *how* to record infants and children in real-time during data collections. Recruitment and data collection are labor intensive, so it is important that participants' time is well used. Moreover, behavior is transient and cannot be restored, so minimizing data loss is imperative. Video captures and preserves behavior with high fidelity, given that the target behavior is recorded entirely and clearly. Thus, researchers need *online feedback* about the quality of the video recordings they are collecting.

A common problem in video recording is that the camera is not pointed toward the body part(s) that exhibits the target behavior. For example, the study of facial expressions requires a clear view of the facial features. Infant behavior is often extremely variable and unpredictable, making it difficult for experimenters to capture behavior clearly from the appropriate viewpoint at all times.

A second common problem arises from methodological differences between researchers. Video can be recorded from many cameras, multiple angles, changing distances, and under varied lighting conditions. Differences in recording practices

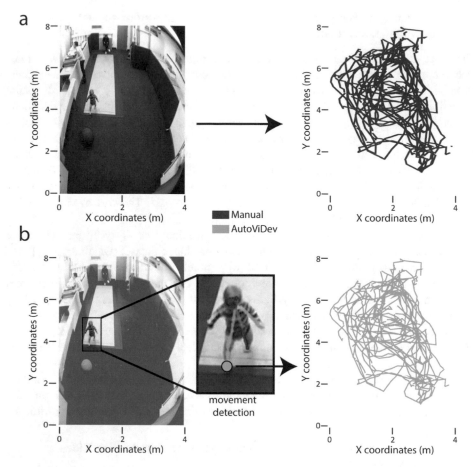

Fig. 2. Using AutoViDev offline mode for automatic detection of infant locomotion. (a) Researchers manually track infants in a video (left) to determine their path over 20 min of free-play. Figure on the right depicts one 20-min path that was manually extracted. (b) AutoViDev can achieve similar tracking by detecting the infant's feet and generating the 20-min path. The AutoViDev tracked path significantly matched the manually digitized path.

among different researchers, labs, and studies can lead to conflicting evidence, failure to replicate, and false scientific conclusions. Video recording practices can be standardized across researchers using a computer-vision-based framework that provides actionable, online feedback to the experimenters while collecting the data.

To address these issues, the feasibility of using AutoViDev was examined (in its online mode) as a real-time automatic instructor/critic for video recordings of faces. To that end, the location of infants' faces was chosen as the body part to track in the UI configuration, with limits to the angle between the camera and the face such that all facial features should be clearly shown (see Fig. 3a for a clear view and a full face detection by AutoViDev). The system was tested on frontal viewpoint recordings, which are widely used in studies of infants watching live or computer displays in the lab or at home over the web [26, 27]. 15 videos (average duration = 2:16 min) of 6-month-old infants was played from a Databrary dataset [28]. To increase the number of 'invalid' viewpoints (i.e., camera location is not informative for analysis), portions of the video were manually manipulated by cropping part of the face in some frames. Therefore, invalid viewpoints were either bad angles between the camera and the face or missing part of the face (Fig. 3b). We then played the videos and used AutoViDev in its online mode to provide real-time feedback about whether the face was shown clearly and entirely. The feedback was based on how many facial features were detected (see white lines in Fig. 3) and the angle between the face and the camera (see colored lines in Fig. 3). For example, if the camera did not detect some of the left facial features (as in Fig. 3b, right), the system instructed the experimenter to move the camera further to the left to get a complete view.

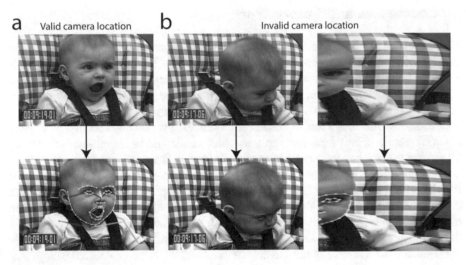

Fig. 3. Using AutoViDev online mode for real-time recording instructions. (a) A valid viewpoint of infant face. Applying AutoViDev in real-time, provided all the relevant features of the face. Calculation of the face angle (shown in colored lines) was valid as well. (b) Invalid viewpoints of the same infant. In the first viewpoint (left), the angle between the camera and the infant face prevents a clear view of the infant face. AutoViDev provided a real-time instruction to move the camera to a bottom-right view. It was calculated based on the angle between the camera and the face (shown in colored line). In the second viewpoint (right), we cropped half the video and tested AutoViDev instructions. AutoViDev detected the missing information (see white line for what AutoViDev succeeded to detect). Based on the false detection of the left part of the face, the system provided instructions to move the camera further to the left (not shown). Average RT for instructions was 2.13 s.

The system worked the same when it detected an angle between the face and the camera that exceeded the limits provided by the user in the configuration (Fig. 3b, left). AutoViDev successfully detected 87.1% of frames with missing information. This was accompanied with accurate graphical instructions on the screen about how to move the camera (e.g., 'move the camera to the right' when the face was cropped as in Fig. 3) to compensate for the missing information. However, AutoViDev's response time (RT) to detect problematic recording practices was slow (mean RT = 2.13 s). This RT should be further reduced to accommodate the high temporal resolution that is required in real-time data collections. Experimenters cannot afford delays of even a few seconds in RT because behavior occurs at the level of milliseconds, so any delay may lead to missing information.

Use case 2 shows the feasibility of automatic quality assurance coaching during video data collection. Future projects should implement this UI as an add-on to video cameras and test it on data across different ages. The tool can be used across researchers, enforcing a joint set of recording rules across the community, and thereby improving the quality, quantity, and reproducibility of behavioral data.

4   Conclusions and Future Work

The current study proposes that computer vision can support the science of human development, thereby accelerating the pace of discovery. We described the AutoViDev system's high level architecture and reported results from two specific use cases, both of which illustrate the potential of this framework. State-of-the-art results were presented on shared developmental video data drawn from the Databrary video library, demonstrating that the positive results for AutoViDev are not specific to carefully curated training or test data, but apply to videos collected in actual research settings. Thus, AutoViDev has the potential to scale widely and rapidly as video data sharing takes off. Future work should test AutoViDev on large-scale datasets (including the PLAY project recordings) and on data from research groups that focus on other developmental domains including perception, emotion, language (by adding auditory-based algorithms), and neuroscience [29]. Furthermore, although our current results for the online mode are limited in accuracy and RT, both can be improved to accommodate the demands for real-time quality assurance feedback in video-centered developmental research.

More broadly, preliminary results from AutoViDev demonstrate that computer-vision tools can reduce the time and cost of human annotation, and human annotations can, in turn, provide training data for computer-vision algorithms. Indeed, by storing and sharing annotations in common data formats on large-scale data repositories like Databrary, computer-provided or human annotations can serve as indices or bookmarks into the video recordings, allowing researchers in both fields to find and select video segments across studies that meet their specific research requirements. Thus, by supporting research on human development and embracing video annotation and sharing tools commonly used in the developmental science community, computer-vision researchers can push their own field forward.

In sum, developmental science has provided meaningful insights about how to build intelligent, learning machines that 'see' with increasing sophistication and power. Those learning machines can now be productively applied to advance understanding of development.

References

1. Vosniadou, S., Brewer, W.F.: Theories of knowledge restructuring in development. Rev. Educ. Res. **57**, 51–67 (1987)
2. Meltzoff, A.N., Kuhl, P.K., Movellan, J., Sejnowski, T.J.: Foundations for a new science of learning. Science **325**, 284–288 (2009)
3. Mitchell, T.M.: The discipline of machine learning. Carnegie Mellon University, School of Computer Science, Machine Learning Department Pittsburgh, PA (2006)
4. Gilmore, R.O., Raudies, F., Jayaraman, S.: What accounts for developmental shifts in optic flow sensitivity? In: 2015 Joint IEEE International Conference on Development and Learning and Epigenetic Robotics (ICDL-EpiRob), pp. 19–25 (2015)
5. Raudies, F., Gilmore, R.O.: Visual motion priors differ for infants and mothers. Neural Comput. **26**, 2652–2668 (2014)
6. Raudies, F., Gilmore, R.O., Kretch, K.S., Franchak, J.M., Adolph, K.E.: Understanding the development of motion processing by characterizing optic flow experienced by infants and their mothers. In: Proceedings of the IEEE Conference on Development and Learning (2012)
7. Smith, B., Yu, C., Yoshida, H., Fausey, C.M.: Contributions of head-mounted cameras to studying the visual environments of infants and young children. J. Cogn. Dev. **16**, 407–419 (2015)
8. Bambach, S., Lee, S., Crandall, D.J., Yu, C.: Lending a hand: detecting hands and recognizing activities in complex egocentric interactions. In: Proceedings of the IEEE International Conference on Computer Vision, pp. 1949–1957 (2015)
9. Schlesinger, M., Amso, D., Johnson, S.P.: The neural basis for visual selective attention in young infants: a computational account. Adapt. Behav. **15**, 135–148 (2007)
10. Ossmy, O., Hoch, J.E., MacAlpine, P., Hasan, S., Stone, P., Adolph, K.E.: Variety wins: Soccer-playing robots and infant walking. Front Neurorobot **12**, 19 (2018)
11. Adolph, K., Gilmore, R.O., Kennedy, J.L.: Video data and documentation will improve psychological science. Psychol. Sci. Agenda (2017)
12. Adolph, K.: Video data and documentation will improve psychological science. APS Observer **29**, 23–25 (2016)
13. Gilmore, R.O., Adolph, K.E.: Video can make behavioral science more reproducible. Nat. Hum. Behav. **1** (2017). s41562-41017
14. Gilmore, R.O., Adolph, K.E., Millman, D.S.: Curating identifiable data for sharing: The databrary project. In: Scientific Data Summit (NYSDS), New York, pp. 1–6 (2016)
15. Gilmore, R.O., Adolph, K.E., Millman, D.S., Gordon, A.S.: Transforming education research through open video data sharing. Adv. Eng. Educ. **5**, 1–17 (2016)
16. Gilmore, R.O., Kennedy, J.L., Adolph, K.E.: Practical solutions for sharing data and materials from psychological research. Adv. Methods Pract. Psychol. Sci. **1**, 121–130 (2018)
17. Adolph, K., Tamis-LeMonda, C.S., Gilmore, R.O., Soska, K.C.: Play & Learning Across a Year (PLAY) Project Summit, 29 June 2018, Philadelphia. Databrary (2018). http://doi.org/2010.17910/B17917.17724. Accessed 30 Aug 2018
18. Ramakrishna, V., Munoz, D., Hebert, M., Bagnell, J.A., Sheikh, Y.: Pose machines: articulated pose estimation via inference machines. In: European Conference on Computer Vision, pp. 33–47. Springer (2014)
19. Simon, T., Joo, H., Matthews, I.A., Sheikh, Y.: Hand keypoint detection in single images using multiview bootstrapping. In: Proceedings of the IEEE Conference on Computer Vision and Pattern Recognition, p. 2 (2017)

20. Wei, S.-E., Ramakrishna, V., Kanade, T., Sheikh, Y.: Convolutional pose machines. In: Proceedings of the IEEE Conference on Computer Vision and Pattern Recognition, pp. 4724–4732. (2016)
21. Cao, Z., Simon, T., Wei, S.-E., Sheikh, Y.: Realtime multi-person 2D pose estimation using part affinity fields. arXiv preprint arXiv:1611.08050 (2016)
22. Hoch, J., O'Grady, S., Adolph, K.: It's the journey, not the destination: Locomotor exploration in infants. Dev. Sci. (2018)
23. Lee, D.K., Cole, W.G., Golenia, L., Adolph, K.E.: The cost of simplifying complex developmental phenomena: a new perspective on learning to walk. Dev. Sci. 21, e12615 (2018)
24. Thurman, S.L., Corbetta, D.: Spatial exploration and changes in infant-mother dyads around transitions in infant locomotion. Dev. Psychol. 53, 1207–1221 (2017)
25. Adolph, K.: It's the journey not the destination: Locomotor exploration in infants (2015). Databrary. http://doi.org/10.17910/B17917.17140
26. Scott, K., Chu, J., Schulz, L.: Lookit (Part 2): assessing the viability of online developmental research, results from three case studies. Open Mind 1, 15–29 (2017)
27. Scott, K., Schulz, L.: Lookit (part 1): a new online platform for developmental research. Open Mind 1, 4–14 (2017)
28. Messinger, D.S.: Facial expressions in 6-month old infants and their parents in the still face paradigm and attachment at 15 months in the strange situation (2014). Databrary. http://doi.org/10.17910/B17059D
29. McNamara, Q., De La Vega, A., Yarkoni, T.: Developing a comprehensive framework for multimodal feature extraction. In: Proceedings of the 23rd ACM SIGKDD International Conference on Knowledge Discovery and Data Mining, pp. 1567–1574 (2017)

Application of Remote Sensing for Automated Litter Detection and Management

Mark Hamill$^{(\boxtimes)}$, Bryan Magee, and Phillip Millar

Built Environment Research Institute, Ulster University,
Shore Road, County Antrim, BT37 OQB Coleraine, Northern Ireland
mj.hamill@ulster.ac.uk

Abstract. The Clean Europe Network (CEN) estimates that cleaning litter in the EU accounts for €10–13 billion of public expenditure every year. The annual budget for managing roadside litter alone, is approximately €1 billion. While local authorities in Northern Ireland and elsewhere have legal requirements to monitor and control litter levels, requirements for compliance are unclear and frequently ignored. Against this background, the overall objective of this research is to develop an integrated management system allowing remote discrimination and quantification of roadside litter. As such, the intention is that local authorities can more effectively meet their statutory requirements with regards to litter management. The research aligns with objectives outlined by the UK Government and CEN in terms of improving litter-related data levels. As plastic containers of type RIC1, Polyethylene terephthalate (PETE), represent one of the most common components of roadside litter, its identification in the natural environment via remote sensing is a key objective. By combining published US Hyperspectral library data and experimental field study results, the initial findings of this research indicate that it is possible to discriminate PETE plastic samples in a grass background using a low-cost multispectral sensor primarily designed for agricultural use. While at an initial phase, the research presented has the potential to have a significant impact on the economic, environmental and statutory implications of roadside litter management. Future work will employ image processing and machine learning techniques to deliver a methodology for automatic identification and quantification of multiple roadside litter types.

Keywords: Image analysis · Multispectral · Litter · Remote sensing · Hyperspectral signatures

1 Introduction

Local environmental charity Keep Northern Ireland Beautiful calculated in 2014 that the average cost to ratepayers for cleaning streets was £38 M per year [1]. The organization stated that 97% of streets in Northern Ireland are littered and recognized plastic litter as a particular issue. The UK Government's primary recommendation [2] relating to litter is that more and better data is required to underpin more accurate management of its collection. Collation of roadside litter data by quantity, type and location is seen as a key component of this recommendation. The widespread physical, environmental and financial impact of littering is well accepted [3] and its monitoring and

© Springer Nature Switzerland AG 2020
K. Arai and S. Kapoor (Eds.): CVC 2019, AISC 944, pp. 157–168, 2020.
https://doi.org/10.1007/978-3-030-17798-0_15

management is a statutory duty of all UK local authorities. However, requirements for compliance under statutory documents such as the Litter (Northern Ireland) Order 1994 are unclear and, as a result, frequently ignored.

Of particular concern, plastic littering has reached such an extent worldwide that it has been recognized an indicator of a distinct geologic era [4]. In the US, for example, plastic litter has increased by 165% since 1969 and according to several studies [4–11], polyethylene terephthalate (PETE) - widely utilised in the food and drink industry - is by far the most common type of discarded plastic product. This plastic is internationally categorized using the Resin Coding Method [12] as Resin Identification Code (RIC) Type 1. Concern surrounding the impact and persistence of plastic litter [5, 6] has generated much research in measuring litter levels in the marine environment. However, despite the terrestrial origin of much plastic litter [7, 8], there appears to be little similar roadside- or built environment-based research.

Measurement of environmental plastic is currently carried out by physical observation, although there have been some attempts to utilise remote sensing. Remote sensing is defined as the detection of reflected electromagnetic radiation emanating from a surface using a wide range of imaging sensor types and imaging techniques such as sonar platforms and aerial cameras [13]. In the past decade, examples of remote sensing application in the built environment include identification of impervious surfaces to interpret urbanization levels [14] and road centerline extraction in support of autonomous driving [15]. Vehicle mounted remote sensing has been successfully deployed to monitor road surface conditions [16]; work that has successfully demonstrated that deterioration of street furniture and road surfaces can be captured in urban environments to high levels of geo-spatial accuracy using vehicle-based sensors. Several other publications consider remote sensing in relation to road surface features and road surface conditions [17] and general sensing of urban surface features [18]. However, the techniques reported have yet to be applied to the challenge of remote sensing of litter on urban and rural road networks.

Two types of remote sensor are principally used to detect reflectance spectra; namely multispectral (MS) and hyperspectral (HS). MS sensors measure reflected energy in typically four or five discreet electromagnetic spectral bands in the green to infra-red range. HS sensors also detect reflectance in discreet contiguous bands, albeit over 200 or more. MS sensors have primarily been used for vegetation management [19] and weed monitoring [20]. While a few examples of MS sensors having been used for litter, and more specifically plastic debris [21], measurement exist in the literature, their focus has been on marine and beach litter [21–23]. In these instances, as the debris being detected is typically large and the background homogenous, debris quantities have been quantifiable by applying brightness thresholds to two-tone images. HS remote sensing has also been considered as a method of discriminating plastics in this environment, with related research [24, 25] considering remote sensing of macro plastics in both visible (VIS) and short-wave infrared (SWIR) spectra. Current HS equipment is most effective in a controlled environment and is commercially deployed in litter recycling facilities to assist with sorting. Serranti et al. [26] first demonstrated effective use of HS reflectance sensing techniques to discriminate specific plastic resin types for litter sorting.

Against this background, the aim of this research is to evaluate the effectiveness of HS and MS sensors to identify RIC Type 1 litter in a roadside environment. The paper presents two distinct phases of research. Phase I initially analyzed HS reflectance profiles held in existing data to compare spectral profiles of common built-environment materials, including RIC Type 1 plastic. HS signatures in four distinct spectral bands, corresponding to MS sensitivity, were examined and the corresponding reflectance compared. The intention was to examine whether sufficient spectral information was contained in these bands to uniquely distinguish materials. Phase II subsequently acquired field data using a commonly available, Parrot Sequoia MS sensor, with data obtained analysed against the Phase I data. The work reported is the first stage of a longitudinal study focused on developing an automated litter monitoring system capable of distinguishing and quantifying roadside litter by type, to assist local authorities meet statutory litter reporting and management.

2 Phase I – Spectral Data Analysis

The research in this section was undertaken to support a hypothesis that litter in the built environment can be remotely identified using a low-cost MS sensor. Data used in Phase I of the study was acquired from the ECOSTRESS Spectral library; a resource created and made freely available by the U.S. Jet Propulsion Laboratory [27]. The library (formally the ASTER spectral library) [28], version 1.0 of which was released in February 2018, comprises more than 2,800 reflective spectra of natural and man-made materials. Spectra in the range 0.4–2.5 µm was used for this study, corresponding to the spectral range of the multispectral sensor employed in Phase II (0.53–0.81 µm). Sampling intervals of 0.001 and 0.004 µm were considered in the 0.4–0.8 and 0.8–2.5 µm ranges respectively. Two spectral samples each, for two common European grasses, Avena fatua and Bromus Diandrus [29] were initially considered (see Fig. 1(a)). Clearly, a simple visual comparison of these profiles indicates a high degree of consistency across the spectrum considered, with strong distinguishing characteristics at several wavelengths.

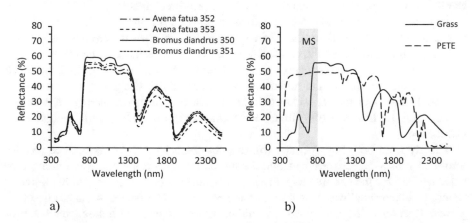

a)   b)

Fig. 1. Comparative HS profiles for: (a) common European grasses, and (b) average value for grass plotted in comparison with PETE (MS spectral range superimposed)

As shown in Fig. 1(b), an averaged grass reflectance profile was then compared to that for PETE. Examination in the MS range (525–800 nm), indicates clear differentiation between profiles in most regions. While the profiles cross at 725 nm and show on 5% reflectance differences in the wavelength range 725–800 nm, in the 500–700 nm range, reflectance diverges significantly, with an average difference of more than 30%. To analyse the unique nature of these spectral profiles more closely in the MS range, specific spectral band ranges corresponding to the capabilities of the MS sensor used in Phase II were then compared. As shown in Fig. 2(a) the discreet bands considered were: (1) Green spectrum (530–570 nm); (2) Red spectrum (640–680 nm); (3) Red Edge spectrum (725–745 nm); and (4) Near Infrared spectrum (770–810 nm).

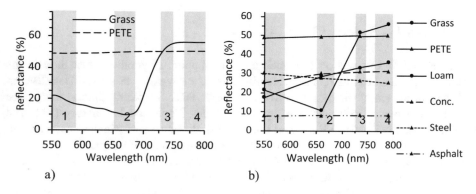

Fig. 2. HS profiles in the 515–815 nm range with: (a) all data points plotted for grass and PETE, and (b) averaged data points across bands 1–4 plotted for grass and PETE in comparison to other common built-environment materials

When comparing PETE to grass across the 525–800 nm spectral range, unique profiles are apparent, with average differences in reflectance of +27.5 and +38.7% in bands 1 and 2 respectively. In bands 3 and 4, the profiles converged, with PETE having slightly lower average reflectance values compared to grass (−1.6 and −5.8% respectively). While the strongest distinction between the two materials is in bands 1 and 2, it is recognized that the distinctive crossover in reflectance in bands 3–4 might play an important role in distinguishing the materials. As shown in Fig. 2(b), this work was extended to compare the reflectance signature of PETE against other materials commonly found in the roadside environment. In this instance, average reflectance values from the ECOSTRESS library were calculated for each material across band widths 1–4.

Clearly from both Fig. 2(a) and (b), distinct characteristics of reflectance profiles across each band range were apparent for the materials considered. This finding provided support for the research hypothesis and confirmed potential for creating numeric material 'fingerprints' based on data considered across MS bands. Against this background, it was decided that the work should progress utilizing MS field-captured sensor data to explore these characteristics further.

3   Phase II - Multispectral Sensor Reflectance Analysis

The equipment used for the MS research was a Parrot Sequoia manufactured by Parrot SAS, France, comprising both sunshine and multispectral sensors. The equipment contained a 16 MP RGB camera with 63.9° × 50.1° field of view (FOV) and four global shutters, 1.2 MP single-band cameras with 61.9° × 48.5° FOV. The assembly was installed in a modified GoPro Hero5 mount with power was supplied by a 5 V, 3A USB battery and supported by firmware version v.1.4.1.

Calibration required the conversion of unprocessed image data to at-sensor radiance values via a linear correlation. No atmospheric corrections were required as the sensor was mounted close to the ground with atmospheric effects deemed negligible [30]. The sensor was mounted on a tripod with the lens assembly normal to the ground surface to reduce distortion and set at a height of 1,200 mm; the minimum focus distance. A sample of calibration images is presented in Fig. 3. Each field survey undertaken commenced with radiometric calibration using a standard Sequoia reference target, labeled 'T' (Fig. 3(a)). This procedure generated reference data for subsequent numerical adjustments or image manipulations to compensate for variations in band sensitivity.

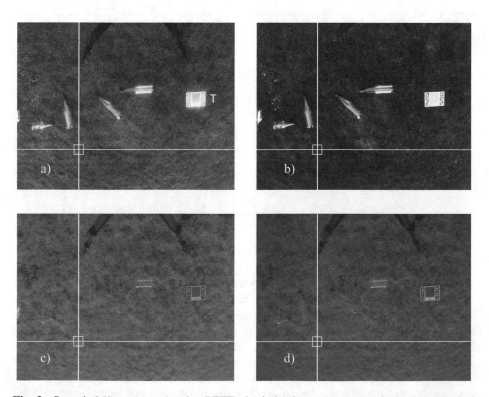

Fig. 3. Sequoia MS captures showing PETE plastic bottles on lawn grass indicating a typical sample point on each frame for: (a) Green, (b) Red, (c) IR Edge & (d) Near-infrared spectrums

3.1   Image Analysis

The Parrott Sequoia MS sensor generated four, 1280×960 pixel greyscale images; one for each spectral band. These images were in 8-bit integer jpeg format giving a range of possible values from 0 to 255 for each pixel, where zero is black and 255 white. For image analysis the GNU Image Manipulation Program (GIMP) was used to capture, report and analyze image pixel values. The software reported pixel values in HSV format where V is a brightness value in the range 0–100. The software contained the appropriate tools to extract individual or sample averaged pixel values. The process allowed reliable and repeatable selection of discrete areas of images and contained tools to allow averaged pixel values to be determined from user-definable kernels. Each image band was separately calibrated to compensate for differing responses in each band. To do this, the reflectance value for the standard grey target was corrected to a reference value of 50%. Three sample points for each grey target were captured using an averaged 10×10 pixel kernel. Table 1 presents a representative example of the sampling process.

Table 1. Example of spectral band reflectance value extraction for lawn grass

Sample	Image reference	X Value	Y Value	R-Value	Average R-value	Target R-Value	R-Value Corr
Lawn Grass 001	180312_165336_0000_GRE	600	450	34.4	37.0	74.9	24.7
		700	450	39.3			
		800	450	37.2			
	180312_165336_0000_RED	600	450	18.1	18.8	70.7	13.3
		700	450	20.6			
		800	450	17.7			
	180312_165336_0000_REG	600	450	30.1	30.6	25.9	59.1
		700	450	30.2			
		800	450	31.6			
	180312_165336_0000_NIR	600	450	39.1	39.1	31.6	61.9
		700	450	38.8			
		800	450	39.4			

Table 1 contains sample data from one set of four multispectral images from a single image capture event as illustrated in Fig. 3. The four image references listed represent the Green, Red, Red-Edge and Near IR spectral bands (top to bottom). In this example, three reflectance measurements are presented for each. By using X and Y values to co-ordinate the sampling, consistency in sample areas across the spectral band images was effectively achieved. The sampling yielded corresponding reflectance R-values, which were averaged for each multispectral band. In this study, six grass samples and six PETE samples were acquired from five reference images, yielding 30 samples for both grass and PETE. A calibration reflectance measurement of the 50% grey calibration target was taken in each frame for the corresponding spectral band (recorded as Target R-Value in Table 1). Radiometric calibration uses Target R-Value to harmonise the reflectance outputs to compensate for the differing sensitivities of the sensor bands as follows:

$$R - Value\ Corr = \frac{50}{Target\ R - value} \times Average\ R - Value$$

The radiometrically correct reflectance values for the samples was calculated for the Average R Values and is shown in Table 1 as R-Value Corr. Data acquired from the sampling described above was then plotted to provide multispectral profiles as shown in Fig. 4, which illustrates the distinct profiles for each material.

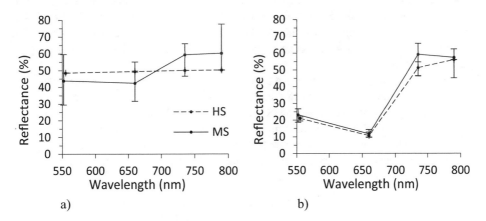

a)   b)

Fig. 4. Comparisons of HS and MS spectral reflectance data for: (a) PETE and (b) grass

3.2 Observations

Figure 4 combines data recovered from the MS sample acquisition described above, compared against raw data from the ECOSTRESS HS library. Values for both PETE plastic and lawn grass were derived from the HS data as described in Sect. 2. The average reflectance values for MS and HS data are presented for both PETE and grass, with the reflectance range in each band being indicated by error bars. The profiles demonstrate that results obtained from the Parrott Sequoia MS sensor were generally as predicted, with a strong correlation between HS and MS reflectance data for both grass and PETE. It was expected that RIC1 plastic, which is optically clear, would transmit background reflectance information and the impact on reflectance profiles can be seen in Fig. 4(a). The study clearly shows that Green and Red spectral bands provide clearest distinction between materials, IR Edge and Near IR the least. This is clearly observable in the grey-scale images in Fig. 3, where PETE plastic bottles are visibly most transparent in the IR bands and least transparent in the visible bands. The narrow variance exhibited in the HS data compared to MS is indicative of the controlled conditions in which the HS data was captured; the HS reflectance being laboratory sampled against plain backgrounds. The wide variation seen in the MS PETE values is symptomatic of the reflective nature and curved surface of plastic bottles; with examples of these reflective highlights being most visible in Fig. 3(a) and (b).

Strong correlation between HS and MS profiles can be observed in all four bands for both materials. The average reflectance values acquired from the MS sensor field trials were in the ranges predicted by the analysis of the HS library data. Very little variation was witnessed in the reflectance values for the HS data, with less than 1% variation in any band for PETE and no more than 8% in any band for grass. This reflects the laboratory-based nature of the acquisition. MS reflectance values for grass showed good consistency in visible bands of no more than 8% and below 20% in the Infra-Red bands. MS reflectance data for PETE had acceptable variance, extending from just under 20% to 30%. The MS reflectance profiles show that there is significant overlap in the IR bands, but despite some variance, no overlap exists in the visible bands. This is as predicted by the HS library data. It can be seen that again each material has a distinguishing, clearly recognizable profile capable of being 'finger-printed' numerically.

Numeric analysis of reflectance data was undertaken as shown in Table 2, defined separately for both the HS and MS profiles. Profiling was based on a simple factoring of values between spectral bands to provide a series of numerical ratios. These ratios describe the mathematical relationship between the values in each band to each other for each material. The relationships are further described in terms of average multipliers and a range reflecting the measured reflectance from the experimental data.

Table 2. Numeric analysis of HS and MS fingerprints

	Logical relationship between spectral bands 1–4			Then spectral profile equates to:
	IF 1:2 =	AND 2:3 =	AND 3:4 =	
PHASE I - Based on data sourced from published hyperspectral libraries	0.440–0.654	3.624–5.663	1.024–1.212	Grass
	1.010–1.023	1.006–1.018	1.002–1.005	PETE
PHASE II – Based on experimental data collected using multispectral camera	0.365–0.763	3.239–6.743	0.685–1.340	Grass
	0.531–1.871	0.845–2.079	0.762–1.666	PETE

This analysis provides a mathematical description of the reflectance curve for each material which has the potential to be used to identify and filter materials from their reflectance data using data-processing techniques. The IF:AND:AND relationship has been trialed in a spreadsheet matrix using 'fingerprints' for several ratios developed from the HS library data. Upper and lower reflectance values for selected common natural and manufactured materials have been used initially to develop a look-up table.

A sample of this is presented in Table 3, in which sample values are entered and subjected to an IF:AND analysis in the form:

$$=IF((AND(H13<\$E\$3, H13>\$E\$4, H14<\$E\$5, H14>\$E\$6, H15<\$E\$7,$$
$$H15>\$E\$8, H16<\$E\$9, H16>\$E\$10)), "Y", " ")$$

Table 3. Sample of IF AND analysis using HS library data

Wavelength (nm)	Spectral Reflectance Library							
	Oak (fresh)	Oak (dry)	Mixed Spruce	Meadow Grass	Lawn Grass	Grass Dry	HDPE TransL	PETE
Band 1 - 550	181.65	204.75	25.2	164.85	101.85	309.75	560.7	510.3
Range	164.35	185.25	22.8	149.15	92.15	280.25	507.3	461.7
Band 2 - 660	106.05	128.1	19.95	274.05	45.15	242.55	534.45	518.7
Range	95.95	115.9	18.05	247.95	40.85	219.45	483.55	469.3
Band 3 - 735	641.55	584.85	95.55	332.85	471.45	286.65	522.9	525
Range	580.45	529.15	86.45	301.15	426.55	259.35	473.1	475
Band 4 - 790	874.65	689.85	130.2	365.4	724.5	309.75	516.6	527.1
Range	791.35	624.15	117.8	330.6	655.5	280.25	467.4	476.9
	Sample values							
Band 1 - 550	23	200	530	160	100	165	291	480
Band 2 - 660	19	120	520	250	42	100	230	480
Band 3 - 735	87	530	480	300	430	600	263	480
Band 4 - 790	120	625	470	335	660	800	285	480
	Results Matrix							
Oak (fresh)				null		Y		
Oak (dry)		Y						
Mixed Spruce	Y							
Meadow Grass								
Lawn Grass					Y			
Grass Dry							Y	
HDPE TransL			Y					
PETE								Y

Where the sample values are within the reflectance range in all four bands for a given material in the library table a 'Y' is shown in the results matrix. Where one or any of the values are beyond the upper and lower values, a null response is returned. The trial IF AND analysis demonstrates the potential of further work focused on automation of a numeric material 'fingerprinting' method.

4   Conclusion

A simple and reliable method to distinguish PETE RIC Type 1 plastic in grassed areas, utilizing inexpensive radiometric survey equipment has been demonstrated in this study. Under field conditions, the research indicates that different materials can be clearly discriminated in the Green and Red spectral bands. Nevertheless, there are some limitations that require deeper investigation and will form the basis of further research. For instance, the limited number of samples presented in this paper is recognized as a limiting factor and further study will address this. In addition, work will extend to collect data from other common materials. Future research will focus on methods such as using automated computer image processing. This will automatically measure and compare the reflectance values generated by the MS sensor allowing profiles for additional materials to be acquired more quickly. It was observed that the capture process benefited from a combination of bright and overcast conditions. Generally, for the trials undertaken, well-lit sunny or hazy but bright conditions were chosen. However, it was also observed that the quality of image capture in early morning or evening could be adversely affected due to low sun generating increased shadow; particularly where surfaces are undulating or uneven. Further trials will be carried out in a range of lighting conditions to establish the impact of these variations.

In terms of practical surveys, the effect of sensor movement on image acquisition quality will be evaluated. The Sequoia multispectral camera is principally designed for aerial drone surveys, where due to its elevation, large overlap between consecutive images exists and high frame rates are not required. The low sensitivity of the sensor means that capture rates up to two frames per second are attainable. Trials are planned to determine performance in vehicle mounted surveys.

Recognizing that this is an early phase in the study, this paper presents an innovative method of litter detection in the built environment using remote sensing. This presents a potentially more economic and flexible solution to litter quantification and qualification than more expensive specialist solutions. The research indicates that in the area of remote sensing, it is possible to use a relatively simple multispectral sensor to distinguish materials. The research will continue with the final aim to develop an economic and reliable method of litter detection and measurement as part of an environmental monitoring scheme. It is anticipated that such a system will have a beneficial impact for those responsible for managing and maintaining the environment. A future phase of the research will involve field trials in collaboration with Mid and East Antrim District Council; a Northern Ireland local authority responsible for a district area of 1046 km^2 comprising a network of major and minor roads. Field trials will initially take place on a representative section of trunk road with the intention of being extended as the project develops.

References

1. Allen, C.: Northern Ireland Litter Survey 2014 [pdf] Keep Northern Ireland Beautiful (2014). http://www.keepnorthernirelandbeautiful.org/keepnorthernirelandbeautiful/documents/0066 55.pdf. Accessed 13 April 2018
2. Communities and Local Government Committee: Litter and fly-tipping in England - Seventh Report of Session 2014–15 [pdf] House of Commons Communities and Local Government Committee (2015). http://www.publications.parliament.uk/pa/cm201415/cmselect/cmcomloc/607/607.pdf. Accessed 13 April 2018

3. Clean Europe Network: Facts and Costs (2018). http://www.cleaneuropenetwork.eu/en/facts-and-costs/aup/. Accessed 21 Dec 2018
4. Zalasiewicz, J., Waters, C.N., Corcoran, P.L., Ivar do Sul, J.A., Barnosky, A.D., Cearreta, A., Edgeworth, M., Gałuszka, A., Jeandel, C., Leinfelder, R., McNeill, J.R., Steffen, W., Summerhayes, C., Wagreich, M.: The geological cycle of plastics and their use as a stratigraphic indicator of the anthropocene. Anthropocene **13**, 4–17 (2016)
5. Eriksen, M., Lebreton, L.C.M., Carson, H.S., Thiel, M., Moore, C.J., Borerro, J.C., Galgani, F., Ryan, P.G., Reisser, J.: Plastic pollution in the World's oceans: more than 5 trillion plastic pieces weighing over 250,000 tons afloat at sea. PLoS ONE **9**(12), e111913 (2014). https://doi.org/10.1371/journal.pone.0111913
6. Andrady, A.L.: Persistence of plastic litter in the oceans. In: Bergmann, M., Gutow, L., Klages, M. (eds.) Marine Anthropogenic Litter, pp. 57–72. Springer, Cham (2015)
7. Sherrington, C.: Plastics in the marine environment. Eunomia (2016). https://www.eunomia.co.uk/reports-tools/plastics-in-the-marine-environment/. Accessed 17 Nov 2018
8. European Environment Agency: Marine litter – a growing threat worldwide [pdf] European Environment Agency (2017). https://www.eea.europa.eu/highlights/marine-litter-2013-a-growing. Accessed 17 Nov 2018
9. Schultz, P.W., Stein, S.R.: Executive summary: litter in America. 2009 National litter research findings and recommendations. [pdf] Keep America Beautiful (2009). https://www.kab.org/news-info/research/litter-america-executive-summary. Accessed 17 Nov 2018
10. House of Commons Environmental Audit Committee: Plastic bottles: Turning Back the Plastic Tide. First Report of Session 2017–19. HC 339 Published on 22 December 2017 by authority of the House of Commons (2017)
11. Derrraik, J.G.B.: The pollution of the marine environment by plastic debris: a review. Mar. Pollut. Bull. **44**(2002), 842–852 (2002)
12. ASTM International: D7611/D7611 M—18, Standard Practice for Coding Plastic Manufactured Articles for Resin Identification (2018). https://compass.astm.org/EDIT/html_annot.cgi?D7611+18. Accessed 28 Feb 2019
13. Lillesand, T.M., Kiefer, R.W., Chipman, J.W.: Remote Sensing and Image Interpretation, 7th edn. Wiley, New York (2015)
14. Weng, Q.: Remote sensing of impervious surfaces in urban areas: requirements, methods, and trends. Remote Sens. Environ. **117**, 34–49 (2012)
15. Cheng, G., Wang, Y., Xu, S., Wang, H., Xiang, S.: Automatic road detection and centerline extraction via cascaded end-to-end convolutional neural network. IEEE Trans. Geosci. Remote Sens. **55**(6), 3322–3337 (2017)
16. Tadic, S., Favenza, A., Kavadias, C., Tsagaris, V.: GHOST: a novel approach to smart city infrastructures monitoring through GNSS precise positioning. In: IEEE International Smart Cities Conference (ISC2). 12–15 September 2016, Trento, Italy (2016)
17. Vaa, T.: Remote sensing of road surface conditions and intelligent transportation systems applications. In: 20th ITS World Congress, 14–18 October 2017, Tokyo, Japan (2013)
18. Le Saux, B., Yokoya, N., Hansch, R., Prasad, S.: Advanced multisource optical remote sensing for urban land use and land cover classification. IEEE Geosci. Remote Sens. Mag. **6**(4), 85–89 (2018)
19. Berni, J.A.J., Zarco-Tejada, P.J., Suares-Barranco, M.D., Fereres-Castel, E.: Thermal and narrowband multispectral remote sensing for vegetation monitoring from an unmanned aerial vehicle. IEEE Trans. Geosci. Remote Sens. **478**(3), 722–738 (2009)

20. Stroppiana, D., Villa, P., Sona, G., Ronchetti, G., Candiani, G., Pepe, M., Brusetto, L., Migliazzi, M., Boschetti, M.: Early season weed mapping in rice crops using multi-spectral UAV data. Int. J. Remote Sens (2017). Received 31 Oct 2017, Accepted 05 Feb 2018, Published online: 21 Feb 2018. https://www.tandfonline.com/doi/abs/10.1080/01431161. 2018.1441569?scroll=top&journalCode=tres20. Accessed 4 Mar 2018

21. Mitchell, K., Driedger, H.D., Van Cappelin, P.: Remote Sensing of Plastic Debris American Geophysical Union Science Policy Conference, Washington, D.C., USA. http://spc.agu.org/ 2013/eposters/eposter/o-05/. https://www.researchgate.net/publication/242651084_Remote_ Sensing_of_Plastic_Debris. Accessed 6 Sept 2018

22. Nakashima, E., Isobe, A., Magome, K.S., Deki, N.: Using aerial photography and in situ measurements to estimate the quantity of macro-litter on beaches. Mar. Pollut. Bull. **62**(4), 762–769 (2011). http://www.sciencedirect.com/science/article/pii/S0025326X11000105. Accessed 20 June 2018

23. Kako, S., Isobe, A., Magome, S.: Sequential monitoring of beach litter using webcams. Mar. Pollut. Bull. **60**(5), 775–779 (2010). http://www.sciencedirect.com/science/article/pii/ S0025326X10000998. Accessed 20 June 2018

24. Goddijn-Murphy, L., Steef, P., Van Sebille, E., James, N.A., Gibb, S.: Concept for a hyperspectral remote sensing algorithm for floating marine macro plastics. Mar. Pollut. Bull. **126**, 255–262 (2018)

25. Asner, G.: Workshop on mission concepts for marine debris sensing, January 19–21,2016, east-west center of the university of Hawaii at Manoa, Honolulu, Hawaii (2016). http://iprc. soest.hawaii.edu/NASA_WS_MD2016/pdf/Asner2016.pdf. Accessed 12 Dec 2018

26. Serranti, S., Gargiulo, A., Bonifazi, G.: Characterization of post-consumer polyolefin wastes by hyperspectral imaging for quality control in recycling processes. Waste Manag **31**(11), 2217–2227 (2011)

27. Meerdink, S.K., Hook, S.J., Abbott, E.A., Roberts, D.A.: The ECOSTRESS Spectral Library 1.0 (in prep). https://speclib.jpl.nasa.gov/. Accessed 4 Mar 2018

28. Baldridge, A.M., Hook, S.J., Grove, C.I., Rivera, G.: The ASTER Spectral Library Version 2.0. Remote Sens. Environ. **113**, 711–715 (2008)

29. Fitter, R., Fitter, F., Farrer, A.: Collins Guide to the Grasses, Sedges, Rushes and Ferns of Britain and Northern Europe. Collins, London (1984)

30. Karpouzli, E., Malthus, T.: The empirical line method for the atmospheric correction of IKONOS imagery. Int. J. Remote Sens. **24**, 1143–1150 (2003). https://doi.org/10.1080/ 01431160210000026779

Design and Evaluation of a Virtual Reality-Based Car Configuration Concept

Marcus Korinth[1], Thomas Sommer-Dittrich[1], Manfred Reichert[2],
and Rüdiger Pryss[2(✉)]

[1] Daimler AG, Ulm, Germany
{marcus.korinth,thomas.sommer-dittrich}@daimler.com
[2] Institute of Databases and Information Systems, Ulm University, Ulm, Germany
{manfred.reichert,ruediger.pryss}@uni-ulm.de

Abstract. The Daimler AG provided a concept preview towards the individualization of interior trim parts at the *International Motor Show* in September 2017, which was named *unleash the color*. At the show, a tablet computer was used to enable the configuration of a car. The configuration output, in turn, could be either directly previewed on the tablet computer or experienced using a virtual reality application. However, as the car configuration procedure is usually performed iteratively, a user experiences frequent context switches of the used software application, which often leads to an embittered perceived user experience and usability. To remedy these drawbacks, one promising approach constitutes the idea to integrate the configuration procedure into a proper virtual reality application. The work at hand presents *Xconcept*, which draws upon various state-of-the-art approaches from the field of human-computer interaction to provide a suitable car configuration procedure based on a virtual reality setting. Among other important factors, one fundamental goal of *Xconcept* constitutes the perceived user experience independently of age, gender, or previous virtual reality experiences. To evaluate whether or not this can be achieved with *Xconcept*, we conducted a study with employees of the Daimler AG. Although the results of the study reveal that with rising age, the rating of the *Xconcept* deteriorates, the overall user experience and usability has been rated positively. Interestingly, gender and previous experiences with virtual reality applications had no significant effect on the rating of the user experience. Altogether, *Xconcept* shows valuable insights to ease the car configuration procedure based on a proper virtual reality setting.

Keywords: Car configuration · Virtual reality · User study

1 Introduction

Today's huge demand for various affordable products can only be satisfied by mass production, which was made possible by automation and their technological advances in the past decades. Especially the mass production of automobiles had

K. Arai and S. Kapoor (Eds.): CVC 2019, AISC 944, pp. 169–189, 2020.
https://doi.org/10.1007/978-3-030-17798-0_16

great impact on the economy and accelerated the growth of multiple new industries. However, one major downside of today's automation is that the diversity
of variants is limited to a given set, i.e., a pre-customizing approach is applied,
which results in a lack of (self-)customization and individualization of a product,
especially a high-tech one, such as a car. The problem intensifies even further
when considering cultural differences between the local markets and the resulting
requirements in terms of customization. One predominant cause among others
are the inflexible standardized processes and methods, which are necessary in the
wake of mass production. However, recent advances in the field of digitalization
might harbor the answer to the customization problem, which could satisfy the
human need for individualization [1].

In the context of individualization demands, in September 2017, the Smart
division of the Daimler AG introduced a new approach, which was named unleash
the color, towards the individualization (and therefore the mass customization)
of interior trim parts at the International Motor Show (IAA). With unleash the
color, which, in turn, was made possible based on an innovative printing process,
the Smart division offers a peek preview at further ways of individualizing a car
on the example of personalized interior trim parts on the basis of a completely
digitalized process [2]. Moreover, the Smart division demonstrated in what way
an entirely digitalized process may look like for a quick and easy realization of
individual small series or parts. For this purpose, a configuration application was
used on a tablet computer to design selected interior trim parts (e.g., air vents,
instrument covers, or multimedia interface panels) with a motif of the customer's
choice. After the visitor completed the configuration process, the entire design
could be viewed immediately on a tablet computer and experienced in a virtual
reality application. As soon as the visitor positively appraised the design, it was
optimized and prepared for printing. Then, the design was printed in color with
tactile effects on the surfaces of the real parts of the car. Therefore, a digital printer
was placed in the direct vicinity of the smart exhibition stand in the Mercedes
FabLab. The visitors could then see how the trim parts were printed and find out
more about the innovative printing process and the technology itself [2].

However, when taking a closer look at the way customers would potentially
like to configure their car of desire, it becomes clear that the process of configuration is usually an iterative one. This is particularly the case when it comes to
a high degree of individualization with a mere endless choice of options as one
might want to test certain design options and their effects. With the innovative
configuration approach used at the IAA, this iterative process still would result
in a continuously occurring context switch (cf. Fig. 1) as customers would have to
remove the head-mounted display and interact with a tablet computer in order
to change the configuration. This loosely coupled process of a car configuration
between the different used devices eventually results in a lack of immersion, a
presumably embittered perceived user experience as well as an overall inefficient
configuration procedure, which could charge a potential customer with negative
emotions and frustration. Therefore, the configuration procedure has to be properly integrated with the virtual reality application. Otherwise, the cause for the
frustration would only be changed compared to the setting without the virtual
reality application.

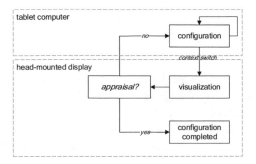

Fig. 1. Abstract configuration process using a Tablet Computer and a Virtual Reality Application

One important aspect, among others, for a satisfying user experience, constitutes the consideration of the age of car buyers in the past and present. The typical customer of the Daimler AG is between 51 and 55 years old. The Daimler AG, on the other, desires to attract younger customers in the same way than senior customers, preferably between the age of 25 and 44. Thus, the considered customer target group for a virtual reality-based car configuration application must consist of any person between the age of 25 and 55+. Based on these considerations, we designed *Xconcept*. The latter is a human-computer interaction concept for a virtual reality-based car configuration application, with a particular focus on the perceived user experience. Thereby, *Xconcept* addresses the (1) aforementioned drawbacks of the existing application shown by the Smart division as well as considers the peculiarities of the (2) illustrated target group. Furthermore, the work at hand presents results of a study that evaluated *Xconcept*. Interestingly, the obtained results show that (1) an approach like *Xconcept* is highly welcome and (2) that selected customer characteristics (e.g., age) should be properly taken into account.

The remainder of this paper involves five sections organized as follows: In Sect. 2, the requirements identified for *Xconcept* are discussed, while Sect. 3 presents *Xconcept*. Section 4, in turn, presents and summarizes the results of the conducted study. Related work is discussed in Sect. 5, whereas Sect. 6 concludes the work with a summary and an outlook.

2   Xconcept Requirements

Prior to the presentation of *Xconcept* requirements, selected background aspects are presented. In general, the design of a virtual reality (VR) environment means to enable a user to interact with that environment under real-time conditions. The interaction inside the environment, in turn, is basically accomplished by the exchange of information between a human and a computer that controls the virtual environment, also known as Human-Computer Interaction (HCI). Therefore, VR can be considered as a kind of HCI, and, hence, *Xconcept* is concerned with the development and evaluation of a human-computer interaction concept for a virtual reality-based car configuration application. In particular, *Xconcept* pursues the following two major objectives:

- Achievement of a positive usability perception: The human-computer interaction concept should be designed in a way that the integrated configuration procedure is perceived as intuitive and easy to use for most of the users, preferably independent of age, sex, or previous virtual reality experiences. In addition, the perceived usability shall be classified as above average in an evaluation design.
- Achievement of a positive user experience: The human-computer interaction concept should be designed in a way that convinces by its pragmatic and its hedonistic qualities, preferably independent of age, sex, or previous virtual reality experiences. In addition, the perceived user experience shall be classified as above average in an evaluation design.

Based on these two major objectives, *Xconcept* describes how a person could interact with a computer by means of a VR system to configure a car. However, in order to ensure a proper definition and selection of the user interface and HCI methodology, it is crucial to define all important requirements beforehand. The *Xconcept* requirements, in turn, are primarily based on the above stated major objectives, the considered target group, and four important interaction archetypes. As the latter are decisive requirements *Xconcept* aims at, they are shortly introduced: Regarding Archetypes 1 (Selection) and 2 (Manipulation), users usually interact with a VR system to select or manipulate virtual objects. Regarding Archetype 3 (Navigation), users want to determine their position as well as viewing direction in a virtual environment. Finally, regarding Archetype 4 (Control), there is a necessity to interact with the VR system itself in order to perform functions outside of the virtual environment at a meta level (e.g., loading a new virtual world).[1]

Based on this, functional as well as non-functional requirements have been elicited for *Xconcept*. The functional requirements are defined by means of UML 2 Use Case Diagrams and User Stories, which are based on the described interaction archetypes. Notably, the *selection* and *manipulation* archetypes are considered together due to the fact that an object usually has to be selected before a manipulation can take place.

2.1   Selection and Manipulation

The use case diagram of the *selection & manipulation* scenario provides an overview of the use cases related to the corresponding interaction archetypes (cf. Fig. 2). The use case of *manipulation* (e.g., change the lacquer finish of a car within the VR simulation) always includes the use case of a *selection* beforehand. In reality, there are usually two subtypes of *manipulation*, which are denoted as

---

[1] Further information can be found at https://dbis.eprints.uni-ulm.de/xconcept.pdf.

local (e.g., close objects, which are within reach) and *remote* (far away objects, which are out of reach); i.e., these two subtypes require a corresponding *selection* interaction archetype.

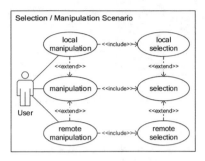

Fig. 2. Use case diagram of the selection & manipulation scenario

Furthermore, Table 1 contains the use cases of the *selection & manipulation* scenario with their corresponding user stories.

Table 1. Selection & manipulation scenario with their corresponding user stories

ID	Use case	User story
F1	Manipulation	As a *User*, I want to manipulate objects (e.g., trim parts of a car) in order to individually customize them
F2	Selection	As a *User*, I want to select objects (e.g., trim parts of a car) in order to manipulate them subsequently
F3	Local manipulation	As a *User*, I want to manipulate objects in direct vicinity in order to individually customize them
F4	Local selection	As a *User*, I want to select objects in direct vicinity in order to manipulate them subsequently
F5	Remote manipulation	As a *User*, I want to manipulate distant objects in order to individually customize them
F6	Remote selection	As a *User*, I want to select distant objects in order to manipulate them subsequently

2.2 Navigation

The use case denoted as navigation can be either extended by *wayfinding* or *traveling* (cf. Fig. 3). In order to find the way through a virtual world, an overview of that world is required, which can be provided by a map. Furthermore, *wayfinding* is the cognitive component of *navigation*[2] (cf. Fig. 3). Therefore, the use case

---

[2] For further explanations see https://dbis.eprints.uni-ulm.de/xconcept.pdf.

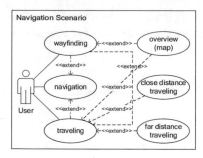

Fig. 3. Use case diagram of the navigation scenario

denoted as *wayfinding* is only extended by the *overview* use case as a map is not necessarily required. The use cases denoted as *wayfinding* and *overview* can be furthermore extended by *traveling* as a user might want to start traveling after he or she is done with the route planning. The use case denoted as *traveling* could be further extended by *close distance traveling* and *far distance traveling*, depending on the distance to the endpoint of the route.

Table 2 contains the use cases of the *navigation* scenario with their corresponding user stories.

Table 2. Navigation scenario with their corresponding user stories

ID	Use Case	User Story
F7	Navigation	As a *User*, I want to navigate through the virtual world in order to get to the points of interest
F8	Wayfinding	As a *User*, I want to be able to find my way through the virtual world in order to know how to get to the points of interest
F9	Overview (map)	As a *User*, I want to be able to get a overview of the virtual world in order to know something about the *where*, *what* and *how*, again in terms of the points of interest
F10	Traveling	As a *User*, I want to be able to travel inside of the virtual world in order to reach the points of interest
F11	Close distance traveling	As a *User*, I want to be able to travel to close points of interest with an appropriate method in order to reach them comfortably
F12	Far distance traveling	As a *User*, I want to be able to travel to distant points of interest with an appropriate method in order to reach them comfortably

2.3   System Control

The use case denoted as *system control* includes always the use case *manipulation*, which, in turn, includes the use case *selection* (cf. Fig. 4).

Table 3 contains the use cases of the *system control* scenario with their corresponding user stories. Although the use cases *manipulation* and *selection* are already described in Table 1, they are again mentioned as they are an integral part of this scenario.

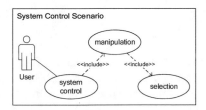

Fig. 4. Use case diagram of the system control

Table 3. System control with their corresponding user stories

ID	Use case	User story
F1	Manipulation	As a *User*, I want to manipulate objects (e.g., trim parts of a car) in order to individually customize them
F2	Selection	As a *User*, I want to select objects (e.g., trim parts of a car) in order to manipulate them subsequently
F13	System control	As a *User*, I want to be able to execute the system control in order to change the various settings of a simulation

Due to space limitations, the non-functional requirements are not discussed in this work.[3]

3   Xconcept

The *Xconcept* aims to provide a properly perceived user experience and usability, independently of age, sex, or previous virtual reality experiences. To meet the stated objectives, *Xconcept* primarily focuses on natural interaction goals. Thereby, *Xconcept* draws upon various fundamentals and state-of-the-art approaches, which are briefly introduced: As input interface, the human hand and body are used, while for the output interface, a head-mounted display and earphones are used. The tracking of the hand is performed by means of an optical

---

[3] See https://dbis.eprints.uni-ulm.de/xconcept.pdf for further information.

markerless inside-out method. The tracking of the body, in turn, is performed by means of an optical markerless outside-in method. Finally, the hand is used to perform the majority of interaction methods, whereas the body is solely used for the purpose of *traveling*. Furthermore, *Xconcept* proposes the utilization of a *hand menu* and a *World-in-Miniature* (WIM) method.

Since the interaction methodology of *Xconcept* constitutes the most important pillar it is based on, this section mainly focuses on relevant aspects of the interaction methodology. According to the presentation of the requirements, the discussion of the interaction methodology is structured along the archetypes.

3.1 Selection

The *selection* interaction archetype allows users to determine semantically relevant objects of the virtual world in order to subsequently interact with them [3,4]. Based on a hand as input interface, it is feasible to use already known 2D interaction methods [5,6]. Additionally, in order to make users aware that they are about to perceive the selection of an object, visual and acoustic indicators are used based on a head-mounted display and earphones as the output interface.

For the purpose of *local selection*, *Xconcept* suggests to utilize a finger (or multiple fingers) to select objects by simply touching them for a specified period of time (*hold gesture*). This is shown in Figs. 5 and 6. Thereby, Fig. 5 shows an interaction scene in which a user is in direct vicinity of a virtual reality car. In the given scenario, it is assumed that the user wants to change the color of the car lacquer finish.[4] Figure 6, in turn, shows the user while performing a *hold gesture* to select the exterior of a car with his or her left index finger. During *selection*, a visual (e.g., displayed in a blue circle around the finger) and an audible indicator are provided.

Fig. 5. The first image of the selection & manipulation interaction scene shows a user in direct vicinity of a VR car. In the given scenario, it is assumed that the user wants to change the color of the car lacquer finish. Drawn by [7].

---

[4] Drawn by [7]. All other figures in this section are also drawn by [7].

For the purpose of *remote selection*, the *Xconcept* uses the *World-in-Miniature* (WIM) method [8]. The usage of the WIM method, in turn, is shown in Figs. 7, 8, 9, and 10. Thereby, the *remote selection* interaction archetype is performed by using the *hold gesture* onto an appropriate object on the WIM map.

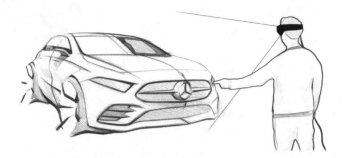

Fig. 6. The second image of the selection & manipulation interaction scene shows the user while performing a *hold gesture* to select the cars exterior with his or her left index finger. During the selection, a visual (e.g., displayed in a blue circle around the finger) and an audible indicator are provided. Drawn by [7].

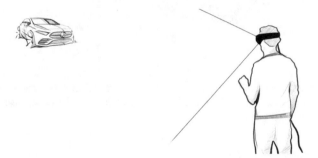

Fig. 7. The first image of the WIM interaction scene shows a user distant to a car, while the user intends to see the car from a closer distance. Drawn by [7].

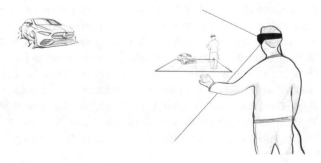

Fig. 8. The second image of the WIM interaction scene shows the user with the open WIM map, where he or she may easily obtain all semantically relevant objects. Drawn by [7].

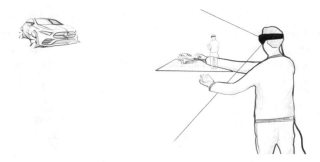

Fig. 9. The third image of the WIM interaction scene shows the user while reaching for the car in order to move it on the map, which will eventually result in a movement in the virtual world. Drawn by [7].

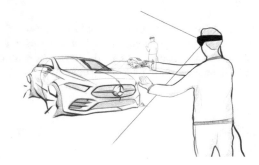

Fig. 10. The fourth and final image of the WIM interaction scene shows the final state of the translation of the car. The car is now in direct vicinity of the user as he or she intended it to be. Drawn by [7].

3.2   Manipulation

The manipulation of car parts is essential for a car configuration procedure and is defined as an interactive change of object parameters [3]. Within *Xconcept*, users can manipulate objects either by means of a menu or gestures, after the *selection* has taken place. Menus spawn after a successful *selection* and are always pointing at their origin (selection point) with the goal that a user knows where the menu is coming from or which object will be manipulated by using it.

An example of a *local manipulation* interaction is shown in Fig. 11. For the interaction with required menus, *Xconcept* uses common 2D interaction methods [5,6]. This can be seen in Fig. 12, in which a user applies the *tap gesture* to select an item in the menu. Furthermore, a virtual hand method [8] is incorporated for the spatial manipulation of smaller objects, e.g., by picking them up and carrying them away.

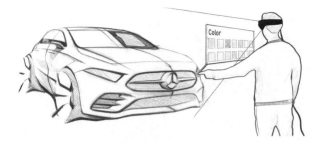

Fig. 11. The third image of the selection & manipulation interaction scene shows the pop-up menu (blue), which spawned after the user finished the selection process. Drawn by [7].

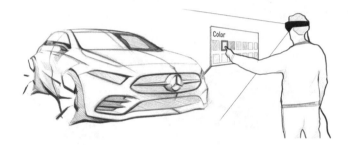

Fig. 12. The fourth and final image of the selection & manipulation interaction scene shows the user interacting with the menu in order to change the color of the car lacquer finish. Drawn by [7].

The *remote manipulation* interaction, in turn, takes place by means of a menu as well. Here, the menu spawns in direct vicinity of the user and points to the object on the WIM map. Gestures can be used after enabling the gestures mode by taping onto the corresponding button within a menu (*menu & module* associated gestures), or the *hand menu* (global gestures). An example of a module associated gesture constitutes the swipe gesture, which allows one to iterate through different object options (e.g., colors). An example of the usage of global gestures constitutes the rotation of a car, which is illustrated in Figs. 13, 14, and 15. Note that these gestures can be also performed locally.

Finally, the gestures supported by *Xconcept* are summarized in Table 4, i.e., with their description and classes (global, menu). Global gestures are used to manipulate the car as an entity and are always performed with both hands. 3D menu gestures, in turn, are used as an alternative menu control, and are always performed with one hand.

Fig. 13. The first image of the gesture interaction scene shows the virtual representation of the hands of a user related to the rotate gesture. The scene is within the user's point of view. Drawn by [7].

Fig. 14. The second image of the gesture interaction scene shows the car and the hands rotating. The car rotates in correspondence to the hands. Drawn by [7].

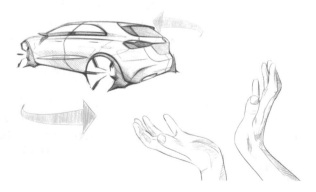

Fig. 15. The third and final image of the gesture interaction scene shows the car and the hands of the user in the final position after the rotation is done. Drawn by [7].

3.3   Navigation

Navigation can simply be described as finding a way to a specified point, which is separated in *wayfinding* and *traveling* (cf. Sect. 2.2). Within *Xconcept, traveling* and *wayfinding* are performed by walking around or using a WIM map (cf. Sect. 3.1).

Regarding *wayfinding* and *overview, Xconcept* supports the process of *wayfinding* by providing a WIM map to get an overview of the virtual world. For the interaction with the WIM map, *Xconcept* uses common 2D interaction methods [5,6]. In order to be able to select a object of desire or to manipulate

Table 4. 3D gestures supported by *Xconcept*

Class	Gesture	Description
Menu	Single hand thumbs up	The thumbs up gesture is performed by simply extending the thumb upwards. This gesture is used within *Xconcept* to approve and save a selected menu item or module
Menu	Single hand horizontal swipe	The horizontal swipe gesture is performed by simply extending all fingers in the direction of the car (thumb facing the ceiling), and then either moving the wrist joint to the left or right. Thereby, swiping left selects the previous menu Item, while swiping right the next respectively
Menu	Single hand clap	The single hand clap gesture is performed by simply clapping with a single hand. Notably, the hand has to face the head. This gesture is used to exit a menu module
Menu	Single hand vertical swipe	The vertical swipe gesture is performed by simply extending all fingers in the direction of the car, while the palm is facing the ceiling or the floor. If the palm is facing the floor while swiping, the previous menu module is selected. If the palm is facing the ceiling while swiping, the next menu module is selected
Global	Dual hand vertical swipe	The vertical swipe gesture is performed equally to the one stated above for the menu interaction. Here, both hands are used in parallel. Furthermore, the hands have to face the car. The gesture is used to change the absolute height of the car (spatial dimension)
Global	Dual hand rotate	The rotate gesture is performed as illustrated above. Here, both hands are brought together, such that their wrists almost touch each other. Furthermore, the hands should be opened in a curved manner. Now, both hands have to be rotated in the same direction in parallel. The rotation of the car follows the rotation of both hands
Global	Dual hand pinch and spread	The dual hand pinch and spread gesture is performed as follows: First, one has to perform the pinch gesture for each hand, which is bringing the thumb and index finger together. In the next step, both pinched finger pairs have to get in touch to each other. This enables the zoom mode. When moving the finger pairs apart, such that the forearms do not cross, then, the car is scaled-up. If the opposite event is the case, the car is scaled-down respectively

the viewed zone, it is possible to perform scaling and rotation actions on the WIM map. Scaling is performed with the *spread and pinch* gesture, while rotation, in turn, is performed with the *rotate* gesture. Finally, the *Drag and Slide* gesture can also be performed on the WIM map in order to move the current zone presently viewed on the map.

Regarding *traveling*, *Xconcept* is performed by physical walking (close distance traveling) or the use of the aforementioned WIM map with teleportation (far distance traveling) capabilities. Within the WIM map, it is possible to teleport objects, including oneself, by simply moving the objects on the map (*Drag and Slide*). An example for far distance traveling is shown in Figs. 7, 8, 9 and 10.

3.4 System Control

The *System Control* allows interaction with the virtual world on a meta level [4], and is realized by means of the *hand menu* method [8]. Due to space limitations, Fig. 16 shows only one selected example for the *system control* interaction. Note that the *hand menu* is opened by facing the palm to the head. *Selection* takes place as described in Sect. 3.1. The aforementioned WIM map is opened via the use of a *hand menu*, for example, by performing the single hand clap gesture with the appropriate hand, which is the one for which the menu is opened at.

Fig. 16. The depicted hand menu interaction scene shows the completed selection of a menu item. Thereby, an indicator shows which item has been selected for a specified period of time. Drawn by [7].

3.5 Discussion

Xconcept is a hand-driven virtual reality-based human-computer interaction concept, which primarily focuses on natural interactions. The hand-driven approach aligns with Heidegger's philosophy of being-in-the-world [9]. The human hand is an already and literally experienced tool and is therefore ready-to-hand for the most basic tasks of our daily life and is therefore familiar, proximal, and directly useful. We naturally know how to touch and grab things with our hands and fingers. Furthermore, most people are already used to two-dimensional touch interaction gestures, known from smartphones and tablet computers. The hand also aligns with Merlau-Ponty's intentional arc, which is to allow oneself to respond to the call of objects by knowing (intuitively) how to interact in a way that coheres with them [10]. This could be emphasized by the example of a user standing in front of a virtual reality car, i.e., within the configuration simulation. Here, the user would probably automatically start to walk around after some time, and would try to interact with the virtual reality car, by reaching the hands or fingers towards parts of the car. An indicator would show the user then instantly that he or she is indeed able to interact in that way. After a while, a menu will spawn, which is self-explanatory as it is tied visually to the car part and people are usually used to be working with menus.

4   Conducted Study and Results

The exploratory research study is particularly concerned with the question of how a user perceives *Xconcept* in regards to pragmatic and hedonistic qualities. For a successful commercial application of *Xconcept*, it is of utmost importance that a user perceives the overall user experience and usability as being positive. To reach the discussed target group, it is vital to provide a positive experience, independently of age, sex, and previous virtual reality experiences.

4.1   Hypotheses

To investigate the research questions, how a user perceives *Xconcept*, the following operationalized hypotheses have been suggested:

First, Question A was raised: *Is the overall user experience and usability of Xconcept being positively perceived?*

1. Hypothesis 1: In an evaluation study, users will rate the overall user experience of *Xconcept* as being positive.
2. Hypothesis 2: In an evaluation study, users will rate the overall usability of *Xconcept* as being positive.

Second, Question B was raised: *Is the perceived user experience and usability of Xconcept independent (no statistical correlation) of age, sex, and previous virtual reality experiences?*

1. Hypothesis 3: In an evaluation study, user experience does not correlate with age.
2. Hypothesis 4: In an evaluation study, the usability evaluation does not correlate with age.
3. Hypothesis 5: In an evaluation study, user experience does not correlate with sex.
4. Hypothesis 6: In an evaluation study, the usability evaluation does not correlate with sex.
5. Hypothesis 7: In an evaluation study, user experience does not correlate with previous virtual reality experiences.
6. Hypothesis 8: In an evaluation study, the usability evaluation does not correlate with previous virtual reality experiences.

4.2   Study Setting

All participants in this presented study were recruited at the Daimler AG at the research and development facility in Ulm, Germany. Prior to participation, the participants were informed about the exclusion criteria, which are VR sickness, motion sickness, sea sickness, and epilepsy. In total, 92 people participated in this study. Four participants failed to respond to all items of the used questionnaire, therefore their data was not included in the analyzes. The final sample consisted of 88 people (73.9% males and 26.1% females), who completed the experiment (N = 88), with an average age of 34.03 years ($SD = 14.15$). Previous experiences with VR applications were reported by 43.2% of the participants.

Due to the fact that there is currently no prototype available for *Xconcept*, its interaction methodology had to be evaluated by means of a substitute. To test the earlier introduced hypotheses, all participants had to engage in two different VR games, which are similar to the features of the *Xconcept*: *Blocks* [11] and *Force-Directed Graph* [12]. Note that Blocks as well as Force-Directed Graph both provide features that represent selection, manipulation, navigation, and system control interaction scenarios similar to *Xconcept*. Furthermore, both applications use a *hand menu* for *system control* and physical walking for *navigation*. Moreover, Blocks provides the use of gestures and the virtual *hand method*, while Force-Directed Graph provides the use of floating menus for the local manipulation. Therefore, these games are used as a basis for the evaluation of the user experience and the usability of *Xconcept*.[5]

In order to provide the two VR games to the participants, a Dell Precision Tower 7910 was used. As the virtual reality output, the head-mounted display HTC VIVE was used, whereas the Leap Motion Controller was used as the input device. For optical body tracking, two common HTC base stations were used.

For study purposes, a questionnaire was used to measure the user experience and usability.[6] The used questionnaire, in turn, mainly comprises the existing questionnaires *AttrakDiff* [13] as well as the *System Usability Scale* [14], plus an additional sociodemographic section, which addresses sex, age, and previous virtual reality experiences.

Regarding the AttrakDiff questionnaire, it captures the perceived user experience and usability of a product. The questionnaire consists of the following scales: attractiveness (ATT), pragmatic quality (PQ), hedonistic quality (HQ), which is further separated in hedonistic quality – identity and hedonistic quality – stimulation. Furthermore, each of the four scales is evaluated by the use of a likert scale.

Regarding the System Usability Scale (SUS) questionnaire, it is essentially a questionnaire used for the reliable, quick, and dirty measurement of the usability. The SUS questionnaire has been chosen as additional measurement as it is considered to be an industry standard as well as provides reliable and valid results for small sample sizes.

---

[5] For a detailed description of the game setting, see https://dbis.eprints.uni-ulm.de/xconcept.pdf.

[6] The questionnaire can be found on Page 97 of https://dbis.eprints.uni-ulm.de/xconcept.pdf.

4.3   Study Procedure

Based on the presented background information, the study was conducted following the *Publication Manual of the American Psychological Association* [15]. To be more precise, the study was conducted in a reasonably distraction free and quiet meeting room, in which a VR system was already pre-installed. The VR room was set to a size of around 2.5 by 5 m. The examination took place with one participant at a time and took about 45 min in total (cf. Fig. 17). On arrival, the participants were repeatedly asked if they suffer of any of the exclusion criteria and were reminded that they could abandon the examination at any time. Next, they received an orally VR related safety instruction, where they were told that they could move freely in the room until a blue line or grid appears, which is marking the end of the area. After the given instructions, participants started with Blocks, followed by Force-Directed Graph. After completing the two games, the participants were debriefed. They were also asked in this context if they could imagine to configure the interior of a car with *Xconcept*. Before they received the questionnaire, they were told that they should only rate the experienced interaction concept and not the games, their realization and design, the technology, including their perceived problems (if any occurred), and the interfaces. On average, participants completed the questionnaire in about 8 min. After returning all of the materials, the participants had the opportunity to ask any questions they have, or to give some personal feedback about the experiment.

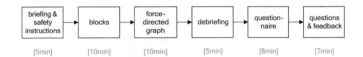

Fig. 17. Procedure of the study. The study started with a briefing and a safety instruction. In the next step, the two games were played; first Blocks, then Force-Directed Graph. Before the participants continued with the questionnaire, they were debriefed. In the final step, they were given the chance to ask questions about the experiment or to provide personal feedback.

4.4   Data Preparation

Regarding data preparation, all statistical calculations were performed with IBM SPSS Statistics 3 on a significance level of $\alpha = .05$. Furthermore, the following necessary presumption was made: From a formal point of view, the rating scales used for the AttrakDiff and SUS questionnaires are ordinal or ranked because it cannot be necessarily assumed that test participants perceive the different answer options as equidistant. However, since the scales are formulated symmetrically, the values can be treated as quasi-metric, and, hence, as an interval scale. Therefore, and due to the fact how the scales have to be calculated, the mean instead of the median will be considered as primary location parameter.

4.5   Study Results

Due to space limitations, not all study results cannot be presented in detail. Therefore, we summarize the results and solely present two selected aspects. Reconsider that two major hypotheses were suggested: Question A: *Is the overall user experience and usability of Xconcept being positively perceived?* and Question B: *Is the perceived user experience and usability of Xconcept independent (no statistical correlation) of age, sex, and previous virtual reality experiences?*. Based on the conducted study, Research Question A could be confirmed by the statistical data, while research Question B could not be confirmed.

Regarding Question A and its Hypothesis 1 that users will rate the overall user experience of *Xconcept* as being positive, this could be confirmed by taking a closer look at the AttrakDiff questionnaire scales. For this purpose, the mean value of the attractiveness (ATT) and hedonistic quality (HQ) AttrakDiff scales have to be equal or greater than 5 per definition to confirm the hypothesis. The overall mean of the ATT scale is $mATT = 5.57$ (Median = 5.57, SD = .71, Min = 3.86, Max = 7.00), and the overall mean of the HQ scale is $mHQ = 5.29$ (Median = 5.29, SD = .53, Min = 4.21, Max = 6.36). Therefore, the hypothesis can be considered as confirmed. The same holds for Hypothesis 2.

Regarding Question B and its Hypothesis that *Xconcept* is independent of age, sex, and previous virtual reality experiences, it was found that with rising age, the overall rating of the *Xconcept* deteriorates. Consider therefore Table 5.

Table 5. Results of the bivariate correlation examination for two identified age groups

			ATT	PQ	HQ	SUS
Age	**Group A** (age < 40) N = 54	Pearson r	-.084	-.036	-.024	.074
		Pearson p	.544	.797	.864	.593
		Spearman ρ	-.103	-.038	.007	.042
		Spearman p	.456	.783	.962	.761
		Kendall τ_a	-.088	-.028	-.002	.021
		Kendall p	.384	.779	.988	.838
	Group B (age ≥ 40) N = 29	Pearson r	-.444*	-.112	-.436*	-.226
		Pearson p	.016	.563	.018	.238
		Spearman ρ	-.443*	-.214	-.389*	-.167
		Spearman p	.016	.264	.037	.386
		Kendall τ_a	-.327*	-.158	-.274*	-.136
		Kendall p	.018	.248	.045	.324

The * annotation indicates that the correlation is significant at the .05 level (2-tailed).
ATT = Attractiveness — PQ = Pragmatic Quality — HQ = Hedonistic Quality
SUS = System Usability Scale

Regarding **Group A**, the results shown in Table 5 suggest that there is no bivariate correlation between age and the user experience relevant scales. For example, the correlation coefficient for the ATT and the HQ scales are relatively close to zero and Pearson's r p-value is way bigger than the confidence level α ($p > .05$; $p > \alpha$).

In contradiction to **Group A**, the results shown in Table 5 for **Group B** illustrate that there is a bivariate correlation between age and the user experience relevant scales. For example, the age variable correlates significantly (2-tailed, .05 level) with the ATT scale.

However, the data revealed that the overall user experience and usability still has been rated above average on the respective scales and therefore being positively perceived. Sex and previous experiences with virtual reality applications, however, had no effects on the rating. Therefore, the results of the study indicate that it is worthwhile to improve *Xconcept* by implementing a prototype as well as further work on its interaction methodology.

4.6   Limitations

The study was conducted at the Daimler AG in Ulm, Germany. All participants were employees at the research and development facility and therefore usually have an university education in a technical or science-related subject. Another fact to keep in mind is that presumably most employees are likely to be positively minded about and interested in the subject of the investigation as they are working in the automotive industry. In addition, most participants had extremely fun while playing Blocks, for which the Halo Effect cannot be excluded. Furthermore, most participants were of German heritage. The findings of the study poses therefore a bias and cannot be applied to the general population as the study was conducted in an extremely homogeneous field.[7]

5   Related Work

In general, less approaches can be found that directly address virtual reality solutions in the context of car configuration procedures. As already discussed in Sect. 1, the Smart division of the Daimler AG introduced an approach [2] that is related to this work. However, we also discussed its limitations, which served as the basis for the work at hand. In addition, works exist that deal with virtual reality applications and how they can be used to experience a car [16]. Again, these approaches do not focus on the car configuration procedure. To utilize a virtual reality application in other related fields like maintenance can be found [17]. Although these approaches show that virtual reality can play an important role to better support maintenance procedures, they are not directly deal with aspects that are important while configuring a car. Moreover, approaches exist

---

[7] All study results can be obtained from https://dbis.eprints.uni-ulm.de/xconcept. pdf.

that are related to selected techniques used for *Xconcept* [3, 4, 11, 12]. However, they focus on other scenarios. In addition, only few approaches introduced study results in this context. On the other, existing prototypes like the one of the Smart division or others [18] show the relevance of virtual reality solutions in the context of the individual configuration of a car.

6   Summary and Outlook

In this paper, a hand-driven VR-based HCI concept, for the purpose of car configuration, named *Xconcept*, has been designed and evaluated. In particular, *Xconcept* aims towards a solution of the stated problem of continuously occurring context switches while configuring a car like shown for the approach of the Smart division. Such context switches eventually result in frustration and therefore often in an embittered user experience and usability. In order to ensure that the reasons for the frustration decrease, *Xconcept* must also deal with the fact that it is designed especially for its target group, which could be any person between 25 and 55+ years. Thus, another goal of *Xconcept* constitutes the provision of a user experience and usability that is independent of age, sex, or previous virtual reality experiences. The paper discussed how *Xconcept* deals with these goals and on top of that it showed results of a study that revealed that *Xconcept* can achieve these goals. However, it is also shown that *Xconcept* is only the first step into a promising direction.

References

1. Coletti, P., Aichner, T.: The need for personalisation. In: Mass Customization, pp. 1–21. Springer (2011)
2. Daimler AG: Smart at the International Motor Show 2017: Experience the Future for Yourself. Press Release (2017). https://media.daimler.com/marsMediaSite/en/instance/ko/smart-at-the-International-Motor-Show-2017-Experience-the-future-for-yourself.xhtml?oid=29182467. Accessed 31 Oct 2018
3. Dörner, R., et al.: Interaktionen in virtuellen Welten. In: Virtual und Augmented Reality (VR/AR), pp. 157–193. Springer (2013)
4. Preim, B., Dachselt, R.: Grundlegende 3D-Interaktionen. In: Interaktive Systeme, pp. 339–397. Springer (2015)
5. Saffer, D.: Designing Gestural Interfaces: Touchscreens and Interactive Devices. O'Reilly Media Inc., Cambridge (2008)
6. Villamor, C., et al.: Touch Gesture Reference Guide. Touch Gesture Reference Guide (2010)
7. Meinzer, J.: Visualization of the Interaction Methodology of the Xconcept in Four Scenes (2018). jannmeinzer@icloud.com
8. Bowman, D., et al.: 3D User Interfaces: Theory and Practice, CourseSmart eTextbook. Addison-Wesley, Boston (2004)
9. Heidegger, M.: Being and Time. Suny Press, Albany (2010)
10. Merleau-Ponty, M.: Phenomenology of Perception. Routledge, Abingdon (2013)
11. leapmotion.com: Blocks (2018). https://gallery.leapmotion.com/blocks/. Accessed 31 Oct 2018

12. leapmotion.com: Force-Directed Graph (2018). https://gallery.leapmotion.com/forcedirected-graph/. Accessed 31 Oct 2018
13. Hassenzahl, M., et al.: Attrakdiff: Ein fragebogen zur messung wahrgenommener hedonischer und pragmatischer qualität. In: Mensch & Computer 2003, pp. 187–196. Springer (2003)
14. Brooke, J.O.: SUS-A quick and dirty usability scale. Usability Eval. Ind. **189**, 4–7 (1996)
15. American Psychological Association: Publication Manual of the American Psychological Association (2018). https://www.apastyle.org/manual. Accessed 31 Oct 2018
16. Klinker, G., et al.: Fata morgana-a presentation system for product design. In: Proceedings of International Symposium on Mixed and Augmented Reality, pp. 76–85. IEEE (2002)
17. Kammerer, K., et al.: Towards context-aware process guidance in cyber-physical systems with augmented reality. In: 4th International Workshop on Requirements Engineering for Self-Adaptive, Collaborative, and Cyber Physical Systems, pp. 44–51. IEEE (2018)
18. demodern.com: Mazda Virtual Reality Car Configurator (2018). https://demodern.com/projects/mazda-vr-experience. Accessed 24 Oct 2018

Using Computer Vision Techniques for Parking Space Detection in Aerial Imagery

Andrew Regester and Vamsi Paruchuri[✉]

University of Central Arkansas, Conway, AR 72034, USA
vparuchuri@uca.edu

Abstract. We propose a pattern recognition algorithm that uses computer vision to analyze and map the location of parking spaces from aerial images of parking lots. The analysis method developed made use of line detection coupled with selective filtering based on the prevalence of line length and angle. The goal of this algorithm was to provide a means of automated detection of regions of interest in parking lot images for further use in collection of parking data. Aerial images that used for development and testing were collected via a quad-copter. The quad-copter was equipped with a camera mounted via a gimbal that maintained a camera angle parallel to the parking lot surface. Video was collected from an altitude of 400 ft and individual frames were selected for content. The images were then split into a development set and a testing set. For analysis, images were converted to gray-scale followed by the application of a binary filter. Line features in the binary images were then detected using a Hough transform. Resulting features were then analyzed iteratively to find recurring line patterns of similar length and angle. After filtering for noise, line end-points and intersections were grouped to estimate individual parking space locations. We performed analysis of the proposed methodology over real spaces and demonstrate good performance results.

Keywords: Parking space detection · Computer vision · Autonomous systems

1 Introduction

Many new businesses and technologies centered around transportation have arisen in the recent past, and as cities grow larger and populations become more dense, it is likely that many cities and business will need solutions to address increasingly dense populated areas; possible solutions such autonomous and shared vehicles have been proposed, but need addition development and work before becoming viable options. In addition, present-day transportation and navigation software available to the public is able to provide directions to a destination, but then leaves the driver who is often in an unfamiliar area to find suitable parking.

© Springer Nature Switzerland AG 2020
K. Arai and S. Kapoor (Eds.): CVC 2019, AISC 944, pp. 190–204, 2020.
https://doi.org/10.1007/978-3-030-17798-0_17

To address these issues, an automated solution using computer vision is proposed to provide parking space analysis and mapping with currently available or easily obtainable aerial images. The information from this solution could provide immediate improvements in current consumer applications as well as provide data for future city planning and the development of transportation.

1.1   Solution Scope

In order to accomplish space identification without any information other than the image contains parking spaces, some assumption and expectations about the parking areas characteristics were required.

First, parking spaces are assumed to be marked using straight lines, and the lines must be marked in a paint that contrasts the parking surface. Second, spaces are assumed to have other similar spaces near them - that is, there are multiple spaces that have parallel lane markings located in the same proximity.

Due to these assumptions and the how the images are analyze, this solution is most ideal for parking areas that are paved and have a larger number of spaces. While this proposed approach would not achieve any meaningful results on unpaved surfaces or smaller parking lots, it would be beneficial for data collection on larger parking areas such as would be found in downtown city centers or commercial parking areas.

While it may be possible to modify or improve on the techniques used for analysis of other surfaces, this approach was developed with larger, paved, and marked parking areas in mind.

1.2   Data Collection

In order to develop software to analyze aerial parking images, it was first necessary to collect some sample and test images. These initial test images were acquired from a GoPro Hero 3 camera. The camera was mounted to a 3D Robotics Iris quadcopter via a Tarot T-2D 2-Axis Brushless Gimbal. The gimbal was used to maintain a camera angle perpendicular to the ground, which helped to ensure consistency between subsequent images.

Test images were collected from a parking area that was chosen based on characteristics that would be commonly shared among many large parking lots. These characteristics include large parking areas (greater than 50 parking spaces), contiguous parking spaces, shared parking lines between adjacent spaces, and multiple rows of parking. An example aerial image is shown in Fig. 1.

Once parking lots were selected, a flight path consisting of a set of longitude, latitude, and elevation way points were uploaded to the UAV Pixhawk flight controller. Once the flight path was determined, the UAV was taken to the launch location. Before launching, the camera was turned on and the gimbal angle set to maintain a constant. Upon completion of the flight path, the flight video was retrieved from the GoPro camera.

Because of the wide angle lens of the camera, the flight video had considerable barrel distortion. To remove the distortion, GoPro Studio software was used with the "Fisheye Removal" option selected on video import. Once imported, relevant frames were selected and exported in JPEG format.

Fig. 1. An example of a test image collected using a UAV.

Initial Processing. The desired approach to achieve parking space mapping was intended to be generic enough that the method could be applied to aerial images of any parking lot that fit the typical set of characteristics. In order to keep the algorithm as universal as possible, a line detection approach was chosen that would attempt to detect the parking lines that separate parking spaces. Once the lines were detected, a pattern detection method would be developed that would be used to identify which detected lines were from parking spaces, and subsequently parking space identification could be achieved from the collection of parking lines.

The initial step of the proposed method required finding all parking lines. Ideally, for best results with this method, space detection would be perfectly achieved with bright white parking lines painted on a uniform dark gray or black asphalt surface. Since this is almost never the case due to surface imperfections, faded paint, varying surface characteristics and paint color, and variable lighting, the first stages of data processing aim to normalize the input images. This involves steps that attempt to remove noise and reduced the lines detected in the subsequent step to only those relevant for parking space detection.

2   Related Works

Various existing works that attempted to detect vehicle or parking spaces within parking lots were referenced while developing this method of parking space detection.

2.1 Parking Lot Occupancy Determination from Lamp Post Camera Images

Delibatlov et al. approached the parking lot analysis using a human defined 3-D space model of the monitored parking stalls from a single camera view [3]. Although the goal is handling occlusions by estimating volume based occupancy from an 2D image, the method used to determine occupancy could also be used to identify parking spaces if it was assumed that a detected vehicle was in a parking space (Fig. 2).

Vehicle detection relied heavily on classifying parking spaces from low camera angles with machine learning using vehicle features paired with Support Vector Machines. Since we will be using aerial images, some of the obstacles faced do not apply, though the challenges faced in regards to environmental variables is valid. Vehicle features and machine learning could also be used to improve accuracy with region of interest identification, though it would increase the complexity.

Fig. 2. Delibatlov et al. combined human-defines spaces from fixed cameras with vehicle detection using a Support Vector Machine.

2.2 Parking Space Detection with OpenCV and a Raspberry Pi

As line detection was a necessity of this project, work done by Fatsi using SimpleCV with canny edge detection to detect the presence of a car in a one parking spot provided useful insights [4]. Using images from a stationary camera and converting the image from RGB to binary, edge detection produced features that showed up as white pixels while non-edges show up as black (see Fig. 3). In the specified region of interest that was human-defined, a higher prevalence of white pixels on the image indicated many edges, which in turn was assumed to increase the likelihood that a car was present in the image.

While the approach used was simple and easy to implement, multiple factors decreased the accuracy of this method and ultimately influenced us to pursue another approach when attempting to find parking lines.

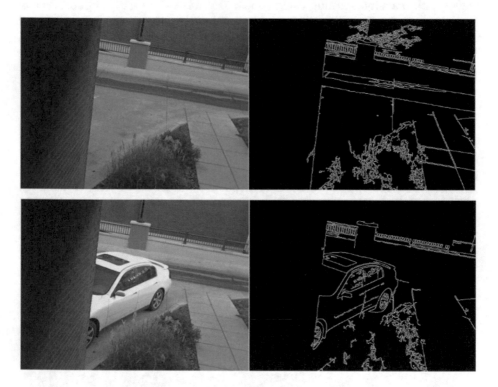

Fig. 3. Fatsi et al. used a simple edge-detection approach to determine the presence of vehicles in a parking space from a fixed camera.

2.3 Vacant Parking Space Detection in Static Images

Another approach to identifying parking spaces featured the use of machine learning and classifiers trained on vehicle features. True acquired pictures from a slightly elevated angle that mimicked images like those that might be collected from a fixed security cameras. These images were analyzed using color histograms with k-Nearest Neighbor classifiers and Support Vector Machines as well as vehicle feature detection algorithms [9]. The color histogram approach produced good results, and it was speculated that combining approaches was suggested as likely producing even better results as shown in Fig. 4.

While this is method made use of more complex techniques, detection of vacant parking spaces was higher than other works that were examined. Ultimately the combination of multiple methods that provide overlapping data could be used in our work to improve accuracy and produce better analysis.

Fig. 4. Combining advanced machine learning techniques showed good accuracy in the method proposed by True.

2.4 Other Related Works

Several other works studied parking lot management in order to determine whether a parking space is vacant using fixed cameras. A 8-class Support Vector Machine (SVM) classifier is used to distinguish parking spaces in [10]. A 3-layer Bayesian hierarchical detection framework (BHDF) for robust parking space detection is proposed in [5]. Multiple cameras were used to build a parking lot management system in order to monitor a wide parking area in outdoor environments in [2]. Features like colors in LUV space, gradient magnitudes and then trained a SVM-based classifier were extracted and used to classify a parking lot into vacant/occupied status in [7]. Amato et al. also used Deep CNNs (Convolutional Neural Networks) were proposed to train and detect parking slots and their status in [1] All the above works attempt to solve the classification problem whether it contains a vehicle or not using fixed cameras.

To the best of our knowledge, there is only one other work that studies a drone-based system where the camera (attached to the drone) is always moving. Deep Neural Networks was used to determine the occupancy status of parking spaces based the three features (vehicle color, local gray-scale variant, and corners) [8]. While in some scenarios the proposed system achieved good accuracy, in several scenarios the precision is only about 80%.

3   Design and Implementation

The desired approach to achieve parking space mapping was intended to be generic enough that the method could be applied to aerial images of any parking lot that fit the typical set of characteristics. In order to keep the algorithm as universal as possible, a line detection approach was chosen that would attempt to detect the parking lines that separate parking spaces. Once the lines were detected, a pattern detection method would be developed that would be used to identify which detected lines were from parking spaces, and subsequently parking space identification could be achieved from the collection of parking lines. An overview of the parking line detection algorithm is shown in Fig. 5.

In order to remove noise that may cause issues later in the analysis, a filtering process is required. This is mostly useful for removing many small or insignificant lines that are not required for use in detecting parking lines: things such as shadow contrast or areas of adjacent features such as a row of bushes or sidewalk expansion joints are examples of things that may produce false lines, and as such an attempt is made to remove them during pre-processing.

After the images are passed through the initial filtering, The initial step of the proposed method required finding all parking lines. Ideally, for best results with this method, space detection would be perfectly achieved with bright white parking lines painted on a uniform dark gray or black asphalt surface. Since this is almost never the case due to surface imperfections, faded paint, varying surface characteristics and paint color, and variable lighting, the first stages of data processing aim to normalize the input images. This involves steps that attempt to remove noise and reduced the lines detected in the subsequent step to only those relevant for parking space detection.

3.1   Normalization and Initial Filtering

Normalization and initial filtering of the input images was accomplished in a three step process:

Gray Scale Conversion. First, input images that were received as RGB images were converted to a gray scale image. In an RGB image, each pixel is represented by a three numeric values between zero and 255, each representing the value of a red, green, and blue component. After converting to gray scale, each pixel is represented by one numeric value between zero and 255 (see Fig. 6).

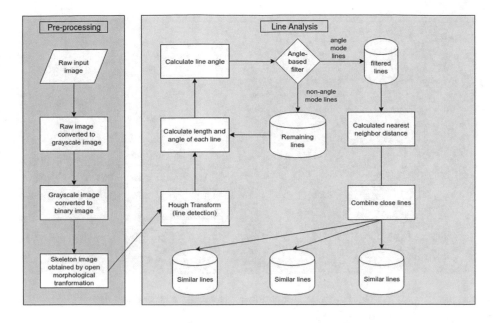

Fig. 5. An overview of the parking line detection algorithm

Binary Filtering. Once the image has been converted to gray scale, it was passed through a binary filter. The filter produced an image of pixels with values of only zero or 255 exclusively. This produces a true black and white image (see Fig. 7)

Many papers detail the specifics of the Hough line detection method, but a brief overview will be included to aid with the description of the approach used in this paper. Since parking are often a light color, typically white or yellow, and parking lot surfaces are usually dark, the contrast can be used with a line detection method to provide initial data as to the location of parking lines.

Images are analyzed individually. Initial preprocessing is done by first converting the image to gray scale. Next, a threshold is defined and applied to isolate pixels from the image that are above a certain value, which would indicate pixels in the initial image that were lighter in color. With only lighter pixels remaining, an open morphological transformation is applied to remove noisy pixels.

Morphological Skeleton. Before passing the binary image to the line detection algorithm, it is necessary to reduce the white pixels of the binary image to their minimum width; this is because images typically will have parking lines that are more than one pixel wide. Due to how line detection works, any lines that are represented by multiple pixels could be detected as multiple lines corresponding to each side of a multi-pixel representation of a parking line. This is achieved through a series of erosions and dilations in OpenCV, called a morphological transformation. In this case, the open morphological transformation is applied in order to produce a skeleton image, with the goal of producing a virtual representation of lines that are a minimum width prior to attempting line detection as seen in Fig. 8.

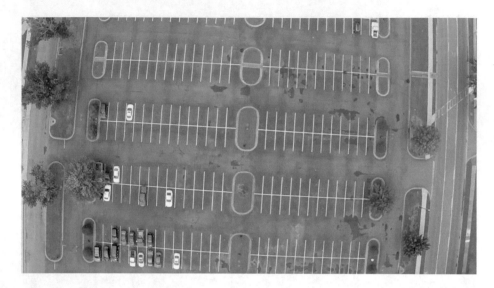

Fig. 6. A test image after gray scale conversion

Line Detection. Once preprocessing is complete, a Hough Transformation is used to detect lines [6]. These lines are then partitioned according to the angle of the line. The angle of each line is placed in a histogram. The angle range which contains the most lines is considered to be most likely parking lines. Further analysis gets the length of each line in the partition and then removes lines that do not fall within a standard deviation set by a percentage of the average line length. The distance to each line's nearest neighbor is also calculated, and those longer or shorter than a percentage of the average length are removed from consideration.

Lines that run perpendicular to the initial detected lines and intersect those lines are found by calculating lines which have an angle that is perpendicular to the angle of the largest group of lines partitioned by angle. If a number of these lines are found to intersect with the initial group of lines, this is a strong indication that a parking area has been found.

3.2 Parking Line Detection and Identifying Regions of Interest

After normalization, all images are binary and can be analyzed using the same method to determine if the lines are valid for use in identifying regions of interest.

Combine Close Lines. Since images and line detection techniques are imperfect and environmental factors can cause breaks or darkening of sections of parking lines, it is common for what would be considered a single parking line to be detected as two or more shorter lines. These broken lines can be combined to

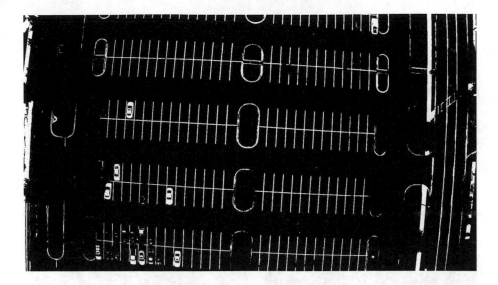

Fig. 7. A test image after the binary filtering step of pre-processing

produce better results by comparing the distances between detected lines. This is accomplished calculating the distance between lines. If either of two lines compared are within a factor of 0.2 of their length, they are considered close lines and subsequently combined.

Majority Parallel Lines. The first requirement is to find the majority parallel lines; these are lines that share similar lengths and angles. In theory, if the image contains a parking lot, then out of all the detected lines, those that have similar length and angle values would likely be parking lines.

Similar Lengths and Angles. To find these lines, the same method is used to isolate lines with similar length and angle values. First, a histogram is used to find the range of line length that is most common as well as the most common line angle. Then, standard deviation is calculated for all lines across both length and angle values.

In the case of the example image (Fig. 9), the most common lines detected are those that separate adjacent parking spaces from side to side. Note that these do not indicate lines marking individual spaces in this instance, but rather they run along two spaces that are oriented with the front of the spaces facing each other. This is not apparent from this image, nor is it able to be ascertained algorithmically at this point due to the exclusion of the longer lines separating the two adjacent rows.

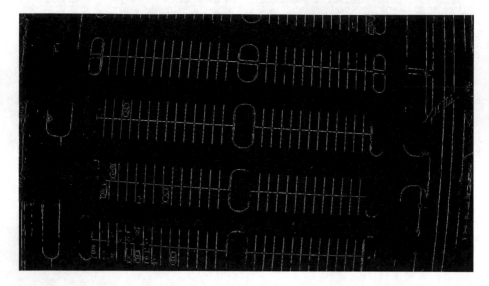

Fig. 8. The binary image undergoes an open morphological transform to reduce virtual line width.

Fig. 9. After the Hough Line detection is performed, the mode of the line length and angle values are used to isolate lines that are the most common in these characteristics. This process is repeated iteratively until all remaining lines fall withing one standard deviation with respect to length and angle.

Additional Parking Lines. It is common that parking areas will have lines that run parallel to the most commonly occurring lines. These are typically painted lines that separate adjacent rows of cars. In the example parking lot, this is demonstrated as sets of lines that divide the most common parking lines to create two adjacent rows of cars in each parking area.

To detect these lines, the same process that was used to find the primary parking lines can be used. Using the same original dataset after removing the already detected most common lines, the algorithm is run again to find lines of similar angles and lengths. This can be repeated until similar lines no longer exist, indicating all lines that share similar properties have been found. All of these similar lines can then be combined and drawn to produce a visual of what the algorithm has detected for parking lines as seen in (Fig. 10).

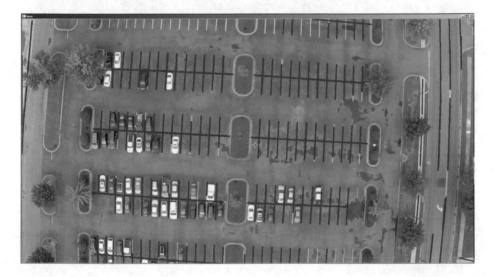

Fig. 10. After being run iteratively against the set of lines remaining after previously detected lines have been removed, it is possible to produce a representation of all similar lines.

Locating Regions of Interest. For further analysis, it is useful to identify "blocks" of parking spaces that are contiguous. Once all parking lines have been detected, it is possible to identify these blocks by using a rotated minimum bounding rectangle. A rotated minimum bounding rectangle will take features, in this case lines, and produce four points around contiguous features that enclose the features using the smallest area possible (see Fig. 11).

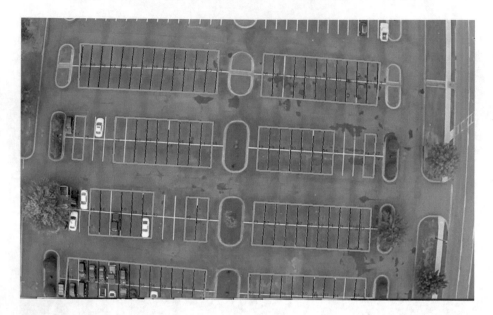

Fig. 11. Once all lines have been detected, a rotated minimum bounding rectangle can be used to identify contiguous line features. This gives a good representation of the parking blocks, or sections of contiguous parking spaces, captured in the image.

4   Experimental Results

To test the effectiveness of the line detection, 10 test images were analyzed. In each image, the total number of lines that were fully contained in the image were counted - partial lines that were cut off were not counted in the total lines. Line detection was divided into three categories: lines detected completely, lines partially detected, and lines not detected. Finally, lines detected that were not parking lines, or false positives, were counted.

The results, as presented in Table 1, showed a parking line detection rate of 76 % for complete lines, 18 % for partially detected lines, and 6 % of lines were not detected. This method proved to provide acceptable results for automated extraction of regions of interest from aerial images for mapping or further analysis, as all parking blocks contained within images were at least partially detected.

5   Future Work and Conclusions

Further improvements could be made to produce better results by removing noise and recover other lines that may have been removed from the set in error. Furthermore, some lines are more difficult to detect because of environmental causes or wear and tear or due to variation in lighting and shadows that obscured

Table 1. Experimental values

Image number	Total parking lines	Lines correctly detected	Lines partially detected	Parking lines not detected	Non-parking lines detected
1	76	67	8	1	9
2	25	20	4	1	4
3	82	68	9	5	9
4	54	37	4	12	9
5	31	24	6	1	5
6	22	17	5	0	7
7	18	14	3	1	4
8	38	34	4	0	5
9	115	92	20	3	7
10	94	83	8	3	9

lines partially or completely. Addition work and further analysis could possibly infer those lines from existing data.

In addition, this algorithm was developed using images of parking lots with only parking lines that were strictly offset in their orientation by 90°. In the future, addition functionality could provide detection for angled parking spaces.

Beyond improvements in accuracy, with accurate detection of additional parking lines, further analysis would be possible to produce detailed maps of parking areas down to individual spaces. Using the current detection of regions of interest, further work could identify "critical points" within the regions of interest: line endpoints line intersections. Endpoints mark the back edges of parking spaces or the end of a parking row, while line intersections mark the front edges and indicate adjacent spaces.

Once individual spaces are able to be differentiated, each space could also be examined to indicate vacancy of spaces. This could be useful for urban areas to determine parking efficiency and utilization. With real-time or near real-time images, vacant parking spaces could be detected and communicated to those using the parking lot.

References

1. Amato, G., Carrara, F., Falchi, F., Gennaro, C., Vairo, C.: Car parking occupancy detection using smart camera networks and deep learning. In: 2016 IEEE Symposium on Computers and Communication (ISCC), pp. 1212–1217, June 2016
2. Chen, Y.-L., Wu, B.-F., Fan, C.-J..: Real-time vision-based multiple vehicle detection and tracking for nighttime traffic surveillance. In: Proceedings of the 2009 IEEE International Conference on Systems, Man and Cybernetics, SMC 2009, pp. 3352–3358, Piscataway, NJ, USA. IEEE Press (2009)

3. Delibaltov, D., Wu, W., Loce, R.P., Bernal, E.A.: Parking lot occupancy determination from lamp-post camera images. In: 2013 16th International IEEE Conference on Intelligent Transportation Systems-(ITSC), pp. 2387–2392. IEEE (2013)
4. Fatsi, E.: Parking space detection with OpenCV and a raspberry Pi (2014)
5. Huang, C.C., Wang, S.J.: A hierarchical bayesian generation framework for vacant parking space detection. IEEE Trans. Circuits. Syst. Video Technol. **20**(12), 1770–1785 (2010)
6. Illingworth, J., Kittler, J.: A survey of the hough transform. Comput. Vis. Graph. image Process. **44**(1), 87–116 (1988)
7. Jensen, T.H.P., Schmidt, H.T., Bodin, N.D., Nasrollahi, K., Moeslund, T.B.: Parking space occupancy verification - improving robustness using a convolutional neural network. In: Proceedings of the 12th International Joint Conference on Computer Vision, Imaging and Computer Graphics Theory and Applications - Volume 5: VISAPP, (VISIGRAPP 2017), pp. 311–318. INSTICC, SciTePress (2017)
8. Peng, C., Hsieh, J., Leu, S., Chuang, C.: Drone-based vacant parking space detection. In: 2018 32nd International Conference on Advanced Information Networking and Applications Workshops (WAINA), pp. 618–622 (2018)
9. True, N.: Vacant parking space detection in static images. University of California, San Diego, 17 (2007)
10. Wu, Q., Huang, C., Wang, S., Chiu, W., Chen, T.: Robust parking space detection considering inter-space correlation. In: 2007 IEEE International Conference on Multimedia and Expo, pp. 659–662, July 2007

Hybrid Navigation Information System for Minimally Invasive Surgery: Offline Sensors Registration

Uddhav Bhattarai and Ali T. Alouani$^{(\boxtimes)}$

Tennessee Technological University, Cookeville, TN 38501, USA
ubhattara42@students.tntech.edu, aalouani@tntech.edu

Abstract. Current Minimally Invasive Surgery (MIS) technology, although advantageous compared to open cavity surgery in many aspects, has limitations that prevents its use for general purpose surgery. This is due to reduced dexterity, cost, and required complex training of the currently practiced technology. The main challenges in reducing cost and amount of training is to have an accurate inner body navigation advisory system to help guide the surgeon to reach the surgery location. As a first step in making minimally invasive surgery affordable and more user friendly, quality images inside the patient as well as the surgical tool location should be provided automatically and accurately in real time in a common reference frame. The objective of this paper is to build a platform to accomplish this goal. It is shown that a set of three heterogeneous asynchronous sensors is a minimum requirement for navigation inside the human body. The sensors have different data rate, different reference frames, and independent time clocks. A prerequisite for successful information fusion is to represent all the sensors data in a common reference frame. The focus of this paper is on off-line calibration of the three sensors, i.e. before the surgical device is inserted in the human body. This is a pre-requisite for real time navigation inside the human body. The proposed off-line sensor registration technique was tested using experimental laboratory data. The result of calibration was promising with an average error of 0.1081 mm and 0.0872 mm along the x and y directions, respectively, in the 2D camera image.

Keywords: Multisensor system · Sensor fusion · Calibration ·
Computer assisted surgery · Biomedical signal processing ·
Image guided treatment

1 Introduction

MIS does not require opening the patient body to perform surgical procedure. This method has distinct merits of faster recovery, shorter hospital stays, less pain, and decreased scarring. However, restricted visualization of operative site, minimal accessibility, and reduced dexterity have increased the challenges of its implementation. Image Guided Surgery (IGS) during MIS will help to solve such problems and improve safety and accuracy to significant level [1].

© Springer Nature Switzerland AG 2020
K. Arai and S. Kapoor (Eds.): CVC 2019, AISC 944, pp. 205–219, 2020.
https://doi.org/10.1007/978-3-030-17798-0_18

Computed Tomography (CT)/Magnetic Resonance Imaging (MRI) provide high quality images of the inside of the patient body [2]. Surgical planning may involve getting insight of patient anatomy, analyzing it, and developing the effective treatment approach [2]. The preoperative images expire within minutes because the intraoperative environment changes continuously due to manipulation by surgeon or organ movement. Hence an intraoperative imaging system with navigation mechanism is required to provide the real-time changes in the map provided by the preoperative data [2]. The process of registration/calibration is needed to transform a point/collection of points from one coordinate frame to another. Thus, one can acquire all the data in a single coordinate frame for the purpose of data fusion in order to achieve more accurate and informative results than what single sensor can provide.

Spatial calibration deals with determination of spatial transformation parameters between the coordinate frames while the temporal calibration is for time synchronization of multiple asynchronous sensor data. Temporal calibration is beyond the scope of this paper, and further reading can be found in [3–5]. The surface based spatial registration between preoperative CT and intraoperative ultrasound in [6, 7] was carried out by using Iterative Closest Point (ICP) algorithm. ICP suffers from being trapped in local minima unless a good initial guess is provided. In addition, it requires computation of closest point pair for operation so it has limited computational speed. Furthermore, the reported accuracy in [7] doesn't provide reliability for clinical application in human body.

The use of Hand-Eye calibration for rigid registration among robotic arm, tracking devices (EMTS/Optical Tracking System (OTS)) and imaging devices (Endoscope/Laparoscopic Ultrasound (LUS)) was reported in [4, 8]. Minimally invasive procedure requires precise tracking and hand eye calibration because of limited view of camera where image may need to be magnified for better interpretation of anatomy [9]. The calibration using optical tracking in proximal end is prone to large tracking error compared to using Electromagnetic Sensor (EMS) near the camera [4, 8, 9]. Furthermore, hand eye calibration requires at least two distinct motions with nonparallel rotation axes. The transformation cannot be obtained if there exist limiting cases such as pure translation or rotation.

In order to calculate optimum transformation parameters, [3, 10–13] implemented linear least square algorithm with OTS as main reference frame. The calibration in [3, 10, 12, 13] were limited to rigid surgical device while MIS require frequent use of flexible surgical device. As the whole distortion correction was based on OTS, the magnetic distortion correction mechanism may provide false correction vector even if Line of Sight (LOS) is blocked for few seconds accidentally for both real time and preoperative correction mechanism [3, 11–13]. Furthermore, the system was modeled for static distortion [3, 11, 12]. Hence, the correction vector would be redundant if the distortion in the vicinity of EMTS changes during surgery.

The Levenberg-Marquardt, iterative method to solve nonlinear least squares problems by minimizing the cost function, was implemented for calibration of LUS probe with tracking devices [4–6, 11, 14–16]. Levenberg-Marquardt algorithm suffers from two complementary problems: slow convergence, and robustness to initial guess [17]. Method implemented to increase the convergence speed yield decreased robustness to the initial guess. Hence user needs to manually adjust the algorithm parameters according to the particular requirement [17].

Researchers in [18, 19] performed fiducial marker/landmark-based calibration between preoperative CT with the tracking device by using Horn's absolute orientation method [20]. The work of [19] provided the contextual information for localizing targets for novice and experienced surgeon. However, the high precision task such as needle placement, ablation require higher accuracy; ultrasound probe itself has tendency to distort the EM tracking measurement. On the other hand, the evaluated accuracy of 24.17 mm in [19] is not acceptable for clinical application.

Researchers in [21] presented the calibration of LRS with OTS. Same fiducial markers were extracted in both coordinate frame for calibration. In addition to requirement of constant line of sight, performance of OTS is widely affected by various lighting condition in room. Furthermore, the optical markers located at handle of surgical pointer require additional fixed transformation between the optical makers and surgical pointer tip. This may induce additional error in the system.

Existing calibration systems use the fusion of Optical Tracking System (OTS) and/or EMTS with intraoperative image such as endoscope, Laparoscopic Ultrasound System (LUS), and preoperative CT/MRI image [3–6, 8, 11, 15, 19]. OTS becomes redundant in scenario crowded with medical device and surgeon in incision-based MIS approach. [22] Ultrasound suffers from shadowing, multiple reflections, low signal to noise ratio, requirement of expertise and training of surgeon. The use of ultrasound within the EMTS field is also responsible for added distortion in EMTS measurement [22]. While implementing two heterogeneous sensors, researchers have fused information from intraoperative images (LUS/Endoscope/DynaCT) with preoperative images (CT/MRI) [23] or with navigation system (EMTS/OTS), [1, 8, 10, 14, 24–26]. The information gathered from two sensors is not sufficient for performing successful MIS.

In order to perform successful MIS, one needs at least three heterogeneous sensors: at least two for preoperative and intraoperative imaging, and one for navigation purpose. The combination of these three heterogeneous sensors provide sufficient information for real time visualization, positional information of the surgical tools, and real time path planning. This paper presents our offline spatial calibration among three heterogeneous sensors. The proposed hybrid system involves EMTS, videoscope, and LRS. In addition to not requiring the LOS, the EMS can be directly inserted to the point of interest without surgical pointer. To the best of our knowledge this is first approach of offline calibration of three heterogeneous sensors which involve LRS, EMTS, and Camera together. LRS is used here to emulate CT/MRI preoperative data, camera for real-time high-quality images, and EMTS for positional information inside the human body. EMTS is the best method of tracking in MIS approaches where line of sight is not available [27]. Up to now no universally acceptable alternative of EMTS has been developed [27].

Each of three heterogeneous sensors used in this work have their own coordinate frame and data rates. To provide the surgeon with useful real time video and positional information of the surgical tool(s), all the sensors data have to be represented in the same coordinate frame with proper synchronization. Hence the calibration process may be classified as problem of spatiotemporal calibration. Off-line calibration is a prerequisite for real time tracking, once the time synchronization problem is resolved. Hence, we have considered temporal calibration of asynchronous sensors and their real-time tracking as future work.

This paper is organized as follows. Section 2 discusses the hardware used to perform the calibration and experimental testing. Section 3 discusses the proposed calibration technique and its justification. Section 4 discusses the accuracy obtained using the proposed calibration technique. Section 5 contains conclusions and discusses future work.

2   Hybrid Tracking Hardware

LRS from Next Engine was used as 3D scanner to imitate the preoperative CT/MRI machine. The LRS consists of four scanning lasers with scanning resolution of 500 DPI (Dots Per Inch) in macro mode and 200 DPI in wide mode [28]. The images from each scanning lasers were processed and fused to give the xyz position and RGB value of a pixel in the LRS coordinate frame. The navigation sensor used was NDI Type-2 6DOF sensor for Aurora EMTS with measurement frequency of 40 Hz [29]. It provides position and orientation information in reference to tabletop field generator. According to NDI, the accuracy is 0.8 mm for position and 0.7degree for orientation for EMTS measurement [29]. The third sensor was the Go 5000C series color camera from JAI Corporation [30], with 5-mega-pixel resolution.

3   Proposed Calibration Technique

The implementation of multimodal display including tracked videoscope along with preoperative data can be potentially helpful to detect and correct possible anatomical shifts. The videoscope data will provide updated information that the surgeon can rely on, while he/she can also benefit from preoperative data with real time view and understanding of anatomy. We have selected LRS as the standard reference frame. Figures 1 and 2(a) show all the coordinate systems involved and the coordinate transformation among them. Registering the data in preoperative images as the absolute coordinate frame allows precise advanced AR visualization as well as therapy delivery [31]. The camera itself consists three coordinate systems as shown in Fig. 2(b): 3D camera focal point coordinates, 2D coordinates of the center of the image plane, and 2D coordinates of the origin of the camera image. The depth information is lost during transformation of focal point to image plane coordinate system.

The selected calibration object involves two planes and each plane with four circular patterns of different colors (Blue, Pink, Red, Purple), Fig. 3(a) [32]. The design of the calibration object satisfies the requirement of Direct Linear Transform (DLT) camera calibration: at least six calibration points located in different planes [33]. The idea of using different color for calibration objects is to simplify the calibration point extraction for LRS and camera coordinate frame by using a color filter algorithm. This work advances the work of [32] toward spatiotemporally calibration of three heterogeneous sensors which are necessary for real time visualization and navigation to perform successful MIS.

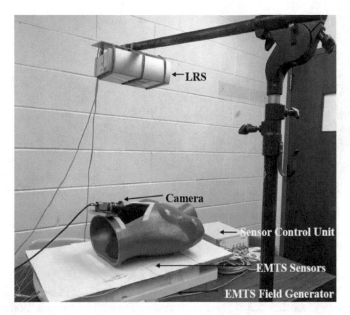

Fig. 1. Heterogeneous sensors for proposed system. EMTS is the tracking system; Camera is for intraoperative while LRS is for preoperative imaging

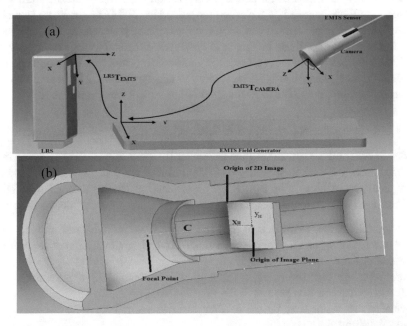

Fig. 2. (a) Transformation among the different coordinate systems. (b) Three coordinate systems (Focal point, Origin of 2D image, Origin of image plane) involved in camera

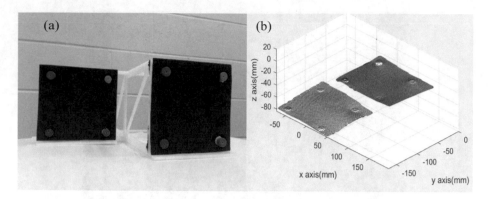

Fig. 3. (a) Calibration object with two planes. Blue, Pink, Red, and Purple calibration objects of radius 5 mm are attached in each plane of the device. (b) 3D point cloud of calibration object after preprocessing.

Let us consider the surgeon needs to identify and reach the target area in minimally invasive fashion. Preoperative imaging provides 3D overview of patient. Surgeon can rely on these high-quality 3D images to diagnose the problem inside body. When the target anatomy is recognized, the same preoperative images can be used for 3D path planning to reach the destination with shortest path facing the minimum obstacle. As there exists fixed transformation between EMTS and CT/MRI reference frames, every point along with the planned path can be recognized in CT/MRI coordinate frame. Once the position of camera, planned path, and the destination point all are in CT/MRI coordinate frame, surgical tool can be driven to destination correctly with real time feedback from EMS attached to the camera. Once the destination is reached, the target anatomy in CT/MRI can be segmented to extract relevant features such that the 2D camera image can be overlaid on the top of 3D image for augmented view within human body.

3.1 Calibration Point Extraction in LRS and EMTS Coordinate Frame

In order to obtain the eight calibration points in LRS coordinate frame, the calibration device was scanned with ScanStudio HD. The point cloud was preprocessed to remove any unnecessary artifacts, Fig. 3(b) After preprocessing the point cloud was fed to an algorithm to extract the eight calibration points. The algorithm works as follows:

1. Extract calibration points in four bins according to their RGB color range. Each bin will contain set of position and RGB value of two circular patterns of same color but located at front and black plane.
2. Determine the centroid along the z axis of point cloud. Z axis centroid acts as reference between two planes.
3. Transfer the calibration points in four bins to eight bins with reference to the centroid along the z axis.
4. Remove any outlier pointed detected beyond the 5 mm radius of circular pattern Each circular pattern has radius of 5 mm. Any point that is detected beyond that range is falsely detected point and should be removed.
5. Finally, calculate eight calibration points (red dots) in LRS frame, Fig. 4(a).

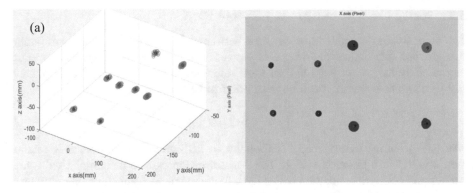

Fig. 4. (a) Calibration point in LRS coordinate frame. Red points, hidden within blue circles, indicate the calibration points. (b) Calibration point extraction in camera image. Labelling from 1 to 8 indicate the calibration points and their measurement order.

3.2 Calibration Point Extraction in Camera Coordinate Frame

In order to extract calibration points in camera coordinate frame, an image processing algorithm was developed. The algorithm performs morphological image processing [34] and removes any background objects in the field of view of camera. The next step is to extract the calibration points and arrange them in specific order. Before calculation of centroids of each connected component, one needs to label them first. During connected component labeling, Image processing toolbox scans objects from top to bottom starting from the leftmost position and ending at rightmost position. This labelling may change according to different position from where the image is taken. In order to solve this problem, the labeled connected components were first arranged as top and bottom components in image. Later the arranged connected components were rearranged in order on the basis of their presence in front and back plane of the calibration object. Once we order the labeling of connected component, we can calculate the calibration points, Fig. 4(b).

In order to calibrate LRS with EMTS, Horn's absolute orientation method based on unit quaternion was implemented [20]. In addition, we implemented Direct Linear Transform (DLT) for camera calibration. Horn's quaternion-based approach and DLT method both provide closed form solution [20, 33]. Both approaches are computationally efficient as they are not iterative. Iterative approaches have tendency to end up in local minima unless a good initial approximation provided. Horn's method provides the efficient solution compared to training based Artificial Neural Network (ANN), Genetic algorithm [32]. The DLT camera calibration can be done by using single image of calibration object unlike planar pattern which require image of at least two different orientations of the object [35]. The process is less prone to error because it doesn't require additional hand-eye calibration to transform camera position and orientation in planar pattern to the EMTS coordinate frame.

3.3   Spatial Calibration Between LRS and EMTS

The problem of coordinate transformation between LRS and EMTS consists of finding rotation and translation matrices using positional information of the same entity measured by the two sensors in their local reference frame. If P_{EMTS} is a 3D point in EMTS coordinate frame, it can be transferred to LRS coordinate frame as

$$P_{LRS} = RP_{EMTS} + T \tag{1}$$

where, P_{LRS} is the transformed EMTS point in LRS coordinate frame, R is rotation matrix and T is translation vector. The transformation can be determined with three perfect non-collinear calibration points [20]. Including more points for calibration leads to overdetermined system with increased accuracy [20]. Maximum accuracy of transformation between LRS and EMTS is achieved by using eight calibration points.

Pseudocode for Horn's Absolute Orientation for transformation from LRS to EMTS.

1. Inputs: Calibration point in LRS (P_{LRS}) and EMTS (P_{EMTS}).
2. Compute: C_{LRS} = Centroid of P_{LRS} and C_{EMTS} = centroid of P_{EMTS}
3. Calculate: $A = P_{LRS} - C_{LRS}$ $B = P_{EMTS} - C_{EMTS}$
4. Calculate M = A*B^T, Such that

$$M = \begin{bmatrix} S_{xx} & S_{xy} & S_{xz} \\ S_{yx} & S_{yy} & S_{yz} \\ S_{zx} & S_{zy} & S_{zz} \end{bmatrix}$$

5. Calculate

$$N = \begin{bmatrix} S_{xx}+S_{yy}+S_{zz} & S_{yz}-S_{zy} & S_{zx}-S_{xz} & S_{xy}-S_{yx} \\ S_{yz}-S_{zy} & S_{xx}-S_{yy}-S_{zz} & S_{xy}+S_{yx} & S_{zx}+S_{xz} \\ S_{zx}-S_{xz} & S_{xy}+S_{yx} & -S_{xx}+S_{yy}-S_{zz} & S_{yz}+S_{zy} \\ S_{xy}-S_{yx} & S_{zx}+S_{xz} & S_{yz}+S_{zy} & -S_{xx}-S_{yy}+S_{zz} \end{bmatrix}$$

6. λ_{max} = Most positive Eigen Value of N
7. v_{max} = Eigen Vector Corresponding to λ_{max}
8. Normalize v_{max} to get unit quaternion representation of rotation $\dot{q} = q_o + iq_x + jq_y + kq_z$
9. Calculate

$$R = \begin{bmatrix} q_o^2+q_x^2-q_y^2-q_z^2 & 2(q_xq_y-q_oq_z) & 2(q_xq_z+q_oq_y) \\ 2(q_xq_y+q_oq_z) & q_o^2-q_x^2+q_y^2-q_z^2 & 2(q_yq_z-q_oq_x) \\ 2(q_xq_z-q_oq_y) & 2(q_yq_z+q_oq_x) & q_o^2-q_x^2-q_y^2+q_z^2 \end{bmatrix}$$

10.

$$T = C_{EMTS} - RC_{LRS}$$

3.4   Spatial Calibration Between EMTS and Camera

Normalized DLT maps any point in world coordinate system to the camera coordinate system. Data normalization involves the translation and scaling of calibration points and it should be carried out before implementation of DLT algorithm [33]. Apart from improved accuracy in result, the result of data normalization will be invariant with respect to the arbitrary choices of scale and coordinate origin [33]. The matrix M in Eq. (2) has eleven unknown parameters. In order to determine the unique solution of these parameters, we need at least 6 points, and all of them should not lie in same plane [33]. Let us consider a point P in EMTS coordinate system is to be transformed to camera sensor coordinate system p both in homogeneous form.

$$
\begin{aligned}
p_{3x1} &= K_{3x3}R_{3x3}[I_{3x3}| - X_{0_{3x1}}]P_{4x1} \\
&= M_{3x4}X_{4x1}
\end{aligned}
\tag{2}
$$

Where, K: 3×3 camera intrinsic parameter matrix: consists 5 intrinsic parameters: camera constant (c), scale difference (m), sheer component (s), transformation between plane coordinate system to sensor coordinate system (x_H, y_H); R: 3×3 rotation matrix; X_0: 3 x 1 translation vector; I: 3×3 identity matrix. Pseudocode for DLT.

1. Inputs: camera image coordinate (p_{cam}) and EMTS coordinate (P_{EMTS}) ($i \geq 6$)

$$
\bar{p}_{cam} = \text{mean}(p_{cam}), \bar{P}_{EMTS} = \text{mean}(P_{EMTS})
$$

3. Shift origin of camera and EMTS data to $\bar{p}_{cam}$, $\bar{P}_{EMTS}$
4. $[p_n \quad P_n]$ = Normalize (p_{cam} P_{EMTS}),
5. Calculate Homography (M)

$$
M = \begin{bmatrix} -X_{n_i} & -Y_{n_i} & -Z_{n_i} & -1 & 0 & 0 & 0 & 0 & x_{n_i}X_{n_i} & x_{n_i}Z_{n_i} & x_{n_i}Z_{n_i} & x_{n_i} \\ 0 & 0 & 0 & 0 & -X_{n_i} & -Y_{n_i} & -Z_{n_i} & -1 & y_{n_i}X_{n_i} & y_{n_i}Y_{n_i} & y_{n_i}Z_{n_i} & x_{n_i} \end{bmatrix}_{2ix12}
$$

6.

$$
[U \quad S \quad V] = \text{SVD}(M)
$$

7. Select eigen vector (v) corresponding to smallest singular value which minimizes error
8. Renormalize and Rearrange (v)
9.

$$
[R,K] = \text{QR\_decomp}(v)
$$

10.

$$
[X_o] = \text{camcenter}(v)
$$

4 Experimental Performance Evaluation

In order to assess the accuracy of the proposed calibration, ten colored circular objects were attached to the surface of an artificial liver available in lab, Fig. 5(a). The centroids were acquired by the LRS, EMTS, and Camera in their respective frames. These circular objects can represent presence of liver tumor. In LRS coordinate frame, the colored objects were first extracted on the basis of color filter algorithm, Fig. 5(b). In camera coordinate system the image was fed to image processing algorithm to remove background, clear border, clear holes, label and arrange the connected components, and determine the centroid of each connected component, Fig. 6(a). EMTS data were acquired by inserting the EMS in colored object.

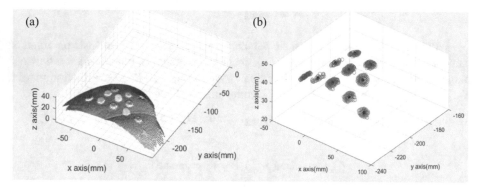

Fig. 5. (a) 3D point cloud of artificial liver after preprocessing in ScanStudio. (b) Extraction of accuracy evaluation in LRS coordinate frame. Red points indicate the extracted points

The calculated centroid from LRS frame was first transformed to EMTS coordinate frame according to calibration parameter, Eq. (3). The performance of two rigid registration algorithms for LRS to EMTS transformation were compared for accuracy evaluation: algorithm proposed by Horn, and the algorithm proposed by Walker et al. [36]. The calibration, and accuracy evaluation were performed in environment without any ferromagnetic material near the EM field generator. Tracking in the electromagnetic field generator is unaffected by the medical-grade stainless steel (300 series), titanium, and aluminum [22, 29]. The tabletop field generator also minimizes distortions produced from the patient table or materials located below it [29]. Once the set of points were transformed from LRS to EMTS they were projected to distortion corrected camera image, Eq. (4). We have tested the accuracy of the calibration for the liver shown in Fig. 5(a) for ten different arrangements. Accuracy was evaluated at least 12 in from the top surface of EMTS field generator so as to provide room for placement of patient table.

$$^{EMTS}X_{LRS} = {}^{EMTS}T_{LRS} \bullet X_{LRS} \qquad (3)$$

$$\begin{bmatrix} {}^{CAM}X_{LRS} \\ 1 \end{bmatrix} = {}^{CAM}T_{EMTS} \begin{bmatrix} {}^{EMTS}X_{LRS} \\ 1 \end{bmatrix} \tag{4}$$

In Eqs. (3) and (4), ${}^{A}T_{B}$ represents the transformation parameter from coordinate system B to coordinate system A, while ${}^{A}X_{B}$ represents the transformed point from B to A.

Figure 6(b) illustrates the registration error in each axis for LRS to EMTS transformation. As both algorithms provide the closed form solution, the average error difference is within millimeter range for LRS to camera coordinate frame, Fig. 7(a). In addition to the accuracy evaluation, the computation time of each algorithm was compared. Both algorithms in [20] and [36] require at least three calibration points spatially located at same place. The computation time for the algorithm proposed by Walker et al. increases drastically compared to the almost constant computation time for the Horn's method with increasing number of calibration points, Fig. 7(b).

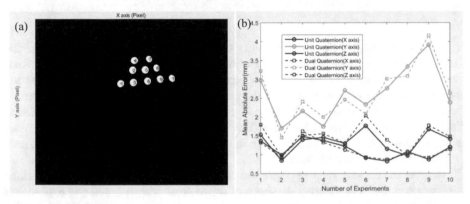

Fig. 6. (a) Extraction of accuracy evaluation point in camera frame. (b) Error of transformation from LRS to EMTS using Horn's method and Walker's method.

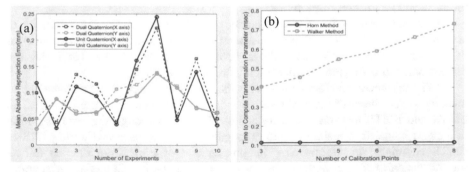

Fig. 7. (a) Error of transformation from transformed LRS points to 2D camera image. (b) Computation time vs number for calibration points [20, 36]

Table 1 summarizes the absolute positional error for coordinate transformation from LRS to EMTS as well as EMTS to camera frame using Horn's, and DLT method respectively. The average error for LRS to EMTS coordinate transformation is minimum along Z axis. The Y coordinate seems to be most affected by error with maximum standard deviation and range. The X and Z coordinate provide more consistent reading compared to largely fluctuating Y coordinate values, Fig. 6(b). There might be two possible reasons for the error.

Table 1. Absolute error of transformation from LRS to EMTS to 2D image (Horn's method)

	LRS to EMTS			EMTS to image	
	X	Y	Z	X	Y
Mean (mm)	1.3511	2.6019	1.1325	0.1081	0.0872
S. D. (mm)	0.9321	1.5239	0.9285	0.0606	0.0298
Range (mm)	4.1423	7.0469	4.2079	0.1802	0.0879

First reason may be the varying ability of LRS to correctly scan and replicate the scanned object at varying distance. According to Feng *et al.* [37] the signal attenuation increases with increase in the distance of scanned surface from LRS. This reduces the ability of scanner to correctly localize the point cloud. If Fig. 3(b) is closely observed there are two separate planes along the Z axis of calibration device at the increasing distance from the scan position of LRS. The Z coordinate in LRS might be transformed to the Y coordinate of EMTS during coordinate transformation. The second reason may be due to error in data collection.

In order to correctly scan an object, the laser beam should be normal to the surface to be scanned [37]. Considering the shape of the scanned liver it might be possible that some surfaces were not perfectly normal to the laser beam and contributed for the system error. Ten experiments were performed to measure the registration error of the two-plane calibration device by moving it to another position. The total registration error was 0.5862 ± 0.3901 mm, 1.1255 ± 0.5850 mm, 0.5815 ± 0.4440 mm along the x, y, and z direction respectively.

The result supports the claim proposed by [37]. Furthermore, the capability to accuracy measure the data with each measurement system also affects the overall error of the hybrid system. For instance, the accuracy in measurement of EMTS is 0.8 mm for position and 0.7degree for orientation as reported by NDI.

EMTS to camera transformation error is the overall error associated with the hybrid tracking system because the evaluated error is the integrated error from the LRS to EMTS and EMTS to camera transformation, Table 1. Although the error is within millimeter range, it is mainly due to propagation of error generated during LRS to EMTS coordinate transformation. The propagation of error from one coordinate transformation to another is the main disadvantages of the hybrid tracking system.

Researcher in [19] calibrated OTS with CT scan using Horn's absolute orientation method with reported overall system error of 24.17 mm. In addition to Fiducial Registration Error (FRE), the transformation error between the optical marker and the surgical pointer contributed the poor performance of the system [19]. Our system is immune to the possible registration error between the optical marker and surgical pointer because of direct insertion of EMS coils to the point of interest. Furthermore, our result show that the EMTS can work as the efficient localization device under non-ferromagnetic condition. This result is also an improvement over the distortion corrected average accuracy of 2.1 ± 0.8 mm for OTS and EMTS calibration reported in [12].

5 Conclusion and Future Work

This paper provided a first step toward building a platform for using a set of asynchronous sensors to make safe navigation inside the human body possible. The LRS is used in this paper to provide the preoperative information that would be given by a CT/MRI in a hospital setting. However, the registration process is still applicable when CT/MRI is used. Furthermore, for laboratory testing, a low cost 2D camera is used. The proposed technique applies to any camera as long as the specific parameters of the camera are provided to the registration algorithm. Laboratory testing using an artificial liver was carried out which showed promising accuracy.

Future work will extend the result of this paper to include temporal and spatial registration of asynchronous heterogeneous sensors. The next phase is necessary because in addition to having data in different spatial coordinate frames, the heterogeneous sensors have different data rate based on independent clocks.

References

1. Thoranaghatte, R.U., et al.: Endoscope-based hybrid navigation system for minimally invasive ventral spine surgeries. Comput. Aided Surg. **10**(5–6), 351–356 (2005)
2. Peters, T., Cleary, K.: Image-Guided Interventions: Technology and Applications. Springer Science and Business Media (2008)
3. Nakada, K. et al.: A rapid method for magnetic tracker calibration using a magneto-optic hybrid tracker, In: International Conference on Medical Image Computing and Computer-Assisted Intervention, pp. 285–293 (2003)
4. Feuerstein, M., et al.: Magneto-optical tracking of flexible laparoscopic ultrasound: model-based online detection and correction of magnetic tracking errors. IEEE Trans. Med. Imaging **28**(6), 951–967 (2009)
5. Nakamoto, M., et al.: Intraoperative magnetic tracker calibration using a magneto-optic hybrid tracker for 3-D ultrasound-based navigation in laparoscopic surgery. IEEE Trans. Med. Imaging **27**(2), 255–270 (2008)
6. Fakhfakh, H.E., et al.: Automatic registration of pre-and intraoperative data for long bones in minimally invasive surgery, In: 2014 36th Annual International Conference of the IEEE Engineering in Medicine and Biology Society (EMBC), pp. 5575–5578 (2014)
7. Martens, V., et al.: LapAssistent—a laparoscopic liver surgery assistance system. In: 4th European Conference of the International Federation for Medical and Biological Engineering, pp. 121–125 (2009)
8. Wengert, C., et al.: Endoscopic navigation for minimally invasive suturing. In: International Conference on Medical Image Computing and Computer-Assisted Intervention, pp. 620–627 (2007)

9. Thompson, S., et al.: Hand–eye calibration for rigid laparoscopes using an invariant point. Int. J. Comput. Assist. Radiol. Surg. **11**(6), 1071–1080 (2016)
10. Solberg, O.V., et al.: Navigated ultrasound in laparoscopic surgery. Minim. Invasive Ther. Allied Technol. **18**(1), 36–53 (2009)
11. Konishi, K., et al.: A real-time navigation system for laparoscopic surgery based on three-dimensional ultrasound using magneto-optic hybrid tracking configuration. Int. J. Comput. Assist. Radiol. Surg. **2**(1), 1–10 (2007)
12. Birkfellner, W., et al.: Calibration of tracking systems in a surgical environment. IEEE Trans. Med. Imaging **17**(5), 737–742 (1998)
13. Nakamoto, M., et al.: Magneto-optic hybrid 3-D sensor for surgical navigation. In: MICCAI, pp. 839–848 (2000)
14. Prager, R.W., et al.: Rapid calibration for 3-D freehand ultrasound. Ultrasound Med. Biol. **24**(6), 855–869 (1998)
15. Sato, Y., et al.: Image guidance of breast cancer surgery using 3-D ultrasound images and augmented reality visualization. IEEE Trans. Med. Imaging **17**(5), 681–693 (1998)
16. Chaoui, J., et al.: Virtual movements-based calibration method of ultrasound probe for computer assisted surgery. In: 2009 IEEE International Symposium on Biomedical Imaging: From Nano to Macro, ISBI 2009 (2009)
17. Transtrum, M.K., Sethna, J. P.: Improvements to the Levenberg-Marquardt algorithm for nonlinear least-squares minimization. arXiv Preprint arXiv:1201.5885 (2012)
18. Pyciński, B., et al.: Image navigation in minimally invasive surgery. Inf. Technol. Biomed. **4**, 25–34 (2014)
19. Jose Estepar, R.S., et al.: Towards scarless surgery: an endoscopic ultrasound navigation system for transgastric access procedures. Comput. Aided Surg. **12**(6), 311–324 (2007)
20. Horn, B.K.: Closed-form solution of absolute orientation using unit quaternions. Josa A **4**(4), 629–642 (1987)
21. Cash, D.M., et al.: Incorporation of a laser range scanner into image-guided liver surgery: surface acquisition, registration, and tracking. Med. Phys. **30**(7), 1671–1682 (2003)
22. Birkfellner, W., et al.: Tracking devices. In: Image-Guided Interventions. Springer (2008)
23. Marami, B., et al.: Dynamic tracking of a deformable tissue based on 3D-2D MR-US image registration, In: SPIE Medical Imaging, International Society for Optics and Photonics, pp. 90360T (2014)
24. Sindram, D., et al.: Novel 3-D laparoscopic magnetic ultrasound image guidance for lesion targeting. Hpb **12**(10), 709–716 (2010)
25. Mercier, L., et al.: A review of calibration techniques for freehand 3-D ultrasound systems. Ultrasound Med. Biol. **31**(2), 143–165 (2005)
26. Ng, C.S., et al.: Hybrid DynaCT-guided electromagnetic navigational bronchoscopic biopsy. Eur. J. Cardiothoracic Surg. **49**(suppl_1), i88 (2015)
27. Franz, A.M., et al.: Electromagnetic tracking in medicine—a review of technology, validation, and applications. IEEE Trans. Med. Imaging **33**(8), 1702–1725 (2014)
28. NextEngine 3D Laser Scanner. http://www.nextengine.com/products/scanner/specs
29. Aurora. https://www.ndigital.com/medical/products/aurora/
30. GO-5000 M-USB/ GO-5000C-USB. https://www.jai.com/products/go-5000c-usb
31. Sauer, F.: Image registration: Enabling technology for image guided surgery and therapy, In: 27th Annual International Conference of the 2005 Engineering in Medicine and Biology Society, IEEE-EMBS 2005, pp. 7242–7245 (2006)
32. Ruehling, D.E.: Development and Testing of a Hybrid Medical Tracking System for Surgical Use. MS Thesis, Tennessee Technological University, Cookeville, TN (2015)
33. Hartley, R., Zisserman, A.: Multiple View Geometry in Computer Vision (2003)
34. Gonzalez, R.C., Woods, R.E.: Digital image processing (2012)

35. Camera Calibration Toolbox for Matlab. http://www.vision.caltech.edu/bouguetj/calib_doc/index.html
36. Walker, M.W., Shao, L., Volz, R.A.: Estimating 3-D location parameters using dual number quaternions. CVGIP: Image Understanding **54**(3), 358–367 (1991)
37. Feng, H., Liu, Y., Xi, F.: Analysis of digitizing errors of a laser scanning system. Precis. Eng. **25**(3), 185–191 (2001)

Seeking Optimum System Settings for Physical Activity Recognition on Smartwatches

Muhammad Ahmad[1,2]([✉]), Adil Khan[1], Manuel Mazzara[1], and Salvatore Distefano[2]

[1] Innopolis University, Innopolis, Kazan, Russia
mahmad00@gmail.com
[2] University of Messina, Messina, Italy

Abstract. Physical activity recognition using wearable devices can provide valued information regarding an individual's degree of functional ability and lifestyle. Smartphone-based physical activity recognition is a well-studied area. However, research on smartwatch-based physical activity recognition, on the other hand, is still in its infancy. Through a large-scale exploratory study, this work aims to investigate the smartwatch-based physical activity recognition domain. A detailed analysis of various feature banks and classification methods are carried out to find the optimum system settings for the best performance of any smartwatch-based physical activity recognition system for both personal and impersonal models in real life scenarios. To further validate our hypothesis for both personal and impersonal models, we tested single subject out cross validation process for smartwatch-based physical activity recognition.

Keywords: Smartwatch · Accelerometer · Magnetometer · Gyroscope · Machine learning · Physical activity recognition · Health care services

1 Introduction

Last few years have witnessed a massive growth in consumer interest in smart devices [1]. These include smartphones, smart TVs, and recently smartwatches. Because of their increased penetration, smartwatches have claimed a handsome market share. A typical smartwatch contains a heart rate monitor, GPS, thermometer, camera and accelerometer. Thus, it can provide a variety of services. These services include temperature and pulse measurements and the number of calories consumed when performing a physical activity, such as walking, running, cycling and so forth [2].

Recently, smartwatches have emerged in our daily lives and like their smartphone counterparts, smartwatches do contain a gyroscope, magnetometer and accelerometer sensors to support similar capabilities and applications. These applications include health applications i.e. calories consumption that require

© Springer Nature Switzerland AG 2020
K. Arai and S. Kapoor (Eds.): CVC 2019, AISC 944, pp. 220–233, 2020.
https://doi.org/10.1007/978-3-030-17798-0_19

sensor based physical activity recognition. A number of calories consumed during physical activities vary from one activity to another. Therefore, the first thing to do when estimating the number of calories consumed, is to first recognize which activity have been performed [3].

This work explores the idea of physical activity recognition on smartwatches using the embedded gyroscope, magnetometer and accelerometer sensors. The gyroscope, magnetometer and accelerometer sensors are ideal for physical activity recognition. We further discussed the detailed analysis of various feature banks and classification methods to seek the best settings for optimum performance on two different types of models such as impersonal and personal model over the smartwatch. Personal models referred as the classifier is built using the data only form one specific user whereas the impersonal model referred as the classifier is built using the data form every user except the one under study. To summarize, the main contributions of this work is towards answering the following questions: (1) Existing smartphone-based physical activity recognition studies have a number of features that were utilized for smartphone-based physical activity recognition; can they be used for smartwatch-based physical activity recognition as well? (2) Which of the most commonly used classifiers from the reference field provides the best recognition accuracy? (3) Does features permutation help to improve the accuracy? (4) Does features normalization help to improve the performance of the classifier? (5) Which model (Personal or Impersonal) works better for smartwatch-based physical activity recognition? (6) Does changing the window size affect the recognition accuracy?

Physical activity recognition in general using motion sensors especially using smartphones has been studied in recent years [4–6] and currently being studied extensively [7,8] but there are a few studies on physical activity recognition using smartwatches [9–18] in which the authors studied the role of the smartwatch in physical activity recognition system for health recommendation purposes.

For instance, Guiry et al. [9] proposed physical activity recognition system to recognize nine different activities using five different classifiers. These activities include walking, standing, cycling, running, stair ascent, stair descent, elevator ascent and descent. However, the authors studied smartphone and smartwatch based activity recognition separately and did not combine the data from both devices. Furthermore, for real-time data collection, the author's used magnetometer, accelerometer and gyroscope sensors for smartphone and accelerometer for a smartwatch. Trost et al. [10] used sensors on hip and wrist to detect physical activities using logistic regression based classification method. In [10], Trost et al. just focused on showing the potential of using the wrist position for physical activity recognition for health recommendation system. Chernbumroong et al. [11] used a single wrist-worn accelerometer to detect five different physical activities for recommendation system. These activities include standing, sitting, lying, running and walking. However, Da-Silva et al. [12] used the same sensor to detect the eight different physical activities for recommendation system. In addition to the activities identified in [11] Da-Silva et al. include the activity of working on a computer too. The works [13] and [14] detect the eating activity using a Hidden Markov Model (HMM) with a wrist-worn accelerometer and gyroscope sensors. Ramos-Garcia identifies the eating activity by dividing it into further sub-activities in [13] which includes resting, drinking, eating, using utensil and others. Meanwhile, Dong et al. [14] proposed to differentiate eating periods from non-eating periods to identify the eating and not-eating activity. Sen et al. [15] used both accelerometer and gyroscope sensor data to recognize the eating and non-eating activities. Furthermore, the authors used smartwatch camera to take images of the food being consumed to analyze what a user is eating.

Scholl et al. [16] presented a feasibility study to detect and identify the smoking activity using accelerometer sensor where Scholl et al. only reported a user-specific accuracy for smoking activity. Whereas, Parade et al. [17] used both accelerometer, magnetometer and gyroscope sensors to recognize smoking puffs where they just try to distinguished smoking activity from other activities. A similar work has been reported in [18] where the authors used support vector machine (SVM) to recognize six different physical activities. These activities include standing, walking, writing, jacks, smoking and jogging. Kim et al. [19] proposed the collaborative classification approach for recognizing daily activities using smartwatches. The authors exploit a single off-the-shelf smartwatch to recognize five daily activities; namely eating, watching TV, sleeping, vacuuming and showering. For fair comparisons Kim et al. suggested using single and multiple sensors-based approaches to compare with their proposed approach with the overall accuracy of 91.5% with the improved recall rate up to 21.5% for each activity. Weiss et al. showed that accelerometers and gyroscopes sensors that sense user's movements and can help identify the user activity [20]. Weiss et al. propose to compare the smartphone and smartwatch-based user activity recognition in which smartwatch have the ability to recognize hand-based activities, namely, eating which is not possible through a smartphone. They showed that the smartwatch is able to recognize the drinking activity with 93.3% accuracy while smartphones achieve only 77.3%.

Nurwanto et al. [21] proposed a daily life light sport exercise activities. These activities include squat jump, push up, sit up and walking. Nurwanto et al. focuses only on accelerometer-based signal captured by left-handed persons and the obtained signals were processed through sliding window approach with the k-nearest neighbor method and dynamic time warping as the main classifier. Furthermore, Nurwanto et al. proved that k-NN together with dynamic time warping method is an efficient method for daily exercise recognition with the accuracy of 76.67% for push up 96.69% for squat jump and 80% for sit up activity. They intentionally exclude the walking activity for their experimental process because of the random patterns. Al-Naffakh et al. [22] proposed the usefulness of using smartwatch motion sensor namely gyroscope and accelerometer to perform user activity recognition. Al-Naffakh et al. proved the natures of the signals captured are sufficiently discriminative to be useful in performing activity recognition.

Different activities can be recognized using motion sensors at the wrist position i.e. drinking, eating, smoking, walking, walking down stairs, walking up stairs, running, jogging, elevator up, elevator down, doing regular exercises, writing etc. However there is no study that discusses these activities all together with the recommendations of optimum system settings. Therefore, to extend the existing works, we evaluate the combination of different feature banks and classification techniques for the reliable recognition of various activities. We also evaluate three different built-in motion sensors. Moreover, unlike the current trends, we also evaluate the effect of window size on each activity performed by the individual user and its impact on the recognition performance of simple and complex activities. Although the effect of window size on physical activity recognition has been studied by Huynh et al. in [23] but the windowing effect on complex activities is yet not fully explored. Moreover, unlike the recent works, we explore the use of both *Personal* and *Impersonal* models individually for each activity as explained: (1) Personal model means the classifier is trained using the data only from one specific individual whereas, the Impersonal model means the classifier is built using the data from every user except the one under study. (2) For both models, one subject cross-validation process is adopted. (3) Finally, our study suggests the optimum system settings for any activity recognition system based on smartwatches.

This work analyzes the effects of combining gyroscope, magnetometer and accelerometer sensors data for physical activity recognition for different size of windows varying from one second to 12 s long window. We further analyze the effect of various sampling rates on the recognition performance in different scenarios. However, in our current study, we still have room for further improvement. For example, our current dataset is not balanced (as shown in Fig. 1) for all performed activities which might lead to the biased results towards majority classes. To handle the said problem to some extent, we used one subject cross-validation process for different models (personal and impersonal) which somehow overcome the class imbalance issue. This study also overcomes the limitations of overlapping while segmenting the raw data which forcefully improve the recognition performance because this overlap cause using some part of data in both training and testing which limiting the use of such techniques in a real-time environment. Moreover, we proposed and suggest some methods which helps to improve the recognition performance for those activities that have low recognition performance.

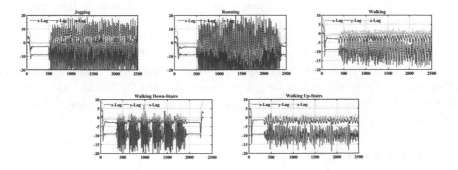

Fig. 1. Raw signal for individual activity.

The rest of the paper has the following organization. Section 2 describes the Material and Methods. Section 3 contains experimental settings. Section 4 presents the experimental results and discussion and finally Sect. 5 concludes the paper with possible future directions.

2   Materials and Methods

Today's smartwatches are equipped with a variety of motion sensors that are useful for monitoring device movements like tilt, rotate and shake. Some of these sensors are the ambient light sensor, accelerometer, compass, gyroscope and GPS sensors. For experimental study we only collected the raw signals using three different sensors such as accelerometer, magnetometer and gyroscope sensor.

2.1   Data Collection

We collected the raw signals for five different activities from six users (three female and three male) having a mean age of twenty-five years old. The criterion for selecting the subjects is based on the gender because different genders exhibit different patterns when performing the same activity. These activities include walking, walking upstairs, walking downstairs, running and jogging. All subjects performed these activities twice on each day for more than a month. Therefore we collected raw data from the same user for same activity but performed on different days.

The participants enrolled in our study are approved by the laboratory head because this is a formal prerequisite because our experiments involved human subjects and there is a negligible risk of injury. The involved subjects then asked to answer few nontechnical questions like gender, age, height, weight, left or right handed etc, which we used to characterize our study. Then the subjects were asked to fastens the smartwatch on their wrist and places a Bluetooth paired smartphone in their pocket. Both devices run a simple custom designed application that controls the data collection process and instructs the participant to first add their name and then select the activity from the list of five different activities and the sensor from three different sensors. Once the initial instructions were done, turn the smartphone screen off and place the smartphone into the pants pocket. The smartphone instructs the smartwatch running our paired data collection application to collect the raw signal at the 20 Hz rate. Each of these sensors generates 3-dimensional signals and appends a time-stamp to the values. Every after five minutes the smartwatch sends the data to the smartphone and after a successful transmission, the smartphone vibrate to notify the user that the data collection process is successfully completed and they can stop the current activity.

2.2   Feature Extraction

The embedded sensor generates three-dimensional time series signals which are highly fluctuating and oscillatory in nature. The oscillation and fluctuations make the physical activity recognition more difficult than other applications in nature. These raw signals are shown in Fig. 1 for individual activity performed by the single user. Therefore, it is compulsory to gather the nontrivial signals from raw data through a feature extraction process. The raw signals are divided into several equal sized windows to control the flow rate and pass fewer data to the system to extract the meaningful information. In [24] Khan et al. have presented a detailed analysis of using different kinds of features with a different number of samples per windows for smartphone-based physical activity recognition.

From these raw signals, the extracted features are divided into two different feature banks. First feature bank include the features obtained by taking; average, median, variance, standard deviation, interquartile, autocorrelation, partial autocorrelation, coefficients of autoregressive model, coefficients of moving average model, coefficients of autoregressive moving average model and wavelet coefficients. Second feature bank includes the features obtained by taking; average acceleration, average absolute difference, standard deviation, average resultant acceleration and the average difference between peaks. We also calculate the binned distribution in which we determine what fraction of reading fall in a 10 equal-sized bins. This function generates 10 features. Every feature except average resultant acceleration is extracted for each axis.

These features are extracted from each axis of the three-dimensional acceleration signal and in total for each window of acceleration data 43 (first-feature-bank) and 70 (second-feature-bank) features are extracted respectively. Prior to the feature extraction process moving average filter of order three is used for noise reduction.

2.3   Classifiers

Khan et al. and Saputri et al. showed that support vector machines (SVM) and artificial neural networks (ANN) are superior to other traditional classification methods for user-independent physical activity recognition system in [24–26]. K-nearest neighbor (KNN) is also one of the most popular classification method used for smartphone-based physical activity recognition and user identification [27]. ELM [28], Naive Bayes [29], Adaptive learning/Ensemble learning (BAG) [30] and Decision tree [31] classifiers are also famous for smartphone-based physical activity recognition system. All these classification methods have shown several advantages in several fields most probably in their respective domains. Once again, given that ours is an exploratory study and the fact that each of these classifiers has shown a good performance in their respective studies. Instead of choosing one we decided to explore the use of all of these classifiers except neural networks in this work.

3   Experimental Settings

The following experimental studies are performed using all the data to create both personal and impersonal models. In our first study, using a fixed size of 75 samples per window the performance of five classifiers are compared to the following three cases on both feature banks and for both personal and impersonal models: (1) Randomly permuted features with normalization. (2) Without random permutation and normalization. (3) Randomly permuted features without normalization.

In the second study, the best setting for each classifier is chosen and tested multiple times while changing the number of samples in a window *i.e.* {*25, 50, 75, 100, 125, 150, 175, 200, 225, 250, 275, 300*} samples. The goal is to measure the effect of changing the window size on the performance of each classifier for both personal and impersonal model respectively. In our both studies, the training and testing data is randomly divided and classification results are obtained using 10-fold cross-validation process. The values used for different parameters of classifiers are as follows where all these parameters are carefully tuned and optimized. SVM is trained with quadratic kernel function is used, KNN with Euclidean distance is used and K is set to 10, Ensemble model (BAG) with tree model and the number of weak classifiers are set to 50 and the numbers of decision tree are set to 85.

4   Experimental Results and Discussion

This section contains the obtained results for smartwatch-based physical activity recognition system. To validate our system we conduct different experiments. Our first experiment is for both personal and impersonal model-based physical activity recognition analyses. In this experiment, we conduct a detailed comparison of two different feature banks and five classifiers for both personal and impersonal models. Our second experiment explains the physical activity recognition process behavior with a different number of samples per window within the best settings obtained in our first experiment. Finally, the third experiment presents the single subject cross-validation recognition for each activity for both personal and impersonal models individually. All these experiments are based on 10-fold cross-validation process and all these experiments are carried out using Matlab R2014b installed on core i5 and 8 GB of RAM machine.

In the introduction section, we stated some important research questions which our explanatory study seeks to provide the basis of answers. The answers to those questions based on the obtained results are as follows: Based on our findings, yes the features used in existing smartphone-based physical activity recognition studies can also be used for the smartwatch-based physical activity recognition. If we consider normalized and random permutation feature space, classifiers perform much better opposed to the case of without normalization and random permutation in several cases. Tables 1, 2, 3 and 4 present the 98% confidence intervals about overall physical activity recognition rate. Using pairwise T-tests between groups of normalized/randomly permuted and unnormalized/non-permuted data at the 98% confidence level.

Table 1. Overall classification accuracy and confidence interval for **impersonal model** with **43 features** extracted using **75 samples per window**. Where $NR = normalized$, $UNR = unnormalized$, $RP = randomly\ permuted$, and $NRP = non\ randomly\ permuted$.

Classifiers	Walking			Walking Up-Stairs			Walking Down-Stairs			Running			Jogging		
	NR-RP	NR-NRP	UNR-RP	NR-RP	NR-NRP	UNR-RP	NR-RP	NR-NRP	UNR-RP	NR-RP	NR-NRP	UNR-RP	NR-RP	NR-NRP	UNR-RP
Decision Tree	**0.90±0.08**	0.89±0.09	0.41±0.26	**0.90±0.07**	**0.90±0.09**	0.46±0.20	0.87±0.12	0.87±0.12	0.60±0.13	0.82±0.18	0.82±0.18	0.61±0.15	**0.93±0.06**	**0.94±0.06**	0.85±0.10
Naive Bayes	**0.86±0.06**	**0.86±0.06**	0.63±0.18	0.78±0.12	0.78±0.12	0.36±0.15	0.76±0.16	0.76±0.16	0.42±0.11	0.55±0.18	0.55±0.18	0.59±0.17	**0.86±0.02**	**0.86±0.02**	0.86±0.04
KNN	**0.86±0.08**	**0.86±0.08**	0.71±0.08	0.62±0.07	0.62±0.07	0.18±0.07	0.70±0.13	0.70±0.13	0.38±0.12	0.61±0.17	0.61±0.17	0.56±0.13	**0.91±0.06**	**0.91±0.06**	0.88±0.04
SVM	**0.94±0.07**	**0.94±0.06**	0.65±0.32	**0.91±0.04**	**0.91±0.04**	0.25±0.27	0.81±0.10	0.81±0.10	0.43±0.20	0.76±0.13	0.76±0.13	0.55±0.16	**0.92±0.03**	**0.92±0.03**	0.88±0.05
BAG	**0.91±0.08**	0.90±0.08	0.43±0.24	**0.89±0.09**	**0.89±0.09**	0.42±0.19	**0.87±0.12**	**0.87±0.12**	0.60±0.13	0.81±0.18	0.81±0.18	0.60±0.16	**0.94±0.05**	0.93±0.06	0.85±0.09

Table 2. Overall classification accuracy and confidence interval for **personal model** with **43 features** extracted using **75 samples per window**

Classifiers	Walking			Walking Up-Stairs			Walking Down-Stairs			Running			Jogging		
	NR-RP	NR-NRP	UNR-RP	NR-RP	NR-NRP	UNR-RP	NR-RP	NR-NRP	UNR-RP	NR-RP	NR-NRP	UNR-RP	NR-RP	NR-NRP	UNR-RP
Decision Tree	**0.98±0.01**	0.88±0.09	**0.94±0.01**	**0.93±0.02**	0.78±0.06	0.69±0.10	**0.98±0.05**	0.75±0.03	0.75±0.07	**0.95±0.01**	0.95±0.01	0.78±0.09	**0.98±0.01**	**0.95±0.06**	0.94±0.01
Naive	**0.91±0.02**	0.72±0.02	**0.87±0.08**	0.77±0.11	0.61±0.09	0.55±0.07	0.81±0.07	0.60±0.08	0.64±0.12	0.61±0.19	0.67±0.13	0.65±0.07	**0.87±0.08**	0.73±0.07	**0.87±0.06**
KNN	**0.92±0.07**	0.79±0.07	0.87±0.07	0.64±0.09	0.45±0.11	0.22±0.05	0.62±0.12	0.53±0.10	0.44±0.10	0.77±0.07	0.72±0.12	0.71±0.08	**0.91±0.08**	0.88±0.09	0.92±0.03
SVM	**0.93±0.07**	0.76±0.01	**0.93±0.01**	0.79±0.09	0.74±0.07	0.56±0.16	**0.90±0.03**	0.68±0.04	0.6±0.10	**0.87±0.06**	0.83±0.09	0.74±0.07	**0.97±0.01**	0.94±0.06	0.94±0.01
BAG	**0.98±0.01**	0.88±0.09	0.87±0.09	0.85±0.09	0.80±0.08	0.66±0.06	**0.89±0.07**	0.75±0.04	0.72±0.09	**0.95±0.02**	0.95±0.02	0.72±0.10	**0.99±0.01**	0.92±0.08	0.94±0.01

Table 3. Overall classification accuracy and confidence interval for **personal model** with **70 features** extracted using **75 samples per window**

Classifiers	Walking			Walking Up-Stairs			Walking Down-Stairs			Running			Jogging		
	NR-RP	NR-NRP	UNR-RP	NR-RP	NR-NRP	UNR-RP	NR-RP	NR-NRP	UNR-RP	NR-RP	NR-NRP	UNR-RP	NR-RP	NR-NRP	UNR-RP
Decision Tree	**0.98±0.01**	0.87±0.09	0.90±0.08	0.84±0.09	0.67±0.04	0.62±0.13	0.79±0.11	0.74±0.12	0.74±0.04	**0.86±0.07**	0.87±0.06	0.72±0.07	**0.97±0.01**	0.94±0.07	0.93±0.01
Naive	**0.88±0.02**	0.75±0.08	0.77±0.09	0.65±0.16	0.51±0.10	0.53±0.12	0.62±0.15	0.47±0.09	0.60±0.08	0.58±0.12	0.55±0.06	0.65±0.14	**0.88±0.02**	0.68±0.02	0.83±0.07
KNN	**0.96± 0.01**	0.93± 0.07	0.79± 0.09	0.60± 0.10	0.46± 0.08	0.19± 0.11	0.49± 0.16	0.46± 0.10	0.17± 0.06	0.58± 0.14	0.59± 0.16	0.26± 0.14	**0.95± 0.02**	0.95± 0.02	0.90± 0.02
SVM	0.98±0.01	0.91±0.08	0.87±0.07	0.89±0.08	0.77±0.08	0.52±0.07	0.86±0.08	0.76±0.08	0.54±0.08	0.91±0.02	0.85±0.09	0.66±0.09	0.97±0.01	0.90±0.10	0.92±0.01
BAG	**0.97±0.01**	0.87±0.09	0.90±0.07	**0.86±0.07**	0.69±0.04	0.68±0.10	0.84±0.05	0.65±0.09	0.65±0.06	**0.90±0.03**	0.86±0.05	0.72±0.08	**0.97±0.01**	0.87±0.10	0.91±0.07

Table 4. Overall classification accuracy and confidence interval for **impersonal model** with **70 features** extracted using **75 samples per window**.

Classifiers	Walking			Walking Up-Stairs			Walking Down-Stairs			Running			Jogging		
	NR-RP	NR-NRP	UNR-RP	NR-RP	NR-NRP	UNR-RP	NR-RP	NR-NRP	UNR-RP	NR-RP	NR-NRP	UNR-RP	NR-RP	NR-NRP	UNR-RP
Decision Tree	**0.91±0.07**	0.91±0.07	0.50±0.21	0.82±0.09	0.81±0.09	0.45±0.19	0.64±0.12	0.64±0.12	0.57±0.10	0.62±0.21	0.63±0.21	0.54±0.18	**0.86±0.07**	0.86±0.07	0.85±0.094
Naive	0.76±0.09	0.73±0.08	0.39±0.11	0.54±0.08	0.14±0.03	0.50±0.07	0.53±0.11	0.13±0.03	0.46±0.07	0.42±0.10	0.16±0.07	0.54±0.17	0.81±0.05	0.80±0.11	0.80±0.08
KNN	**0.89±0.05**	0.89±0.05	0.73±0.08	0.54±0.08	0.54±0.08	0.14±0.03	0.42±0.13	0.42±0.13	0.13±0.03	0.31±0.11	0.31±0.11	0.16±0.07	**0.86±0.06**	0.88±0.09	0.80±0.11
SVM	**0.97±0.02**	0.97±0.02	0.84±0.12	**0.90±0.04**	**0.90±0.04**	0.12±0.07	0.82±0.06	0.82±0.06	0.50±0.1	0.80±0.09	0.80±0.09	0.60±0.06	**0.94±0.03**	0.94±0.03	0.87±0.06
BAG	0.90±0.07	0.90±0.08	0.51±0.20	0.80±0.09	0.80±0.10	0.44±0.18	0.63±0.12	0.63±0.12	0.57±0.09	0.61±0.21	0.61±0.21	0.54±0.18	**00.86±.07**	0.86±0.07	0.85±0.09

Statistically significant results are in boldface, shows decision trees and SVM classifiers are statistically better when normalized and randomly permuted features are used. This increase was as much as almost 50% (from 40% to 90% for the impersonal model and 94% to 98% for the personal model) looking at Tables 1 and 2 with 43 features. Other classifiers are also statistically better using normalized features in a number of cases. For 70 number of feature-bank we again used pairwise T-tests between groups of normalized/randomly permuted and unnormalized/non-permuted data at 97%–98% confidence level.

Statistically significant results are in boldface shows in all cases again decision trees and SVM classifiers are statistically better when normalized features are used. This increase was as much as almost 10% (from 87% to 97% for the personal model) looking at Tables 3 and 4. Tables 5, 6, 7 and 8 shows single subject cross-validation process for each activity recognition.

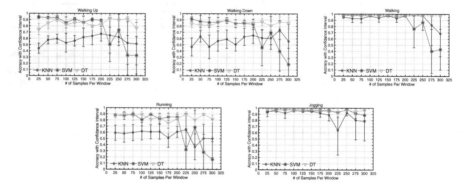

Fig. 2. Overall recognition accuracy for different activities with different number of samples per window.

Examining the overall and single subject cross-validation accuracy of all classifiers, each personal and the impersonal model becomes statistically significant in a number of cases. This leads us to prefer the use of normalization and random permutation in future applications of this method and in practical implementations. All of our chosen classifiers provide acceptable performance but decision tree and SVM are found to be better than other classifiers and have a number of other attractive features to their applicability.

In terms of the applicability of the smartwatch system, decision tree and SVM have smaller confidence intervals implying that they have more reliability in their training than other classification models. In terms of the applicability of the smartwatch system, the personal model has smaller confidence intervals implying that personal model has more reliability overall accuracies for both numbers of feature banks.

Table 5. Single subject cross-validation-based accuracy analysis for **impersonal model with 43 features extracted using 75 samples per window.**

Users	Walking			Walking Up-Stairs			Walking Down-Stairs			Running			Jogging		
	NR-RP	NR-NRP	UNR-RP	NR-RP	NR-NRP	UNR-RP	NR-RP	NR-NRP	UNR-RP	NR-RP	NR-NRP	UNR-RP	NR-RP	NR-NRP	UNR-RP
Decision Trees Classifier															
User 1	0.8706	0.8617	0.0595	0.9714	0.9689	0.4646	0.7909	0.7942	0.8123	0.6807	0.6922	0.5766	0.9759	0.9765	0.9175
User 2	0.7145	0.6961	0.0839	0.9768	0.9794	0.7112	0.9944	0.9940	0.7330	0.9724	0.9688	0.7371	0.9844	0.9871	0.8945
User 3	0.9757	0.9761	0.3730	0.7431	0.7000	0.7931	0.9497	0.9492	0.4230	0.9274	0.9259	0.7513	0.8691	0.8864	0.6500
User 4	0.9935	0.9894	0.5847	0.9380	0.9390	0.2432	0.9277	0.9312	0.4050	0.4287	0.4201	0.2616	0.9911	0.9941	0.9410
User 5	0.8998	0.8825	0.8987	0.9500	0.9442	0.2238	0.9690	0.9665	0.6171	0.9164	0.9126	0.7327	0.9687	0.9646	0.9400
User 6	0.9563	0.9518	0.4594	0.8450	0.8462	0.2972	0.6004	0.6025	0.6017	0.9804	0.9812	0.5938	0.8039	0.7993	0.7626
Naive Bayes Classifier															
User 1	0.7597	0.7597	0.2070	0.8883	0.8883	0.3495	0.6883	0.6883	0.5390	0.2448	0.2448	0.4323	0.8522	0.8522	0.8595
User 2	0.7979	0.7979	0.6905	0.9270	0.9270	0.3863	0.9742	0.9742	0.6137	0.7647	0.7647	0.7235	0.8855	0.8855	0.8909
User 3	0.9179	0.9179	0.6589	0.6373	0.6373	0.6961	0.8907	0.8907	0.3060	0.7411	0.7411	0.7614	0.8287	0.8287	0.7832
User 4	0.9369	0.9369	0.8342	0.8545	0.8545	0.2723	0.7525	0.7525	0.2624	0.3171	0.3171	0.2317	0.8614	0.8614	0.9242
User 5	0.8667	0.8667	0.5657	0.7791	0.7791	0.2500	0.8734	0.8734	0.4430	0.6918	0.6918	0.7673	0.9038	0.9038	0.8308
User 6	0.8847	0.8847	0.8176	0.5663	0.5663	0.1847	0.4017	0.4017	0.3473	0.5134	0.5134	0.6205	0.8256	0.8256	0.8479
KNN Classifier															
User 1	0.7523	0.7523	0.6728	0.5680	0.5680	0.1359	0.6299	0.6299	0.4416	0.4688	0.4688	0.5365	0.9179	0.9179	0.8942
User 2	0.7370	0.7370	0.6279	0.7854	0.7854	0.2017	0.8970	0.8970	0.3605	0.7765	0.7765	0.6882	0.9655	0.9655	0.9091
User 3	0.9661	0.9661	0.8268	0.6618	0.6618	0.3235	0.7814	0.7814	0.1913	0.7310	0.7310	0.6853	0.8759	0.8759	0.8444
User 4	0.8901	0.8901	0.6811	0.6150	0.6150	0.1878	0.7030	0.7030	0.2327	0.2317	0.2317	0.2378	0.9778	0.9778	0.9298
User 5	0.9371	0.9371	0.8495	0.5698	0.5698	0.0698	0.7848	0.7848	0.5190	0.6667	0.6667	0.5723	0.9615	0.9615	0.9269
User 6	0.8985	0.8985	0.6127	0.5422	0.5422	0.1606	0.4184	0.4184	0.5565	0.7813	0.7813	0.6339	0.7658	0.7658	0.8051
SVM Classifier															
User 1	0.9760	0.9760	0.1257	0.9223	0.9223	0.3204	0.8377	0.9657	0.6883	0.4740	0.4740	0.4792	0.9398	0.9380	0.9033
User 2	0.7710	0.7710	0.8497	0.9742	0.9742	0.0172	0.9657	0.8361	0.6695	0.8706	0.8706	0.7059	0.9164	0.9164	0.9073
User 3	0.9679	0.9679	0.1304	0.8137	0.8137	0.8971	0.8361	0.7277	0.2787	0.8985	0.8985	0.6954	0.8654	0.8654	0.7815
User 4	0.9874	0.9874	0.9550	0.9061	0.9061	0.0329	0.7277	0.8608	0.2772	0.7378	0.7378	0.1707	0.9464	0.9464	0.9575
User 5	0.9600	0.9600	0.9448	0.9302	0.9302	0.0814	0.8608	0.6109	0.5823	0.6792	0.6792	0.6604	0.9788	0.9788	0.9192
User 6	0.9776	0.9776	0.8864	0.9197	0.9197	0.1365	0.6109	0.9657	0.0669	0.8839	0.8839	0.5848	0.8991	0.8974	0.8308
Ensemble Learning-BAG Classifier															
User 1	0.8784	0.8656	0.0980	0.9670	0.9670	0.3529	0.7805	0.7864	0.7968	0.6734	0.6786	0.5661	0.9763	0.9748	0.9224
User 2	0.7086	0.7038	0.1243	0.9712	0.9721	0.6236	0.9944	0.9953	0.7416	0.9641	0.9647	0.7429	0.9827	0.9811	0.8982
User 3	0.9757	0.9759	0.3900	0.6730	0.6985	0.7902	0.9475	0.9497	0.4240	0.9279	0.9228	0.7472	0.8692	0.8764	0.6458
User 4	0.9894	0.9874	0.5782	0.9376	0.9366	0.2531	0.9292	0.9233	0.4183	0.3988	0.4073	0.2317	0.9930	0.9904	0.9442
User 5	0.8989	0.8970	0.9006	0.9517	0.9448	0.2017	0.9614	0.9608	0.6253	0.8994	0.9031	0.7409	0.9677	0.9662	0.9448
User 6	0.9546	0.9497	0.4305	0.8454	0.8426	0.2924	0.6167	0.6079	0.6004	0.9772	0.9777	0.5960	0.8373	0.8113	0.7576

Table 6. Single subject cross-validation-based accuracy analysis for **personal model with 43 features extracted using 75 samples per window.**

Users	Walking			Walking Up-Stairs			Walking Down-Stairs			Running			Jogging		
	NR-RP	NR-NRP	UNR-RP	NR-RP	NR-NRP	UNR-RP	NR-RP	NR-NRP	UNR-RP	NR-RP	NR-NRP	UNR-RP	NR-RP	NR-NRP	UNR-RP
Decision Trees Classifier															
User 1	0.9815	0.7889	0.9500	0.9073	0.7366	0.8000	0.9067	0.7067	0.5867	0.9632	0.9368	0.7895	0.9908	0.9945	0.9633
User 2	0.9820	0.9802	0.9604	0.9604	0.7348	0.8217	0.7870	0.7826	0.8043	0.9471	0.9647	0.8765	0.9836	0.9873	0.9382
User 3	0.9786	0.7768	0.9214	0.9450	0.7500	0.6750	0.9278	0.7167	0.7611	0.9333	0.9128	0.5487	0.9772	0.9579	0.9193
User 4	0.9874	0.9946	0.9568	0.9714	0.7619	0.5000	0.9600	0.7400	0.8100	0.9500	0.9625	0.7875	0.9833	0.9815	0.9500
User 5	0.9867	0.9905	0.9314	0.9235	0.7412	0.7176	0.9226	0.7290	0.8129	0.9548	0.9484	0.8581	0.9865	0.7885	0.9519
User 6	0.9741	0.7741	0.9102	0.9306	0.5959	0.5959	0.8894	0.8000	0.6979	0.9636	0.9682	0.7955	0.9829	0.9846	0.9282
Naive Bayes Classifier															
User 1	0.9019	0.6981	0.9000	0.8049	0.6146	0.4732	0.6800	0.3467	0.3467	0.5000	0.6474	0.5421	0.8789	0.7046	0.8844
User 2	0.9099	0.7171	0.9027	0.7087	0.5087	0.5000	0.8174	0.6478	0.7391	0.8412	0.8412	0.6118	0.9109	0.9036	0.8982
User 3	0.9518	0.7571	0.6589	0.9250	0.7300	0.6950	0.8889	0.6778	0.7500	0.7846	0.5949	0.7897	0.8912	0.6421	0.8930
User 4	0.9279	0.7423	0.9261	0.9429	0.7381	0.4952	0.8750	0.6600	0.6050	0.7500	0.7312	0.7250	0.9296	0.7315	0.7278
User 5	0.9219	0.7200	0.9105	0.6176	0.6176	0.6294	0.8903	0.6710	0.7355	0.6194	0.8194	0.6258	0.9308	0.7231	0.9269
User 6	0.8741	0.6862	0.8931	0.6490	0.4735	0.4776	0.7319	0.4809	0.6723	0.1955	0.3955	0.6273	0.6735	0.6803	0.8906
KNN Classifier															
User 1	0.7333	0.7389	0.6852	0.7122	0.4732	0.2585	0.5800	0.3467	0.3667	0.8316	0.8211	0.6895	0.9450	0.9541	0.9376
User 2	0.9604	0.7477	0.9045	0.5217	0.2783	0.1826	0.8391	0.6087	0.4435	0.8059	0.8412	0.8059	0.9691	0.9691	0.9164
User 3	0.9643	0.7571	0.9054	0.5400	0.5900	0.2050	0.7222	0.5222	0.3389	0.8051	0.6103	0.7436	0.9561	0.9491	0.9316
User 4	0.9568	0.9658	0.9351	0.8333	0.6333	0.1333	0.4450	0.6250	0.6350	0.6813	0.4813	0.7688	0.8741	0.6889	0.8481
User 5	0.9638	0.7600	0.9219	0.6765	0.4118	0.3000	0.6452	0.6581	0.5032	0.6387	0.6645	0.7355	0.9769	0.7769	0.9558
User 6	0.9397	0.7466	0.8759	0.5673	0.3143	0.2408	0.5064	0.4340	0.3234	0.8773	0.8773	0.5318	0.7316	0.9350	0.9009
SVM Classifier															
User 1	0.7481	0.7370	0.9315	0.8049	0.6732	0.6000	0.8400	0.6200	0.7333	0.7263	0.9158	0.7474	0.9688	0.9761	0.9431
User 2	0.9622	0.7658	0.9279	0.6522	0.6609	0.8261	0.9435	0.7565	0.7696	0.8824	0.6882	0.8059	0.9855	0.9855	0.9327
User 3	0.9714	0.7679	0.9089	0.7500	0.7250	0.7450	0.8556	0.6333	0.6889	0.8974	0.6769	0.7538	0.9719	0.9579	0.9246
User 4	0.9838	0.7892	0.9640	0.9238	0.7286	0.3095	0.9350	0.7150	0.7500	0.8500	0.8375	0.5938	0.9685	0.9685	0.9537
User 5	0.9811	0.7562	0.9067	0.9333	0.7235	0.5000	0.9226	0.6968	0.7484	0.8968	0.9161	0.8194	0.9673	0.7769	0.9596
User 6	0.9552	0.7638	0.9414	0.6939	0.9020	0.4082	0.8851	0.6638	0.4511	0.9636	0.9364	0.7045	0.9573	0.9607	0.9162
Ensemble Learning-BAG Classifier															
User 1	0.9870	0.7852	0.9481	0.7122	0.7317	0.7707	0.9133	0.7067	0.5933	0.9632	0.9421	0.7842	0.9872	0.7963	0.9486
User 2	0.9830	0.9874	0.9604	0.8870	0.7261	0.5652	0.9783	0.7826	0.8174	0.9588	0.9647	0.8706	0.9873	0.9909	0.9491
User 3	0.9786	0.7732	0.9179	0.9450	0.9500	0.6400	0.9278	0.7111	0.5556	0.9077	0.9026	0.8154	0.9807	0.9561	0.9105
User 4	0.9874	0.9964	0.9495	0.9619	0.7714	0.6762	0.9650	0.7400	0.8350	0.9375	0.9563	0.6000	0.9778	0.9778	0.9444
User 5	0.9867	0.9886	0.7276	0.9059	0.7235	0.7118	0.7290	0.7161	0.7871	0.9677	0.9548	0.6452	0.9942	0.7846	0.9462
User 6	0.9741	0.7724	0.7138	0.6980	0.9224	0.5918	0.8553	0.8255	0.7064	0.9727	0.9591	0.5955	0.9846	0.9897	0.9197

Looking at Tables 1 to 4, KNN, Decision trees and SVM perform much better in the personal model for both number of features with normalization and for overall accuracy prospect; decision trees outperformed then KNN and SVM. The highest obtained accuracies are in boldface. For further analysis the best-recommended system settings are 70 features normalized and randomly permuted used by Decision tree and SVM classifier to recognize the individual's activity on a smartwatch. Given a fixed amount of data, the performance of KNN and SVM decreases significantly when the window size is increased as it reduces the number of training samples but this does not affect decision tree classifier. Since decision tree, KNN and SVM classifiers trained using 70 normalized and randomly permuted features happens to be the best setting for smartwatch based PAR system.

Results are summarized in Fig. 2 which presents the overall classification accuracy with 98% confidence interval. According to these results, decision tree classifier shows higher accuracy for every size of the window. The performance of SVM and KNN through good but is relatively less than that of the decision tree in all cases. Overall, decision trees performance happens to be best followed by SVM as their performance did not much degraded when changing the size of samples per window. On the other hand, the performance of KNN reduced significantly as the window size is increased. We think that it is because we have a fixed amount of data. So a bigger window size results in a smaller number of training samples. To confirm this, we repeated the same experiment for each activity individually. The entire results are summarized in Tables 1 to 4 which confirms our hypothesis. The additional experiments on single subject cross-validations are presented in Tables 5, 6, 7 and 8 for both feature banks and personal and impersonal models, respectively.

Table 7. Single subject cross-validation-based accuracy analysis for **personal model with 70 features extracted using 75 samples per window**.

Users	Walking			Walking Up-Stairs			Walking Down-Stairs			Running			Jogging		
	NR-RP	NR-NRP	UNR-RP	NR-RP	NR-NRP	UNR-RP	NR-RP	NR-NRP	UNR-RP	NR-RP	NR-NRP	UNR-RP	NR-RP	NR-NRP	UNR-RP
Decision Trees Classifier															
User 1	0.9593	0.9574	0.7074	0.6780	0.6780	0.7171	0.5800	0.5467	0.7000	0.6947	0.9053	0.7632	0.9908	0.9908	0.9541
User 2	0.9748	0.7640	0.9405	0.8913	0.7217	0.6130	0.9217	0.9522	0.7739	0.9235	0.7235	0.8176	0.9909	0.9873	0.9218
User 3	0.9821	0.7821	0.9357	0.9550	0.7500	0.4950	0.9389	0.7111	0.7167	0.8769	0.8872	0.7487	0.9421	0.9246	0.9105
User 4	0.9820	0.9856	0.9586	0.9286	0.7143	0.6762	0.8200	0.5950	0.8100	0.8562	0.9000	0.6062	0.9778	0.7741	0.9389
User 5	0.9790	0.7733	0.9257	0.8412	0.6235	0.8294	0.6903	0.8710	0.7548	0.9290	0.9419	0.6129	0.9750	0.9731	0.9462
User 6	0.9741	0.9810	0.9172	0.7265	0.7102	0.3673	0.7830	0.7489	0.6894	0.8727	0.8773	0.7773	0.9641	0.9675	0.9248
Naive Bayes Classifier															
User 1	0.8593	0.6333	0.7870	0.5073	0.4634	0.6537	0.5467	0.3733	0.6200	0.5368	0.5632	0.7211	0.8789	0.6642	0.8917
User 2	0.8757	0.8793	0.6505	0.3609	0.3826	0.5696	0.7478	0.5739	0.7739	0.7176	0.5529	0.7941	0.8636	0.6727	0.7000
User 3	0.9161	0.7179	0.6071	0.8750	0.6650	0.6300	0.7889	0.5722	0.4667	0.5333	0.6462	0.7590	0.8561	0.6614	0.8842
User 4	0.8955	0.6991	0.8541	0.8381	0.6524	0.6000	0.5050	0.4550	0.5400	0.6000	0.4063	0.4625	0.9130	0.7148	0.7333
User 5	0.8857	0.8800	0.8781	0.5765	0.3647	0.2588	0.8065	0.5613	0.6065	0.7613	0.5871	0.7677	0.9096	0.7135	0.9154
User 6	0.8638	0.6879	0.8448	0.7224	0.5429	0.4612	0.3362	0.3064	0.5745	0.3273	0.5318	0.3955	0.8598	0.6547	0.8838
KNN Classifier															
User 1	0.9648	0.9426	0.8944	0.4390	0.3951	0.1268	0.4267	0.4667	0.1600	0.7053	0.7158	0.1842	0.9835	0.9725	0.9138
User 2	0.9532	0.9532	0.8685	0.6565	0.5087	0.1391	0.5174	0.5174	0.2217	0.5118	0.7118	0.2000	0.9545	0.9600	0.8927
User 3	0.9625	0.9732	0.8536	0.7700	0.5950	0.4650	0.7389	0.5278	0.1444	0.4718	0.5795	0.2974	0.9263	0.9070	0.8526
User 4	0.9550	0.9568	0.7964	0.5048	0.4810	0.2048	0.3450	0.5200	0.2700	0.3250	0.2562	0.0938	0.9481	0.9574	0.9204
User 5	0.9771	0.7638	0.7048	0.5471	0.3176	0.1353	0.7161	0.5484	0.1613	0.7742	0.7935	0.5935	0.9635	0.9654	0.9058
User 6	0.9603	0.9759	0.6190	0.6898	0.4612	0.0980	0.2213	0.2043	0.0596	0.6955	0.4727	0.2182	0.9675	0.9556	0.8940
SVM Classifier															
User 1	0.9778	0.7759	0.8852	0.8634	0.7317	0.4000	0.8067	0.8067	0.5600	0.9000	0.9263	0.5105	0.9706	0.9725	0.9193
User 2	0.9946	0.9910	0.9297	0.9522	0.7652	0.4826	0.9348	0.7435	0.4870	0.9000	0.7294	0.5353	0.9909	0.9855	0.9200
User 3	0.9804	0.9839	0.8732	0.9150	0.7200	0.6150	0.9333	0.9389	0.4333	0.8718	0.8769	0.7231	0.9439	0.7175	0.9123
User 4	0.9820	0.9748	0.9063	0.9667	0.7619	0.6048	0.9100	0.7000	0.7100	0.9313	0.9437	0.7375	0.9667	0.7778	0.9241
User 5	0.9829	0.9752	0.6990	0.6882	0.6588	0.4765	0.8968	0.7032	0.4516	0.9097	0.6968	0.7677	0.9827	0.9808	0.9346
User 6	0.9759	0.7862	0.8983	0.9510	0.9633	0.5673	0.6851	0.6638	0.5872	0.9273	0.9545	0.7091	0.9675	0.9641	0.8872
Ensemble Learning-BAG Classifier															
User 1	0.9556	0.9500	0.9241	0.8780	0.6829	0.7317	0.7333	0.5133	0.6467	0.8947	0.8842	0.8000	0.9908	0.9890	0.9596
User 2	0.9658	0.7730	0.9405	0.8826	0.7087	0.8261	0.9087	0.7174	0.5696	0.9353	0.7529	0.8000	0.9873	0.9873	0.9364
User 3	0.9821	0.9800	0.9268	0.9500	0.7400	0.4700	0.8944	0.7056	0.5500	0.8667	0.8564	0.7487	0.9386	0.7228	0.9211
User 4	0.9874	0.9802	0.9532	0.6905	0.6952	0.6333	0.8250	0.5150	0.6650	0.8625	0.9000	0.6000	0.9741	0.7741	0.9444
User 5	0.9619	0.7695	0.9390	0.8235	0.6000	0.7941	0.8645	0.6903	0.7613	0.9032	0.9226	0.8065	0.9750	0.7750	0.9519
User 6	0.9741	0.9724	0.7241	0.9061	0.6939	0.6367	0.8170	0.7872	0.6936	0.9500	0.8409	0.5864	0.9761	0.9658	0.7231

Table 8. Single subject cross-validation-based accuracy analysis for **impersonal model with 70 features extracted using 75 samples per window**.

Users	Walking			Walking Up-Stairs			Walking Down-Stairs			Running			Jogging		
	NR-RP	NR-NRP	UNR-RP	NR-RP	NR-NRP	UNR-RP	NR-RP	NR-NRP	UNR-RP	NR-RP	NR-NRP	UNR-RP	NR-RP	NR-NRP	UNR-RP
Decision Trees Classifier															
User 1	0.7455	0.7412	0.3338	0.9296	0.9286	0.3888	0.7896	0.7896	0.7383	0.3563	0.3677	0.3885	0.8648	0.8739	0.9086
User 2	0.9370	0.9458	0.1308	0.9682	0.9665	0.8129	0.7210	0.7223	0.6605	0.8600	0.8524	0.7094	0.8445	0.8320	0.8367
User 3	0.9846	0.9832	0.5609	0.7505	0.7471	0.6240	0.6268	0.6311	0.5557	0.8437	0.8508	0.7061	0.7346	0.7392	0.6330
User 4	0.9524	0.9532	0.6688	0.7906	0.7850	0.4028	0.6485	0.6470	0.3876	0.2348	0.2274	0.1835	0.9762	0.9773	0.9486
User 5	0.9617	0.9636	0.8874	0.8023	0.7849	0.1552	0.6791	0.6747	0.5766	0.7264	0.7157	0.7535	0.9442	0.9456	0.9133
User 6	0.8523	0.8491	0.4219	0.6711	0.6643	0.3040	0.3448	0.3552	0.5130	0.7254	0.7513	0.4955	0.7814	0.7879	0.8374
Naive Bayes Classifier															
User 1	0.6192	0.8429	0.4455	0.7087	0.1408	0.3883	0.5649	0.1039	0.5649	0.2396	0.1719	0.3646	0.7646	0.8102	0.8266
User 2	0.9123	0.7728	0.3953	0.5536	0.1330	0.5880	0.6609	0.1116	0.5536	0.5118	0.1294	0.7118	0.8582	0.8873	0.8273
User 3	0.8286	0.6982	0.5643	0.4216	0.1961	0.5392	0.5683	0.2022	0.4481	0.5736	0.2792	0.6954	0.7902	0.7115	0.7413
User 4	0.8342	0.5802	0.4432	0.5164	0.0845	0.5164	0.5149	0.1535	0.3960	0.3415	0.0183	0.2378	0.8965	0.9279	0.9205
User 5	0.6724	0.8114	0.1410	0.5291	0.1279	0.5756	0.6076	0.1266	0.4873	0.4969	0.1698	0.7610	0.7404	0.8769	0.6308
User 6	0.7126	0.6713	0.3270	0.4859	0.1767	0.3735	0.2594	0.0879	0.3347	0.3527	0.2188	0.4955	0.8222	0.5573	0.8393
KNN Classifier															
User 1	0.8096	0.8096	0.8429	0.6408	0.6408	0.1408	0.4026	0.4026	0.1039	0.2448	0.2448	0.1719	0.8029	0.8029	0.8102
User 2	0.9141	0.9141	0.7728	0.6438	0.6438	0.1330	0.5794	0.5794	0.1116	0.3235	0.3235	0.1294	0.9182	0.9182	0.8873
User 3	0.9143	0.9143	0.6982	0.5931	0.5931	0.1961	0.5683	0.5683	0.2022	0.5381	0.5381	0.2792	0.8759	0.8759	0.7115
User 4	0.9387	0.9387	0.5802	0.4131	0.4131	0.0845	0.2574	0.2574	0.1535	0.1037	0.1037	0.0183	0.9593	0.9593	0.9279
User 5	0.9733	0.9733	0.8114	0.4767	0.4767	0.1279	0.5380	0.5380	0.1266	0.3019	0.3019	0.1698	0.9231	0.9231	0.8769
User 6	0.8176	0.8176	0.6713	0.4618	0.4618	0.1767	0.2008	0.2008	0.0879	0.3705	0.3705	0.2188	0.7726	0.7726	0.5573
SVM Classifier															
User 1	0.9372	0.9372	0.6007	0.9417	0.9417	0.2233	0.7857	0.7857	0.5909	0.6146	0.6146	0.5313	0.9197	0.9197	0.8832
User 2	0.9517	0.9517	0.9660	0.9828	0.9828	0.0258	0.8712	0.8712	0.6266	0.6059	0.6059	0.6882	0.9655	0.9655	0.8909
User 3	0.9875	0.9875	0.8750	0.8578	0.8578	0.1716	0.8852	0.8852	0.5355	0.9137	0.9137	0.6802	0.8776	0.8776	0.7483
User 4	0.9928	0.9928	0.6883	0.8685	0.8685	0.2066	0.8861	0.8861	0.4208	0.8049	0.8049	0.5183	0.9778	0.9778	0.9482
User 5	0.9867	0.9867	0.9543	0.8663	0.8663	0.0291	0.8291	0.8291	0.5886	0.7358	0.7358	0.5849	0.9654	0.9654	0.9000
User 6	0.9639	0.9639	0.9466	0.8996	0.8996	0.0723	0.6820	0.6820	0.2176	0.8527	0.8527	0.5670	0.9573	0.9573	0.8256
Ensemble Learning-BAG Classifier															
User 1	0.7423	0.7240	0.3532	0.9194	0.9301	0.3888	0.7721	0.7825	0.7123	0.3219	0.3255	0.3906	0.8746	0.8611	0.9111
User 2	0.9331	0.9392	0.1644	0.9665	0.9597	0.7931	0.7120	0.7206	0.6652	0.8288	0.8253	0.7141	0.8393	0.8480	0.8349
User 3	0.9832	0.9852	0.5695	0.7333	0.7216	0.6039	0.6186	0.6235	0.5322	0.8376	0.8365	0.7015	0.7362	0.7392	0.6451
User 4	0.9450	0.9505	0.6701	0.7690	0.7643	0.3911	0.6416	0.6391	0.3856	0.2305	0.2329	0.1951	0.9758	0.9747	0.9444
User 5	0.9598	0.9577	0.8851	0.7628	0.7756	0.1651	0.6918	0.6842	0.5766	0.7182	0.7170	0.7440	0.9415	0.9463	0.9137
User 6	0.8522	0.8515	0.4389	0.6627	0.6514	0.2892	0.3477	0.3485	0.5188	0.7201	0.7076	0.4906	0.7805	0.7701	0.8487

5 Conclusion

Feature banks created by different smartphone-based physical activity recognition studies can be used for physical activity recognition on smartwatches because smartphone also uses accelerometer, gyroscope and magnetometer sensors and hence the same set of features is acceptable for a smartwatch. Decision trees, SVM and KNN showed the best results with a minor difference. Additionally, Naive Bayes and ensemble learning with Bag classifiers also produce a good performance. Furthermore, feature normalization and random permutation processes significantly improve the classification performance in several cases. In addition to above, without any doubt, changing window size affects the recognition accuracy.

From results, we observed that every activity and each classifier's results of recognition is highly correlated with window size. In general, each classifier performs well when features are extracted using 25–125 samples per window. Personal models perform better for smartwatch based physical activity recognition providing an average accuracy of 98%. Furthermore, the impersonal model produces 94% and 90% accuracy respectively with the same settings of the personal model. Finally, the recommended combination for smartwatch based physical activity recognition system is decision tree classifier trained using 25–125 samples per window of normalized and randomly permuted features. In our future work, we will investigate the effects of several linear/non-linear supervised and unsupervised feature selection methods.

References

1. Ahmad, M., Khan, A.M., Brown, J.A., Protasov, S., Khattak, A.M.: Gait fingerprinting-based user identification on smartphones. In: Proceedings of the IEEE International Joint Conference on Neural Networks (IJCNN), in Conjunction with World Congress on Computational Intelligence (WCCI), Canada, pp. 3060–3067 (2016)
2. Ahmad, M., Alqarni, M.A., Khan, A., et al.: Smartwatch-based legitimate user identification for cloud-based secure services. Mob. Inf. Syst. **2018**, 14 (2018). https://doi.org/10.1155/2018/5107024. Article ID 5107024
3. Khan, A.M., Lee, Y.K., Lee, S.Y., Kim, T.S.: A triaxial accelerometer-based physical-activity recognition via augmented-signal features and a hierarchical recognizer. IEEE Trans. Inf. Technol. Biomed. **14**(5), 1166–1172 (2010)
4. Incel, O.D., Kose, M., Ersoy, C.: A review and taxonomy of activity recognition on mobile phones. BioNanoSci. **3**, 145–171 (2013)
5. Khan, W.Z., Xiang, Y., Aalsalem, M.Y., Arshad, Q.: Mobile phone sensing systems: a survey. IEEE Commun. Surv. Tutor. **15**, 402–427 (2013)
6. Bulling, A., Blanke, U., Schiele, B.: A tutorial on human activity recognition using body-worn inertial sensors. ACM J. Comput. Surv. **46**(3), 33:1–33:33 (2014). Article No. 33
7. Lane, N.D., Miluzzo, E., Lu, H., Peebles, D., Choudhury, T., Campbell, A.T.: A survey of mobile phone sensing. IEEE Commun. Mag. **48**, 140–150 (2010)
8. Shoaib, M.: Human activity recognition using heterogeneous sensors. In: Proceedings of the Adjunct Proceedings of the ACM Conference on Ubiquitous Computing, pp. 8–12 (2013)
9. Guiry, J.J., Van de Ven, P., Nelson, J.: Multi-sensor fusion for enhanced contextual awareness of everyday activities with ubiquitous devices. J. Sens. **14**, 5687–5701 (2014)
10. Trost, S.G., Zheng, Y., Wong, W.K.: Machine learning for activity recognition: hip versus wrist data. Physiol. Meas. **35**, 2183–2189 (2014)
11. Chernbumroong, S., Atkins, A.S., Yu, H.: Activity classification using a single wrist-worn accelerometer. In: Proceedings of the 5th IEEE International Conference on Software, Knowledge Information, Industrial Management and Applications (SKIMA), pp. 1–6 (2011)
12. Da Silva, F.G., Galeazzo, E.: Accelerometer-based intelligent system for human movement recognition. In: Proceedings of the 5th IEEE International Workshop on Advances in Sensors and Interfaces (IWASI), pp. 20–24 (2013)
13. Ramos-Garcia, R.L., Hoover, A.W.: A study of temporal action sequencing during consumption of a meal. In: Proceedings of the ACM International Conference on Bioinformatics, Computational Biology and Biomedical Informatics, p. 68 (2013)
14. Dong, Y., Scisco, J., Wilson, M., Muth, E., Hoover, A.: Detecting periods of eating during free-living by tracking wrist motion. IEEE J. Biomed. Health Inf. **18**, 1253–1260 (2013)
15. Sen, S., Subbaraju, V., Misra, A., Balan, R., Lee, Y.: The case for smartwatch-based diet monitoring. In: Proceedings of the IEEE International Conference on Pervasive Computing and Communication Workshops (PerCom), pp. 585–590 (2015)
16. Scholl, P.M., Van Laerhoven, K.: A feasibility study of wrist-worn accelerometer based detection of smoking habits. In: Proceedings of the 6th IEEE International Conference on Innovative Mobile and Internet Services in Ubiquitous Computing (IMIS), pp. 886–891 (2012)

17. Parade, A., Chiu, M.C., Chadowitz, C., Ganesan, D., Kalogerakis, E.: Recognizing smoking gestures with inertial sensors on a wristband. In: Proceedings of the 12th Annual International Conference on Mobile Systems, Applications, and Services, pp. 149–161 (2014)
18. Varkey, J.P., Pompili, D., Walls, T.A.: Human motion recognition using a wireless sensor-based wearable system. Pers. Ubiquit. Comput. **16**, 897–910 (2012)
19. Kim, H., Shin, J., Kim, S., Ko, Y., Lee, K., Cha, H., Hahm, S.-I., Kwon, T.: Collaborative classification for daily activity recognition with a smartwatch. In: Proceedings of the IEEE International Conference on Systems, Man, and Cybernetics (SMC), pp. 3707–3712 (2016)
20. Weiss, G.M., Timko, J.L., Gallagher, C.M., Yoneda, K., Schreiber, A.J.: Smartwatch-based activity recognition: a machine learning approach. In: Proceedings of the IEEE International Conference on Biomedical and Health Informatics (BHI), pp. 426–429 (2016)
21. Nurwanto, F., Ardiyanto, I., Wibirama, S.: Light sports exercise detection based on smartwatch and smartphone using k-nearest neighbor and dynamic time warping algorithm. In: Proceedings of the 8th IEEE International Conference on Information Technology and Electrical Engineering (ICITEE), pp. 1–5 (2016)
22. Al-Naffakh, N., Clarke, N., Dowland, P., Li, F.: Activity recognition using wearable computing. In: Proceedings of the IEEE 11th International Conference on Internet Technology and Secured Transactions (ICITST), pp. 189–195 (2016)
23. Huynh, T., Schiele, B.: Analyzing features for activity recognition. In: Proceedings of the Joint Conference on Smart Objects and Ambient Intelligence: Innovative Context-Aware Services: Usages and Technologies, pp. 159–163 (2005)
24. Khan, A.M., Siddiqi, H.M., Lee, S.W.: Exploratory data analysis of acceleration signals to select light-weight and accurate features for real-time activity recognition on smartphones. J. Sens. **13**(10), 13099–13122 (2013)
25. Saputri, T.R.D., Khan, A.M., Lee, S.W.: User-independent activity recognition via three-stage GA-based feature selection. Int. J. Distrib. Sens. Netw. **2014**, 15 (2014)
26. Khan, A.M., Lee, Y.-K., Kim, T.-S.: Accelerometer signal-based human activity recognition using augmented autoregressive model coefficients and artificial neural nets. In: Proceedings of the 30th Annual International Conference of the IEEE Engineering in Medicine and Biology Society, pp. 5172–5175 (2008)
27. Su, X., Tong, H., Ji, P.: Activity recognition with smartphone sensors. Tsinghua Sci. Technol. Int. J. Inf. Sci. **19**(3), 235–249 (2014)
28. Ahmad, M., Khan, A.M., Mazzara, M., Distefano, S., Ali, A., Tufail, A.: Extended sammon projection and wavelet kernel extreme learning machine for gait-based legitimate user identification. In: Proceedings of the 34th ACM/SIGAPP Symposium On Applied Computing, SAC 2019 (2019)
29. Al Jeroudi, Y.: Online sequential extreme learning machine algorithm based human activity recognition using inertial data. In: Proceedings of the IEEE 10th Control Conference (ASCC), pp. 1–6 (2015)
30. Shoaib, M., Bosch, S., Incel, O.D., Scholten, H., Havinga, P.J.M.: Complex human activity recognition using smartphone and wrist-worn motion sensors. J. Sens. **16**, 1–24 (2016)
31. Shoaib, M., Bosch, S., Incel, O.D., Scholten, H., Havinga, P.J.M.: A survey of online activity recognition using mobile phones. J. Sens. **15**(1), 2059–2085 (2015)

SURF Based Copy Move Forgery Detection Using kNN Mapping

Kelvin Harrison Paul[1], K. R. Akshatha[1(✉)], A. K. Karunakar[2], and Sharan Seshadri[1]

[1] Department of Electronics and Communication Engineering, Manipal Institute of Technology, Manipal Academy of Higher Education, Manipal 576104, Karnataka, India
kelvin.harrison@learner.manipal.edu,
akshatha.kr@manipal.edu, sharan2510@gmail.com
[2] Department of Computer Applications, Manipal Institute of Technology, Manipal Academy of Higher Education, Manipal 576104, Karnataka, India
karunakar.ak@manipal.edu

Abstract. Digital images can be edited with the help of photo editing tools to improve or enhance the image quality. On the other hand, digital images can also be subject to manipulations which can alter the visual information being conveyed by the image. The forged images can also be used to spread false information through various media platforms and in some cases may be surreptitiously used as false evidence in a court of law. Therefore, it is crucial to test the authenticity of such images and ensure that it does not spread falsified information. One of the most common types of forgery being used today is copy-move forgery in which one part of the image is copied and placed over another part of the same image in order to either conceal certain details or multiply certain features seen in the original image. This work introduces a method of detecting copy-move forgery in digital images using speeded-up robust features (SURF) to extract keypoints from the image and then uses k-nearest neighbor (kNN) training and mapping to yield accurate matches. The SURF algorithm is capable of performing equally or even exceed the more widely accepted SIFT-based counterparts in terms of ensuring distinctive features, reproducibility, and robustness. As a result, this technique ensures a robust detection of copy move forgery while ensuring lower computational costs compared to the SIFT-based techniques used for the same purpose.

Keywords: Copy move forgery detection · CMFD · SURF · kNN mapping · Image forensics

1 Introduction

In the past few decades, digital images have become a widely accepted form of visual media. One of the main reasons for its drastic rise to popularity is that digital images can be easily transferred from one device to another using peripheral electronic device such as compact discs and portable USB drives. Additionally, the past two decades have witnessed a rapid increase in the number of images being circulated through the

© Springer Nature Switzerland AG 2020
K. Arai and S. Kapoor (Eds.): CVC 2019, AISC 944, pp. 234–245, 2020.
https://doi.org/10.1007/978-3-030-17798-0_20

internet which include social media, news platforms, and the entertainment industry. With the rampant increase in the popularity of digital images, those who are trying to convey false information have taken up manipulating digital images to achieve their purpose. Even though some forgeries can be easily identified by the average person, other manipulations can be very subtle and escape the scrutiny of the human eye. These manipulations can be easily made by those who are well versed with using sophisticated photo-editing tools which are available in the market.

As more people have become familiar with the use of sophisticated photo-editing tools, it is not surprising to see that the number of tampered images being published in social media, news articles, academic research, and lawful evidence has also increased over the past few years. The forged images often have a negative influence on the people viewing them as it can lead them to wrong interpretations of events and people. Tampered images are also frequently presented as evidence in courts of law. In this scenario, the image has the potential to sway the judgment of the people involved and thereby hinder the process of carrying out justice in modern societies. Therefore, it is crucial to determine the authenticity of digital images before they can have a negative impact on people and their communities.

More recently, copy-move forgery has become a very common method for manipulating digital images. This form of forgery involves copying a section of the image and pasting it in other sections of the same image. The forged regions of the image are made more subtle by incorporating scaling and rotational transformations. Blurring is also often carried out on these regions to make it difficult for human scrutiny. In order to detect such forgeries in digital images, researchers have proposed various novel techniques which can be categorized as either block-based techniques or keypoint-based techniques.

The block-based techniques involve dividing the image being tested into overlapping blocks of fixed dimensions, each block is then characterized using different feature algorithms. This method ensures robustness to brightness adjustment, addition of noise, and lossy compression of images. But these techniques are often found to have high computational cost and are less accurate when scaling and rotation are involved in the test image. On the other hand, keypoint based techniques are robust when scaling and rotation are present in the test image. Keypoint-based techniques involve extracting keypoints from the test image, this process is often called keypoint detection. The extracted keypoints are then matched to other keypoints in the image; this step is often denoted as keypoint matching. As a result, the keypoint-based techniques would reveal the copy-moved regions in the test image.

This paper will focus on a keypoint-based method of identifying copy-moved regions in the test image. The proposed method makes use of the Speeded-Up Robust Features (SURF) algorithm, introduced by Bay et al. [1] for computer vision applications, with k-nearest neighbor (kNN) mapping to yield improved matching of keypoints in the test image. The SURF algorithm has been proven as a technique which is lower in computational costs compared to the widely accepted Scale-Invariant Feature Transform (SIFT) algorithm which has been frequently used in keypoint-based techniques for copy move forgery detection [1]. But the state of the art SIFT based methods are often quite heavy on computational cost and need to be modified to be able to handle images of larger sizes. On the other hand, the SURF technique is just as robust as the SIFT technique and even outperforms SIFT in terms of reproducibility and extracting distinctive features, making it a more intriguing technique for researchers in the computer vision and information sciences domains. The proposed method exploits the robustness of the SURF algorithm and uses the kNN mapping algorithm to improve the detection of copy-moved regions in the test image.

The remaining sections of the paper are organized as follows. The second section will briefly highlight the various works and techniques which have already been introduced to detect copy-move forgery in digital images. The third section will explain the proposed method in detail. Section four will display the results of the proposed technique. And lastly, the fifth section serves as the conclusion for the paper.

2   Literature Survey

The existing method of copy move forgery detection can be categorized as either block-based techniques or keypoint-based techniques as stated in the introduction. This section will give the readers an overview of the existing methods in each category and state the merits and demerits of each method.

2.1   Block-Based Techniques

The block-based technique was introduced by Fridrich et al. [4] whose work involved the detection of copy-move forgery by dividing the test image into overlapping blocks or tiles of fixed dimensions and then implemented discrete cosine transform (DCT) to distinguish each block. The matching was then performed using lexicographic sorting which ensured reduced computational complexity. Following this work, several researchers further developed the technique and introduced modifications such as intensity-based methods [5], frequency-domain based methods [6], and moment-based methods to distinguish each block [7]. In order to reduce the time taken in computation for matching the blocks, Cozzolino et al. [8] proposed an adaptation of the Patch-Match algorithm. Even though these techniques yield improved detection, they are associated with high computational cost and lack robustness when scaling and rotational transformations are involved.

2.2   Keypoint-Based Techniques

In contrast to the block-based techniques, the keypoint-based techniques are robust to scaling operation, rotational transformations, and blurring. As a result, keypoint-based techniques can be easily adapted to detect copy-move forgeries in digital images. The Scale-Invariant Feature Transform (SIFT) introduced by Lowe [13] has been a widely accepted keypoint-based algorithm to detect copy-move forgeries. Using the SIFT algorithm to detect image forgeries was introduced by Huang et al. [9]. In order to enable better matching, Amerini et al. [3] used the generalized 2-Nearest Neighbor (g2NN) matching method, agglomerative hierarchical clustering was then used to make clusters of the matched pairs which resolved the issue of detection problem of the multiple cloned regions. The random sample consensus (RANSAC) algorithm was used by Pan et al. [10] to estimate the affine transformation matrix among the original and copy-moved regions, the duplicated regions were determined by using region correlation mapping. The two techniques mentioned previously were combined by Christlein et al. [2] whose work established a foundation for the existing keypoint-based methods: feature extraction followed by keypoint matching, clustering matches, estimation of affine transformations (which involves ensuring rotation invariance), and visually distinguishing the copy-moved regions.

Most of the research work, following Christlein's work [2], involves improving the processes of feature extraction and forming clusters. For example, SURF has been used as an alternate method to extract keypoints from the test image. Another example involves the combination of Harris Corner and SIFT descriptors. But the use of Harris Corner yields lower robustness to scaling operations [9]. On the other hand, researchers have also experimented with extracting keypoints from different color spaces. For example, Jin et al. [14] has proposed a method to use Opponent-SIFT and extracting keypoints based on smaller regions from the test image in order to successfully detect smooth-tampered regions. Opponent-SURF has also been used to extract feature vectors from the opponent color space of the test image. In order to improve the process of clustering so as to account for copy move regions which are spatially close to the original region, researchers have used the J-Linkage algorithm [15]. But as the time complexity of the J-Linkage algorithm is related to the number of matched pairs by a quadratic function, the time taken in the process of clustering increases by a great degree as the number of matched pairs increases [15].

More recently, researchers have also proposed techniques which combine block-based and keypoint-based methods to detect copy-move forgeries. For example, the method proposed by Silva et al. [11] focuses on using multi-scale analysis followed by a voting process to determine the susceptible regions using SURF feature matching and clustering. The block-based method was then used to detect the tampered regions in the multi-scale space. Another example can be observed from the work of Li et al. [12] who have implemented image segmentation; where the regions which are suspected of copy-move forgery are determined and then an iterative nearest neighbor algorithm is used to continually improve the precision of the detected regions. But these methods involve high computational cost and are not feasible for larger images.

As it is evident from the preceding paragraphs, the keypoint-based techniques yield more accurate results when modifications are made in the process of extracting key-points and clustering the keypoints to localize the copy-moved regions in the test image. However, these modification often increase the computational cost of the algorithm and thereby makes the method impractical for larger images. This paper proposes a keypoint-based method for copy-move forgery detection which involves low computational cost while ensuring robustness in detecting the copy-moved regions in the test image. The lower computational cost of the SURF algorithm ensures that the proposed method can be easily adapted to detect copy-move forgeries in larger images when compared to the SIFT techniques. Additionally, the proposed method makes use of kNN mapping to match the keypoints to ensure lower numbers of false matches.

3 Proposed Method

This paper proposes a method for copy move forgery detection which is robust and involves lower computational complexity, making this method more feasible for the detection of copy-move forgery in larger images. Additionally, the computational cost involved in clustering is bypassed by using the kNN mapping algorithm. The proposed method can be viewed in two main steps. The first step is the extraction of keypoins from the entire test image and its suspected regions using the SURF algorithm [1]. The second step is the matching and classification of keypoints using kNN mapping.

3.1 Keypoint Extraction Using SURF

Before executing the SURF algorithm to extract keypoints, the suspected region is isolated by cropping it out from the test image. This may be more time consuming and may seem ad-hoc for the user but it ensures that more number of keypoints are extracted from the suspected region as compared to having the SURF algorithm to extract keypoints from the entire test image and then matching the keypoints in the same image. The increase in the number of extracted keypoints from the suspected region also ensures higher accuracy in detecting copy-moved regions of smaller scale in the test image. The isolated suspected region and the test images are converted from color to grayscale as color information is not required in the accurate extraction and matching of the descriptors. Intensity and contrast are the primary deciding factors for keypoint extraction. As a result, including the color information in the SURF algorithm would only make the process more computationally complex for the SURF algorithm.

Following the isolation of the suspected region from the test image, the second stage of the method involves the extraction of keypoints from the suspected region and the test image using the Speeded-Up Robust Features (SURF) algorithm introduced by Bay et al. [1] The SURF algorithm defines a process of extracting keypoints and calculating descriptors. The keypoints are acquired by using a Fast-Hessian Detector. After successfully extracting the keypoints, descriptors are generated for each keypoint. These descriptors are then matched using the kNN algorithm. The following paragraphs will briefly explain the SURF algorithm.

Fast Detection of Keypoints
The process of extracting the keypoints from the test image and the suspected region begins with acquiring the Hessian matrices. The Hessian matrix, $H(p, \sigma)$, at point p with the coordinates (x,y) in the image with the scale σ is defined in the Eq. 1.

$$H(p, \sigma) = \begin{bmatrix} L_{xx}(p, \sigma) L_{xy}(p, \sigma) \\ L_{xy}(p, \sigma) L_{yy}(p, \sigma) \end{bmatrix} \tag{1}$$

In the above equation, L_{xx}, L_{xy}, and L_{yy} are the convolution of the Gaussian-Second Order Derivative with the image in the point p over x, x and y, and y, respectively. This involves a great degree of complexity but an approximation of the Gaussian-Second Order Derivative is achieved with the help of box filters: D_{xx}, D_{xy}, and D_{yy} over x, x and y, and y respectively. The illustrations of the Gaussian Partial Derivative in y and xy-direction and the approximation for the second order Gaussian partial derivative in y-direction (D_{yy}) and xy-direction (D_{xy}) are seen in Fig. 1. Integral images are used to calculate the box filters; integral images are capable of calculating the sum of intensities inside any rectangular dimension in a fixed time, independent of its size. The integral images enable the fast computation of box type convolution filter [1]. The entry of an integral image at a specific location $x = (x,y)^T$ represents the sum of all pixels in the image I within a rectangular dimension formed by the origin and x.

$$I\sum(x) = \sum_{i=0}^{i \leq x} \left(\sum_{j=0}^{j \leq y} I(i,j)\right) \tag{2}$$

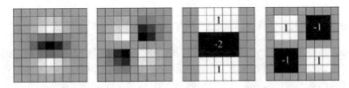

Fig. 1. From left to right: The illustration (discretised and cropped) of Gaussian-Second Order Partial Derivative in y-(Lyy), xy-direction (Lxy); the approximation using box filters of the second order Gaussian partial derivative in y-(Dyy), and xy-direction (Dxy). The grey regions are equated to zero values in the box filters.

Therefore, once the integral image is acquired, only three additions are needed to calculate the sum of intensities over any upright, rectangular area. As a result, the calculation time is not dependent of its size [1]. Consequently, this enables for big filter sizes which can work on larger images as well. The weights of the box filter can be chosen such that it adequately estimates the different intensities of the image. This also allows for a good estimation of the Hessian Determinant as shown in the equation below.

$$det\left(H_{approx}\right) = D_{xx}D_{yy} - \left(0.9D_{xy}\right)^2 \tag{3}$$

The determinant of the Hessian matrix helps in selecting the location and scale of the keypoints. As a final step of the extraction of keypoints, a non-maximum suppression in a fixed neighborhood is applied to localize keypoints in the image. To acquire a more complete idea on the keypoint extraction in the SURF algorithm, one can refer to the work by Bay et al. [1].

Keypoint Descriptors
Once the keypoints have been extracted from the image, the keypoints are assigned a dominant orientation in order to render rotation invariance to the image. For each keypoint, the dominant rotation is estimated by the summation of the horizontal and vertical Haar wavelet responses within a circular sector which has a span of $\pi/3$ radian in the wavelet response space. This process determines the maximum which is used to describe the orientation of the keyoint's descriptor. The keypoint descriptors are then constructed by defining square regions around the keypoints and are oriented according

to the dominant orientation. Each of these square regions are divided into 4 by 4 sub-sections and each of these sub-sections are defined by a vector quantity V and given below [1].

$$V = \left[\sum dx, \sum dy, \sum |dx|, \sum |dy| \right] \qquad (4)$$

In the equation, $\sum dx$ is the sum of the wavelet responses in the horizontal direction, and $\sum dy$ is a similar summation for the vertical direction. Additionally, the absolute value summations in the x and y direction are used to acquire information about the polarity of the image intensity changes. The final keypoint descriptor vector for all the 4 by 4 sub-sections is of length 64.

3.2 Matching of Keypoints Using kNN Mapping

The matching is done using the k-Nearest Neighbor (kNN) Mapping algorithm which is commonly used as a method of classifying new or unknown data-points to the existing classified data-points. This algorithm is a comparatively faster method of classifying the keypoint descriptors of the image while ensuring accuracy in matching the keypoints. As the classification is done in comparison to an existing data-set, training is a crucial step for the algorithm. The keypoint descriptors extracted from the test image are used as the test data or training data to which the unclassified keypoint-descriptors are compared to for successful matching.

Another factor which is crucial for ensuring successful classification of the key-point descriptors is the value of k. The parameter k denotes the number of nearest neighbors to be considered in classifying the keypoint descriptor. For example, k = 2 will compare the unmatched keypoint descriptor to two of the nearest keypoint descriptors in the training set, which can also be considered as the number of neighbors from which the algorithm takes the vote from to determine if the keypoint descriptor from the suspected region can be matched to those in the complete test image. A successful match is determined when the keypoints from the test image and its k-neighbors are matched to that of the suspected region. Upon successful matching, the algorithm will then highlight the region with matching keypoint.

4   Results and Discussion

The proposed method was implemented using the OpenCV platform. The performance of the algorithm was tested with the MICC-F8 Dataset which consists of a set of images which involve copy move forgeries in each image. The copy moved regions have been scaled and rotated to other parts of the same image. In this implementation, a k-value of 3 is used in the kNN mapping process to yield successful matches. The results of the OpenCV program execution results for two images from the image dataset are displayed here. The isolated suspected region from the test image are shown in Figs. 2(a) and (b). The extracted keypoints for each of these suspected regions are shown with highlighted points in Figs. 3(a) and (b). Figures 4(a) and (b) display the input test image to the algorithm. Lastly, Figs. 5(a) and (b) display the final output of the algorithm which is the test image with the matched keypoints.

(a)   **(b)**

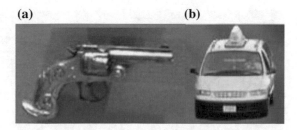

Fig. 2. (a, b) The isolated suspected copy-moved regions from each test image.

(a)   **(b)**

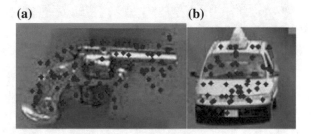

Fig. 3. (a, b) The extracted keypoints from the isolated suspected region.

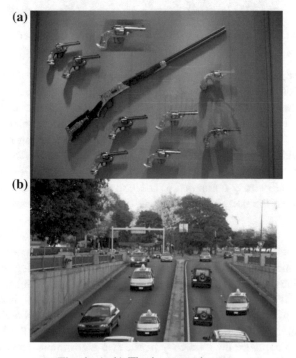

(a)

(b)

Fig. 4. (a, b) The input test images.

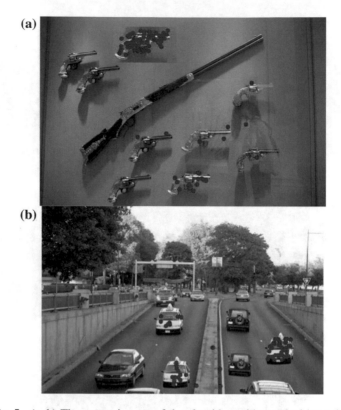

Fig. 5. (a, b) The output images of the algorithm with matched keypoints.

From the output images (Figs. 5(a) and (b)), the reader can also notice the high spatial density of matched keypoints in the copy-moved regions. The copy-moved region which was isolated from the test image has the highest density of matching keypoints in the test image whereas the copy-moved regions which differ in scale and rotation from the isolated region have a lower density of matching keypoints. But the density is high enough to conclude that the region is a copy-moved region. The outliers found in the result images are successful keypoint matches but these matches are of much lower spatial density compared to the copy-moved regions. Therefore, one can visually distinguish the copy-moved regions by observing the density of the matched keypoints.

In order to highlight the improved performance of the proposed method over the SIFT algorithm, we have executed the SIFT algorithm on the same test images. Similar to the proposed method from the previous section, the suspected region for copy-move forgery was isolated from the test image to ensure that optimal number of keypoints are extracted. The SIFT algorithm also involves converting the color image to grayscale prior to extracting keypoints from the test image and the suspected region. The Hessian Threshold is maintained at the same value for both SIFT and SURF implementations; this enables us to compare the efficiency of the two algorithms. The images below are the output windows of the SIFT algorithm on the same OpenCV platform. Please note that, the SIFT algorithm implemented here does not involve hierarchical clustering or extraction of keypoints based on region. This is done in order to serve as a more direct comparison of the proposed method to the SIFT algorithm. While techniques which involve region based extraction and hierarchical clustering yield improved results compared to the pure SIFT implementation, these steps are associated with increasing the computational cost of the algorithm. Additionally, the SIFT algorithm has already been distinguished as the one involving higher computational cost compared to the SURF algorithm. As a result, computational cost is compromised and the algorithm is rendered ineffective for larger images. Paulo et al. [16] has proven the low computational cost of the SURF algorithm over the SIFT algorithm. The results of his work displayed that the time taken for both processes of keypoint detection and generating keypoint descriptors are significantly lower for the SURF algorithm. Moreover, as the number of keypoints increases, the time taken for execution also increases for both algorithms but SIFT still takes more time for obtaining descriptors.

The output of the SIFT implementation (Figs. 6(a) and (b)) reveals that fewer keypoints are extracted from the isolated suspected region compared to the proposed method. The reader would also observe that algorithm yields fewer matches on the test image compared to the proposed method. Neither output window of the SIFT implementation shows more than four keypoint matches for the copy-moved regions. These factors would make the pure SIFT algorithm more ineffective when detecting copy-moved regions which are made in much smaller scale. Table 1 gives a comparison of

Fig. 6. (a, b) The output windows of the SIFT algorithm implementation. The isolated suspected region is seen at the top-left corner of each output window. The successful matches of keypoints are illustrated with the colored lines.

the performance parameters of the proposed algorithm as well as the SIFT implementation. From these values, it can be noted that the proposed algorithm yields better results compared to the SIFT implementation.

Table 1. Comparing results of SURF and SIFT extractors

Image	Image of guns	Image of traffic
Number of Keypoints extracted from isolated suspected region	SIFT: 21 SURF: 70	SIFT: 27 SURF: 58
Average number of successful Keypoint matches with the test image (excluding the isolated region)	SIFT: 4 SURF: 12	SIFT: 4 SURF: 11
Number of False Keypoint Matches	SIFT: 11 SURF: 9	SIFT: 13 SURF: 4

5 Conclusion

The proposed algorithm has proven to be an accurate method for distinguishing the copy-moved regions in the test image. To summarize, the method involved two major steps: the extraction of keypoints with the help of the SURF algorithm and the matching of the keypoint descriptors using the kNN mapping algorithm. Using the SURF algorithm in order to detect copy move forgery doesn't only ensure invariance to scaling and rotational operations but also ensures reduced computational cost, making the algorithm more suitable for larger images. As copy move forgery can be easily implemented on much larger images, it is crucial that the algorithm would be able to detect the copy-moved regions with acceptable accuracy and reduced computational cost to meet the demands of the future. The results obtained from comparing the proposed method against the standard SIFT implementation proves that the proposed method is capable of extracting a greater number of keypoints from the suspected regions, yields more successful keypoint matches, and results with fewer false matches compared to the SIFT algorithm.

References

1. Bay, H., Ess, A., Tutyelaars, T., Van Gool, L.: Speeded-up robust features (SURF). Comput. Vis. Image Underst. **110**, 346–359 (2008)
2. Christlein, V., Riess, C., Jordan, J., Angelopoulou, E.: An evaluation of popular copy-move forgery detection approaches. IEEE Trans. Inf. Forensics Secur. **7**(6), 1841–1854 (2012)
3. Amerini, I., Ballan, L., Caldelli, R., Del Bimbo, A., Serra, G.: A SIFT-based forensic method for copy move attack detection and transformation recovery. IEEE Trans. Inf. Forensics Secur. **6**(3), 1099–1110 (2011)
4. Fridrich, A.J., Soukal, B.D., Lukáš, A.J.: Detection of copy-move forgery in digital images. In: Digital Forensic Research Workshop, DFRWS. Citeseer (2003)

5. Wang, J., Liu, G., Li, H., Dai, Y., Wang, Z.: Detection of image region duplication forgery using model with circle block. In: International Conference on Multimedia Information Networking and Security, Hubei, China. MINES, vol. 1, pp. 25–29. IEEE (2009)
6. Bayram, S., Sencar, H.T., Memon, N.: An efficient and robust method for detecting copy-move forgery. In: International Conference on Acoustics, Speech and Signal Processing, ICASSP, Taipei, Taiwan, pp. 1053–1056. IEEE (2009)
7. Ryu, S.-J., Lee, M.-J., Lee, H.-K.: Detection of copy-rotate-move forgery using Zernike moments. In: Information Hiding, pp. 51–65. Springer, Heidelberg (2010)
8. Cozzolino, D., Poggi, G., Verdoliva, L.: Copy-move forgery detection based on patch match. In: International Conference on Image Processing, ICIP, Paris, France, pp. 5312–5316. IEEE (2014)
9. Huang, H., Guo, W., Zhang, Y.: Detection of copy-move forgery in digital images using SIFT algorithm. In: Pacific-Asia Workshop on Computational Intelligence and Industrial Application, Wuhan, China. PACIIA, vol. 2, pp. 272–276. IEEE (2008)
10. Pan, X., Lyu, S.: Region duplication detection using image feature matching. IEEE Trans. Inf. Forensics Secur. 5(4), 857–867 (2010)
11. Silva, E., Carvalho, T., Ferreira, A., Rocha, A.: Going deeper into copy-move forgery detection: exploring image telltales via multi-scale analysis and voting processes. J. Vis. Commun. Image Represent. 29, 16–32 (2015)
12. Li, J., Li, X., Yang, B., Sun, X.: Segmentation-based image copy-move forgery detection scheme. IEEE Trans. Inf. Forensics Secur. 10(3), 507–518 (2015)
13. Lowe, D.G.: Distinctive image features from scale-invariant keypoints. Int. J. Comput. Vis. 60(2), 91–110 (2004)
14. Jin, G., Wan, X.: An Improved method for SIFT-based copy-move forgery detection using non-maximum value suppression and optimized J-Linkage. Signal Process. Image Commun. 57, 113–125 (2017)
15. Amerini, I., Ballan, L., Caldelli, R., Del Bimbo, A., Del Tongo, L., Serra, G.: Copy-move forgery detection and localization by means of robust clustering with J-Linkage. Signal Process. Image Commun. 28(6), 659–669 (2013)
16. Drews Jr., P., de Bem, R., de Melo, A.: Analyzing and exploring feature detectors in images (2011)

A Preliminary Approach to Using PRNU Based Transfer Learning for Camera Identification

Sharan Seshadri[1], K. R. Akshatha[1]([✉]), A. K. Karunakar[2],
and Kelvin Harrison Paul[1]

[1] Department of Electronics and Communication Engineering,
Manipal Institute of Technology, Manipal Academy of Higher Education,
Manipal 576104, Karnataka, India
sharan2510@gmail.com, akshatha.kr@manipal.edu,
kelvin.harrison@learner.manipal.edu
[2] Department of Computer Applications, Manipal Institute of Technology,
Manipal Academy of Higher Education, Manipal 576104, Karnataka, India
karunakar.ak@manipal.edu

Abstract. In this paper, a preliminary method of source camera identification using transfer learning methods is proposed. Every image taken by a camera has a unique artefact, known as photo response non-uniformity (PRNU) noise that is manifested in the image, because of irregularities during the manufacturing process of the image sensor present in that camera. A convolutional neural network (CNN) is used here to extract the features pertaining to the noise residual of a specific camera, which can be later passed to a classifier. In this paper, we use a CNN architecture that has weights that have been pre-trained to extract the residual features. This method is called transfer learning and is used after denoising the images in the dataset. The noise residual features are extracted by the transfer learning model and are then passed to a Support Vector Machine (SVM) classifier. Despite the simplicity of this approach, the results obtained are favorable compared to the existing methods, especially considering the small dataset used.

Keywords: Image forensics · Convolutional neural networks · Transfer learning · Deep learning · PRNU noise · Source camera identification

1 Introduction

In today's increasingly connected world, visual data such as images can reach millions of people at a time. With the increase in importance of social media in our lives and the boom in the number of devices used for such data acquisition, the methods for duplicating, forging, or in general, falsifying such images are also becoming increasingly advanced.

In response to these problems, the scientific community has been working on techniques to investigate the authenticity of digital images [1, 2]. The collective name given to the techniques used for verifying the veracity of digital images is called Digital Image Forensics. An important subclass of digital image forensics is Source Camera

© Springer Nature Switzerland AG 2020
K. Arai and S. Kapoor (Eds.): CVC 2019, AISC 944, pp. 246–255, 2020.
https://doi.org/10.1007/978-3-030-17798-0_21

Identification. This deals with the process of identifying the camera that has been used to capture an image or a set of images. The implications of this field are far reaching in security and legal applications as evidence. Sophisticated techniques can also be used to pinpoint creators or forgers of illegal material such as terrorism scenes, child pornography etc. Over the years there have been several approaches for source camera identification, all of which aim to obtain a unique fingerprint for correlating each image to its source camera.

1.1   Lens Aberration-Based Methods

The main focus of lens aberration based methods, primarily researched by Choi et al. [3], is to take advantage of the fact that camera manufacturers design lens systems that have unique radial distortion. This approach fails to work if there is no straight line in the image as the distortion is measure using the straight line method.

1.2   Color Filter Array-Based Methods

The color filter array is an important part of a digital camera, whose role is to filter the incoming light before it reaches the sensor. The demosaicing process, which is one of the functions of the color filter, leaves a unique fingerprint that can be detected. Bayram et al. [4] worked on a method to use the color filter array interpolation process to determine a correlation structure present in each color band for image classification. This method fails because cameras of the same model mostly share the same CFA filter pattern and the interpolation algorithm. This method is also does not work for compressed images.

1.3   Photo Response Non Uniformity-Based Approaches

Lukas et al. [5] based their approach on Photo Response Non-Uniformity (PRNU) noise. This is an image artifact that is the product of the camera's image sensor deformity. At that time, Lukas' work failed to predict the source camera of cropped images, as the sensor pattern noise estimation was distorted. This deformity is unique to each camera, as it is brought about during the manufacturing process. This approach finds a way to obtain the image noise brought about by the deformity, and feed this to a correlation-based model to identify the source camera.

1.4   Feature-Based Approaches

Another class of source camera identification methods that came into existence were those based on features of the images being considered. The first approach by Kharrazi et al. [6] dealt with identifying what components the image was affected by and using these as a basis to obtain features. Gloe et al. [7] built on Kharrazi's work and developed a method to identify more relevant features. Celikutan et al. [8] used a small part of Kharrazi's features and used binary similarity measures to classify the images accordingly, but their approach failed with increasing number of cameras.

These methods required the features to be extracted manually before being passed to the classifier. The disadvantage with this approach is that it is difficult to judge if the manual features identified are accurate representations of the image. Thus, we use unsupervised learning to automatically identify features that best represent the image so that we can use these to classify the image correctly.

2 Literature Review

2.1 Using Pre-determined Features

Kharrazi et al. [6] were among the first researchers to work on feature-based approaches for source camera identification. Their initial approach was to use a pre-determined set of features for each image. They identified that the image was affected by two components:

(a) Colour filter array demosaicing algorithm
(b) Colour processing of the image

Based on the color processing criteria, the signal bands of the RGB components exhibit certain traits regardless of the camera used. They defined a set of features that governed the denoised image, which were then to a support vector machine (SVM) classifier to get trained. Two cameras were initially used for the training of the algorithm. This method achieved high rates of accuracy with 2 cameras and 300 images. However, it was noticed that with the increase in cameras, the accuracy rate decreases.

Gloe et al. [7] used Kharrazi's feature sets with extended color features to identify camera models. One of these extended features was wavelet statistics, where Gloe used the Daubechies-8 wavelet to obtain wavelet statistics of each color channel. Based on the feature sets extracted, they further employed feature selection using a sequential forward floating search method for obtaining the most relevant features.

The number of cameras used in this approach number around 15, which was more than that used in the previous approach. Here, due to the use of the extended wavelet statistics and the feature selection metric, the change in accuracy was negligible with the increase in the number of cameras used. The features extracted are then passed to a support vector machine which classifies each feature point with the appropriate source camera. One additional study done here was the comparison between different camera models, and different devices of the same model. It was observed that the limitation of this method was that it could accurately predict inter-camera model differences, but not intra-model differences.

2.2 Using Binary Similarity Measures

For the task of identifying cell phone source cameras, Celikutan et al. [8] implemented a set of measures that assumed that CFA interpolation leaves unique correlations across the bit planes that can be represented by these measures. 108 binary similarity measures were obtained, and the end goal of this approach was to measure the similarity between the bit planes of the image. This method, however, does not work well with an increase in the number of cameras used.

2.3   Using Convolutional Neural Networks

Tuama et al. [9] implemented what was one of the preliminary approaches that used a convolutional neural network (CNN) [10, 11] to extract the relevant features from the images. The approach used by Tuama et al. [9] makes use of a valuable property that is used to identify cameras based on their Photo-Response non-uniformity noise (PRNU), as written about by Lukas et al. The image is denoised using a famous wavelet denoising filter, developed by Mihcak et al. [12]. The PRNU components for each image are then passed to a CNN for feature extraction. The CNN then obtains the features and passes it on to the Support Vector machine classifier. In addition to these, the authors developed their own network architecture, the results of which were comparable to the existing CNN architectures, but a little more inefficient. The limitations of this problem include the large dataset required per camera to evaluate the features. In addition to the dataset, the computation time required is extremely large and would require industrial grade graphics processing units, which may be expensive to purchase.

Bondi et al. [13] also used a CNN for feature extraction. However, these authors propose a more efficient approach, which does not require as much data as the previous existing approaches. The idea used here is to split the images into patches and extract relevant features from the patches. The classifier used here is also several one versus all SVMs, the results of which are then passed to a majority voting function. This does not affect the performance of the CNN, as the hypothesis is that in different patches of different images in same cameras, there will be unique identifying features.

3   Proposed Approach

The objective of the proposed approach is twofold; the first is to lessen the influence of the dataset used, without much compromise in the accuracy of identification and the second is to reduce the effort used in training the dataset by implementing a pre-trained convolutional neural network. To reduce the dataset creation effort and to optimize the computation required for such a data driven approach, transfer learning [14] is used here. This is a method where the filter weights of the convolutional neural network are predetermined, and only the fully connected layers are changed based on the local identification problem at hand.

The algorithm described comprises of three parts, which are explained in more detail in the following sub-sections:

(a) Wavelet denoising. In this section, the PRNU noise component is extracted with the use of an empirically designed filter by Mihcak et al. [12].
(b) Feature extraction using Transfer Learning. The noise residuals obtained from the denoising process are then analysed using a pre-trained CNN which extracts features that contribute significantly to the image identification process.
(c) Support vector machine classifier to map the features obtained to the camera model.

3.1   Wavelet-Based Denoising

The method used to obtain the PRNU here is the wavelet-denoising based approach used by Lukas et al. [5] and developed by Mihcak et al. [12]. This wavelet denoising sequence is illustrated in Fig. 1. This approach uses a spatially adaptive statistical model for the wavelet image representation. The value taken for the noise variance here is 5, and the wavelet used is the Daubechies-8 wavelet. The coefficients containing the high frequency content, where the PRNU is located, can be modelled using a union of a stationary Gaussian noise signal and a local stationary independent identical signal, with a zero mean.

Fig. 1. Wavelet denoising sequence. These steps are repeated for sub-bands of each decomposition.

This wavelet based pre-processing method is implemented in 2 steps:

(a) The image variance is first estimated
(b) A Weiner filter is used for the denoised image estimation

This is used to extract a significant portion of the PRNU from each test image. The PRNU pattern obtained for each image are then passed to the transfer learning algorithm.

3.2   Transfer Learning

Conventional machine learning approaches to source camera identification have made use of convolutional neural networks, for their ability to effectively extract the distinguishing features from an image and representing it as a vector.

There are several layers in a CNN, including the pooling layer, the activation layer, and the softmax layer, all of which have different functions. The convolutional filter values used in a CNN are required to be trained from scratch with the images used. One major obstacle faced while using a data-driven machine learning based approach to source camera identification is the need for a large, versatile dataset. In the approach most recently used by Bondi et al. [9], nearly 1000 shots from each camera model were required to train the convolutional neural network (CNN) to obtain satisfactory results.

The idea behind transfer learning is that the earlier convolution layers need not be changed as they involve extracting generic features, but the later layers are trained to extract features specifically pertaining to the classes contained in the original dataset. The transfer learning model used here is the well-known AlexNet, initially implemented by Krizhevsky et al. [15], and this is trained on the ImageNet dataset. AlexNet is used on the noise residuals after the wavelet transform based pre-processing, and this leads to unsupervised feature extraction.

3.3   Support Vector Machine Classifier

The noise residual features that are extracted by the transfer learning model are then passed to a Support Vector Machine (SVM) Classifier, which is trained based on the features extracted and the class of the camera.

4   Experiment Setting

4.1   Source Cameras

In this preliminary study, four cameras have been used, with around 300 images per camera. The dataset from where the camera specific images have been downloaded is the Dresden Image Database [16]. A scene is defined here as the combination of both a location and a viewpoint. Here, 12 of the same scenes have been used for each camera, to make the training of the dataset less prone to dependence on the scene itself, but on the pattern of the PRNU noise.

This initial approach is extended to four cameras-Casio EXZ150, Kodak M1063, Olympus MJU, and the Praktica DCZ 5.9. The details of each camera have been elaborated in Table 1.

Table 1. Camera details

Camera model	Resolution	Sensor size	Focal length (mm)
Casio EXZ150	3264 × 2448	1/2.5"	5.8–17.4
Kodak M1063	3664 × 2748	1/2.33"	5.7–17.1
Olympus MJU	3648 × 2736	1/2.33"	6.7–20.1
Praktica DCZ 5.9	2560 × 1920	1/2.5"	5.4–16.2

4.2   Transfer Learning Model

The model used is AlexNet [15]. As shown in Fig. 2, this model consists of 5 convolutional layers and 3 fully connected layers, and it was a hugely popular model developed in 2012. The networks architecture is such that the initial layers have weights that have been pre-trained for other multiple image classes.

The model has been pretrained on the extensively used ImageNet Dataset, making it useful for feature extraction tasks in this scenario. A block diagram-based illustration of transfer learning has been shown in Fig. 3. The only training that is required is the SVM classifier, which is trained based on the features extracted and the output cameras.

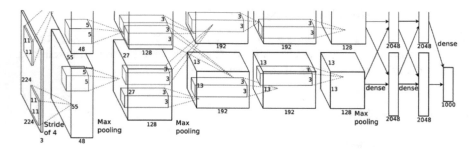

Fig. 2. AlexNet architecture

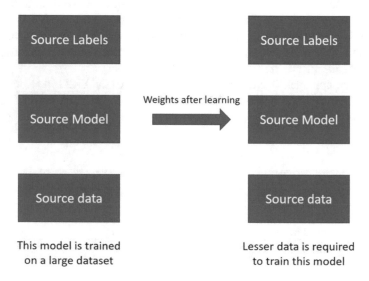

Fig. 3. Transfer learning block diagram

5 Evaluations

The dataset of images is divided into training and testing data, of which 70% is used to train the model and 30% is used to test the model. Figure 4 shows the comparison between images of two cameras, the Casio EX-Z150 and the Kodak M1063. About 80 images per camera are used to test the model, and the results are depicted in Table 2. Overfitting is handled by implementing dropout after the first and second fully connected layers.

The overall results after running this model is 97.97%, with 807 training labels and 345 test labels. The transfer learning algorithm extracted 4096 features which were then passed to the classifier for identification of the source camera.

a) Casio EX-Z150   b) Kodak M1063

Fig. 4. Same scenes captured by the Kodak and the Casio cameras

Table 2. Output results

Actual labels	Predicted labels	Praktica DCZ 5.9	Olympus MJU	Kodak M1063	Casio EXZ150
	Praktica DCZ 5.9	80	0	0	0
	Olympus MJU	1	87	0	0
	Kodak M1063	0	0	90	0
	Casio EXZ150	4	2	0	81

5.1 Discussion

Data driven approaches to source camera identification have been increasing, but the use of a pre-trained network to identify the models is an idea that is new. In comparison to all previous feature extraction-based methods, it is shown that data driven methods generalize much better and more accurately. Earlier approaches like the ones used by Tuama et al. [9], and Bondi et al. [13] require the network to be trained from scratch for feature extraction. The main contribution of this approach is the implementation of a scheme that has a pre-trained convolutional neural network through transfer learning of the AlexNet network, and passing the features extracted by this algorithm to a support vector machine classifier.

Though the accuracy rates are high in both previous approaches, we achieve a similar accuracy metric using a pre-trained network which extracts PRNU features. This significantly reduces computation costs that would have otherwise been extremely high for training the convolutional neural network for feature extraction from scratch, and also reduces the effort taken to obtain the dataset, as this approach is able to generalize well for four cameras without an extensive data set of training images.

6 Conclusion and Future Work

The use of PRNU based denoising and classification is not a new one, as there have been several past approaches to working on this type of image fingerprint extraction. Implementing a data-driven approach was also initially seen using a convolutional neural network, as done by Tuama et al. [9]. However, while their method required the feature extraction network to be trained from scratch, we use an already pre-trained network to reduce the training time and the computation complexity. The only part of the pipeline that is going through training is the support vector machine, which is trained to classify the extracted PRNU pattern with the respective camera.

Future work could include design of a more extensive and sophisticated neural network for feature extraction, though it remains to be seen how much of a trade-off this would be in terms of computing ability. The PRNU extraction method used here is based on the initial approach by Lukas et al. [5], so work in developing an algorithm that enables a more accurate extraction of the PRNU fingerprint could be done.

References

1. Stamm, M.C., Wu, M., Liu, K.J.R.: Information forensics: an overview of the first decade. IEEE Access **1**, 167–200 (2013)
2. Piva, A.: An overview on image forensics. ISRN Signal Process. **2013**, 22 (2013)
3. Choi, K., Lam, E., Wong, K.: Source camera identification using footprints from lens aberration. In: Proceedings of SPIE, Digital Photography II, vol. 6069, no. 1, p. 60690J (2006)
4. Bayram, S., Sencar, H., Memon, N.: Improvements on source camera model identification based on CFA interpolation. In: Advances in Digital Forensics II, IFIP International Conference on Digital Forensics, Orlando Florida, pp. 289–299 (2006)
5. Lukas, J., Fridrich, J., Goljan, M.: Digital camera identification from sensor pattern noise. IEEE Trans. Inf. Forensics Secur. **1**(2), 205–214 (2006)
6. Kharrazi, M., Sencar H., Memon, N.: Blind source camera identification. In: IEEE International Conference on Image Processing, ICIP 2004, vol. 1, pp. 709–712 (2004)
7. Gloe, T.: Feature-based forensic camera model identification. In: Shi, Y.Q., Katzenbeisser, S. (eds.) Transactions on Data Hiding and Multimedia Security VIII. LNCS, vol. 7228, pp. 42–62. Springer, Heidelberg (2012)
8. Celiktutan, O., Sankur, B., Avcibas, I.: Blind identification of source cell-phone model. IEEE Trans. Inf. Forensics Secur. **3**(3), 553–566 (2008)
9. Tuama, A., Comby, F., Chaumont, M.: Camera model identification based on machine learning approach with high order statistics features. In: Proceedings of the IEEE European Signal Processing Conference, Budapest, Hungary, pp. 1183–1187, August 2016
10. Le Cun, Y., Bengio, Y.: Convolutional networks for images, speech, and time series. In: Arbib, M.A. (ed.) The Handbook of Brain Theory and Neural Networks. MIT Press, Cambridge (1998)
11. Bengio, Y.: Learning deep architectures for AI. Found. Trends Mach. Learn. **2**, 1–127 (2009)

12. Mihcak, M.K., Kozintsev, I., Ramchandran, K.: Spatially adaptive statistical modeling of wavelet image coefficients and its application to denoising. In: Proceedings of IEEE International Conference on Acoustics, Speech, and Signal Processing, Phoenix, AZ, vol. 6, pp. 3253–3256, March 1999
13. Bondi, L., Baroffio, L., Guera, D., Bestagini, P., Delp, E.J., Tubaro, S.: First steps toward camera model identification with convolutional neural networks. IEEE Signal Process. Lett. **24**(3), 259–263 (2017)
14. Ben-David, S., Blitzer, J., Cramme, K., Kulesza, A., Pereira, F., Wortman Vaughan, J.: A theory of learning from different domains. Mach. Learn. **79**, 151 (2009)
15. Krizhevsky, A., Sutskever, I., Hinton, G.E.: Imagenet classification with deep convolutional neural networks. In: Advances in Neural Information Processing Systems, vol. 25, pp. 1097–1105. Curran Associates Inc. (2012)
16. Gloe, T., Böhme, R.: The Dresden image database for benchmarking digital image forensics. J. Digit. Forensic Pract. **3**(2–4), 150–159 (2010)

Accident Recognition via 3D CNNs for Automated Traffic Monitoring in Smart Cities

Mikhail Bortnikov[1], Adil Khan[1(✉)], Asad Masood Khattak[2], and Muhammad Ahmad[1]

[1] Innopolis University, Innopolis, Russia
{m.bortnikov,a.khan}@innopolis.ru, mahmad00@gmail.com
[2] Zayed University, Dubai, United Arab Emirates
asad.khattak@zu.ac.ae

Abstract. Automatic recognition of road accidents in traffic videos can improve road safety. Smart cities can deploy accident recognition systems to promote urban traffic safety and efficiency. This work reviews existing approaches for automatic accident detection and highlights a number of challenges that make accident detection a difficult task. Furthermore, we propose to implement a 3D Convolutional Neural Network (CNN) based accident detection system. We customize a video game to generate road traffic video data in a variety of weather and lighting conditions. The generated data is preprocessed using optical flow method and injected with noise to focus only on motion and introduce further variations in the data, respectively. The resulting data is used to train the model, which was then tested on real-life traffic videos from YouTube. The experiments demonstrate that the performance of the proposed algorithm is comparable to that of the existing models, but unlike them, it is not dependent on a large volume of real-life video data for training and does not require manual tuning of any thresholds.

Keywords: Machine learning · Deep learning · Computer vision · 3D convolutional neural networks · Accident recognition

1 Introduction

In recent years cities have become increasingly smart. A smart city is a combination of hard infrastructure such as roads and buildings with information technologies. Such a combination is considered as a driver for economic growth [2].

Smart surveillance algorithms are an important part of the smart cities and in particular Intelligent Transportation Systems (ITS). One of the tasks, the solution of which can significantly improve people's lives, is creation of automatic car accident recognition algorithm for traffic videos. Faster detection and processing information on car accidents can enable emergency services to arrive

© Springer Nature Switzerland AG 2020
K. Arai and S. Kapoor (Eds.): CVC 2019, AISC 944, pp. 256–264, 2020.
https://doi.org/10.1007/978-3-030-17798-0_22

to the accident site within the golden hour [7] and increase the chances to save lives.

An accident detection system will decrease the response time of emergency services. Ambulances will arrive faster and traffic jams due to the accident will be less harmful. More accurate accident statistics will be collected. Such an algorithm could also be applied on the data which have been recorded on the server, to increase the volume of the information available for the analysis in further training of more complex models using real data.

Unfortunately, car accident detection problem still does not have reliable solutions. The existing algorithms of classical computer vision are based on hand-crafted features, which are adapted for normal lighting condition and therefore work not so well with non-standard lighting. Figure 1 shows a frame with illumination noise that can make it difficult for an automated system to analyze the video. Furthermore, accidents usually happen on streets, which are often crowded and may present a strong change of weather conditions such as snow, rain, sun, and time of the day. Also, traditional approaches do not work well with shadows. For example, Fig. 2 shows the internal representation of a background selection system, where it wrongly took the shadows from the clouds for a moving object. Even under normal conditions it is not always possible to accurately recognize accidents as collisions can occur in different ways. These are some of the reasons that make automatic road accident detection a difficult task.

Fig. 1. Examples of challenges: glare from the cars headlights and dirty camera lens.

At the same time, videos are becoming ubiquitous on the internet, which is driving a rapid development of new algorithms that can automatically analyze and understand semantics hidden in videos. Since 2013, for the task of video recognition and understanding, the community began to use algorithms based on 3D convolutional neural networks (CNN) [5,6]. It is a type of neural network that can work directly with raw input data. It can extract both the spatial and the temporal features by performing 3D convolutions on sequences of frames. Since then, the approaches have not changed fundamentally.

Fig. 2. Background Subtraction System detect the shadow from the clouds as a moving object.

As mentioned earlier, recognizing accidents in real life traffic videos suffers from various problems, and traditional computer vision based algorithms do not generalize well to be used in different conditions. Given their excellent generalization ability, it is, therefore, reasonable to explore the use of CNNs for the task of accident detection in videos. However, such models usually require large amounts of training data. Unfortunately, there are no standard datasets of car accidents, and existing works have used their private datasets.

In this work, we provide a solution to this problem. Firstly, we use video games to generate training data under a variety of weather and scene conditions. Secondly, we implement a 3D CNN to train on the generated data and test it against real-life traffic videos from YouTube. The state-of-the-art research in action recognition in videos inspires our CNN. Thirdly, we test our CNN with two different loss functions, with/without optical flow to minimize the influence of illumination and background changes. Finally, we compare our results with a state-of-the-art traditional computer vision based accident recognition system to show that the proposed system works better even though it is solely training on computer generated data.

2 Related Works

Car accident detection is a part of the more general field of action recognition on videos. Therefore, we consider here how car accident detection is done in literature, and which state-of-the-art algorithms from action recognition can be applied to the accident detection.

Traditional computer vision algorithms are built around vehicle track analysis. Moving objects are detected on an image, then their track is recorded and from it some features are extracted and classified by hand-crafted rules or some simple machine learning models.

One of the best result with a recall of 0.91 was achieved by Maaloul et al. [8]. They created non-machine learning approach which needs less fine tuning and is more robust. As features for accident detection, they extract velocity, orientation, acceleration, vehicle area, and traffic trajectory. Next, they calculate several statistics and compare it with predetermined thresholds. Strength of their algorithm is in the use of Optical Flow (OF) pre-processing. It allows their method to ignore the background and lighting of the scene. The disadvantage of their algorithm is that it is necessary to manually configure thresholds for classification on each new camera installation location. Besides, there is no collision video from this camera to tune thresholds and evaluate performance.

Lack of the training data is usually solved using open data from YouTube [8]. However, YouTube videos vary a lot; only a few of them are taken from the correct angle, and most of them contain just normal traffic. As a result, training dataset becomes unbalanced. To overcome this limitation Aköoz, and Karsligil collected their own dataset using toy cars [1]. Total length of training videos is slightly more than 11 min. They achieved a recall of 0.85.

Idea of using OF pre-processing combined with Scale Invariant Feature Transformation (SIFT) was introduced by Chen et al. [3]. Usage of OF-SIFT allows to capture and extract dynamic motion information for motion detection without using the static state information of moving objects. Next, they apply the idea of Bag of Feature (BOF) to store spatial information. Finally, as a classifier they used Extreme Learning Machine (ELM) which is known for it's fast generalization. In different experiments they achieved a recall in the range of 0.5 to 1.0, which depends on the type of collision.

In recent years, the deep learning approach has been applied in more and more new areas, including ITS. Singh and Chalavadi used Stacked Autoencoders to detect road accidents [9]. They showed that convolutional auto-encoders for features representation outperforms the existing hand-crafted features based approaches.

3   Methodology

This section provides technical details of our use of CNNs for road accident recognition. We had a three-fold objective: first, to build a supervised 3D CNN model for accident recognition. Second, to generate a large road video dataset with diverse lighting conditions and varying surrounding infrastructure and environments to train the CNN. Thirdly, to improve the accuracy of the model by integrating an optical flow estimation into the proposed pipeline.

3.1   Convolutional Neural Networks

CNNs have been been shown to provide state-of-the-art results for action recognition [5,6]. Convolution is an operation of multiplying input data tensor by a convolutional kernel. Based on convolution operation it is possible to build multilayer neural network which can learn how to extract features directly from the data. Such a network is complemented with a densely connected layer with activation function to make class prediction.

3D convolutions allow to extract not only spatial, but also temporal features. To do so, it is needed to create an input tensor from several consecutive video frames. After end to end learning, model will be able to extract processes features in time, which includes car accidents in videos.

The final model in our work was found empirically, which consists of four 3d convolution layers alternating with max pooling and dropout layers. Max pooling is the operation of extracting small windows from the input feature maps and outputting the max value. Dropout is a technique to fight overfitting. Dropout randomly sets to zero a number of output features of the layer during the training. The number of convolutional filters doubles for each new layer, that is, it goes from 16 to 128. After all the convolutions and poolings, the time dimension of a tensor decreases from 15 to 1. Next, comes the flattening layer, whose output is finally fed to a dense layer to learn the classification model.

The performance of a machine learning model is linked with the choice of the loss function, which is a feedback signal for learning the weights of the model. Loss is usually calculated as the distance between ground-truth distribution and the model predictions. In this work, for the loss function, we used hinge loss. It is quite unusual because with CNNS it is typical to use categorical or binary cross-entropy loss. However, in our case, we believed that hinge loss would give better performance because of the binary type of the classification task.

As for the input to the network we used bags of short 15-frames clips, which were cut from the training videos.

3.2 Data Generation

Next comes the training and test data for fitting the model and its evaluation, respectively. As mentioned earlier, there are no particular datasets for car accident detection. So, as other researchers did, we also created our dataset. However, in our case, the data came from two sources: first is the training data that was generated using video game GTA V; whereas the test data came from YouTube videos with car accidents which we used to evaluate the performance of the network.

Video game GTA V provides real-life looking graphics, and has a rich variety of environments and realistic traffic models. Therefore, it can be used to generate the training data that we needed. To turn GTA into a traffic simulator, we added free-range camera mod, which gives the ability to set the point of view as we needed and also changed bot cars AI to make cars more or less prone to dangerous driving. Then the game screen was recorded and each part of the captured video was tagged with labels – containing an accident or not. For accidents, we considered a sequence of frames where the first frame is the frame on which it is already clear that the collision cannot be prevented, and the last frame is that when the cars have just stopped. In total, we collected and labeled 5 h of video generated from GTA V for training and YouTube for evaluation. Figure 3 shows a frame generated with GTA V.

Fig. 3. Frame from generated training video.

3.3   Performance Improvement

We add optical flow pre-processing to the network in order to increase its ability to extract temporal features and make it not sensitive to background and lighting changes. Optical flow is the idea of the extraction of vectors of pixels' movement from a video. As its output, we have a tensor which contains amplitudes and directions of each pixel's movement vector. Figure 4 shows a frame in optical flow and in raw data.

Fig. 4. Farneback optical flow on the left and Raw RGB image on the right.

In this work, we used Farneback [4] optical flow algorithm, which returns dense flow feature set, in comparison with the Lucas-Kanade algorithm, which returns only sparse feature set and therefore it is not suitable to capture fast movements such as an accident.

Finally, to further improve the accuracy, we examined the idea of adding LSTM (Long Short-Term Memory) layer and using noise injection. LSTM can extract the temporal information embedded in the data, which can further improve the accuracy of the accident detection. Noise injection works similar to data augmentation. It expands feature space and therefore helps to deal with overfitting.

4   Experiments and Results

In this work, we have used F1-score as the evaluation metric and performed four experiments.

In the first experiment, we run a series of trials with 3D CNNs. In each trial, we varied one or more of the following attributes: the type of input (raw frames or optical flow), the loss functions (cross entropy or hinge), and training data (generated data from the game or real life videos from YouTube). The same real-life test data were used in each trial. The results of these trials are summarized in Table 1. The results showed that (i) the hinge loss function performed better in all experiments, and (ii) the use of the game-generated data for training demonstrated a better performance than using the real-life data for training. Therefore, the model trained on game-generated data with optical flow pre-processing and hinge loss function outperformed other models.

Table 1. F_1-scores for a series of trails that we run in the first experiment.

Input	Loss function	Training data	F1-Score
Raw	Cross-entropy	Youtube	0.49
Raw	Cross-entropy	Game	0.64
Raw	Hinge	Youtube	0.49
Raw	Hinge	Game	0.64
OF	Cross-entropy	Youtube	0.55
OF	Cross-entropy	Game	0.63
OF	Hinge	Youtube	0.60
OF	Hinge	Game	0.71

In the second experiment, we took the best model from Table 1, and retrained it after injecting the data with noise. The goal was to explore the use of noise as a data augmentation technique (a means of adding randomness to the data with an intention to increase the generalization power of the model) to improve the performance of the system. We got an F1-score of 0.68, which showed that noise injection was not a suitable method for this task, likely due to the use of optical flow pre-processing. We plan to investigate this further in our future work.

In the third experiment, we took the best model from Table 1, and retrained it by replacing the convolutional units with LSTM units. The goal was to explore the use of LSTM units to improve the performance of the system. This model received a F1-score of 0.63, which showed that for short 15 frames sequences simpler 3D convolutional model is able to extract more significant temporal data than LSTM.

5 Conclusion

We developed 3D CNN based algorithm for automated car accidents recognition in videos. This approach demonstrated good performance on different types of roads, including those, which were not used for the training purposes. We also demonstrated that game-generated data can be successfully used for network training, especially in the case of insufficient availability of real-life data.

A comparison between our algorithm and traditional approaches, such as car track analysis, showed that convolutional neural networks are more robust in operation and need less setting during camera installation. On the other hand, an application of deep learning algorithm requires more computing power for training and operation and, therefore, can be challenging for a low-budget projects. Nevertheless, relatively inexpensive single board computers which are capable of running CNNs in real-time, such as Nvidia Jetson, are already available on the market.

It is difficult to examine the proposed algorithm against existing methods since a direct comparison requires the use of identical datasets for car accident detection. Its performance on the game-generated data was proven to be more accurate than in previously developed models, therefore proving the feasibility of employing the proposed approach for the task of car accident detection.

Acknowledgment. This research work was funded by Zayed University Cluster Research Award R18038. This research was also supported by Innosoft (https:// innosoft.pro). We thank Anton Trantin and Vyacheslav Lukin for providing us access to their GPUs for performing our experiments.

References

1. Aköoz, Ö., Karsligil, M.: Severity detection of traffic accidents at intersections based on vehicle motion analysis and multiphase linear regression. In: 2010 13th International IEEE Conference on Intelligent Transportation Systems (ITSC), pp. 474–479. IEEE (2010)
2. Caragliu, A., Del Bo, C., Nijkamp, P.: Smart cities in Europe. J. Urban Technol. **18**(2), 65–82 (2011)
3. Chen, Y., Yu, Y., Li, T.: A vision based traffic accident detection method using extreme learning machine. In: International Conference on Advanced Robotics and Mechatronics (ICARM), pp. 567–572. IEEE (2016)
4. Farnebäck, G.: Two-frame motion estimation based on polynomial expansion. In: Scandinavian Conference on Image Analysis. pp. 363–370. Springer, Heidelberg (2003)

5. Ji, S., Xu, W., Yang, M., Yu, K.: 3d convolutional neural networks for human action recognition. IEEE Trans. Pattern Anal. Mach. Intell. **35**(1), 221–231 (2013)
6. Karpathy, A., Toderici, G., Shetty, S., Leung, T., Sukthankar, R., Fei-Fei, L.: Large-scale video classification with convolutional neural networks. In: Proceedings of the IEEE Conference on Computer Vision and Pattern Recognition, pp. 1725–1732 (2014)
7. Kotwal, R.S., Howard, J.T., Orman, J.A., Tarpey, B.W., Bailey, J.A., Champion, H.R., Mabry, R.L., Holcomb, J.B., Gross, K.R.: The effect of a golden hour policy on the morbidity and mortality of combat casualties. JAMA Surg. **151**(1), 15–24 (2016)
8. Maaloul, B., Taleb-Ahmed, A., Niar, S., Harb, N., Valderrama, C.: Adaptive video-based algorithm for accident detection on highways. In: 2017 12th IEEE International Symposium on Industrial Embedded Systems (SIES), pp. 1–6. IEEE (2017)
9. Singh, D., Mohan, C.K.: Deep spatio-temporal representation for detection of road accidents using stacked autoencoder. IEEE Trans. Intell. Transp. Syst. (2018)

On Image Based Enhancement for 3D Dense Reconstruction of Low Light Aerial Visual Inspected Environments

Christoforos Kanellakis, Petros Karvelis, and George Nikolakopoulos[✉]

Robotics Group, Department of Computer, Electrical and Space Engineering,
Luleå University of Technology, Luleå, Sweden
{chrkan,petkar,geonik}@ltu.se

Abstract. Micro Aerial Vehicles (MAV)s have been distinguished, in the last decade, for their potential to inspect infrastructures in an active manner and provide critical information to the asset owners. Inspired by this trend, the mining industry is lately focusing to incorporate MAVs in their production cycles. Towards this direction, this article proposes a novel method to enhance 3D reconstruction of low-light environments, like underground tunnels, by using image processing. More specifically, the main idea is to enhance the low light resolution of the collected images, captured onboard an aerial platform, before inserting them to the reconstruction pipeline. The proposed method is based on the Contrast Limited Adaptive Histogram Equalization (CLAHE) algorithm that limits the noise, while amplifies the contrast of the image. The overall efficiency and improvement achieved of the novel architecture has been extensively and successfully evaluated by utilizing data sets captured from real scale underground tunnels using a quadrotor.

Keywords: Low-illumination image processing · 3D reconstruction · MAVs

1 Introduction

Lately, the Micro Aerial Vehicles (MAVs) have received increased research attention within the robotics community. These platforms are mechanically simple and by providing agile navigation capabilities have increased their popularity in the society, since they are able to fly in different modes, aggressively, smoothly, hover close to a target and perform advanced maneuvers. Until now the majority of consumer grade MAVs have been directed towards the photography-cinematography industry, taking advantage of their payload capacity and stable flight characteristics. Lately, there is an increasing interest from other industries targeting autonomous infrastructure inspection in close proximity with the Region of Inter-

This work has received funding from the European Unions Horizon 2020 Research and Innovation Program under the Grant Agreement No. 730302, SIMS.

© Springer Nature Switzerland AG 2020
K. Arai and S. Kapoor (Eds.): CVC 2019, AISC 944, pp. 265–279, 2020.
https://doi.org/10.1007/978-3-030-17798-0_23

est (ROI). A characteristic example of this trend is the underground mines that have the potential to deploy MAVs (Fig. 1) in their operation cycles and by that to reduce the operating costs, increase the productivity, while allowing for an overall increase of the human safety in underground challenging mining conditions [1].

Fig. 1. Left: MAV endowed with a visual sensor and a light bar for artificial illumination. Right: Typical surrounding in a dark mining environment.

These attributes will have an imminent impact to change the common practices currently utilized in the mining sector such as at: (1) overall mine operation, (2) mine production and (3) safety in the operations. So far, there have been developed multiple systems, limited to remotely operate in open pit mines above ground and assist in stockpile surveying, 3D pit model build, facility monitoring, security inspection and environmental assessment of the mine sites. In all these application scenarios, multiple challenges arise before employing autonomous MAVS, mainly towards the planning and control of these vehicles, such as the narrow passages, reduced visibility due to rock falls, dust, uncertainty in localization, wind gusts and lack of proper illumination.

Within the related literature of MAVs in underground mine operations, few research efforts have been reported trying to address these challenging tasks. In [2] a visual inertial navigation framework has been proposed, to implement position tracking control of the platform. In this case, the MAV was controlled to follow obstacle free paths, while the system was experimentally evaluated in a real scale tunnel environment, simulating a coal mine, where the illumination challenge was assumed solved. In [3] a more realistic approach, compared to [2] regarding underground localization, has been performed. More specifically, a hexacopter equipped with a Visual Inertial (VI) sensor and a laser scanner was manually guided across a vertical mine shaft to collect data for post-processing. The extracted information from the measurements have been utilized to create a 3D mesh of the environment and localize the vehicle. Finally, in [4] the estimation, navigation, mapping and control capabilities for autonomous inspection of penstocks and tunnels using aerial vehicles has been studied, using IMUs, cameras and lidar sensors.

This article focuses on the perception task and aims in proposing novel methods for enhancing the 3D dense reconstructions of low light environments by using MAVs. More specifically, video data recorded on board of the MAV during the mission can be post processed to provide a detailed 3D model of the visited area. The 3D model of the area of interest provides an actual comprehensive visual and geometric information for the asset owner for further analysis, or the mine inspectors to contextualize the location of the damages found during the inspection task, while the 3D information further facilities the evaluation of defects relative to the neighboring areas [5]. The combination of small scale and agile robotic platforms with advanced computer vision algorithms have the potential to create a powerful tool that is able to address complex tasks and provide better visual data and subsequently enabling a better decision making around the inspected infrastructure.

On the other hand, the quality of an image or a video sequence largely depends on the light conditions of the environment. For example strong light produces images with a wash out effect and weak light produce images that are not visible due to the darkness. For both of the cases, the contrast of the two images is extremely low and needs further modification in order to reveal the details of the images. These typical low light conditions occur in mines, usually due to the lack of proper illumination and in other underground or indoor dark environments, e.g. factories. In all these cases, even if it was possible and realistic to utilize a light on board of the MAV in order to illuminate the surrounding area of interest, this would have drawbacks for the mission design due to: (a) the limited weight that the MAV can lift and (b) the limited power supply that the MAV can support with the corresponding impact on the flight duration. A characteristic example of such a low light image capturing conditions acquired from a camera on a MAV is displayed in Fig. 2.

Fig. 2. Image of low light conditions in a mine captured by the camera on board of a MAV.

The solution to the described low light capturing conditions will be to employ image processing methods in order to contrast enhance the captured image since higher contrast level images display more detailed images and color differences, when compared to the lower contrast level ones. Contrast enhancement techniques [6] have received great attention mostly because of their simplicity as also their effectiveness. Even now days in the era of Deep Learning (DL), efforts are being made to enhance images especially in the case of low light conditions [7].

This work aligned with the vision of deploying aerial robotic platforms in the underground mine, focuses on the development of elaborate perception capabilities for MAVs flying in low light environments. The contribution of the proposed work is three-folded. Initially, this article proposes a method for 3D dense reconstruction by using a low cost aerial platform equipped with a single camera that can be considered as a consumable and easily replaceable. Secondly the paper focuses on employing image enhancement methods that are best suitable for low light enhancement methods. Finally, the third contribution stems from the fact that this work is among the few in the field that reports experimental trials on real scale tunnels, demonstrating the concept of enhancing 3D reconstruction of low light areas.

The rest of this article is organized as it follows. In Sect. 2 the image processing method for low light image enhancement is introduced and also the methods used for the 3D reconstruction are described for a generic clarity. In Sects. 3 and 4 the data collection procedure and the results of the proposed novel method are described respectively. Finally, the concluding remarks and future work are presented in Sect. 5.

2   Methodology

2.1   Contrast Limited Adaptive Histogram Equalization

One of the most prominent and simple techniques like the Histogram Equalization (HE) [8] has been employed to enhance images in low light conditions. This method alters the histogram of the original image in such a way that the resulting image will have a constant histogram. Another method that is based on locally equalizing regions or blocks of the image is the Adaptive histogram Equalization (AHE) [9,10] method which has the advantage that is adaptive to local information of the image. However both of these methods suffer from the fact that they actually enhance noise, particularly in homogeneous regions of the image, since the histogram in such regions is highly concentrated.

Contrast Limited AHE or (CLAHE) [11] overcomes such problems and the histogram of the new image is clipped in a way that the clipped pixels are reassigned to each gray level. This is the case in homogeneous or uniform regions of the image where high peaks of the histogram are present. In this case both methods, the AHE or HE enhance the image noise, since a very narrow range of input intensities will be mapped to a wider range of output intensity values. CLAHE on the other hand will enforce a maximum on the counts of the output histogram, thus limiting the amount of contrast enhancement.

The key parameters of the CLAHE method are two: the Block Size (BS) of the region, which the method will be employed and the Clip Limit (CL). As the threshold CL is increased, the resulting image will be brighter because larger CL will produce a flatter histogram. Finally, as the BS will be larger the dynamic range will become larger and the contrast of the image is also increased. The CLAHE method is composed of a number of steps that for clarity will be briefly described in the sequel:

1. Divide the original image into a number of blocks of size $M \times N$, where M and N are the number of pixels in the x and y direction respectively, and for each region compute:

$$N_{aver} = \frac{M \cdot N}{N_{gray}}, \tag{1}$$

   where N_{gray} is the number of gray levels in the region.
2. The actual clip limit N_{CL} will be computed as:

$$N_{CL} = N_{clip} \cdot N_{aver}, \tag{2}$$

   where N_{clip} is the normalized CL in the range of $[0,1]$.
3. The total number of clipped pixels will be defined as $S_{clipped}$ with the average number of pixels (N_{anp}) having to distribute uniformly to all gray levels is defined as:

$$N_{anp} = \frac{N_{anp}}{N_{gray}}. \tag{3}$$

   The contrast limited histogram of the contextual region can be calculated with the following set of rules:

$$H_{clipped}(i) = \left\{ \begin{array}{ll} N_{CL}, & \text{if } (i) > N_{CL} \\ N_{CL}, & \text{if } H(i) > N_{CL} + N_{anp} \\ H(i) + N_{CL}, & \text{otherwise} \end{array} \right\}, \tag{4}$$

   where H and $H_{clipped}$ are the initial histogram and clipped histogram, respectively.
4. Redistribute the remaining pixels N_{rp} by searching from the minimum to the maximum gray level with the following step: $step = \frac{N_{gray}}{N_{rp}}$. One pixel will be distributed to the gray level if the number of pixels in the gray level is less than N_{CL}. The procedure is repeated until all remaining pixels are all distributed to the new histogram.
5. Apply Histogram Equalization (HE) to the resulting histogram $H_{clipped}$.
6. In order to reduce some abruptly changes in the resulting histogram apply linear contrast stretch [8].

An example of the CLAHE method applied to an image of the tunnel is shown in the following Fig. 3.

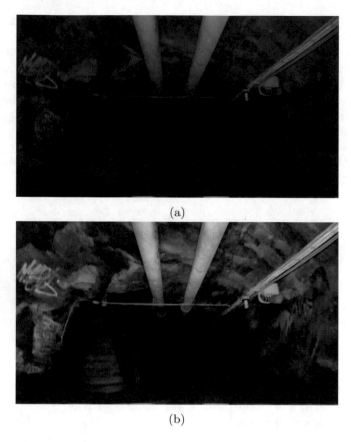

(a)

(b)

Fig. 3. CLAHE method employed to a tunnel image (a) before CLAHE and (b) after CLAHE.

2.2   Tunnel Reconstruction

The application scenario considered in this work targets the inspection of underground mine tunnels by utilizing aerial robotic platforms. The final outcome of the inspection mission will be a high fidelity 3D model of the inspected surface by post processing the collected visual data using the Structure from Motion technique (SfM) [12]. The reconstruction can be further analyzed by inspection experts to detect abnormalities or other type of defects speeding up and facilitating the maintenance task, or even overall updates on the a priori mining map libraries.

In the SfM process, different camera viewpoints are used offline to reconstruct the 3D structure. The process starts with the correspondence search step, which identifies overlapping scene parts among input images. During this stage, feature extraction and matching algorithms between frames are performed to extract information about the image scene coverage. Next, it follows the geometric verification using the epipolar geometry [13] to remove false matches.

In this approach, it is crucial to select an initial image pair I_1 and I_2 with enough parallax to perform two-view reconstruction, before incrementally registering new frames. Firstly, the algorithm recovers the sets of matched features f_1 and f_2 in both images and in the sequel estimates the camera extrinsics for I_1 and I_2 using the 5-point algorithm [14]. Afterwards, it decomposes the resulting Essential matrix with Singular Value Decomposition (SVD) and finally builds the projection matrices $P_i = [R_i|t_i]$ that contain the estimated rotation and translation for each frame. In the final step, by utilizing the relative pose information, the identified features are triangulated to recover their 3D position X^{3D}.

Afterwards, the two-frame Bundle Adjustment [15] refines the initial set of 3D points by minimizing the re-projection error and the remaining images are incrementally registered in the current camera and point sets. More specifically, the frames that observe the largest amount of the recovered 3D points are processed by the Perspective-n-Point (PnP) [16] algorithm that uses 2D feature correspondences to 3D points to extract their pose. Furthermore, the newly registered images will extend the existing set of the 3D scene (X^{3D}) using multi-view triangulation and in the end a global Bundle Adjustment is performed in the entire model to correct drifts in the process.

3   Data Collection

The proposed novel methodology for 3D dense reconstruction in dark environments has been initially evaluated by using datasets collected from actual flights of a custom designed aerial platform (Fig. 1) inside a tunnel under Mjölkuddsberget mountain located at Luleå, Sweden. The selected environment, which resembles an underground mine tunnel, was pitch dark without any external illumination, while the tunnel surfaces consisted by uneven rock formations. The dimensions of the testing tunnel area were $100 \times 2.5 \times 3\,\mathrm{m}^3$, capturing the camera sequences, while the MAV was following the path along the tunnel. Furthermore, the tunnel lack the presence of strong magnetic fields, while small particles were floating in the air during the flights. The aerial platform has been equipped with a LED light bar pointing towards the field of view of the camera to illuminate its surroundings. In more details, this light bar was set to different illumination levels (luminous flux per unit area or lux) of 270 lux and 230 lux. The illumination levels have been recorded from $1\,\mathrm{m}$ distance from the light source using the ILM-01 digital light meter[1]. The camera used in the dataset sequences was the FOXEER Box[2] that was recording with a resolution of 1920×1080 at 60 Frames Per Second (FPS), and with a diagonal field of view of $155°$, while Fig. 4 depicts snapshots of the field trials during the dataset collection, where the dominating darkness in the surrounding environment is evident.

---

[1] https://docs-emea.rs-online.com/webdocs/156e/0900766b8156e2ba.pdf.
[2] http://foxeer.com/Foxeer-4K-Box-Action-Camera-SuperVision-g-22.

Fig. 4. Snapshots of the MAV flight within the tunnel for the dataset collection from Mjölkuddsberget mountain at Luleå Sweden.

The datasets collected from the field trials have been divided in two cases capturing different locations of the tunnel environment and different illuminations. More specifically, *dataset1* includes 150 image frames with 230 lux illumination level, while *dataset2* includes 500 image frames with 270 lux illumination level. Figures 5, and 6 depict some sample image frames from the two datasets.

Fig. 5. Representative snapshots from the *dataset1* collected during the MAV flight. On the left column the original images and on the right column the CLAHE enhanced images.

Fig. 6. Representative snapshots from the *dataset2* collected during the MAV flight. On the left column the original images and on the right column the CLAHE enhanced images.

4   Experimental Results

The described experimental part was designed to demonstrate the performance of the proposed enhanced image based 3D reconstruction scheme for underground tunnel inspection, using the datasets discussed in Sect. 3. The presented evaluation includes quantitative and qualitative results from the aspects of the 3D reconstruction and image processing. The main goal is to demonstrate the ability to enhance the images fed to the reconstruction pipeline and increase the information that can be extracted, towards a more detailed 3D model generation. The evaluation considers the comparison of the CLAHE enhanced images with the original images by using the state-of-the-art SfM software Colmap [12]. In both reconstruction cases the same parameters have been selected regarding the feature extraction, matching as well as the sparse and dense reconstruction, as shown in Table 1.

4.1   Dataset1

Initially for the case of the original images, the reconstruction pipeline provided two separate pointclouds. Based on the software documentation COLMAP attempts to reconstruct multiple models if not all images are registered into the same model. Therefore in the original images the SfM pipeline is not able to place the collected images in the same model. The first pointcloud processed 31

Table 1. Colmap parameters used for *dataset1* and *dataset2*

	Dataset1	Dataset2
Camera model	*Simple radial fisheye*	*Simple radial fisheye*
Feature matching	*Exhaustive matching*	*Exhaustive matching*
−DenseStereo.window_radius	15	15
−StereoFusion.min_num_pixels	5	15
−PoissonMeshing.trim	7	7
−PoissonMeshing.depth	20	20

image frames and resulted in 300564 points, while the second pointcloud processed 58 image frames and provided 587098 points. Figure 7(a) depicts the 2 resulting pointclouds from the processing of the original images.

For the case of the CLAHE enhanced images the reconstruction pipeline provided a complete pointcloud processing in total 117 image frames, while resulting in 1090067 points. Figure 7(b) depicts the resulting pointcloud from the processing of the enhanced images.

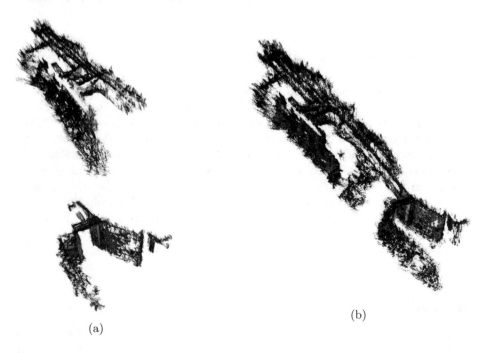

(a)

(b)

Fig. 7. Dense reconstruction of *dataset1* using COLMAP. (a) Original images, (b) CLAHE enhanced images

Table 2 presents the total number of points for each of the generated point-clouds as well as the total image frames processed for each case. The proposed method provides a 3D model with an increase of 22.08% compared to the point-clouds generated from the original images. Moreover, the proposed method was able to use 24% more image frames compared to the original images. Based on the quantitative results the proposed method is able to enrich the image content and improve the reconstruction outcome. Generally, the original images fail to provide a single pointcloud as a result and were able only to reconstruct parts of the dataset where the illumination conditions where substantial.

Table 2. Original vs Enhanced images pointcloud summary

	Original		Enhanced
	Pointcloud1	Pointcloud2	Pointcloud
#points	300564	587098	1090067
#frames	31	58	117

The generated models have been converted also to a 3D mesh using Poisson surface reconstruction method [17]. The mesh resulting from the proposed method is characterized by improved texture, with brighter colors, compared to the same location in the mesh resulting from the original images. Thus, another merit of the proposed method considers the texture enhancement of the 3D mesh. Figure 8 visualizes the different meshes generated from original and CLAHE enhanced images.

4.2 Dataset2

Regarding this case, the reconstruction pipeline processed 110 image frames and resulted in 208525 points, Fig. 7(a) depicts the 2 resulting pointclouds from the processing of the original images.

For the case of the CLAHE enhanced images the reconstruction pipeline provided a complete pointcloud processing in total 408 image frames, while resulting in 3562538 points. Figure 9(b) depicts the resulting pointcloud from the processing of the enhanced images.

Table 3 presents the total number of points for each of the generated point-clouds, as well as the total image frames processed for each case. The proposed method provides a 3D model with 16× more points compared to the pointclouds generated from the original images. Moreover, the proposed method was able to use 3× more image frames compared to the original images. Based on the quantitative results the proposed method is able to enrich the image content and improve the reconstruction outcome. Similarly to *dataset1*, the original images provide a single pointcloud as a results only in the parts of the scene where the illumination conditions were substantial.

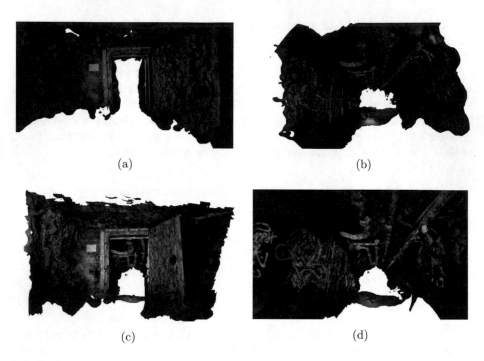

Fig. 8. Instances of the generated 3D meshes using *dataset1*. (a) and (b) based on original image, while (c) and (d) based on CLAHE enhanced images

Fig. 9. Dense reconstruction of *dataset2* using COLMAP. (a) Original images, (b) CLAHE enhanced images

Table 3. Original vs Enhanced images pointcloud summary

	Original	Enhanced
#points	208525	3562538
#frames	110	408

(a)   (b)

Fig. 10. Instances of the generated 3D meshes using *dataset2*. (a) based on original image and (b) based on CLAHE enhanced images

Similarly to Sect. 4.1 the 3D mesh for each case has been generated and is depicted in Fig. 10. In this scenario the original images provide slightly smoother mesh but only from the areas with sufficient illumination, whereas the proposed method was able to provide a mesh including bigger part of the inspected tunnel, trading off completeness and accuracy.

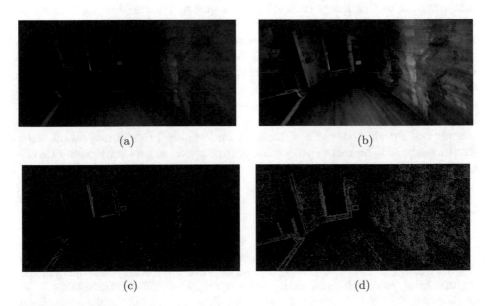

Fig. 11. Effect of the Sobel edge detection (threshold $= 0.02$) on the processed image. (a) Original image, (b) CLAHE enhanced image, (c) Edge detection on the original image and (d) Edge detection on the enhanced image.

Based on the results from the datasets presented in this work, the proposed method is able to enhance the information extracted from the image and used by the reconstruction pipeline. The critical part that emphasizes the importance of this study is that it focuses on low cost solutions for 3D model generation applied in underground tunnel environments. This system can be the basis of a robust inspection system structured around aerial robotics and visual sensors.

4.3   Edge Detection Comparison

In order to evaluate the effectiveness of the proposed method the Sobel edge detection method [8] has been used, with a threshold value of 0.02.

Below a representative example of a tunnel image is depicted with its detected edges before and after the application of the CLAHE method. As one can see significant edge information is absent from the original image (Fig. 7(c)) instead of the processed image where the edge information is significant more (Fig. 11(d)).

5   Conclusions

Despite the vast advances and amount of methods for feature extraction and matching methods that are used in the 3D reconstruction pipeline when it comes to tunnel images these methods fail due to the lack of light in the images. This article investigated the effect of the CLAHE method to boost the hidden details in these type of images. The experimental results showed that a significant improvement in image contrast enhancement of tunnel images is achieved through the CLAHE method, which later significantly enhances the quality of the 3D reconstruction. Aerial platforms will have a major role in the upcoming years in the underground mining and the proposed system can be considered among the first experimental step to address the challenging problem of lacking illumination when using visual sensors in such environments. Future work will focus on further examination of low light image enhancement methods as also Deep learning techniques aiming to employ them in real time localization and mapping of autonomous aerial vehicles in underground tunnel inspection tasks. Additionally, can be merged to online mapping techniques for coarse obstacle avoidance tasks.

References

1. Attard, L., Debono, C.J., Valentino, G., Castro, M.D.: Tunnel inspection using photogrammetric techniques and image processing: a review. ISPRS J. Photogramm. Remote Sens. **144**, 180–188 (2018). https://doi.org/10.1016/j.isprsjprs.2018.07. 010. http://www.sciencedirect.com/science/article/pii/S0924271618302028
2. Schmid, K., Lutz, P., Tomić, T., Mair, E., Hirschmüller, H.: Autonomous vision-based micro air vehicle for indoor and outdoor navigation. J. Field Robot. **31**(4), 537–570 (2014)

3. Gohl, P., Burri, M., Omari, S., Rehder, J., Nikolic, J., Achtelik, M., Siegwart, R.: Towards autonomous mine inspection. In: 2014 3rd International Conference on Applied Robotics for the Power Industry (CARPI), pp. 1–6. IEEE (2014)
4. Özaslan, T., Loianno, G., Keller, J., Taylor, C.J., Kumar, V., Wozencraft, J.M., Hood, T.: Autonomous navigation and mapping for inspection of penstocks and tunnels with MAVs. IEEE Robot. Autom. Lett. **2**(3), 1740–1747 (2017)
5. Karpowicz, J.: UAVs in civil infrastructure. http://www.expouav.com/wp-content/uploads/2016/04/free-report-uavs-in-civil-infrastructure.pdf?_ga=2.149425910.1944537166.1536777582-1263854739.1536777582
6. Lal, S., Narasimhadhan, A., Kumar, R.: Automatic method for contrast enhancement of natural color images. J. Electr. Eng. Technol. **10**(3), 1233–1243 (2015)
7. Chen, C., Chen, Q., Xu, J., Koltun, V.: Learning to see in the dark. arXiv preprint arXiv:1805.01934 (2018)
8. Gonzalez, R.C., Woods, R.E.: Digital Image Processing. Pearson, London (2018)
9. Pizer, S.M., Zimmerman, J.B., Staab, E.V.: Adaptive grey level assignment in CT scan display. J. Comput. Assist. Tomogr. **8**(2), 300–305 (1984)
10. Zimmerman, J.B., Pizer, S.M., Staab, E.V., Perry, J.R., McCartney, W., Brenton, B.C.: An evaluation of the effectiveness of adaptive histogram equalization for contrast enhancement. IEEE Trans. Med. Imaging **7**(4), 304–312 (1988)
11. Pizer, S.M., Amburn, E.P., Austin, J.D., Cromartie, R., Geselowitz, A., Greer, T., ter Haar Romeny, B., Zimmerman, J.B., Zuiderveld, K.: Adaptive histogram equalization and its variations. Comput. Vis. Graph. Image Process. **39**(3), 355–368 (1987)
12. Schonberger, J.L., Frahm, J.M.: Structure-from-motion revisited. In: Proceedings of the IEEE Conference on Computer Vision and Pattern Recognition, pp. 4104–4113 (2016)
13. Hartley, R., Zisserman, A.: Multiple View Geometry in Computer Vision. Cambridge University Press, Cambridge (2003)
14. Nistér, D.: An efficient solution to the five-point relative pose problem. IEEE Trans. Pattern Anal. Mach. Intell. **26**(6), 756–770 (2004)
15. Triggs, B., McLauchlan, P.F., Hartley, R.I., Fitzgibbon, A.W.: Bundle adjustment—a modern synthesis. In: International Workshop on Vision Algorithms, pp. 298–372. Springer (1999)
16. Gao, X.S., Hou, X.R., Tang, J., Cheng, H.F.: Complete solution classification for the perspective-three-point problem. IEEE Trans. Pattern Anal. Mach. Intell. **25**(8), 930–943 (2003)
17. Kazhdan, M., Hoppe, H.: Screened poisson surface reconstruction. ACM Trans. Graph. (ToG) **32**(3), 29 (2013)

Robots in Healthcare: A Survey

Arshia Khan[⊠] and Yumna Anwar[⊠]

University of Minnesota Duluth, Duluth, USA
{akhan, anwar033}@d.umn.edu

Abstract. Advances in robotic technology is stimulating growth in new treatment mechanism by enhancing patient outcomes and helping reduce healthcare costs, while providing alternate care apparatus. Provision of care by assistive therapeutic robots has increasingly grown in the past decade. Although the healthcare industry has been lagging in the use of assistive robots; the use of assistive robots in the manufacturing industry has been a norm for a long time. The vulnerable population of patients with illnesses, cognition challenges, and disabilities are some of the causes for the delay in the use of assistive therapeutic robots in healthcare. In this paper we explore the various types of assistive robots and their use in the healthcare industry.

Keywords: Robots · Healthcare · Dementia · Surgical robots ·
Robot classification · Therapy robots

1 Introduction

Robots have been used in the industry as early as the 1960s [9, 15]. The first manufacturing robot was built by General Motors in 1958, called *Unimate* [47]. The concept of the robots came in the awareness after the play "Rossum's Universal Robots" [47, 48]. Robotics was first introduced to the Healthcare industry through the surgical robots that assisted in surgery. To this day the surgical robots are the most popular and approved robots in the healthcare industry. The first healthcare robot was a surgical robot in 1985 [47]. The need for assistive service robots in healthcare was reviewed in 2017 and the results revealed a scarcity of service robots in the hospitals and housing services [3]. The drive to develop assistive robots in healthcare is stemmed off of the rising healthcare costs and the need for caregivers. There is a clearly identified need for robotics in healthcare. Societal drivers of improvement in healthcare, reduction in healthcare costs, shorter recovery time, increased access to healthcare, revolutionizing and invention of new treatment mechanisms, improvement in patient outcomes, reduction in caregiver stress and workload are only a few of many idealized outcomes [20].

1.1 Robot Definition

Robots can be defined as artificially intelligent physical system that is capable of interacting with the environment. The robots can be mobile or not. A robot can also be defined as a set of sensors paired with a set of algorithms, configured to communicate data and organized in some physical form to provide autonomous operation.

© Springer Nature Switzerland AG 2020
K. Arai and S. Kapoor (Eds.): CVC 2019, AISC 944, pp. 280–292, 2020.
https://doi.org/10.1007/978-3-030-17798-0_24

Robots have been classified in multiple domains based on the needs, abilities, performance and technology. The term "robot" was coined from the Czech word, "robota", which means serf or laborer [47].

1.2 Robot Acceptance and Implementation

The spectrum of robotic use in the healthcare industry spans the operating room through the living area. Robots have been utilized in multiple areas to assist humans with tasks that can be repetitive, or require significant risk, or require specific precision or some form of sophisticated complex ability. In healthcare robots have been used in Europe and a few other countries effectively. For example, as of 2010, approximately 1,500 Paro the seal, a mental commitment robot were sold across the world. Of these 1,500 robots about 1,300 were sold in Japan, about 100 were sold in Denmark, and 100 were sold in other countries.

Robots have been sparingly implemented in the healthcare system in the United States although the results from robotic usage in other parts of the world as well as the USA have shown encouraging positive results. Research in assistive robotics is slowly building and progressively providing evidence that therapeutic assistive robots are helpful in providing care for the elderly [15–19].

2 Robot Classification

Robots can be categorized [3] based on their type such as humanoid (human like), animoid (animal like), machine like, pet like, screen only or a combination of these. Robots can also be classified as mobile, stationary or a combination of the two. Robots have been used as social companions, as surgery bots, as mechanical devices, as machines or in a humanoid form. There is always a question of acceptability of these robots especially when it comes to the healthcare industry where the safety and security of the individuals, HIPAA laws, and the costs pose restrictions. In this paper we will first explore the use of assistive therapeutic robots classified as assistive robots in the areas of nursing, companionship and surgery, rehab aid providing robots, assistive robots for disabled, assistance providers to cognitively impaired, motivational robots, robots in telemedicine, medication management robots and meal delivery robots [7].

2.1 Therapy Robots

This is a new emerging field of assistive robots in healthcare where robots are used as therapy companions in place of animals. This kind of therapy has shown promise and positive results encouraging further research in this area. Psychologists and researchers are equally intrigued by this therapy mechanism [9]. The use of animals in therapy has shown to have improved health outcomes and reduced mortality. Findings from various studies have confirmed that social affiliations and companionships improve health [34]. Some challenges with real animals used for therapy are that there is a chance that the patient or the caregiver may be allergic to animals or the care of an animal may be restricting the therapy, plus a live animal needs to be looked after and his/her needs.

These scenarios encourage the use of robot animal therapy. Most animoid robots are equipped with sensors to gauge touch such as petting, scratching motions etc. These motions help interact and engage the animals. The therapy robot is designed to have a sense of touch, and to respond and mimic the real live animal. Its design involves responding by displaying gestures such as cat purring or a dog wagging its tail. These interactions have had positive impact on the health of individuals and offered a means to engage the patients [35].

2.2   Mental Commitment Robots

The mental commitment robots are used to not only provide companionship, but these robots are designed with a psychological component where the robots work towards providing psychological enrichment and mental stimulation to its users. These robots are much more than a therapy robot and assist by providing guidance, education on certain subject matters, encouragement and enablement of communication, in addition to providing assistance [9]. Robots of this type are not meant to offer any physical support. These robots are specifically designed to stimulate, arouse and engender emotions and mental effects. These robots are also perceived to receive stimulation from the humans they interact with, thus demonstrating human like feelings [15]. Paro is an excellent example for this type of robot as seen in Fig. 1.

Fig. 1. Paro the seal robot [15]

2.3   Socially Assistive Robots

Socially assistive robots (SAR) are those that are both assistive and socially interactive. They are an amalgamation of robotics, medicine and psychology and been widely used in mental health applications [1].

2.4   Surgical Assistance Robots

Surgical robots have been routine used in surgery currently. Surgical robotics is advancing much faster than the other types of robotics in healthcare. The drive to overcome human limitations, transference of information to action, increase in the effectiveness of the procedure, the ability to maneuver between robot control and human control and finally the ability to use advanced enhanced precision based surgical instruments are some of the drivers for the growth in surgical robots. The current surgical robots are offering increased dexterity to the surgeons while giving them the ability to tele-operate, freeing the surgeon to also lookup information as the surgery proceeds [20].

Magnetic Resistance (MR) Guided Robotic Surgery

This type of robot assisted surgery can be used in minimally invasive surgeries such as catheter biopsies, where the surgeon can get assistance in positioning and directing the catheter. Due to the large size of the MR equipment it was placed over head of the surgical area, while the arms extend to the surgical workspace. The robot itself was made with paramagnetic material and driven by non-magnetic ultrasonic motors. This system is currently in clinical trials [39].

Miniature Wireless in Vivo Robots

These miniature robots can be inserted into the peritoneal cavity to assist in laparoscopic surgeries. The benefit of these miniature robots is that they are not only less expensive, but they are also easily transportable due to their size, making them more desirable [40].

Endoluminal Mobile Robot

This orifice transgastric endoscopic surgery robot has the potential to eliminate skin incisions while reducing postoperative discomfort and pain by utilizing miniature robots that can explore the gastric cavity instead of inserting an endoscope [43].

SpineAssist Surgical Robot

This type of robotic surgery assisted with spinal implant insertions such as screw insertions [42].

2.5   Robot Assisted Telesurgery

Is one useful implementation where the surgeon can be in a different location from the patient. This type of surgery can benefit the Military as well as the rural areas where healthcare is not easily accessible.

Transatlantic Robot Assisted Telesurgery

Transatlantic robot assisted tele-surgery not only improved dexterity but also enabled surgery to be performed from a distance. The ZEUS robotic system was used to perform a full laparoscopic cholecystectomy from New York while the patient was in Strasbourg, France. The total time for surgery was 54 min. The challenge in this type of robot assisted surgery is the time delay in data transmission and in the ZEUS system the delay was somewhat overcome by using fiber optic cables [38].

Da Vinci Robotic System
Arobot assisted tele-surgery was successfully performed with the surgeon in a different location than the patient. The dexterity of the surgeon was greatly enhanced by the da Vinci robotic system in addition to the ability to perform a teleoperation.

Video Assisted Surgery
Video assisted surgery allows the surgeon to be able to perform thoracic surgical lobectomy. This is another da Vinci robotic system [41].

2.6   Rehabilitation Robots

Rehabilitation robots is a domain in robotics to help physically impaired restore motor control in limbs using robots. One common and widely researched upon is Stroke rehabilitation. Studies have identified that robot aided therapy is better for improvement in motor control of a stroke patient than conventional physiotherapy [4]. In a systematic review [4] of seventeen clinical trials to investigate robot aided therapies for patients affected by stroke, all these studies showed that most trials resulted in a short and long-term improvement of the motor control in shoulder and elbow of the patients.

MIT-Manus system [5], and ARM Guide [6] are two of the rehabilitation robots used in studies for movement impairment. MIT-Manus system, unlike commercial robots, is smooth, gentle, low mechanical impedance, designed specifically for neuro-rehabilitation [5]. Robot aided therapy has proven to be favorable for improvement in sensorimotor and reducing the impairment in hemiparetic patients. The research study [5], investigated 76 stroke patients using MIT-Manus system for physiotherapy.

2.7   Robots to Aid Elderly

Increasing number of older adults are greatly strained due to the lack of healthcare services. There are a wide range of needs that the aging population requires assistance in, like performing and reminding of day to day tasks, mobility, navigation, and most importantly companionship and someone to spend their time with. The need for assistance is the reason why they are admitted to the nursing homes. The drive to delay admission into the nursing homes reasons promote and encourage the advancements towards service and companion robots for the elderly [14].

Service Robots for Assisting Elderly in Mobility and Navigation
Service robots for aging adults assist the users in performing a specific task. One such service Personal Aid for Mobility and Monitoring (PAMM) system [10], has a walker based and a cane-based design to help provide physical support, health monitoring by recording basic vitals via health sensors on the handle, acoustic sensors and camera in the front to detect obstacles and provide guidance. Cane is for elderly adults with minor physical impairment and walker for people with severe impairment.

Most of the visually impaired people are elderly of age seventy plus [11], this poses a greater risk in their mobility. The guide dog for visually impaired elderly adults is not always suitable especially when they are not very active, and on the other hand canes has limitations and many risks associated with it. Due to this there has been a lot of research and development in the area of mobility aid. One such research in the late 20th century was on personal adaptive mobility AID (PAM-AID) [11], the mechanical design of this robot was in a form of a walking support frame. One of the main challenges in developing such a robot is designing the user interface, as the robot aids blind users. In PAM-AID the direction of the push on the handles determined the user's indication for the direction of the movement of the robot. The information related to the command and or any obstacle that the sensors detect were delivered to the user as an audio feedback. The system architecture of this robot includes dynamic decision-making measures to keep user balance and also taking into account the short-term memory of most of the elderly population the interface of the robot is kept easy to use and understand and requires inherent knowledge. Robotic Travel Aid (RoTA) [12] and Care-o-bot [13] are other similar work on physical support and navigational guide for visually impaired people.

Physical Coaching Robots

There has also been some research on assistive robots to coach and engage the elderly in physical exercises. One such SAR robot as discussed in [2], motivates the user to engage in physical exercise using game sessions. Additionally, the robot interacts with the elderly and gives feedback on their actions. The system architecture is composed of facial expressions, arm movement, speech, vision and user communication [2].

Companionship Robots for Elderly in Nursing Homes

Research on companion robots for elderly is aimed at improving psychological health and address their loneliness and depression. Pearl, Homie and Paro, as discussed in the later section, are some of the Companion robots specifically designed and developed for elderly, to keep them company, relieve stress, reduce depression, remind them of daily tasks like taking medications, and even reading out a text message or making a call to their loved ones for them [21, 26, 36].

2.8   Some Common Existing Companion Robots

Aibo

Aibo (as seen in Fig. 2) is a dog-like robot developed by Sony, costing around USD $1000. Though it's an entertainment robot, for example it has been programmed to play in the robocup league [45], it has also been programmed as a companion in many studies to interact with the elderly to decrease stress and loneliness [21], some studies have also used it as a therapeutic tool for dementia patients, concluding positive outcomes of its effectiveness and usefulness [46].

Homie

Is a dog-like robot companion (as seen in Fig. 3) for elderly, which can be personalized by giving a name, and can show emotions with movement of its eyebrows, mouth, tail and other parts. The expressions are mainly focused on positive emotions. Emotions for sadness and anger are only for when the battery is low or there is some technical difficulty. The soft skin complemented by the heart beat feeling of homie gives a sense of warmth [36].

Fig. 2. Sony Aibo robot

Fig. 3. Homie [36]

Homie is equipped to perform medication management. It has a box with pills inside the stomach so that they are available at the prescribed time, and the dogcollar-cum-bracelet is a medium to transfer data, such as blood pressure and pulse rate, from the user to homie. For emergencies the bracelet also contains GPS, and the data can also be sent to and fro the clinician and Homie. The robot also has a microphone and audio feature to receive commands and read out information or messages for the elderly. Homie is especially designed for aging population and like any other robots for elderly, is motivated to reduce depression and loneliness [36].

Huggable

Huggable is a bear-like robot as seen in Fig. 4. The main idea is to use it in place of animal therapy as real animals pose many restrictions. This soft bear has a set of sensors for touch, visual, audio, temperature, force and audio output. Huggable is being used as an interactive companion as well as to send sensory data for health monitoring of an individual [35].

Pearl

Is a Nursebot for assistance of the elderly developed by Carnegie Mellon University [26]. It aids elderly by reminding and assisting in their daily tasks and also helps with navigation.

Fig. 4. Huggable robot [35]

Paro

Paro is a seal-like robot as seen in Fig. 1 and is developed by National Institute of Advanced Industrial Science and Technology (AIST) in Japan, specifically, for elderly [21]. It has touch sensors spread out over the whole body, with visual sensors, light sensors and has been used widely as a social therapeutic robot for elderly and dementia patients [23, 24]. There are many researches involving Paro to interact, relieve stress and calm the dementia patients. Studies have also shown patients are willing to interact with Paro [24].

iCat

iCat is a cat-like robot developed by Philips (as seen in Fig. 5). It has a user interface that displays different facial expressions such as angry, happy etc. and is mainly used to study human-robot interaction [21, 22].

Fig. 5. Philips iCat robot [22]

2.9   Use of Robots for Autism Spectrum Disorders

Robot assisted therapy [RAT] has shown promise specifically in the area of assisting children with autism spectrum disorder. An argument can be made to provide additional autonomy and training/machine learning ability where the robot, while being under the supervisor control can still learn from individual children and adapt to their needs [32]. There is sufficient research that suggests that RAT has a positive impact on children affected with autism spectrum disorder. The effects are demonstrated in the form of reactions and behaviors that are positive and encouraging towards robots when compared to human caregivers and therapists [33]. These results are reassuring and encourage further research in RAT technologies.

Use of robots in autism therapy is intended to improve the social and communication skills of the children and to engage them, grasp their attention and improve eye-contact [8, 37]. One such robot that has been used in autism therapy is 'Kaspar/ as shown in Fig. 6. It has a human like face with a less complex overall design as simplicity is preferred in SAR systems for Autism [37].

Fig. 6. Kasper robot for ASD [37]

3   Robot Acceptance in the Healthcare Industry

While robots have been used in the industry since 1960s and been widely adopted due to the support and assistance they provide in terms of speeding up the manufacturing process, offering assistance where human reach is restricted, offering improved precision and assistance in areas where security is challenged or in dangerous situations; robots in healthcare are less accepted and scrutinized by not only the patients but the caregivers and the clinicians. Healthcare robots face challenges that are not so pronounced in the manufacturing industry. Challenges such as working in an environment that is surrounded by vulnerable individuals in terms of physical, mental and other weaknesses further reduce the acceptance of healthcare robots. Some factors identified that are unique to the healthcare industry are Usability and Acceptability, Safety and Reliability, Capability and Function, Clinical Effectiveness, and Cost Effectiveness. Usability in terms of ease of use and knowledge and skills required to use the robot can potentially reduce the usability. Occasionally there is a societal stigma associated with the use of a robot besides the cognition abilities that pose a challenge [25]. A study found approximately 75% of the hand rehabilitation robots were never used and tested [27]. Issues of lack of technical skills and knowledge and or fear of technology, specifically robotic technology can further reduce the robot usability and acceptability [28].

3.1 Functionality and Capabilities as Pros and Cons

The capabilities and functionalities of the robots are restricted by the advancements in the robotics domain. Some areas of technology are not sufficiently advanced to provide the functionality necessary for the required task. Although there are great advancements in this area most of the research is set up as a prototype and is used under supervised demonstration purposes only. The healthcare field is one field that varies individual to individual in terms of needs and restrictions. It is challenging to offer a assistive robot that can satisfy the needs of multiple type of patients, in other words the robots developed are very unique to specific individual needs [29, 30]. This leads to another major concern of acceptability which is safety and reliability.

3.2 Safety and Reliability

Safety around patients with respect to the use of robots is questioned due to the patient population's vulnerabilities. This population that has many issues such as cognition, elderly population, patients with physical, mental and other disabilities. Children and patients with cognition issues are especially more vulnerable requiring additional safety measures to be taken into consideration when working with robots. Reliability of a robot poses some challenges as well. Another issue is the fairness and ethical considerations in terms of deception with cognitively impaired and elderly or disabled patients [31]. Another issue that relates to the safety and reliability is the ethical considerations in the use of the assistive robots in healthcare.

3.3 Ethics Consideration in Robots in Healthcare

Robotics has shown great promise in terms of improving the quality of care, the quality of life and even improvement in patient outcomes. One aspect that needs to be taken into consideration is the ethical implications of these developments from the perspective of the patient, the caregiver/family member, care provider, clinician and the hospital [44]. Often times the robots are designed and developed by engineers who are not thinking about the ethical considerations. It is important to involve ethicists in the design of assistive robots so the ethical considers can be examined from the perspectives of various stakeholders.

4 Conclusion

In conclusion this paper has explored multiple types of healthcare robots from miniature and magnetic resonance large surgical robots to robots in living rooms. Therapy robot, mental commitment robots, socially assistive robots, surgical robots, tele-surgery robots, companionship robots, rehabilitation robots, blind mobility assistance robots, assistance with autism spectrum robots, robots for elderly are some of the more prominently used robots. Robots are created in all sizes from miniature robots that can be sent into the abdominal cavity to large magnetic robots that hang overhead in surgeries to mobile robots that can move freely or robots that can have conversations with individuals. This wide spectrum of robots provides assistance mentally, physically, psychologically and emotionally. The main purpose of these robots is to enhance care delivery, improve patient outcomes, improve recovery time, reduce pain and discomfort, provide companionship and provide emotional and mental support while educating the patient. The stakeholders in this field are the patients, the caregivers, the family members, the clinicians and the care providers.

References

1. Rabbitt, S.M., Kazdin, A.E., Scassellati, B.: Integrating socially assistive robotics into mental healthcare interventions: applications and recommendations for expanded use. Clin. Psychol. Rev. **35**, 35–46 (2015). https://doi.org/10.1016/j.cpr.2014.07.001. ISSN 0272-7358
2. Fasola, J., Mataric, M.: A socially assistive robot exercise coach for the elderly. J. Hum. Robot Interact. **2**(2), 3–32 (2013)
3. Vänni, K.J., Salin, S.E.: A need for service robots among health care professionals in hospitals and housing services. In: International Conference on Social Robotics. Springer, Cham (2017)
4. Prange, G.B., Jannink, M.J., Groothuis-Oudshoorn, C.G., Hermens, H.J., Ijzerman, M.J.: Systematic review of the effect of robot-aided therapy on recovery of the hemiparetic arm after stroke. J. Rehabil. Res. Dev. **43**, 171–183 (2006)
5. Krebs, H.I., Volpe, B.T., Aisen, M.L., Hogan, N.: Increasing productivity and quality of care: robot-aided neuro-rehabilitation. J. Rehabil. Res. Dev. **37**(6), 639–652 (2000)
6. Reinkensmeyer, D.J., Kahn, L.E., Averbuch, M., McKenna-Cole, A., Schmit, B.D., Rymer, W.Z.: Understanding and treating arm movement impairment after chronic brain injury: progress with the ARM guide. J. Rehabil. Res. Dev. **37**(6), 653–662 (2000)
7. Broadbent, E., Stafford, R., MacDonald, B.: Acceptance of healthcare robots for the older population: Review and future directions. Int. J. Soc. Robot. **1**(4), 319 (2009)
8. Diehl, J.J., Schmitt, L.M., Villano, M., Crowell, C.R.: The clinical use of robots for individuals with autism spectrum disorders: a critical review. Res. Autism Spectr. Disord. **6**(1), 249–262 (2012). https://doi.org/10.1016/j.rasd.2011.05.006. ISSN 1750-9467
9. Shibata, T., Wada, K.: Robot therapy: a new approach for mental healthcare of the elderly–a mini-review. Gerontology **57**(4), 378–386 (2011)
10. Dubowsky, S., Genot, F., Godding, S., Kozono, H., Skwersky, A., Yu, H., Shen Yu, L.: PAMM - a robotic aid to the elderly for mobility assistance and monitoring: a "helping-hand" for the elderly. In: Proceedings of the IEEE International Conference on Robotics and Automation, vol. 1, pp. 570–576 (2000). https://doi.org/10.1109/ROBOT.2000.844114
11. Lacey, G., Dawson-Howe, K.M.: The application of robotics to a mobility aid for the elderly blind. Robot. Auton. Syst. **23**, 245–252 (1998)
12. Mori, H., Kotani, S.: A robotic travel aid for the blind – attention and custom for safe behavior. In: International Symposium on Robotics Research, pp. 237–245. Springer, London (1998)
13. Schraft, R.D., Schaeffer, C., May, T.: Care-O-bot(tm): the concept of a system for assisting elderly or disabled persons in home environments. In: IECON: Proceedings of the IEEE 24th Annual Conference, vol. 4, pp. 2476–2481 (1998)
14. Robinson, H., Macdonald, B., Broadbent, E.: The role of healthcare robots for older people at home: a review. Int. J. Soc. Robot. **6**, 575–591 (2014). https://doi.org/10.1007/s12369-014-0242-2

15. Shibata, T., et al.: Human interactive robot for psychological enrichment and therapy. In: Proceedings of AISB, vol. 5 (2005)
16. Sabelli, A.M., Kanda, T., Hagita, N.: A conversational robot in an elderly care center: an ethnographic study. In: 2011 6th ACM/IEEE International Conference on Human-Robot Interaction (HRI). IEEE (2011)
17. Broekens, J., Heerink, M., Rosendal, H.: Assistive social robots in elderly care: a review. Gerontechnology 8(2), 94–103 (2009)
18. Harmo, P., et al.: Needs and solutions-home automation and service robots for the elderly and disabled. In: 2005 IEEE/RSJ International Conference on Intelligent Robots and Systems (IROS 2005). IEEE (2005)
19. Flandorfer, P.: Population ageing and socially assistive robots for elderly persons: the importance of sociodemographic factors for user acceptance. Int. J. Popul. Res. 2012, 13 (2012)
20. Okamura, A.M., Mataric, M.J., Christensen, H.I.: Medical and health-care robotics. IEEE Robot. Autom. Mag. 17(3), 26–37 (2010)
21. Broekens, J., Heerink, M., Rosendal, H.: Assistive social robots in elderly care: a review. Gerontechnology 8, 94–103 (2009)
22. van Breemen, A., Yan, X., Meerbeek, B.: iCat: an animated user-interface robot with personality, 143–144 (2005). https://doi.org/10.1145/1082473.1082823
23. Mcglynn, S., Snook, B., Kemple, S., Mitzner, T., Rogers, W.: Therapeutic robots for older adults: investigating the potential of Paro. In: ACM/IEEE International Conference on Human-Robot Interaction, pp. 246–247 (2014). https://doi.org/10.1145/2559636.2559846
24. Šabanović, S., Bennett, C.C., Chang, W., Huber, L.: PARO robot affects diverse interaction modalities in group sensory therapy for older adults with dementia. In: 2013 IEEE 13th International Conference on Rehabilitation Robotics (ICORR), Seattle, WA, pp. 1–6 (2013). https://doi.org/10.1109/icorr.2013.6650427
25. Riek, L.D.: Healthcare robotics. Commun. ACM 60(11), 68–78 (2017)
26. Pollack, M.E., Brown, L., Colbry, D., Orosz, C., Peintner, B., Ramakrishnan, S., Engberg, S., Matthews, J., Dunbar-Jacob, J., Mccarthy, C.E., Montemerlo, M., Pineau, J., Roy, N.: Pearl: a mobile robotic assistant for the elderly (2002)
27. Balasubramanian, S., Klein, J., Burdet, E.: Robot-assisted rehabilitation of hand function. Curr. Opin. Neurol. 23(6), 661–670 (2010)
28. Lluch, M.: Healthcare professionals' organisational barriers to health information technologies—a literature review. Int. J. Med. Inform. 80(12), 849–862 (2011)
29. Gonzales, M.J., Cheung, V.C., Riek, L.D.: Designing collaborative healthcare technology for the acute care workflow. In: 9th International Conference on Pervasive Computing Technologies for Healthcare (Pervasive Health) (2015)
30. Riek, L.D.: The social co-robotics problem space: six key challenges. In: Robotics Challenges and Vision (RCV 2013) (2014)
31. Hartzog, W.: Unfair and deceptive robots. Md. L. Rev. 74, 785 (2014)
32. Thill, S., et al.: Robot-assisted therapy for autism spectrum disorders with (partially) autonomous control: challenges and outlook. Paladyn 3(4), 209–217 (2012)
33. Ricks, D.J., Colton, M.B.: Trends and considerations in robot-assisted autism therapy. In: 2010 IEEE International Conference on Robotics and Automation (ICRA). IEEE (2010)
34. Friedmann, E., et al.: Animal companions and one-year survival of patients after discharge from a coronary care unit. Public Health Rep. 95(4), 307 (1980)
35. Stiehl, W.D., et al.: The huggable: a therapeutic robotic companion for relational, affective touch. In: ACM SIGGRAPH 2006 Emerging Technologies. ACM (2006)

36. Kriglstein, S., Wallner, G.: HOMIE: an artificial companion for elderly people. In: CHI 2005 Extended Abstracts on Human Factors in Computing Systems, pp. 2094–2098 (2005). https://doi.org/10.1145/1056808.1057106
37. Scassellati, B., Admoni, H., Mataric, M.J.: Robots for use in autism research. Ann. Rev. Biomed. Eng. **14**, 275–294 (2012)
38. Marescaux, J., et al.: Transatlantic robot-assisted telesurgery. Nature **413**(6854), 379 (2001)
39. Chinzei, K., et al.: Surgical assist robot for the active navigation in the intraoperative MRI: Hardware design issues. In: Proceedings of the 2000 IEEE/RSJ International Conference on Intelligent Robots and Systems (IROS 2000), vol. 1. IEEE (2000)
40. Hawks, J.A., et al.: Towards an in vivo wireless mobile robot for surgical assistance. Stud. Health Technol. Inform. **132**, 153–158 (2008)
41. Park, B.J., Flores, R.M., Rusch, V.W.: Robotic assistance for video-assisted thoracic surgical lobectomy: technique and initial results. J. Thorac. Cardiovasc. Surg. **131**(1), 54–59 (2006)
42. Devito, D.P., et al.: Clinical acceptance and accuracy assessment of spinal implants guided with SpineAssist surgical robot: retrospective study. Spine **35**(24), 2109–2115 (2010)
43. Rentschler, M.E., et al.: Natural orifice surgery with an endoluminal mobile robot. Surg. Endosc. **21**(7), 1212–1215 (2007)
44. Bouazzaouri, S., Witherow, M.A., Castelle, K.M.: Ethics and robotics. In: Proceedings of the International Annual Conference of the American Society for Engineering Management. American Society for Engineering Management (ASEM) (2016)
45. Quinlan, M.J., Chalup, S.K., Middleton, R.H.: Techniques for improving vision and locomotion on the sony AIBO robot (2004)
46. Tamura, T., Yonemitsu, S., Itoh, A., Oikawa, D., Kawakami, A., Higashi, Y., Fujimooto, T., Nakajima, K.: Is an entertainment robot useful in the care of elderly people with severe dementia? J. Gerontol. Ser. A **59**(1), M83–M85 (2004). https://doi.org/10.1093/gerona/59.1.M83
47. Hockstein, N.G., et al.: A history of robots: from science fiction to surgical robotics. J. Robot. Surg. **1**(2), 113–118 (2007). PMC. Web: 1 October 2018
48. Capek, K.: R.U.R. (Rossum's Universal Robots). Penguin Group, New York (2004)

An Efficient Approach for Detecting Moving Objects and Deriving Their Positions and Velocities

Andreas Gustavsson[(✉)]

Mälardalen University, Högskoleplan 1, 722 20 Västerås, Sweden
andrgust@gmail.com

Abstract. Well-functioning autonomous robot solutions heavily rely on the availability of fast and correct navigation solutions. The presence of dynamic/moving objects in the environment poses a challenge because the risk of collision increases. In order to derive the best and most foreseeing re-routing solutions for cases where the planned route suddenly involves the risk of colliding with a moving object, the robot's navigation system must be provided with information about such objects' positions and velocities.

Based on sensor readings providing either 2-dimensional polar range scan or 3-dimensional point cloud data streams, we present an efficient and effective method which detects objects in the environment and derives their positions and velocities. The method has been implemented, based on the Robot Operating System (ROS), and we also present an evaluation of it. It was found that the method results in good accuracy in the position and velocity calculations, a small memory footprint and low CPU usage requirements.

Keywords: Computer vision algorithms · Object detection · Robotics · ROS

1 Introduction

Fast and correct navigation, including object avoidance, is a crucial ability for any autonomous agent (e.g. autonomous robot, human, etc.) working in a dynamic environment. Navigation is often based on a global (i.e. known and static) map of the environment. When the environment might be dynamically changing, account is usually also taken to sensor readings, which are used to create a local (i.e. temporarily valid) map, on top of the global map, to avoid collisions with dynamic objects.

Based on the idea of using a global and a local map, a novel method, introducing navigation based on magnetic flow fields, has been proposed [24]. This method relies on the existence of a base flow field (cf. a global map), showing a collision-free path (if any) from the start point to the goal point. In case objects can move, or be moved, they cannot simply be static parts of the base flow field.

© Springer Nature Switzerland AG 2020
K. Arai and S. Kapoor (Eds.): CVC 2019, AISC 944, pp. 293–313, 2020.
https://doi.org/10.1007/978-3-030-17798-0_25

The method therefore relies on information about the position and velocity of such objects. Each object is seen as having a dipole flow field, whose magnitude is proportional to the speed of the object and whose direction is aligned with its *velocity*. The total flow field, and thus the path, is updated in accordance to how the *position* of each detected object affects it. This strategy provides a very foreseeing and effective solution to the object avoidance re-routing problem. The proposed method [24] is destined to be implemented within the Robot Operating System, ROS [22].

In this paper, we present an effective and efficient method for detecting objects and deriving the position and velocity of these. The method relies on receiving a sequence of either 2-dimensional (2D) polar range scans or 3-dimensional (3D) point clouds. Our method thus serves to satisfy the prerequisite knowledge about moving objects for the mentioned path-planning algorithm, and for a multitude of other systems relying on the information provided by our system.

An implementation of the method as a ROS library and a few executable ROS nodes is provided open source [13]. Two of the nodes take as input a stream of `LaserScan` [2] messages and a stream of `PointCloud2` [5] messages, respectively, and use the library for detecting objects and deriving their positions and velocities. The position and velocity of each detected object are derived in several ROS coordinate systems (from here on referred to as *frames*) [18], including `map`, so that an agent's findings can be shared with other agents in a globally interpretable way. Thus, an agent could easily be made to benefit from other agents' findings. The library also provides the possibility for the user to define a confidence value for each found object, if desired. The implemented method has been verified and evaluated, both using simulations and physical experiments.

The rest of this paper is organized as follows. Section 2 details the foundations of our approach for finding moving objects, based on a single sensor data stream, while Sect. 3 presents our approach for utilizing several sensor data streams. Section 4 briefly explains the ROS-based implementation of our method. Section 5 explains the experiment setup we have used to perform the evaluation (based on the implementation) which is presented in Sect. 6. Some related research is presented in Sect. 7, and Sect. 8 concludes the paper and states some future tracks for our research.

2   Finding Moving Objects in 2D Polar Range Scan Data

This section presents our method for finding the positions and velocities of detected moving objects (no object classification is performed). The method is founded on the data having the form of consecutive 2D polar range scans of the environment (cf. Fig. 1). The sensory source is assumed always to produce the same number, n, of scan point ranges. Each range should represent the distance to the closest obstructing object at a given and known angle, around the sensor z-axis and relative the sensor x-axis (cf. the data produced by a 2D laser range scanner). Note, however, that the above assumptions only represent the

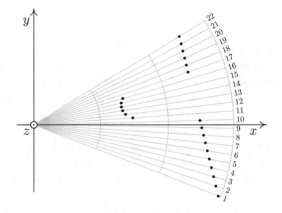

Fig. 1. 2D polar range scan example—a circular object in front of a wall. Scan rays are numbered.

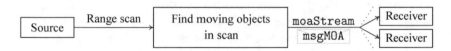

Fig. 2. Architectural description.

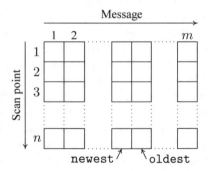

Fig. 3. The bank of m consecutive messages, each containing n scan points.

internal structure of our method and that an actual sensory source could deliver for example 3D data and have the z-axis pointing forward instead of the x-axis (cf. an *optical* frame [11]), as will become apparent in Sect. 4. The frame of the sensor is assumed to be Cartesian and right-handed.

Each scan is assumed to have an attached timestamp, representing the time at which the scan was acquired. The scans are assumed to be streamed in messages (an actual system need not have this type of setup, this is just the terminology we use to describe our method), from a *single* source (i.e. sensor). The data related to the set of found moving objects is, in turn, streamed in a message, to a set of receivers (cf. Fig. 2).

The main structure of our method is outlined in Algorithm 1. We keep a history of data scans in a bank, `bank` (cf. Fig. 3), which is similar to a circular buffer, except that incoming data is immediately handled. Setting the size, m, of the bank appropriately allows us to derive accurate velocities for the found objects, as discussed below. The oldest and newest scan messages in the bank are pointed to by `oldest` and `newest`, respectively. When a message is received on the stream, `dataStream`, the (timestamped) range data of that message replaces the oldest range data in the bank. To allow for reducing the influence of high-frequent noise, the received range data can be adapted using exponential moving average (EMA; row 7). In Algorithm 1, the parameter $\alpha \in [0, 1]$ represents the degree of weighting decrease—a higher value of α decreases the influence of old range data faster.

Objects are detected in the incoming scan data message by iterating through, comparing and grouping consecutive range values. First, the next valid scan point is found (rows 11–15). To be valid, its range value must lie within $[\text{th}_{\text{dist}}^{\min}, \text{th}_{\text{dist}}^{\max}]$. Second, all consecutive valid scan points, whose ranges are not pair-wise separated by more than $\text{th}_{\delta_{\text{edge}}}^{\max}$, are counted (rows 16–24). The derived scan points are considered to constitute one unique object. If the considered object includes the scan point with the lowest index and the sensor delivering the messages is viewing 360°, then the high end of the scan points must also be accounted for, since they might also belong to this object (rows 25–34). If the found object is constituted by less than $\text{th}_{\text{pts}}^{\min}$ points, then the object is discarded and the search continues. Otherwise, the found object is further considered.

Since the angle, relative the sensor, for each scan point is known, it is straightforward to calculate the coordinates, in the frame of the sensor, of the closest point and the estimated center point of the found object. It is also straightforward, using the law of cosine, to calculate the seen width of the object.

The velocity of the found object is calculated based on how it has moved, relative the sensor (more on this in Sect. 4), from its position in the oldest scan data message in the bank to its position in the newest scan data, and on the duration of this movement. However, for this calculation to be possible, the object must exist in the oldest scan data. Therefore, we try to track the object through each scan data message stored in the bank, from the newest to the oldest (cf. Fig. 4). The basic assumption while tracking the object is that its center scan point in the newer data covers some part of the object in the older data. There are several unrelated reasons for not succeeding in tracking an object: the center scan point in the newer data tracks to a scan point in the older data whose range value lies outside the interval given by $[\text{th}_{\text{dist}}^{\min}, \text{th}_{\text{dist}}^{\max}]$; the center scan point in the newer data tracks to an object constituted by less than $\text{th}_{\text{pts}}^{\min}$ scan points (given $\text{th}_{\delta_{\text{edge}}}^{\max}$); the distance to the object (from the sensor) differs more than $\text{th}_{\delta_{\text{dist}}}^{\max}$ between two consecutive scans; or the difference in seen width of the object between two consecutive scans is larger than $\text{th}_{\delta_{\text{width}}}^{\max}$.

Algorithm 1. Find moving objects

Input: dataStream, bank, oldest, newest, α, $\text{th}^{\max}_{\delta_{\text{edge}}}$, $\text{th}^{\min}_{\text{pts}}$, $\text{th}^{\min}_{\text{dist}}$, $\text{th}^{\max}_{\text{dist}}$, $\text{th}^{\max}_{\delta_{\text{dist}}}$, $\text{th}^{\max}_{\delta_{\text{width}}}$, $\text{th}^{\min}_{\text{conf}}$

Output: msgMOA on moaStream

```
 1 procedure
 2   loop
 3     Create new msgMOA
 4     Receive msg with timestamp, timestamp, and array of distances, data, from dataStream
 5     bank[oldest].timestamp ← msg.timestamp
 6     for all i iterating msg.data do                   ▷ Parallel operation—replace oldest data (EMA:ed)
 7     |   bank[oldest].data[i] ← α × msg.data[i] + (1 − α) × bank[newest].data[i]
 8     Update oldest and newest pointers
 9     i ← 1
10     limit ← |bank[newest].data|
11     while i ≤ limit do                                            ▷ Sequential operation
12     |   range_i ← bank[newest].data[i]                            ▷ First point of object
13     |   if range_i < th^min_dist or th^max_dist < range_i then
14     |   |   i ← i + 1
15     |   |   continue
16     |   nrObjectPoints ← 1                                 ▷ Object consists of scan point i so far
17     |   rangePrev ← range_i
18     |   for j ← i + 1 ; j ≤ |bank[newest].data| ; j ← j + 1 do    ▷ Count object scan points
19     |   |   range_j ← bank[newest].data[j]
20     |   |   if th^min_dist ≤ range_j ≤ th^max_dist and |rangePrev − range_j| ≤ th^max_{δ_edge}  then
21     |   |   |   nrObjectPoints ← nrObjectPoints + 1
22     |   |   else
23     |   |   |   break
24     |   |   rangePrev ← range_j
25     |   if i = 1 and sensor views 360° then   ▷ Valid first point, must also account high end of scan
26     |   |   rangePrev ← range_i
27     |   |   for k ← |bank[newest].data| ; j < k ; k ← k − 1 do
28     |   |   |   range_k ← bank[newest].data[k]
29     |   |   |   if th^min_dist ≤ range_k ≤ th^max_dist and |rangePrev − range_k| ≤ th^max_{δ_edge} then
30     |   |   |   |   nrObjectPoints ← nrObjectPoints + 1
31     |   |   |   |   limit ← limit − 1                  ▷ Do not account for this point from lower end
32     |   |   |   else
33     |   |   |   |   break
34     |   |   |   rangePrev ← range_k
35     |   if th^min_pts ≤ nrObjectPoints then
36     |   |   Calculate position, position_new, and width of object
37     |   |   Try to track object through bank given th^min_dist, th^max_dist, th^max_{δ_edge} th^max_{δ_dist}, th^min_pts, th^max_{δ_width}
38     |   |   if object could be tracked then
39     |   |   |   Calculate position, position_old, in oldest scans
40     |   |   |   δ_t ← bank[newest].timestamp − bank[oldest]).timestamp
41     |   |   |   velocity ← (position_new − position_old)/δ_t
42     |   |   |   Calculate confidence
43     |   |   |   if th^min_conf ≤ confidence then
44     |   |   |   |   Create MO from timestamp, position_new, velocity and confidence
45     |   |   |   |   Add MO to msgMOA
46     |   i ← j                                     ▷ Skip already-considered scan points
47     Publish msgMOA on moaStream if we found at least one object, otherwise discard msgMOA
48 end procedure
```

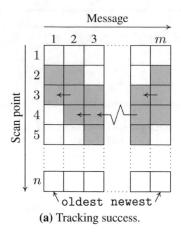 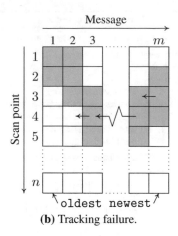

(a) Tracking success.   (b) Tracking failure.

Fig. 4. Tracking an object through the bank—gray scan points indicate actual location.

The size of the bank can, clearly, affect the accuracy of our method. For example, a relatively small size, in combination with a sensor which is producing scans with a relatively high frequency, means that $\mathtt{position_{new}} - \mathtt{position_{old}}$ and δ_t are, most likely, too small in relation to the noise in the data, to accurately estimate the current velocity of the object. On the other hand, a relatively large size, in combination with a sensor which is producing scans with a relatively low frequency, means that $\mathtt{position_{new}} - \mathtt{position_{old}}$ and δ_t, most likely, do not cover the instantaneous changes in movement well enough. For the latter case, it is easy to see that the estimated velocity of a constantly accelerating (or retarding) object could be very imprecise—the change in position between two consecutive scans is larger (or smaller) for the newer scans than for the older scans. Thus, to yield the best result, the bank size should be adapted to the scan rate of the sensor and the environmental context, if possible. (We, of course, also have the case where the sensor per se is not fast enough to satisfactory handle the given environmental context—e.g. if the environment contains objects moving very rapidly—but, this issue is very difficult to solve by means of adapting the bank size.)

If we are confident enough about our calculations, then the object is added to `msgMOA`, which is published on `moaStream`, provided that at least one object was found, after the entire newest scan data has been searched for objects. How the confidence value, `confidence`, should be calculated depends on the given context: how fast objects in the environment move, how fast our robot moves, with what rate the source produces scans, how large the bank is etc. This is further discussed in Sects. 4 and 5.

The details of a found object are stored as a `MO` object. `MO` should be viewed as containing four data fields: a timestamp, `timestamp`; a position, `position`; a velocity, `velocity`; and a confidence value, `confidence`. The `msgMOA` should be viewed as an array, list or the like, containing `MO` objects.

3   Handling Several Sources

As previously stated, the method, outlined in Algorithm 1, can receive input from a single source only. If we desire to use s sensory sources for finding moving objects, then one instance of the method must be applied for each source (cf. Fig. 5).

Whenever the method is instantiated for several sources, and we know how the sources are physically located in relation to each other, then we can determine whether an object seen by one source is the same object which is seen by another source. If several sources see the same object, then that object should be assigned a higher confidence value (cf. Fig. 5). This section presents a method for achieving this.

Our strategy is architecturally visualized in Fig. 5 and outlined in Algorithm 2. moaStream$_1$,..., moaStream$_s$ are the streams on which msgMOA messages are received. We keep a cache containing the latest message from each source. For each incoming msgMOA, and each MO it contains, a matching object is searched for in the cached message from all other sources. An object is considered to match another object if the two objects' timestamps, positions and velocities do not differ by more than $\text{th}_{\delta_t}^{\max}$, $\text{th}_{\delta_{\text{pos}}}^{\max}$ and $\text{th}_{\delta_{\text{vel}}}^{\max}$, respectively. The considered object's confidence value is increased by the average confidence of all objects, seen by other sources, found to match it (but upper-bounded by 1). All objects in msgMOA, with possibly increased confidence values, are then further reported on moaStream.

Algorithm 2. Enhance confidence of MO

Input: moaStream$_1$,..., moaStream$_s$, $\text{th}_{\delta_t}^{\max}$, $\text{th}_{\delta_{\text{pos}}}^{\max}$, $\text{th}_{\delta_{\text{vel}}}^{\max}$

Output: msgMOA on moaStream

```
 1 procedure
 2 │  Create cache of s messages, one per source
 3 │  loop
 4 │  │   Receive msgMOA from moaStreamᵢ, where i ∈ {1, ..., s}
 5 │  │   Replace cached message for source i with msgMOA
 6 │  │   for all MO in msgMOA do                                    ▷ Parallel operation
 7 │  │   │   confidenceSum ← 0
 8 │  │   │   matchingSources ← 0
 9 │  │   │   for all sources, j, such that j ≠ i do                 ▷ Parallel operation
10 │  │   │   │   for all MOⱼ in j's cached message do               ▷ Parallel operation
11 │  │   │   │   │   if |MO.timestamp − MOⱼ.timestamp| ≤ th^max_{δt} and
   │  │   │   │   │      |MO.position − MOⱼ.position| ≤ th^max_{δpos} and
   │  │   │   │   │      |MO.velocity − MOⱼ.velocity| ≤ th^max_{δvel} then
12 │  │   │   │   │   │   confidenceSum ← confidenceSum + MOⱼ.confidence
13 │  │   │   │   │   │   matchingSources ← matchingSources + 1
14 │  │   │   │   │   │   break
15 │  │   │   if 0 < matchingSources then
16 │  │   │   │   MO.confidence ← min(1, MO.confidence + confidenceSum/matchingSources)
17 │  │   Publish msgMOA (with possibly updated MOs) on moaStream
18 end procedure
```

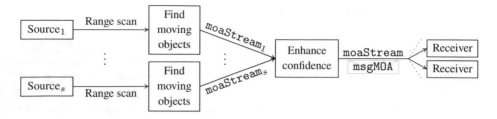

Fig. 5. Enhancing the confidence by using s sources.

4   Implementation

The confidence-enhancing method of Algorithm 2 has been implemented [13] in C++ as a ROS node [22]. Since ROS provides a communications layer, structured the way assumed by our algorithms, the implementation of the confidence enhancer node is very straightforward.

The functionality of the bank (i.e. storing a history of incoming range scan data, and deriving the position and velocity of objects found based on that data), as outlined in Algorithm 1, has been implemented [13] as a ROS C++ library. The ROS transformation system, `tf` [10], is used for transforming the object position and velocity from the sensor frame into the `base_link`, `odom` and `map` frames [18]. This automatically accounts for how the sensor has moved, as long as such information is available. The output messages (i.e. `msgMOA`) are of the type `MovingObjectArray.msg` [13]. This message type contains the name of the sending node and an array of `MovingObject.msg` [13] messages, where each such message hence corresponds to a `MO` object.

The bank can be told to publish messages meant for visualization in RViz [15]. We use several types of "visualization" messages: a `LaserScan` [2], which visualizes each scan point range, adapted using EMA, from the newest received message (the scan points of found objects have an intensity distinct from scan points not belonging to a found object); a `LaserScan`, which marks the point on each found object closest to the sensor; a `MarkerArray` [4], which adds an arrow (using a `Marker` [3] message) for each found object, where the tail of the arrow marks the position of the object and the direction and length of the arrow shows the velocity of the object; a `MarkerArray`, which adds a line (using a `Marker` message) for each found object, showing the change in position (from the oldest scan to the newest scan in the bank) for that object; and a `MarkerArray`, which adds a line (using a `Marker` message) for each found object, showing the width of the object (in the newest scan). The velocities can be shown in any of the four available frames but, the default frame is `map`.

The bank implementation provides some functionality for handling `LaserScan` or `PointCloud2` [5] input messages in a convenient manner (the user of the library can simply provide the bank with a message of any of these types and the data of that message is added to the bank automatically). An instance of the bank should, of course, only be used for a single source, sending only one of these message types, as discussed in Sects. 2 (cf. Fig. 2) and 3 (cf. Fig. 5).

The ranges in a `LaserScan` message can be handled without modification, since the bank expects a 2D polar structure. On the other hand, each point in a `PointCloud2` message must be extracted and projected, from a 3D volume onto a 2D plane, before it can be stored in the bank. The projection, in our implementation, is made from (x, y, z) volume coordinates to (x, y) plane coordinates. A 2D polar range scan structure (as expected by the bank) is achieved, by mapping each resulting (x, y) coordinate onto one or several (based on a given voxel grid leaf size) scan point(s) of the bank data structure. If several (x, y) coordinates map to the same scan point(s), then the chosen range value for that scan point is the distance to the coordinate closest to the sensor. All this, of course, makes the handling of `PointCloud2` messages more costly.

A sensor-interpreting application, using our bank library, is required to implement a function for calculating the confidence values for the found objects. The function should take as input: the information derived for that object; the arguments to the bank (such as the bank size, thresholds etc.); the difference in time between the newest and oldest data in the bank; and whether transformations of the object position and velocity, from the sensor frame into the `base_link`, `odom` and `map` frames, were possible. This allows for the user to make a context-based implementation of the confidence calculation. Note that all found objects can be reported, using the calculation `confidence` ← 1.

Along with the bank library, we provide [13] two sensor-interpreting executable ROS nodes, one for handling `LaserScan` message streams and one for handling `PointCloud2` message streams, which use the bank for finding moving objects. Each of these two nodes can take any data stream of its handled type as input which allows for them to be used on data streams from a large variety of sensors. The size of the bank (i.e. the number of scan messages it stores) can be automatically calculated by the nodes, based on measuring the rate of received scan messages and a desired time interval which the messages in the bank should cover.

We also provide [13] several ROS launch files for running our sensor-interpreting nodes. The sensor-interpreting nodes can be launched to run in a live setting, with actual physical sensors providing the data streams. Or, they can be launched in a simulated setting, with the sensor data coming from a provided recording of a 360° LIDAR and a depth camera (more about these sensors in the next section).

5   Experiment Setup and Verification

The moving-object-finding algorithm and the confidence-enhancing algorithm have been evaluated on a research laptop receiving input from a Slamtec RPLIDAR A2M8 [6] 360° LIDAR (providing a `LaserScan` [2] message stream) and an Intel RealSense D435 [1] depth camera (providing a `PointCloud2` [5] message stream) over USB 3.0 ports. The sensors were physically mounted close together with over-lapping fields of view. The laptop had an Intel i7 quad-core processor clocked at 2.9 GHz and 32 GB of DDR4 (2.4 GHz) Random Access Memory (RAM).

The evaluation environment consisted of an outdoor area with varying sun-light conditions. An object, about 70 cm wide and moving with speeds between $0\,\text{ms}^{-1}$ and $2\,\text{ms}^{-1}$, was used for detection purposes.

Experiments showed that when the bank covered a time period of 0.5 s, good accuracy in the position and velocity calculations resulted. The bank size was therefore automatically calculated by the sensor-interpreting nodes to cover a time period of 0.5 s. This resulted in a bank size of 6 messages for the `LaserScan` interpreter and 10 messages for the `PointCloud2` interpreter (the rate of received messages on the two streams were about 11 Hz and 18 Hz, the latter after voxel-grid-filtering, respectively).

The points of the `PointCloud2` messages were mapped onto a 2D polar range scan structure with a field of view of 180° and a total of 360 scan points.

The confidence value for each found object, MO, was derived using Eq. 1. Here, the resulting confidence value is cropped to lie within $[0, 1]$; α is the degree of EMA weighting decrease; $expr_1$? $expr_2$: $expr_3$ is $expr_2$ if $expr_1$, and $expr_3$ otherwise; `transformSucces` is a boolean value representing whether we could transform MO's position and velocity from the sensor frame into the `base_link`, `odom` and `map` frames [18]; δ_t is the difference between the timestamps of the newest and oldest scans in the bank; `width(MO)` is the current seen width of MO and `widthOld(MO)` is the width of MO as found in the oldest scan in the bank; and `baseConfidence` is a measure for how well we trust the given sensor (we used 0.3 for the LIDAR and 0.4 for the camera).

$$
\begin{aligned}
\max(0, \min(1, \alpha \times (&(\texttt{transformSucces ? 0.5 : 0.0}) \\
&+ \tfrac{-10}{3}(\delta_t - 0.2)(\delta_t - 0.8) \\
&- 5.0 \times |\texttt{width(MO)} - \texttt{widthOld(MO)}| \\
&+ \texttt{baseConfidence})))
\end{aligned}
\tag{1}
$$

The expression involving δ_t in Eq. 1 is visualized in Fig. 6 and results in a maximum confidence increase when $\delta_t = 0.5\,\text{s}$, which is the time period that the bank is expected to cover. If $\delta_t < 0.2\,\text{s}$ or $0.8\,\text{s} < \delta_t$, then the resulting confidence value is decreased. Thus, successful transformations of the object's position and velocity, a well-adapted bank size in combination with a well-functioning data stream from a trustworthy sensory source without much high-frequent noise, and a similar old and current object width result in high confidence.

To verify the correctness of Algorithms 1 and 2, we logged all found objects and verified the correctness of the derived position and velocity in the different frames, for both sensors. We also used RViz [15] to visually verify the correctness of the derived information. The used EMA weighting decrease factor and thresholds for the two different cases of using Algorithm 1 are presented in Table 1 and the used thresholds for Algorithm 2 are presented in Table 2.

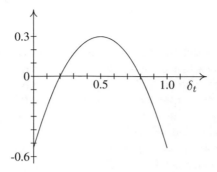

Fig. 6. The effect of δ_t on the confidence: $\frac{-10}{3}(\delta_t - 0.2)(\delta_t - 0.8)$

Table 1. Used EMA weighting decrease factor and thresholds for Algorithm 1.

LaserScan		PointCloud2	
α	1.0	α	1.0
$\text{th}_{\delta_{\text{edge}}}^{\text{max}}$	0.2 m	$\text{th}_{\delta_{\text{edge}}}^{\text{max}}$	0.15 m
$\text{th}_{\text{pts}}^{\text{min}}$	4 points	$\text{th}_{\text{pts}}^{\text{min}}$	4 points
$\text{th}_{\text{dist}}^{\text{min}}$	0.01 m	$\text{th}_{\text{dist}}^{\text{min}}$	0.01 m
$\text{th}_{\text{dist}}^{\text{max}}$	8 m	$\text{th}_{\text{dist}}^{\text{max}}$	6.5 m
$\text{th}_{\delta_{\text{dist}}}^{\text{max}}$	0.4 m	$\text{th}_{\delta_{\text{dist}}}^{\text{max}}$	0.4 m
$\text{th}_{\delta_{\text{width}}}^{\text{max}}$	20 points	$\text{th}_{\delta_{\text{width}}}^{\text{max}}$	50 points
$\text{th}_{\text{conf}}^{\text{min}}$	0.5	$\text{th}_{\text{conf}}^{\text{min}}$	0.5

Table 2. Used thresholds for Algorithm 2.

Confidence enhancer	
$\text{th}_{\delta_t}^{\text{max}}$	0.1 s
$\text{th}_{\delta_{\text{pos}}}^{\text{max}}$	0.1 m
$\text{th}_{\delta_{\text{vel}}}^{\text{max}}$	0.1 ms^{-1}

All experiments showed good accuracy in the position and velocity calculations. The Valgrind [20] heap profiler, Massif, was used for some long-running experiments to successfully verify the lack of memory leaks in our implementation. It is hence our opinion that the algorithms and implementations can be considered sufficiently and successfully verified.

6   Evaluation

To evaluate the presented method, we have measured several quantities and qualities. The first such is the resident set size, i.e. memory usage (the RAM page size was 4 kB). For the LaserScan-interpreting node, this was constant and showed 11 896 kB in our experiments. For the PointCloud2-interpreting node, this varied, depending on the size of the incoming messages (voxel-grid-filtered point clouds vary in size depending on how close objects are to the sensor) but never exceeded 40 048 kB in our experiments. For the confidence-enhancing node, this was constant and showed 10 436 kB in our experiments. Thus, we find the memory footprints of these different nodes to be of reasonable size and provide a good opportunity for achieving a high-performing system.

The second measured quantity is the average CPU usage. For the LaserScan-interpreting node, this was about 1.3%. For the PointCloud2-interpreting node, this was about 25%. For the confidence-enhancing node, this was about 0.2%. The CPU usage is of course dependent on the rate of the incoming messages and the complexity of the presented algorithms. However, the LaserScan messages arrive at a rate of about 11 Hz (see the previous section) and are of constant size, so the CPU usage required by the implementation of Algorithm 1 to find objects seems to be quite low. The higher CPU usage visible for the PointCloud2 part thus seems to stem from reading and mapping the points in each received cloud. The requirements of the confidence-enhancing node is clearly dependent on how many instants of Algorithm 1 are sending moving object information to it, and on how many objects each source has found. In our experiments, the requirements have thus shown to be very low while successfully handling the two incoming message streams. There is also a risk of the confidence-enhancing node becoming a bottleneck, due to the architecture of the system (cf. Fig. 5). For our setup, the experiments show that this is not the case, however.

The third measured quantity is the output (MovingObjectArray) message rates. For the LaserScan-interpreting node, this was about 11 Hz. For the PointCloud2-interpreting node, this was about 17 Hz. Comparing these rates to the rates of the incoming messages, it is obvious that the LaserScan-interpreting node seems to be able to handle the incoming message stream without having to drop any messages. For the PointCloud2-interpreting node, it seems that a few incoming messages have to be dropped. This conclusion is rather vague, however, because of the constantly (to a small extent) varying message rates. For the confidence-enhancing node, the rate of the outgoing message stream seemed to be equal to the total rate of the incoming messages. Thus, it seems like the confidence-enhancing node was able to handle the incoming message streams without having to drop any messages.

The fourth measured quality is how well the system is able to correctly calculate the position and velocity of objects moving in different directions, relative the sensors. The results of the position and velocity calculations (arrows; as discussed in Sect. 4), along with the width of the detected object (green/gray lines),

(a) Toward.   (b) Away from.

Fig. 7. Object moving toward and away from the sensors.

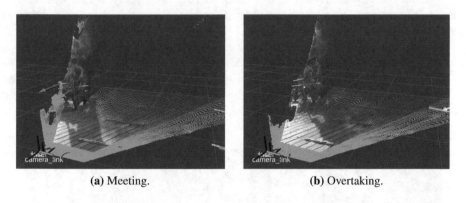

(a) Meeting.   (b) Overtaking.

Fig. 8. Object meeting and overtaking the sensors on their left.

are visualized in Figs. 7, 8, 9, 10, 11, 12, 13, 14 and 15 for an occurrence within the following cases. Note that the point cloud sometimes covers (parts of) the arrows and lines in the figures. For clarity, only the output of the sensor-interpreting nodes is shown in the figures. For the first five cases, the sensors remain in a static position (i.e. do not move).

The first case is when the object is moving toward/away from the sensors along their depth-viewing axes (Fig. 7), and the second case is when the object is meeting/overtaking the sensors on one side (Fig. 8).

The third case is when the object is moving across the field of view of the sensors (Figs. 9, 10, 11 and 12). This case poses a challenge, because the object is moving across the sensors'/banks' scan rays, not along them, which makes tracking the object from the newest scan in the bank to the oldest more difficult. When the object is crossing the field of view of the sensors at a distance of 1 m with a speed of $2\,\text{ms}^{-1}$, the system mostly fails to correctly calculate its position and velocity. For speeds around $1\,\text{ms}^{-1}$, the `PointCloud2`-interpreting node successfully calculates the position and velocity at a distance of $1\,\text{m}$, though (cf. Fig. 9). For speeds around $0.5\,\text{ms}^{-1}$ and below, both interpreting nodes successfully calculate the position and velocity at a distance of $1\,\text{m}$. When the object is crossing the field of view of the sensors at a distance of 2–4 m with a speed of $2\,\text{ms}^{-1}$, the `PointCloud2`-interpreting node is able to derive the position and velocity of the object, while the `LaserScan`-interpreting node mostly fails to do so. For speeds around $1\,\text{ms}^{-1}$ and below and at distances of 2–4 m, the `LaserScan`-interpreting node is able to successfully calculate the position and velocity of the object, though. Thus, we have established somewhat of a soft limit for at what distances and speeds the sensor interpreters fail to correctly calculate the position and speeds of the object because the rate at which the sensor produces scans is too low.

The fourth case is when the object is moving diagonally across the field of view of the sensors (Fig. 13), while the fifth case is when the object is moving in an arc in front of the sensors (Fig. 14).

The sixth case is when the *sensors are moving* forward while the object is moving perpendicularly across their path (Fig. 15).

All cases (except for the mentioned exceptions) show good accuracy in the calculation of the position and velocity of the object, and the rates of the `MovingObjectArray` output message streams are very stable.

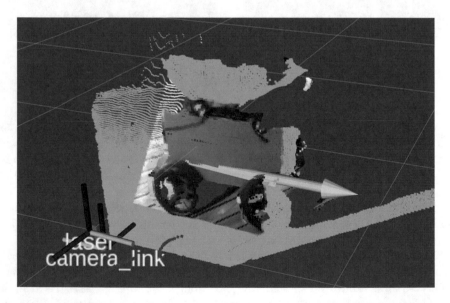

Fig. 9. Object crossing the sensors' field of view with a speed of $1\,\text{ms}^{-1}$ at a distance of 1 m.

 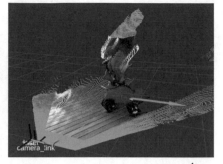

(a) Crossing with a speed of $0.5\,\mathrm{m\,s^{-1}}$.   (b) Crossing with a speed of $2\,\mathrm{m\,s^{-1}}$.

Fig. 10. Object crossing the field of view of the sensors at a distance of 2 m.

(a) Crossing with a speed of $0.5\,\mathrm{m\,s^{-1}}$.   (b) Crossing with a speed of $2\,\mathrm{m\,s^{-1}}$.

Fig. 11. Object crossing the field of view of the sensors at a distance of 3 m.

(a) Crossing with a speed of $0.5\,\mathrm{m\,s^{-1}}$.   (b) Crossing with a speed of $2\,\mathrm{m\,s^{-1}}$.

Fig. 12. Object crossing the field of view of the sensors at a distance of 4 m.

(a) 30°.   (b) 60°.

Fig. 13. Object moving diagonally across the field of view of the sensors.

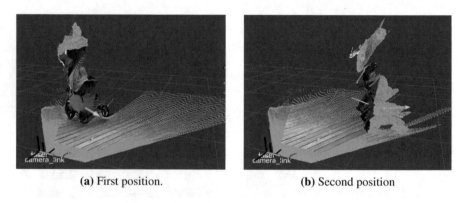

(a) First position.   (b) Second position

Fig. 14. Object moving in a convex arc in front of the sensors.

The fifth measured quality is how well instantaneous changes in how the object moves (e.g. rapidly stopping/starting or changing direction) are interpreted. Both the LaserScan- and PointCloud2-interpreting nodes handle the instantaneous changes very well when the object is moving at speeds up to about $1\,\mathrm{ms}^{-1}$. For speeds higher than this, the interpretation of the instantaneous changes becomes more difficult, but the result can still be seen as quite satisfactory. The best result was achieved for speeds between 0.5 and $1\,\mathrm{ms}^{-1}$. The confidence-enhancing node could successfully handle its two incoming data streams very well for all speeds and movement patterns.

The overall results are hence: good accuracy in the position and velocity calculations for reasonable object speeds at reasonable object distances; small memory footprints; and low CPU usage requirements for the implementations of Algorithms 1 and 2.

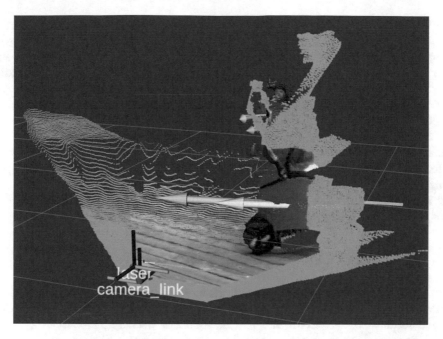

Fig. 15. Sensors moving forward at $0.5\,\mathrm{ms}^{-1}$ and object crossing their path at $1\,\mathrm{ms}^{-1}$.

7   Related Work

To the best of the author's knowledge, there is no other work trying to achieve the exact same results as those presented herein. However, this work has some similarities to work on localization and map building [8,12,17] (e.g. simultaneous localization and mapping—SLAM [7,9,17]) and on detecting and tracking moving objects [14,16,19,21,23,25,26].

Many localization and map-building solutions rely on feature-to-feature, point-to-feature or point-to-point matching techniques for determining changes in consecutive environment scans. Our method could be said to rely on a very simple feature-to-feature, or rather segment-to-segment, technique for tracking an object through the bank. The flat structure of the bank, only consisting of range readings at known directions, makes our solution quite effective and efficient.

We do not focus on tracking the detected moving objects in the environment, but only provide an instantaneous view of the surroundings. One could argue that the tracking instead will be done "automatically" by the navigation system when fed with information by our system. Like most approaches to tracking objects [26], our approach works the best for rather smooth motions of the objects in the environment (i.e. an object is assumed not to move too far between two consecutive sensor readings). Some techniques (applicable to for example street surveillance applications) rely on a static sensor with a known default/background image input, then the deviations from this known default image and the current input image represent objects.

We use partly the same strategies as those presented by Diosi and Kleeman in [8] where they try to estimate the motion of a laser range scanner using its range readings (thus, they are primarily targeting localization/odometry- and mapping-related problems but the technique could also be applied for tracking moving objects). Like Diosi and Kleeman, we too work in a polar coordinate system and two consecutive scan points are considered to belong to the same object if their ranges do not differ more than a specific threshold value. However, we do not group points whose ranges differ more than this threshold, but which lie on a straight line, into objects. The latter grouping of scan points is a very effective strategy when trying to detect walls [8] and other large plane-shaped surfaces. We do not specialize on detecting walls, however, but try to be quite general in our object detection and focus on moving objects. Therefore, we do not include such a specialized condition.

8   Discussion, Conclusions and Future Work

In this paper, we have presented an efficient and effective method for detecting moving objects and deriving their positions and velocities, based on 2D polar range scans of the environment. The method has been implemented as a C++ ROS library along with two executable nodes using the library/method. One of the nodes takes a (2D) LaserScan [2] message stream as input, while the other node takes a (3D) PointCloud2 [5] message stream as input. Thus, great flexibility in which type of sensor can be used to feed our method with a data stream is provided.

We have also provided a very efficient method for increasing the confidence value for objects which are seen/detected by several sensors. Note that this method can be used to easily perform sensor fusion with the goal of providing information about moving objects (since this method only relies on the output of the first method, and the first method can be applied to a great variety of sensor types).

In this paper, the presented methods were evaluated using a 360° LIDAR from Slamtec (RPLIDAR A2M8) [6] and a depth camera from Intel (RealSense D435) [1]. It was found that the simplicity of our approaches resulted in good accuracy in the position and velocity calculations for reasonable object speeds at reasonable object distances, and small memory footprints and low CPU usage requirements for the implementations of the presented methods.

The complexity of Algorithm 1 is dependent on the number of messages that are stored in the bank, m, and on the size of each such message, n. The complexity of caching the range values of an incoming message, and performing EMA on them, is $\mathcal{O}(n)$. The object detection (rows 11–34) loops through the range values of the newly arrived message only once. Thus, the object detection complexity is $\mathcal{O}(n)$. Tracking a detected object through the messages in the bank (row 37) is of $\mathcal{O}(mn)$ in the worst case (i.e. when all scan points constitute the object in all cached messages). The rest of the operations are of constant complexity. Thus, the total complexity is $\mathcal{O}(mn^2)$ in the worst case. However, setting the thresholds used by the algorithm appropriately will drastically reduce this worst-case complexity, like our evaluation experiments show.

The complexity of Algorithm 2 is dependent on the number of sources, s, and on the number of objects detected by each source, which is n_i, where $i \in \{1, \ldots, s\}$, in the worst case. When receiving a message from source i, each of its detected objects is compared to each object detected by all other sources. Thus, the complexity is $\mathcal{O}(n_i \prod_{j \in \{1,\ldots,s\} \setminus \{i\}} n_j) = \mathcal{O}(\prod_{j=1}^{s} n_j)$ in the worst case. For our evaluation setup, the worst-case complexity is hence $\mathcal{O}(n_1 n_2)$. Again, setting the thresholds used by the algorithms appropriately will drastically reduce this worst-case complexity, since then it will not be that each scan point for each source represents an object. This is evident in our evaluation experiments.

We consider making parallel implementations of some strategically chosen parts of the algorithms and compare their performance to the sequential versions presented in this paper. It should be noted that there are several pitfalls to consider, though. Since the number of scan points, n, in an incoming polar scan message is typically quite small, it is expected that the overhead from using several threads might not be covered by the parallel computations. It should be noted that detecting objects and tracking them through the bank are inherently sequential operations in our approach. The detection could be done in parallel by several threads, but that would require special handling of the threads' respective search boundaries. This approach is not expected to out-perform the sequential approach for realistic values of n. The objects could be tracked through the bank in parallel if changing the structure of the algorithm to first identify all objects in an incoming message and then tracking them as a subsequent operation.

Since the number of scan points might be too small for an efficient parallel implementation of Algorithm 1, it should also be that the number of found objects in an incoming msgMOA to the confidence-enhancing node is too small as well. In case we have a very large number of sources, though, then it might be efficient to compare an incoming message to the cached messages in parallel.

Acknowledgments. The research presented herein was funded by the Knowledge Foundation through the research profile "DPAC - Dependable Platforms for Autonomous systems and Control".

References

1. Intel® RealSense™ Depth Camera D435. https://click.intel.com/intelr-realsensetm-depth-camera-d435.html. Accessed 29 Sept 2018
2. LaserScan message. http://docs.ros.org/api/sensor_msgs/html/msg/LaserScan.html. Accessed 29 Sept 2018
3. Marker message. http://docs.ros.org/api/visualization_msgs/html/msg/Marker.html. Accessed 29 Sept 2018
4. MarkerArray message. http://docs.ros.org/api/visualization_msgs/html/msg/MarkerArray.html. Accessed 29 Sept 2018
5. PointCloud2 message. http://docs.ros.org/api/sensor_msgs/html/msg/PointCloud2.html. Accessed 29 Sept 2018

6. RPLIDAR A2. https://www.slamtec.com/en/Lidar/A2. Accessed 29 Sept 2018
7. Bailey, T., Durrant-Whyte, H.: Simultaneous localization and mapping (SLAM): part II. IEEE Robot. Autom. Mag. **13**(3), 108–117 (2006)
8. Diosi, A., Kleeman, L.: Fast laser scan matching using polar coordinates. Int. J. Robot. Res. **26**(10), 1125–1153 (2007)
9. Durrant-Whyte, H., Bailey, T.: Simultaneous localization and mapping: part I. IEEE Robot. Autom. Mag. **13**(2), 99–110 (2006)
10. Foote, T.: tf: the transform library. In: International Conference on Technologies for Practical Robot Applications (TePRA), pp. 1–6. IEEE (2013)
11. Foote, T., Purvis, M.: REP 103: Standard Units of Measure and Coordinate Conventions (2010). http://www.ros.org/reps/rep-0103.html. Accessed 29 Sept 2018
12. Gonzalez, J., Gutierrez, R.: Direct motion estimation from a range scan sequence. J. Robot. Syst. **16**(2), 73–80 (1999)
13. Gustavsson, A.: Find moving objects repository on GitHub. https://github.com/andreasgustavsson/find_moving_objects. Accessed 9 Nov 2018
14. Hue, C., Le Cadre, J.P., Pérez, P.: Tracking multiple objects with particle filtering. IEEE Trans. Aerosp. Electron. Syst. **38**(3), 791–812 (2002)
15. Kam, H.R., Lee, S.H., Park, T., Kim, C.H.: RViz: a toolkit for real domain data visualization. Telecommun. Syst. **60**(2), 337–345 (2015). https://doi.org/10.1007/s11235-015-0034-5
16. Khan, Z., Balch, T., Dellaert, F.: MCMC-based particle filtering for tracking a variable number of interacting targets. IEEE Trans. Pattern Anal. Mach. Intell. **27**(11), 1805–1819 (2005)
17. Kohlbrecher, S., Von Stryk, O., Meyer, J., Klingauf, U.: A flexible and scalable SLAM system with full 3D motion estimation. In: 2011 IEEE International Symposium on Safety, Security, and Rescue Robotics (SSRR), pp. 155–160. IEEE (2011)
18. Meeussen, W.: REP 105: Coordinate Frames for Mobile Platforms (2010). http://www.ros.org/reps/rep-0105.html. Accessed 29 Sept 2018
19. Montesano, L., Minguez, J., Montano, L.: Modeling the static and the dynamic parts of the environment to improve sensor-based navigation. In: Proceedings of the International Conference on Robotics and Automation (ICRA), pp. 4556–4562. IEEE (2005)
20. Nethercote, N., Seward, J.: Valgrind: a framework for heavyweight dynamic binary instrumentation. In: ACM SIGPLAN Notices, vol. 42, pp. 89–100. ACM (2007)
21. Pu, S., Rutzinger, M., Vosselman, G., Elberink, S.O.: Recognizing basic structures from mobile laser scanning data for road inventory studies. ISPRS J. Photogram. Remote Sens. **66**(6), S28–S39 (2011)
22. Quigley, M., Conley, K., Gerkey, B., Faust, J., Foote, T., Leibs, J., Wheeler, R., Ng, A.Y.: ROS: an open-source robot operating system. In: ICRA Workshop on Open Source Software, Kobe, Japan, vol. 3, p. 5 (2009)
23. Schulz, D., Burgard, W., Fox, D., Cremers, A.B.: Tracking multiple moving targets with a mobile robot using particle filters and statistical data association. In: Proceedings of the International Conference on Robotics and Automation (ICRA), vol. 2, pp. 1665–1670. IEEE (2001)
24. Trinh, L., Ekström, M., Çürüklü, B.: Toward shared working space of human and robotic agents through dipole flow field for dependable path planning. Front. Neurorobotics **1**, 1–24 (2018). http://www.es.mdh.se/publications/5128-

25. Wang, C.C., Thorpe, C., Suppe, A.: Ladar-based detection and tracking of moving objects from a ground vehicle at high speeds. In: Proceedings of the Intelligent Vehicles Symposium, pp. 416–421. IEEE (2003)
26. Yilmaz, A., Javed, O., Shah, M.: Object tracking: a survey. ACM Comput. Surv. (CSUR), **38**(4) (2006). http://doi.acm.org.ep.bib.mdh.se/10.1145/1177352.1177355

Adaptive Fusion of Sub-band Particle Filters for Robust Tracking of Multiple Objects in Video

Ahmed Mahmoud and Sherif S. Sherif[✉]

Department of Electrical and Computer Engineering, University of Manitoba,
75 Chancellor's Circle, Winnipeg, MB R3T 5V6, Canada
Sherif.Sherif@umanitoba.ca

Abstract. Video tracking is a relevant research topic because of its many surveillance, robotics, and biomedical applications. Although remarkable progress was made on this topic the capability to track objects precisely in video frames that contain difficult conditions, such as an abrupt variation in scene illumination, incomplete object camouflage, background motion and shadow, presence of objects with distinct sizes and contrasts, and presence of noise in the video frame, is still considered a vital research problem. To overcome the presence of these difficult conditions, we proposed a robust multi-scale tracker that used different sub-bands frame in the wavelet domain to express a captured video frame. Then N independent particle filters are employed to a selected subset of these sub-bands, where the selection of this wavelet sub-bands varies with every captured frame. Finally, the output position paths of these N independent particle filters were fused to obtain more precise position paths for moving objects in the video. To show the robustness of the proposed multi-scale video tracker, we employed it to various example videos that have different challenges. Opposed to a standard full-resolution particle filter-based tracker and a single wavelet sub-band $(LL)_2$ based tracker, the proposed multi-scale tracker shows greater tracking performance.

Keywords: Robust tracking · Particle filter · Wavelet transform

1 Introduction

Video tracking is a vital research topic because of its broad variety of applications that include biomedical, industrial and security applications. The purpose of video tracking is to evaluate the states of moving objects, e.g., position, velocity, and acceleration, in a video sequence. Over the past few decades, significant progress was made on video tracking. However robust video tracking remains an active research topic [1, 2]. Robust video tracking relates to the capability to avoid tracking failures [3] and to track objects precisely across video frames that have difficult conditions [4]. These problematic conditions could comprise the presence of (1) background motion and object shadows; (2) simultaneously the presence of objects with distinct sizes and contrasts; (3) presence noise in the video frame; (4) abrupt variation in scene illumination; and (5) incomplete object camouflage.

© Springer Nature Switzerland AG 2020
K. Arai and S. Kapoor (Eds.): CVC 2019, AISC 944, pp. 314–328, 2020.
https://doi.org/10.1007/978-3-030-17798-0_26

In this paper, a multi-scale visual tracker is proposed, it uses different sub-bands in the wavelet domain to express a captured video frame [4, 5]. Then N independent Particle filters (PFs) are employed to a selected subset of these sub-bands, where the selection of this wavelet sub-bands varies with every captured video frame. Finally, the output position paths of these N independent PFs were fused to obtain more accurate position paths for moving objects in the video sequence.

2 Related Work

Based on a Sequential Bayesian Framework (SBF) and Data Fusion (DF), numerous video trackers were developed to address the above problematic conditions [4, 5]. A SBF, e.g., a Kalman Filter (KF) or a PF could be considered as information fuser [6] because of their capability to integrate observation data and a motion model of the object into a single mathematical framework, this ability could be extended to provide a consistent framework for DF [4]. That is why they are attractive, although they can be expensive computationally with an increased number of tracked objects [7]. Fusion is a well-known way to accomplished the robustness of a video tracker [8]. Fusion could be performed through fusing (1) different measurements from independent sensors [9], (2) multiple features from a video frame [10], or (3) tracking results from cooperating or independent trackers [11–13].

Based on the fusion of multiple visual cues work in [14] represented a robust tracker to manipulate the variation in illumination. The authors concurrently combined color, depth cues, and optical flow in one framework to reach a robust tracker. They used the optical flow cue to obtain a coarse estimation of the object position, then the accurate location of the object is attained by using color and depth statistics.

Another robust video tracker based on the fusion of multiple visual cues was developed in [15] to address the existence of objects with distinct sizes or contrast levels, and the existence of incomplete object camouflage. The likelihood function of its Bayesian model represented human motion using multiple visual cues including color and edge shape. Because the models for visual cues were generated off-line before the tracking started, this tracker was however limited to tracking a single type of object, e.g., a human body.

Another work in [16], the authors represented a tracker that used a sequential detection framework, this framework contains both detection and tracking steps. In the detection step, DPM, *deformable part-based model*, [17] was used to get several candidates Although, in the tracking step, a GBM, a *group behaviour model*, was employed to predict potential locations of the moving objects also to handle complex interactions between object.

Another potential approach for robust video tracking is Domain Transformation because the effects of both noise and abrupt variations in illumination can be suppressed. For example the introduced work in [18], a video tracker was proposed based on comparing the energy of biorthogonal wavelet coefficients among the next frames. In the beginning, the user picks a region R that is filled by an object, then the energy of the wavelet coefficients in this region is computed. Considering the size of the moving object would not vary among adjacent frames, the evaluated energy is matched with the other energies for the potential object's region in the following frame. The area with the most similar match would consider the new object's area.

To handle the background motion and variation in illumination difficulties, the work in [19] represented a video frame by a single low-resolution sub-band that obtained by applying a two-dimensional discrete wavelet transform (2D-DWT). However, for robust tracking of objects with various contrasts and sizes would need information from every level in the wavelet domain [20]. Therefore, robust video tracking in video sequence containing such difficulties would be unachievable. To detect the edges map of a moving object at time t, another work in [21] used the dual-tree complex wavelet transform (DT-CWT) to represent three consecutive video frames $(Z_{t-1}, Z_t,$ and $Z_{t+1})$. Then two sets of difference-frames were computed based on the difference between the corresponding sub-bands at time $t-1$ and t, and the corresponding sub-bands at time t and $t+1$. Then a merger between corresponding difference-frames of the two sets was used to obtain the final edge map of the moving object.

3   Bayesian Visual Tracking

A conventional Bayesian method for video tracking is standard PF that applies to the full-resolution video frame. The PF provides an approximate solution to the states of the moving objects by a point mass function as a summation of weighted of random particles that are usually known as a sample [22]. This conventional method could have several shortcomings when tracking multiple objects across video frames that have difficult conditions, including:

1. Background motion and object shadow could create fake objects, i.e., a fake mode in the likelihood function.
2. Residence of objects with distinct sizes or contrast could give rise to a dominant likelihood problem [23, 24], in which the particle filter's posterior distribution contains multi-modes that represent different objects, but it would be biased to one object (likelihood mode) that has the most significant size or intensity.
3. Residence of noise in the video frames could generate fake objects, i.e., modes in the likelihood.
4. Abrupt variations in illumination could cause abrupt variation in the likelihood function.
5. Residence of incomplete object camouflage could produce abrupt alterations in the likelihood function.

4 Benefits of Wavelet Domain for Multiple Objects Tracking

Multiple objects tracking in the wavelet domain has several benefits comprising:

1. The wavelet transform allows a *multi-resolution analysis* (MRA). Therefore, it is suitable for tracking objects of distinct size or contrast that may be simultaneously present in a video frame. Since a coarse-resolution view is proper for tracking great size or contrast objects, while a high-resolution view is appropriate for tracking low size objects or contrast objects [20];
2. The wavelet transform is a conventional edge detector, where objects' boundaries could be detected in different directions, e.g., edges along horizontal, vertical and diagonal directions;
3. The wavelet domain is a simple approach for image denoising, generally accomplished by thresholding techniques of the wavelet coefficients [25].

5 Implementation of Our Visual Tracker

To start visual tracking, we constructed a *background frame* from the full-resolution video sequence. Then we applied the *discrete wavelet transform* to both background and current frames to generate the sub-band frames. Then we subtracted the sub-bands of the background frame from their corresponding sub-bands of the current frame to produce sub-band *difference frames*. We then applied three independent PFs to three adaptively chosen sub-bands of the difference frame. We obtained our final position path by fusing the position paths that resulted from our three sub-band particle filters (SPFs).

5.1 Background Extraction and Update

We detected moving objects across the video frames by constructing a frame that represents the Background Frame (BF). Long Term Average Background Modeling (LTABM) technique was employed for background extraction [26], as it satisfies the real-time requirement for video tracking. We constructed the initial background frame, B_0, by averaging the first few frames of the video sequence. Then we transformed B_0 to wavelet domain to get, B_0^l, the initial sub-band BFs at various l levels. At every time point, we updated the sub-band BFs, B_t^l, as described in [26].

5.2 Generation of Sub-band Frames

We generated the sub-band frames, Z_t^l, at different scales l using the *dual-tree complex wavelet* transform. This transform produced two low-frequency sub-bands, and six high-frequency sub-bands that represent detected edges at various orientations. The level l refers to the set of all sub-band frames in first and second levels of the complex wavelet tree.

5.3   Generation and Selection of Sub-band Difference Frames

We generated sub-band difference frames, D_t^s, by obtaining the differences between the current sub-band BFs, B_t^s, and the current sub-band frames, Z_t^s. As an approximation of the energy in a sub-band, we calculated the l_1 norm for all D_t^s. Then we retained only three sub-band frames possessing the greatest l_1 norm values and got rid of the others. We note that discarding the sub-bands having the lowest energies would result in denoising.

5.4   Generation of Sub-band Binary Frames and Labeling of Objects

Pixels were classified as foreground if their values were beyond a positive threshold. Therefore, the candidate moving objects were expressed by the white pixels in these sub-band binary frames. After pixel classification, morphological operations, e.g., dilation and fill operations, were applied to improve the shapes of the candidate objects.

5.5   Development of the Sub-band Particle Filters

Three independent SPFs were developed, where each SPF applied to a sub-band labeled frame. These SPFs continuously updated the kinematic states of the moving objects. The employed linear motion model in SPFs is similar to the one described in [27]. Also, the employed measurement model in SPFs is a motion cue that is analogous to the one used in [28].

5.6   Fusion of the Resulting Position Paths

SPFs generated three sets of the potential position paths for moving objects. Two fusion steps were applied to formulate the final potion path. First object confirmation step, it contains voting for the presence of an object in a predefined area. Second averaging step, it involves calculation of the average of the position paths of confirmed objects.

To correlate an object i in the previous frame with an object j in the current frame, we used one or maybe two inter-frame data association steps. First, we used a position-gating method described in [29] which imposes a distance constrain to associated object i with an object, j. If position-gating method failed, we would resort to gray-scale histogram comparison that is similar to the one described in [28].

6   Evaluation of Tracking Performance

To show the enhanced performance of the proposed multi-scale tracker, compared to a standard full-resolution particle filter based tracker (PF tracker), and to a low-resolution $(LL)_2$ tracker (LL tracker) [30], we used several video sequences that contain the above mentioned problematic conditions.

6.1   Example Demonstrating, Incomplete Object Camouflage

We used "ATCS" a Visor database video sequence. To evaluate the tracking performance of the three video trackers, we described a *detection-frame* of a particular object as a frame where the object was detected rightly by the three trackers.

As presented in Table 1, we remark that object 1 in this video emerged in 385 detection frames, with Cumulative Track Errors (CTEs) of 2177 pixels, 5320 pixels, 3893 pixels utilizing (1) PF tracker, (2) LL tracker, and (3) proposed multi-scale tracker, respectively. Object 2 emerged in 156 detection frames, with CTEs of 1446 pixels, 2456 pixels, 1892 pixels utilizing (1) PF tracker, (2) LL tracker, and (3) the proposed multi-scale tracker, respectively. Object 3 emerged in 289 detection frames, with CTEs of 1621 pixels, 3730 pixels, 2463 pixels utilizing (1) PF tracker, (2) LL tracker, and (3) the proposed multi-scale tracker, respectively. In addition to the minimum CTE values achieved by proposed multi-scale tracker, it also gave the minimum standard deviation of track errors.

6.1.1   Demonstrating Challenging Video Conditions

Incomplete Object Camouflage: Figure 1(a), (b), and (c) shows the binarization outputs produced from the 1036th frame using the full-resolution frame, using sub-band $(LL)_2$, and sub-band $(HL)_2$, which is a selected sub-bands for this 1036th frame in the proposed multi-scale tracker, respectively.

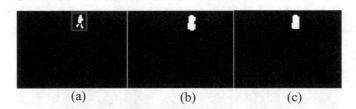

(a)   (b)   (c)

Fig. 1. Binarization outputs produced from the 1036th frame utilizing (a) the full-resolution frame (b) sub-band $(LL)_2$ (c) sub-band $(HL)_2$

(a)   (b)   (c)

Fig. 2. Tracking outcomes for 1036th frame utilizing (a) PF tracker (b) LL tracker, and (c) proposed multi-scale tracker

We remark that the red box in Fig. 1(a) refers to the division of an object because of incomplete object camouflage problem. Figure 2(a), (b) and (c) show tracking outcomes for the 1036th frame, produced by PF tracker, LL tracker, and the proposed multi-scale tracker, respectively. We note that the PF tracker creates phantom object due to the presence of incomplete object camouflage problem, while the proposed multi-scale tracker handled the existence of this problem.

6.2   Example Demonstrating, Background Motion and Shadow and Incomplete Object Camouflage

We used "OneLeaveShopReenter2front" a Caviar database video sequence. We remark that object 1 emerged in 58 detection frames, with CTEs of 381 pixels, 474 pixels, 309 pixels utilizing (1) PF tracker, (2) LL tracker, and (3) the proposed multi-scale tracker, respectively. Object 2 emerged in 470 detection frames, with CTEs of 1876 pixels, 3407 pixels, 2149 pixels utilizing (1) PF tracker, (2) LL tracker, and (3) the proposed multi-scale tracker, respectively. Object 3 emerged in 121 detection frames, CTEs of 1595 pixels, 1510 pixels, 805 pixels utilizing (1) PF tracker, (2) LL tracker, and (3) the proposed multi-scale tracker, respectively.

6.2.1   Demonstrating Challenging Video Conditions

Incomplete Object Camouflage: Figure 3(a), (b), and (c) shows the binarization outputs produced from the 88th frame utilizing the full-resolution frame, sub-band $(LL)_2$, and sub-band $(LL)_1$, which is a selected sub-bands for this 88th frame in the proposed tracker, respectively.

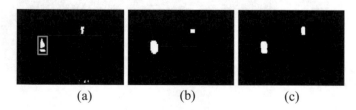

(a)   (b)   (c)

Fig. 3. Binarization outputs produced from the 88th frame utilizing: (a) the full-resolution frame; (b) sub-band $(LL)_2$; (c) sub-band $(LL)_1$

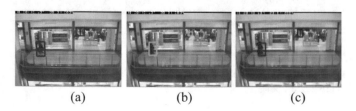

(a)   (b)   (c)

Fig. 4. Tracking outcomes for the 88th frame utilizing: (a) PF tracker (b) LL based tracker, and (c) proposed multi-scale tracker

We remark that the green box in Fig. 3(a) refers to the division of an object because of incomplete object camouflage problem. Figure 4(a), (b) and (c) represent the tracking outcomes for the 88th frame, produced by the PF tracker, LL tracker, and the proposed multi-scale tracker, respectively. We note that the PF tracker generated a phantom object because of the object division in Fig. 3(a), while the proposed multi-scale tracker and LL tracker handled this problem.

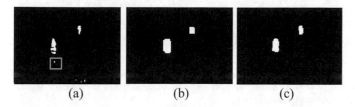

(a)   (b)   (c)

Fig. 5. Binarization outputs produced from the 116th frame utilizing: (a) the full-resolution frame; (b) sub-band $(LL)_2$; (c) sub-band $(HL)_2$

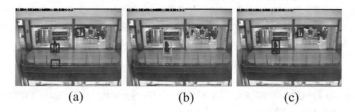

(a)   (b)   (c)

Fig. 6. Tracking outcomes for the 116th frame utilizing: (a) PF tracker (b) LL tracker, and (c) proposed multi-scale tracker

Object Shadow: Figure 5(a), (b) and (c) shows the binarization outputs produced from the 116th frame utilizing the full-resolution frame, sub-band $(LL)_2$, and sub-band $(HL)_2$, which is a selected sub-bands for this 116th frame in the proposed - tracker, respectively.

Figure 6(a), (b) and (c) represent the tracking outcomes for the 116th frame, produced by the PF tracker, LL tracker, and the proposed multi-scale tracker, respectively. We remark that the PF tracker created a phantom object because of the artifact in Fig. 5 (a), while the proposed multi-scale tracker and LL tracker cope with this problem.

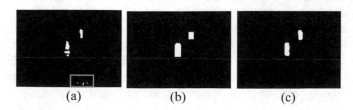

(a)   (b)   (c)

Fig. 7. Binarization outputs produced from the 138th frame utilizing: (a) the full-resolution frame; (b) sub-band $(LL)_2$; (c) sub-band $(LL)_1$

Background Motion: Figure 7(a), (b), and (c) shows the binarization outputs produced from the 138th frame utilizing the full-resolution frame, sub-band $(LL)_2$, and sub-band $(LL)_1$, which is a selected sub-bands for this 138th frame in the proposed tracker, respectively.

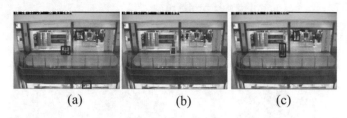

(a)   (b)   (c)

Fig. 8. Tracking outcomes for the 138th frame utilizing: (a) PF tracker (b) LL tracker, and (c) proposed multi-scale tracker

We remark that the green box in Fig. 7(a) refers to an artifact due to the background motion. Figure 8(a), (b), and (c) show the tracking outcomes for the 138th frame, produced by the PF tracker, LL tracker, and the proposed tracker, respectively. We also note that the PF tracker produced a phantom object because of the artifact in Fig. 7(a), while the proposed multi-scale tracker and PF tracker cope with the presence of background motion.

6.3   Example Demonstrating, Illumination Variation, Objects of Different Sizes and Incomplete Object Camouflage

We used in this example *"Meet_WalkTogether2"* a Caviar database video sequence. We remark that object 1 emerged in 109 detection frames, with CTEs of 721 pixels, 738 pixels, 547 pixels utilizing (1) PF tracker, (2) LL tracker, and (3) the proposed multi-scale tracker, respectively. Object 2 emerged in 8 detection frames, with CTEs of 72 pixels, 101 pixels, 80 pixels using (1) PF tracker, (2) LL tracker, and (3) the proposed multi-scale tracker, respectively. Object 3 emerged in 60 detection frames, with CTEs of 718 pixels, 1023 pixels, 653 pixels using (1) PF tracker, (2) LL tracker, and (3) the proposed multi-scale tracker, respectively.

6.3.1   Demonstrating Challenging Video Conditions

The Presence of Abrupt Illumination Variation: Figure 9(a), (b), and (c) represents the binarization outputs produced from the 67th frame utilizing the full-resolution frame, sub-band $(LL)_2$, and sub-band $(HL)_2$, which is a selected sub-bands for this 67th frame in the proposed tracker, respectively. We remark that the green box in Fig. 9(a) refers to artifacts because of abrupt illumination variation in this frame. Figure 10(a), (b), and (c) introduce the tracking outcomes for the 67th frame, produced by a PF tracker, LL tracker, and the proposed multi-scale tracker, respectively.

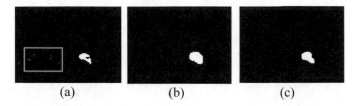

(a)   (b)   (c)

Fig. 9. Binarization outputs produced from the 67th frame utilizing: (a) the full-resolution frame; (b) sub-band $(LL)_2$; (c) sub-band $(LH)_2$

We also remark that the PF tracker produced a phantom object because of the artifact in Fig. 9(a), while the proposed multi-scale tracker and LL tracker handle the effect of abrupt illumination variation in this 67th frame.

The Presence of Objects of Distinct Sizes and Incomplete Object Camouflage: Figure 11(a), (b), and (c) shows the binarization outputs produced from the 202nd frame utilizing the full-resolution frame, sub-band $(LL)_2$, and sub-band $(HL)_2$, which is a selected sub-bands for this 202nd frame in proposed multi-scale PF based tracker, respectively.

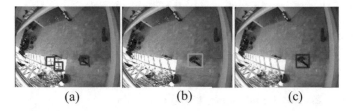

(a)   (b)   (c)

Fig. 10. Tracking outcomes for the 67th frame utilizing: (a) PF tracker (b) LL tracker, and (c) proposed multi-scale tracker

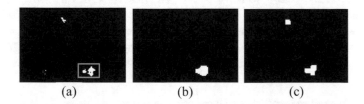

(a)   (b)   (c)

Fig. 11. Binarization outputs produced from the 202nd frame utilizing: (a) the full-resolution frame; (b) sub-band $(LL)_2$; (c) sub-band $(HL)_2$

We remark that (1) the object sizes in Fig. 11(b) and (c) are comparable to each other than the size in Fig. 11(a); and (2) the green box in (a) refers to the division of an object because of the incomplete object camouflage. Figure 12(a), (b) and (c) introduce the tracking outcomes for the 202nd frame, produced by the PF tracker and the

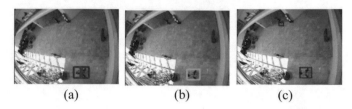

(a)   (b)   (c)

Fig. 12. Tracking outcomes for the 202nd frame utilizing: (a) PF tracker (b) LL tracker, and (c) proposed multi-scale tracker

proposed multi-scale trackers, respectively. We remark that, because of the existence of a large object, the PF tracker failed to track the smaller object. Moreover, it also suffered from an incomplete camouflage problem. Also, the LL tracker was unable to track small object due to using only one sub-band in a fixed scale. However, the proposed multi-scale tracker, cope with these problems.

6.4   Example Demonstrating Object Shadow and Incomplete Object Camouflage

In this example, we utilized *"Intelligentroom_raw"* a VISOR database video sequence. We remark that object 1 emerged in 214 detection frames, with CTEs of 982 pixels, 2024 pixels, 1025 pixels using (1) PF tracker, (2) LL tracker, and (3) the proposed multi-scale tracker, respectively. Also, we remark that PF tracker created 90 phantom objects, and the LL tracker created 11 phantom objects, while the proposed multi-scale tracker created no phantom objects.

6.4.1   Demonstrating Challenging Video Conditions

Incomplete Object Camouflage: Figure 13(a), (b), and (c) shows the binarization outputs produced from the 267th frame utilizing the full-resolution frame, sub-band $(LL)_2$, and sub-band $(HL)_1$, which is a selected sub-bands for this 267th frame in the proposed tracker, respectively. We remark that the green box in Figure 13(a) and (b) refers to the division of the object because of incomplete object camouflage problem. Figure 14(a), (b) and (c) shows the tracking outcomes for the 267th frame, produced by a PF tracker, LL tracker, and the proposed tracker, respectively.

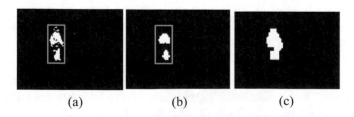

(a)   (b)   (c)

Fig. 13. Binarization outputs produced from the 267th frame utilizing: (a) the full-resolution frame; (b) sub-band $(LL)_2$; (c) sub-band $(HL)_1$

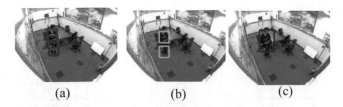

(a)   (b)   (c)

Fig. 14. Tracking outcomes for the 267th frame utilizing: (a) PF tracker (b) LL tracker, and (c) proposed multi-scale tracker

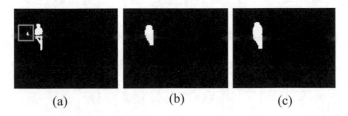

(a)   (b)   (c)

Fig. 15. Binarization outputs produced from the 240th frame utilizing: (a) the full-resolution frame; (b) sub-band $(LL)_2$; (c) sub-band $(HL)_2$

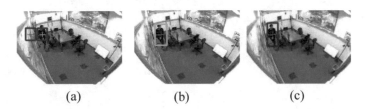

(a)   (b)   (c)

Fig. 16. Tracking outcomes for the 240th frame utilizing: (a) PF tracker (b) LL tracker, and (c) proposed multi-scale tracker

Object Shadow: Figure 15(a), (b), and (c) show the binarization outputs produced from the 240th frame using the full-resolution frame, sub-band $(LL)_2$, and sub-band $(HL)_2$, a selected sub-bands for this 240th frame in the proposed multi-scale tracker, respectively. The green box in Fig. 15(a) refers to an artifact because of the object shadow problem. Figure 16(a), (b) and (c) represented the tracking outcomes for the 240th frame, produced by the PF tracker, LL tracker, and the proposed multi-scale tracker, respectively. We remark that the PF tracker created a phantom object because of the artifact in Fig. 15(a), while the proposed multi-scale tracker and LL tracker coped successfully with the of object shadow problem.

Table 1. Number of missed object events, average position track errors, and number of phantom object events

Video seq.	Visual tracker type	Missed object (event/video frames)	Average position track error (pixel/*detection frame*)			Standard deviation of track errors			Phantom object (event/video frames)
			Obj.1	Obj.2	Obj.3	Obj.1	Obj.2	Obj.3	
ATCS	PF tracker	83	5.65	7.34	5.61	2.95	3.60	4.53	31
	LL tracker	34	13.81	15.76	12.90	5.03	4.59	3.90	0
	Multi-scale tracker	15	10.11	12.13	8.52	3.91	3.49	3.26	0
OneLeaveShop Reenter2front	PF tracker	55	6.56	3.99	13.18	3.6	3.12	5.58	469
	LL tracker	80	8.17	7.24	12.47	3.84	2.97	3.41	2
	Multi-scale tracker	23	5.32	4.57	6.65	2.68	2.37	2.82	0
Meet_Walk Together2	PF tracker	81	6.6	9.04	12	7.94	3.91	3.19	122
	LL tracker	62	6.7	12.7	17.05	451	2.45	3.70	0
	Multi-scale tracker	45	5	10.05	10.8	3.52	0.69	3.39	0
Intelligent room_raw	PF tracker	0	4.635	0	4.635	5.5	-	-	90
	LL tracker	2	9.547	2	9.547	5.8	-	-	11
	Multi-scale tracker	0	4.8	0	4.8	2.86	-	-	0

7 Conclusions

We described a multi-scale video tracker that showed a robust tracking performance. It used different sub-bands frame in the wavelet domain to express a captured frame. Then N independent PFs were applied to selected sub-bands, where this selection of this wavelet sub-bands varies with every captured frame. Finally, the output position paths of these N independent PFs were fused to obtain more precise position paths for moving objects in the video sequence. To show the robustness of the proposed multi-scale video tracker, we employed it to various example videos that have different kinds of difficulties. Opposed to a standard full-resolution particle filter-based tracker and a single low resolution $(LL)_2$ based tracker, the proposed multi-scale tracker showed significantly greater tracking performance.

References

1. Yang, H., Shao, L., Zheng, F., Wang, L., Song, Z.: Recent advances and trends in visual tracking: a review. Neurocomputing **74**(18), 3823–3831 (2011)
2. Zhang, B., Li, Z., Perina, A., Del Bue, A., Murino, V., Liu, J.: Adaptive local movement modeling for robust object tracking. IEEE Trans. Circuits Syst. Video Technol. **27**(7), 1515–1526 (2017)
3. Biresaw, T.A., Cavallaro, A., Regazzoni, C.S.: Tracker-level fusion for robust bayesian visual tracking. IEEE Trans. Circuits Syst. Video Technol. **25**(5), 776–789 (2015)
4. Zheng, N., Xue, J.: Statistical Learning and Pattern Analysis for Image and Video Processing. Springer, London (2009)
5. Bar-Shalom, Y., Li, X.-R.: Multitarget-Multisensor Tracking: Principles and Techniques. University of Connecticut, Storrs (1995)
6. Raol, J.R.: Data Fusion Mathematics: Theory and Practice. CRC Press, Boca Raton (2015)
7. Rao, G.M., Satyanarayana, C.: Visual object target tracking using particle filter: a survey. Int. J. Image Graph. Signal Process. **5**(6), 1250 (2013)
8. Maggio, E., Cavallaro, A.: Video Tracking: Theory and Practice. Wiley, Hoboken (2011)
9. Vadakkepat, P., Jing, L.: Improved particle filter in sensor fusion for tracking randomly moving object. IEEE Trans. Instrum. Meas. **55**(5), 1823–1832 (2006)
10. Hu, M., Liu, Z., Zhang, J., Zhang, G.: Robust object tracking via multi-cue fusion. Signal Process. **139**, 86–95 (2017)
11. Leang, I., Herbin, S., Girard, B., Droulez, J.: On-line fusion of trackers for single-object tracking. Pattern Recogn. **74**, 459–473 (2018)
12. Bailer, C., Pagani, A., Stricker, D.: A superior tracking approach: building a strong tracker through fusion. In: European Conference on Computer Vision 2014, pp. 170–185. Springer (2014)
13. Kwon, J., Lee, K.M.: Tracking by sampling trackers. In: 2011 IEEE International Conference on Computer Vision, ICCV, pp. 1195–1202. IEEE (2011)
14. Wang, Q., Fang, J., Yuan, Y.: Multi-cue based tracking. Neurocomputing **131**, 227–236 (2014)
15. Islam, M.Z., Oh, C.-M., Lee, C.W.: An efficient multiple cues synthesis for human tracking using a particle filtering framework. Int. J. Innov. Comput. Inf. Control **7**(6), 3379–3393 (2011)
16. Yuan, Y., Lu, Y., Wang, Q.: Tracking as a whole: multi-target tracking by modeling group behavior with sequential detection. IEEE Trans. Intell. Transp. Syst. **18**(12), 3339–3349 (2017)
17. Felzenszwalb, P.F., Girshick, R.B., McAllester, D., Ramanan, D.: Object detection with discriminatively trained part-based models. IEEE Trans. Pattern Anal. Mach. Intell. **32**(9), 1627–1645 (2010)
18. Prakash, O., Khare, A.: Tracking of moving object using energy of biorthogonal wavelet transform. Chiang Mai J. Sci. **42**(3), 783–795 (2015)
19. Cheng, F.-H., Chen, Y.-L.: Real time multiple objects tracking and identification based on discrete wavelet transform. Pattern Recogn. **39**(6), 1126–1139 (2006)
20. Gonzalez, R., Woods, R.: Digital Image Processing. Pearson Prentice Hall, Upper Saddle River (2008)
21. Celik, T., Ma, K.-K.: Moving video object edge detection using complex wavelets. In: Advances in Multimedia Information Processing, PCM 2008, pp. 259–268 (2008)
22. Dunn, W.L., Shultis, J.K.: Exploring Monte Carlo Methods. Elsevier, Amsterdam (2011)

23. Khan, Z., Balch, T., Dellaert, F.: An MCMC-based particle filter for tracking multiple interacting targets. In: Computer Vision, ECCV 2004, pp. 279–290. Springer (2004)
24. Tao, H., Sawhney, H.S., Kumar, R.: A sampling algorithm for tracking multiple objects. In: International Workshop on Vision Algorithms, pp. 53–68. Springer (1999)
25. Lang, M., Guo, H., Odegard, J.E., Burrus, C.S., Wells, R.O.: Noise reduction using an undecimated discrete wavelet transform. IEEE Signal Process. Lett. 3(1), 10–12 (1996)
26. Hassanpour, H., Sedighi, M., Manashty, A.R.: Video frame's background modeling: reviewing the techniques. J. Signal Inf. Process. 2(02), 72 (2011)
27. Rowe, D., Huerta, I., Gonzàlez, J., Villanueva, J.J.: Robust multiple-people tracking using colour-based particle filters. In: Iberian Conference on Pattern Recognition and Image Analysis, pp. 113–120. Springer (2007)
28. Pantrigo, J.J., Hernández, J., Sánchez, A.: Multiple and variable target visual tracking for video-surveillance applications. Pattern Recogn. Lett. 31(12), 1577–1590 (2010)
29. Amditis, A., Thomaidis, G., Karaseitanidis, G., Lytrivis, P., Maroudis, P.: Multiple hypothesis tracking implementation. INTECH Open Access Publisher (2012)
30. Hsia, C.-H., Chiang, J.-S., Guo, J.-M.: Multiple moving objects detection and tracking using discrete wavelet transform. INTECH Open Access Publisher (2011)

Human Tracking for Facility Surveillance

Shin-Yi Wen, Yu Yen, and Albert Y. Chen[✉]

National Taiwan University, No. 1, Sec. 4, Roosevelt Road, Taipei 10617, Taiwan
albertchen@ntu.edu.tw

Abstract. This research provides two main changes based on Detect-And-Track. To improve the Multi-Object Tracking Accuracy (MOTA) while keeping the lightweight of the original approach, this paper proposes a gradient approach to obtain higher MOTA. We use the location of two previous frames of the same identified person to calculate the gradient for the location prediction of the current frame. Then, the predicted and the detected locations are compared. We also compare the current and previous detections. With a weighted combination for matching, we increase the MOTA score and improve the results of Detect-And-Track. Moreover, this research replaces cosine distance, the original feature extractor, with Euclidean distance. By doing so, feature extraction can match Intersection over Union (IoU) better. The weighted combination, which consists of IoU and Euclidean distance, provides a better MOTA than Detect-And-Track. In addition, a greedy approach facilitates a higher MOTA when implement with IoU and Euclidean distance. This weighted combination utility is superior than the combination of IoU and cosine distance, achieving 56.1% MOTA in total on the validation data of PoseTrack ICCV'17 dataset.

Keywords: Human tracking · PoseTrack · Detect-And-Track · Mask R-CNN

1 Introduction

As computer vision technology advances rapidly on object detection and tracking, the vision community has pushed human tracking technique to a higher level than ever before. In general, the achievement is driven by a powerful detection facility and an adequate matching algorithm. With the huge improvement of detection techniques over the recent years, it is time now to consider how to enhance the efficiency of matching algorithms for real-time object tracking. Kalman Filter is one of the choices [1], it can predict the current location of the specific object based on the object's moving behaviors in the past. However, the process of Kalman Filter costs time while computing the current behavior and updating the parameters inside. To save computing time of prediction, we present a lightweight method to predict the current location of an object by using a gradient approach. This method directly using the previous behaviors to predict the current location of an object to save time. The details are described in Sect. 3.

© Springer Nature Switzerland AG 2020
K. Arai and S. Kapoor (Eds.): CVC 2019, AISC 944, pp. 329–338, 2020.
https://doi.org/10.1007/978-3-030-17798-0_27

2   Related Work

In this section, we reviewed the current literature on multi-person detection and tracking for the identification of the research gap for improvement.

2.1   Multi-person Detection

Convolutional neural network (CNN) based feature-extracting models have enabled very efficient multi-object detection, e.g., ResNet [7] and VGG [14].

The R-CNN [5] implemented the selective search method [15] to attend candidate object regions, and used CNN and the support vector machine [2] for object recognition. To mitigate the double counting problem in R-CNN, Fast R-CNN [4] extented Region of Interest Pooling (RoIPooling) after the region proposal. The extension lead Fast R-CNN to better accuracy and efficiency. Faster R-CNN [13] optimized Fast R-CNN by using a network which provided region proposals based on the feature map.

Being different from the previous object detection systems, You Only Look Once (YOLO) [11] used a single neural network to produce bounding boxes and to classify objects. YOLO9000 [12] is an improved version of the original YOLO. The model can detect objects in higher accuracy and better efficiency and is able to recognize more than 9000 kinds of objects.

Mask R-CNN is a framework extended from Faster R-CNN [13]. Mask R-CNN [6] took inspiration from Faster R-CNN. It contains a ResNet [7] as the backbone of the model. By adding a branch neural network, Mask R-CNN not only predicted the bounding boxes of objects, but also estimated object masks of each objects and detected object keypoints as well. Mask R-CNN is able to estimate mask/pose of human in images.

3D Mask R-CNN is a modified version of original Mask R-CNN. It extends its 2D convolutional kernels to 3D, by adding the third dimension – time to Mask R-CNN. More precisely, 3D Mask R-CNN computes the detection results not by considering one frame, but by considering a set of adjacent frames instead. According to Detect-and-Track [3], the 3D Mask R-CNN with ResNet-18 backbone outperforms the 2D mask R-CNN with the same feature-extracting model.

2.2   Multi-object Tracking

The multi-object tracking problem is about paring the same object across a set of continuous frames in a video. As a result, most of the multi-object tracking methods are composed of two steps: object detection and object matching.

Simple Online and Realtime Tracking (SORT) [1] implemented Faster R-CNN [13] for detecting human in each frame. For the object matching step, the Kalman Filter [1] and the Hungarian method [10] were employed to match the predicted and newly detected bounding boxes.

PoseTrack [9] transferred body joint detections from a video into a spatio-temporal graph. The tracking approach utilized integer linear programming to find the trackers of each person and each person's keypoints.

Detect-and-Track [3] compared two object detection models and several object paring methods in the literature. For object detection, 3D Mask R-CNN was introduced to compare with Mask R-CNN [6]. For the object matching step, cosine similarity, IoU, and human pose similarity were implemented and compared with each other.

3 Method

The approach of multi-person tracking is described in this section. Taking inspiration from Detect-and-Track [3], this work also employed a two-step approach. In the first step, multi-person detection was implemented, while in the second stage, different matching methods were employed to pair each detection into tracks.

3.1 Multi-person Detection

We have adapted two frame-based human pose estimation approaches, Mask R-CNN [6] and 3D Mask R-CNN [3], for human detection in each frame before matching for tracks. Both 3D and 2D Mask R-CNN with various feature-extracting model were compared.

3.2 Multi-object Tracking

As similar as the synopsis of Detect-And-Track [3], the second-stage of the approach is to link every detection pair for tracking. The multi-person detection results are predicted by the method mentioned in 2.1, and each of those detections need to be associated with the corresponding detections in previous frames. The problem is treated as a data association problem. Since Hungarian algorithm and the greedy method performed well and very closely in Detect-And-Track [3], this research also adapted those two matching algorithms. However, the main difference between this research and Detect-And-Track [3] is the way objects are compared for association. Detect-And-Track [3] compared the detection of the previous and current frames. In contrast, this research compares the current frame with a predicted detection. With this method, this research yields a better performance while tracking pedestrians.

As mentioned in the previous paragraph, this research mainly adopts two approaches to deal with data association, including IoU, cosine distance and Euclidean distance to be parts of the matching utility. Before the step, this research first predicts the current detections' locations.

The Current Detections' Location Prediction. Given one of the detections of the current frame as d_n^{id} , where id is the detection, this research exploits the detection of previous frame d_{n-1}^{id} and also d_{n-2}^{id} to predict d_n^{id} (see Fig. 1).

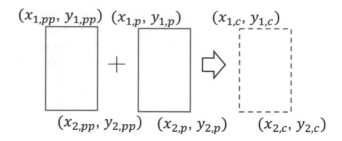

$(x_{1,pp}, y_{1,pp})$ $(x_{1,p}, y_{1,p})$ $(x_{1,c}, y_{1,c})$

$(x_{2,pp}, y_{2,pp})$ $(x_{2,p}, y_{2,p})$ $(x_{2,c}, y_{2,c})$

Fig. 1. The current detection's location

Detect the Current Detection and Compare with the Previous Detection. Applying the detection process, which is the detection mechanism in Detect-And-Track [3], it predicts a set of detection results in the previous frame and those in the current frame.

Matching Utility. This research considers two similarities for matching: (1) Location similarity and (2) Feature similarity. Location similarity is represented by IoU; Feature similarity is described by cosine distance and Euclidean distance. The matching utility is formed with a weighted combination of these similarity terms. This research does not use PCKh (Percentage of Correct Keypoints based on head) as similarity as Detect-And-Track [3] does since it is not significant in contributing to tracking performance. Please see Table 4 for details.

Base Networks. We have employed a CNN to extract features from an image. This research tests many state-of-the-art CNNs, as shown in Table 1. Each one has its advantages and disadvantages, such as high computing speed or high tracking performance.

Table 1. State-of-the-art CNNs.

Number	CNN
1	VGG-19 [14]
2	ResNet-18 [6]
3	ResNet-50 [6]
4	ResNet-101 [6]
5	ResNet-152 [6]
6	SqueezeNet1.0 [8]

Matching. Once we have the matching utility, the Hungarian Algorithm is adopted to accomplish linkage of each detection in different frames.

The weighted combination is inspired by Detect-And-Track [3] which made use of IoU, PCKh and cosine distance, for their combination as the similarity cost. Detect-And-Track [3] indicated that by only using location similarity may not be enough to yield good tracking performance. Therefore, this research applies a weighted combination the costs. However, instead of including PCKh, this paper uses the combination of IoU and cosine distance or IoU and Euclidean distance. That is because the contribution of PCKh is not significant. Moreover, our research includes Euclidean distance to be a part of the cost. Hence, Eqs. (1) and (2) are the weight combination of IoU-cosine similarity and IoU-Euclidean, respectively.

$$Cost_{IoU-cosine} = w_{IoU} \times Cost_{IoU} + w_{cosine} \times Cost_{cosine} \qquad (1)$$

$$Cost_{IoU-Euclidean} = w_{IoU} \times Cost_{IoU} + w_{Euclidean} \times Cost_{Euclidean} \qquad (2)$$

Furthermore, we adopt another cost made up of IoU comparison between the predicted detection location and the current detection location, as Eq. (3).

$$Cost_{IoU-pr} = 1.0 \times Cost_{IoU-pc} \qquad (3)$$

We consider three types of weighted utilities. The first two, shown in Eqs. (4) and (5), are discussed as before; however, since using IoU can yield a faster result, we also consider using Eq. (6) to be the matching utility.

$$Cost_{all} = w_{IoU-pr} \times Cost_{IoU-pr} + w_{IoU-cosine} \times Cost_{IoU-cosine} \qquad (4)$$

$$Cost_{all} = w_{IoU-pr} \times Cost_{IoU-pr} + w_{IoU-Euclidean} \times Cost_{IoU-Euclidean} \qquad (5)$$

$$Cost_{all} = w_{IoU-pr} \times Cost_{IoU-pr} + w_{IoU} \times Cost_{IoU} \qquad (6)$$

Matching Algorithms. This paper experiments with two matching methods: Hungarian algorithm and a greedy algorithm. These two algorithms are followed from Detect-And-Track [3]. As Detect-And-Track's experiment [3], Hungarian algorithm could yield slightly higher score in MOTA (Multi-Object Tracking Accuracy) than a greedy algorithm. Thus, we compare these two algorithms as well.

4 Results

4.1 Dataset

We have used the ICCV'17 PoseTrack Challenge dataset in this work. PoseTrack consists of human body keypoints and tracking bounding boxes. The dataset has 514 video streams, 66,374 frames in total. Moreover, the dataset is split into 300 training videos, 50 validation videos. In the dataset, images contain large crowd of pedestrians or moving people with many occlusions.

4.2   Detection Method Comparison

This paper compares 2D Mask R-CNN and 3D Mask R-CNN performance by their mean Average Precision (mAP) results. According to the evaluation from Detect-And-Track who compared 2D Mask R-CNN and 3D Mask R-CNN, both are ResNet-18 based network. In this comparison, 3D Mask R-CNN seizes a better mAP performance. The 3D Mask R-CNN with ResNet-101 captures a higher mAP score than 2D Mask R-CNN with the same network. However, because of GPU memory limitations, we compare a ResNet-101 2D Mask-RCNN architecture with a ResNet-18 3D Mask-RCNN. The mAP results are shown in Table 2.

Table 2. Comparison of mAP between 2D and 3D Mask R-CNN (Unit: %)

Dimension	Network	mAP Head	mAP Shou	mAP Elbo	mAP Wri	mAP Hip	mAP Knee	mAP Ankl	mAP Total	MOTA Total
2D	ResNet-101	68.3	70.9	62.3	52.1	62.0	59.2	50.3	61.2	55.9
3D	ResNet-18	28.0	24.8	17.1	10.6	23.8	16.7	11.4	19.5	13.9

As shown in Table 2, a ResNet-101 2D Mask-RCNN architecture seizes even higher detection performance. Furthermore, the detection performance affects tracking results which results to MOTA of 55.9. Therefore, this paper uses ResNet-101 based 2D Mask R-CNN for the following experiments.

Tracking Performance. To find an appropriate CNN for image feature extracting, this paper compares computing speed and tracking performance of CNNs mentioned in Table 1. After choosing an adequate CNN, the matching utility is calculated. Finally, either Hungarian algorithm or a greedy algorithm will be selected to be the final matching algorithm, according to their tracking performance.

CNNs. We tests on CNNs including VGG-19, ResNet-18, ResNet-50, ResNet-101, ResNet-152 and SqueezeNet1.0 for image feature extracting.

Time Cost. A 5-s video data is utilized which has 71 frames (see Fig. 2). It is called TUD-campus which is one of video in Multi-Object Tracking (MOT) Challenge dataset.

Table 3 shows the computing time with different cost combination. Moreover, since Detect-And-Track [3] adopted cosine distance to be the feature extractor, we also use their cost function to be the baseline cost.

As shown in Table 3, Squeezenet1.0 has the quickest computing time. In the following, we compare tracking performance of every CNNs.

Fig. 2. Some frames of the 5-s video.

Table 3. Computing a 5-s video time consuming with different cost functions. (Unit: second)

CNN/Cost combination	cosine distance	cosine distance + IoU	cosine distance + PCKh	All combined
VGG-19	7.2386	7.2557	7.2217	7.2597
ResNet-18	**4.6718**	**4.6993**	**4.6802**	**4.7455**
ResNet-50	7.2630	7.3532	7.3153	7.3789
ResNet-101	11.1276	11.1410	11.1489	11.1885
ResNet-152	14.6793	14.7076	14.8085	14.7191
Squeezenet1.0	**4.2109**	**4.1899**	**4.1769**	**4.2184**

Table 4. The tracking performance of dealing with a 5-s video. (Unit: %)

CNN/Cost combination	cosine distance	cosine distance + IoU	cosine distance + PCKh	All combined
VGG-19	15.2	55.9	54.8	55.9
ResNet-18	56.0	56.0	55.9	55.9
ResNet-50	56.0	56.0	56.0	56.0
ResNet-101	56.0	56.0	56.0	56.0
ResNet-152	56.0	56.0	56.0	56.0
Squeezenet1.0	55.9	55.9	54.8	55.9

Tracking Performance. Table 4 reveals the scores of every used CNNs in this work. Table 4 shows tracking performance is irrelevant to computing speed. In addition, ResNet-18 captures decent MOTA scores while using every cost combination. Therefore, ResNet-18 is selected to be the following experiments architecture.

Furthermore, as long as IoU is part of the cost function, we can obtain generally decent MOTA results. That means IoU's contribution is important. In Table 4, except VGG-19, the MOTA differences of different CNNs are subtle. Hence, in the following analysis, we round off MOTA to four decimal digits.

Similarity. This paper compares the computing speed and tracking performance.

Computing Speed. As shown in Table 5, using Euclidean distance to extract image feature is more efficient.

Table 5. The computing time of dealing with a 5-s video. (Unit: second)

Cost combination	x or y	$(x$ or $y) + IoU$
$x =$ cosine distance	4.6718	4.6993
$y =$ Euclidean distance	**4.6538**	**4.6735**

Tracking Performance. We have compared the tracking performance of Eqs. (1) and (2), as shown in Table 6. Except the last item, the rest results show IoU+Euclidean distance is better than IoU+cosine distance. Besides, The best MOTA score falls into $0.6 \times Cost_{IoU} + 0.4 \times Cost_{Euclidean}$. It means that to track human, we may need to believe in location more a bit and also use advantages of image features to match detection results.

Table 6. Tracking performance comparison between IoU+cosine and IoU+Euclidean.

w_{IoU}	w_{cosine}	$w_{Euclidean}$	MOTA (%)	Difference (E-C)
0.0	1.0	0.0	55.9437	0.0384
0.0	0.0	1.0	55.9821	
0.5	0.5	0.0	55.9831	0.0567
0.5	0.0	0.5	56.0389	
0.6	0.4	0.0	55.9638	0.0841
0.6	0.0	0.4	**56.0479**	
0.4	0.6	0.0	56.0020	−0.0089
0.4	0.0	0.6	55.9931	

On the other hand, this paper recommends some parameters of Eqs. (3), (4) and (5) to obtain decent tracking performance. As shown in Table 7, compared to the results of Detect-And-Track, our parameter can seize higher MOTA scores.

Table 7. Parameter recommendation of Eqs. (3), (4) and (5) with the results of Detect-And-Track (DAT). $w = w_{IoU-Euclidean}$ or $w_{IoU-cosine}$ or w_{IoU}.

Approach	w_{IoU}	w_{cosine}	$w_{Euclidean}$	w_{IoU-pr}	w	MOTA (%)
DAT	1.00	0.00	0.00	0.00	1.00	55.9437
Ours				0.80	0.20	**55.9782**
DAT	0.00	1.00	0.00	0.00	1.00	55.9935
Ours				0.40	0.60	**56.0245**
DAT	0.15	0.85	0.00	0.00	1.00	56.0384
Ours	0.60	0.00	0.40	0.00	1.00	**56.0479**

Matching Algorithm. We also compare two matching algorithms: the Hungarian algorithm and the greedy method. According to Detect-And-Track, the greedy method obtains a little lower MOTA scores than Hungarian approach. This paper also compares tracking performance of using these two algorithms.

Tracking Performance. As shown in Table 8, the greedy method can assist to achieve slightly higher MOTA scores. By taking a look at the second item (IoU+Euclidean distance), the greedy method can obtain MOTA of 56.1.

Table 8. Tracking performance comparison: Hungarian algorithm v.s. greedy method.

w_{IoU}	w_{cosine}	$w_{Euclidean}$	Hungarian (%)	Greedy (%)
0.6	0.4	0.0	55.9638	55.9572
0.6	0.0	0.4	**56.0479**	**56.0549**

Using the parameters of greedy method with the MOTA score 56.1%, the tracking results are shown in Fig. 3.

Fig. 3. A continuous photographic record with tracking results.

5   Discussion and Conclusions

The contribution of this paper is in two folds. First, we provide an amiable method based on the Detect-And-Track approach to improve the MOTA score. Since the time and space between the previous frame and the current frame is short, a person cannot move in an unreasonable way. In other words, we can use the previous two frames of the same person to predict that person's movement in the current frame. The results show better MOTA scores, compare to Detect-And-Track results.

To distill image features, this paper changes cosine distance to Euclidean distance. This process successfully increases MOTA as well. In addition, using greedy method can obtain higher MOTA.

While comparing to the results of Detect-And-Track, the outcome of this paper shows comparable, if not superior, MOTA scores. In summary, this paper proposes and tests two adjustments, which are sufficient to compete with the state-of-the-art.

References

1. Bishop, G., Welch, G.: An Introduction to the Kalman Filter, p. 80 (2001)
2. Cortes, C., Vapnik, V.: Support-vector networks. Mach. Learn. **20**, 273–297 (1995)
3. Girdhar, R., Gkioxari, G., Torresani, L., Paluri, M., Tran, D.: Detect-and-track: efficient pose estimation in videos. In: CVPR (2018)
4. Girshick, R.: Fast R-CNN. In: Proceedings of the IEEE International Conference on Computer Vision (2015)
5. Girshick, R., Donahue, J., Darrell, T., Malik, J.: Rich feature hierarchies for accurate object detection and semantic segmentation. In: Proceedings of the IEEE Computer Society Conference on Computer Vision and Pattern Recognition (2014)
6. He, K., Gkioxari, G., Dollar, P., Girshick, R.: Mask R-CNN. In: Proceedings of the IEEE International Conference on Computer Vision (2017)
7. He, K., Zhang, X., Ren, S., Sun, J.: Deep residual learning for image recognition. In: 2016 IEEE Conference on Computer Vision and Pattern Recognition (CVPR) (2016)
8. Iandola, F.N., Han, S., Moskewicz, M.W., Ashraf, K., Dally, W.J., Keutzer, K.: SqueezeNet: alexnet-level accuracy with $50\times$ fewer parameters and < 0.5 Mb model size. In: ICLR (2017)
9. Iqbal, U., Milan, A., Gall, J.: PoseTrack: joint multi-person pose estimation and tracking. In: IEEE Conference on Computer Vision and Pattern Recognition (CVPR) (2017). https://arxiv.org/abs/1611.07727
10. Kuhn, H.W.: The Hungarian method for the assignment problem. Nav. Res. Logist. Q. **2**(1–2), 83–97 (1955). https://doi.org/10.1002/nav.3800020109
11. Redmon, J., Divvala, S., Girshick, R., Farhadi, A.: 2016 YOLO You only look once: unified, real-time object detection. In: CVPR (2016)
12. Redmon, J., Farhadi, A.: YOLO9000: Better, faster, stronger. In: Proceedings - 30th IEEE Conference on Computer Vision and Pattern Recognition, CVPR 2017 (2017)
13. Ren, S., He, K., Girshick, R., Sun, J.: Faster R-CNN: towards real-time object detection with region proposal networks. IEEE Trans. Pattern Anal. Mach. Intell. (2017)
14. Simonyan, K., Zisserman, A.: Very deep convolutional networks for large-scale image recognition. In: International Conference on Learning Representations (ICRL) (2015)
15. Uijlings, J.R., Van De Sande, K.E., Gevers, T., Smeulders, A.W.: Selective search for object recognition. Int. J. Comput. Vis. **104**, 154–171 (2013)

Near Real-Time Robotic Grasping
of Novel Objects in Cluttered Scenes

Amirhossein Jabalameli, Nabil Ettehadi, and Aman Behal[✉]

University of Central Florida, Orlando, FL 32826, USA
amir.jabal@knights.ucf.edu, abehal@ucf.edu

Abstract. In this paper, we investigate the problem of grasping novel objects in unstructured environments. Object geometry, reachability, and force closure analysis are considered to address this problem. A framework is proposed for grasping unknown objects by localizing contact regions on the contours formed by a set of depth edges generated from a single view 2D depth image. Specifically, contact regions are determined based on edge geometric features derived from analysis of the depth map data. Finally, the performance of the approach is successfully validated by applying it to the scenes with both single and multiple objects, in both MATLAB simulation and experiments using a Kinect One sensor and a Baxter manipulator.

Keywords: Grasping · Robotics vision · Object manipulation ·
Depth map · Force closure · Depth edge · Single view · Novel object ·
Stable grasp

1 Introduction

A crucial problem in robotics is interacting with known or novel objects in unstructured environments. Among several emerging applications, assistive robotic manipulators seek approaches to assist users to perform a desired object motion in a partial or fully autonomous system. While the convergence of a multitude of research advances is required to address this problem, our goal is to describe a method that employs the robot's visual perception to identify and execute an appropriate grasp to pick and place novel objects. Finding a grasp configuration relevant to a specific task has been an active topic in robotics for the past three decades. In a recent article by Bohg et al. [1], grasp synthesis algorithms are categorized into two main groups, viz., analytical and data-driven.

This study was funded in part by NIDILRR grant #H133G120275, and in part by NSF grant numbers IIS-1409823 and IIS-1527794. However, these contents do not necessarily represent the policy of the aforementioned funding agencies, and you should not assume endorsement by the Federal Government.
A. Jabalameli and N. Ettehadi are graduate students with the Electrical and Computer Engineering Dept. at the University of Central Florida (UCF), Orlando, FL 32816.
A. Behal is with ECE and NanoScience Technology Center at UCF, Orlando, FL.

© Springer Nature Switzerland AG 2020
K. Arai and S. Kapoor (Eds.): CVC 2019, AISC 944, pp. 339–354, 2020.
https://doi.org/10.1007/978-3-030-17798-0_28

Analytical approaches explore for solutions through kinematic and dynamic formulations [2]. On the other hand, data-driven methods retrieve grasps according to their prior knowledge of either the target object, human experience, or through information obtained from acquired data. In line with this definition, Bohg *et al.* [1] classified data-driven approaches based on the encountered object being considered known, familiar, or unknown to the method. Thus, the main issues relate to how the query object is recognized and then compared with or evaluated by the algorithm's existing knowledge. As an example, [4–6] assume that all the objects can be modeled by a set of shape primitives such as boxes, cylinders and cones. During the off-line phase, they assign a desired grasp for each shape while, during the on-line phase, these approaches are only supposed to match sensed objects to one of the shape primitives and pick the corresponding grasp. A group of methods considers the encountered object as a familiar object and employs 2D and/or 3D object features to measure the similarities in shape or texture properties [1]. In [7], a logistic regression model is trained based on labeled data sets and then grasp points for the query object are detected based on the extracted feature vector from a 2D image. The last group of methods, in data-driven approaches, introduce and examine features and heuristics which directly map the acquired data to a set of candidate grasps [1]. The authors in [8] propose an approach that takes 3D point cloud and hand geometric parameters as the input, then search for grasp configurations within a lower dimensional space satisfying defined geometric necessary conditions. Another approach to grasp planning problem can be performed through object segmentation algorithms to find surface patches [9, 10]. One of the main challenges for data-driven approaches is data preparation and specifically background elimination. This matter forces some of the methods to make simplifying assumptions about an object's situation, *e.g.*, [11] is only validated for objects standing on a planar surface. Finding a feasible grasp configuration subject to the given task and user constraints is required for a group of applications. As discussed by [12–14], suggesting desired grasp configurations, in assistive human-robot interaction, results in increasing the users' engagement and easing the manipulator trajectory adaptation.

This paper is organized as follows. The grasp problem is stated in Sect. 2. In Sect. 3, we propose a framework based on the supporting principle that potential contacting regions for a stable grasp can be found by searching for (i) sharp discontinuities and (ii) regions of locally maximal principal curvature in the depth map. Details of algorithm implementation are provided in Sect. 4 and indicate that the algorithm run with low computational costs and in near real-time. In Sect. 5, we validate our proposed approach by considering different scenarios for grasping objects, using a Kinect One sensor and a Baxter robot as a 7-DOF arm manipulator. Finally, Sect. 6 concludes the paper.

2   Problem Definition

The problem addressed in this paper is to find contacting regions for grasping unknown objects in a cluttered scene. The obtained grasp needs to exhibit force closure, be reachable, and also feasible under the specifications of a given end-effector. Partial depth information of the object, which is sensed by an RGBD camera, is the only input through this process and the proposed approach assumes that the manipulated objects have rigid and non-deformable shapes. In practice, we do not utilize objects with transparent and reflective surfaces since they cannot be sensed by the employed sensor technology.

3   Approach

In this section, first we present an object representation and investigate its geometric features based on the scene depth map; then a grasp model for the end-effector is provided. In the end, pursuant to the development, we draw a relationship between an object's depth edges and force closure conditions. Finally, we specify contact location and end-effector pose to grasp the target object.

3.1   2D Object Representation

Generally, 3D scanning approaches require multiple-view scans to construct complete object models. In this work, we restrict our framework to only utilize partial information captured from a single view and represent objects in a 2-dimensional space. As previously stated, our main premise is that potential contacting regions for a stable grasp can be found by looking for (i) sharp discontinuities or (ii) regions of locally maximal principal curvatures in the depth map. A depth image can be shown by a 2D array of values which is described by an operator $d(.)$:

$$z = d(r, c), \ d(.) : R^2 \rightarrow R$$

where z denotes the depth value (distance to the camera) of a pixel positioned at coordinates (r, c) in the depth image (I_d). Mathematically speaking, our principle suggests a search for regions holding *high gradient property* in depth or depth direction values. Gradient image, gradient magnitude image, and gradient direction image are defined as follows:

$$
\begin{aligned}
&\text{Depth Image:} && I_d = [d(r_i, c_i)] \\
&\text{Image Gradient:} && \triangledown I = (\frac{\partial I_d}{\partial x}, \frac{\partial I_d}{\partial y})^T \\
&\text{Gradient Magnitude Image:} && I_M = [\sqrt{(\frac{\partial I_d}{\partial x})^2 + (\frac{\partial I_d}{\partial y})^2}] \\
&\text{Gradient Direction Image:} && I_\theta = [\tan^{-1}((\frac{\partial I_d}{\partial y})/(\frac{\partial I_d}{\partial x}))]
\end{aligned}
\tag{1}
$$

where gradient magnitude image pixels describe the change in depth values in both horizontal and vertical directions. Similarly, each pixel of gradient direction image demonstrates the direction of largest depth value increase. In our proposed terminology, a *depth edge* is defined as a 2-dimensional collection of points in the image plane which form a simple curve and satisfy the *high gradient property*.

To expound on the kinds of depth edges and what they offer to the grasping problem, we investigate their properties in depth map. All the depth edges are categorized into two main groups: 1. Depth Discontinuity (DD) edges and 2. Curvature Discontinuity (CD) edges. A DD edge is created by high gradient in depth values or a significant depth value difference between its two sides in the 2D depth map (I_d). It intimates a free-space between its belonged surface and its surroundings along the edge. A CD edge emerges from the directional change of depth values (I_θ) although it holds a continuous change in depth values on its sides. Note that the directional change of depth values is equivalent to surface orientation in 3D. In fact, a CD depth edge illustrates intersection of surfaces with different orientation characteristics in 3D. CD edges are also divided into two types, namely, *concave* and *convex*. A CD edge is called convex if the following inequality is satisfied for any points in its local neighborhood.

$$\forall j_1, j_2 \in J, \forall t \in [0, 1] :$$
$$D(t j_1 + (1 - t) j_2) \le t D(j_1) + (1 - t) D(j_2) \tag{2}$$

Otherwise, it is considered as a concave edge. Simply speaking, the outer surface of the object curves like the interior of a circle at concave edges and curves like the circle's exterior at convex edges. Moreover, each surface segment in the image plane is the projection of an object's face. Particularly, projection of a flat/planar surface maps all the belonged points to the corresponding surface segment while, in a case of curved/non-planar face, the corresponding surface segment includes that subset of the face, which is visible in the viewpoint. Assume that operator $\lambda : R^2 \to R^3$ maps 2D pixels to their real 3D coordinates. In Fig. 1, the S_i show 2D surface segments and A_i indicate collections of 3D points. It is clear that S_1 represents a flat face of the cube and $\lambda(S_1) = A$ while the surface segment S_2 implies only a subset of the cylinder's lateral surface in 3D bounded between e_2 and e_4 such that $\lambda(S_2) \subseteq A_2$. Hence, a depth edge in the image plane may or may not represent an actual edge of the object in 3-dimensional space. Thus, edge type determination in the proposed framework, relies on the viewpoint. While a concave CD edge holds its type in all the viewpoints, a convex CD edge may switch to DD edge and vice versa by changing the point of view.

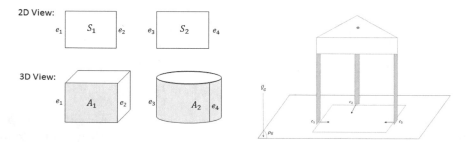

Fig. 1. (a) Geometric interpretation of a surface segment for a cube and a cylinder. (b) Grasp representation for a planar shape

3.2   Grasp and Contact Model

Generally, a precision grasp is addressed by end-effector and fingertips poses with respect to a fixed coordinate system. According to terminology adopted from [15], referring to an end-effector E with n_E fingers and n_θ joints with the fingertips contacting an object's surface, a grasp, G, is defined as follows:

$$G = (p_G, \ \theta_G, \ C_G)$$

where p_G is the end-effector pose (position and orientation) relative to the object, $\theta_G = (\theta_1, \theta_2, ..., \theta_{n\theta})$ indicates the end-effector's joint configuration, and $C_G = \{c_i \in S(O)\}_{i=1}^{n_E}$ determines a set of point contacts on the object's surface. The contact locations on the end-effector's fingers is $C_E = \{\bar{c}_i \in S(E)\}_{i=1}^{n_E}$ and defined by a forward kinematics transform from the end-effector pose p_G. Throughout this paper, we make an assumption regarding the end-effector during the interaction with the object. Each fingertip applies force in the direction of its normal and the exerted forces by all fingertips lie on a plane. We refer to this plane and its normal direction as end-effector's approach plane, ρ_G, and approach direction, $\mathbf{V}_G$. As a result, the contact points between the object and fingers will be located on this plane. In addition, some of the end-effector geometric features can be described according to how they appear on the approach plane. For instance, grasp representation, finger's opening-closing range, finger's shape and width cast on a 2D plane is shown for a 3-finger end-effector in Fig. 1.

3.3   Edge-Level Grasping

To this point, we discussed how to extract depth edges and form closed contours based on available partial information. In other words, objects are captured through 2D shapes formed by depth edges. Experiments show human tendency to grasp the objects by contacting its edges and corners [3]. The main reason is that edges provide a larger wrench convex and accordingly a greater capability to apply necessary force and torque directions. In this part, we aim to evaluate existence of grasps for each of the obtained closed contours as a way to contact/approach an object. For this matter, we use contours as the input for planar grasp synthesis process. The output grasp will satisfy reachability, force closure and feasibility with respect to end-effector geometric properties. Next, we analyze the conversion of a planar grasp to a executable 3D grasp. Finally, we point out the emerging ambiguity and uncertainties due to the 2D representation. If we assume corresponded 3D coordinates of a closed contour locate on a plane, planar grasp helps us to find appropriate force directions lie on this virtual plane. In addition, edge type determination guides us to evaluate the feasibility of applying the force directions in 3D. *Reachability* of a depth edge is measured by the availability of wrench convex lied in the plane of interest. A convex CD edge provides wrench convexes for possible contacting of two virtual planes. While, a concave CD edge is not reachable for a planar grasp. Exerting force to a DD edge, which also points to object interior, is just possible from one side. Therefore, DD and convex CD edges are remarked as reachable edges, while concave CD edges are not considered as the available points for the planar contact. For the purpose of simplicity in the analysis and without loss of generality, we approximate curved edges by a set of line segments. As a result, all 2D contours turn into polygonal shapes. To obtain the planar grasp on a polygon, the force closure construction is reduced to evaluation of all combinations of reachable edges subject to the following tests:

- Angle test: Testing if the combination of the edge wrench convexes makes the force closure possible.
- Overlapping test: Checking the existence of a region on each edge providing contact locations subject to overlapping.

The output of the angle test for 2 opposing fingers is considered valid for a combination of edges if the angle made by two edges is less than twice the friction angle. The angle test for a 3 finger end-effector is passed for a set of three contacts such that a wrench from the first contact with opposite direction overlaps with any positive combination of the other two contacts' provided wrenches. Overlapping test is also validated if there exists a contact region corresponds to each edge. To find overlapping area and corresponding region on edges, the following steps are required:

1. Form orthogonal projection area (H_i) for each edge e_i.
2. Find the intersection of projection areas by the candidate edges (overlapping area $\bar{H}$).
3. Assign the edge contact region by back-projecting overlapped region on each edge ($\bar{e}_i$).

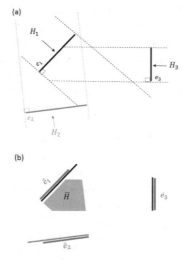

Fig. 2. Overlapping test. (a) shows intersection of orthogonal projection for three edges (b) indicates overlapped region and edge contact regions.

In fact, in this process, the planar force closure test is applied on the possible combinations of reachable edges (polygon sides) with desired number of contacts belong to a certain contour. Figure 2 illustrates the overlapping test for three edges.

To evaluate the feasibility of the obtained planar grasp with respect to the employed end-effector, we extract 3D coordinates of the pixels of interest. By accessing the 3D coordinates, we are able to recognize the Cartesian equation of a plane passes through the edge contact regions ($\bar{e}_i$). According to Sect. 3.2, the obtained plane determines end-effector approach plane (ρ_G), and approach direction ($\mathbf{V}_G$) at the grasping moment.

4 Implementation

In this section, we describe the implementation steps to process a depth image as the input and identify appropriate grasps. Notice that the current implementation focuses on finding grasps for a two-opposing finger gripper. Therefore, we employ the described algorithm in Sect. 3.3 to construct a grasp based on forming combination of two edges to indicate a pair of contact locations. A set of pixel-wise techniques is utilized to achieve the regions of interest in a 2D image and eventually obtain the desired 3D grasp. In addition, to cope with noise effects of edge detection step in the algorithm, we utilize a tweaked procedure to follow the approach steps. In fact, we skip contour formation process in the third step of the approach and directly look for the pairs that meet the discussed conditions. Thus, if an edge is missed in the detection step, we do not lose the whole contour and its corresponding edge pairs. However, the emerging complication is expansion of the pair formation search space. Later in this section, we introduce constraints to restrict this search space.

4.1   Edge Detection and Line Segmentation

According to Sect. 3.1, depth edges appear in depth image and gradient direction image. Due to the discontinuity existing by traveling in the orthogonal direction of a DD edge in depth image (I_d), the pixels belonging to the edge are local maxima of I_M (magnitude of the gradient image $\bigtriangledown I$). Alongside, a CD edge demonstrates a discontinuity in gradient direction image (I_θ) values, which illustrate a sudden change in normal directions corresponding to the edge neighborhood. Thus, in the first step, an edge detection method is required to be applied to I_d and I_θ to capture all the DD and CD edges, respectively. We selected Canny edge detection method [16] that outputs the most satisfying results with our collected data. Generally, the output of an edge detection method is a 2D binary image. Imperfect measurement in depth image yields appearance of artifacts and distorted texture in the output binary images. For instance, an ideal edge is marked out with one pixel-width. However, practically there exist non-uniform thickness along the detected edges. In order to reduce such effects and enhance the output of edge detection, a set of morphological operations is applied to the binary

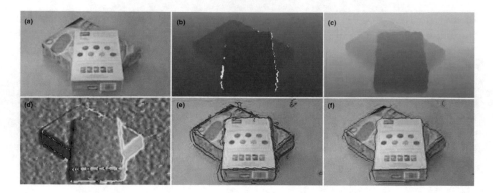

Fig. 3. Applied edge detection on an acquired depth map. (a) RGB image of the scene, I_c (b) Color map of the raw depth map. White pixels imply non-returned values from the sensor (depth shadows) (c) Color map of the processed depth map, I_d (d) Color map of computed gradient direction image, I_θ (e) Detected edges before applying the morphological operations (f) Detected edges after the morphological process, I_{DE}.

images. In coordination with the aforementioned attempt, logical OR operation is used to integrate all the marked pixels corresponding to depth edges from I_d and I_θ in a single binary image called detected depth image I_{DE}. Figure 3 shows the output of edge detection step for an acquired depth image from the Object Segmentation Dataset [17]. Note that, the only input through the whole algorithm is I_d, and color image is merely used to visualize the obtained results. For the purpose of visualizing, a range of colors is also assigned to the values of I_d and I_θ. Improvements made by the morphological operations is noticeable in Fig. 3(d). To perform further processing, a procedure is required to distinguish edges by a 2D representation in the obtained binary image (I_{DE}). Considering a 2D image with the origin on the left bottom corner, each pixel is addressed by a pair of positive integers. We employed a method proposed in [18] to cluster binary pixels into groups and then represent them by start and end points. Given I_{DE}, we first congregate the marked pixels into connected pixel arrays such that each pixel in an array is connected only to its 8 immediate neighbor pixels of the same array. Next, an iterative line fitting algorithm is utilized to divide the pixel arrays into segments such that each segment is indicated by its two end-points. The pixels belong to a segment, satisfy an allowable deviation threshold from the 2D line formed by the end-points. As a result, pixels corresponding to a straight edge are represented by one line segment while, curved edges are captured by a set of line segments. Operator $|L_i|$ computes pixel-length of line segment and $\angle(L_i)$ measures the angle which is made by the line segment and the positive direction of horizontal axis. Assuming the line segment always points out and counter clockwise as the positive orientation, $\angle(.)$ outputs an angle in the range of $[0°\tilde{\ } + 180°)$.

4.2   Edge Feature Extraction and Pair Formation

At the end of the previous step, a set of pixel groups, indicated by a corresponding set of line segments, are provided. In this part, we aim to form pairs of line segments subject to mentioned constraints in Sect. 3.3. We define local and mutual geometry features extracted from edge neighborhoods even though, mathematical relations of features rely on single pixels. We create 2D masks enclosing the line segment. Consider operator $h(.)$ locates the region of interest. A parallelogram binary mask can be obtained by

$$h(\mathbf{L}_i, \mathbf{W}) \equiv h(\mathbf{L}_i, (w, \gamma))$$

where $\mathbf{L}_i$ and $\mathbf{W}$ are the sides. In the equivalent operator representation, w shows pixel-length of the line segment $\mathbf{W}$, while γ denotes the angle between sides $\mathbf{W}$ and L_i in the range of $[-180° : +180°)$. In a similar way, we provide the following predefined masks for a line segment:

$$H^0(L_i) = h(\mathbf{L}_i, (1, \ +90))$$
$$H^+(L_i) = h(\mathbf{L}_i, (w_0, \ +90))$$
$$H^-(L_i) = h(\mathbf{L}_i, (w_0, \ -90))$$

Applying kernels build upon $h(.)$ to depth image and the other constructed images help to make the feature identification process robust. First, we evaluate reachability of each line segment and existence of a wrench convex for it. To do so, the line segments have to be assigned with an edge type label. Comparison of binary masks $H^0(L_i)$ applied to I_d and I_θ images results in distinguishing DD and CD line segments from one another. In addition, a line segment divides its local region in two sides. Therefore, the object is posed either with a positive orientation w.r.t. the line segment or a negative orientation. As discussed earlier, the wrench convex(es) is available in certain side(s) for each line segment. Note that depth value of DD edge sides hint at object relative pose with respect to the line segment. As a result, the side with lower depth value implies object (foreground) while the side with greater depth value points out the background; correspondingly available wrench is suggested. Likewise, evaluating the sides and line segment average depth values based on Eq. (1) specifies convexity/concavity of a CD edge. Mathematically speaking, edge type feature is determined for a DD line segment L_i and a CD line segment L_j as follows:

$$\begin{cases} \text{if} : \bar{d}(H^+(L_i)) < \bar{d}(H^-(L_i)) \\ \text{then} : L_i \text{ is } DD^- \\ \text{otherwise} : L_i \text{ is } DD^+ \end{cases}$$

$$\begin{cases} \text{if} : 1/2[\bar{d}(H^-(L_j)) + \bar{d}(H^+(L_j))] > \bar{d}(H^0(L_j) \\ \text{then} : L_j \text{ is } CD^\pm \\ \text{otherwise} : L_j \text{ is } CD^0 \end{cases}$$

such that $(\pm, +, -, 0)$ signs indicate availability of wrench convex w.r.t the line segment and $\bar{d}(.)$ operator takes average of depth values over the specified region.

The pair (L_i, L_j) represents a planar force closure grasp for 2-opposing fingers, if line segments have opposite wrench signs and satisfy the following conditions obtained from Sect. 3.3:

$$\begin{cases} |\measuredangle(L_i) - \measuredangle(L_j)| < 2\alpha_f \\ \bar{H}_\beta \neq \phi \end{cases}$$

where α_f is determined by the friction coefficient. The $\bar{H}_\beta$ mask is the pair overlapping area which is captured by intersection of edges projections and acquired by the following relations:

$$\begin{aligned} \bar{H}(\beta) &= H_\beta(L_i) \cap H_\beta(L_j) \\ H_\beta(L_i) &= \begin{cases} h(\mathbf{L}_i, (w_{\max}, \beta)) \text{ if } DD^- \text{ or } CD^- \\ h(\mathbf{L}_i, (w_{\max}, -\beta)) \text{ if } DD^+ \text{ or } CD^+ \end{cases} \\ \beta &= 1/2 \times |180 - |\measuredangle(L_i) - \measuredangle(L_j)|| \end{aligned}$$

such that $H_\beta(L_i)$ addresses projection area made by line segment L_i with the angle of β. In fact, β implies orthogonal direction of the bisector. Assuming existence of the overlapping area, edge contact regions, L_i^* and L_j^* are parts of the line segments which are enclosed by the $\bar{H}(\beta)$ mask. To this point, planar reachability and force closure features are assessed. As the final step, we check if the pair is feasible under the employed gripper constraints. We assume $P_i = \lambda(L_i^*)$ is the set corresponding all the 3D points located on L_i^* region. Euclidean distance between the average points of two sets P_i and P_j is required to satisfy:

$$\epsilon_{\min} < ||\bar{P}_i - \bar{P}_j||_2 < \epsilon_{\max}$$

where ϵ denotes the width range of the gripper. In addition, to assure that P_i and P_j posed on a plane, we fit plane model to the data. Throughout the current implementation, we utilized RANSAC method to estimate the plane parameters. The advantage of RANSAC is its ability to reject the outlier points resulting from noise. If a point holds greater distance from the plane than an allowable threshold $(t_{\max})$, it is considered an outlier point. The output plane and the normal unit vector pointing in the plane are referred as ρ_R and $\mathbf{V}_R$. Note that, for further processes, sets P_i and P_j are also replaced with corresponding sets excluding the outliers.

4.3   3D Grasp Specification

We desire to calculate grasp parameters based on the presented model in Sect. 3.2. To reduce the effects of uncertainties, we pick the centroid of the edge contact regions (P_i) as the safest contact points. As stated by [19], a key factor to improve the grasp quality is orthogonality of the end-effector approach direction to the object surface. In addition, the fingers of a parallel-finger gripper can only move toward each other. Hence, according to the employed grasp policy, the gripper holds a certain pose such that the gripper approach direction is aligned with normal of the extracted plane. Subsequently, closing the fingers yields contact with the object at the desired contact points. Thus, for a graspable pair, grasp parameters are described by:

$$G(L_i, L_j) = (p_G, \theta_G, C_G)$$
$$= \begin{cases} p_G = (P_G, \mathbf{R}_G) \\ \theta_G = \{\theta_1, \theta_2\} \\ C_G = \{c_1, c_2\} = \{\bar{P}_i, \bar{P}_j\} \end{cases}$$

where 3D vector P_G and rotation matrix $\mathbf{R}_G$ indicate the gripper pose. We adjust θ_G such that fingers have maximum width before contacting and width equals to $\|\bar{P}_i - \bar{P}_j\|_2$ during the contact. If length of the fingers are equal to ϵ_d and the fingers direction closure is defined by the unit vector $\mathbf{V}_c = (\bar{P}_i - \bar{P}_j)/\|\bar{P}_i - \bar{P}_j\|_2$, then we can obtain:

$$\begin{cases} P_G = 1/2 \times (\bar{P}_i + \bar{P}_j) - \epsilon_d \mathbf{V}_G \\ \mathbf{R}_G = \mathbf{R}_{o_1}^{o_2} \\ \mathbf{V}_G = \mathbf{V}_R \end{cases}$$

The matrix $\mathbf{R}_{o_1}^{o_2}$ represents a rotation from the world coordinate frame o_1 to the coordinate frame o_2 which is captured by the three orthogonal axes $[\mathbf{V}_R; \mathbf{V}_c \times \mathbf{V}_G; \mathbf{V}_G]$.

5   Results

In this section, we first evaluate the performance of detection step of grasp planning algorithm and then conduct experiments to test the overall grasping performance using the 7-DOF Baxter arm manipulator. A standard data set named Object Segmentation Database (OSD) [17] is adopted for the simulation. Besides, we collected our own data set using Microsoft Kinect One sensor for real world experiments. The data sets include a variety of unknown objects from the aspects of shape, size, and pose. In both cases, the objects are placed on a table inside the camera view and data set provides RGBD image. The depth image is fed in the grasp planning pipeline and RGB image is just used to visualize the obtained results. Note that all the computations are performed in MATLAB.

5.1   Simulation-Based Results

In this part, to validate our method, we focus on the output results of detection step in a simulation-based environment, *i.e.*, edge detection, line segmentation, and pair evaluation. To do so, we chose 8 images from OSD dataset including different object shapes and cluttered scenes. Figure 4 shows provided scenes. To specify the ground truth, we manually mark all the reachable edges (DD and Convex CD) for the existing objects and consider them as *graspable edges*. If each graspable edge is detected with correct features, it is counted as a *detected edge*. Assuming there are no gripper constraints, a *graspable surface segment* is determined if it provides at least one planar force closure grasp in the camera view. In a similar way, detected surface segment, graspable object, and detected object are specified. Table 1 shows the obtained results by applying the proposed approach on the data set.

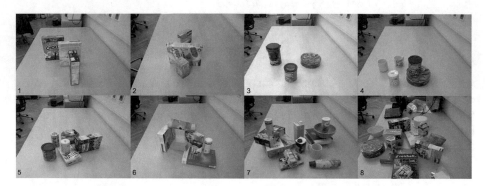

Fig. 4. Utilized images for obtaining simulation results.

Table 1. Simulation section results. Columns describe the number of (G)raspable and (D)etected objects, surface segments, and edges for 8 different scenes. The last row indicates average accuracy rates of detection at object level, surface-level and edge-level.

Scene	Objects	G. Object	D. Object	G. Surface	D. Surface	G. Edge	D. Edge
No.1	Boxes	3	3	6	6	17	14
No.2	Boxes	3	3	8	8	20	17
No.3	Cylinders	3	3	6	5	12	10
No.4	Cylinders	5	5	10	9	20	19
No.5	Mixed - low cluttered	6	6	13	9	28	21
No.6	Mixed - low cluttered	7	7	13	9	28	22
No.7	Mixed - high cluttered	11	11	24	17	55	42
No.8	Mixed - high cluttered	12	10	22	16	49	33
Average detection accuracy rate		97%		81%		80%	

According to the provided results, although 20% of the graspable edges are missed in the detection steps, 97% of the existed objects are detected and represented by at least one of their graspable surface segments. This emphasizes how skipping the contour formation step has positive effects through the grasp planning. Obtained results also indicate that the efficiency of the proposed approach decreases as the scene becomes more cluttered. Addressing how exactly the performance of these pixel-wise techniques, such as edge detection and morphology operations, affect the efficiency of the our approach is complex. Output quality and setting of these methods strongly depend on characteristics of the image view and scene. Therefore here, we only analyze edge length effects and avoid detailing other effective parameters. In fact, an edge appearing longer in a 2D image is composed of a greater number of pixels. Thus, it has a smaller chance of being missed in the detection step. In addition, since there is uncertainty in the measured data, a longer 2D edge signifies more reliable information in the grasp extraction step. On the other hand, appearance of an edge in the image relies on the distance and orientation of the object w.r.t. the camera view. Thus, depth pixel density of an object in 2D image affects the detection performance and reliability of its corresponding grasp.

5.2   Robot Experiments

For the real world experiments, the approach is run in two phases, namely grasp planing and grasp execution. In the first phase, the proposed approach is applied to the sensed data and extracted grasping options are presented to the user by displaying the candidate pairs of contact regions. Based on the selected candidate, a 3D grasp is computed for the execution phase and the grasp strategy is performed. During all the experiments, arm manipulator, RGBD camera, and the computer station are connected through a ROS network. The right arm of Baxter is fitted out with a parallel gripper. The gripper is controlled with two modes, in its "open mode" fingers distance is manually adjusted, $\epsilon_{max} = 7\,cm$ based on the size of the utilized objects. During the "closed mode", fingers take either minimum distance, $\epsilon_{min} = 2\,cm$ or hold a certain force value in the case of contacting. The grasp strategy is described for the end-effector by taking the following steps:

Step (1) Move from an initial pose to the planned pre-grasp pose.
Step (2) Wend through a straight line from pre-grasp pose to final grasp pose with fingers in the open-mode.
Step (3) Contact the object by switching the fingers to the close-mode.
Step (4) Lift the object and move to post-grasp pose.

In the current implementation, pre-grasp and post-grasp poses have the same orientation as the final grasp pose while they take a height 20 cm above the final grasp position. In this way, the end-effector approaches the object while holding a fixed orientation. Consequently, the fingers are prevented from colliding with the object prior to the grasp. Please note that a motion planner is utilized to find feasible trajectories for the arm joints. The scenario to examine algorithm's overall performance is a single object setup. In all the experiments, we assume target object is placed in the camera field of view, there exists at least one feasible grasp based on the employed gripper configuration, and planned grasps are in the workspace of the robot. An attempt is considered as a *successful grasp*, if the robot could grasp the target object and hold it for a 5 s duration after elevating.

In single object experiments, objects are in an isolated arrangement on a table in front of the robot. Four iterations are performed, covering different positions and orientations for each object. The grasp is planned by the algorithm provided in the previous section followed by robot carrying out the execution strategy to approach the object. Prior to conducting each experiment, relative finger position of the Baxter gripper are set to be wide. Table 2 shows the obtained results in single object experiment.

Table 2. Single object experiment results. Four attempts for each object are performed. "L" indicates the large size and "S" indicates small size objects.

Object	% Succ	Object	% Succ
Toothpaste Box	100	L Box	100
S Blue Box	75	L Paper Cup	100
Banana	100	L Plastic Cup	100
S Paper Cup	75	Green Cylinder	100
Apple Charger	75	L Pill Container	100
Tropicana Bottle	100	Chips Container	75
S Pill Container	100	Smoothie Bottle	100
Mouse	50	Fruit Can	100
Average: 90.62%			

According to the provided rates, 90% of the robot attempts were successful for the entire set where 11 objects were grasped successfully in all 4 iterations, 4 objects failed to be grasped successfully in 1 out of 4 iterations, while one object (mouse) had 2 successful and 2 unsuccessful attempts. In the unsuccessful attempts, the inappropriate orientation of the gripper during approaching moment is observed as the main reason of failure (4 out of 6) preventing the fingers from forming force closure on the desired contact regions. Basically, this relates performance of plane extraction from the detected contact regions. Observations during the experiments illustrate high sensitivity of the plane retrieval step to existence of unreliable data in the case of curved shape objects. For instance, in grasping the toothpaste box, although estimated normal direction ($\mathbf{V}_R$) made a 19° angel with the expected normal direction (actual normal of the surface), the object was lifted successfully. However, a 9° error resulted in failure to grasp the mouse. Impact of force closure uncertainties on the mouse case is also noticeable. For the other 2 unsuccessful attempts in the single object experiment, inaccurate positioning of the gripper was the main reason for the failure. For grasping the apple charger, gripper could not contact the planned regions, due to noisy values retrieved from low number of pixels on the object edges. The video of the robot executing the tasks can be found on-line [https://youtu.be/J87oKvFQjAE].

6   Conclusions

We have proposed an approach to grasp novel objects in an unstructured scene. Our algorithm estimates reliable regions on the contours (formed by a set of depth edges) to contact the object based on geometric features extracted from a captured single view depth map. The proposed algorithm leads to a force-closure grasp. Real world experiments demonstrate the ability of the proposed method to successfully grasp a variety of objects of different shapes, sizes, and colors.

References

1. Bohg, J., Morales, A., Asfour, T., Kragic, D.: Data-driven grasp synthesis–a survey. IEEE Trans. Robot. **30**(2), 289–309 (2014)
2. Sahbani, A., El-Khoury, S., Bidaud, P.: An overview of 3D object grasp synthesis algorithms. Robot. Auton. Syst. **60**(3), 326–336 (2012)
3. Nguyen, V.D.: Constructing force-closure grasps. In: Proceedings of the 1986 IEEE International Conference on Robotics and Automation, pp. 1368–1373 (1986)
4. Miller, A.T., Knoop, S., Christensen, H.I., Allen, P.K.: Automatic grasp planning using shape primitives. In: Proceedings of the IEEE International Conference Robotics and Automation, pp. 1824–1829 (2003)
5. Hübner, K., Kragic, D.: Selection of robot pre-grasps using box-based shape approximation. In: Proceedings of the IEEE/RSJ International Conference Intelligent Robots System, pp. 1765–1770 (2008)
6. Przybylski, M., Asfour, T., Dillmann, R.: Planning grasps for robotic hands using a novel object representation based on the medial axis transform. In: Proceedings IEEE/RSJ International Conference on Intelligent Robots and Systems, pp. 1781–1788, September 2011
7. Saxena, A., Driemeyer, J., Ng, A.Y.: Robotic grasping of novel objects using vision. Int. J. Robot. Res. **27**(2), 157–173 (2008)
8. ten Pas, A., Platt, R.: Using geometry to detect grasp poses in 3D point clouds. In: International Symposium on Robotics Research (2015)
9. Teng, Z., Xiao, J.: Surface-based detection and 6-DoF pose estimation of 3-D objects in cluttered scenes. IEEE Trans. Robot. **32**(6), 1347–1361 (2016)
10. Ückermann, A., Haschke, R., Ritter, H.: Realtime 3D segmentation for human-robot interaction. In: 2013 IEEE/RSJ International Conference on Intelligent Robots and Systems, Tokyo, pp. 2136–2143 (2013)
11. Suzuki, T., Oka, T.: Grasping of unknown objects on a planar surface using a single depth image. In: 2016 IEEE International Conference on Advanced Intelligent Mechatronics (AIM), Banff, AB, pp. 572–577 (2016)
12. Parkhurst, E.L., Rupp, M.A., Jabalameli, A., Behal, A., Smither, J.A.: Compensations for an assistive robotic interface. Proc. Hum. Factors Ergon. Soc. Annu. Meet. **61**(1), 1793–1793 (2017)
13. Rahmatizadeh, R., Abolghasemi, P., Boloni, L., Jabalameli, A., Behal, A.: Trajectory adaptation of robot arms for head-pose dependent assistive tasks. In: FLAIRS Conference (2016)
14. Ettehadi, N., Behal, A.: Implementation of feeding task via learning from demonstration. In: 2018 Second IEEE International Conference on Robotic Computing (IRC), Laguna Hills, CA, pp. 274–277 (2018)
15. Stork, A.: Representation and learning for robotic grasping, caging, and planning. Ph.D. dissertation, Stockholm (2016)
16. Canny, J.: A computational approach to edge detection. IEEE Trans. Pattern Anal. Mach. Intell. **PAMI–8**(6), 679–698 (1986)

17. Richtsfeld, A., Mörwald, T., Prankl, J., Zillich, M., Vincze, M.: Segmentation of unknown objects in indoor environments. In: 2012 IEEE/RSJ International Conference on Intelligent Robots and Systems, Vilamoura, pp. 4791–4796 (2012)
18. Kovesi, P.D.: MATLAB and Octave functions for computer vision and image processing (2000). http://www.peterkovesi.com/matlabfns
19. Balasubramanian, R., Xu, L., Brook, P.D., Smith, J.R., Matsuoka, Y.: Human-guided grasp measures improve grasp robustness on physical robot. In: 2010 IEEE International Conference on Robotics and Automation, Anchorage, AK, pp. 2294–2301 (2010)

Identifying Emerging Trends and Temporal Patterns About Self-driving Cars in Scientific Literature

Workneh Y. Ayele$^{(\boxtimes)}$ and Imran Akram

Department of Computer and Systems Science, Stockholm University,
Stockholm, Sweden
workneh@dsv.su.se, imraan.akram@gmail.com

Abstract. Self-driving is an emerging technology which has several benefits such as improved quality of life, crash reductions, and fuel efficiency. There are however concerns regarding the utilization of self-driving technology such as affordability, safety, control, and liabilities. There is an increased effort in research centers, academia, and the industry to advance every sphere of science and technology yet it is getting harder to find innovative ideas. However, there is untapped potential to analyze the increasing research results using visual analytics, scientometrics, and machine learning. In this paper, we used scientific literature database, Scopus to collect relevant dataset and applied a visual analytics tool, CiteSpace, to conduct co-citation clustering, term burst detection, time series analysis to identify emerging trends, and analysis of global impacts and collaboration. Also, we applied unsupervised topic modeling, Latent Dirichlet Allocation (LDA) to identify hidden topics for gaining more insight about topics regarding self-driving technology. The results show emerging trends relevant to self-driving technology and global and regional collaboration between countries. Moreover, the result form the LDA shows that standard topic modeling reveals hidden topics without trend information. We believe that the result of this study indicates key technological areas and research domains which are the hot spots of the technology. For the future, we plan to include dynamic topic modeling to identify trends.

Keywords: Self-driving car · Clustering · Term burst detection ·
Time series analysis · Visual analytics

1 Introduction

Self-driving or autonomous driving is an emerging automotive technology [1], and it has been an interesting topic since the 1920s [2]. The evolution of self-driving vehicles has progressed from radio-based to vision and electronic-based systems. Self-driving cars were controlled in the 1920s by radio, 1960s by electronic, and 1980s by vision & electronic systems [2]. The emergence of self-driving technology may presumably indicate concerns, increasing interests in the use of technology, and market potential. There are increasing concerns such as the impact of self-driving technologies on urban

© Springer Nature Switzerland AG 2020
K. Arai and S. Kapoor (Eds.): CVC 2019, AISC 944, pp. 355–372, 2020.
https://doi.org/10.1007/978-3-030-17798-0_29

planning [3], ethical issues [4], transportation system, affordability, safety, control and liabilities [5]. Despite these concerns, self-driving technologies have several benefits [3].

There are several benefits of utilizing self-driving cars such as promoting safety and improving quality of life [5], efficient parking, optimum driving time, crash reductions and fuel efficiency [6]. Hence, as a disruptive technology, research regarding self-driving cars and self-driven vehicles is valuable as the technology touches a broader sphere of research domains. Also, stakeholders engaged in research, prototyping and manufacturing benefit by collaboration with universities and research centers. Collaboration between academia and industry, despite being mutually beneficial, is often challenging, not only because of varying pace but also because of the different goals of these stakeholders [7]. In addition to that, there is also a communication gap between academia and the industry [8]. Brijs argues that there is no easy way to bridge the communication gap [8]. However, technology incubators, research centers, and technology transfer agents play an important role in bridging this communication gap [9]. Furthermore, despite the exponential growth of research results from increased research efforts, it is getting harder to find ideas [10].

Trend analysis can be used for forecasting trends in technology [11]. Analyzing and identifying emerging trends can be beneficial for the decision makers and stakeholders in academia and the industry such as financiers, universities, companies and academic publishers [12]. For example, identification of emerging trends of topics in science and technology is essential for making decisions [13]. Therefore, in this study, we aim to identify emerging trends about self-driving cars using academic research findings by applying visual analytics and machine learning.

The research question is **"What are emerging trends in auto-industry, in particular, self-driving cars?"** To answer this research question, we applied visual analytics on a scientific literature dataset by applying co-citation clustering, time series analysis of clusters, time-series analysis of term burst detection based on [14], Kleinberg's, algorithm. Temporal patterns are measured using citation burstiness [15]. Also, we analyzed the visualization of global collaboration on Google Earth using CiteSpace generated geospatial data. A corpus containing a total of 3323 was collected from Scopus on September 20, 2018. The collected data spans from 1976 to 2019 of which 18 articles were obtained that are accepted for publication in 2019. We chose Scopus because it covers most recent scientific journals [16] and larger databases than Web of Science [17].

Finally, we used the same dataset and run topic identification using Latent Dirichlet Allocation (LDA) and RStudio[1] and identified 20 hidden topics following the use case presented by [3]. We then compared these 20 topics with trending topics identified by using CiteSpace. The comparison shows that unless temporal, dynamic topic modeling is used, standard topic modeling techniques do not reveal trends of topics. Therefore, for the future, we plan to use dynamic topic modeling to identify temporal trends and compare them with the results of CiteSpace. In this paper, we present six sections.

---

[1] https://www.rstudio.com/products/rstudio/download/

In the second section, we present the A Review of Trend Identification. From the third to the Sixth Sections we present Methodology, Results, Discussion and Future Directions, and Conclusions.

2 A Review of Trend Identification: Visualization Analytics, Scientometric, Text Mining, Clustering of Co-citation Networks

In this section, a brief review of methods that are applicable to identify emerging trends such as text mining, co-citation analysis, visual analytics, and scientometrics are presented.

2.1 Visual Analytics

The first computer visualization of citation network was done by Yemish in 1975 [18]. Visual analytics applied on scientific literature enables to track the emergence of trends and to identify critical evidence in repeatable, flexible, timely, and valuable approach [19]. Keim et al. define visual analytics as a set of techniques that combines automated analysis with interactive visualizations for an adequate understanding, reasoning & decision making based on the analysis of large and complex datasets [20]. Visual analytics can be applied to analyze human mobility to understand mobility patterns and movement behaviors to support human perception, cognition, and reasoning [21]. Also, it is possible to depict geographical collaboration of authors using visual analytics, see Sect. 5.3.

2.2 Scientometrics

The study of academic literature research performance measurement and evaluation of quantitative aspects of communication of science as a system is referred to as scientometrics [22]. Trend analysis using co-citation and co-word detection for either policy or academic purposes could be used to detect research trends [23]. In addition to visualization functionalities, text mining functionalities are becoming common in scientometric tools such as VOSviewer [24]. Scientometric provides the analysis and identification of hot trends by providing insights and reflection for the past and the future [25].

2.3 Text Mining

Text mining is a subfield of computer science that combines techniques of natural language processing, knowledge management, data mining, information retrieval, and, machine learning [26]. Moreover, the techniques and tools used in text mining could be used to analyze social media textual data for commercial and research purposes [27]. Natural Language Processing (NLP) covers any natural language manipulation using computers [28]. NLP can use machine learning for linguistic analysis [29]. Capturing emerging trend is also possible using topic modeling [30].

Topic modeling is an evolving technique in machine learning which enables the identification of hidden topics from collections of textual datasets [31]. Blei et al. introduced LDA, a topic modeling technique that uses probabilistic generative model to generate topics by clustering co-occurring terms [32]. Topic detection methods such as dynamic topic models are also applicable to identify and capture emerging trends [30].

2.4   Trend Detection

Trend detection and analysis is the application of computer science and statistics to predict emerging topics [33]. Trend forecasting is the result of an analysis of data from diverse sources. For example, using interviews trends can be forecasted in early warning systems [34]. Similarly, customer rating or word of mouth can be used to assess the market potential of products [35]. It is also possible to use patent analysis for forecasting, for example, trends in technology [11].

Trend analysis can be done using a wide variety of data sources for various applications. For example, trend analysis on social media data is applied for marketing [36], predicting the future of technology [37], emergency management [38, 39], and monitoring diseases propagation [40]. Also, trend analysis can be applied in emergency management and sensing trending topics [41].

2.5   Clustering of Co-citation Networks: CiteSpace

CiteSpace[2] is a freely available tool developed by Drexel University (U.S.A.) which is used to analyze temporal and structural patterns in scientific literature databases [42]. CiteSpace clusters dataset of literature using co-citation network analysis [42]. Co-citation analysis is the main method to analyze structures of scientific works empirically [43]. CiteSpace provides the use of noun phrases of titles **T**, keywords **K** and abstracts **A** of articles to label clusters identified as suggested by [44]. The labeling of clusters is inspired by three algorithms such as LSI (Latent Semantic Indexing), LLR (Log-Likelihood Ratio) and MI (Mutual Information) [45]. LLR often gives the best results concerning coverage and uniqueness [44]. LSI tend to extract clusters to eliciting implicit semantic relationship across the dataset. However, LLR and MI tend to extract a unique aspect of a cluster [45].

Also, Kleinberg's bust-detection algorithm, which is originally designed to detect bursts of single words [14], is used to detect multiword-term bursts and time series analysis of documents [42]. In CiteSpace, the detection of trending terms is identified using burst detection as proposed by [14] which is based on relative frequency of terms. Finally, the visualization of co-citation makes results easily understandable, as the objective of visualization is the portrayal of meaningful structures [18].

Some of the major metrics used in CiteSpace are centrality measures, burstiness of citations, silhouette and modularity. **Centrality measure,** ranges between 0 and 1, measures betweenness score that indicates how different clusters are connected in CiteSpace [42]. **Burstiness of citations** are surges of citation which enables the

---

[2] http://cluster.ischool.drexel.edu/ ~ cchen/citespace/download/

identification of developing interests of a research area in academia, the higher the value the higher is the intensity of the burst [42]. **Silhouette** value ranges from −1 to 1 and shows the homogeneity of a cluster. Higher values of silhouette represent more consistency among cluster members given that the clusters in comparison are similar in size [44]. **Modularity** measures the extent to which a network can be decomposed into independent blocks or modules. This ranges between 0 and 1 where low modularity suggests that the network cannot be reduced to clusters with clear boundaries, whereas high modularity represents a well-structured network [46].

2.6 Related Research

Marçal et al. used Web of Science datasets and scientometric research approach to identify gaps in prospective studies of self-driving vehicles [47]. Review of trend identification techniques using text mining was done 15 years ago [48]. These trend detection techniques were used in diverse domains of research. Also, there are plenty of academic research findings in the literature that uses visualization analytics, sciento-metrics, and text mining such as topic modeling to identify trends. However, trend detection research regarding self-driving cars is rare.

3   Methodology

The data was extracted from Scopus, and CiteSpace was used on the dataset to generate co-citation clustering, visualizations, labeling of clusters, geographical visualization and analysis and interpretation following [42]. We used Kleinberg's burst detection imple-mentation [14] of CiteSpace to identify temporal patterns. A burst detection algorithm determines whether short interval variations of frequency fluctuation as determined by its frequency functions are statistically significant or not during the overall period [15]. Additionally, the R^3 statistical tool was used to generate topic modeling to identify hidden topics following [32] and using the use case described in [3].

For identifying trends, it is possible to use tools and algorithms through sciento-metrics, topic modeling, time-series-analysis of co-citation and co-occurring terms as described in Sect. 2. We choose CiteSpace because it is freely available and suitable for visualizing trends and patterns. On the other hand, IN-SPIRE could be used for the identification of trends, yet it is commercially available, and other software tools are more focused on other features [49]. Besides, CiteSpace is currently updated fre-quently, for example, it was updated six times from March to August 2018. Also, topic modeling is applicable to explore trends and themes from a collection of scientific literature [50]. LDA is the most widely used unsupervised technique for topic modeling [51]. In this study, we used standard non-dynamic LDA to complement CiteSpace by identifying topics without considering trends; and to compare trendy topics generated by CiteSpace with non-trendy topics generated by LDA.

---

[3] https://www.r-project.org

3.1 Computing Tools Used

A 64-bit operating system, Windows 10 Enterprise, running on an x64-based processor with 2.6 GHz Eight-Core Processor, 20 GB RAM was used. For merging datasets and processing Notepad and Microsoft Excel were used. The analysis of the dataset was done using CiteSpace. In this study, we also used R to generate topics using standard LDA.

3.2 Dataset Extraction

We carefully applied a search query that enabled us to refine our search to the most relevant dataset. Without a carefully formulated query, we would end up having tens of thousands of irrelevant data, so we used synonym terms in the query. For example, the term self-driving has many forms as illustrated in the query below. According to [44] irrelevant data is imminent no matter how we formulate our query, yet CiteSpace visualizes irrelevant data in isolated clusters in most cases [44].

The dataset was extracted from Scopus on September 20, 2018, using the query illustrated below. A corpus with a total of 3323 was obtained. The collected data spans from 1976 to 2019 of which 18 articles were obtained from accepted and added to Scopus before they are published in 2019.

Query Used:
((car OR automobile) AND (selfdrive OR selfdriven OR selfdriving OR self-drive OR self-driving OR self-driven OR "self drive" OR "self driven" OR "self driving" OR autopilot OR autonomous OR ("autonomous driving") OR "autonomous drive" OR "autonomous driven")) AND (LIMIT-TO (LANGUAGE,"English "))

4 Results

The clustering of co-citation shows that there are trends indicating most hot areas of research and development related to self-driving cars, clustering without time-series visualization is illustrated in Fig. 1 with nodes of 747 and 1546 connections. We run CiteSpace with a combination of thresholds of Top 60. These means we used top 60 citations per slice, where the number of years per slice is one. The resulting network has a Silhouette value of 0.39 and Modularity of 0.89. The quality of clustering configuration is measured by silhouette value while modularity measures the soundness of network structure. Here the modularity is sound while the silhouette is a little lower than 0.5. This is a fair clustering as it is recommended to balance modularity and silhouette values simultaneously [15]. A total of 186 Clusters where identified and 22 most significant clusters are listed as illustrated in Table 1 below. Out of 186, 22 clusters are more relevant as illustrated in the figures, Figs. 2, 3 and 4 below.

Clustering of co-citation and time series analysis was done in two ways. The first clustering was done using node-type references. The clusters are labeled based on titles as illustrated in Sect. 4.1 to identify research themes and trends based on themes or research titles. The second analysis as illustrated in Sect. 4.2 was done using node-type references and abstract terms. The clusters were labeled based on abstract terms to identify relevant terminologies. Finally, analysis of global collaboration, term burst detection, and topic modeling are presented in Sects. 4.3, 4.4 and 4.5, respectively.

4.1 Clustering Node Type Reference: Labeling Using Title and Abstract Terms

The identified clusters were labeled using the **Titles** of the manuscripts in the collection of documents as illustrated in Fig. 2. and cluster labeling by abstract terms as illustrated in Fig. 3 is presented, and all the three labeling algorithms presented in Sect. 2.5 are used.

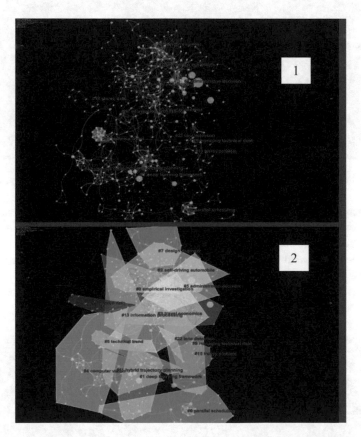

Fig. 1. Co-citation network is colored and labeled based on years, and coloring shows the age of clusters where blue is oldest and greenish and yellowish colors represent newest (1), also (2) each cluster is depicted by a rounded hull colored.

The purpose of analyzing Clusters by title is to identify themes that are commonly discussed among academia from 1976 to 2019. The clusters labeled by title indicates that the most common research themes are an empirical investigation, deep learning framework, travel economics, and computer vision, see Fig. 2 below as:

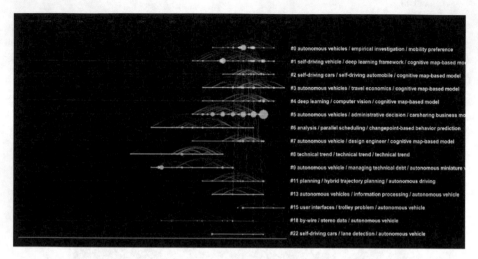

Fig. 2. A timeline analysis of clusters using Titles as cluster labels and node type Reference.

The purpose of analyzing clusters by terms, see Fig. 3 is to identify terms that are representatives of the abstracts between 1976 and 2019. The clusters labeled by terms indicate that the most common research terms are new trip, car detection, transportation network companies, real-world road network, and adversarial attack are the top terms. It is also visible that government agencies are increasingly becoming active participants in self-driving technologies. Also, trolley problem and real-world road networks are currently trending.

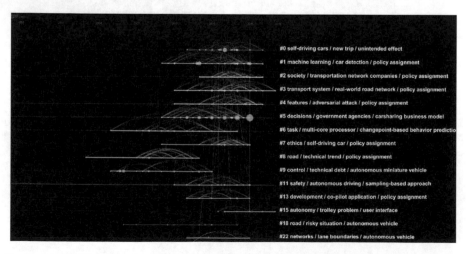

Fig. 3. A timeline analysis of clusters using terms as cluster labels and node type Reference.

4.2   Clustering with Node Type Reference and Abstract: Labeling Using Abstract Terms

In this Section, we present a visualization of a network of clusters generated using node type **Reference** and **Abstract** see Fig. 4 below. The result shows that there is only one difference on the label of the 11th cluster. When cluster label **Abstract** is selected, it becomes "collision checking accuracy" instead of "autonomous driving." The term "collision checking accuracy" is more interesting since "autonomous driving" is part of the search query.

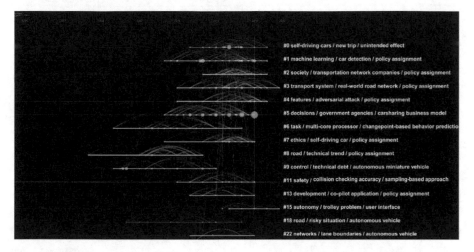

Fig. 4. A timeline analysis of clusters using terms as cluster labels and node type Reference and Term.

To automatically label clusters CiteSpace provides three labeling algorithms such as LSI, LLR, and MI [45] as discussed in Sect. 2.5. Summary of clusters is presented in Table 1 below. The summary of clusters shows that machine learning, computer vision such as car detection, transport network companies, real-world road network, safety, control, ethics, and other important terms. In this study, LSI and LLR cluster labeling give better and interpretable labeling than MI, see Table 1. The silhouette values of most clusters are greater than 0.5 and are closer to 1 which shows that clusters have more consistency as suggested in [44]. The Mean Year indicates most of the biggest clusters are within the range of five years period and hence they are recent.

Table 1. Summary of term clusters sorted by size and labeling using clustering algorithms LSI, LLR, and MI.

Cluster ID	Size	Silhouette	Mean (Year)	Label (LSI)	Label (LLR)	Label (MI)
0	45	0.775	2014	Self-driving cars	New trip	User preference
1	42	0.875	2014	Machine learning	Car detection	Policy assignment
2	41	0.854	2015	Society	Transportation network companies	Policy assignment
3	33	0.831	2014	Transport system	Real-world road network	Policy assignment
4	31	0.902	2015	Features	Adversarial attack	Policy assignment
5	31	0.875	2013	Decisions	Government agencies	Allowing car use
6	29	0.992	2011	Task	Multi-core processor	Policy assignment
7	27	0.901	2014	Ethics	Self-driving car	Policy assignment
8	19	0.981	2008	Road	Autonomous driving	Policy assignment
9	19	0.947	2008	Control	Technical debt	Actor layer
11	18	0.943	2013	Safety	Collision checking accuracy	Sampling-based approach
13	16	0.928	2013	Development	Co-pilot application	Policy assignment
15	14	0.978	2015	Autonomy	Trolley problem	Current finding
18	11	0.96	2011	Road	Risky situation	Policy assignment
22	7	0.989	2014	Networks	Lane boundaries	Autonomous vehicle

4.3   Analysis of Global Collaboration: Google Earth Visualization

Google earth application is used along with a file generated from CiteSpace to explore authors' geographical location and associations to their collaborators as illustrated below. The red links are recent links, and the green ones are old links. As shown in Fig. 5 the general development in self-driving cars has become an emerging technological phenomenon since most connections are reddish.

The U.S.A. and Brazil are active in self-driving cars research. With the U.S.A. being the most active country in undertaking and collaboration with other continentals in the region. Europe has more geographically concentrated self-driving car research and collaboration than any other continent. EU member states collaborate within the EU and at the continental level. Countries such as Germany, the U.K., Ireland, France, Sweden, Netherlands, Denmark, Austria, Switzerland, Spain, Italy, Romania, and Poland are

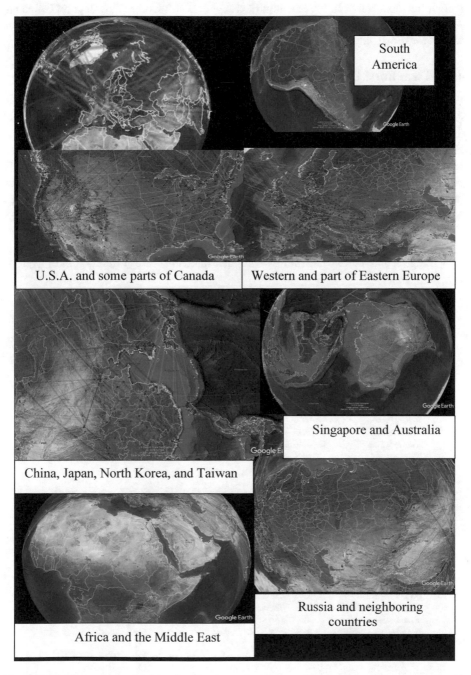

Fig. 5. Google earth worldwide visualization of research collaboration on the topic of self-driving cars.

among the most active European countries. Japan, South Korea, Taiwan, and China are also actively working on self-driving technology. However, Africa, South America except for Brazil, and the Middle East, in general, are inactive in self-driving technology.

4.4   Term Burst Detection

We used the default configuration of burst detection function values of CiteSpace. Additionally, we selected Node Types **Term** and **Keyword** to identify busty terms and their variation. We selected the top 65 terms per a year slice with Top N% = 100% and G-Index = 7. This resulted in 87 burst terms as illustrated below in Fig. 6. The burst terms indicate technological terms and their trends through the years 1976 to 2019. Hence it shows which terms are currently active in academia, see Fig. 6. For a full list of term-bursts refer to Appendix 1. The result shows that deep learning, convolutional neural network, object detection, neural network, learning system, Internet of Things (IoT), deep neural network, automobile safety device, behavioral research, digital storage, network security, semantics, advanced driver assistance, decision making, traffic sign, image segmentation, and human-computer interaction are currently trending terms.

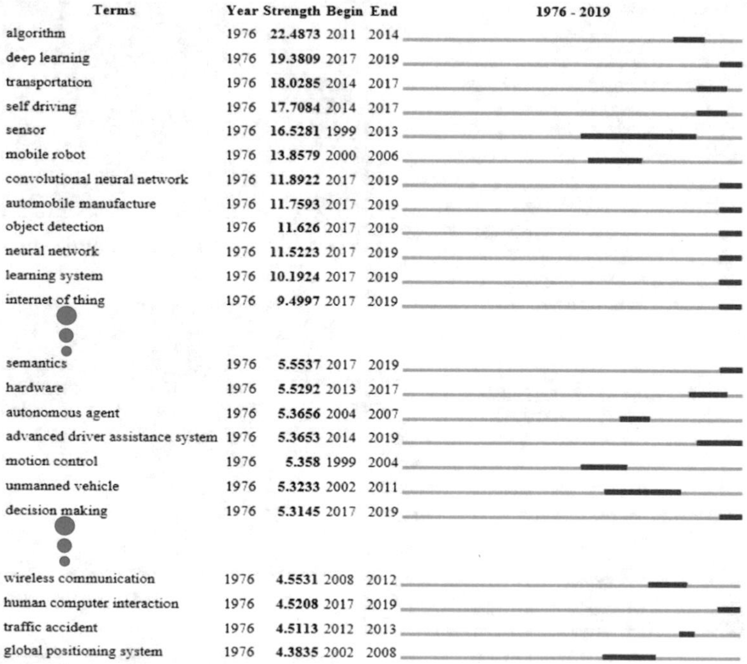

Terms	Year	Strength	Begin	End	1976 - 2019
algorithm	1976	22.4873	2011	2014	
deep learning	1976	19.3809	2017	2019	
transportation	1976	18.0285	2014	2017	
self driving	1976	17.7084	2014	2017	
sensor	1976	16.5281	1999	2013	
mobile robot	1976	13.8579	2000	2006	
convolutional neural network	1976	11.8922	2017	2019	
automobile manufacture	1976	11.7593	2017	2019	
object detection	1976	11.626	2017	2019	
neural network	1976	11.5223	2017	2019	
learning system	1976	10.1924	2017	2019	
internet of thing	1976	9.4997	2017	2019	
semantics	1976	5.5537	2017	2019	
hardware	1976	5.5292	2013	2017	
autonomous agent	1976	5.3656	2004	2007	
advanced driver assistance system	1976	5.3653	2014	2019	
motion control	1976	5.358	1999	2004	
unmanned vehicle	1976	5.3233	2002	2011	
decision making	1976	5.3145	2017	2019	
wireless communication	1976	4.5531	2008	2012	
human computer interaction	1976	4.5208	2017	2019	
traffic accident	1976	4.5113	2012	2013	
global positioning system	1976	4.3835	2002	2008	

Fig. 6. Term and keyword burst detection and trends from 1976 to 2019.

4.5   Topic Modeling

Topic modeling to identify hidden topic was run using R, and we identified 20 topics. After obtaining the dataset text pre-processing such as removing stopwords, stemming & stem completion, remove special characters, removing numbers & punctuations, change cases to lower case, visualization of frequency to check the validity of frequent terms, and conversion of major vocabulary into US English from British English for consistency was done. After preprocessing, we chose the optimum topic number to be generated as described in [3]. The LDA topic model was validated applying a hold out of 10-fold cross validation using 75% training-set and 25% test-set using candidate number of topics, and the evaluation shows that the generated LDA model is suitable if greater than 20 topics are chosen. Some of the major topics identified are safety, car human interaction, computer vision & pattern recognition, urban planning to accommodate self-driving cars, navigation systems, and machine learning. However, these topics do not reflect any trend as time is not considered in standard topic modeling algorithms. Hence, these generated topics are all identified by CiteSpace in trending and non-trending topics.

5   Discussion and Future Directions

The purpose of this study is to identify emerging trends in self-driving cars. The result of this study enabled the visualization of existing trends regarding self-driving cars. Recent academic findings indicate that there are concerns related to using self-driving vehicles – for example, control and security [52], land management policy and marketing [53]. The result of the study also suggests that there are concerns related to safety, control, security, urban traffic management, and others.

In this study, CiteSpace is used because it has more visualization options than most exiting visualization tools reviewed. The limitation of this study is that due to time limitation the results are not compared with other visualization tools. For the future, we plan to compare other visualization tools and make recommendations regarding the utilization of existing tools. Also, since LDA only shows hidden topics covered in the discourse of the scientific literature disregarding temporal patterns we plan to apply dynamic topic modeling techniques to identify trending topics and compare them with visual analytics and scientometric tools.

6   Conclusions

Knowledge processing through the utilization of the growing academic research findings and patents serves many purposes. There is an exponential growth of research findings, and it has become difficult to find innovative ideas [10]. The problem with finding innovative ideas can be a result of lack of communication between academia and the industry. For example, Brijs argue that there is a communication gap between the academic sector and the industry and there is no easy way to bridge this gap [8]. To circumvent the communication gap which hinders the implementation and commercialization of new technology it is possible to analyze the growing academic research findings to identify emerging trends. This means the industry could utilize the results of the academic realm all over the world in a systematic way through generated insights and trends.

Identification of new technological advancement enables the elicitation of valuable ideas. Technology incubators, research centers, and technology transfer agents are dedicated to bridging the communication barriers hindering industry-academia collaboration [9]. Therefore, the identification of emerging trends illustrated in this paper can serve technology incubators, research centers, and technology transfer agents.

From the results of this study we arrive at the following conclusions, the most important and trending areas of research are:

1. Machine learning
   a. Neural network - deep neural network and convolutional neural network
2. Computer vision, robotics and IoT
   a. Sensor
   b. Object detection
   c. Image segmentation
   d. Network security
   e. Human-computer interaction
   f. Control and safety
3. Social and human aspects
   a. Trolley problems, ethics, and behavioral research
   b. Safety and security
4. Urban planning and policy
   a. Transport network companies
   b. Traffic line management
   c. Government agencies

Finally, from the analysis of global collaboration, we can deduce that African, the Middle East, South American, and some Asian countries are behind the rest of the world. Also, most European countries collaborate on self-driving technology with other EU member states and with other countries mostly with the United States. Some Asian countries such as Japan, South Korea, Taiwan, China, and Singapore are also active in self-driving research. Also, potential regions where self-driving cars could be tested, and their applicability could be investigated could be African countries, the Middle East, and most Latin America countries. Brazil, however, is more active than most South American countries.

Appendix 1: Top 87 Terms with the Strongest Citation Bursts

Terms	Strength	Begin	End	1976 - 2019
algorithm	22.4873	2011	2014	
deep learning	19.3809	2017	2019	
transportation	18.0285	2014	2017	
self driving	17.7084	2014	2017	
sensor	16.5281	1999	2013	
mobile robot	13.8579	2000	2006	
convolutional neural network	11.8922	2017	2019	
automobile manufacture	11.7593	2017	2019	
object detection	11.626	2017	2019	
neural network	11.5223	2017	2019	
learning system	10.1924	2017	2019	
internet of thing	9.4997	2017	2019	
mathematical model	9.4557	1998	2006	
navigation	8.923	2007	2011	
fuzzy control	8.73	2003	2009	
accident	8.6202	2012	2013	
convolution	8.4386	2017	2019	
deep neural network	8.4386	2017	2019	
intelligent robot	6.3478	2010	2011	
autonomous navigation	6.2944	2006	2014	
steering	6.2261	2002	2012	
cyber physical systems (cpss)	6.2261	2013	2014	
parking	6.1613	2004	2012	
network security	6.0563	2017	2019	
automobile electronic equipment	6.0549	2004	2010	
multi agent system	6.0296	2005	2013	
human	5.9664	2012	2017	
design	5.8713	2009	2014	
computer software	5.8158	2002	2009	
computer simulation	5.7242	2010	2013	
car	5.6677	2014	2017	
semantics	5.5537	2017	2019	
hardware	5.5292	2013	2017	
autonomous agent	5.3656	2004	2007	
advanced driver	5.3653	2014	2019	

model	**4.2763** 2010	2013	
accident prevention	**4.2355** 2001	2006	
brake	**4.188** 2011	2012	
electric vehicle	**4.1741** 2011	2013	
lane detection	**4.1278** 2012	2013	
robotics	**4.125** 2000	2007	
signal processing	**4.1123** 2012	2014	
safety	**4.0855** 2014	2017	
fuzzy system	**4.0755** 2008	2010	
maneuverability	**4.0004** 2003	2006	
pedestrian safety	**3.9206** 2012	2013	
intelligent control	**3.91** 2007	2013	
reinforcement learning	**3.8925** 2011	2013	

References

1. Davidson, P., Spinoulas, A.: Autonomous vehicles: what could this mean for the future of transport. In: Australian Institute of Traffic Planning and Management (AITPM) National Conference, Brisbane, Queensland (2015)
2. Bimbraw, K.: Autonomous cars: past, present and future a review of the developments in the last century, the present scenario and the expected future of autonomous vehicle technology. In: 12th International Conference Informatics in Control, Automation and Robotics, ICINCO, vol. 1, pp. 191–198. IEEE (2015)
3. Ayele, W.Y., Juell-Skielse, G.: Unveiling topics from scientific literature on the subject of self-driving cars using latent Dirichlet allocation. In: 2018 IEEE 9th Annual Information Technology, Electronics and Mobile Communication Conference, IEMCON, pp. 1113–1119. IEEE (2018)
4. Lin, P.: Why ethics matters for autonomous cars. In: Autonomous Driving, pp. 69–85. Springer, Heidelberg (2016)
5. Howard, D., Dai, D.: Public perceptions of self-driving cars: the case of Berkeley, California. In: Transportation Research Board 93rd Annual Meeting, vol. 14, no. 4502 (2014)
6. Fagnant, D.J., Kockelman, K.: Preparing a nation for autonomous vehicles: opportunities, barriers and policy recommendations. Transp. Res. Part A: Policy Practice **77**, 167–181 (2015)
7. Sandberg, A.B., Crnkovic, I.: Meeting industry-academia research collaboration challenges with agile methodologies. In: ICSE-SEIP IEEE/ACM 39th International Conference Software Engineering: Software Engineering in Practice Track, pp. 73–82. IEEE (2017)
8. Brijs, K.: Collaboration between Academia and Industry: KU Leuven. Cereal Foods World **62**(6), 264–266 (2017)
9. Villani, E., Rasmussen, E., Grimaldi, R.: How intermediary organizations facilitate university–industry technology transfer: a proximity approach. Technol. Forecasting Soc. Change **114**, 86–102 (2017)
10. Bloom, N., Jones, C.I., Van Reenen, J., Webb, M.: Are ideas getting harder to find? (No. w23782). National Bureau of Economic Research (2017)

11. You, H., Li, M., Hipel, K.W., Jiang, J., Ge, B., Duan, H.: Development trend forecasting for coherent light generator technology based on patent citation network analysis. Scientometrics **111**(1), 297–315 (2017)
12. Salatino, A., Osborne, F., Motta, E.: AUGUR: forecasting the emergence of new research topics. In: The 18th ACM/IEEE Joint Conference on Digital Libraries, JCDL 2018. ACM, New York (2018)
13. Small, H., Boyack, K.W., Klavans, R.: Identifying emerging topics in science and technology. Res. Policy **43**(8), 1450–1467 (2014)
14. Kleinberg, J.: Bursty and hierarchical structure in streams. Data Mining Knowl. Discov. **7**(4), 373–397 (2003)
15. Chen, C., Ibekwe-SanJuan, F., Hou, J.: The structure and dynamics of cocitation clusters: a multiple-perspective cocitation analysis. J. Am. Soc. Inf. Sci. Technol. **61**(7), 1386–1409 (2010)
16. Aghaei, C.A., Salehi H., Yunus, M., Farhadi, H., Fooladi, M., Farhadi, M., Ale, E.N.: A comparison between two main academic literature collections: web of science and scopus databases (2013)
17. Mongeon, P., Paul-Hus, A.: The journal coverage of web of science and scopus: a comparative analysis. Scientometrics **106**(1), 213–228 (2016)
18. Small, H.: Visualizing science by citation mapping. J. Am. Soc. Inf. Sci. **50**(9), 799–813 (1999)
19. Chen, C., Hu, Z., Liu, S., Tseng, H.: Emerging trends in regenerative medicine: a scientometric analysis in CiteSpace. Expert Opin. Biol. Therapy **12**(5), 593–608 (2012)
20. Keim, D., Andrienko, G., Fekete, J.D., Görg, C., Kohlhammer, J., Melançon, G.: Visual analytics: definition, process, and challenges. In: Information Visualization, pp. 154–175. Springer, Heidelberg (2008)
21. Andrienko, G., Andrienko, N., Wrobel, S.: Visual analytics tools for analysis of movement data. ACM SIGKDD Explor. Newslett. **9**(2), 38–46 (2007)
22. Mingers, J., Leydesdorff, L.: A review of theory and practice in scientometrics. Eur. J. Oper. Res. **246**(1), 1–19 (2015)
23. Zitt, M., Bassecoulard, E.: Development of a method for detection and trend analysis of research fronts built by lexical or cocitation analysis. Scientometrics **30**, 333–351 (1994)
24. Van Eck, N.J., Waltman, L.: Text mining and visualization using VOSviewer. arXiv preprint arXiv:1109.2058 (2011)
25. Tseng, Y.H., Lin, Y.I., Lee, Y.Y., Hung, W.C., Lee, C.H.: A comparison of methods for detecting hot topics. Scientometrics **81**(1), 73–90 (2009)
26. Feldman, R., Sanger, J.: The Text Mining Handbook: Advanced Approaches in Analyzing Unstructured Data. Cambridge University Press, Cambridge (2007)
27. Hu, X., Liu, H.: Text analytics in social media. In: Mining Text Data, pp. 385–414 (2012)
28. Bird, S., Klein, E., Loper, E.: Natural Language Processing with Python: Analyzing Text with the Natural Language Toolkit. O'Reilly Media, Inc., Sebastopol (2009)
29. Sidorov, G., Velasquez, F., Stamatatos, E., Gelbukh, A., Chanona-Hernández, L.: Syntactic n-grams as machine learning features for natural language processing. Expert Syst. Appl. **41**(3), 853–860 (2014)
30. AlSumait, L., Barbará, D., Domeniconi, C.: On-line lda: adaptive topic models for mining text streams with applications to topic detection and tracking. In: Eighth IEEE International Conference Data Mining, ICDM 2008, pp. 3–12. IEEE (2008)
31. Blei, D.M.: Probabilistic topic models. Commun. ACM **55**(4), 77–84 (2012)
32. Blei, D.M., Ng, A.Y., Jordan, M.I.: Latent Dirichlet allocation. J. Mach. Learn. Res. **3**, 993–1022 (2003)
33. Kataria, D.: A review on social media analytics. Int. J. Adv. Res. Ideas Innov. Technol. **3**(2), 695–698 (2017)

34. Hando, J., Darke, S., O'brien, S., Maher, L., Hall, W.: The development of an early warning system to detect trends in illicit drug use in Australia: the illicit drug reporting system. Addict. Res. **6**(2), 97–113 (1998)
35. Moe, W.W., Trusov, M.: The value of social dynamics in online product ratings forums. J. Market. Res. **48**(3), 444–456 (2011)
36. He, W., Zha, S., Li, L.: Social media competitive analysis and text mining: a case study in the pizza industry. Int. J. Inf. Manag. **33**(3), 464–472 (2013)
37. Asur, S., Huberman, B.A.: Predicting the future with social media. In: IEEE/WIC/ACM International Conference Web Intelligence and Intelligent Agent Technology (WI-IAT), vol. 1, pp. 492–499. IEEE (2010)
38. Yin, J., Karimi, S., Lampert, A., Cameron, M., Robinson, B., Power, R.: Using social media to enhance emergency situation awareness. In: Proceedings of the 24th International Conference on Artificial Intelligence, pp. 4234–4238. AAAI Press (2015)
39. Pohl, D., Bouchachia, A., Hellwagner, H.: Automatic sub-event detection in emergency management using social media. In: Proceedings of the 21st International Conference on World Wide Web, pp. 683–686. ACM (2012)
40. Corley, C., Mikler, A.R., Singh, K.P., Cook, D.J.: Monitoring influenza trends through mining social media. In: BIOCOMP, pp. 340–346 (2009)
41. Ayele, W.Y., Juell-Skielse, G.: Social media analytics and internet of things: survey. In: Proceedings of the 1st International Conference on Internet of Things and Machine Learning, p. 53. ACM (2017)
42. Chen, C.: CiteSpace II: detecting and visualizing emerging trends and transient patterns in scientific literature. J. Am. Soc. Inf. Sci. Technol. **57**(3), 359–377 (2006)
43. Gmür, M.: Co-citation analysis and the search for invisible colleges: a methodological evaluation. Scientometrics **57**(1), 27–57 (2003)
44. Chen, C.: The CiteSpace Manual (Version 0.65). http://cluster.ischool.drexel.edu/-cchen/citespace/CitespaceManual.pdf (2014). Accessed 06 Apr 2014
45. Zhu, Y., Kim, M.C., Chen, C.: An investigation of the intellectual structure of opinion mining research. Inf. Res.: Int. Electron. J. **22**(1), n1 (2017)
46. Chen, C., Chen, Y., Horowitz, M., Hou, H., Liu, Z., Pellegrino, D.: Towards an explanatory and computational theory of scientific discovery. J. Informetrics **3**(3), 191–209 (2009)
47. Marçal, R., Antonialli, F., Habib, B., Neto, A.D.M., de Lima, D.A., Yutaka, J., Luiz, A., Nicolaï, I.: Autonomous Vehicles: scientometric and bibliometric studies. In: 25th International Colloquium of Gerpisa-R/Evolutions. New technologies and Services in the Automotive Industry (2017)
48. Kontostathis, A., Galitsky, L.M., Pottenger, W.M., Roy, S., Phelps, D.J.: A survey of emerging trend detection in textual data mining. In: Survey of Text Mining, pp. 185–224. Springer, New York (2004)
49. Cobo, M.J., López-Herrera, A.G., Herrera-Viedma, E., Herrera, F.: Science mapping software tools: review, analysis, and cooperative study among tools. J. Am. Soc. Inf. Sci. Technol. **62**(7), 1382–1402 (2011)
50. Youssef, A., Rich, A.: Exploring trends and themes in bioinformatics literature using topic modeling and temporal analysis. In: Systems, Applications and Technology Conference (LISAT), pp. 1–6. IEEE, Long Island (2018)
51. Gerlach, M., Peixoto, T.P., Altmann, E.G.: A network approach to topic models. Sci. Adv. **4**(7), eaaq1360 (2018)
52. Schoettle, B., Sivak, M.: A survey of public opinion about autonomous and self-driving vehicles in the US, the UK, and Australia (2014)
53. Bansal, K.M., Kockelman, P.: Are we ready to embrace connected and self-driving vehicles? A case study of Texans. Transportation **45**(2), 641–675 (2018)

Making a Simple Game in Unreal™ Engine: Survival Knight

Aamir Jamal and Sudhanshu Kumar Semwal$^{(\boxtimes)}$

Department of Computer Science, University of Colorado,
Colorado Springs, CO, USA
{aamirjam, ssemwal}@uccs.edu

Abstract. Videogames have been around for more than 50 years now. As technology manufacturers move towards more efficient methods to create and distribute interactive experiences, we find further proof that videogames are here to stay. Initially introduced in the entertainment sector, they have begun to branch out in the education arena as well. By researching about game development using visual scripting within a game engine, we will focus our attention on the development of a third person top-down wave-based game which is the main output of our work. We use Unreal™ engine to develop a game called Survival Knight. Details of our effort are discussed.

Keywords: Unreal · Action-adventure · Game design

1 Introduction

Survival adventure games invoke a sense of curiosity and panic in players when they are left with no choice but to strategize and find the best possible way to sustain and succeed. Games like *PlayerUnknown's Battlegrounds, Fortnite* use a battle-royale formula in which multiple players are dropped onto a map that they can explore, while gathering useful items for their inventory that will help them survive in battle against the other players. The addition of different twists and themes gives survival games an elevated level of variability in terms of interactive experiences and combinations of playthroughs. No two rounds are the same, where every player has merely one thought on his mind – staying alive throughout the duration of the round. Other games which rely on horror as a medium of evoking emotions during survival have been very well received too, since survival horror games like *Resident Evil, Amnesia, Outlast* and the like have solidified their positions as some of the most memorable games with this theme to be released in recent years. Videogames are an art form recognized by millions of players from across the planet as one of the most engaging activities to partake in. Numerous events are organized to showcase the finest examples of game development happening at various locations in different countries like Game Development Conferences, Major League Gaming contests and many other gaming competitions. Games of today have become highly complex, requiring the player to learn and memorize various patterns of controls and movement over many hours. In short, video games are the medium of the computer representing the most polished, powerful, and thoroughly digital learning experiences known to man. There has been a lot of debate on whether

© Springer Nature Switzerland AG 2020
K. Arai and S. Kapoor (Eds.): CVC 2019, AISC 944, pp. 373–384, 2020.
https://doi.org/10.1007/978-3-030-17798-0_30

the video games industry has had a profound positive effect on the (mental) health of young individuals [1, 2]. With technology shifting its focus towards more immersive experiences as time goes by, we find ourselves surrounded by a myriad of videogames belonging to various genres on different platforms which have a thing or two to teach if the player is willing to invest time in understanding the fundamental mechanics of interaction of various objects in a 3D environment governed by a set of rules. Video-games have also been prescribed as an effective treatment method for children diag-nosed with Attention Deficit Hyperactivity Disorder (ADHD) [3].

2 Literature Review

Unreal Engine provides us with a visual scripting centric approach to game develop-ment, where in-stead of typing out actual pieces of code, it is possible for a designer to use logic-based nodes to set up game prototypes. In his book, Nicola Valcasara [2] explains the pros of utilizing Unreal Engine's built in Blueprints editor for game creation. Third person action-adventure games come in a variety of flavors, and plat-formers are a major category of these type of games. Using the tools provided within this framework, it is possible for us to create an environment from scratch, make the obstacles that one would face in the level and also create the player models and enemy types that would populate the game environment. What is it that makes a video game so appealing some may ask. The answer to this question varies from person to person, and genre to genre. But the common subset of all those answers is that fact that videogames provide a fun and engaging factor which is generally not found in any other medium of interactive entertainment.

A. Survival Games

Technically speaking, most of the games that have existed since the early days of game development are survival games. Even in games like Pong which was developed by Allan Alcorn in 1972, the main aim of the game is to bounce a ball between your paddle and the opponent's paddle and ensure that you score more than your opponent which is basically a fight for survival. Games from different subgenres like fighting, sports and story driven adventures are also examples of survival games, where the objective of each game is to make it through the entire level without having the 'Game Over' screen show up. All the different variables that go into making such games ensure that the player has a reasonable level of difficulty to overcome with the given constraints.

Minecraft, perhaps one of the most popular games on the planet which focuses on resource gathering and building a shelter using those resources in order to make it through the night when the monsters show up in probably one of the best examples of survival games in recent times.

B. Action-Adventure Games

Action-adventure games come in a variety of distinct categories. The central theme of an action game is traversing the world environment by solving puzzles and defeating enemies that spawn along the player's path. Action Adventure titles generally offer a storyline, dialogues and interactions between the player and Non-Player Characters (NPCs), an inventory system to name a few things. Some of the most successful titles in this category include the Tomb Raider series, where the objective is to travel through immensely large levels consisting of caves and tombs in dense forests and complete timed puzzles to proceed to the next area of play. The Prince of Persia series, which while similar to Tomb Raider in certain aspects, introduced more platformer mechanics and a middle eastern theme to entice the player to time his jumps and movement to solve in-credibly baffling puzzles. Uncharted is yet another shining example of this genre, which was very widely received by the international audience on the Playstation platform.

3 Motivation

By paying attention to the genres of games listed above, we took inspiration for the development of our game by combining play styles from the survival and the action-adventure subcategory. The initial goal was to create a game that randomly spawns enemies at specific intervals in waves. The complexity of enemy behavior increases with time. As soon as an enemy is spawned in the environment, it starts looking for the main player character and uses a set of actions to try and take down the player.

4 Methodology

Unreal Engine™ was utilized in the creation of the base level design using built in tools [5]. Using the development options provided to us, we created a game that would have a different playthrough each time the game was started. All the assets were placed at defined locations to mold the level. There is one arena which stays the same throughout the duration of the game but the enemies that appear on the scene and their behavior changes with their location relative to items on the map and the main player character.

A. Concept and Story

The core concept behind this project was to in-corporate swarming or clustering in enemy AI so that as soon as an enemy popped up on the level, it would direct its attention towards items of interest (the player) within a set overlap radius. Continuing development based on our previous work, where we had a randomly generated point of interest where 3D agents would flock towards in a group, we have tried incorporating similar behavior but with a player-controlled target this time.

Our game is centered around the protagonist, who finds herself in an arena, in her quest to find the fountain of immortality. The player finds out that in reality, it is a powerful artifact disguised as a fountain, which clones anyone who comes in contact with it, thereby creating evil alter-egos of them. Cursed by the player's own quest, the player now battles waves of evil clones who will stop at nothing to eliminate her from the world.

The game's atmosphere is set in the ruins, where all the action takes place. It builds upon art styles where we have massive columns and structures. Although the theme of the game is somewhat dark, we took inspiration from one of the most popular mobile games on iOS, Infinity Blade as the base for developing our environment and added bright, sunny elements to it to make the environment more visually appealing.

The player must constantly fight against waves of enemies who spawn at randomized locations at intervals based on the number of enemies remaining within a wave.

5   Level Design

A. Material Creation

The most basic aspect of level design is the creation of terrain where the rest of the assets for the level will be placed. Our first area of attention was the creation of a material that would wrap over sculpted terrain. For this project, we created two materials, one for dirt and the other for grass. The dirt terrain consisted of rocks and sand while the grass terrain included mud and short green grass. Using material textures available on the internet, we set up the dirt material to have brighter values for any rocks or twigs present on the surface while the sand was set to a contrasting, dark color. We combined normal maps and texture maps to output a visually realistic representation of the dirt terrain. For the grassy material, we followed a similar procedure but tuned it in such a way that the blades of grass, which were slightly above the ground would be highlighted in lighter colors while the mud would have a contrasting darker color. Blending the normal and texture maps together, we were presented with a favorable output which would become the canvas for our game. Setting a reasonably high value for tile size, we used fuzzy shading to ensure that viewing the terrain from afar and from close-up would make it look aesthetically pleasing. By default, we aimed at keeping the dirt terrain a shade of brown while the grass terrain would have a green hue to it.

B. Adding Tessellation

The next step in making the terrain seamless was to add tessellation to the scene. Tessellation grants us the ability to add texture to the materials. What this means is that rocks and grass would have raised feel to them as compared to the flat dirt and mud. Unreal Engine provides us with the ability to incorporate this feature using blueprints and tweak the distribution and distance of tessellation for optimum visual performance. We can choose to have a denser distribution, which would require more system resources to render at runtime or keep it less dense but spread out over a larger area in order to lower resource utilization for rendering the play area.

C. Setting up Foliage

We used a pre-built grass patch acquired from the asset store to set up our foliage. Within Unreal, we can use an asset and set up a 'foliage brush' that allows us to paint foliage over the landmass. We adjusted density of grass over different areas to make the environment fairly randomized, without it looking computer generated (Fig. 1).

Fig. 1. Foliage and Tessellation up close using Unreal™ Engine Game Platform.

D. Landscaping

We chose to create a landmass plane consisting of 63 × 63 quads. Using the landscape editor tools present within Unreal, we used a combination of sculpt, flatten and erosion presets to elevate and give shape to the flat piece of land we had defined in the initial stages of our development (Fig. 1). Once the desired result was achieved, we then proceeded to paint the landscape with our dirt and grass textures to get started. Our base landscape resembles two concentric circles, interconnected with 4 pathways that make a cross in the middle. All of this is surrounded by high hills, with a waterfall pouring in water in the scene. The lower areas of the map are filled with water which is inaccessible by the player or the enemies.

E. Adding Assets

Once the base landscape was sculpted, we focused our attention towards adding elements to the game which would make up our level (Fig. 2). For this purpose, we utilized freely available assets on the Epic Games asset store. The crossway connecting the two concentric circles is made up of tiles that seamlessly fit together. Paths that lead onto the circle landscape have stairs and ramps which have been rotated and scaled to match the elevation of the path and the depth of the landmass. This was done in order to ensure that the player would be able to traverse the path and not get stuck while moving from a lower area to a higher area in the scene. After the pathways were added. We added a few columns, arches and statues that surround our fountain placed at the center of the map. We made sure that the paths had obstructive railings so that the player or the enemies would not fall into the water as it would not be possible for them to get back to land as the elevation is very steep. We then proceeded to add some elements that would give atmosphere to the game, which includes some barrels, some broken floors, walls, tree roots and other items.

Fig. 2. Level Design showing water reflections using Unreal™ Engine Game Platform.

F. Lighting and Post Processing

The last step in our level design involved getting the lighting in place. Once the lighting has been built, moving any asset within the environment would render our lighting useless, so we made sure that all the assets we wanted to place in the level were just right as per our requirement. Unreal allows us to use a feature called LightmassImpotanceVolume, which engulfs the entire play area and allows us to apply bloom effects and light shafts. We made further adjustments to the direction of the light and intensity to match our needs. We also placed an unbound PostProcessVolume segment in our scene which allowed us to add post processing lighting effects to the level and change the Ambient Occlusion settings to add a different tone to the entire scene. The lighting temperature was increased to make our level have a golden glow, instead of a bluish initial tint (Fig. 3).

Fig. 3. Light Shafts passing through assets using Unreal™ Engine Game Platform.

6 Player Character

Once the initial level design was complete and the light was baked into the scene, it was time to work on the main player character. We acquired assets, such as a knight character, with many raw animation states, which would then be processed to make meaningful animation montages and merge those animations with interactions within the game world using Animation State Machines (Fig. 4).

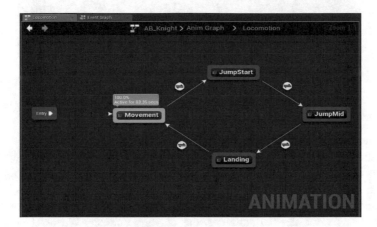

Fig. 4. Jump Animation States using Unreal™ Engine Game Platform.

A. Character Animation State Machines

For this purpose, the most important aspect at this stage was our focus on Blendspaces. Blendspaces allow us to blend animations between two or more states, which helps in computing approximate animation sequences. The easiest example for this would be this – if we have an animation for moving forward and another animation for moving to the right, we can utilize blendspaces to compute what the animation sequence would be while moving forward and right at the same time. We Also divided animation states based on whether they would affect only the upper body (attacks) or full body (movement).

B. Character Movement AI

To set the traversal limits on the main character and any additional agents within the scene, we utilized a feature within Unreal called Nav Mesh Bounds Volume in the scene, which calculates the navigable area for an AI based on the boundaries set by the creator of the asset. We had to ensure that the paths reachable by any agent had properly defined collision boundaries and there were no gaps so that the agents would not get stuck at a point from where moving back to the play area would be impossible. Once the navigable area has been calculated, we attached a wander function to the player character which would choose a point within the area at regular intervals and move to that point to test whether movement to that point was possible or not (Fig. 5).

Fig. 5. Navigable Area using Unreal™ Engine Game Platform.

C. Player Movement Input/Controls

After player navigation limits were set, we defined keyboard bindings for player movement, which were then communicated to the player character pawn through blueprints that had the logic defined for each button press. All control mappings were defined in a separate Controller Blueprint, which had a basic pattern for each action – get the position of the player in the world, get input from the user through keyboard or mouse and apply the desired action to the player character. We used positive and negative Y-axis for up and down movement and positive and negative X-axis for right and left movement. Spacebar was assigned for jump, while left mouse click was attached to melee attack and right mouse click triggered a projectile magic attack. The player also faces the direction of the mouse pointer in the game world.

D. Attacks – Melee and Ranged

The attacks call a sequence of events playing simultaneously. The structure of the attack can be described as getting user input, determining whether the player is already performing another action and if not – triggering two things: playing animation associated with the attack and dealing damage to player in range of the attack. We created two types of attacks that the player and enemies could use, a melee attack using a sword attached to the player and a ranged attack, which launches a particle effects projectile in the direction of the mouse. The attacks by themselves do not make any difference to other enemies unless the logic to deal damage has been associated with the animations. These have been implemented in a separate blueprint called PlayerController.

On a left click of the mouse, a melee attack is invoked. In order to make the melee attack effective, we first created a sword and applied a reflective material to it to recreate real life reflections. As soon as the game begins, we spawn the sword and attach it to the player's right hand, which is mapped to an attack animation montage that plays when the left mouse button is clicked. A single click plays one part of the animation whereas two simultaneous clicks play an attack combo and the player swings the sword twice.

For the ranged attack, we first created a particle system that would be used as a projectile and thrown in the direction of the mouse click. A simple material was created, which was replicated multiple times in a particle system whose gravity was centered at a focal point. The life of each particle was set to expire randomly within a set limit. A right click would trigger a magic throw animation montage, summon the particle system and fling it in the direction of the mouse pointer. The magic projectile was given bounciness to make it ricochet off global collision meshes and deal damage to any enemy that encountered it.

E. Health Points and Damage Dealt

The base player character has a health of 100 (arbitrarily chosen) points and so do the enemies. A melee attack deals 50 (1/2 of what was chosen) damage and a range attack deals 50 damage as well. This was done because when the enemies swarm towards the player, they often become overwhelming and a lower attack damage would make the game even more difficult than it is at this point in development. Health can be replenished by walking over green health pickups that spawn randomly over the map. Since the player and enemy attacks deal equal amount of damage, it is important to prioritize picking up health to survive against the enemy because as soon as the enemies get hit, they prioritize picking up health before the player which puts the player at a disadvantage.

7 Gameplay Algorithm

The enemies that spawn in the level use a basic seek and destroy algorithm. There are two sub-parts to the seek algorithm – wandering and zeroing in. As soon as an enemy is created, it scans for a random point in the navigable radius and tries to move towards that point. While moving towards the location, it triggers another scan at every 2 s interval. If the scan sphere overlaps the player during locomotion, the enemy AI prioritizes on moving towards the player by setting it as a focal point. While moving towards the player, the AI again checks if the player is still alive and continues to reposition and rush towards the player. When the enemy comes within an acceptable attack radius, it invokes the attack function which is either a ranged magic attack if there is a distance between the two or a melee attack if the enemy AI is in direct contact with the player. This loop runs as long as the player is still alive. Once the player has been killed, the loop ends, and the enemy goes back into wander mode. While these functions are being called, if the enemy loses health points, it automatically scans the area for any health pickups that may help replenish its health and changes its priority on picking up health instead (Fig. 6).

Every wave is a struct that holds an array of queues. A wave consists of a mixture of two enemy types, one that uses just melee attack while another that uses both melee and magic attacks. Enemy actors may either spawn all at one location or distributed locations. This logic is randomized and does not have a set pattern, which means enemy approach is unpredictable and thus the game lends re-playability through the element of surprise.

Fig. 6. Enemy AI swarming towards player using Unreal™ Engine Game Platform.

8 Software and Packages Used

A. Game Engine

We chose to implement our idea using Unreal Engine instead of Unity3D as Unreal provides us more flexible level creation and asset management options as compared to Unity3D, in our opinion, and a visual scripting based approach towards game design. The ability to 'paint' terrain and textures on top of that terrain was a bonus that assisted us in the creation of the game map which we found useful. Visual fidelity of the engine, lighting and post processing effects are handled in a more detailed manner as compared to other game engines.

B. Assets

Assets from the Unreal Engine Marketplace were utilized to populate the world, which come in a large variety. The rocks, arches and walls were pre-fabricated pieces that were each given distinct adjustments to fit the requirements of our level design.

C. Controls

Unreal Engine has in-built options to define button mappings to assign controls to various actions within our game. These are located under Project Settings > Engine > Input > Bindings.

D. Visual Scripting

Possibly the most powerful feature within Unreal, Blueprints visual scripting offers an extensive array of features that can be combined in multiple ways to create a project. We can define game logic, player input control, animation sequences, AI algorithms all using the power of diverse types of nodes and attaching them using wires. It is C++ logic that is represented visually [4] (Fig. 7).

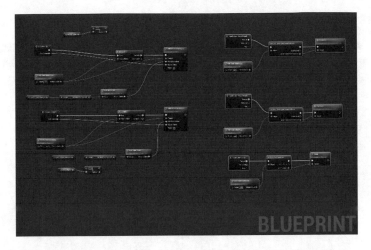

Fig. 7. Player Controls Blueprint using Unreal™ Engine Game Platform.

9   Conclusions and Future Research

With the help of videogames, it is possible to impart knowledge where conventional methods of teaching may not be as effective, including programming itself and problem solving. As noted earlier, games have transformed from a means of entertainment to a means of education and we firmly believe that they can be utilized to promote effective learning in children and young adults alike. Our game is an attempt to utilize readily available game engine tools to craft an experience that is challenging and enjoyable at the same time. Creation of videogames requires logical thinking ability which can be translated to various events that are called every time a certain trigger is activated. We have attempted to blend different sub-genres of games into one third person top-down survival action-adventure wave-based game.

Acknowledgments. The authors would like to thank Epic Games, the Unreal™ community and gratefully acknowledge that several existing functions from Unreal Engine were used to implement the Figures shown in this paper.

We also used several tutorials from YouTube™ including: (a) Painting Our Terrain! - #22 Creating A Survival Horror (Unreal Engine 4), (b) Unreal Engine 4 Beginner Tutorial Series - #28 Using Particle Systems, (c) UE4 Tutorial - Create a Movable Player Character From Scratch, (d) Unreal Engine 4 Beginner Tutorial Series - #17 Terrain Materials, (e) How To Create AI And Enemy Basics - #42 Unreal Engine 4 Beginner Tutorial Series, (f) How To Use Lighting And Post Processing In Unreal Engine 4 [Tutorial], (g) Unreal Engine 4 Tutorial - Basic AI Navigation.

References

1. Squire, K.: Video games and education: designing learning systems for an interactive age. Educ. Technol. **48**, 17 (2008)
2. Valcasara, N.: Unreal Engine Game Development Blueprints. Packet Publishing, Birmingham (2015)

3. Anderson, P.: Video game may help treat ADHD. Medscape, October 2015
4. Blueprints Visual Scripting
5. Jamal, A.: Making a simple game in Unreal™ engine: SurvivalKnight. MS Report, Department of Computer Science, University of Colorado, Colorado Springs, USA, pp. 1–10 (2018). Advisor: S. K. Semwal

Towards Approximate Sphere Packing Solutions Using Distance Transformations and Dynamic Logic

Sudhanshu Kumar Semwal[1]([⊠]), Mike Janzen[1], John Promersbeger[1], and Leonid Perlovsky[2]

[1] University of Colorado, Colorado Springs, USA
{ssemwal,mjanzen,jpromers}@uccs.edu
[2] Northeastern University, Boston, USA
lperl@rcn.com

Abstract. Stereotactic Radiosurgery (SRS) treatment and surgical planning could involve approximating the affected region with spheres of unequal size. This quickly turns into an optimization problem of finding minimum number of unequal size spheres to cover a bounded region. Since this optimization problem is NP-Hard, search for approximate algorithms is of immense important specially to support such SRS treatment. Our goal is to provide fast computational tools using well known Distance Transformations (DTs) and rarely used Dynamic Logic (DL). These two techniques are on the opposite ends of solving the approximate sphere packing problem: DT is bottoms up and DL in top down. Our research is significant in the sense that DL has not been previously used towards developing a sphere packing algorithm. We plan to implement both techniques in 3D on volume data sets in future.

Keywords: Stereotactic Radiosurgery (SRS) treatment ·
Distance Transformations · Dynamic Logic

1 Introduction

Treatment of tumors require precise surgical planning using focused radiation beam during Stereotactic Radiosurgery (SRS) [1, 12–15, 24]. Gamma rays [23] delivers a suitable dose, called a shot, to the targeted area. In clinical applications, the shot is usually prescribed at 50% ISL (isdose level). The physical distribution of this shot at 50% ISL or higher can be approximately as a solid sphere. Wang [15] maps these shots to finding a minimum number of non-elastic solid 3D spheres, so that these unequal and non-overlapping spheres can be packed into a 3D bounded region. This optimization problem is NP-hard [10].

Major benefit of SRS is that it is considered non-invasive. The basic idea of SRS is to focus multiple beams onto a tumor to deliver a targeted dose, called a shot, yet leaving the tissue it passes through with only little effect. The disadvantage of SRS is that side effects such as fatigue, hair loss, nausea, scalp issues such as redness may result. Variety of techniques are used: Linear accelerator (LINAC) which uses X-rays;

K. Arai and S. Kapoor (Eds.): CVC 2019, AISC 944, pp. 385–398, 2020.
https://doi.org/10.1007/978-3-030-17798-0_31

Gamma Knife [23] uses hundreds of Gamma rays; and proton beams or charged particle radiosurgery treats tumors in multiple sessions leading to fractionated SRS. Gamma Knife radiation treatment and its connection to packing spheres was also described by Lim [24]. 3D Slicer [26] is an open source image processing software which works on multiple platforms.

Sphere Packing considers a 3D bounded region R with volume V_R. Let S be the set of spheres representing 50% ISL shots and V_S be the total volume of all spheres in S. Then the coverage or package density $P_D = V_S/V_R$ represents how much of the cancerous tumor is covered by the treatment. Clinical applications require that the P_D is .9 or more [15]. During SRS, physical collimators can use relatively different dosimetric weights to change the shot *diameter*. Usually one sphere shape may not be enough to guarantee packing density of .9 and above, so multiple shots at different locations are needed to cover the volume V_R. Thomas Hales proved that the Kepler's conjecture of 1611 AD, that no packing of spheres can be denser that face-centered-cubic lattice [21, 22, 25]. This also meant that the maximum value of packing density can only be .74048 if equal spheres are used [15, 16]. We must use spheres of unequal sizes because package density of .9 or more is required for SRS planning. Because present solution is to use beams of rays, overlap is necessary yet needs to be minimum to prevent damage to healthy cells. SRS planners can also provide location constraints so that a shot must be placed on exactly the sub-region of interest. We also require that spheres do not overlap to avoid unnecessary *increase* in ISdose Level (ISL) intensity due to overlapping doses.

The input to our algorithms can be 2D images and 3D data volume data sets [1–9]. In this paper, we provide only solutions to 2D. We are working on planning to extend these ideas to 3D as well. As in [15], we will also assume that area of interest, e.g. tumor, is a bounded region. We will also require that region of interest R is well defined binary data set. Our goal is to find minimum number of spheres to minimize the number of shots administered to the patient. Earlier solutions were to fill a region with maximally inscribed spheres (balls) (MIBs) [15, 20]. The locus of the centers of such MIBs is called a skeleton or medial-axis for that region. Sphere packing has received much attention recently in [22, 25]. Using similar framework as in Wang [15], our research formulation can be described as follows:

Formulation of the Research
Input: A 3D region R, a positive integer V with $V < V_R$. A set of m spheres D = (D_1, D_2, ..., D_m) representing the shot-sizes, and location constraints L for placement of spheres.

Constraints: A packing of R using inscribed spheres from S such that: (1) Packing density (pD = V_S/V_R) is .9 or more. (2) Location constraint L is satisfied. (3) Let k be the minimum number of spheres used for packings of R using spheres of D that satisfy constraints 1 and 2. (4) The number of points (N_P) on the contour of R touched by spheres is maximum among all packing of R that satisfy statements 1–3. (5) Packing density pD is such that p1 $\leq$ pD $\leq$ p2. Value of p1 $\geq$ .9 and p2 = 1 will imply that at least 90% of the region is covered by the spheres (6) Measure (M_S) of overlap by spheres: Every data point (2D pixel or 3D voxel) should be covered by at least one and at most one shot, this may not be possible for our application due to nature of applications as rays must pass through healthy tissues to reach the cancerous cells area. If a voxel is covered by multiple doses (shots) then this number needs to be minimized as it will indicate overdose of radiation for healthy tissues along the path of the beam. (7) Measure of effectiveness of coverage: Both sets of false positives (F_P) and false negatives (F_N) should be minimized during planning. False positive is the number of those data points (2D pixels or 3D voxels) which are covered by the spheres but are not inside the given region R. False negatives are those data points which are covered by the region but are not in the coverage provided by the spheres. If we assign high penalty to false positives and false negatives based on the classification of voxels (tissue, bones, etc.) then this penalty should be minimized both during the planning, and after operation, and will provide a measure of efficiency of the coverage.

2 Finding Inscribed Spheres

Finding minimum number of inscribed spheres is NP-Hard [10]. When we consider only the first four constraints from our research formulation then the optimization problem becomes the Min-Max Sphere Packing (Min-Max SP) from Wang in [15]. The first three constraints define the optimization problem Minimum Sphere Packing (Min-SP) in [15]. The corresponding decision problem of Min-SP, called Sphere Packing (SP) [15], is also shown to be NP-Complete by transforming the 3PARTION to SP. This means that our research formulation where we are trying to find unequal spheres packed into the region is also part of NP-Hard problems because it contains the Min-Max SP in its formulation. We also note here that the corresponding Cube packing problems are similarly NP-Hard [10]. Such spheres are defined as maximally inscribed balls (MIBs) in the literature, and the locus of center of MIBs is the skeleton for the region R [15]. Then recursively applying the same idea after subtracting the volume of the sphere from the region R. This situation also encourages the search for *approximate* algorithms [10], where combinations are more manageable, particularly the one we are focusing on this research by implementing Algorithms using Distance Transformations (DTs) and Dynamic Logic (DL).

2.1 Distance Transformations

Distance transformations [17] have been suggested for both the 2D/3D applications as a means to use distance values as radius of spheres, and use the maximum such value as explained in Algorithm A below.

   Algorithm A:

1. Calculate City-Block distance transformation.
2. For those voxels on the skeleton, find the maximum distance value voxel, and use this voxel to place a Shot as representing a sphere (S) centered at that voxel with radius defined by the maximum city-block distance.
3. Subtract sphere S from Region R, and Repeat steps 1–2 until necessary coverage V/V_R found.

The city-block distance transformation can be found by using two scans for 2D data, and six scans for the 3D data. So, steps 1 and 2 are O(N) where N is the number of voxels in region R. Step 3 repeats steps 1–2 k times if k spheres are identified in step 3. So, the Algorithm A is O(kN).

2.2   Dynamic Logic

DL is a process-logic designed to overcome Gödelian dimensionality limitation, and has been applied towards efficient algorithms for many applications. The applications include detection, pattern recognition, data mining, clustering, tracking, swarm intelligence and sensor fusion, including prediction and regression, and has been known to terminate fast [11]. We proposed to improve the state of the art by applying DL to isolate set of spheres for SRS planning, and possibly find minimum number of spheres. DL is a technique which reduces the complexity of the problem by working with models, nudging the iterative process from vagueness to crispness [11] as parameters of the models are refined at every iteration. The parameters of the models are refined based on a *similarity* measure, which is explained later in this document. Once this measure does not change significantly the iteration terminates. In many examples, the Dynamic Logic based algorithms converge in a few iterations (*c* in *algorithm DL* mentioned earlier could be low e.g. 5 to 17).

Algorithm DL is presented below where simultaneously k_{DL} models for spherical regions are started during step 3 which is cN where N is the number of voxels, and c is *usually* a constant representing iterations during step 3. Step 3 will be $k_{DL}cN$ or O(N). Step 3 is further explained in the next section. Step 4 will be O(N). It is projected that *c*, the number of times the iteration will occur is very small relative to other methods, e.g. compared to *p*, the number of times steps 2–4 are repeated in Algorithm C. Our research will verify our claim/conjecture as we will compare all four algorithms (A, B, C, DL) and evaluate these algorithms based of measures explained earlier.

Algorithm DL (Dynamic Logic)

1. Use Dynamic Logic to obtain a sphere. Details of DL are explained later in this document using an iterative method. The benefit of DL is that step 1 is expected to take approximately k steps, where k typically is 5–20 iterations.
2. Subtract area defined by the sphere from step 1, and Repeat steps 1 until necessary coverage V/V_R found. Let the number of iterations for repeating steps 2–4 be D which means there were D spheres found defining the necessary coverage.

   Algorithm DL has time complexity O(kD).

2.3   Dynamic Logic Theory

In 1968, Minsky and Papert exposed the limits of perceptron learning. At the same time, statistical approaches were being developed to first identifying features in the patterns, and then a classification space. Usually classifiers such as planes, or more complex surface, are used to separate the data in classification space. Other methods use a variation of near-neighbor, or a kernel method, where kernel functions, bell shaped curve, or Gaussians are used. In [11], Perlovsky, Deming and Ilin explain that these methods suffer from *dimensionality of the feature space* as the number of training samples grow *exponentially* with number of features or dimensions. Real world problems can also have many *new* changes [11] many of them not considered by learning because those situations did not exist before. If there are N signals and the number of models is M then the complexity of number of combinations in which the system can be is M^N [11]. Formal logic excludes the middle, i.e. things are either true or false and nothing is in between. Rule based system lead to exponential or worse combinatorial explosion. Training based system also suffer similar fate because number of examples become exponential for complex situations. Multivalued logic and fuzzy logic tried to overcome the situation of excluded middle, leading to the degree of fuzziness [11] that is difficult to capture, once again leading to search for such fuzziness in exponential combinatorial space.

Dynamic logic combines dynamics and structure (logic), and can be considered a gradient descent along the variables [11]. Model based learning algorithms maximize a similarity between incoming signals and internal model representation of the world [11]. DL is an *unsupervised* learning technique and is model based. There are three steps: (a) A model is chosen, and there are several options and ways to incorporate existing expertise in choosing a model. (b) A similarity measure is defined. These measures could be simple measures such as least mean squares. (c) Finally, the similarity measure is maximized *iteratively*, in this case association of model with input signal is involved. Let M_m be m models where m = 1 to M. Each model (denoted by boldface) M_m is characterized by its parameters S_m, these parameters are usually unknown, and learning actually means refinement of parameter values, reminding us of Poincare's work on dynamical systems. Task of associating models with signals is complex and explained in detail [11].

2.4   2D Implementation of City-Block

The basic city-block algorithm [17] treats the input image as a grid of points where each point is either occupied (part of the feature of interest) or unoccupied (not part of the feature). Two masks are then applied which will compute the shortest distance from each point to the feature. Because starting at each point and scanning to a feature would be prohibitively time consuming, the forward mask is applied from the upper left corner to the lower right corner, and the backward mask is applied from the lower right corner to the upper left corner. Each application will produce independent numbers which will combine to produce the truth data.

Tables 1 and 2 show the masks that are applied to the image. Each cell (i, j) in the image is assigned a value that is the minimum of its current value and the propagated distance value. This is done to ensure that the value in the cell is the distance to the closest feature since it is entirely possible that the image contains more than one feature. This relationship is defined by the following equation:

$$v_{i,j} = minimum\left(v_{i,j},\ M[k, l] + v_{i+k,j+l}\right)$$

The values of k and l change depending on either the forward mask or the backward mask is being applied. For the forward mask, we use $(k, l) = (0, -1)$ and for the

Table 1. Forward mask

	$j-1$	$j+0$
$i-1$	∞	1
$i+0$	1	0

Table 2. Backward mask

	$j+0$	$j+1$
$i-1$	0	1
$i+0$	1	∞

backward mask we use $(k, l) = (0, 1)$. As an example, we will apply the forward mask to the cell at (i, j) whose value is 5. The cell to be used in the comparison is at $(i+k, j+l)$, which is actually $(i, j-1)$ and has a value of 2. We add the value of that cell to the value of the mask at location $(0, -1)$ which is 1 and yields a total distance of 3. We then compare that distance to the distance of the cell of interest, which is 5, and assign the minimum value to the cell of interest. In this case, the 5 would be replaced by 3, and the interpretation of that change is that either a shorter route to the same feature was discovered or there is a second feature which is closer to that point.

2.5   Finding Internal Distances Representing Radius of Spheres

In order to apply this algorithm to find distances inside the feature, we must reinterpret the data. Instead of treating the feature as having a distance of 0, we treat every cell that is not part of the feature as 0. In the same way, the feature is treated as being populated with ∞ (an arbitrarily large number) so that application of the masks will assign the correct distance values to the cells.

Figure 1 shows the original image on the left and a gradient-shaded image on the right that demonstrates the distance of a given cell from the boundary of the feature. Each cell that is colored green has a distance that is the maximum distance from the edge of the feature. Due to the self-correcting nature of the forward and backward masks, we are guaranteed to not exceed the feature boundaries if we draw a circle whose radius is the distance value of the cell. While it is beyond the scope of this project, follow-on development could demonstrate the ability to remove circles from the image.

Fig. 1.   Left and right images show the implementation of 2D city-block distance transformation, with the radius (distance transformation) values shown so that our first circle could be easily approximated and then we can look for another circle to fill the region of the left images with circles.

Our choice to utilize Processing Language (PL) [27] to implement our particular alteration of the city-block algorithm provided several powerful aids. Processing Language essentially acts as a wrapper around the Java™ language to provide significant graphical improvements.

One of the more useful features of PL is the ability to load an image as an array of pixels, each with its own color that is defined as RGB values ranging from 0 to 255. We can then produce a weighted brightness for each pixel as an average of each RGB value by comparing each brightness to a threshold defined as half the maximum brightness (127). This technique converts the original into an array where 0 means we will not compute the distance and an arbitrarily large number (INFINITY) means that we will compute a distance. As a side note, INFINITY is defined to be the image width times the image height which guarantees that no matter what the size of the input image is, computing distances will always be possible.

Once the array of distances has been created, the city block algorithm is applied which produces the gradient image. Due to the reinterpretation of the input data, it is unnecessary to alter the established algorithm. During the rendering phase, we do pad the computed gradient by a constant amount in order to produce an image that is more easily viewed. Lastly, we determine the cells with the maximum distance and render those in green to highlight the best candidate locations for the center of a circle that could be removed from the image.

3   Finding Circle Using Dynamic Logic Interpretation on Images

The goal of this study was to determine if we could use dynamic logic to find a single, simple, smooth-edged shape; a two-dimensional Gaussian distribution, represented by the following equation. This was designated as model 1.

$$\ell(1, x, y) \; = \; \frac{r(1)}{2\pi s^2} e^{-\left[\frac{(x-x_0)^2 + (y-y_0)^2}{s^2}\right]}$$

For simulation of a realistic input dataset, the distribution was modified by allowing the edges to be squared by removing the edge data so it looks as in Fig. 2 (right).

A second model, $\ell(2, x, y)$, was combined with X(n). This model was a uniform distribution, shown by the following model.

$$\ell(2, x, y) \; = \; \frac{1 - r(1)}{x * y}$$

This model implemented distributed a uniform value across every index in the array with dimensions $[x, y]$. Any number of models, $m = 1 \ldots M$ can be implemented and compared simultaneously. The goal is that through unsupervised learning, the system will determine which model is determined to have the highest confidence when evaluated against the data.

Fig. 2. Our baseline image (left), X(n), modified to be contained within the rectangle (right).

The first goal is to determine a confidence for each of our models, m_1 (defined by $\ell(1, x, y)$) and m_2 (defined by $\ell(2, x, y)$). The confidence is noted as f(m|n), which is the confidence, or similarity of a model m of n models and must be calculated for all models that are being evaluated. The actual values are unknown, so the first step is to make an estimate for each parameter of each model. This conditional likelihood must be calculated every iteration and is calculated by the following equation.

$$f(m|n) = \frac{r_m \ell(m|n)}{\sum_{m \in M} r_m \ell(m|n)}$$

If we choose a 2D matrix size of [1000, 1000], we would have n 1000 × 1000 matrices, each providing an estimation of what our model could be. Our models could look like the following model calculations where r_1 is a randomly assigned probably, for example, .7 or 70%.

$$f(m_1|n) = \frac{r_1 \ell(1, x, y)}{r_1 \ell(1, x, y) + r_2 \ell(2, x, y)}$$

$$f(m_2|n) = \frac{r_2 \ell(2, x, y)}{r_1 \ell(1, x, y) + r_2 \ell(2, x, y)}$$

If a mathematical *and* is performed on the original image, X(n), by multiplying each x, y index by each model estimate, f(m|n), we end up with an array, or image, that contains the intersection of the two datasets, the truth data and the model estimate, as shown in Fig. 2. Absolutely essential for this implementation (although others may not) is to locating the model within the original image is an overlap within the two datasets, X(n) and our model estimate. Without an overlap, the mathematical *and* of the two datasets, would provide an extremely low confidence that we had detected any part of the model within X(n).

By using the derivative of the model, $\frac{r(1)}{2\pi s^2} e^{-\left[\frac{(x-x_0)^2 + (y-y_0)^2}{s^2}\right]}$, for each parameter, (r, x, y, s), we can determine the optimal direction and distance to modify our model's estimated parameters to find the instance of our desired model within $X(n)$. By iterating through each index and summing the intersection of the dataset using an And operation with the probably of the estimated model (Fig. 3), we can make a new estimate for each parameter that defines our model that is contained within the original dataset, $X(n)$. Since a mathematical 'and' is performed, any indices outside of the intersection will be 0 or negligible and will not contribute towards future predictions. This allows us to make an estimate for the instance of the model within $X(n)$, by using the following equation to make an estimate of each parameter of the instance, in this case, the x location of the instance.

$$\Delta x = \sum_{x=0}^{x*y} abs(X(n)) * f(m|n) * \frac{2(x - x_0)}{s_0}$$

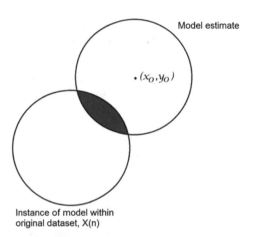

Model estimate

$\bullet (x_0, y_0)$

Instance of model within
original dataset, X(n)

Fig. 3. Intersecting datasets

By summing the value of each index within the model in relation to our model's estimated x location contained within x_0, we can determine the direction that our estimate must move to find the actual x location of the model within $X(n)$. All intersecting indices left of x_0 will accumulate negative values and intersecting indices to the right of x_0 will accumulate positive values. A negative Δx value will be generated, which modifies our model's estimate x to shift in the negative from its current position towards a balance point, where the intersecting area between $X(n)$ and f(m|n) is balanced on x_0, allowing x_0 to be equal to the true x position (Fig. 4).

Since dynamic logic is an iterative process, the process is repeated, taking the step changes from the previous iteration, allowing a new f(m|n) to be created with new parameters (x_0, y_0, s_0, r_0) until the Δx value is 0. Each model must be refined every iteration based upon the generated parameters from the previous iteration. Since the change within the individual parameter is being determined, the final parameter x for model m in instance S for each iteration i, is solved by adding the step to the previous iterations value, as shown by the following equation.

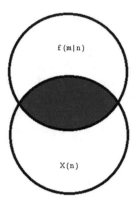

Fig. 4. Intersecting datasets balanced on x_0

$$S_m^{i+1} = S_m^i + \sum_{x=0}^{y*n} abs(X(n)) * f(m|n) * \frac{2(x - x_0)}{s_0}$$

Since y and $y0$ can be a direct replacement for x and x_0, respectively, we can move forward and see how the dynamic logic would handle the radius, parameter s.

By solving for $\frac{ds_0}{dt}$, we can see that $\frac{ds_0}{dt} = \frac{-2}{s_0} - \frac{(x-x_0)^2 + (y-y_0)^2}{s_0^3}$, which doesn't give a lot of insight, but with a little manipulation yields an equation form which starts to explain how dynamic logic handles the radius parameter.

$$\frac{ds_0}{dt} = \frac{-2(s_0^2 - (x - x_0)^2 - (y - y_0)^2)}{s_0^3}$$

The derivation again creates an equation that is trying find the optimal direction and step to change to radius in an effort to balance the intersecting areas based upon the (s, x, y) relationship. It is fascinating to see that this equation is performing a comparison between the distance from the estimate center, (x_0, y_0), to each index and the model's estimated radius, s_0. The comparison shows that when the distance to an index is less than s_0 from (x_0, y_0), the data points inside the estimated radius s_0 is negative, data with a distance equal to s_0 is 0, and data indices that are outside of s_0 are positive.

To visually understand one instance, a brief example is demonstrated by Fig. 5. In this case, the model estimate complete encompasses the model instance with $X(n)$. Since the values that dominantly contribute to the sum of indices are from the intersected area between the estimated model and the model instance within the truth dataset, all the points of interest are closer to x_0 than the estimated radius, generating a cumulative negative radius step, Δs.

This example will generate negative increments from each iteration to be added to the estimated radius, forcing the estimated radius, s_0, to decrease until it approaches the true radius. Once the iteration's parameter values are calculated for each model, the iterative process repeats by calculating the confidence based on new predictions from the last iteration.

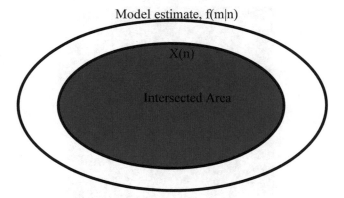

Fig. 5. Solving for correct radius

For this simple example in our study, the results appeared promising. Within the first three iterations, the models were making correct estimates and incrementally heading directly for the instance of the model within the baseline image X(n). As long as the overlap of the model prediction and the model within the baseline image existed. The estimate headed directly towards the center of the model within the baseline image. As the overlap of the two increased, the confidence of model 1 increased and the estimates moved closer, but then slowed down as it tried to center on the target location and slow to the precise center x, y coordinates. Moments of instability arose, seeming due to the interdependency of the parameters as the models approached the target, allowing the radius to increase dramatically and then slow return to realistic sizes. After the best estimate was calculated next iterations take the solution further away as shown in Figs. 6(a, b, c) where we show the iterations 4–6 of our algorithm. In this case, we can terminate with k = 6 which for our O(kD) algorithm where D is the number of circles found in the image, by subtracting the circle found, and iteratively proceeding to find D such spheres. Ofcourse k could change from iteration to iteration, but a constant upper limit of k could be imposed for any implementation if desired.

3.1   Implementation Summary

Our goal is to provide fast computational tools using Dynamic Logic (DL) iterative method, and Distance Transformations. We have also used Processing language which work on multi-platforms. It is important to reduce both false positive and false negative cases, as we want to treat unhealthy tissue, and keep healthy tissues intact during surgery. The goal in future is to interactive, create (near) real-time solutions towards spatial and temporal analysis of medical data generated from real-time functional imaging applications. Since packing spheres map to provable NP-complete and NP-Hard problems, approximate solutions are usually sought.

Fig. 6. (a) Top left, (b) top right and (c) Bottom left show Dynamic Logic iterative algorithm's iteration 4, 5, 6 where a red-circle shows how the data is being captured as the both the radius and center of the circle move towards data of interest. Top image encloses the area, Middle provides a better circle approximation, and bottom image (c) shows how the solution moves away from a better solution image (b) in the next iteration.

4 Conclusions and Further Research

It is well documented that some of the NP-Complete algorithms, under special cases, can be solved in polynomial time for average case or in some instances. Some examples are: Hamiltonian Path [18] and GRAPH 3-Colorability [19]. We consider both DT and DL algorithms to provide approximate sphere packing solutions. Our implementations used the Processing language which is considered well established and is available across most major computing platforms. Processing also has a very light footprint. Our research has the possibilities of opening new pathways for better computational tools for future image guided surgeries or other treatment planning specially where we need to fill areas with circles or spheres for SRS treatment.

Acknowledgments. We want to acknowledge the role Slicer3D Community has played in our research directions. Although we have implemented the code using Processing Language, we want to implement the DL and Distance Transformations algorithms using 3D Slicer platform in future. The first author of this paper is also grateful to Dr. Arcady Godin who introduced Dr. Leonid Perlovsky, and Dr. Perlovsky's work on Dynamic Logic to the first author during conversations in Summer of 2016 when Dr. Godin was visiting Colorado.

References

1. 3D Slicer. https://www.3DSlicer.org/
2. Semwal, S.K., Chandrashekher, K.: 3D morphing for volume data. In: The 18th Conference in Central Europe, on Computer Graphics, Visualization, and Computer Vision, WSCG 2005 Conference, pp. 1–7, January 2005
3. Lorenson, W.E., Cline, H.E.: Marching cubes: a high resolution 3D surface reconstruction algorithm. In: Proceedings of Computer Graphics, SIGGRAPH 1987, vol. 21, pp. 163–169 (1987)
4. Levoy, M.: Display of surfaces from volume data. IEEE Comput. Graph. Appl. **8**, 29–37 (1988)
5. Upson, C., Keeler, M.: V-buffer: visible volume rendering. In: Proceedings of Computer Graphics, SIGGRAPH 1988, vol. 22, pp. 59–64 (1988)
6. Drebin, R., Carpenter, L., Hanrahan, P.: Volume rendering. In: Proceedings of Computer Graphics, SIGGRAPH 1988, vol. 22, pp. 65–74 (1988)
7. Spitzer, V.M., Whitlock, D.G.: High resolution electronic imaging of the human body. Biol. Photogr. **60**, 167–172 (1992)
8. Buchanan, D.L., Semwal, S.K.: Front to back technique for volume rendering. In: Chua, T. S., Kunii, T.L. (eds.) Computer Graphics International 1990, Computer Graphics Around the World Singapore, pp. 149–174. Springer, Tokyo (1990)
9. [VisbleHumanProject]. http://www.nlm.nih.gov/research/visible/
10. Garey, M.R., Johnson, D.S.: Computers and Intractability – A Guide to Theory of NP-completeness. WH Freeman and Company, San Francisco, p. 202 (1979). ISBN 0-7167-1044-7, Subgraph Isomorphism [GT48]
11. Perlovsky, L., Deming, R., Ilin, R.: Emotional Cognitive Neural Algorithms with Engineering Applications Dynamic Logic: From Vague to Crisp. Studies in Computational Intelligence, vol. 371, pp. 1–198. Springer, Heidelberg (2011)

12. Fedorov, A., Beichel, R., Kalpathy-Cramer, J., Finet, J., Fillion-Robin, J.-C., Pujol, S., Bauer, C., Jennings, D., Fennessy, F., Sonka, M., Buatti, J., Aylward, S.R., Miller, J.V., Pieper, S., Kikinis, R.: 3D Slicer as an Image Computing Platform for the Quantitative Imaging Network. Magnetic Resonance Imaging. **30**(9), 1323–1341 (2012). PMID: 22770690
13. Ungi, T., Gauvin, G., Lasso, A., Yeo, C.T., Pezeshki, P., Vaughan, T., Carter, K., Rudan, J., Engel, C.J., Fichtinger, G.: Navigated breast tumor excision using electromagnetically tracked ultrasound and surgical instruments. IEEE Trans. Biomed. Eng. 1–7 (2015). http://www.slicerigt.org/wp/breast-cancer-surgery/
14. Mayo Clinic Explains Stereotactic Radiosurgery (SRS). http://www.mayoclinic.org/tests-procedures/stereotactic-radiosurgery/home/ovc-20130212. Accessed 20 Dec 2016
15. Wang, J.: Packing of unequal spheres and automated radiosurgical treatment planning. J. Comb. Optim. **3**, 453–463 (1999). Supported in part by NSF Grant CCR-9424164 Kluwer Academic Publishers
16. Rogers, C.A.: Packing and Covering. Cambridge University Press, Cambridge (1964)
17. Borgefors, G.: Distance transformation in digital images. Comput. Vis. Graph. Image Process. **34**, 344–371 (1986)
18. Gurevich, Y., Shelah, S.: Expected computation time for Hamiltonian path problem. SIAM J. Comput. **16**(3), 486–502 (1987)
19. Bender, E.A., Wilf, H.S.: A theoretical analysis of backtracking in the graph coloring problem. J. Algorithms **6**, 275–282 (1985)
20. Bergin, E.G., Sobral, F.N.C.: Minimizing the object dimensions in circle and sphere packing problems. Comput. OR **35**(7), 2357–2375 (2008)
21. Schuermann, A.: On packing spheres into containers (about Kepler's finite sphere packing problem). Documenta Mathematica **11**, 393–406 (2006)
22. Sloane, N.J.A.: Kepler's conjecture confirmed. Nature **395**, 435–436 (1998). News and Views Thomas C Hales announced that he had proved Kepler's conjecture of 1611 that no packing of spheres can be denser that face-centered-cubic lattice
23. Lea, M.S., Mara, M.T., Zhand, Q.: Gamma Knife Treatment Planning, pp. 1–6 (2010). http://sites.williams.edu/fws1/files/2011/10/GammaKnife.pdf
24. Lim, J.: Optimization in radiation treatment planning, pp. 1–179, Ph.D. Industrial Engineering, University of Wisconsin (2002)
25. Hales, T.: A proof of Kepler's conjecture. Ann. Math. Second Ser. **162**(3), 1065–1183 (2005)
26. Whiteside, A.: Isolating bon and gray matter in MRI images using 3DSlicer, pp. 1–10, Summer 2016 Independent Study Report working with Dr. SK Semwal, Department of Computer Science, University of Colorado, Colorado Springs (2016)
27. Open Source Processing Language. https://processing.org/

Automatic Nucleus Segmentation with Mask-RCNN

Jeremiah W. Johnson[✉]

University of New Hampshire, Manchester, NH 03101, USA
jeremiah.johnson@unh.edu
https://mypages.unh.edu/jwjohnson

Abstract. Mask-RCNN is a recently proposed state-of-the-art algorithm for object detection, object localization, and object instance segmentation of natural images. In this paper, it is demonstrated that Mask-RCNN can be used to perform highly effective and efficient automatic segmentations of a wide range of microscopy images of cell nuclei, for a variety of cells acquired under a variety of conditions. In addition, it is shown that a cyclic learning rate regime allows effective training of a Mask-RCNN model without any need to finetune the learning rate, thereby eliminating a manual and time-consuming aspect of the training procedure. The results presented here will be of interest to those in the medical imaging field and to computer vision researchers more generally.

Keywords: Deep learning · Neural networks · Microscopy · Instance segmentation · Detection · Mask-RCNN

1 Introduction

In the last few years, algorithms based on convolutional neural networks (CNNs) have led to dramatic advances in the state of the art for fundamental problems in computer vision, such as object detection, object localization, semantic segmentation, and object instance segmentation [1–4]. This has led to increased interest in the applicability of convolutional neural network-based methods for problems in medical image analysis. Recent work has shown promising results on tasks as diverse as automated diagnosis of diabetic retinopathy, automatic diagnosis of melanoma, precise measurement of a patient's cardiovascular ejection fraction, segmentation of liver and tumor 3D volumes, segmentation of mammogram images, and 3D knee cartilage segmentation [5–11].

Semantic segmentation of natural images is a long-standing and not fully solved computer vision problem, and in the past few years progress in this area has been almost exclusively driven by CNN-based models. Notable developments in recent years include the development of the Region-CNN, or R-CNN, model in 2014, fully convolutional neural networks in 2015, and the development shortly thereafter of the Fast-RCNN and Faster-RCNN models [12–15]. These algorithms were designed with semantic segmentation, object localization, and

© Springer Nature Switzerland AG 2020
K. Arai and S. Kapoor (Eds.): CVC 2019, AISC 944, pp. 399–407, 2020.
https://doi.org/10.1007/978-3-030-17798-0_32

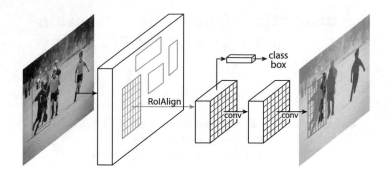

Fig. 1. The Mask-RCNN model. Image from [26]. Used with permission.

object instance segmentation of natural images in mind. Recently there has been an explosion of development in this area, with convolutional neural network based architectures such as Feature Pyramid Networks, SegNets, RefineNets, DilatedNets, and Retinanets developed all pushing benchmarks for this task forward [16–20]. In addition, single-shot models such as YOLOV3 and SSD have enabled object detection to occur at speeds up to 100–1000 times faster than region proposal based algorithms [21, 22].

In 2015, the U-Net architecture was developed explicitly with the segmentation of medical images in mind, and used to produce state-of-the-art results on the ISBI challenge for segmentation of neuronal structures in electron microscopic stacks as well as the ISBI cell tracking challenge 2015 [23]. U-Net architectures have since been adapted and used for a wide range of tasks in medical image analysis including volumetric segmentation of 3D structures and sparse-view CT reconstructions [24, 25].

1.1   The Mask-RCNN Model

The Mask-RCNN model was developed in 2017 and extends the Faster-RCNN model for semantic segmentation, object localization, and object instance segmentation of natural images [26]. Mask-RCNN is described by the authors as providing a 'simple, flexible and general framework for object instance segmentation'. Mask-RCNN was used to outperform all existing single-model entries on every task in the 2016 COCO Challenge, a large-scale object detection, segmentation, and captioning challenge [27].

Many modern algorithms for image segmentation fall into one of two classes: those that rely on a region proposal algorithm and those that do not. U-Net, for instance, is an example of a segmentation algorithm that does not rely on a region proposal algorithm; rather, U-Net uses an encoder-decoder framework in which a convolutional neural network learns, or encodes, a representation of the content of the image and a second network, such as a deconvolutional neural network, constructs the desired segmentation mask from the learned representation produced by the encoder (note that a deconvolutional neural network may also be referred to as a fully convolutional neural network, a transposed convolutional neural network or a fractionally-strided convolutional neural network in the literature) [13,28]. Encoder-decoder architectures have been used in machine learning for a variety of tasks outside of object detection or segmentation for some time, such as denoising images or generating images [29,30].

Mask-RCNN, in contrast, relies on a region proposals which are generated via a region proposal network. Mask-RCNN follows the Faster-RCNN model of a feature extractor followed by this region proposal network, followed by an operation known as ROI-Pooling to produce standard-sized outputs suitable for input to a classifier, with three important modifications. First, Mask-RCNN replaces the somewhat imprecise ROI-Pooling operation used in Faster-RCNN with an operation called ROI-Align that allows very accurate instance segmentation masks to be constructed; and second, Mask-RCNN adds a network head (a small fully convolutional neural network) to produce the desired instance segmentations; c.f. Fig. 1. Finally, mask and class predictions are decoupled; the mask network head predicts the mask independently from the network head predicting the class. This entails the use of a multitask loss function $L = L_{cls} + L_{bbox} + L_{mask}$. For additional details, we refer interested readers to [26].

Mask-RCNN is built on a backbone convolutional neural network architecture for feature extraction [12,15]. In principle, the backbone network could be any convolutional neural network designed for images analysis, such as a residual neural network (ResNet) [3]. However, it has been shown that using a feature pyramid network (FPN) as the Mask-RCNN backbone gives gains in both accuracy and speed [26]. A feature pyramid network takes advantage of the inherent hierarchical and multi-scale nature of convolutional neural networks to derive useful features for object detection, semantic segmentation, and instance segmentation at many different scales. Feature pyramid network models all require a backbone network themselves, from which the feature pyramid is constructed. Again, the backbone model is typically chosen to be a convolutional neural network known for high performance at object detection, and may be pretrained [16].

Although the properties of natural images will in general differ significantly from medical images, given the effectiveness of Mask-RCNN at general-purpose object instance segmentation, it is a reasonable candidate for use in automated segmentation of medical images. In addition, several high-quality opensource and liberally licensed implementations of Mask-RCNN have been released, including the Detectron implementation by the original authors of the algorithm, as well as the Matterport Inc. implementation [31,32]. Here, the efficacy of a Mask-RCNN model at detecting nuclei in microscopy images is investigated.

1.2 Cyclic Learning Rates

Traditionally, neural networks are trained using a gradient based optimization method such as stochastic gradient descent [33]. The performance of these methods is highly dependent on the learning rate parameter, and significant time and attention must be spent to find an ideal learning rate. Once a learning rate is chosen, the network is trained at that learning rate for some number of epochs,

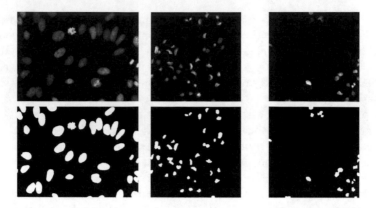

Fig. 2. Sample nuclei images and masks. For each image, an individual mask is provided for each nucleus detected. To generate the masks in the bottom row, all of the masks provided for each image have been merged into a single mask. Note that images vary widely, including in size.

after which the learning rate is reduced, and training continues for some additional epochs. The reduction and continued training may occur several times, and may take place as learning plateaus, or may occur according to a predetermined schedule.

Recent research using convolutional neural networks for object recognition tasks has highlighted the effectiveness of cyclic learning rates for accelerating the training process and preventing overfitting [34,35]. In this regime, both a maximum and a minimum learning rate are set and the learning rate cycles linearly back and forth between these values during the training process. These values can be selected automatically using a learning rate range test, and the length of each cycle can be set heuristically based on the training time [34]. Little research has been done on the effectiveness of a cyclic learning rate regime for an image segmentation task; therefore, we investigate such a regime here.

2   Nuclear Segmentation

Nuclear segmentation is a challenging problem in computational pathology. Effective nuclear segmentation has many applications, such as the extraction of high-quality features for nuclear morphometrics. However, conventional methods of automated analysis tend to perform poorly on challenging cases, such as images that exhibit crowding [36]. In addition, while conventional methods may be used at times to produce state-of-the-art results on a particular tissue morphology, such methods often fail to generalize [36]. Many datasets used for detecting and segmenting nuclei are often small, due to the difficulty and expense of collecting and hand-annotating large numbers of images, or lack the requisite information for training an automated segmentation algorithm [37,38].

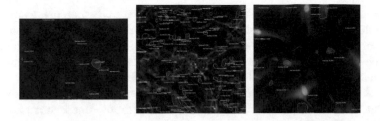

Fig. 3. Sample detections from the ResNet-50-FPN model.

The obvious need for larger annotated datasets on which overparamater-ized algorithms with many millions of parameters, such as Mask-RCNN, can be trained has led to the release of much larger annotated datasets in recent years, such as that described in [36]. To facilitate the creation of larger datasets for nuclei segmentation, often heterogeneous data will be collected from a variety of sources and combined. The datasets may include nuclei from a variety of different patients and organs, and in a variety of different disease states.

3 Methodology and Results

3.1 The Dataset

The dataset utilized for these experiments is image set BBBC038v1, available from the Broad Bioimage Benchmark Collection [39]. These data were used for Stage 1 of the 2018 Data Science Bowl, an annual competition sponsored by Booz Allen Hamilton and hosted on the data science website kaggle.com. The data consist of 729 microscopy images and corresponding annotations for each individual nucleus detected by an expert in each image; c.f. Fig. 2. The nuclei in the images are derived from a wide range of organisms including humans, mice, and flies. Furthermore, the nuclei in the images have been imaged and treated in variety of conditions and appear in a variety of contexts and states, including tissues and embryos, and cell division and genotoxic stress. 610 of the images are grayscale images. The remaining 119 images are color RGB images. The images vary in size also, from a minimum height of 256 pixels to a maximum height of 1040 pixels and a minimum width of 161 pixels to a maximum height of 1388 pixels. The variety present in the dataset provides a significant additional challenge as convolutional neural networks approaches can be expected to perform best in general when the input data is as uniform and standardized as possible. This includes standardization in terms of color, contrast, scale, and class balance.

Of these 729 images in the dataset, 664 images were used for training and validating the model and 65 images were held out for testing.

3.2 Experiments

For all experiments described here, a Mask-RCNN model with a feature pyra-mid network backbone is used. This model is based on an existing open-source implementation by Matterport Inc., which is written in Python using the libraries Keras and Tensorflow [32,40,41]. This implementation is released under an MIT License and is well-documented and easy to extend. For these experiments, the backbone networks used were feature pyramid networks derived from residual neural networks with either 50 or 101 layers (ResNet-50-FPN, ResNet-101-FPN). Note that the model with ResNet-50-FPN backbone has a somewhat lower computational load than that with a ResNet-101 backbone, but in our experiments the ResNet-101-FPN gives significantly improved results with no other changes to the model or training procedure. Rather than training the network end-to-end from the start, the model is initialized using weights obtained from pretraining on the MSCOCO dataset [27], after which the layers are trained in three stages: first, training only the network heads, which are randomly initialized, then training the upper layers of the network (from stage 4 and up in the ResNet model), and then training end to end. Two training strategies are considered: first, a learning rate schedule consisting of a learning rate of 0.001 for the first two stages of training followed by a learning rate of 0.0001 for the remaining 60 epochs, and second, a cyclic learning rate strategy in which the learning rate cycles linearly from 0.0001 to 0.001 and back over a period of 4000 minibatches. The cyclic learning rate strategy performed nearly as well as the learning rate schedule, but required significantly less finetuning. In total the models were trained for 100 epochs using stochastic gradient descent with momentum of 0.9, starting with a learning rate of 0.001 and ending with a learning rate of 0.0001. Numerous experiments were conducted using longer and shorter training times, but additional training did not lead to noticeable improvement and using fewer epochs led to underfit. A batch size of 6 was used on a single NVIDIA Titan Xp GPU. Gradients were clipped to 5.0 and weights were decayed by 0.0001 each epoch. Additional experiments with other learning rate schedules were conducted, but they gave no noticeable additional improvements to the schedule described above and thus aren't reported here.

For all of these experiments, image preprocessing was kept to a minimum. Images were upsampled by a factor of two and the channel means were normalized. Multiple experiments on mirroring the image edges by various numbers of pixels to improve detection of small nuclei at the edges as described in [23] were conducted, but no noticeable improvement from doing so was obtained, so this augmentation step was omitted from the final algorithm. However, to help avoid overfitting, the dataset was augmented during training with random crops, random rotations, Gaussian blurring, and random horizontal and vertical flips [42].

The model with ResNet-50 backbone and parameters as described above obtains an average mask intersection over union (IoU) of 66.98% on the validation dataset. The mean average precision at thresholds 0.5 to 0.95 by steps of size 0.05 as defined as the primary metric for the MSCOCO challenge (AP) is 56.06% for this model [27]. The model with ResNet-101 backbone and the same parameters and training procedures as described above obtains an average mask IoU of 70.54% and a mean average precision as defined for the MSCOCO

Table 1. Instance segmentation mask results on validation data. All results are single-model results. ResNet-XXX FPN denotes a feature pyramid network built using a residual neural network with XXX layers as backbone

Backbone	AP	Mask average IoU	Learning rate
ResNet-50 FPN	56.06	66.98	Scheduled
ResNet-50 FPN	55.64	66.70	Cyclic
ResNet-100 FPN	**59.40**	**70.54**	**Scheduled**
ResNet-100 FPN	59.09	69.61	Cyclic

challenge of 59.40%. These results are summarized in Table 1. Several sample detections are illustrated in Fig. 3.

4   Conclusions and Future Work

In this paper, it has been demonstrated that the Mask-RCNN model, while primarily designed with object detection, object localization, and instance segmentation of natural images in mind, can be used to produce high quality results for the challenging task of segmentation of nuclei in widely varying microscopy images. Furthermore, Mask-RCNN can be adapted to this task with limited modification. There are several similar tasks in medical image analysis for which it is likely that a Mask-RCNN based model could also easily be adapted to improve performance without extensive modification or customization. Examples of this include the task of segmentation of the left ventricle of the heart, where accurate segmentations can be used to estimate a cardiac patient's ejection fraction and improve their outcomes, or liver and tumor segmentation as described in [9]. Exploring the efficacy and performance of Mask-RCNN-based models for a range of such tasks will be the subject of further work.

Acknowledgments. The author would like to thank NVIDIA Corp. for GPU donation to support this research.

References

1. Krizhevsky, A., Sutskever, I., Hinton, G.E.: ImageNet classification with deep convolutional neural networks. In: Proceedings of the 25th International Conference on Neural Information Processing Systems, NIPS 2012, vol. 1, pp. 1097–1105. Curran Associates Inc., USA (2012)
2. Simonyan, K., Zisserman, A.: Very deep convolutional networks for large-scale image recognition. CoRR, abs/1409.1556 (2014)
3. He, K., Zhang, X., Ren, S., Sun, J.: Deep residual learning for image recognition. CoRR, abs/1512.03385 (2015)
4. Zagoruyko, S., Lerer, A., Lin, T.-Y., Pinheiro, P.H.O., Gross, S., Chintala, S., Dollár, P.: A multipath network for object detection. CoRR, abs/1604.02135 (2016)

5. Pratt, H., Coenen, F., Broadbent, D.M., Harding, S.P., Zheng, Y.: Convolutional neural networks for diabetic retinopathy. Procedia Comput. Sci. **90**, 200–205 (2016). 20th Conference on Medical Image Understanding and Analysis (MIUA 2016)

6. Esteva, A., Kuprel, B., Novoa, R.A., Ko, J., Swetter, S.M., Blau, H.M., Thrun, S.: Dermatologist-level classification of skin cancer with deep neural networks. Nature **542**, 115–118 (2017)

7. Prasoon, A., Petersen, K., Igel, C., Lauze, F., Dam, E., Nielsen, M.: Deep Feature Learning for Knee Cartilage Segmentation Using a Triplanar Convolutional Neural Network, pp. 246–253. Springer, Heidelberg (2013)

8. Tran, P.V.: A fully convolutional neural network for cardiac segmentation in short-axis MRI. CoRR, abs/1604.00494 (2016)

9. Christ, P.F., Ettlinger, F., Grün, F., Elshaer, M.E.A., Lipková, J., Schlecht, S., Ahmaddy, F., Tatavarty, S., Bickel, M., Bilic, P., Rempfler, M., Hofmann, F., D'Anastasi, M., Ahmadi, S.-A., Kaissis, G., Holch, J., Sommer, W.H., Braren, R., Heinemann, V., Menze, B.H.: Automatic liver and tumor segmentation of CT and MRI volumes using cascaded fully convolutional neural networks. CoRR, abs/1702.05970 (2017)

10. Zhu, W., Xiang, X., Tran, T.D., Hager, G.D., Xie, X.: Adversarial deep structured nets for mass segmentation from mammograms. CoRR, abs/1710.09288 (2017)

11. Kabani, A., El-Sakka, M.R.: Ejection fraction estimation using a wide convolutional neural network. In: Karray, F., Campilho, A., Cheriet, F., (eds.) Image Analysis and Recognition, pp. 87–96. Springer, Cham (2017)

12. Girshick, R., Donahue, J., Darrell, T., Malik, J.: Rich feature hierarchies for accurate object detection and semantic segmentation. In: Computer Vision and Pattern Recognition (2014)

13. Shelhamer, E., Long, J., Darrell, T.: Fully convolutional networks for semantic segmentation. IEEE Trans. Pattern Anal. Mach. Intell. **39**(4), 640–651 (2017)

14. Girshick, R.: Fast R-CNN. arXiv preprint arXiv:1504.08083 (2015)

15. Ren, S., He, K., Girshick, R.B., Sun, J.: Faster R-CNN: towards real-time object detection with region proposal networks. CoRR, abs/1506.01497 (2015)

16. Lin, T.-Y., Dollár, P., Girshick, R.B., He, K., Hariharan, B., Belongie, S.J.: Feature pyramid networks for object detection. CoRR, abs/1612.03144 (2016)

17. Badrinarayanan, V., Kendall, A., Cipolla, R.: SegNet: a deep convolutional encoder-decoder architecture for image segmentation. In: IEEE Transactions on Pattern Analysis and Machine Intelligence (2017)

18. Lin, G., Milan, A., Shen, C., Reid, I.: RefineNet: multi-path refinement networks for high-resolution semantic segmentation. In: CVPR, July 2017

19. Lin, T.-Y., Goyal, P., Girshick, R.B., He, K., Dollár, P.: Focal loss for dense object detection. CoRR, abs/1708.02002 (2017)

20. Yu, F., Koltun, V.: Multi-scale context aggregation by dilated convolutions. In: ICLR (2016)

21. Redmon, J., Farhadi, A.: Yolov3: an incremental improvement. arXiv (2018)

22. Liu, W., Anguelov, D., Erhan, D., Szegedy, C., Reed, S., Fu, C.-Y., Berg, A.C.: SSD: single shot multibox detector (2016), to appear

23. Ronneberger, O., Fischer, P., Brox, T.: U-net: convolutional networks for biomedical image segmentation. In: Medical Image Computing and Computer-Assisted Intervention (MICCAI). LNCS, vol. 9351, pp. 234–241. Springer (2015)

24. Han, Y., Ye, J.C.: Framing U-Net via deep convolutional framelets: application to sparse-view CT. CoRR, abs/1708.08333 (2017)

25. Çiçek, Ö., Abdulkadir, A., Lienkamp, S.S., Brox, T., Ronneberger, O.: 3D U-Net: learning dense volumetric segmentation from sparse annotation. In: Medical Image Computing and Computer-Assisted Intervention–MICCAI 2016, pp. 424–432. Springer, Cham (2016)

26. He, K., Gkioxari, G., Dollár, P., Girshick, R.B.: Mask R-CNN. CoRR, abs/1703.06870 (2017)

27. Lin, T.-Y., Maire, M., Belongie, S., Hays, J., Perona, J., Ramanan, D., Dollár, P., Lawrence Zitnick, C.: Microsoft COCO: common objects in context. In: Fleet, D., Pajdla, T., Schiele, B., Tuytelaars, T. (eds.) Computer Vision–ECCV 2014, pp. 740–755. Springer, Cham (2014)

28. Johnson, J., Alahi, A., Fei-Fei, L.: Perceptual losses for real-time style transfer and super-resolution. In: European Conference on Computer Vision, pp. 694–711. Springer (2016)

29. Kingma, D.P., Welling, M.: Auto-encoding variational bayes. arXiv preprint arXiv:1312.6114 (2013)

30. Vincent, P., Larochelle, H., Lajoie, I., Bengio, Y., Manzagol, P.-A.: Stacked denoising autoencoders: learning useful representations in a deep network with a local denoising criterion. J. Mach. Learn. Res. 11(Dec), 3371–3408 (2010)

31. Girshick, R., Radosavovic, I., Gkioxari, G., Dollár, P., He, K.: Detectron (2018). https://github.com/facebookresearch/detectron

32. Mask-RCNN. https://github.com/matterport/Mask_RCNN. Accessed 27 Apr 2018

33. Goodfellow, I., Bengio, Y., Courville, A.: Deep Learning. MIT Press (2016). http://www.deeplearningbook.org

34. Smith, L.N.: No more pesky learning rate guessing games. CoRR, abs/1506.01186 (2015)

35. Smith, L.N.: A disciplined approach to neural network hyper-parameters: part 1-learning rate, batch size, momentum, and weight decay. CoRR, abs/1803.09820 (2018)

36. Kumar, N., Verma, R., Sharma, S., Bhargava, S., Vahadane, A., Sethi, A.: A dataset and a technique for generalized nuclear segmentation for computational pathology. IEEE Trans. Med. Imaging 36, 03 (2017)

37. Gelasca, E.D., Obara, B., Fedorov, D., Kvilekval, K., Manjunath, B.S.: A biosegmentation benchmark for evaluation of bioimage analysis methods. BMC Bioinf. 10(1), 368 (2009)

38. Cheng, J., Rajapakse, J.C., et al.: Segmentation of clustered nuclei with shape markers and marking function. IEEE Trans. Biomed. Eng. 56(3), 741–748 (2009)

39. Ljosa, V., Sokolnicki, K.L., Carpenter, A.E.: Annotated high-throughput microscopy image sets for validation. Nat. Methods 9, 637 (2012)

40. Abadi, M., Agarwal, A., Barham, P., Brevdo, E., Chen, Z., Citro, C., Corrado, G.S., Davis, A., Dean, J., Devin, M., Ghemawat, S., Goodfellow, I., Harp, A., Irving, G., Isard, M., Jia, Y., Jozefowicz, R., Kaiser, L., Kudlur, M., Levenberg, J., Mané, D., Monga, R., Moore, S., Murray, D., Olah, C., Schuster, M., Shlens, J., Steiner, B., Sutskever, I., Talwar, K., Tucker, P., Vanhoucke, V., Vasudevan, V., Viégas, F., Vinyals, O., Warden, P., Wattenberg, M., Wicke, M., Yu, Y., Zheng, X.: TensorFlow: large-scale machine learning on heterogeneous systems (2015). Software available from tensorflow.org

41. Chollet, F., et al.: Keras (2015). https://keras.io

42. Perez, L., Wang, J.: The effectiveness of data augmentation in image classification using deep learning. arXiv preprint arXiv:1712.04621 (2017)

Stereo Vision Based Object Detection Using V-Disparity and 3D Density-Based Clustering

Shubham Shrivastava$^{(\boxtimes)}$

Renesas Electronics America, Farmington Hills, MI 48331, USA
Shubham.Shrivastava@renesas.com

Abstract. In recent years, autonomous driving has inexorably progressed from the domain of science fiction to reality. For a self-driving car, it is of utmost importance that it knows its surroundings. Several sensors like RADARs, LiDARs, and Cameras have been primarily used to sense the environment and make a judgment on the next course of action. Object detection is of a great significance in Autonomous Driving wherein the self-driving car needs to identify the objects around it and must take necessary actions to avoid a collision. Several perception-based methods like classical Computer Vision techniques and Convolutional Neural Networks (CNN) exist today which detects and classifies an object. This paper discusses an object detection technique based on Stereo Vision. One challenge in this process though is to eliminate regions of the image which are insignificant for the detection, like unoccupied road and buildings far ahead. This paper proposes a method to first get rid of such regions using V-Disparity and then detect objects using 3D density-based clustering. Results given in this paper show that the proposed system can detect objects on the road very accurately and robustly.

Keywords: Stereo vision · Object detection · Modified DBSCAN · Clustering

1 Introduction

Research and development of self-driving vehicles (also known as autonomous vehicles, or driverless vehicles) have increased significantly over the last decade. These cars use numerous perception techniques to detect their surroundings such as object detection, object classification, 3D position estimation, and localization. A diverse array of on-board sensors like LiDARs, RADARs, Cameras, and GPS are used to achieve this. While cameras can sense most objects in sight, it still cannot *'see'* the world in 3D. One of the cost-effective ways to sense the surroundings for 3D perception of objects is by making use of *Stereo Vision*. Stereo Vision provides dense 3D data (and intensity/color) while having enough

© Springer Nature Switzerland AG 2020
K. Arai and S. Kapoor (Eds.): CVC 2019, AISC 944, pp. 408–419, 2020.
https://doi.org/10.1007/978-3-030-17798-0_33

accuracy for most driving assistance applications. In addition, these data can be further exploited to detect objects.

In stereo vision, two cameras, displaced horizontally from one another are used to obtain two differing views on a scene. The stereo vision algorithms utilize the *disparity* of the two corresponding pixels of stereo images to compute the *depth map*. Values in depth map are inversely proportional to the values in *disparity map*. This depth map can further be utilized to perceive 3D surroundings like road surfaces, objects and thus provide a higher sense of localization for self-driving cars.

This paper is composed of the following sections: Sect. 2 describes the Stereo Vision technology and briefly discusses methods for computing disparity and depth map. Section 3 proposes a combination of V-Disparity and 3D density-based clustering method for object detection. Section 4 provides some results of the proposed method of object detection on testing urban dataset [10]. Section 5 provides a brief conclusion based on the experiment conducted.

2 Stereo Vision

Stereo Vision is the process of extracting 3D information from two or more 2D images. *Stereo vision* can reconstruct the complete 3D space, but due to some factors, such as *occlusion* and the complexity of the imaging environment, the imaging of the 3D space is prone to regional hollow or only partial information [14]. During the projection process of real world onto a two-dimensional picture, we lose one dimension. This third-dimensional information can be recovered by means of a pair of synchronized camera images which *epipolar lines* [6] are aligned. A simple stereo geometry is shown in Fig. 1.

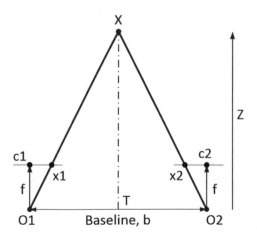

Fig. 1. Basic stereo geometry

In the above figure, $c1$ and $c2$ are the centers of left and right lenses respectively, f is the focal length, Z is the distance in Z direction (depth), $x1$ is the image coordinate in the left image, *x2* is the image coordinate in the right camera, and *b* is the stereo *baseline* or distance between the centers of the projection *O1* and *O2*. For this system, the depth *Z* of an object at real-world coordinate *X* can be determined [2] as shown below:

$$\frac{b + x1 + x2}{Z - f} = \frac{b}{Z}$$

$$or, \quad Z = f \cdot \left(\frac{b}{x1 + x2}\right) \tag{1}$$

Where, *(x1 + x2)* is the *disparity* which is the difference in image location of the same 3D point when projected under perspective to two different cameras. For a given stereo system, the focal length *f* and baseline *b* are fixed and known. If the disparity *(x1 + x2)* can be computed from left and right images, the depth can be determined from Eq. 1. This provides the basis of so-called *depth map* which contains the real-world depth for each pixel in the *disparity map*. There is a wide range of stereo matching algorithms, each with its own advantages and disadvantages [3–5]. For any of these algorithms to work, it is necessary that the stereo vision system satisfies *epipolar constraint* [6]. However, if the two image planes are not *coplanar*, then the images can be warped such that both images are on the *epipolar plane*. This process is often referred to as *Rectification* [7,15].

Disparity Map also forms the basis of *U-Disparity* and *V-Disparity* maps. *U-Disparity* is essentially the histogram of each column of the *disparity map* and *V-Disparity* is the histogram of each row of the *disparity map*.

The v-disparity image computation can be seen as a function *H* on the disparity map *D* such that $H(D) = D_v$ wherein, *H* accumulates the points in the *disparity map* with the same disparity value *d* in each row. The abscissa u_p of a point *p* in the *v-disparity* image corresponds to the disparity value d_p and its gray value $g(d_p) = r_p$ to the number of points with the same disparity in row *r*.

$$r_p = \sum_{P \in D} \delta_{v_p, r} \cdot \delta_{d_p, d_p} \tag{2}$$

Where $\delta_{i,j}$ is Kronecker delta and is defined as follows:

$$\delta_{i,j} = \begin{cases} 1, & for\ i = j \\ 0, & for\ i \neq j \end{cases} \tag{3}$$

Similarly, u-disparity image can be constructed by accumulating the points in the disparity map with the same disparity value d column-wise.

3 Object Detection

Object detection has been an active and very significant research area in the automotive industry for quite a long time now. Detecting on-road objects such as vehicles, pedestrians is a major challenge for self-driving cars and some research suggests using *u-v-disparity* for stereo vision based object detection [1,12].

This paper proposes a different method for object detection using three-dimensional density-based clustering combined with V-Disparity. Extracting unoccupied road surface information can give us a major advantage towards object detection. *Stereo Vision* provides a very robust method for extracting unoccupied road surface and free space boundary and is explained in the following subsection.

3.1   Unoccupied Road Surface Extraction

Road surface can be modelled as a succession of parts of oblique planes [8]. In case of a flat road geometry, the plane of the road is projected as a straight line on the lower portion of *v-disparity map* as shown in Fig. 2(d). This straight line can very easily be found by first performing *canny edge detection* and then applying *Hough Transforms* [9]. Each pixel which is a part of this *lower envelope* then provides us the disparity value corresponding to unoccupied road surface area for each row of *disparity map*. If the road was modelled as a non-flat earth road geometry, then we will need to extract a *piecewise linear curve* rather than a straight line and [11] suggests the use of *steerable filter* to achieve this.

Figure 2(a) and (b) shows a stereo image pair *(Source: Daimler Urban Segmentation Dataset [10])*, for which the computed *disparity map*, *v-disparity*, and *u-disparity* are shown in Fig. 2(c), (d), and (f) respectively. The lower envelope of *v-disparity* is further extracted using *Hough Transforms* and is shown in Fig. 2(e).

The lower envelope of *v-disparity* can be used to generate a *Road Mask*. This can be achieved by considering each disparity value d in a row r of *disparity map* as road, if the row r and column d of Fig. 3(e) is 'TRUE'. Please note that Fig. 3(e) is logical and each pixel has a value of either 'TRUE' or 'FALSE'. The *road mask* thus computed is shown in Fig. 3(b). This is a logical image which pixel values are 'FALSE' for the region classified as *unoccupied road*, and 'TRUE' otherwise. Applying this mask to the *disparity map* results in a *masked disparity map* which is free from unoccupied road surface as shown in Fig. 3(c).

3.2   Proposed Method for Object Detection

There has been some research on stereo vision based object detection and most of it is based on *u-v-disparity*. Each horizontal line in *u-disparity* has a corresponding vertical line in *v-disparity* and represents an object. Horizontal boundaries of the object can be found from column boundaries of the line in *u-disparity* and Vertical boundaries of the object can be found from row boundaries of the line in *v-disparity*. Some results of this approach are given in [1,12]. In some cases when an object is represented by multiple disparity values (e.g a vehicle parked on the side of the road for which pixels can have many depth levels), it causes that object to be represented by multiple lines in *u-disparity* and *v-disparity*. The only way to classify all these pixels as one object is by grouping the lines in *u-disparity* which are close to each other. However, this can result in misclassification of vehicles that are at a far distance and close to each other because they will end up being classified as one big object.

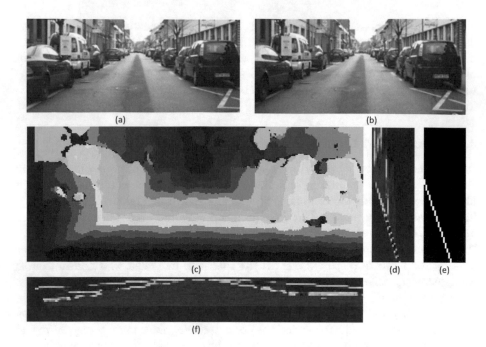

Fig. 2. (a) Left Stereo Image (b) Right Stereo Image (c) Disparity Map (d) V-Disparity Map (e) V-Disparity lower envelope extracted using Hough Transforms (f) U-Disparity Map

This paper approaches this problem by looking at all the pixels in three-dimensions and classifies a cluster of pixels that are close to each other as an object. The basis for this three-dimensional density-based clustering is a very well-known algorithm, 'DBSCAN' [13] which is slightly modified here to work with stereo vision. This new 'Modified DBSCAN' algorithm works with a dataset which essentially is the *depth map* computed using *disparity map, focal length*, and *baseline* as given by Eq. 1. The difference between traditional *DBSCAN* and *Modified DBSCAN* is in the fact that (1) the database for *Modified DBSCAN* is a two-dimensional array with each value representing depth of the pixel such that DATA[row, column] represents a pixel at the 3D coordinate (column, row, DATA[row, column]) in the (x, y, z) directions respectively, (2) the cost function here is computed using 3D distance between points, and (3) the *Modified DBSCAN* discussed in this paper also takes in a 'mask' such as the *road mask* to perform 'Region Query'. A very high-level flowchart of *Modified DBSCAN* is shown in Fig. 4. This paper does not go in detail of the algorithm but addresses the modifications from implementation point of view in the following subsections.

Initialization of Cluster Labels. Modified DBSCAN proposed here initializes labels for each pixel as either 'Unclassified' or 'Unoccupied Road' as opposed to the traditional *DBSCAN* which initializes labels for all the points in dataset to 'Unclassified'.

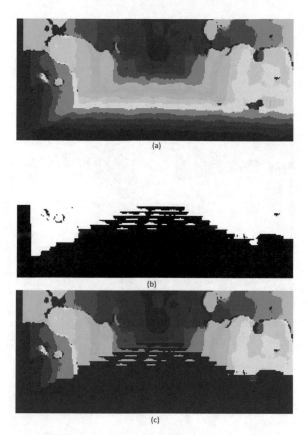

Fig. 3. (a) Disparity Map (b) Road Mask generated from the lower envelope of V-Disparity (c) Disparity Map as a result of applying the Road Mask

```
1   FOR i FROM 1 TO depthMap.numberOfRows
2       FOR j FROM 1 TO depthMap.numberOfColumns
3           IF roadMask.data(i, j) IS FALSE THEN
4               labels(i, j) = UNOCCUPIED_ROAD
5           ELSE
6               labels(i, j) = UNCLASSIFIED
7           END IF
8       END FOR
9   END FOR
```

The algorithm then performs 'Region Query' and 'Expand Cluster' for all the pixels which label is marked as 'Unclassified'.

Region Query. Region Query finds pixels in the *Eps-Neighborhood* [13] of points, meaning that it returns a list of pixels within a distance of 'epsilon' from the current pixel in consideration. This distance can be found by defining

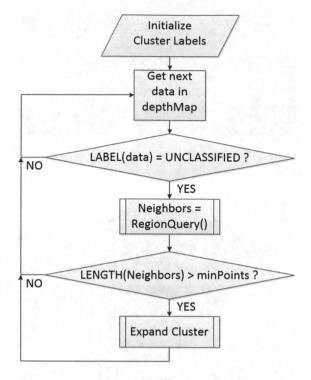

Fig. 4. High-Level flowchart of Modified DBSCAN

a *cost function* which computes the cost to reach each pixel in three-dimensions. *Modified DBSCAN* excludes pixels labelled as 'Unoccupied Road' for querying the pixels in *dataset*.

```
1   RegionQuery(pixelValue, labels, epsilon):
2       FOR i FROM 1 TO depthMap.numberOfRows
3           FOR j FROM 1 TO depthMap.numberOfColumns
4               IF labels(i, j) IS NOT UNOCCUPIED_ROAD THEN
5                   IF Cost(pixelValue, depthMap.data(i, j)) <
                        epsilon THEN
6                       ADD depthMap.data(i, j) TO NEIGHBOR_LIST
7                   END IF
8               END IF
9           END FOR
10      END FOR
11      RETURN NEIGHBOR_LIST
12  END FUNCTION
```

Cost Function. *Region query* uses a *cost function* to determine the distance between two points. This *cost function* can be defined to either return the absolute three-dimensional distance or it can allow putting more weight in one dimension depending on the type of *dataset*.

$$Cost = X_{weight} \cdot (point1_x - point2_x)^2$$
$$+Y_{weight} \cdot (point1_y - point2_y)^2$$
$$+Z_{weight} \cdot (point1_z - point2_z)^2 \tag{4}$$

Where, $pointn_x$ is the x or *column* location of nth pixel in *depthMap*, $pointn_y$ is the y or *row* location of nth pixel in *depthMap*, $pointn_z$ is the value of the pixel at `depthMap(row, column)`. X_{weight}, Y_{weight}, and Z_{weight} are the weights assigned to distance between pixels in x, y, and z directions respectively.

Expand Cluster. The *Modified DBSCAN* method discussed here thus finds clusters satisfying *'epsilon'* and *'minPoints'* criteria [13] and provides us three-dimensional density based clusters. Further filtering can be applied to exclude clusters which are either at a far distance or clusters forming at the top of frame. Such clusters most likely will be sky, buildings and other objects that self-driving cars do not care about. Remaining clusters can then be identified as a potential object on the road. Figure 5(a) shows these clusters. Figure 5(b) maps the bounding-box found from clusters, back on the left stereo image.

Point-Cloud data can be constructed based on the depth map obtained from disparity map. Furthermore, the pixel intensities can also be mapped from the stereo image to each point in the point-cloud data. Figures 6 and 7 demonstrates one such point-cloud reconstructed from stereo image pair. 3D bounding-boxes are also drawn to identify detected objects.

4 Results

To evaluate the proposed object detection method, this algorithm was applied to multiple stereo image pairs and the results were evaluated manually. The proposed method of object detection showed very promising results. It was found that the vehicles, cyclist and traffic signs on the road were detected accurately. The dataset used for testing were obtained from 'Daimler Urban Segmentation Dataset' [10] which contains a huge set of stereo image pairs collected on urban roads. Figure 8 shows the result of object detection on few of these stereo image pairs. It can be seen in Fig. 8 that the objects on road are detected accurately and the system robustly estimates their distance from the camera.

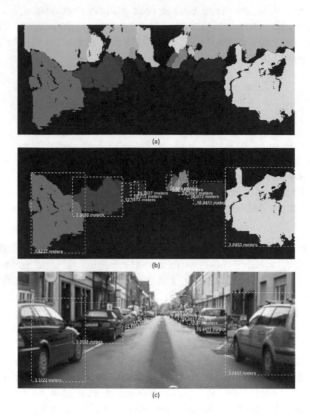

Fig. 5. (a) Clusters found as a result of the Modified DBSCAN (b) Filtered Clusters (c) Cluster bounding box mapped on the left stereo image.

Fig. 6. Point-Cloud data reconstructed from stereo image pair with objects 3D bounding boxes (Front View).

Fig. 7. Point-Cloud data reconstructed from stereo image pair with objects 3D bounding boxes (Top View).

Fig. 8. Experimental results of object detection on urban road test dataset.

5   Conclusion

In this paper, an accurate and robust method of object detection is proposed. The detection process is based on extraction of unoccupied road surface by constructing *v-disparity* map and finding its lower envelope. After identifying the road surface, *Modified DBSCAN* method is proposed for three-dimensional density based clustering which extracts clusters of objects. Furthermore, stereo vision based object detection also provides an accurate distance to each object which can help self-driving cars localize themselves and avoid obstacles on the road. Using sensors such as LiDARs and RADARs can also perform object detection and provide an accurate distance as well, however, they are computationally expensive. *Stereo Vision* can be used to reconstruct 3D space from 2D images and provide self-driving cars a way to *'see'* the world in 3D with a very low computational cost.

References

1. Benacer, I., Hamissi, A., Khouas, A.: A novel stereovision algorithm for obstacles detection based on U-V-disparity approach. In: 2015 IEEE International Symposium on Circuits and Systems (ISCAS), Lisbon, pp. 369–372 (2015). https://doi.org/10.1109/ISCAS.2015.7168647
2. Shah, M.: Fundamentals of Computer Vision (1997)
3. Lazaros, N., Sirakoulis, G.C., Gasteratos, A.: Review of stereo vision algorithms: from software to hardware. Int. J. Optomechatron. **2**, 435–462 (2008). https://doi.org/10.1080/15599610802438680
4. Kumari, D., Kaur, K.: A survey on stereo matching techniques for 3D vision in image processing. Int. J. Eng. Manuf. (IJEM) **6**(4), 40–49 (2016). https://doi.org/10.5815/ijem.2016.04.05
5. Kuhl, A.: Comparison of Stereo Matching Algorithms for Mobile Robots (2005)
6. Hata, K., Savarese, S.: Epipolar geometry. In: Computer Vision, From 3D Reconstruction to Recognition (2018)
7. Kudryavtsev, A.V., Dembélé, S., Piat, N.: Stereo-image rectification for dense 3D reconstruction in scanning electron microscope. In: 2017 International Conference on Manipulation, Automation and Robotics at Small Scales (MARSS), Montreal, pp. 1–6 (2017). https://doi.org/10.1109/MARSS.2017.8001905
8. Labayrade, R., Aubert, D., Tarel, J.P.: Real time obstacle detection in stereovision on non flat road geometry through v-disparity representation. In: IEEE Intelligent Vehicle Symposium, vol. 2, pp. 646–651 (2002)
9. Pao, D.C.W., Li, H.F., Jayakumar, R.: Shapes recognition using the straight line Hough transform: theory and generalization. IEEE Trans. Pattern Anal. Mach. Intell. **14**(11), 1076–1089 (1992). https://doi.org/10.1109/34.166622
10. Scharwächter, T., Enzweiler, M., Roth, S., Franke, U.: Stixmantics: a medium-level model for real-time semantic scene understanding. In: European Conference on Computer Vision (ECCV) (2014)
11. Gao, Y., Ai, X., Wang, Y., Rarity, J., Dahnoun, N.: U-V-disparity based obstacle detection with 3D camera and steerable filter. In: 2011 IEEE Intelligent Vehicles Symposium (IV), Baden-Baden, pp. 957–962 (2011). https://doi.org/10.1109/IVS.2011.5940425
12. Zhu, K., Li, J., Zhang, H.: Stereo vision based road scene segment and vehicle detection. In: Proceedings of 2nd International Conference on Information Technology and Electronic Commerce (2014). https://doi.org/10.1109/ICITEC.2014.7105591
13. Ester, M., Kriegel, H.P., Sander, J., Xu, X.: A density-based algorithm for discovering clusters in large spatial databases with noise. In: Proceedings of KDD 1996 (1996)

14. Yang, J., Chen, H.: The 3D reconstruction of face model with active structured light and stereo vision fusion. In: 2017 3rd IEEE International Conference on Computer and Communications (ICCC), Chengdu, China, pp. 1902–1906 (2017). https://doi.org/10.1109/CompComm.2017.8322869

15. Liu, X., Li, D., Liu, X., Wang, Q.: A method of stereo images rectification and its application in stereo vision measurement. In: 2010 Second IITA International Conference on Geoscience and Remote Sensing, Qingdao, pp. 169–172 (2010). https://doi.org/10.1109/IITA-GRS.2010.5602989

Building a Weighted Graph to Avoid Obstacles from an Image of the Environment

Kevin G. Prehn$^{(\boxtimes)}$ and John M. Jeffrey$^{(\boxtimes)}$

Elmhurst College, Elmhurst, IL 60102, USA
kevin.prehn@365.elmhurst.edu, johnj@elmhurst.edu

Abstract. In robot pathfinding problems, the accuracy and efficiency of the solution path is dependent on the representation of the environment. While many local navigation systems describe pathfinding with limited information, there has been a lack of research on developing a global representation of an environment with image processing techniques. The described algorithm constructs a bidirectional weighted graph representation from an aerial input image designed to not intersect with obstacles in that image. The algorithm is structured as a four-stage pipeline, where each stage uses existing or modified algorithms. To the best of our knowledge, this algorithm is unique in the construction of a well understood representation of an environment from an aerial input image. Future work could improve algorithm runtime and overall accuracy of the output graph.

Keywords: Image processing · Robot navigation ·
Bidirectional weighted graph

1   Introduction

In robot navigation, the pathfinding problem is to reach a target point from an initial point after avoiding obstacles. Bidirectional weighted graphs are commonly used to represent a navigable space in video games and robotics applications. Much is already known about navigation using this representation, and many shortest-path graph algorithms like *Dijkstra's* and A^* already use this representation [2]. Previous studies do not discuss constructing this representation from image data.

According to Algfoor et al., the first stage of robot navigation is extracting information about the environment [2]. The difficulty is how to construct an accurate representation of a given environment; once the environment representation is created, many algorithms of the literature are able to find the path. However, analyzing an aerial image for this purpose is challenging, since it may be difficult to discern what is navigable space and what is not. In this paper, we assume that non-salient regions are navigable spaces.

© Springer Nature Switzerland AG 2020
K. Arai and S. Kapoor (Eds.): CVC 2019, AISC 944, pp. 420–435, 2020.
https://doi.org/10.1007/978-3-030-17798-0_34

This paper introduces an algorithm which creates a representation of a given environment in the form of a bidirectional weighted graph. We evaluate two variations of the algorithm's image segmentation stage to determine whether a slower yet more accurate segmentation method produces a better output graph.

The outline of this paper is as follows. In Sect. 2, we introduce related work and existing algorithms. We describe the algorithm and time complexity in Sects. 3 and 4, respectively. In Sect. 5, we describe experiments and present results. Finally, Sects. 6 and 7 we discuss the results and future work.

2 Background

This section is organized as follows: Subsect. 2.1 reviews related robot navigation methods and solutions; Subsect. 2.2 covers image segmentation; Subsect. 2.3 discusses salient regions; finally, Subsect. 2.4 briefly discusses the convex hull and visibility graphs.

2.1 Robot Navigation

Robot Navigation can be loosely described as a navigation task where a system uses a representation of the environment to navigate from some start location to a desired target location. The environment can be represented either locally or globally, depending on how much information a robot has about its surroundings. If a robot only has information on its immediate surroundings, then it is local navigation. A robot using global navigation would initially have all information about its environment.

Local navigation systems have limited information. In [7], Kiy describes a local robot navigation system where the robot has a front-mounted camera. He assumes objects with higher contrast are obstacles and navigates to avoid them. Limited to a one-directional view, the robot may not be able to find the best path around obstacles. Despite this simple assumption, it was able to identify obstacles within the environment.

Global navigation systems have all information of an environment at the start. In [5], Gehrig and Stein modeled the path from a start and target point as an elastic band with obstacles "pushing" the elastic band away from themselves. This required the system to know the entire environment before finding a path. Pathfinding with a weighted graph representation of the environment is already well understood and is a common solution for robots and video games [2].

2.2 Image Segmentation

Image processing algorithms refer to algorithms meant to extract information from images. Image segmentation is a common first step in image processing which involves separating all of the pixels in an image to a set of disjoint sets, where each set represents one region within the image [11].

A more primitive image processing method related to segmentation is binarization or thresholding where the output image will have each pixel from the input image either included or excluded, represented as a boolean value [9]. This is especially common in document analysis. Binarization may also be viewed as segmenting the image into only two regions. While these methods work great in some problem domains because of their simplicity, efficiency, and accuracy, they are not enough for more complex problems.

In [11], Yu segments images using a genetic algorithm by partitioning the set of all colors in the image into k clusters, where each cluster is defined by a single color as its centroid. All other colors belong to whichever cluster whose centroid color is closest to that color. For the genetic algorithm, each population represented a set of centroids. His algorithm optimizes the centroids to maximize cluster dissimilarity.

While digital images are often taken and stored as a set of pixels where each pixel has a single color, an improved representation can help with segmentation and image classification. By definition, the texture describes the qualities of some set of pixels, usually a subimage of a larger image, represented as a vector of different features such as brightness, color variance, and average color [10]. Textures can be compared and classified using the euclidean distance between their feature vectors [6].

In [10], Sidorova segments aerial image using texture instead of only color with improved results. Because k-means clustering only works with points and not regions, Sidorova maps each pixel in an image to a texture, where the texture is the feature vector of the window around that pixel. The image is segmented by partitioning the set of textures in the image, rather than the set of colors as done by Yu.

2.3 Saliency Regions

Intuitively, salient regions are the regions in an image that stand out, such as a red flower in a field of green grass. Determining salient regions is useful in image processing because one may want to focus on certain subregions of an image [8]. Scharfenberger et al. calculated the saliency value of a region by comparing the distance of the feature vectors of a region to all other regions in the image, making it a global calculation. More salient regions have higher saliency values.

2.4 Convex Hull and Visibility Graphs

The convex hull of a set of vertices is a convex polygon (or n-gon) such that every point in the set is within the bounds of that polygon. An algorithm to efficiently compute this is described in [3].

A visibility graph is a weighted bidirectional graph where all nodes in the graph have an edge between them if and only if the edge does not intersect with any lines or polygons in some obstacle set [4]. This only makes sense when the graph is in Euclidean space and if every node has a position in this space. In this paper, the vertices, polygons, and corresponding nodes and edges of graphs are assumed to be in 2-dimensional Euclidean space.

3 Algorithm Description

This section is organized as follows: Sect. 3.1 details the image segmentation by texture stage, Sect. 3.2 describes the vertex grouping stage, Sect. 3.3 describes the polygon construction stage, and finally Sect. 3.4 briefly describes the weighted graph construction.

3.1 Image Segmentation

The first stage attempts to categorize each pixel in the image as an obstacle or background. It is assumed that obstacles are non-navigable, while the background is navigable. All regions which are not salient regions are assumed to be background.

The following is the formal representation of images and the image segmentation process is similar to the ones used in [9,11]. For this specific problem, the image is restricted to two dimensions.

Let I be a 2D image with width w and height h, comprised of pixels P, where $(x, y) \in P$ where $0 \leq x < w$ and $0 \leq y < h$ are the pixel coordinates such that each pixel has a unique coordinate. Each pixel has an RGB color value represented by $color : P \rightarrow C \times C \times C$ where C is a set of integers ranging from 0 to 255. $color(p)$ then gives the color for pixel $p \in P$.

Before segmentation, a textural representation of the image must be calculated. The method of calculating textures is very similar to what is done in [8,10].

Given an input image I with pixels P, let a vector $v_p \in \mathbb{R}^a$ be the feature vector associated with pixel $p \in P$ with a feature values. Given a window size $w \in \mathbb{Z}$:

$$texture : P \times \mathbb{Z} \rightarrow \mathbb{R}^a$$

Where $texture(p, w) = v_p$. $texture$ computes the feature vector of p by computing the feature vector of the region in the window around p. The algorithm computes the multiset of textures T such that:

$$T = \{v_p \mid \forall p \in P, v_p = texture(p, w)\}$$

Each pixel in P has a corresponding feature vector in T such that $|P| = |T|$. The feature vectors are partitioned into k disjoint sets, representing the different clusters. The similarity of two feature vectors $v_{p1}, v_{p2} \in T$ is the Euclidean distance between them. This function is then used to cluster regions by texture with k-means clustering. k is a user defined value, and we found $k = 5$ to be sufficient. Figure 1 shows the result of this segmentation.

Once the textures have been partitioned, the algorithm computes the saliency value of each cluster. The saliency value is the average distance of the feature vectors in one region compared to all other feature vectors in all other regions [1]. The more dissimilar a region is from the rest of the image, the higher its saliency value.

The regions are sorted by their saliency values, and the h regions with the highest values are labeled as obstacle, while the remaining regions are labeled as background. h is a user-defined value, and we found $h = 2$ to be sufficient.

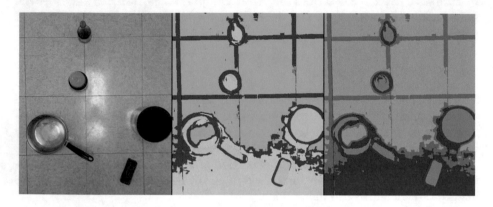

Fig. 1. Result of segmenting the image with clustering textures and selecting salient regions. The left shows the original image. The middle shows the segmented image. The right highlights the h ($h = 2$) regions that are the most salient. Gray is labeled as background and red is labeled as obstacle.

All pixels associated with the features vectors in the h most salient clusters are considered obstacles and added to the set $P_{obst} \subseteq P$ if and only if the pixel is on an edge of a region labeled as obstacle. These pixels are transformed to a set of vertices V_{obst} defined as:

$$V_{obst} = \{(x, y) \mid \forall (x_p, y_p) \in P_{obst}, x = x_p, y = y_p\}$$

3.2   Vertex Grouping

The set V_{obst} is separated to different groups so the polygons can be constructed. Groups are sets of vertices that are assumed to be part of the same object. V_{obst} is partitioned and reduced to a set of sets G where $\forall g \in G, g \subseteq V_{obst}$. Algorithm 1 describes the procedure for grouping vertices. Initially, G is set to a set of sets where each set is unique and contains one vertex $v \in V_{obst}$ so that $|G| = |V_{obst}|$. These groups are continuously combined until no further combining is possible. Two groups $g_a, g_b \in G$ may be combined if any two vertices from the two groups are at least d_{min} close but not closer than d_{max}. d_{min} and d_{max} are user defined values. Once combining is complete, all groups which have fewer than a minimum number of vertices (we made $size_{min} = 10$ the minimum) are removed. Additionally, we remove any group where the standard deviation of all points' x or y coordinates is less than some threshold (we used $\sigma_{min} = 3$ as the threshold). This removes insignificant groups which may have resulted from noise in the input image. Figure 2 shows the result of this stage.

Data: A set of vertices V_{obst}
Result: A set of sets of vertices G
$G \leftarrow$ A set of sets, where each set has exactly one vertex from V_{obst} ;
$combine \leftarrow true$;
while $combine$ **do**
    $combine \leftarrow false$;
    for $g_i \in G$ **do**
        for $g_j \in G$ **do**
            $min \leftarrow d_{min} + 1$;
            $max \leftarrow d_{max} - 1$;
            for $g_h \in G$ **do**
                for $g_k \in G$ **do**
                    if $dist(g_h, g_k) < min$ **then**
                        $min \leftarrow dist(g_h, g_k)$;
                    end
                    if $dist(g_h, g_k) > max$ **then**
                        $max \leftarrow dist(g_h, g_k)$;
                    end
                end
            end
            if $min < d_{min}$ **and** $max < d_{max}$ **then**
                $g_i \leftarrow g_i \cup g_j$;
                remove g_j from G ;
                combine = true ;
            end
        end
    end
end
for $g_i \in G$ **do**
    if $|g_i| < size_{min}$ **then**
        remove g_i from g ;
    else
        $\sigma_x \leftarrow$ standard deviation of all x positions in g_i ;
        $\sigma_y \leftarrow$ standard deviation of all y positions in g_i ;
        if $\sigma_x < \sigma_{min}$ **or** $\sigma_y < \sigma_{min}$ **then**
            remove g_i from g ;
        end
    end
end

Algorithm 1. Vertex Grouping

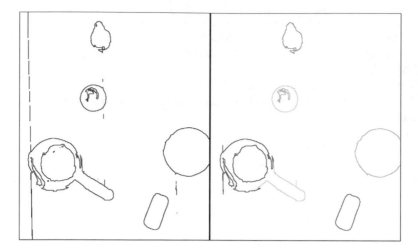

Fig. 2. On the left are the edge pixels of the selected salient clusters. On the right are the resulting groups of points.

3.3 Polygon Construction

This stage transforms the set G to a set of polygons N, where each group $g \in G$ will correspond to one polygon $n \in N$. Each polygon is the convex hull of the set of points belonging to the corresponding group. This stage uses the convex hull algorithm from [3]. Once the convex hull polygons are constructed, we remove any polygons that are completely contained inside of any other polygon, since these are not needed. Consequently $|N| \leq |G|$.

Data: A set of convex polygons N
Result: A set of polygons with fewer vertices
for $n \in N$ **do**
    for $v_i \in n$ **do**
        $\mathbf{a} \leftarrow v_i - v_{i-1}$;
        $\mathbf{b} \leftarrow v_{i+1} - v_{i-1}$;
        $\theta \leftarrow$ angle between $\mathbf{a}$ and $\mathbf{b}$;
        if $\theta < \theta_{pmin}$ **and** $dist(v_{i-1}, v_i) < d_{pmin}$ **then**
            | remove v_i from n ;
        end
    end
end

Algorithm 2. Polygon reduction

As described in Algorithm 2, polygons are also reduced where vertices are removed if and if only if removing them would not likely impact the overall structure of that polygon. If two consecutive vertices v_i and v_{i+1} are within a user defined distance d_{pmin} from each other and the angle between the first line intersecting v_{i-1} and v_i and the second line intersecting v_{i-1} and v_{i+1} is below some user defined threshold θ_{pmin}, then v_i is removed from the polygon. This helps reduce the number of nodes in the output graph while not impacting its overall structure. Figure 3 shows the result of this stage.

3.4 Weighted Graph Construction

The problem of constructing a weighted graph which does not collide with any obstacle polygons is analogous to the problem of constructing a visibility graph. Therefore, the algorithm in [4] may be used to construct the final output graph. In order to give the navigating robot space, all polygons are first scaled up from their center point to create padding around the obstacle as done in [5]. Every vertex in each polygon becomes a nodes in the visibility graph, and the edges are constructed by the algorithm creating the graph.

As seen in Fig. 3, the output is a bidirectional weighted graph whose edges should not collide with obstacles in the input image.

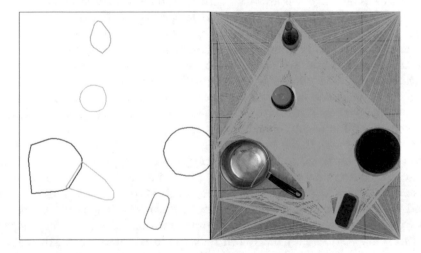

Fig. 3. On the left is the result of finding the convex hull of every vertex group. On the right is the result of finding the visibility graph with the convex hulls.

4   Analysis

Every stage of the algorithm runs independently and sequentially, so the time complexity may be analyzed separately. However, because it is a pipeline architecture, the output of the previous stage will also affect the time complexity of the next stage.

The first stage is calculating the texture value for each pixel in the input image. Each texture has a values in its feature vector, and each one of these values uses at least every pixel within the region for its calculation, so the worst-case time for each texture is $O(a \times w^2)$. Given that the image has $|P|$ pixels, the total time for finding all of the textures is then $O(|P| \times a \times w^2)$.

Once the textures are calculated, the image may be segmented and the salient regions may be computed. The image segmentation stage uses a k-means clustering algorithm on the textures. The initial cluster centers are selected randomly, and the number of iterations required for k-means clustering is therefore non-deterministic. The analysis for the time of this clustering process is too complex for the purposes of this paper.

To find the saliency values of the regions, the similarity of each texture value in each region is calculated with each texture value in every other region. The average similarity of a given region compared to every other region is the average of this sum. Since every pixel's texture must be compared with each other texture for each of the k clusters, the time of this algorithm is $O(|P|^2 \times k)$ in the worst case. Next, every pixel on the edge of a region is transformed into a point for the next stage, creating n_g points. In the worst case, every pixel is an obstacle, so the worst case for the number of points is $n_g = O(|P|)$.

The grouping stage will initialize with n_g groups, and continuously combine these groups as pairs until they may no longer be combined. Therefore, each iteration will reduce n_g groups to up to $n_g/2$ groups. Since every group may be compared to every other group, this process has a time of $O(n_g^2)$, and there will be at most $\log n_g$ iterations. Therefore, the worst case time of the grouping stage is $O(n_g^2 \times \log n_g)$. The number of groups produced will depend on the complexity of the input image and how many groups were constructed in the previous stage.

In the polygon construction stage, the convex hull of each group is calculated; each convex hull is constructed in $O(|g| \times \log |g|)$ time, where $g \in G$ from the point grouping stage. If the largest group contains n_{maxg} points, then the time complexity of constructing the initial polygons is $O(|g| \times n_{maxg} \times \log n_{maxg})$. Polygons are also removed if they overlap, meaning each one must be compared to each other one, requiring $O(|g|^2)$ time. Each vertex in each polygon is also tested with the next and previous vertex and removed if it is not expected to affect the overall structure of the polygon. Since this iterates over every vertex, but some vertices may have been removed in previous states, the time can reasonably be defined as $O(|g| \cdot n_{maxg})$. The polygon construction stage has an estimated time of $O(|g|^2 + |g| \times n_{maxg} \times \log n_{maxg})$.

In the final stage, an existing algorithm which constructs a weighted bidirectional graph may be used. The naive one we used simply created an edge between two nodes of that edge did not intersect with any other polygon. If so, then the line was not valid; otherwise, the algorithm adds an edge between the two nodes with a weight equal to the distance in pixels between them. The nodes of the graph are the vertices of the polygons from the previous stage with padding. Assume all polygons from the previous stage had n_v vertices and n_e edges between those vertices. The naive method of calculating a weighted bidirectional graph must compare every vertex with every other vertex, and for each comparison compare one line with every other polygon, making the worst case time $O(n_v^2 \times n_e \times |g|)$.

Because $n_v \leq n_{maxg} \leq |P|$ and $|g| \leq n_g$ we can use $|P|$ for n_v and n_{maxg} and use n_g for $|g|$. Since the graph has $n_v \leq |P|$ nodes, the maximum number of edges n_e is $n_v \times (n_v - 1)$ or $O(|P|^2)$. Therefore, the estimated overall time complexity is:

$$O(n_v^2 \times n_e \times |g| + |g|^2 + |g| \times n_{maxg} \times \log n_{maxg}) =$$
$$O(|P|^4 \times |g| + |g|^2 + |g| \times |P|) =$$
$$O(|P|^4 \times n_g + n_g^2)$$

As previously mentioned, $n_g = O(|P|)$, so $|P|$ may replace n_g, making the final overall time complexity $O(|P|^5)$.

5   Experiment and Results

To evaluate the results of the algorithm, we tested different input images depicting simple and complex aerial scenes where the camera is perpendicular to the ground. The experiments used both researcher-created images and satellite images. We recorded the runtime of each stage for each test image and the number of nodes and edges in the output. Because the runtime complexity for many pathfinding algorithms which use bidirectional weighted graphs depend on the number of nodes and edges, a more efficient graph would have fewer nodes and edges.

For each of the Figs. 4, 5, 6, 7 and 8, the top row is the ground-truth image, the middle row shows the results of segmenting by color, and the bottom row shows the results of segmenting by texture. The images shown are, from right

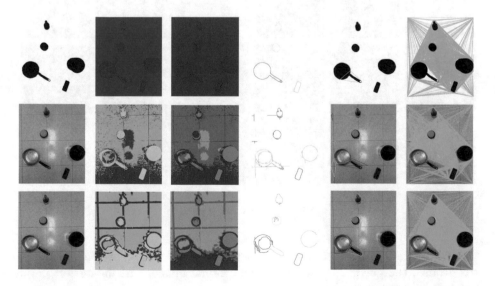

Fig. 4. Sample on a floor setup

Fig. 5. Sample of vegetables

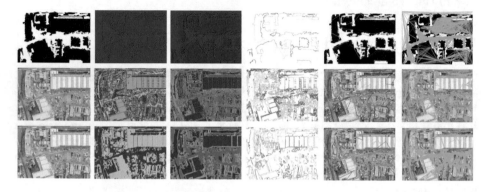

Fig. 6. Sample of the warehouse

Fig. 7. Sample of a park

to left: The original image, the clustered images, the salient regions, the vertex groups, the polygons overlaid on the original image, and the output graph overlaid on the original image.

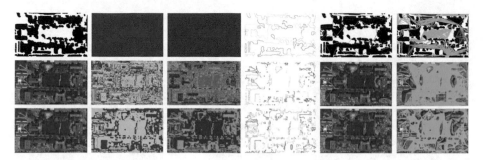

Fig. 8. Sample of a college campus

Table 1. Number of groups, polygons, nodes, and edges for each image with each segmenting method.

Figures	Method	Quantity of entity in applicable stage			
		Groups	Polygons	Nodes	Edges
4	Ground Truth	6	6	168	5,116
	Color	13	12	179	3,822
	Texture	8	6	156	4,350
5	Ground Truth	31	31	468	5,447
	Color	75	67	748	7,311
	Texture	66	55	703	6,993
6	Ground Truth	49	47	541	7,129
	Color	213	160	1,563	11,849
	Texture	168	149	1,515	20,506
7	Ground Truth	143	136	1,478	29,499
	Color	264	218	2,193	18,332
	Texture	156	133	1,410	14,019
8	Ground Truth	52	52	654	7,393
	Color	108	103	892	25,155
	Texture	174	159	1,540	31,617

We hypothesize that higher accuracy image segmentation will result in a higher quality output graph. Clustering pixels based on their texture is slower than clustering pixels only on their color. To determine whether a slower but more accurate image segmentation method is worthwhile, we compared the number of nodes and edges in the output graphs of different images using different segmentation techniques. These were also compared to manually created ground-truth images.

Table 1 shows the number of vertex groups, polygons, and the number of nodes and edges in the final output graph. The simpler images like Figs. 4 and 5 have fewer groups, polygons, nodes, and edges than the more complex images. Segmenting by texture resulted in fewer nodes and edges in the output graph in

Table 2. Runtime of each stage for each figure with every segmenting method. All units are in milliseconds. The Texture, Cluster, and Salient region times are not applicable for the Ground Truth Image, and the Texture time is not applicable for the Color clustering segmenting method.

Figures	Method	Quantity					
		Textures	Clusters	Salient regions	Grouping	Polygons	Graph
4	Ground Truth	N/A	N/A	N/A	383	16	78
	Color	N/A	1,906	344	2,375	15	94
	Texture	5,093	2,766	422	859	16	47
5	Ground Truth	N/A	N/A	N/A	6,359	219	219
	Color	N/A	4,734	610	12,2872	78	640
	Texture	7,325	8,202	766	31,667	47	546
6	Ground Truth	N/A	N/A	N/A	14,640	47	312
	Color	N/A	2,968	672	1,340,410	109	4,641
	Texture	8,874	4,734	782	225,189	109	4,422
7	Ground Truth	N/A	N/A	N/A	31,722	78	4,312
	Color	N/A	3,218	656	1,500,551	125	11,515
	Texture	8,203	6,530	750	146,707	78	3,547
8	Ground Truth	N/A	N/A	N/A	12,499	93	516
	Color	N/A	3,845	613	66,014	58	1,210
	Texture	7,429	4,352	603	78,281	93	6,307

every image except Fig. 8. A possible explanation for this might be that Fig. 8 appears to have fewer colors and shading, making color clustering sufficient.

Table 2 shows the runtime of each stage on each input image for each segmenting method in milliseconds. The grouping stage clearly impacts the runtime the greatest, especially in more complicated images.

6   Discussion

The algorithm succeeds in building a graph, but does not perform satisfactory in images which do not have a clear contrast of colors or textures of the foreground obstacles and background. There could be a more accurate way to distinguish between navigable and non-navigable spaces given an image in the general case. Future algorithms could be interactive, requiring a user to select a few points in the navigable and non-navigable space to more accurately segment and label the image.

We had assumed that non-salient regions are navigable and salient regions are obstacles. Therefore, when the input image had a greater color and texture contrast between the background and obstacles, the resulting weighted graph appeared more accurate. In Fig. 6, the warehouse, there were two different shades of navigable space; the road and the dirt. The algorithm sometimes labeled highlighted areas of the dirt and road as obstacles, failing to recognize it as part of the background and creating an inaccurate graph. This also happens in Fig. 5 with the vegetables. The wrong salient regions were chosen as obstacles, although the segmented image seemed to determine the outline of most vegetables fairly accurately. This reduced the quality of the output graph.

Assuming non-salient regions are background regions may be a poor method of selecting navigable space. The output graph appeared most accurate with images that had the greatest color and textural contrast between the foreground objects and the background. The algorithm was unable to accurately and consistently determine what was navigable space running on images closer to practical application, which have more complex shapes, textures, and variously colored backgrounds.

The runtime is not practically applicable, especially considering that most aerial and satellite images are much larger than the test input used in this paper. Although every image we tested was slower when segmenting by texture than by color, this usually resulted in a graph with fewer nodes and edges. Based on these results, the overhead of calculating textures and clustering with them is worth-while, since it will make the output more efficient for systems that use it. However, this claim should not be overgeneralized because of the small number of test images used.

Based on the empirical runtime results, the grouping stage runtime needs the most improvement, since it takes up most of the overall runtime for many of the test image. One reasonable solution would be to develop parallel versions of the existing series algorithms used in this paper.

The output graph, especially those containing thousands of nodes, may have too many nodes for practical pathfinding applications. Future work could investigate reducing the navigation graphs while still properly representing the environment. This would increase the efficiency of some navigation system using the output graph.

The complexity of the output graph is directly related to the complexity and number of vertices in the polygons resulting from the polygon construction stage. Further improving the polygon vertex reduction algorithm would improve the output graph, since polygons with fewer vertices would result in a graph with fewer nodes, which in turn would result in more efficient pathfinding. Because the polygons are convex, complex images often had poor area coverage, even for some ground-truth images. This could be improved by using an improved algorithm that can more accurately construct a polygon surrounding a set of vertices.

The output graph and subsequent pathfinding solution outputs could be transformed into a robot instruction language. This instruction set can then allow a robot to follow the shortest path calculated using the representation produced by this algorithm, creating an automated navigation system starting with input image data and information about the target point.

One possible implication of this algorithm includes using aerial images from satellite and Unmanned aerial vehicles to construct weighted graph representations of environments they photograph. This could make automated navigation over areas without roads more possible. However, the runtime of this algorithm is currently impractical for this application.

Because the overall algorithm relies on many different algorithms, it could be improved as the algorithms it uses are further improved.

7   Conclusion

The goal of this paper was to develop an algorithm to construct a bidirectional weighted graph usable for navigating through an environment using an aerial image of that environment. We investigated whether the image segmentation method would meaningfully impact the output, finding that the output graph often had fewer nodes and edges with a better but slower segmentation method. However, because of the small number of images tested, these results can not be overgeneralized. Given the greater access and availability of satellite and aerial images, it would be beneficial to investigate further practical uses for this data.

Acknowledgment. We would like to thank Elmhurst College for the opportunity to conduct this research project.

References

1. Achanta, R., Estrada, F., Wils, P., Süsstrunk, S.: Salient region detection and segmentation. In: International conference on Computer Vision Systems, pp. 66–75. Springer, Heidelberg (2008). https://doi.org/10.1007/978-3-540-79547-6_7
2. Algfoor, Z.A., Sunar, M.S., Kolivand, H.: A comprehensive study on pathfinding techniques for robotics and video games. Int. J. Comput. Games Technol. **2015**, 1148–1148. Hindawi Publishing Corp. (2015). https://doi.org/10.1155/2015/736138
3. Andrew, A.: Another efficient algorithm for convex hulls in two dimension. Inf. Process. Lett. **9**(5), 216–219. Elsevier (1979). https://doi.org/10.1016/0020-0190(79)90072-3
4. Chen, D., Wang, H.: A new algorithm for computing visibility graphs of polygonal obstacles in the plane. J. Comput. Geom. **6**(1), 316–345 (2015). https://doi.org/10.20382/jocg.v6i1a14
5. Gehrig, K.S., Stein, J.F.: Collision avoidance using elastic bands for an autonomous car. Research Institute DaimlerChrysler AG (2003). http://www.lehre.dhbw-stuttgart.de/~sgehrig/resume/papers/ias.pdf
6. Liu, L., Fieguth, P., Kuang, G.: Sorted random projections for robust texture classification. In: International Conference on Computer Vision, pp. 391–398. IEEE (2011)
7. Kiy, K.I.: Segmentation and detection of contrast objects and their application in robot navigation. In: Pattern Recognition and Image Analysis, vol. 25, no. 2, pp. 338–346. Springer, Heidelberg (2015). https://doi.org/10.1134/S1054661815020145
8. Scharfenberger, C., Chung, A.G., Wong, A., Clausi, D.A.: Salient region detection using self-guided statistical non-redundancy in natural images. In: IEEE Access, vol. 4, pp. 48–60. IEEE (2015). https://doi.org/10.1109/ACCESS.2015.2502842

9. Shafait, F., Keysers, D., Breuel, T.M.: Efficient implementation of local adaptive thresholding techniques using integral images. In: Proceedings of the SPIE Document Recognition and Retrieval XV, vol. 6815. SPIE (2008). https://doi.org/10.1117/12.767755

10. Sidorova, V.S.: Global segmentation of textural images on the basis of hierarchical clusterization by predetermined cluster separability. In: Pattern Recognition and Image Analysis, vol. 25, no. 3, pp. 541–546. Springer, Heidelberg (2015). https://doi.org/10.1134/S1054661815030232

11. Yu. L.: Vehicle extraction using histogram and genetic algorithm based fuzzy image segmentation from high resolution UAV aerial imagery. In: The International Archives of the Photogrammetry, Remote Sensing and Spatial Information Sciences, vol. 37. Citeseer (2008)

Performance Evaluation of Autoencoders for One-Shot Classification of Infectious Chlamydospore

Raphael B. Alampay$^{(\boxtimes)}$, Josh Daniel Ong,
Ma. Regina Justina E. Estuar, and Patricia Angela R. Abu

Information Systems and Computer Science, Ateneo de Manila University,
Quezon City, Philippines
{ralampay, restuar, pabu}@ateneo.edu,
joshua.ong@obf.ateneo.edu

Abstract. In the Philippines, there is a growing need for the protection of banana plantation from various diseases that directly affects the livelihood of farmers, markets and overall ecosystem. One such fatal disease is *Fusarium oxysporum cubense* (TR4 Chlamydospores) which allows growth of such fungi in banana crops that permanently damages the soil for further fertility. As of this writing, there is very small visual distinction between TR4 Chlamydospores and non-infectious Chlamydospores. This paper proposes the use of autoencoders to engineer relevant features in order to distinguish Fusarium Oxysporum from similar fungi or other artifacts present in the soil. Furthermore, the paper tries to address the problem with minimal data available for supervised learning as opposed to traditional methods that require thousands of data points for classification. The purpose of the experiments presented here will aid towards the creation of more sophisticated models to visually discriminate *Fusarium Oxysporum*.

Keywords: Autoencoders · Fusarium oxysporum · One-shot learning · Computer vision

1 Introduction

The problem of Fusarium Oxysporum (TR4) is widely known in the Philippines, Indo-malay countries as well as in other countries wherein agriculture is the majority of the locals' livelihood. One of the major effects of such a disease is the wilting of crops and making the underlying land infertile for further growth thus directly affecting the livelihood of farmers and the community. There have been numerous studies on its widespread [6] in order to develop regulatory policies and possible prevention [8]. Unfortunately, once the presence of the said fungi has been verified, it is often considered to be in a terminal stage since the wilting has reached the leaves of the crop. The only successful way to prevent its spread is through burning the entire affected area leaving the land infertile resulting in as well in the loss of livelihood [7].

This paper examines the use of machine learning models for feature engineering in order to attempt to create a pre-emptive mechanism through visual identification of early

© Springer Nature Switzerland AG 2020
K. Arai and S. Kapoor (Eds.): CVC 2019, AISC 944, pp. 436–446, 2020.
https://doi.org/10.1007/978-3-030-17798-0_35

stages of the disease. To do this, soil samples are garnered from the field as well as synthesized samples to examine the behavior of early *Fusarium Oxysporum* and its characteristics. The theory is that distinctive visual representation of the fungi can be extracted from microscopic images of the samples to trigger necessary steps for its prevention without permanently damaging the land. The challenges presented in this study can be elaborated in two major areas. First, studies suggest that there are very little morphological features to distinguish TR4 from other fungi [9]. False detection will affect the succeeding actions to be taken if this is the case. Second, there is very little data available on visual representation of said fungi to apply traditional machine learning models for visual classification. To address these, autoencoders are used to automatically engineer and extract the relevant features present and evaluate its performance with minimal data available. The end goal of this research is to be able to create a robust model that can be applied to the automation of early detection of *Fusarium Oxysporum*.

2 Related Literature

2.1 Applications of MLP for Classification

Multilayer perceptron networks (MLP) is the classical form of neural networks used for classification. This model makes use of mathematical operations applied to input vectors (often termed as nodes representing feature descriptors) through a series of sequential layers (hidden layers) with varying dimensionality. These layers of nodes are connected through matrices of weight values which determine the relevance of the computed values previous to it making it non-parametric since no previous assumptions are made as to the distribution of data (i.e. Gaussian). These hidden layers allow modeling of non-linear dependencies for more complex datasets by means of non-linear activation functions. Such an approach has become advantageous for non-trivial or highly unpredictable data such as classification of similar samples of different categories [4, 5].

Neural network models, specifically MLP, has been widely used as a tool for image classification in agriculture such as that of [2]. In such approaches, labeled image samples measuring 32×32 pixels were extracted from crop images. From these samples, performance evaluation was examined against three main feature extraction methods namely gray level co-occurrence matrix (GLCM), discrete wavelet transform (DWT) and discrete cosine transform (DCT). These features achieved an accuracy of 86%, 98.27% and 98.002% respectively [1]. These features allowed a more concrete representation of the data as opposed to using all 32×32 pixel values as descriptors.

Multilayer perceptron networks have also been used in the field of biomedical engineering to classify microscopic samples similar to this paper's image dataset. In [3], it has been used to evaluate the classification of segmented regions of interested for automated processing.

The methods used in previous literature are said to be hand crafted features as they measure a specific type of visual representation of the image such as texture descriptors. This study adapts the same approach of using multilayer perceptron networks for classification however the approach to feature engineering is via the use of autoencoders to allow the data to dictate relevant features using minimal samples.

2.2   Applications of MLP for Classification

Autoencoders are a type of neural network that attempts to learn the identity function as seen in Eq. (1).

$$f(x) = x \tag{1}$$

This approach in machine learning is to be able to replicate the input of some vector x, and reconstruct it as close as possible. To do this, the model seeks to learn a compressed representation of data by capturing relevant information from the given samples in an unsupervised fashion. Typically, this is done by connecting the input layer to a hidden layer with lower dimensionality which is referred to as the latent layer (and its neurons referred to as latent variables denoted as z) before being fully connected to the output which as the same dimensionality as x as illustrated in (Fig. 1). The latent variables represent a generalization of the input. It is because of this that autoencoders are also considered to be a type of feature reduction technique which attempts to capture relevant information about the given data. After training the network, the latent variables (z) are then collected as representations of the sample data. They are then used as input for other classification models such as MLP, support vector machines or bayesian models [14]. The following figure illustrates a basic representation of an autoencoder. In this case, nodes at the output layer y are set to be the same values of input vector x during training. The latent variables z represents the compressed version of x.

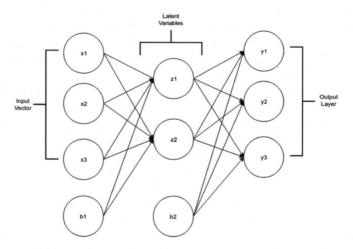

Fig. 1.   An illustration of the autoencoder model

Being a type of neural network, autoencoders share the same properties of a typical feed forward neural network. It can be trained using back propagation techniques and is subject to the same optimization approaches for neural networks such as drop-out rates and bias terms as well as taking advantage of non-linear activation functions like sigmoid and hyperbolic tangent. One such activation function that has proven to be effective in previous studies is the rectified linear unit (ReLU) as seen in Eq. (2).

$$f(x) = max(x, 0) \tag{2}$$

The derivative of ReLU used for back propagation can be used using the form as seen in Eq. (3).

$$f(x) = 1 \text{ if } x > 0 \text{ } else \text{ } 0 \tag{3}$$

Such approaches have been proven to be successful in similar cases wherein there was a significant increase in performance from the use of features given by autoencoders despite having a small dataset for training. This study attempts to see if such a feature reduction technique will be viable for classification [15, 16].

2.3   One-Shot Learning

Most machine learning methods require large labeled datasets in order to achieve an effective result for classification. In our case however, there is very few image data available captured from the field for proper study of Fusarium oxysporum in a visual manner. Thus there has been growing interest in the domain of one-shot learning in the past decade where the goal was to use one or few data samples to be able to create an accurate model for classification [10, 11]. In recent studies, one-shot learning has generally been successful with models that involve either a form of neural networks such as siamese networks [12] as well as the bayesian models [13]. This study will be taking the approach of applying one-shot learning in a neural network approach. It is theorized that since there are very little morphological or visual elements that distinguishes Fusarium oxysporum from normal fungi in a soil sample, a neural network approach such as autoencoders allows us to let the data samples themselves determine these integral features for differentiation.

3   Methodology

This section presents the methods used in the extraction of information relating to Fusarium Oxysporum and how autoencoders were applied to extract the relevant features for classification as illustrated in (Fig. 2).

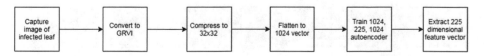

Fig. 2. Feature extraction and classification pipeline

3.1   Data Extraction, Labelling and Preprocessing

The microscopic images for testing were taken in two approaches and categorized in two resolutions. The first context was brightfield wherein the specimen is illuminated from below to be observed from above. Second was fluorescence wherein the specimen was introduced with a substance for better illumination. For each context, there were two fields of view that were used namely low power optics (LPO) and high power optics (HPO). 10 regions of interest were extracted that encloses an example of Fusarium Oxysporum as well as 10 regions of interest that encloses an example of non-infected fungi or artifacts in the image that are similar to Fusarium Oxysporum. The infected images are referred to as positive samples and non-infected images as negative examples. This would allow performance evaluation on four types of experiments. First would be for positive vs negative images under brightfield context with HPO as illustrated in Figs. 3 and 4. Second would be positive vs negative images under brightfield context with LPO as illustrated in Figs. 5 and 6. Third would be positive vs negative images under fluorescent context with HPO as illustrated in Figs. 7 and 8. Finally, last would be positive vs negative images under fluorescent context with LPO as illustrated in Figs. 9 and 10.

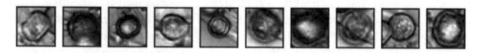

Fig. 3. Brightfield HPO positive

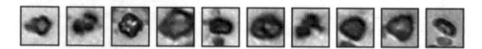

Fig. 4. Brightfield HPO negative

Fig. 5. Brightfield LPO positive

Fig. 6. Brightfield LPO negative

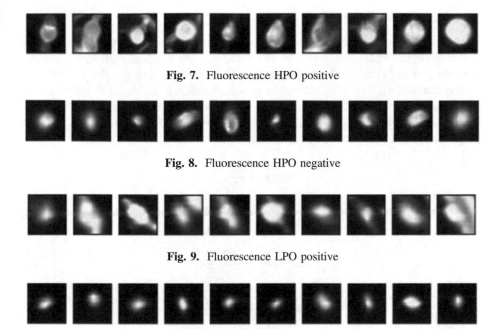

Fig. 7. Fluorescence HPO positive

Fig. 8. Fluorescence HPO negative

Fig. 9. Fluorescence LPO positive

Fig. 10. Fluorescence LPO negative

3.2 Feature Extraction with Autoencoders

Given the labelled information, the data was then vectorized/flattened for feature extraction. The 28 × 28 scaled version of the images would yield a 784 dimensional vector. The training data was then fed to the three-layer autoencoder whose latent variables yielded a 250 dimensional vector which was a third of the original vector. After the model has been trained, its weights were used to create the reduced version of both the training and validation set (each datapoint in both sets now measuring 250 dimensions). This was done by encoding the original vector and performing a single feed forward pass without back propagation. The activated values (latent variables z) in the hidden layer were then used as the compressed version of sample data. The topology of the autoencoder model was then 784 dimensions for the input layer, 250 for the hidden layer and 784 for the output layer (the output layer is then disregarded). The activation function used in the hidden layer was ReLU whereas sigmoid was used as the activation function for the output layer with a standard value of 1 for the bias and $\alpha = 0.05$ for the learning rate.

3.3 Classification with MLP

Once the derived datapoints have been extracted (250 dimensional latent representations of the data), it was trained with a fully connected multilayer perceptron neural network for classification. The topology of the MLP neural network was 250 dimensions for the input layer, 125 for the hidden layer and 2 for the output layer (one neuron representing positive and the other neuron representing the negative probability of the input being TR4).

3.4   Performance Evaluation

To evaluate the performance of the experiments, four measurements were used namely accuracy as seen in Eq. (4), precision as seen in Eq. (5), recall as seen in Eq. (6) and f1-score as seen in Eq. (7). They are computed as follows (where positive P is referred to as infectious, negative N is referred to as noninfectious samples, TP is referred to as true positives or correctly classified infectious samples, TN is referred to as true negatives or correctly classified non-infectious samples, FP is referred to as false positives and FN is referred to as false negatives).

$$\text{Accuracy} = \frac{\text{TP} + \text{TN}}{\text{Validation}} \tag{4}$$

$$\text{Precision} = \frac{\text{TP} \cap \text{Validation}}{\text{Validation}} \tag{5}$$

$$\text{Recall} = \frac{\text{TP}}{(\text{TP} + \text{FN})} \tag{6}$$

$$\text{F1 Score} = \frac{2}{\frac{1}{\text{Recall}} + \frac{1}{\text{Precision}}} \tag{7}$$

4   Results and Analysis

After training the autoencoder, it was possible to see the visual representation of latent variables by attempting to reconstruct some samples with the trained weights. The example representation of reconstructed infectious samples depicts the features learned by the neural network model as illustrated in Figs. 11 and 12. There was a slight variation with the reconstructed results suggesting a generalized set of features detected by the autoencoder.

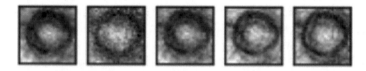

Fig. 11. Brightfield reconstructed positive examples

Fig. 12. Fluorescence reconstructed positive examples

The summary of results for the four experiments (HPO and LPO infectious vs non-infectious for brightfield and fluorescence) displays the performance of each context as illustrated in Fig. 13.

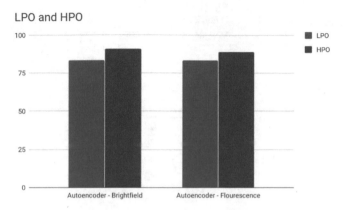

Fig. 13. Summary of results

The following tables show the results of the different measurements from the four experiments. LPO measurements seem to show low precision scores which is presented in Tables 1 and 3. For LPO brightfield experiments, it can be seen that results are largely deviated from the other measurements. This suggests that in an LPO perspective, there is a lot of variability and that the model might generate features that perform poorly regardless of an increase in sample size.

Table 1. LPO brightfield measurements

Measurement	Result
Accuracy	80%
F1 Score	83.3%
Precision	71.4%
Recall	99.99%

Table 2. HPO brightfield measurements

Measurement	Result
Accuracy	90%
F1 Score	90.9%
Precision	83.3%
Recall	99.99%

Table 3. LPO fluorescence measurements

Measurement	Result
Accuracy	80%
F1 Score	83.3%
Precision	71.4%
Recall	99.99%

Table 4. HPO fluorescence measurements

Measurement	Result
Accuracy	90%
F1 Score	88.9%
Precision	99.99%
Recall	80%

A general trend is shown wherein the HPO context yields better results across majority of the measurements compared to LPO which logically follows since a higher magnification of optics allows capturing finer features of the specimen as presented in Tables 2 and 4. There are still some performance measurements that fall below 90% which indicative of high tolerance in differentiating infectious and non-infectious fungi which is possibly Fusarium oxysporum. This also proves the theory in a biological aspect that there are similar characteristics between Fusarium and fungi or other artifacts in the scene which comprises some of the negative samples (visually similar but are not necessarily fungi) used in training. It is also seen that using non-parametric, automated feature engineering models such as autoencoders allowed the extraction of general features that aren't explicitly introduced in the system. This case proves that using such a model will allow the data itself to dictate how a category or label should be characterized even if there is no prior knowledge as what features should be sought after. The model itself will generate the needed information that might not be evident to the practitioner at first glance or based on previous studies.

The measurement that is interesting to take a look at however is the F1 score. This allows us to see how accurate the model is to take into account false positives and false negatives which will have a direct impact in the application of the model. As mentioned in literature, false positives might lead to the burning of land making it infertile for further growth of crops. Having a high F1 score indicates that the model lowers the risk of false positives. For both the HPO and LPO experiments, it can be seen that it generally tends to stay high relative to the volatility of precision and accuracy. This might suggest that the features extracted from the autoencoder are mostly relevant to its category/label for visual distinction.

5   Conclusion

Using autoencoders, the identification of prominent features for the classification of Fusarium oxysporum using image samples can be auto-engineered and extracted from data presented. This study serves as the starting point to allow the automation of feature engineering for fungi classification which is believed to have very little morphological and visual properties that distinguishes it from other fungi. Automated feature engineering techniques such as autoencoders proves to show some promise in progressing towards modeling early detection model of the infectious fungi to aid in the prevention of the loss livelihood of farmers and agricultural communities affected by this phenomenon. Such a preprocessing step will allow better understanding of relevant features for specimens with close visual resemblance. For succeeding research and experiments, it is recommended to evaluate the approach with the context of measuring possible overfitting. It is also recommended to evaluate the claim of usage of minimal data by comparing it with classical machine learning approaches which make use of as much training samples as needed. As mentioned in the results and analysis section, precision is volatile with the change of magnification. It is then recommended to evaluate if this still holds true with larger samples. The same holds true for the performance of the model's f1-score. If the results still hold true for a larger sample size, then it might strengthen the hypothesis that autoencoder models allow relevant feature extraction for one-shot learning classification.

Acknowledgment. We would like to thank the Philippine California Advanced Research Institute (PCARI) and the Commission on Higher Education (CHED), Philippines, University of California Berkeley Bioengineering and the Ateneo de Manila University for the guidance, funding and support.

References

1. Yanga, C., Everitt, J.H., Murden, D.: Evaluating high resolution SPOT 5 satellite imagery for crop identification (2011)
2. Qayyum, Z.U., Akhtar, A., Sarwar, S., Ramzan, M.: Optimal feature extraction technique for crop classification using aerial imagery. In: 2013 International Conference on Information Science and Applications (ICISA). ISBN 978-1-4799-0604-8
3. Fasihi, M.S., Mikhael, W.B.: Overview of current biomedical image segmentation methods. In: 2016 International Conference on Computational Science and Computational Intelligence. ISBN 978-1-5090-5510-4
4. Schmidt, M.: Automatic brain tumor segmentation. Masters thesis. University of Alberta (2005)
5. Sharma, M., Mukharjee, S.: Brain tumor segmentation using genetic algorithm and artificial neural network fuzzy inference system (ANFIS). In: Advances in Computing and Information Technology 2013, pp. 329–339., Springer, Berlin, Heidelberg (2013)
6. Mostert, D., Molina, A.B., Daniells, J., Fourie, G., Hermanto, C., Chao, C.-P., et al.: The distribution and host range of the banana Fusarium wilt fungus, Fusarium oxysporum f. sp. cubense, in Asia. PLoS ONE **12**(7), e0181630 (2017)

7. Jendoubi, W., Bouhadida, M., Boukteb, A., Bji, M., Kharrat, M.: Fusarium wilt affecting chickpea crop. Agriculture **7**, 23 (2017)
8. Aquino, A.P., Bandoles, G.G., Lim, V.A.A.: R&D and policy directions for effective control of Fusarium wilt disease of Cavendish banana in the Asia-Pacific region (2013)
9. Fourie, G., Ploetz, R., Steenkamp, E., Viljoen, A.: Current status of the taxonomic position of Fusarium oxysporum formae specialis cubense within the Fusarium oxysporum complex **11**(3), 533–542 (2011). https://doi.org/10.1016/j.meegid.2011.01.012. Epub 21 Jan 2011
10. Mocanu, D.C., Mocanu, E.: One-shot learning using mixture of variational autoencoders: a generalization learning approach, April 2018. arXiv:1804.07645v1
11. Wong, A., Yuille, A.L.: One shot learning via compositions of meaningful patches. In: Proceedings of the IEEE International Conference on Computer Vision, pp. 1197–1205 (2015)
12. Vinyals, O., Blundell, C., Lillicrap, T.P., Kavukcuoglu, K., Wierstra, D.: Matching networks for one shot learning, pp. 3630–3638 (2016)
13. Fei-Fei, L., Fergus, R., Perona, P.: A Bayesian approach to unsupervised one-shot learning of object categories. In: IEEE Ninth International Conference on Computer Vision, 10 2003
14. Coates, A., Ng, A.Y., Lee, H.: An analysis of single-layer networks in unsupervised feature learning. In: International Conference on Artificial Intelligence and Statistics (2011)
15. Kavukcuoglu, K., Sermanet, P., Boureau, Y.-L., Gregor, K., Mathieu, M., LeCun, Y.: Learning convolutional feature hierarchies for visual recognition. In: NIPS, vol. 1, p. 5 (2010)
16. Makhzani, A., Frey, B.: k-sparse autoencoders. In: International Conference on Learning Representations, ICLR (2014)

Ursa: A Neural Network for Unordered Point Clouds Using Constellations

Mark B. Skouson[(⊠)], Brett J. Borghetti, and Robert C. Leishman

Department of Electrical and Computer Engineering,
United States Air Force Institute of Technology,
Wright-Patterson Air Force Base,
Dayton, OH 45433, USA
mark.skouson@afit.edu

Abstract. This paper describes a neural network layer, named Ursa, that uses a constellation of points to learn classification information from point cloud data. Unlike other machine learning classification problems where the task is to classify an individual high-dimensional observation, in a point-cloud classification problem the goal is to classify a set of d-dimensional observations. Because a point cloud is a set, there is no ordering to the collection of points in a point-cloud classification problem. Thus, the challenge of classifying point clouds inputs is in building a classifier which is agnostic to the ordering of the observations, yet preserves the d-dimensional information of each point in the set. This research presents Ursa, a new layer type for an artificial neural network which achieves these two properties. Similar to new methods for this task, this architecture works directly on d-dimensional points rather than first converting the points to a d-dimensional volume. The Ursa layer is followed by a series of dense layers to classify 2D and 3D objects from point clouds. Experiments on ModelNet40 and MNIST data show classification results comparable with current methods, while reducing the training parameters by over 50%.

Keywords: Point cloud classification · Point sets · 3D vision · Machine learning · Deep learning

1 Introduction

A large bulk of the recent computer vision research has focused on applying artificial neural networks to 2D images. More recently, a growing research area focuses on applying neural networks to 3D physical scenes. Point clouds or point sets are a common format for representing 3D data since some sensors, including laser-based systems, collect scene data directly as point clouds. Voxelization is a straightforward way of applying powerful deep convolutional neural network techniques to point clouds, as is done in VoxNet [1] and 3DShapeNets [2]. Voxelization, however, is not always desirable because point clouds can, in

© Springer Nature Switzerland AG 2020
K. Arai and S. Kapoor (Eds.): CVC 2019, AISC 944, pp. 447–457, 2020.
https://doi.org/10.1007/978-3-030-17798-0_36

many cases, represent structural information more compactly and more accurately than voxelized alternatives.

In contrast to voxelization methods, PointNet [3] and others have developed architectures that operate directly on point clouds, including ECC [4], Kd-Net [5], DGCNN [6], and KCNet [7]. This research adds to the growing body of knowledge about learning on point sets.

This paper describes a neural network layer (Ursa layer) that accepts a point cloud as input and efficiently yields a single feature vector, which is both agnostic to the ordering of the points in the point cloud and encodes the dimensional features of every point. This output feature vector is an efficient representation of the entire point cloud - an observation which can be used for classification (or other machine learning tasks) in later portions of the network. The layer's trainable parameters are centroids, and each centroid has the same dimension as a point in the point cloud. For the remainder of this paper, in order to distinguish the centroids from the points in the point cloud, the centroids will be referred to as stars, and the collection of stars in the Ursa layer will be referred to as a constellation. The output of the layer is a feature vector with length equal to the number of constellation stars, which us used in the later layers of the neural network to inform the classification output. Another important characteristic of this approach is that it does not require a preprocessing step - the Ursa layer is trained as part of the overall network structure using backpropagation and gradient descent learning.

The Ursa layer is invariant to the ordering of the input points. The Ursa layer is not inherently invariant to shift, scale, or rotation; rather, it relies on demonstrations of those types of variations (possibly through data augmentation) in the input data during training to learn these variations. The output of the Ursa layer is a global shape descriptor of the point cloud that is fed to later layers in the classifier to classify the point cloud.

Experiments on this architecture show the classification accuracy is comparable to current point cloud-based classifiers, but with a significantly smaller model size. The experiments tested the Ursa architecture with various distance functions and various numbers of constellation stars using MNIST (2D) data and ModelNet40 (3D) data. Experimentally, the best distance measure was dependent on the data set. For both data sets, too few or too many stars generally degraded performance. Performance gains leveled off with 256 or more stars and, in some cases, more stars led to worse performance.

2   Selected Related Works

The work presented herein is informed by the PointNet [3] research and architecture. In [3], Qi et al., introduce the concept of symmetric functions for unordered points. A symmetric function aggregates the information from each point and outputs a new vector that is invariant to the input order. Example symmetric operators are summation, multiplication, maximum, and minimum. Alternatives to a symmetric function for point order invariance would be to sort the input into a canonical order or augment the training data with all kinds of permutations. PointNet uses a 5-layer multi-layer perceptron (MLP) to convert the input points to a higher-dimensional space, then uses max pooling as the symmetric function to generate a single global feature, which is then fed through 3-layer MLP for classification. The architecture experimented on in this paper replaces PointNet's first 5 MLP layers and max pooling layer with a single Ursa layer. The Ursa layer generates a global feature and, as in PointNet, uses a 3-layer MLP for classification. As with PointNet, the Ursa layer's output is invariant to the order of the input data.

This work is also closely related to the KCNet architecture [7]. KCNet uses a concept similar to the Ursa constellation layer, which they call kernel correlation. Kernel correlation has been used for point set registration, including by [8]. Whereas [8] attempts to find a transformation between two sets of points to align them, Ursa and KCNet allow each point in the constellation (or kernel) to freely move and adjust during training. The KCNet architecture maintains all the layers of PointNet, and augments them by concatenating kernel correlation information to the intermediate vectors within the 5 layers of MLP. In a forward pass in PointNet, each input point is treated independently of all other points until the global max pooling layer, but that is not the case for KCNet. KCNet uses a set of kernels that operate on local subsets of the input points using a K-nearest neighbor approach. The kernels are trained to learn local feature structures important for classification and segmentation. Thus, KCNet improves on PointNet by adding additional local geometric structure and feature information prior to global max pooling.

There are several difference between the KCNet and the Ursa-based architecture used for this paper. The KCNet kernel correlation produces a scalar value while the Ursa layer produces a vector. KCNet uses several kernels at the local level to augment the PointNet architecture. The Ursa architecture uses a single star constellation at the global level to replace the first several layers of PointNet.

Other deep learning methods that operate on point clouds include Dynamic Graph Convolutional Neural Networks (DGCNNs) [6], Edge-Conditioned Convolution (ECC) [4], Kd-Networks [5], and OctNet [9]. These methods organize the data into graphs. In the cases of [6] and [4], the graphs are based on a vertex for each point and edges that define a relationship between the vertex and near neighbors, and weighted sum operations that operate on vertices and edges of the graph. The DGCNN architecture in [6] is quite similar to the PointNet structure, but where the multi-layer perceptron layers are replaced with Edge Convolution Layers. Both [5] and [9] use non-uniform spatial structure to partition the input space, and they also used weighted sum operations. In contrast to these methods, the learning in the Ursa layer is not stored in the weights of a weighted sum operation. Instead the learning is stored in the locations of a set of constellation stars as will be described in the next section.

This work explores the use of radial basis functions (RBFs) in point cloud classification and so bears some commonality with RBF networks [10–13]. RBFs are able to project an input space into a higher-dimensional space. It does this through a radial function, which varies with distance from a central point, and a set of vectors known as RBF centers. In common usage, a function $f(x)$ of an input vector x can be modeled as a weighted sum of a radial basis function, ϕ, of the distances between the input vector and m RBF centers, q_i:

$$f(x) \approx \sum_{i=1}^{m} \omega_i \left(\phi(||x - q_i||) \right) \tag{1}$$

where $|| \cdot ||$ is the L2 norm.

In general, the ω_i, the q_i, and ϕ can be selected or trained to fit the RBF network to the function. In practice, ϕ and the q_i are usually first selected, then the ω_i are adjusted or trained to fit the data. While the work herein does explore the use of RBFs, it differs from RBF networks. Rather than computing a weighted sum of RBF outputs to determine a classification of a single input vector (a point cloud), Ursa uses an RBF to transform a set of input vectors to a higher-dimensional feature vector as will be described below.

3   Method

This section describes the Ursa layer and a neural network architecture that uses a Ursa layer to classify objects. The overall classification architecture is shown in Fig. 1. It is an Ursa layer followed by a three-layer fully connected (dense) multi-layer perceptron classifier. Compared to the PointNet architecture used by several researchers [3,6,7,14], the Ursa layer replaces the first 5 MLP layers of PointNet architecture, while maintaining the last three-layer MLP portion. Maintaining an end structure similar to other methods aids in comparison.

During training, the neural network model makes use of data augmentation at the input and data dropout just prior to the final MLP layer. Because the final layers are straightforward, the remainder of this section will focus on only the the Ursa layer. Parameters used during implementation are discussed in Sect. 4.

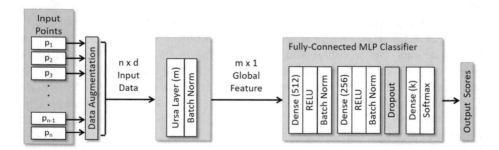

Fig. 1. The Ursa architecture. The first hidden layer is an Ursa layer. The remaining three hidden layers are fully-connected (dense) layers. Data augmentation and dropout (the gray boxes) are used only during training.

To define the Ursa layer, consider a set of n d-dimensional input points in $\Re^d$ that make up a point cloud $P = \{p_1, ..., p_n\} \subset \Re^d$. P is the input to the Ursa layer. Within the layer is a constellation of m stars, with the same dimensionality as the input points, $Q = \{q_1, ..., q_m\} \subset \Re^d$. The output of the layer is an $m \times 1$ vector $V = \{v_1, ..., v_m\} \subset \Re$. The Ursa layer converts a set of n d-dimensional points into an m-dimensional feature vector. There are $m \times d$ trainable parameters within the layer.

As mentioned earlier, RBFs have the ability to convert vectors to a higher-dimensional feature space. The Ursa layer makes use of RBFs for this purpose. In fact, the experiments investigated the use of three different candidate RBFs and compared their effectiveness. The first function explored is the Gaussian RBF, $\phi(\cdot) = \exp(-(\cdot)^2/(2\sigma^2))$, with which the relationship between P, Q, and V is

$$v_i = \sum_{j=1}^{n} \exp\left(-\frac{\|p_j - q_i\|^2}{2\sigma^2}\right), \qquad i = 1, ..., m \qquad (2)$$

where σ controls the "width" of the function. In this paper, σ is considered a user-selected hyper-parameter, potentially tunable in cross-validation. So, each input point's contribution to the i-th entry in the output vector is a function its distance to the i-th constellation star, with the contribution decreasing according to the Gaussian function as the point is farther away. In other words, v_i accumulates into a single scalar value the weighted distance information between the ith star and each point in the point cloud. The summation provides the symmetry that makes the output of the layer invariant to the ordering of the input points.

The second function explored was an exponential decaying RBF, applied according to Eq. 3, where the hyper parameter λ controls the width of the function similar to σ in Eq. 2.

$$v_i = \sum_{j=1}^{n} \exp\left(-\lambda \|p_j - q_i\|\right), \qquad i = 1, ..., m \qquad (3)$$

The exponential decay has the effect of more rapidly depreciating the contribution of each point to a star's feature output the further they are from the star's location, as can be seen in Fig. 2.

Finally, the experimentation explored a linear RBF applied according to Eq. 4.

$$v_i = \min_{1 \le j \le n} \|p_j - q_i\|, \qquad i = 1, ..., m \qquad (4)$$

In this case, the symmetry is provided by the minimum function. The effect of Eq. 4 is that v_i is the distance from constellation star q_i to its nearest point from the point cloud. This is the RBF that provides the most efficient computation of the three investigated. The relative effectiveness of all three measures is shown in Sect. 4. This paper refers to Eqs. 2, 3, and 4 as the Gaussian distance function, the exponential distance function, and the minimum distance function, respectively, throughout this paper.

The Ursa layer is followed by a three-layer MLP. The non-linearity for each MLP layer is the ReLU function, except for the final layer, which uses softmax. Each ReLU is followed by a batch normalization. The Ursa layer does not

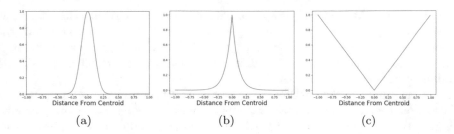

(a)   (b)   (c)

Fig. 2. Visual comparison of RBFs, where the x-axis is the distance from the centroid. (a) Gaussian RBF, (b) exponential decay RBF, (c) RBF for the minimum distance function.

require a ReLU function afterward because the v_i in Eqs. 2, 3, and 4 are already always non-negative. Additionally, the Ursa distance measures defined by these equations are not matrix multiplies, which require a separate non-linearity afterward to enable the network to emulate non-linear function of the input; the L2 norm within the computations provides an inherent nonlinear component to the layer. All three distance measures are differentiable and are trained as part of the overall back-propagation of the entire network. For this implementation, the gradient was computed using the standard tensorflow gradient calculations.

4   Experiments

A series of experiments evaluated the described Ursa network architecture for classification of 3D and 2D objects. For 3D data, the ModelNet40 shape database [2] was selected. For 2D data, the MNIST handwritten character recognition database [15] was converted to 2D point clouds and used.

For the ModelNet40 data, 2048 points per object were evenly sampled on mesh faces and normalized into the unit sphere as provided by [14]. To convert an MNIST image to a 2D point cloud, the coordinates of all pixels with values larger than 128 were used. The maximum number of pixels greater than 128 for any MNIST image was 312. For those images with fewer than 312 points, the available points from the set were randomly repeated to reach 312 points.

During training for both 3D and 2D data, the data was augmented by scaling the shape to between 0.8 and 1.25 of the unit-sphere size with a random uniform distribution; rotating the shape between -0.18 and 0.18 radians along each angular axis with a random normal distribution (clipped) with standard deviation 0.06; shifting the shape in every dimension between -0.1 and 0.1 away from its original position with a random uniform distribution; and adding jitter between -0.05 and 0.05 to each point according to a random normal distribution (clipped) with standard deviation 0.01. Also during training a dropout layer was used with a dropout rate of 0.3 just before the last dense layer. The value for σ in Eq. 2 was chosen to be 0.1, and λ in Eq. 3 was chosen to be 10. Additional tuning of these hyper-parameters may improve performance.

The trainable parameters within the Ursa layer were initialized by randomizing them according to a uniform distribution between ±1 in each dimension. The trainable parameters for the dense layers were initialized using the glorot uniform method. Future research may consider other Ursa layer initialization methods such as uniformly distributing across the space [12] or using information from the input data, e.g. k-means clustering techniques.

Experiments explored the three feature space transforms in Eqs. 2, 3, and 4 for several values of m, the number of constellation stars. Classification performance was evaluated for $m = 32, 64, 128, 256, 512,$ and 1024. Ten independent tests were conducted with each of the distance measures at each value of m. The average accuracy is plotted in Figs. 3 and 4.

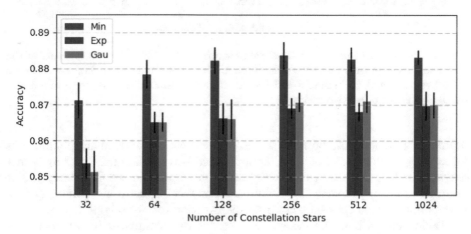

Fig. 3. The performance of the Ursa architecture on ModelNet40 for each of the distance measures with respect to the number of constellations stars. The y-axis limits have been selected to highlight the small differences. The minimum distance measure slightly outperforms the other two methods.

5   Results and Discussion

An analysis of Figs. 3 and 4 shows a general trend of performance improving steadily as the constellation grows to 256 stars, but leveling out or perhaps worsening beyond 256 stars. It is interesting that the experiments did not show any clear difference in the number of stars needed based on the dimensionality of the data (3D vs 2D), number of points per shape (2048 vs 312), or number of possible classes (40 vs 10). An in-depth analysis revealed that when the number of stars increased over 512, many of the constellation stars were effectively unused; they were pushed to the edges of the 2- or 3-dimensional space. The data suggest a good starting point for other data sets may be 256 or 512 stars.

The minimum distance function performed slightly better on the Model-Net40 data, while the Gaussian distance function performed slightly better on the MNIST data. This may be a function of the number of points per shape. The

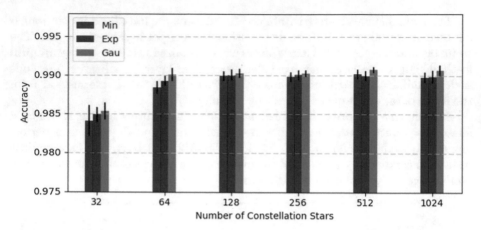

Fig. 4. The performance of the Ursa architecture on the MNIST data for each of the distance measures with respect to the number of constellations stars. The y-axis limits have been selected to highlight differences. The Gaussian distance measure slightly outperforms the other two methods.

minimum function may be more effective in situations with many points. It is interesting to note that PointNet [3] performed better using max pooling rather average pooling or an attention weighted sum to provide the symmetry (point order invariance) property. The minimum distance measure has similarity to a max pooling, while the Gaussian distance function uses a weighted sum.

Table 1 shows how the evaluated Ursa network compares to other classification methods on the same data sets. Ursa achieved impressive results, especially considering the Ursa model uses far fewer parameters than any other method compared. While some more sophisticated methods outperformed Ursa, this paper demonstrates the viability, effectiveness, and potential of the Ursa layer and the constellation approach to pattern learning.

The 2D data was used to demonstrate the movement over time of the Ursa constellation stars during training. Figures 5 and 6 show the constellation stars adjusting over time during training to span the space for the minimum distance measure of Eq. 4 and the Gaussian distance measure of Eq. 2, respectively. The constellation resulting from the minimum distance measure appears more compact in the center and the stars are more spread out toward the outside. On the other hand, the Gaussian distance measure constellation is more uniformly distributed throughout the space. Also, the resulting range is larger for the minimum distance measure than for the Gaussian distance measure.

As mentioned in Sect. 2, The Ursa layer computes a single global feature for a set of points, so it is limited in its ability to recognize local structures and patterns. Future research should explore a hierarchical network architecture based on the Ursa layer that evaluates object structure at varying levels. The current architecture is also not invariant to shifts, scales, and rotations of the input data. This is another area for future research.

Table 1. Classification results and model size comparisons for various methods. Accuracy results for the Ursa method are the mean of 10 independent runs for each data set. Best Ursa single run performances are in parentheses. Accuracy results for the other methods are adapted from [7] and [6]. Model size for Ursa is based on 512 stars.

	ModelNet40 accuracy (in percent)	MNIST accuracy (in percent)	Model size (x1M params)
LeNet5 [15]	–	99.2	–
3DShapeNets [2]	84.7	–	–
VoxNet [1]	85.9	–	–
Subvolume [16]	89.2	–	–
ECC [2]	87.4	99.4	–
PointNet (baseline) [3]	87.2	98.7	0.8
PointNet [3]	89.2	99.2	3.5
PointNet++ [14]	90.7	99.5	1.0
KCNet [7]	90.0	99.3	0.9
Kd-Net [5]	91.8	99.1	2.0
DGCNN [6]	92.2	–	1.8
Ursa (ours)	88.2 (89.0)	99.1 (99.2)	0.4

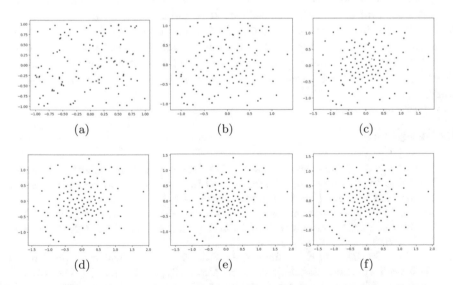

(a)   (b)   (c)

(d)   (e)   (f)

Fig. 5. Depiction of the Ursa constellation stars and their positions during training for m = 128 and using the minimum distance measure. Sub-figures show (a) random initialization, (b) after 10 training epochs, (c) after 100 epochs, (d) after 200 epochs, (e) after 300 epochs, and (f) after 500 epochs.

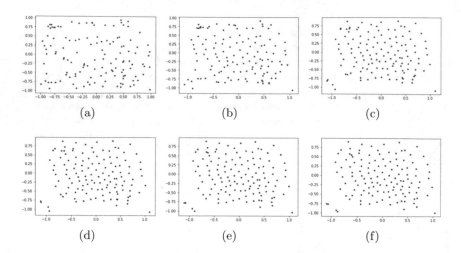

Fig. 6. Depiction of the Ursa constellation stars and their positions during training for m = 128 and using the Gaussian distance measure. Sub-figures show (a) random initialization, (b) after 10 training epochs, (c) after 100 epochs, (d) after 200 epochs, (e) after 300 epochs, and (f) after 500 epochs.

6 Conclusion

This paper has presented an Ursa neural network layer and demonstrated its effectiveness and viability for classification of point cloud data. The Ursa layer stores information in the form of constellation points, rather than a set of multiplicative weights in a matrix. While other more sophisticated methods achieved higher classification rates on the data sets used in testing, all other methods compared used at least twice the model parameters of the baseline Ursa network.

References

1. Maturana, D., Scherer, S.: VoxNet: a 3D convolutional neural network for real-time object recognition. In: 2015 IEEE/RSJ International Conference on Intelligent Robots and Systems (IROS), pp. 922–928. IEEE, September 2015
2. Wu, Z.: 3D ShapeNets: a deep representation for volumetric shapes. In: IEEE Conference on Computer Vision and Pattern Recognition, CVPR 2015, pp. 1912–1920 (2015)
3. Qi, C.R., Su, H., Mo, K., Guibas, L.J.: PointNet: deep learning on point sets for 3D classification and segmentation. In: Proceedings - 30th IEEE Conference on Computer Vision and Pattern Recognition, CVPR 2017, January 2017, pp. 77–85 (2017)
4. Simonovsky, M., Komodakis, N.: Dynamic edge-conditioned filters in convolutional neural networks on graphs. In: Proceedings - 30th IEEE Conference on Computer Vision and Pattern Recognition, CVPR 2017, pp. 29–38 (2017)

5. Klokov, R., Lempitsky, V.: Escape from cells: deep Kd-networks for the recognition of 3D point cloud models. In: Proceedings of the IEEE International Conference on Computer Vision, October 2017, pp. 863–872 (2017)
6. Wang, Y., Sun, Y., Liu, Z., Sarma, S.E., Bronstein, M.M., Solomon, J.M.: Dynamic graph CNN for learning on point clouds. arXiv:1801.07829 (2018)
7. Shen, Y., Feng, C., Yang, Y., Tian, D.: Mining point cloud local structures by kernel correlation and graph pooling. In: IEEE Conference on Computer Vision and Pattern Recognition (2018)
8. Tsin, Y., Kanade, T.: A correlation-based approach to robust point set registration. In: European Conference on Computer Vision (2004)
9. Riegler, G., Ulusoy, A.O., Geiger, A.: OctNet: learning deep 3D representations at high resolutions. In: Proceedings - 30th IEEE Conference on Computer Vision and Pattern Recognition, CVPR 2017, January 2017, pp. 6620–6629 (2017)
10. Buhmann, M.D.: Radial Basis Functions: Theory and Implementations. Cambridge University Press, Cambridge (2003)
11. Orr, M.: Introduction to radial basis function networks (1996)
12. Broomhead, D.S., Lowe, D.: Multivariable functional interpolation and adaptive networks. Complex Syst. **2**, 3221–355 (1988)
13. Chen, S., Cowan, C., Grant, P.: Orthogonal least squares learning algorithm for radial basis function networks. IEEE Trans. Neural Netw. **2**, 302–309 (1991)
14. Qi, C.R., Yi, L., Su, H., Guibas, L.J.: PointNet++: deep hierarchical feature learning on point sets in a metric space. In: Neural Information Processing Systems Conference (2017)
15. Lecun, Y., Bottou, L., Bengio, Y., Haffner, P.: Gradient-based learning applied to document recognition. Proc. IEEE **86**(11), 2278–2324 (1998)
16. Qi, C.R., Su, H., Niessner, M., Dai, A., Yan, M., Guibas, L.J.: Volumetric and multi-view CNNs for object classification on 3D data. In: Proceedings - 29th IEEE Conference on Computer Vision and Pattern Recognition, CVPR (2016)

Persistent Homology for Detection of Objects from Mobile LiDAR Point Cloud Data in Autonomous Vehicles

Meirman Syzdykbayev[✉] and Hassan A. Karimi[✉]

Geoinformatics Laboratory, School of Computing and Information,
University of Pittsburgh, 135 North Bellefield Avenue, Pittsburgh, PA, USA
{mis180, hkarimi}@pitt.edu
http://gis.sis.pitt.edu/

Abstract. Recently, researchers have paid significant attention to problems related to object detection and computer vision for autonomous vehicles. Such vehicles offer many benefits, including their ability to help address transportation-related issues such as safety concerns, traffic jams, and overall mobility. Multi-beam 'light detection and ranging' (LiDAR) is one of the main sensors that is used to sense and detect objects by creating a point cloud data map of the surrounding environment. Current object detection tasks that use only mobile LiDAR data divide the entire area into cubes and employ image object detection methods. Such an approach poses challenges due to the third dimension that increases computational time, which thus requires a tradeoff between performance and time optimization. In this paper, we propose a new approach to detect objects using point cloud data by investigating the shapes of the objects. To this end, we developed a method based on topological data analysis achieved via persistent homology to analyze the qualitative properties of the data. To the best of our knowledge, our work is the first to develop topological data analysis for real-world mobile LiDAR point cloud data exploration. The evaluation result shows a high accuracy classification result using features extracted from barcodes.

Keywords: Mobile LiDAR point cloud data · Persistent homology · Autonomous vehicles · Object detection

1 Introduction

In recent years, automobile manufacturers have made significant progress in bringing automation and computerization to driving. Cars currently have various automated features such as adaptive cruise control, parking assist, distance control, and blind spot detection. Some manufacturers have gone further and designed autonomous vehicles (AVs) that can drive and navigate without a driver [1], thereby allowing passengers to travel on demand. A transportation network of AVs could offer numerous economic benefits to society, including congestion relief, increased fuel efficiency, and improved passenger safety [2].

© Springer Nature Switzerland AG 2020
K. Arai and S. Kapoor (Eds.): CVC 2019, AISC 944, pp. 458–472, 2020.
https://doi.org/10.1007/978-3-030-17798-0_37

For tasks such as localization, perception, planning, and control, AVs use a complex system that is equipped with various sensors [3]. LiDAR is one of the main sensors used in current AVs. LiDAR is used to create a three-dimensional (3D) point cloud of the surrounding environment for localization and perception (object detection) tasks. For the perception task, LiDAR has an advantage over two-dimensional (2D) images, because the spatial coordinates of the 3D point cloud make it easier to obtain the shape and position or orientation of the detected objects. However, no efficient algorithms are currently available that can compute mobile LiDAR point cloud data. Also, processing 3D point cloud data requires much more computational power than processing 2D images. So, due to their high computational and financial costs, the use of LiDAR sensors and mobile LiDAR point cloud data is not widespread in comparison with traditional cameras. Nonetheless, the increasing availability and use of mobile LiDAR sensors should begin to reduce the sensors' price and lead to the creation of efficient algorithms to store and compute large point cloud datasets [4]. For example, a high-definition map can be created using mobile LiDAR point cloud data and serve as a base map for AVs to perform various tasks, such as localization [3].

For the object detection task, current AVs use a sensor fusion method where the advantages of each sensor are combined with those of the other sensors to perform a task as no single sensor can provide all the required data or required accuracy. Despite its advantages, however, sensor fusion in AVs increases computational complexity and makes sensors highly reliant on each other to accomplish a specific task. For example, the effect of direct sunlight on a camera image may cause the task to fail to recognize specific objects.

One challenge in object detection using 3D point cloud data is that, unlike 2D images, LiDAR-driven 3D data do not have clear texture information and do not capture the color of objects. Another challenge is the presence of noise in point cloud data. The third challenge is the higher degree of freedom for objects in 3D space compared to 2D space [5]. Evaluation results obtained from the KITTI Vision Benchmark Suite [6] show that car detection accuracy using a method based on point cloud data is close to 71% for moderate objects. This percentage calls for new and more accurate algorithms to detect objects using mobile LiDAR point cloud data.

In this paper, we propose a method that addresses the challenges of detecting cars using 3D point cloud data. We used the geometric and topological features of objects to obtain reliable results. The geometric features that we used are the size of the object and the distance between the object and the sensor. The topological features are the numbers of connected components, circles, and voids imposed on the object. The proposed approach computes persistent homology (PH) that falls within the scope of topological data analysis and generates persistent barcodes. These barcodes can later be modified and integrated with common data analysis and machine learning workflows. We also compared our results with state-of-the-art object detection algorithms, with considerable success.

To the best of our knowledge, this work is the first to use both geometric and topological features of objects captured by mobile LiDAR point cloud data for object detection. The contributions of this paper can be summarized as follows:

- A method is developed to identify the set of points that belongs to a particular class (cars) from mobile LiDAR point cloud data using PH and a machine learning approach.
- A ground segmentation algorithm is proposed and implemented that can separate ground points from the rest of the points for the ground segmentation task.
- Suitable classifiers for classification are compared.

The rest of the paper is organized as follows: Sect. 2 provides fundamental information about PH and LIDAR as well as some related works. Section 3 elucidates the proposed methods. Section 4 presents the experiment and steps for creating object detection methods from topological features. Section 5 discusses the results. Section 6 provides concluding remarks and ideas for future work.

2   Background and Related Works

2.1   Background

Persistent Homology. Topological data analysis (TDA) has attracted the attention of researchers and developers in recent years as means of analyzing data, in particular image data, and a complementary framework to the machine learning techniques. Persistent homology (PH) is one of the foundational methods for TDA that can be used to study qualitative features of complex data in multiple scales. Applying PH method to point cloud datasets obtained through LiDAR technology is increasingly gaining interest due to the widespread availability and use of the technology.

Homology is associated with measuring/counting the number of holes in an object [7]. In other words, homology is an invariant of topological data that are computable, general, and simple. The homology of a space gives a sequence of vector spaces and counts various types of holes and voids in space, depending on the dimension. To compute homology algorithmically, simplicial complexes are employed; these are combinatorial structures that approximate topological space and are used because topological space is hard to compute. A simplicial complex is a space that is created from a union of points, edges, triangles, tetrahedra, and higher-dimensional polytopes. It is inherently linear-algebraic and serves as the inspiration for topological algebra. Homological methods are robust and do not rely on precise coordinates or careful estimates for efficacy; thus, these methods are most useful in conditions where geometric precision fails. PH is an algebraic tool for recording the history of the emergence and vanishing of topological features of the space such as number of connected components, holes, and voids [8]. These features can be presented in one of the three forms: barcode, persistent diagram (PD), or persistence landscape. The barcode is a shape descriptor for point cloud data that has its base in algebraic topology and differential geometry [9]. It uses a finite collection of intervals to present lifetime of the features (Fig. 1a). The PD presents lifetime of the features on an extended graph with X and Y coordinates, where X represents the birth of the hole and Y is the death. In a PD, each point corresponds to a feature, and its significance can be defined by its distance from the diagonal line [10] (Fig. 1b). Persistence landscape maps the space of PD to a vector space allowing the performance of statistical analysis and machine learning to the data as well as faster computation [11] (Fig. 1c).

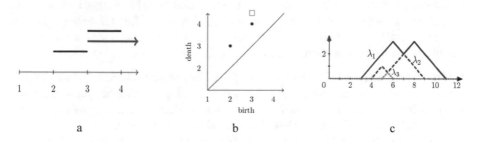

a   b   c

Fig. 1. The output of persistent homology represented as (a) barcode, (b) persistent diagram and (c) persistence landscape.

LiDAR. Over the last several decades, LiDAR has become a common distance measurement technology and is now used in many fields. LiDAR measures distance by sending pulsed laser light to an object and measuring the reflected pulses from it [12]. LiDAR sends rapid pulses to the surrounding environment; then, the sensor receives the reflected pulses and measures the time that it takes for each pulse to bounce back. As a result, a 3D representation of an object can be created by using the difference between the laser send and return times and the wavelength, as light moves at a constant speed that allows the distance between the sensor and reflected object to be calculated. The quality of the point cloud is affected by different scanning methods that are available for different purposes. Also, LiDAR consists of two major components: a scanner and receiver. The two types of receivers are silicon avalanche photodiodes and photomultipliers. The LiDAR sensor model commonly used for AVs is a Velodyne HDL-64. The laser in this model generates a wavelength of 905 nm and 5-nm pulse. In addition, it fires approximately 2 million pulses per second up to 120 m and has an accuracy of 2 cm [3].

2.2   Related Works

Two areas of related work are discussed object detection approaches using mobile LiDAR point cloud data and PH implementation in computer vision applications.

Object detection using mobile LiDAR point cloud data. One of the early approaches to organizing point cloud data was to segment the points into clusters. Early reasonable segmentation was generated by removing data from the ground [13, 14]. Better segmentation results were obtained by forming graphs on the point cloud [15] and by implementing hierarchical segmentation and mixture of bag-of-words approaches [16].

With the addition of machine learning to object detection methods, a feature-centric voting algorithm that takes a sliding window approach was introduced by [17]. This algorithm considers only the occupied grids, which helps decrease computational time significantly. End-to-end fully convolutional network and bounding boxes for object detection were used by [18] for data which was presented in a 2D point map. Researchers have employed other similar concepts that also utilize a convolutional network point cloud object detection approach [19, 20]. In addition, VoxelNet, a 3D detection network that can process feature extraction in a single stage and bounding box prediction, both in a single step, was proposed by [21]. This network divides the space into equally spaced voxels and transforms a group of points from each voxel into a unified feature representation, and then computes only the occupied voxels.

Persistent Homology for Computer Vision. A framework for object (3D meshes, 2D gesture contours, and texture images) recognition using topological persistence was presented by [22] and showed that PDs can serve as informative descriptors for shapes and images. Multiscale kernel for PDs that provides a stable summary representation of topological features was designed by [23]. The results were shown in two datasets for 3D shape classification/retrieval and texture recognition. Different surface textures were classified using topological descriptors by conducting a comprehensive study of the topological descriptors and by comparing them with other current computer vision approaches [24].

The current methods use full 3D representation of objects, whereas our proposed approach uses real-world mobile LiDAR point cloud data that consist of only the points that are returned from the objects. The sparsity of the points varies depending on the distance between the sensor and the object.

3 Proposed Method

3.1 Point Cloud Data Segmentation

Ground detection and filtering play a significant role in point cloud segmentation [14]. The simplest and fastest way to use ground segmentation is to assume that the terrain is flat and to set a fixed threshold at a specific height. However, this method is not very accurate. A grid-based approach was introduced by [25] where grid cells are either ground-based or not ground-based, depending on the range between the heights of the points inside the cell. The purpose of this algorithm is to detect non-drivable terrain at a sufficient range. Another method involves building a Markov random field and inference that implements loopy belief propagation [26]. This method, although it provides results with accuracy, is computationally intensive. A new method that provides the same degree of accuracy as [26] but with less computational time was proposed by [27].

In this paper, we introduce a new ground detection algorithm that separates points that return from the ground using other points. Our proposed ground segmentation algorithm can separate ground points with a high degree of accuracy. Unlike airborne

LiDAR sensors, mobile LiDAR sensors are set up vertically and collect return points from the top and side of an object. The algorithm extracts the extreme maximum and minimum points from the whole dataset. Extreme maximum points count as an object and extreme minimum points count as a ground. Thus, in taking out these extreme points, we can collect all of the points from the top of the object. For the remaining points, we created a box around each point and selected a range of elevation (Z) values in that box. Thus, the points with small range values inside the box are from the ground, and the points with large range values are from the object. Next, we set a threshold for the elevation values (Algorithm 1). The difference between the proposed algorithm and the algorithm by [25], in which the focus is on the ground and on detecting non-drivable terrain, is that our algorithm focuses on points and detecting non-ground points. The limitation of the proposed algorithm is that it assumes the terrain of the area is not extreme while setting the minimum and maximum points.

Algorithm 1. Ground segmentation

Input:
X, Y, and Z coordinates for all point cloud data, size of box, elevation range threshold
Output: X, Y, and Z coordinates of objects
1: Select min and max points.
2: **for** i in points
3: Create a box around point.
4: **for** j in points in the box
5: Check elevation range.
6: **if** range > threshold then
7: i = point from object.
8: **return** j
9: **end if**
10: **end for**
11: **end for**

After separating the ground from the objects, we needed to separate the objects from each other. To do so, we implemented a DBCAN unsupervised clustering algorithm. This algorithm expands clusters from core samples of high-density point cloud data and requires two variables: the maximum distance between two samples and the minimum number of samples [28]. If the maximum distance between two samples is great, then two close objects can be clustered as a single object. If the maximum distance is too small, then each point can be presented as a cluster.

3.2   Computation of Persistent Homology

Developing an efficient algorithm to compute PH is currently an active research direction. Such algorithms have been implemented in many software programs in different programming languages. For example, a pipeline for the computation of PH was proposed by [29] (Fig. 2).

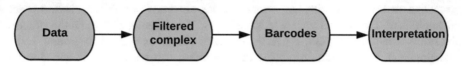

Fig. 2. Pipeline for computation of persistent homology [29].

Data are a subset of points that are segmented to form mobile LiDAR point cloud data. As mentioned in Sect. 2, the computation of PH requires building simplicial complexes. Among the several types of simplicial complexes, we selected alpha complexes to build our simplicial complex to compute PH. The alpha complexes are similar to the Cech complex and use Voronoi decomposition but differ by having natural geometric realization. Due to known acceleration that uses duality between 0-dimensional and 1-dimensional persistence, alpha complexes are well suited for point data. In barcode form, the filtered complexes need to be associated with a boundary matrix that stores information about the faces of each simplex. Then, this boundary matrix needs to be reduced using Gaussian elimination [29].

3.3   Feature Extraction from Persistent Diagrams

Features retrieved from PH and presented as barcodes contain firm, valuable information about the structure of a point cloud. However, implementing machine learning algorithms for barcodes is not possible due to the absence of an inner product structure. Different approaches that combine machine learning algorithms with PH can be taken to extract features from barcodes, but none has been implemented for object detection using mobile LiDAR point cloud data. Examples of the feature extraction methods include: (i) clustering between two global positioning system (GPS) paths of birds using features such as the number of holes, maximum hole lifetime, number of relevant holes, average lifetime of all holes, and sum of all lifetimes [30]; (ii) classification of black and white images of numbers by extracting features from barcodes [31] and selecting the four most important among the features; and (iii) extraction of features from barcodes for fingerprint classification [32]. In this last example, five different metrics for the barcodes and then 552 features for each fingerprint were identified by [32]. Based on these examples of the feature extraction methods, clearly the science of choosing the best feature from barcodes is still in its early stages.

In this work, we used the feature extraction method to transform the output of PH into a standard vectorized format by assigning a real vector to the barcode. We identified 54 features from [32], with 18 features for each dimension. These features are:

$f_1 = \sum_{i=1}^{n} (y_i - x_i)$	$f_7 = mean\{x_i\}$	$f_{13} = median\{y_{max} - y_i\}$
$f_2 = \sum_{i=1}^{n} n(y_i - x_i)$	$f_8 = mean\{y_i\}$	$f_{14} = median\{y_i - x_i\}$
$f_3 = \sum_{i=1}^{n} (y_{max} - y_i)(y_i - x_i)$	$f_9 = mean\{y_{max} - y_i\}$	$f_{15} = SD\{x_i\}$
$f_4 = \sum_{i=1}^{n} n(y_{max} - y_i)(y_i - x_i)$	$f_{10} = mean\{y_i - x_i\}$	$f_{16} = SD\{y_i\}$
$f_5 = \sum_{i=1}^{n} (y_{max} - y_i)^2 (y_i - x_i)^2$	$f_{11} = median\{x_i\}$	$f_{17} = SD\{y_{max} - y_i\}$
	$f_{12} = median\{y_i\}$	$f_{18} = SD\{y_i - x_i\}$

where n denotes the number of bars in each dimension, x_i denotes the left endpoint of the barcode or birth time, y_i denotes the right endpoint of the barcode or the death time of a homological feature, and y_{max} denotes the right-most endpoint of the bar in the barcode.

4   Experiments

The mobile LiDAR point cloud data contained a return point from an environment, including the ground. Although our objective was object detection, we did not need any returns from the ground. Thus, we set our first task as ground segmentation followed by the second task, implementing a DBSCAN unsupervised clustering algorithm to obtain a subset of adjacent points. The advantage of using mobile LiDAR point cloud data over 2D images is that we were able to analyze the precise 3D size of the object. In the

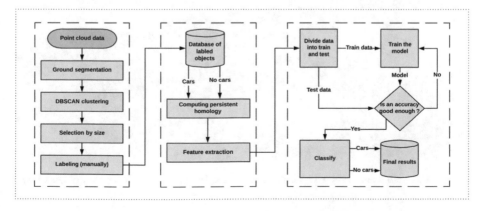

Fig. 3. Tasks of the proposed object detection method.

third task, we selected subsets of the point clouds with a total size that resembled the size of a car. In the fourth task, we trained the model to classify objects. We manually labeled selected objects and stored them in a database. We then implemented computation and feature extraction for each selected object barcode. In addition to topological features, we calculated geometric features such as the average distance between the sensor and object. A machine learning classification model requires labeled data for the learning process. Hence, we manually labeled 114 (43 'car' and 71 'no car') observations and stored them in a database. Figure 3 shows the tasks of the conducted experiments. We conducted our experiments using a workstation with 3.2-GHz Intel Core i5 PC with 16 GB of RAM.

4.1   Dataset

Mobile LiDAR point cloud data can be obtained from different datasets, such as the Ford Campus Vision and Lidar Data Set [33], the KITTI Vision Benchmark Suite dataset [6], and the Visual Localization Data Set [34]. All these datasets, in addition to the X, Y, and Z coordinates, contain data from an inertial measurement unit (IMU). These datasets are available for different cities, each with unique road infrastructures, in both the United States and Germany. In this work, we used the dataset from the KITTI Vision Benchmark Suite. This dataset contains data from sensors such as LiDAR, high-resolution color and grayscale video cameras, and IMU. The data were collected while driving in the city of Karlsruhe, Germany, both on highways and in rural areas. In addition to the raw data, a benchmark, an evaluation metric, and an evaluation website are all available from this data source. We assigned each object in the dataset to one of three categories: easy, moderate, and hard. Easy objects have a minimum bounding box height of 40 pixels and are clearly visible. Moderate and hard objects have a minimum bounding box height of 25 pixels and are not clearly visible. In this work we used only easy objects. Figure 4 shows raw data from the KITTI Vision Benchmark Suite dataset with six easy objects (cars).

Fig. 4. Raw data from the KITTI Vision Benchmark Suite dataset with six easy objects (cars).

4.2   Persistent Homology

We utilized the R-package topological data analysis component, which contains tools for computing barcodes [35]. The output of this algorithm is a table with three columns, where the first column represents the dimension, the second column represents the birth time, and the third column represents the death time of a homological feature. The dimension contains three distinct values of 0, 1, and 2. The first value corresponds to the number of connected components. The second value corresponds to the number of circles, and the third value corresponds to the number of voids. Figure 5 shows an example of a PD for a set of points that has been returned both from the car and from an object that is not a car.

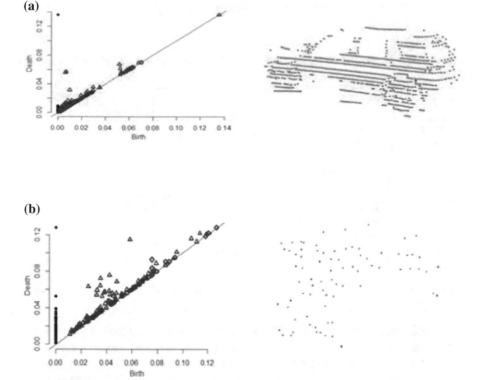

Fig. 5. (a) Persistent diagram and return point cloud data from car and (b) persistent diagram and return point cloud data from object that is not car.

4.3   Classification

Machine learning was used to classify whether a selected subset of point cloud data belongs to the class "car" or to the class "not car" using features derived from a collection of barcodes. For each selected subset of a point cloud, we computed persistent homology and created a barcode. Using a feature extraction method, we retrieved 54 features from these barcodes for machine learning.

We used Python's Scikit-Learn package to apply supervised machine learning to PH. To find a suitable classifier in our study, we compared five well-established classifiers (support vector machine, logistic regression, decision tree, neural network, and AdaBoost) for accuracy, precision, and recall using various parameters in order to select the best one for our labeled data with 55 features. In addition to the 54 defined features, we identified one feature as the distance between an object and a sensor. For each classifier, we divided the dataset into training and testing. We did not divide the dataset further into a train and validation set because our dataset was small, and the number of features exceeded the number of observations in the dataset.

To extract features, we implemented principal component analysis (PCA), a linear dimensionality reduction technique that uses a singular value decomposition of the data to project in the lower dimensional space. Table 1 shows the accuracy rates that we obtained for the five classifiers with and without using PCA.

Table 1. Accuracy rate obtained from different classifiers with and without principal component analysis.

Classifier	Parameters	Accuracy without PCA	Accuracy using PCA with 2 components
Support vector machine	Kernel = 'rbf' C = 10 Gamma = 10	0.862	0.827
Logistic regression	Penalty = 'l2' C = 1000	0.846	0.79
Neural network	Hidden layers = [5, 10] Activation = 'relu' Solver = 'lbfgs'	0.86	0.758
Decision tree	Max depth = 2	0.896	0.827
AdaBoost	Number of estimators = 500	1	0.48

The evaluation website of the KITTI Vision Benchmark Suite [6] states that KITTI uses pattern analysis, statistical modeling, and computational learning (PASCAL) criteria to evaluate 3D object detection performance. For object detection experiments, these criteria require a bounding box overlap of at least 70% [36]. However, our approach identifies exact point clouds that belong to cars. In order to evaluate our proposed approach, as mentioned in Sect. 3, we manually labeled the subset of point cloud data that belongs to car as 'car' and 'no car' otherwise.

5   Results and Discussion

The aim of the proposed object detection method using 3D point cloud data is to identify objects in order to create a bounding box around the identified objects. The best classifier (AdaBoost) provided 100% accuracy using features that were extracted from the barcode. Table 2 shows the accuracy, precision, recall, and confusion matrix for each classifier. The 100% accuracy is due to the fact that we used only easy objects in the experiment. These objects (cars) are clearly visible and have distinct holes (windows) which were captured using persistent homology. These holes played a significant role in the classification task.

Table 2. Accuracy, precision, recall, and confusion matrix obtained from different classifiers.

Classifier	Accuracy	Precision	Recall	Confusion matrix
Support vector machine	0.862	0.785	0.9166	[[14 3] [1 11]]
Logistic regression	0.846	0.705	1	[[12 5] [0 12]]
Neural network	0.86	0.896	0.862	[[13 4] [0 12]]
Decision tree	0.896	0.909	0.896	[[16 1] [2 10]]
AdaBoost	1	1	1	[[17 0] [0 12]]

The second-best classifier was the decision tree classifier, which is understandable given that the AdaBoost classifier uses a decision tree classifier algorithm as its base estimator. Hence, AdaBoost, with a low number of estimators, provides the same result as the decision tree classifier. The support vector machine and neural network classifiers were tied for third. The difference between a support vector machine and neural network classifiers is the number of true negatives and false positives.

The main reasons that we found low accuracy for the classifiers (with the exception of AdaBoost) may be the low number and skewed nature of the observations. We labeled and stored a total of 114 (43 'car' and 71 'no car') observations; that is, the number of observations that were not a car is approximately twice the number of observations that were a car. The dataset is unbalanced and thus may be a cause of the low accuracy. Another reason for low accuracy may be the number of features in the classification process. In this paper, we used all the features in our classification to show our results. To achieve greater accuracy, we need to identify and implement additional significant features. Feature selection is important because the features may be highly correlated with each other.

6 Conclusions and Future Work

In this paper, we have shown that objects can be detected from mobile LiDAR point cloud data using the geometric and topological features of the objects. We used cars as target objects because they clearly play a critical role in the tasks that AVs must perform. We used the size of the target object and the distance between the sensor and the object as geometric features. By taking the PH approach and computing the barcodes of the object, we acquired the topological features of the target objects.

In the ground segmentation task, we created and implemented an algorithm that separates return points from the ground and from the objects. After clustering adjacent points using DBSCAN unsupervised clustering, we set the size of the objects to the maximum values of the manually labeled objects and a minimum value of 1 m. We selected a subset of point cloud data that was bigger than 1 m and less than approximately 5 m for further classification. Then, we computed barcodes for each selected object and extracted 55 features for each object. We later classified these features through five machine learning classifiers.

When using AdaBoost as the classifier, the classification accuracy with 114 (43 'car' and 71 'no-car') observations was 100% without altering features. This perfect result indicates that topological features of objects for computer vision applications can serve as powerful tools. Providing reliable results is vital when working with AVs. However, time performance was not good enough for real-world deployment. For example, the ground segmentation algorithm takes approximately 180 s to separate the points of one frame (100,000 points), and a frame is collected every 0.1 s.

We believe that PH can be a significant tool for computer vision applications that use mobile LiDAR point cloud data. PH is a new method in the data analysis field, and mobile LiDAR is a new technology for collecting data about the environment. Using this current method, PH shows that it can help identify objects (cars) from point cloud data with high accuracy.

The proposed method performs well on easy objects, but we do not know how it will perform when complex objects are used. Evaluating the proposed method on moderate and hard objects is one of our future research works. Considering that the proposed method assumes point cloud data with no, or little, noise, we will evaluate the method with noisy data such as point cloud data in presence of rain or snow. As we used limited features and specific objects (cars) in this work, but we plan to include additional features and evaluate the new method by using other objects such as pedestrians and bicycles. To improve the time performance of our method, our future research works include development of efficient algorithms and implementation of a parallel processing approach.

References

1. Fagnant, D.J., Kockelman, K.: Preparing a nation for autonomous vehicles: opportunities, barriers and policy recommendations. Transp. Res. Part A: Policy Pract. **77**, 167–181 (2015)
2. Howard, D., Dai, D.: Public perceptions of self-driving cars: the case of Berkeley, California. In: Transportation Research Board 93rd Annual Meeting, vol. 14, no. 4502 (2014)

3. Liu, S., Li, L., Tang, J., Wu, S., Gaudiot, J.L.: Creating autonomous vehicle systems. Synth. Lect. Comput. Sci. **6**(1), i–186 (2017)
4. Velodyne Lidar Price Reduction – Self-Driving Cars – Medium. https://medium.com/self-driving-cars/velodyne-lidarprice-reduction-d358f245f086. Accessed 14 Apr 2018
5. Huang, J., You, S.: Detecting objects in scene point cloud: a combinational approach. In: 3D Vision-3DV 2013, pp. 175–182. IEEE (2013)
6. Geiger, A., Lenz, P., Stiller, C., Urtasun, R.: The KITTI vision benchmark suite (2015)
7. Ghrist, R.: Homological algebra and data. preprint (2017)
8. Zomorodian, A., Carlsson, G.: Computing persistent homology. Discrete Comput. Geom. **33** (2), 249–274 (2005)
9. Collins, A., Zomorodian, A., Carlsson, G., Guibas, L.J.: A barcode shape descriptor for curve point cloud data. Comput. Graph. **28**(6), 881–894 (2004)
10. Edelsbrunner, H., Letscher, D., Zomorodian, A.: Topological persistence and simplification. In: Proceedings of the 41st Annual Symposium on Foundations of Computer Science, pp. 454–463. IEEE (2000)
11. Bubenik, P.: Statistical topological data analysis using persistence landscapes. J. Mach. Learn. Res. **16**(1), 77–102 (2015)
12. Schwarz, B.: LIDAR: mapping the world in 3D. Nat. Photonics **4**(7), 429 (2010)
13. Himmelsbach, M., Hundelshausen, F.V., Wuensche, H.J.: Fast segmentation of 3D point clouds for ground vehicles. In: Intelligent Vehicles Symposium (IV), pp. 560–565. IEEE (2010)
14. Douillard, B., Underwood, J., Kuntz, N., Vlaskine, V., Quadros, A., Morton, P., Frenkel, A.: On the segmentation of 3D LIDAR point clouds. In: 2011 IEEE International Conference on Robotics and Automation, ICRA, pp. 2798–2805. IEEE (2011)
15. Wang, D.Z., Posner, I., Newman, P.: What could move? Finding cars, pedestrians and bicyclists in 3D laser data. In: 2012 IEEE International Conference on Robotics and Automation, ICRA, pp. 4038–4044. IEEE (2012)
16. Behley, J., Steinhage, V., Cremers, A.B.: Laser-based segment classification using a mixture of bag-of-words. In: 2013 IEEE/RSJ International Conference on Intelligent Robots and Systems, IROS, pp. 4195–4200. IEEE (2013)
17. Wang, D.Z., Posner, I.: Voting for voting in online point cloud object detection. In: Robotics: Science and Systems, vol. 1, p. 5 (2015)
18. Li, B., Zhang, T., Xia, T.: Vehicle detection from 3D lidar using fully convolutional network. arXiv preprint arXiv:1608.07916 (2016)
19. Maturana, D., Scherer, S.: 3D convolutional neural networks for landing zone detection from lidar. In: 2015 IEEE International Conference on Robotics and Automation, ICRA, pp. 3471–3478. IEEE (2015)
20. Graham, B.: Sparse 3D convolutional neural networks. arXiv preprint arXiv:1505.02890 (2015)
21. Zhou, Y., Tuzel, O.: VoxelNet: end-to-end learning for point cloud based 3D object detection. arXiv preprint arXiv:1711.06396 (2017)
22. Li, C., Ovsjanikov, M., Chazal, F.: Persistence-based structural recognition. In: Proceedings of the IEEE Conference on Computer Vision and Pattern Recognition, pp. 1995–2002. IEEE (2014)
23. Reininghaus, J., Huber, S., Bauer, U., Kwitt, R.: A stable multi-scale kernel for topological machine learning. In: Proceedings of the IEEE Conference on Computer Vision and Pattern Recognition, pp. 4741–4748. IEEE (2015)
24. Zeppelzauer, M., Zieliński, B., Juda, M., Seidl, M.: Topological descriptors for 3D surface analysis. In: International Workshop on Computational Topology in Image Context, pp. 77–87. Springer, Cham (2016)

25. Thrun, S., Montemerlo, M., Dahlkamp, H., Stavens, D., Aron, A., Diebel, J., Fong, P., Gale, J., Halpenny, M., Hoffmann, G., Lau, K.: Stanley: the robot that won the DARPA Grand Challenge. J. Field Robot. **23**(9), 661–692 (2006)

26. Zhang, M., Morris, D.D., Fu, R.: Ground segmentation based on loopy belief propagation for sparse 3D point clouds. In: 2015 International Conference on 3D Vision, 3DV, pp. 615–622. IEEE (2015)

27. Velas, M., Spanel, M., Hradis, M., Herout, A.: CNN for very fast ground segmentation in Velodyne lidar data. arXiv preprint arXiv:1709.02128 (2017)

28. Ester, M., Kriegel, H. P., Sander, J., Xu, X.: A density-based algorithm for discovering clusters in large spatial databases with noise. In: KDD, vol. 96, no. 34, pp. 226–231 (1996)

29. Otter, N., Porter, M.A., Tillmann, U., Grindrod, P., Harrington, H.A.: A roadmap for the computation of persistent homology. EPJ Data Sci. **6**(1), 17 (2017)

30. Pereira, C.M., de Mello, R.F.: Persistent homology for time series and spatial data clustering. Expert Syst. Appl. **42**(15–16), 6026–6038 (2015)

31. Adcock, A., Carlsson, E., Carlsson, G.: The ring of algebraic functions on persistence bar codes. arXiv preprint arXiv:1304.0530 (2013)

32. Giansiracusa, N., Giansiracusa, R., Moon, C.: Persistent homology machine learning for fingerprint classification. arXiv preprint arXiv:1711.09158 (2017)

33. Pandey, G., McBride, J.R., Eustice, R.M.: Ford campus vision and lidar data set. Int. J. Robot. Res. **30**(13), 1543–1552 (2011)

34. Badino, H., Huber, D., Kanade, T.: The CMU visual localization data set. Computer Vision Group (2011)

35. Fasy, B. T., Kim, J., Lecci, F., Maria, C.: Introduction to the R package TDA. arXiv preprint arXiv:1411.1830 (2014)

36. Everingham, M., Van Gool, L., Williams, C.K., Winn, J., Zisserman, A.: The Pascal visual object classes (VOC) challenge. Int. J. Comput. Vis. **88**(2), 303–338 (2010)

Context-Based Object Recognition: Indoor Versus Outdoor Environments

Ali Alameer[1,2](✉), Patrick Degenaar[1,3], and Kianoush Nazarpour[1,3](✉)

[1] School of Engineering, Newcastle University, Newcastle NE1 7RU, UK
{a.m.a.alameer,kianoush.nazarpour}@newcastle.ac.uk
[2] School of Natural and Environmental Sciences, Newcastle University,
Newcastle Upon Tyne NE1 7RU, UK
[3] Institute of Neuroscience, Newcastle University, Newcastle NE2 4HH, UK

Abstract. Object recognition is a challenging problem in high-level vision. Models that perform well for the outdoor domain, perform poorly in the indoor domain and the reverse is also true. This is due to the dramatic discrepancies of the global properties of each environment, for instance, backgrounds and lighting conditions. Here, we show that inferring the environment before or during the recognition process can dramatically enhance the recognition performance. We used a combination of deep and shallow models for object and scene recognition, respectively. Also, we used three novel topologies that can provide a trade-off between classification accuracy and decision sensitivity. We achieved a classification accuracy of 97.91%, outperforming the performance of a single GoogLeNet by 13%. In another experiment, we achieved an accuracy of 95% to categorise indoor and outdoor scenes by inference.

Keywords: Indoor and outdoor classification · Deep learning ·
Object recognition · Scene recognition

1 Introduction

Over the last decade, machine vision algorithms have reached the level of reliability required in complex environments. In particular, many approaches were advanced for object recognition [1–4]. These approaches share common basis which stands on sifting the input images through a large number of filters to extract features. The extracted features attempt to provide an invariant representation of the object.

Many approaches were developed to achieve invariance to the transformation of objects [5–7]. One successful approach is based on stacking convolutional layers and pooling layers together in a hierarchical structure. Extracting high-level features throughout the advanced layers of this hierarchy has proven successful to achieve invariance. The literature shows many examples of hierarchical structures for object recognition. For instance, the Hierarchical MAX model (HMAX) [8], consisted of only two stages of convolution/pooling layers. It was inspired by the

© Springer Nature Switzerland AG 2020
K. Arai and S. Kapoor (Eds.): CVC 2019, AISC 944, pp. 473–490, 2020.
https://doi.org/10.1007/978-3-030-17798-0_38

primate visual cortex. The HMAX model attempts to mimic the processing in the first 100ms of the primates visual cortex, that include processing the visual data through the ventral stream pathway. In this pathway, informative information about the shape and texture of objects that account for rapid categorisation are extracted. Recently, the number of the convolutional/pooling layers have dramatically increased [9–11]. Increasing the depth of models has increasingly enhanced the classification performance.

Recent studies have shown that models that function well in an indoor environment, perform poorly in an outdoor environment and the reverse is also true [12,13]. This is due to the stark difference in local and global properties of both environments. The daily life environment, such as living-rooms and city streets, comprises a large number of objects. The nature of these objects depends on the context in which they can be found. Current algorithms of object recognition are trained to recognise objects regardless of their context, dismissing all the information in the backgrounds. This poses a great difficulty for these models to make logical decisions.

Scene understanding is a necessary stage that provides important information about the possible object identity. Identifying the scene can dramatically reduce the probabilities of the object identity and therefore increasing the recognition chance level. For example, outside in a desert, it is more likely to expect a camel than a microscope. We believe that context-based object recognition that depends equally on the environment can characterise the recognition process and therefore enhance the recognition performance.

Hybrid intelligent systems, in particular combining classifiers, can offer a practical solution to handle increasingly complex problems. It allows the use of a priori knowledge to inspire the solution. The concept of hybrid intelligent systems was applied in handwriting recognition, where several neural networks were aligned and a voting process was applied for decision making [14,15]. It was also shown that averaging the output of an infinite number of independent classifiers can produce an optimal performance [16,17]. However, the literature has not witnessed utilising hybrid intelligent systems for context-based object recognition, for instance, indoor and outdoor environment.

In this work, we propose three topologies for context-based object recognition. The common factor in all topologies is identifying the environment prior/during the object classification stage. This prior knowledge, i.e., environment type, has given the topologies an advantage in performing classification on a diverse object dataset. We used an object dataset that comprises objects that are likely to be found in an indoor environment. Similar criteria were applied to the outdoor object dataset. We formed three topologies to perform object recognition.

The proposed topologies have the following advantages:

1. It enhances the classification accuracy significantly.
2. It provides more decision confidence. Each decision is based on the posterior probability of more than one classifier (topology-B and topology-C).
3. In topology-C, further to the enhanced classification performance, the object category, i.e., indoor and outdoor, was inferred with an accuracy of 90%.

2   Method

We selected six models of object recognition to form our topologies. We tested them in a challenging diverse visual environment. Below are short descriptions of the models used in this work.

2.1   Shallow Models

Here, we refer to the models that consist of five convolutional layers or less as shallow models. Also, we refer to models with more than five convolutional layers as deep models.

HMAX. It was inspired by the simple and complex cells hierarchy of the primate visual cortex [2,8,18,19]. It was developed to extract invariant features to object transformations, such as, scaling and translation. It consists of four stages that comprise pooling and convolutional layers. The combinations of convolution and pooling layers are believed to extract a high-level representation of objects.

En-HMAX. In the En-HMAX model [20–22], the number of layers was increased. It comprised three convolutional layers and three pooling layers. Additionally, sparse coding and independent component analysis (ICA) were introduced [23,24]. The ICA method was used to extract Gabor-like filters from natural images in the first simple layer (S_1). Sparse coding was used to train dictionaries in both S_2 and S_3 layers of the En-HMAX model.

AlexNet. The AlexNet model [9] is a convolutional neural network (CNN) that consists of five convolutional layers, three pooling layers, and two fully connected layers. It comprises 60 million parameters to fine-tune. It transforms objects in the input images into distinctive features. The AlexNet model operates in a similar fashion to the HMAX model. They share similar hierarchal structure and the same classic alternation of convolutional and pooling layers. Across shallow models, it achieved the highest performances on many datasets [25]. The success of AlexNet has attracted the attention of researchers of computer vision towards CNNs. Due to its simplicity and good performance, in this work, we consider AlexNet, pre-trained with Places dataset [26], as our default model for indoor versus outdoor categorisation task.

2.2   Deep Models

Here, we used the following three well-known deep learning CNN models as a major platform to form the topologies. GoogLeNet that comprises 57 convolutional layers is the deepest network used in this work.

VGG16 and VGG 19. The VGGNet architecture introduced in [10], is designed to significantly increase the depth of the existing CNN architectures with 16 or 19 convolutional layers. The last three layers of both versions, i.e., VGG16 and VGG19, are the following layers:

- Fully connected layer: In this layer, the input data is multiplied by the weight matrix and then adds a bias vector.
- Softmax layer: In this layer, a softmax function is used for classification purposes. It is considered as the multi-class generalisation of the logistic sigmoid function, also known as the normalised exponential layer.
- Classification layer: In this layer, the output predicted label is generated. It is formed by cross-entropy loss function that defines the pre-existed trained classes.

GoogLeNet. The GoogLeNet model [11], also known as the inception model, is significantly deeper than the previously explained CNN models. It comprises 57 convolution layers with 5 million parameters to fine-tune. A key feature in the design of GoogLeNet is applying the network in network architecture introduced in [27], in the form of inception modules. Inception module uses a set of parallel convolution layers with a MAX pooling stage along each module. A concatenating layer is used to concatenate the responses of each individual module. In this work, the used version of GoogLeNet comprises a total of 9 inception modules. A more detailed overview of GoogLeNet architecture can be found in [11].

3   Transfer Learning

Transfer learning is increasingly becoming a powerful tool in the field of machine learning [28]. It involves utilising the stored knowledge of a model acquired for solving a particular task and applying it to solve a different problem. For instance, the knowledge acquired while learning to distinguish between different types of trucks could be utilised to recognise different types of cars.

Fine-tuning a network with randomly initialized weights is extremely complicated and time-consuming task. Here, we used networks that were pre-trained with scene images (Places dataset [26]) and object images (ImageNet dataset [25]) depending on the classification task. The CNN models were then adjusted to the new datasets' configurations. To retrain a CNN model on a particular dataset, we froze the weights of earlier layers and only retrained the weights of the advanced layers.

4   Posterior Probability

The posterior probability is the conditional probability that is computed after an occurrence of a relevant event. In the field of pattern recognition [29], the posterior probability indicates the uncertainty of assessing a particular class of images. The posterior probability is produced when a generative model makes

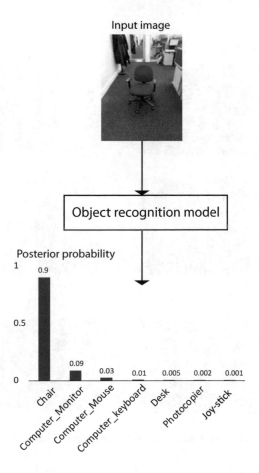

Fig. 1. The distribution of posterior probabilities of an input image. It can be seen that in this example, the classifier is 90% confident that the object in this image is a chair.

a decision [30]. Higher posterior probabilities indicate higher confidence of the classifier's decision. Figure 1 shows an example of how an indoor classifier distributes posterior probabilities for a given input image. Usually, the maximum posterior probability is used to determine the class label. In this work, the maximum posterior probability was utilised to indicate the confidence in the classifier. A threshold for each classifier was set and accordingly, the classifiers made decisions based on their confidence. The threshold was set based on the average posterior probability of all the testing dataset.

5   Datasets

The image classes were collected from ImageNet dataset [25], Caltech 101 dataset [31] and Caltech 256 dataset [32]. These classes were categorised into two uncorrelated set of images: outdoor and indoor. The outdoor image subset does not contain classes of the indoor image subset and the reverse is also true.

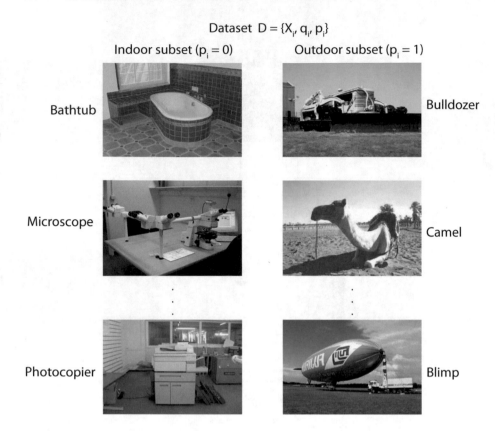

Fig. 2. Selected indoor and outdoor images from our dataset.

Figure 2 shows six examples of the dataset, reflecting the richness of the dataset in terms of the variety of objects and their backgrounds.

6 Classification

In this work, the classification settings are briefly explained. In this section, for all classification scenarios, the extracted features were classified using a linear support vector machine (SVM) [33]. In each of the experiments, 50% of the dataset was allocated for testing the classifier. In addition, to ensure that the classification scores were not biased by the random choice of training samples, the classification was repeated for 20 runs where the random selection in each round is independent of the other. The average classification score and the standard deviation are reported.

7 Proposed Topologies

The hierarchical topologies developed in this section are designed to achieve an improved classification performance over the existing methods of object recognition. Additionally, providing higher confidence level and decision sensitivity. In this section, a detailed description of the proposed topologies is provided. The method and the architecture of each topology are explained. The designed topologies obtain the environment in which the object is found as an essential component of the recognition process. Furthermore, the designed topologies comprise a decision-making stage that can be tuned to increase the confidence or the decision sensitivity for the process of object recognition.

Topology-A and topology-B consist of three different models for object and scene recognition. They comprise one shallow model for recognising the environment and two deep models for object recognition. Topology-C, however, consists of only two models for object recognition. The environment type, whether indoor or outdoor, in topology-C, is categorised by inference. In this topology, the identity of the environment does not directly contribute to the object recognition process and only computed as an external label.

The architecture of topology-A was inspired by the human visual system, where scenes are rapidly categorised in a small time of 50 ms which give a clear information about the identity of the objects within [34]. However, topology-B and topology-C are purely computational with less relevance to biology. Topology-B was designed to minimise the error chance in the first stage of topology-A, the scene recognition stage. The scene recognition stage was designed in-parallel to other stages of object recognition with a different mechanism in the decision-making stage. Topology-C was designed to minimise the number of models in topology-A and topology-B. Only two models for object recognition are used in topology-C for understanding the environment and for identifying objects. Finally, each of the below topologies have several advantages and disadvantages. The below subsections will discuss these in more details.

7.1 Topology-A

Figure 3 shows the basic structure of topology-A. In the used dataset $\mathbb{D} = \{\mathbf{X}_i, q_i, p_i\}_{i=1}^{N}$, each image $\mathbf{X}_i$ has class label q_i (for example: chair) and category label p_i (for example: indoor). The indoor category is denoted by using $p_i = 0$ and the outdoor category by using $p_i = 1$. For a given image, q_i* denotes the predicted class label and p_i* denotes the predicted category label. The confusion matrix of the indoor versus outdoor classifier $C_M = \{c_{ij}\}_{i,j=1}^{2}$ was used to calculate the ratio of the correctly classified images. Using the total probability theorem, the overall accuracy in topology-A can be calculated as shown below:

$$Accuracy(\%) = \frac{100}{\sum_{i,j} c_{ij}} \left[c_{11} \, \mathbb{P}(q* = q \mid p* = p = 0) + \right.$$
$$+ c_{22} \, \mathbb{P}(q* = q \mid p* = p = 1) + \tag{1}$$
$$+ c_{12} \, \mathbb{P}(q* = q \mid p* = 1, p = 0) +$$
$$\left. + c_{21} \, \mathbb{P}(q* = q \mid p* = 0, p = 1) \right]$$

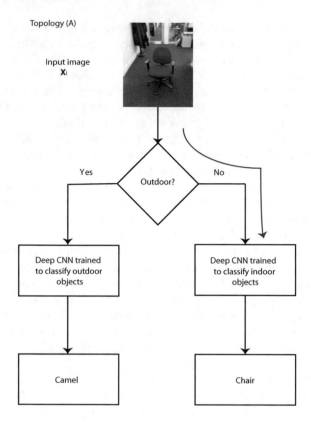

Fig. 3. The structure of topology-A. The input image is first categorised (i.e., indoor and outdoor) then classified (i.e., chair, microscope).

7.2   Topology-B

In topology-B, shown in Fig. 4, the three classifiers operate in parallel to identify an object in an input image. The object identity depends on the decision of all three classifiers. The three classifiers have an equal influence in making the final decision. Making an incorrect decision in any of the stages does not guarantee an incorrect class label in the final stage. The posterior probability is used to quantify the reliability of the classifiers. Classifiers with higher confidence level have more influence on making the final class label decision.

In the experiments performed in this work, the mean of the posterior probabilities of the whole testing data $\mathbb{D}$ was set as a confidence threshold. However, an optimal confidence threshold can be tuned differently depending on the classification context. The final decision is based on the posterior probabilities of all three classifiers as shown in Table 1.

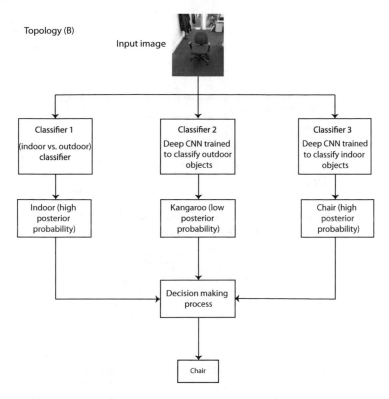

Fig. 4. The structure of topology-B. In topology-B, the classifier that categorises indoor versus outdoor images operates in parallel with other classifiers.

7.3   Topology-C

In this topology, shown in Fig. 5, only two classifiers were used to predict the class label and the category label. Table 2 shows the scenarios in which this topology make the final decision.

In this work, the collected image dataset has two separate image subsets. The image classes of the indoor subset do not correlate with the image classes of the outdoor subset. This suggests that when an indoor classifier is used, classes from the outdoor subset tend to give lower posterior probabilities than classes from the indoor subset. Figure 6 shows an analysis of the average posterior probability for both the indoor classifier and the outdoor classifier. In this analysis, GoogLeNet was used to produce the figures. As expected, in both scenarios, i.e., indoor classifier and outdoor classifier, testing a classifier with unseen images within the same training categories produced a significantly higher posterior probability than testing it with different image categories. For the indoor classifier, the Mann-Whitney U test, with a risk $\alpha = 0.05$, shows that the posterior probabilities for indoor test images (M $= 87.6$, SD $= 18.9$) were significantly higher than that of outdoor test images (M $= 41.7$, SD $= 21.5$); Z-score $= 22.3$,

Table 1. The decision-making process of topology-B. The table shows only 2 possible scenarios of the 16th possible combinations. In all other scenarios, a no-decision state will be produced. The ✓ marker denotes higher confidence, X marker denotes lower confidence and d denotes the "do not care status".

		Confidence	
Indoor classifier (1)		✓	X
Outdoor classifier (2)		X	✓
Indoor versus	Indoor decision	✓	d
Outdoor classifier	Outdoor decision	d	✓
Classifier selection		1	2

Table 2. The decision-making process of topology-c. The ✓ marker denotes higher confidence and X marker denotes lower confidence

	Confidence			
Indoor classifier (1)	✓	X	✓	X
Outdoor classifier (2)	X	✓	✓	X
Classifier selection	1	2	No-decision	

p-value < 0.05. Similarly, for the outdoor classifier, the above test shows that the posterior probabilities of the outdoor test images (M $= 74.0$, SD $= 26.4$) were significantly higher than that of indoor test images (M $= 31.2$, SD $= 18.0$); Z-score $= 20.9$, p-value < 0.05. The data above comprises unpaired non-parametric samples. Therefore, we used Mann-Whitney U method to test for significance. Therefore, we hypothesised that the posterior probability can give a notion of the image category, i.e., indoor versus outdoor.

8   Results

The below subsections display the results for the discussed topologies in the previous sections.

8.1   Indoor Versus Outdoor

Models of object recognition tend to produce higher performances in a binary classification scheme. The chance level in binary classification scenarios is 50%. In this work, shallow models were utilised for categorising indoor and outdoor scenes. Figure 7 shows a comparison in classification performance between these models. It can be noticed that AlexNet (pre-trained with scene images) outperforms other shallow models for the categorisation task, with a high accuracy of 99.46%. The En-HMAX model achieves higher scores of 87.96%, however, it is still far less than the performance of AlexNet. This is due to the large

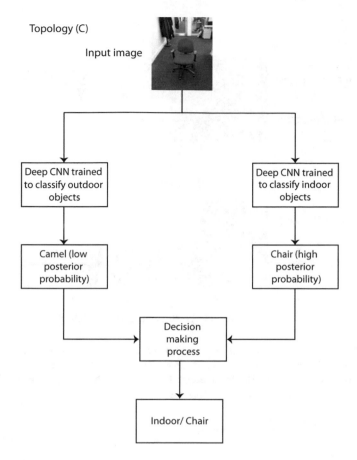

Fig. 5. The structure of topology-C. In topology-C, no classifier is used to categorise the environment (indoor and outdoor), however, it is able to categorise the environment by inference.

size of the image data, in which the En-HMAX model cannot handle efficiently due to its abstract architecture. The same applies to the HMAX model, where 75.03% of classification accuracy is achieved. Therefore, AlexNet was elected as a default model with regard to all indoor versus outdoor categorisation schemes, i.e., topology-A and topology-B.

In topology-A, AlexNet spread the images to either the indoor classifier or the outdoor classifier. Although AlexNet has a very high classification performance, the few incorrect decisions it makes lead to failure in the output stage. This is due to the uncorrelated image data used in both classifiers. In another word, the indoor classifier knows nothing about the outdoor environment and the reverse is also true. Therefore, when an outdoor image passes the indoor classifier, an incorrect class label will be guaranteed.

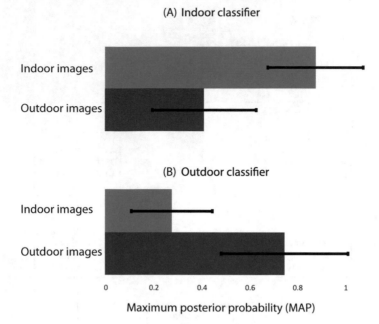

Fig. 6. An example of the average posterior probability of the indoor and the outdoor classifiers using GoogLeNet. (A) Indoor classifier. (B) Outdoor classifier. This chart illustrates the decorrelation in the average posterior probability between the indoor classifier and the outdoor classifier of topology-C.

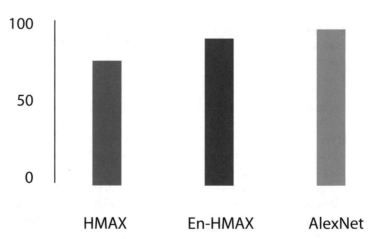

Fig. 7. Results of categorising indoor and outdoor images.

In topology-B, the decision of AlexNet has less impact on the final class label due to the structure of the topology. An incorrect decision at any stage does not guarantee an incorrect class label. In topology-C, however, no shallow network is used to categorise the scene type. The scene type is inferred from the indoor and the outdoor classifier.

8.2   Classification Scores Using Topology-A

In Fig. 8, AlexNet, VGG16, VGG19 and GoogLeNet were utilised as the main platforms to quantify the performance of topology-A. To compute the classification accuracy of the whole classification task, the above models were used individually. In particular, all the image dataset was used without segregating it into an indoor subset and an outdoor subset. This process was repeated for each of the above models separately. As a result, the classification accuracy of each of the above models was quantified for the comparison with topology-A. A similar process was performed for topology-B and topology-C.

Finally, topology-A scores were compared with the above scores. For completeness, the comparisons are only performed between a certain classification model and the topology that is formed within the same model, for instance, the VGG19 network results are compared with topology-A that is formed by only the VGG19 models.

For all used models, topology-A outperformed the original models. For example, in AlexNet, an increased classification performance of 7% is achieved. The difference is constantly decreased for deeper models.

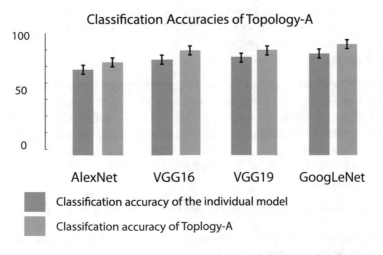

Fig. 8. Results of topology-A. AlexNet is used as a default model for categorising indoor and outdoor images. The classification accuracies in the second-row represent the performance of below models to individually classify the whole dataset.

This is particularly interesting because deeper models are capable of understanding large data. Therefore, using a bigger abject dataset is believed to increase the above differences dramatically.

Topology-A has the following advantages:

1. Advanced performance over using a single network.
2. Only two models can operate to recognise each input image.

The disadvantages of topology-A can be summarised as the followings:

1. It involves three different classifiers that require more memory in terms of implementation.
2. An incorrect decision in the first stage guarantees an incorrect class label. The first stage (indoor versus outdoor classifier) has more power in making the final decision.

8.3   Classification Scores Using Topology-B

Figure 9 shows the classification scores of using topology-B. It also shows the percentages of the no-decision state. In line with topology-A, similar models were used in this experiment to form this topology. AlexNet was used to categorise the indoor and outdoor images in all scenarios. In the above calculations, the no-decision state is considered as a correct classification. It can be noticed that deeper models such as GoogLeNet and VGG19 do not outperform other models when using this topology. The performances are more balanced. However, the topology formed by VGG19 tends to make more decisions than other models. The decision-making conditions can be tuned using an optimised threshold. In this experiment, the mean posterior probability of all the testing images was used as a threshold of confidence.

Topology-B has the following advantages:

1. The decision-making process depends equally on all three classifiers.
2. It achieves the highest performance among the other topologies.
3. It is designed to make no decisions when a lower confidence level is obtained. The confidence threshold can be tuned depending on the allocated task. Applications with higher risks, for instance, autonomous cars, need higher confidence threshold. The "no-decision" state is an important measure in such applications.

The disadvantages of topology-B can be summarised as the followings:

1. It requires more memory in terms of implementation because of the three classifiers in its architecture.
2. It is more computationally expensive than the other topologies because it needs all three classifiers to operate simultaneously.

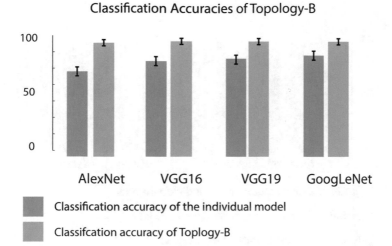

Classification Accuracies of Topology-B

■ Classification accuracy of the individual model

■ Classifcation accuracy of Toplogy-B

Fig. 9. Results of topology-B. AlexNet is used as a default model for categorising indoor and outdoor images for all the below calculations.

8.4   Classification Scores Using Topology-C

In topology-C, the objects are classified using only two classifiers as shown in Fig. 5. Similar to topology-A and topology-B, the same previously explained models were used to form topology-C. Furthermore, the classification scores were reported in a similar fashion. Unlike topology-B, there was no allocated classifier for categorising the indoor and the outdoor environments. Instead, the category label was inferred throughout the process of recognising an object. Figure 10 shows the categorisation and classification scores of topology-C. A high categorisation accuracy of 95% was achieved using VGG19. This is particularly interesting because this score is achieved without using a specific classifier for the task. In this topology, the percentages of the no-decision state are less than that of topology-B. However, the classification accuracies are slightly decreased. Interestingly, VGG19 performs slightly better than other models using this topology.

Topology-C has the following advantages:

1. It involves only two classifiers for the recognition process.
2. It infers the category label without using a specific classifier, i.e., indoor versus outdoor classifier.
3. It makes no decision when a lower confidence level is obtained.

The disadvantages of topology-C can be summarised as the followings:

1. It provides reduced performance comparing to the other topologies due to the decreased number of the classifiers in its architecture.
2. It shows lower decision frequency than other topologies, due to the limited number of input parameters in the decision-making stage.

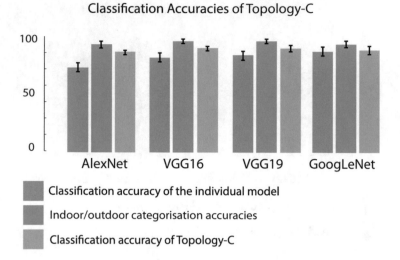

Classification Accuracies of Topology-C

- Classification accuracy of the individual model
- Indoor/outdoor categorisation accuracies
- Classification accuracy of Topology-C

Fig. 10. The results of topology-C

9   Conclusions

The architectures presented in this work provide three essential elements for image classification: classification accuracy, decision sensitivity, and computational complexity. In topology-A, two models can operate to recognise objects for each input image. The categorisation stage filters the input images to either the indoor classifier or the outdoor classifier. This topology is less complex than other topologies. However, an incorrect decision at the first stage may guarantee an incorrect image class label. In topology-B, we overcome the problems of topology-A by electing the decision via all classifiers. All three classifiers operate simultaneously and a voting decides the final decision. This topology is computationally complex, as it needs three classifiers to operate simultaneously for each input image. However, it provides higher classification accuracies. Topology-C provides the advantages of topology-A and topology-B. The voting includes only two classifiers to infer the image category and class. This topology also offers to control the sensitivity of the decision making. Results show that with the proposed topologies, the performance of GoogLeNet can be improved by 13%.

Acknowledgments. The work of A. Alameer was supported by the Higher Committee for Education Development, Iraq (HCED, D1201017). The work of K. Nazarpour was supported by the Engineering and Physical Sciences Research Council, U.K., grants EP/M025977/1 and EP/M025594/1.

References

1. Li, Z., Wang, Y., Yu, J., Guo, Y., Cao, W.: Deep learning based radiomics (DLR) and its usage in noninvasive IDH1 prediction for low grade glioma. Sci. Rep. **7**(11), 5467 (2017)
2. Hu, X., Zhang, J., Li, J., Zhang, B.: Sparsity-regularized hmax for visual recognition. PloS One **9**(1), 215–243 (2014)
3. Krizhevsky, A., Sutskever, I., Hinton, G.E.: ImageNet classification with deep convolutional neural networks. In: Advances in Neural Information Processing Systems, pp. 1097–1105 (2012)
4. Jia, Y., Shelhamer, E., Donahue, J., Karayev, S., Long, J., Girshick, R., Guadarrama, S., Darrell, T.: Caffe: convolutional architecture for fast feature embedding. In: Proceedings of the 22nd ACM International Conference on Multimedia, pp. 675–678 (2014)
5. Ghazaei, G., Alameer, A., Degenaar, P., Morgan, G., Nazarpour, K.: Deep learning-based artificial vision for grasp classification in myoelectric hands. J. Neural Eng. **14**(3), 036025 (2017)
6. Abolghasemi, V., Chen, M., Alameer, A., Ferdowsi, S., Chambers, J., Nazarpour, K.: Incoherent dictionary pair learning: application to a novel open-source database of chinese numbers. IEEE Sig. Process. Lett. **25**(4), 472–476 (2018)
7. Ghazaei, G., Alameer, A., Degenaar, P., Morgan, G., Nazarpour, K.: An exploratory study on the use of convolutional neural networks for object grasp classification. In: Proceedings of the 2nd IET International Conference on Processing Intelligent Signal Processing (ISP), pp. 5–8 (2015)
8. Riesenhuber, M., Poggio, T.: Hierarchical models of object recognition in cortex. Nat. Neurosci. **2**(11), 1019–1025 (1999)
9. Krizhevsky, A., Sutskever, I., Hinton, G.E.: ImageNet classification with deep convolutional neural networks. In: Advances in Neural Information Processing Systems 25, pp. 1097–1105 (2012)
10. Simonyan, K., Zisserman, A.: Very deep convolutional networks for large-scale image recognition. arXiv preprint arXiv:1409.1556, vol. 9, no. 1 (2014)
11. Szegedy, C., Liu, W., Jia, Y., Sermanet, P., Reed, S., Anguelov, D., Erhan, D., Vanhoucke, V., Rabinovich, A.: Going deeper with convolutions. In: Proceedings of the IEEE Conference on Computer Vision and Pattern Recognition, pp. 1–9 (2015)
12. Quattoni, A., Torralba, A.: Recognizing indoor scenes. In: IEEE Conference on Computer Vision and Pattern Recognition, pp. 413–420 (2009)
13. Alameer, A., Degenaar, P., Nazarpour, K.: Biologically-inspired object recognition system for recognizing natural scene categories. In: International Conference for Students on Applied Engineering (ICSAE), pp. 129–132. IEEE (2016)
14. Hansen, L.K., Salamon, P.: Neural network ensembles. IEEE Trans. Pattern Anal. Mach. Intell. **12**(10), 993–1001 (1990)
15. Xu, L., Krzyzak, A., Suen, C.Y.: Methods of combining multiple classifiers and their applications to handwriting recognition. IEEE Trans. Syst. Man Cybern. **22**(3), 418–435 (1992)
16. Tumer, K., Ghosh, J.: Analysis of decision boundaries in linearly combined neural classifiers. Pattern Recogn. **29**(2), 341–348 (1996)
17. Ho, T.K., Hull, J.J., Srihari, S.N.: Decision combination in multiple classifier systems. IEEE Trans. Pattern Anal. Mach. Intell. **16**(1), 66–75 (1994)

18. Serre, T., Oliva, A., Poggio, T.: A feedforward architecture accounts for rapid categorization. Proc. Natl. Acad. Sci. **104**(15), 6424–6429 (2007)
19. Hubel, D.H., Wiesel, T.N.: Receptive fields, binocular interaction and functional architecture in the cat's visual cortex. J. Physiol. **160**(1), 106–154 (1962)
20. Alameer, A., Ghazaei, G., Degenaar, P., Nazarpour, K.: An elastic net-regularized HMAX model of visual processing. In: Proceedings of the 2nd IET International Conference on Processing Intelligent Signal Processing (ISP), pp. 1–4 (2015)
21. Alameer, A., Ghazaei, G., Degenaar, P., Chambers, J.A., Nazarpour, K.: Object recognition with an elastic net-regularized hierarchical MAX model of the visual cortex. IEEE Sig. Process. Lett. **23**(8), 1062–1066 (2016)
22. Alameer, A., Degenaar, P., Nazarpour, K.: Processing occlusions using elastic-net hierarchical max model of the visual cortex. In: IEEE International Conference on Innovations in Intelligent SysTems and Applications (INISTA), pp. 163–167. IEEE (2017)
23. Shen, B., Liu, B.-D., Wang, Q.: Elastic net regularized dictionary learning for image classification. Multimedia Tools Appl. **75**, 1–14 (2014)
24. Hyvärinen, A., Gutmann, M., Hoyer, P.O.: Statistical model of natural stimuli predicts edge-like pooling of spatial frequency channels in V2. BMC Neurosci. **6**(1), 12 (2005)
25. Deng, J., Dong, W., Socher, R., Li, L.-J., Li, K., Fei-Fei, L.: ImageNet: a large-scale hierarchical image database. In: IEEE Conference on Computer Vision and Pattern Recognition, CVPR 2009, pp. 248–255 (2009)
26. Zhou, B., Lapedriza, A., Xiao, J., Torralba, A., Oliva, A.: Learning deep features for scene recognition using places database. In: Advances in Neural Information Processing Systems, pp. 487–495 (2014)
27. Lin, M., Chen, Q., Yan, S.: Network in network. arXiv preprint arXiv:1312.4400, vol. 6, no. 11, pp. 1019–1025 (2013)
28. Pan, S.J., Yang, Q.: A survey on transfer learning. IEEE Trans. Knowl. Data Eng. **22**(10), 1345–1359 (2010)
29. Alameer, A., Akkar, H.A.: ECG signal diagnoses using intelligent systems based on FPGA. Eng. Technol. J. **31**(7), 1351–1364 (2013). Part (A) Engineering
30. Ronquist, F., Huelsenbeck, J.P.: MrBayes 3: Bayesian phylogenetic inference under mixed models. Bioinformatics **19**(12), 1572–1574 (2003)
31. Fei-Fei, L., Fergus, R., Perona, P.: Learning generative visual models from few training examples: an incremental bayesian approach tested on 101 object categories. Comput. Vis. Image Underst. **106**(1), 59–70 (2007)
32. Griffin, G., Holub, A., Perona, P.: Caltech-256 object category dataset (2007)
33. Vapnik, V.: Support-vector networks. Mach. Learn. **20**(3), 273–297 (1995)
34. Joubert, O.R., Rousselet, G.A., Fize, D., Fabre-Thorpe, M.: Processing scene context: fast categorization and object interference. Vis. Res. **47**(26), 3286–3297 (2007)

Comparison of Machine Learning Algorithms for Classification Problems

Boran Sekeroglu[1], Shakar Sherwan Hasan[2], and Saman Mirza Abdullah[3(✉)]

[1] Department of Information Systems Engineering, Near East University,
TRNC, Mersin 10, Turkey
boran.sekeroglu@neu.edu.tr
[2] Department of Software Engineering, Near East University,
TRNC, Mersin 10, Turkey
20168731@std.neu.edu.tr
[3] Department of Software Engineering, Koya University,
University Park, Danielle Mitterrand Boulevard, KOY45, Koysinjaq, Iraq
saman.mirza@koyauniversity.org

Abstract. Machine learning algorithms become wide tools that are used for classification and clustering of data. Several algorithms were proposed and implemented for different applications in multi-disciplinary areas. However, diversity of these algorithms makes the selection of effective algorithm difficult for specific application. Thus, comparison of benchmark algorithms is required. This paper presents preliminary results of the comparison of three different types of machine learning algorithms; Backpropagation Neural Network, Radial Basis Function Neural Network and Support Vector Machine using several numerical datasets for classification problems. Comparison is performed by considering the performance of these algorithms using obtained accuracy rates. The results show that Radial Basis Function Neural Network is superior to other considered algorithms for classification of numerical data.

Keywords: Machine learning · Backpropagation ·
Radial Basis Function Neural Network · Support Vector Machine

1 Introduction

Machines are able to learn and improve their performances by gathering more data and experience [1,2]. Machine learning (ML) is a part of Artificial Intelligence (AI) and usually, its objective is to recognize and fit the statistics into models. It enables computers in building models from sample data according to automate processes making decision, based on data inputs. Several methods are proposed and used for ML in multi-disciplinary fields and topics. In recent years, researchers in different fields and for a variety of purposes, moved towards utilizing machine learning approach to cluster, classify, and predict cases and events [3–5].

© Springer Nature Switzerland AG 2020
K. Arai and S. Kapoor (Eds.): CVC 2019, AISC 944, pp. 491–499, 2020.
https://doi.org/10.1007/978-3-030-17798-0_39

Even unsupervised methods are effective in machine learning, supervised methods are more popular for the classification of data because of making relation between inputs and target data. Though supervised learning, between the input and correct given output, there is an association type to be conducted. Some supervised machine learning algorithms are known as Logistic Regression, Naïve Bayes, Linear Regression, Decision Tree, k-Nearest Neighbor algorithm (kNN), Support Vector Machines (SVM) and several Artificial Neural Networks (ANN) algorithms.

There are many areas where machine learning techniques have been utilized [6]. Most applications perform computational learning, natural language processing (NLP) and pattern recognition (PR). Machine learning techniques are used to make computers capable of solving complex real-life problems with non-explicit programming. Systems are able to construct with the algorithms which able to learn from the data then can make driven-data predictions and decisions. Usage of machine learning improves the artificial intelligence (AI)- human interaction level.

Several comparisons had been performed in order to determine optimum classifier for a specific application. Recently, Dutta et al. [7] compared three models; Gaussian Process for Regression (GPR), Multi Adaptive Regression Spline (MARS) and Minimax Probability Machine Regression (MPMR) to predict compressive strength of concrete. Zeng et al. [8] compared three techniques; partial least square regression, SVM, and deep-learning for estimating soil salinity from hyperspectral data. SVM, Random Forest, and Extreme Learning Machine techniques are compared by [9] for intrusion detection. Deist et al. [10] used outcome prediction in chemo-radiotherapy to compare decision tree, random forest, ANN, SVM, elastic net logistic regression, LogitBoost. Random Forest (RF), SVM, and kNN are compared by [11] in the problem of natural forest mapping. Yan and Fenzhen [12] compared a decision tree (CART), backpropagation neural network (BPNN) and SVM for the classification of OLI images. Additionally, SVM was compared by Extreme Learning Machine (ELM) in [13] and Multilayer Perceptron, Random Forest and Simple Logistic algorithm were compared by [14].

All researchers concluded different comparison results dependent on the applications and it is obvious that, determining optimum machine learning technique is not possible for all kind of applications in different areas. Therefore, it is necessary to test and analyze the performances of those ML techniques with different datasets. In this research, common machine learning techniques of supervised learning are discussed, and common algorithmic approaches as Backpropagation Neural Network (BPNN), Radial Basis Function Neural Network (RBFNN) and Support Vector Machine (SVM) are considered for analysis using four numerical datasets for classification problems.

The rest of the paper is organized as: Sect. 2 briefly explains the considered machine learning algorithms, Sects. 3 and 4 presents experimental design and obtained results respectively. Finally, Sect. 5 concludes the experimental results obtained in this research.

2   Machine Learning Algorithms

In this research Support Vector Machine, Radial Basis Function Neural Network and Backpropagation Neural Network are considered for the comparison. Following subsections briefly explain these algorithms.

2.1   Backpropagation Neural Networks

During the few past decades ANNs have become a dynamic research and implementation area [15–17]. One of the most popular supervised NN algorithm is backpropagation [18]. Several real-life implementations used BPNN effectively in classification [5,19], control [3], and prediction [4] problems. Basic BPNN architecture can be seen in Fig. 1.

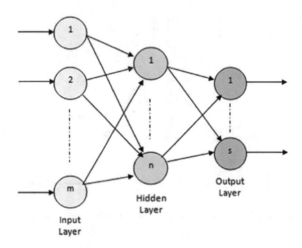

Fig. 1. General architecture of backpropagation neural network

BPNN is based on propagating the error signals obtained by comparing actual and target outputs, through previous layers and to adjust weights according to gradient-descent algorithm [20]. Weights update is performed by:

$$w_j^{k+1} = w_j^k + \eta(y_i - \hat{y}_i^{(k)})x_{ij} \tag{1}$$

where $w^{(k)}$ is the weight associated with the ith input link after the kth iteration, η is the learning rate, and w_{ij} is the value of the jth attribute of the training parameter x_i. Activation function of layers are generally used as Sigmoid activation function:

$$\frac{1}{(1 + exp^{(-w^T x)})} \tag{2}$$

2.2 Support Vector Machine

Support Vector Machine is a non-probabilistic machine learning algorithm that makes a model learns by classifying points in the space features [21]. It finds the optimal hyperplane for the separation of the classes by maximizing the margin to the support vectors. For nonlinearly separable problems, it transforms problem into higher dimensional space using proper basis functions, where the problem becomes linear (dougherty). Linear, polynomials, and Gaussians (radial basis functions) are the basic kernel functions for SVM. Several real life implementations were performed using SVM in risk evaluation [22], medical image classification [23] and data classification [24]. Figure 2 shows basic support vectors and classification margin for linear problems.

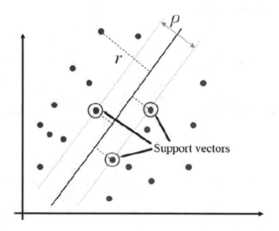

Fig. 2. SVM support vectors and classification margin [20]

Separating hyperplane can be defined as:

$$r = (w^T x + b)/||w|| \tag{3}$$

where w and b are parameters of the model, the distance from an example, x_i.

2.3 Radial Basis Function Neural Network

It is a simple 2-layer type of ANN uses radial basis functions as activation function. Combination of the linear outputs from radial basis functions provides the output of the RBF neural network. It is frequently used for prediction [25], regression [26] and classification [27]. Basic architecture of RBF neural network can be seen in Fig. 3. Radial Basis Function is defined as:

$$h(x) = \phi((x - c)^T R^{-1}(x - c)) \tag{4}$$

where ϕ is the function, x is the input, c and R is the center and metric, respectively.

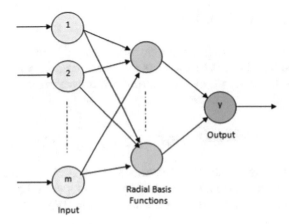

Fig. 3. Basic architecture of Radial Basis Function Neural Network

3   Experimental Design

Four different numerical datasets; Wines [28], Sonar [29], Spect Heart [30] and Iris [31], which are used for classification, have been considered for the experiments in order to analyze and compare the efficiencies of considered ML techniques. Table 1 shows characteristics of datasets.

Wine dataset consists 3 classes of wines with 13 chemical constituents (attributes) of corresponding classes with 178 instances. Sonar Dataset was created by obtaining bouncing sonar signals to classify objects as Rocks or Mines. It consists 60 attributes and 208 instances of signals with different angles under different conditions.

Spect Heart dataset contains features of Single Proton Emission Computed Tomography (SPECT) images of patients and has 2 classes as normal and abnormal. Iris dataset contains 3 classes; Iris Setosa, Iris Versicolour and Virginica and 50 instances for each attribute; sepal length, sepal width, petal length and petal width.

50% of instances for each dataset is used for training and the rest is used for testing. Analysis and comparison is performed by using instances considered for testing of each machine learning technique. Parameters for BPNN, SVM and RBFNN are determined by trial and error.

Table 1. Characteristics of datasets

Dataset	No. of instances	No. of attributes	No. of classes
Wines	178	13	3
Sonar	208	60	2
Spect Heart	278	22	2
Iris	147	14	3

4   Experimental Results

This section presents the results of all datasets using BPNN, SVM and RBFNN. As it was mentioned above, four datasets are considered for the experiments and obtained results are analyzed by considering accuracy rates for each experiment. Accuracy is calculated as follows:

$$Accuracy = \frac{(TP + TN)}{(TP + TN + FP + FN)} \qquad (5)$$

where TP and TN are True Positive and True Negative values, and FP and FN are False Positive and False Negative values, respectively.

Tables 2, 3 and 4 shows the results of BPNN, SVM and RBFNN for all considered datasets.

Table 2. Results of backpropagation neural network

Dataset	Accuracy
Wines	0.98
Sonar	0.90
Spect Heart	0.81
Iris	0.97

Table 3. Results of Support Vector Machine

Dataset	Accuracy
Wines	0.97
Sonar	0.98
Spect Heart	0.81
Iris	0.96

Table 4. Results of Radial Basis Function Neural Network

Dataset	Accuracy
Wines	1.00
Sonar	1.00
Spect Heart	1.00
Iris	1.00

Table 5 shows the combined accuracy results for all machine learning algorithms considered in this research.

Table 5. Accuracy results for all datasets

Dataset	BPNN	SVM	RBFNN
Wines	0.98	0.97	1.00
Sonar	0.90	0.98	1.00
Spect Heart	0.81	0.81	1.00
Iris	0.97	0.96	1.00

5   Conclusions

The main aim of machine learning is to solve problems that humans are not capable or insufficient. This helps humans to improve decision making or simulating improved skills of humans where it is impossible to applied by them. In this paper, three benchmark machine learning algorithms Backpropagation neural network, Support Vector Machine and Radial Basis Function Neural Network is implemented to compare the efficiency of each for classification problems. Four numerical datasets are considered for experiments and accuracy rates are used for the comparison of these machine learning algorithms.

Preliminary results show that, Radial Basis Function Neural Network is superior to Backpropagation neural network and Support Vector Machine by classifying all data effectively for all datasets. However, increment of number of datasets and the usage of big data may increase this accuracy during the testing. Similar results are obtained by Support Vector Machine and Backpropagation Neural Network, however, Backpropagation Neural Network achieved better results than Support Vector Machine even with little differences for datasets with three classes. For datasets with two classes, Support Vector Machine achieved better results than Backpropagation, that means increment of the classes decreases the efficiency of Support Vector Machine.

Future work will include the implementation of more machine learning algorithms with more numerical datasets and by adding image-based datasets. Also big data will be considered for the comparison of more machine learning algorithms.

References

1. Ding, W., Tong, Y.: Image and video quality assessment using neural network and SVM. IEEE 112–116 (2008)
2. Drouhard, J.P., Sabourin, R., Godbout, M.: A neural network approach to off-line signature verification using directional PDF. Pattern Recogn. **29**, 415–424 (1996)
3. Rubio, J.J.: Modified optimal control with a backpropagation network for robotic arms. IET Control Theory Appl. **6**(14), 2216–2225 (2012)
4. Kumar, J., Singh, A.K.: Workload prediction in cloud using artificial neural network and adaptive differential evolution. Future Gener. Comput. Syst. **81**, 41–52 (2018)

5. Sekeroglu, B.: Classification of sonar images using back propagation neural network. In: IEEE Geoscience and Remote Sensing Society Symposium, pp. 3092–3095 (2004)
6. Rashid, T.A., Abdullah, S.M.: A hybrid of Artificial Bee Colony, Genetic Algorithm, and Neural Network for Diabetic Mellitus Diagnosing. ARO Sci. J. Koya Univ. **6**(1), 55–64 (2018)
7. Dutta, S., Samui, P., Kim, D.: Comparison of machine learning techniques to predict compressive strength of concrete. Comput. Concr. **21**, 463–470 (2018). https://doi.org/10.12989/cac.2018.21.4.463
8. Zeng, W., Zhang, D., Fang, Y., Wu, J., Huang, J.: Comparison of partial least square regression, support vector machine, and deep-learning techniques for estimating soil salinity from hyperspectral data. J. Appl. Remote Sens. **12** (2018). https://doi.org/10.1117/1.JRS.12.022204
9. Ahmad, I., Basheri, M., Iqbal, M.J., Rahim, A.: Performance comparison of support vector machine, random forest, and extreme learning machine for intrusion detection. IEEE Access **6**, 33789–33795 (2018). https://doi.org/10.1109/ACCESS.2018.2841987
10. Deist, T.M., Dankers, F.J.W.M., Valdes, G., Wijsman, R., Hsu, I.C., Oberije, C., van Lustberg, T., Soest, J., Hoebers, F., Jochems, A., et al.: Machine learning algorithms for outcome prediction in (chemo)radiotherapy: an empirical comparison of classifiers. Med. Phys. **45**, 3449–3459 (2018). https://doi.org/10.1002/mp.12967
11. Isuhuaylas, L.A.V., Hirata, Y., Santos, L.C.V., Torobeo, N.S.: Natural forest mapping in the Andes (Peru): a comparison of the performance of machine-learning algorithms. Remote Sens. **10**, 782 (2018). https://doi.org/10.3390/rs10050782
12. Yan, G., Fenzhen, Z.: Study on machine learning classifications based on OLI images. In: 2013 International Conference on Mechatronic Sciences, Electric Engineering and Computer (MEC), China, pp. 1472–1476 (2013)
13. Bucurica, M., Dogaru, R., Dogaru, I.: A comparison of extreme learning machine and support vector machine classifiers, In: 2015 IEEE International Conference on Intelligent Computer Communication and Processing (ICCP), Cluj-Napoca, pp. 471–474 (2015). https://doi.org/10.1109/ICCP.2015.7312705
14. Uysal, E., Ozturk, A.: Comparison of machine learning algorithms on different datasets. In: 26th Signal Processing and Communications Applications Conference (SIU), Izmir, pp. 1–4 (2018). https://doi.org/10.1109/SIU.2018.8404193
15. Ghiasi, M.: Complexity revisited. In: 9th International Conference on Application of Information and Communication Technologies (AICT), pp. 553–557 (2015)
16. Wildes, R.P.: Iris recognition: an emerging biometric technology. Proc. IEEE **85**, 1348–1363 (1997)
17. Moghaddam, B., Yang, M.Y.: Gender classification with support vector machines. In: 4th IEEE International Conference on Automatic Face and Gesture Recognition, pp. 306–311 (2000). https://doi.org/10.1109/AFGR.2000.840651
18. Khashman, A., Sekeroglu, B.: Document image binarisation using a supervised neural network. Int. J. Neural Syst. **18**, 405–418 (2008)
19. Singh, K.R., Chaudhury, S.: Efficient technique for rice grain classification using back-propagation neural network and wavelet decomposition. IET Comput. Vis. **10**, 780–787 (2016)
20. Dougherty, G.: Pattern Recognition and Classification. Springer, Germany (2013)
21. Kashyap, K., Yadav, M.: Fingerprint matching using neural network training. Int. J. Eng. Comput. Sci. 2041–2044 (2013)
22. Jianga, H., Ching, W.K., Yiu, K.F.C., Qiu, Y.: Stationary Mahalanobis kernel SVM for credit risk evaluation. Appl. Soft Comput. **71**, 407–417 (2018)

23. Sekeroglu, B., Emirzade, E.: A computer aided diagnosis system for lung cancer detection using support vector machine. In: Third International Workshop on Pattern Recognition (2018). https://doi.org/10.1117/12.2502010

24. Li, H., Chung, F.L., Wanga, S.: A SVM based classification method for homogeneous data. Appl. Soft Comput. **36**, 228–235 (2015)

25. Wang, J., Zhang, W., Wang, J., Han, T., Kong, L.: A novel hybrid approach for wind speed prediction. Inf. Sci. **273**, 304–318 (2014)

26. Fidencio, P.H., Poppi, R.J., Andrade, J.C.: Determination of organic matter in soils using radial basis function networks and near infrared spectroscopy. Anal. Chim. Acta **453**, 125–134 (2002). https://doi.org/10.1016/S0003-2670(01)01506-9

27. Joutsijoki, H., Meissner, K., Gabbouj, M., et al.: Evaluating the performance of artificial neural networks for the classification of freshwater benthic macroinvertebrates. Ecol. Inf. **20**, 1–12 (2014). https://doi.org/10.1016/j.ecoinf.2014.01.004

28. Forina, M., Leardi, R., Armanino, C., Lanteri, S.: PARVUS - an extendible package for data exploration, classification and correlation. Institute of Pharmaceutical and Food Analysis and Technologies, Via Brigata Salerno, 16147 Genoa, Italy (1988)

29. Gorman, R.P., Sejnowski, T.J.: Analysis of hidden units in a layered network trained to classify sonar targets. Neural Netw. **1**, 75–89 (1988)

30. Kurgan, L.A., Cios, K.J., Tadeusiewicz, R., Ogiela, M., Goodenday, L.S.: Knowledge discovery approach to automated cardiac SPECT diagnosis. Artif. Intell. Med. **23**, 149–169 (2001)

31. Fisher, R.A.: The use of multiple measurements in taxonomic problems. Annu. Eugenics **7**, 179–188 (1936)

Automatic Recognition System for Dysarthric Speech Based on MFCC's, PNCC's, JITTER and SHIMMER Coefficients

Brahim-Fares Zaidi[1]([⊠]), Malika Boudraa[1], Sid-Ahmed Selouani[2], Djamel Addou[1], and Mohammed Sidi Yakoub[2]

[1] Laboratory of Speech Communication and Signal Processing (LSCSP), U.S.T.H.B University, Algiers, Algeria
fzaidi@usthb.dz
[2] Laboratory of Research in Human-System Interaction (LARHSI), University of Moncton, Shippagan Campus, Moncton, Canada

Abstract. The aim of this work is to improve the automatic recognition of the dysarthria speech. In this context, we have compared two techniques of speech parameterization; these two techniques are based on the recently proposed coefficients Power Normalized Cepstral Coefficients and Mel-Frequency Cepstral Coefficients. In this paper we have concatenate several variants of JITTER and SHIMMER with the techniques of speech parameterization to improve an automatic recognition of the dysarthric word system. The aim is to help the fragile persons having speech problems (dysarthric voice) and the doctor to make a first diagnosis about the patient's disease. For this, an Automatic Acknowledgment of Continuous Pathological Speech System has been developed based on the Hidden Models of Markov and the Hidden Markov Model Toolkit. For our tests, we used the Nemours Database which contains 11 speakers representing dysarthric voices.

Keywords: Automatic recognition system of continuous pathological speech (ARSCPS) · Hidden Markov Model Toolkit (HTK) ·
Hidden Models of Markov (HMM) ·
Power Normalized Cepstral Coefficients (PNCC) ·
Mel-Frequency Cepstral Coefficients (MFCC)

1 Introduction

Dysarthria is a difficulty to speak, mainly due to a dysfunction of the organs that allow the formation of the words in the mouth [9] and not caused by a problem of phonation (the voice). In a person with dysarthria, it is difficult to use or control the muscles of the mouth, tongue, larynx or vocal cords, which allow to hold a speech. Dysarthria can be caused by diseases that affect nerves and muscles. So in a person with dysarthria, the latter can isolate himself from his entourage and people, compromising his employability and his social relations.

To cope with this disease we have made an interface of an automatic recognition of dysarthria speech system [4] to help not only the patient but also the doctor to make a primary diagnosis.

© Springer Nature Switzerland AG 2020
K. Arai and S. Kapoor (Eds.): CVC 2019, AISC 944, pp. 500–510, 2020.
https://doi.org/10.1007/978-3-030-17798-0_40

The aim of this paper is to improve this automatic recognition of speech dysarthric system based on the HMM and HTK [5–7]. To do that, we calculate several variants (parameters) of JITTER and SHIMMER [3] then Combine these parameters with the MFCC's and PNCC's coefficients [1, 2]. Finally, we compared the results obtained in order to obtain the most relevant parameter for the recognition automatic dysarthric speech.

To our knowledge, this work is the first which proposes and applies the NEMOURS database [8] for the improvement of an automatic recognition of the speech dysarthric system with the combination of the two parameters JITTER and SHIMMER. The latter is a promising approach to improve communication between people with speech disorders and normal speakers.

This paper is outlined as follows; Sect. 2 is for JITTER and SHIMMER. Section 3 is for proposed technique.

2   Jitter and Shimmer

2.1   Jitter Variants

We define jitter as a quantification of cycle-to-cycle $F0$ perturbations (small deviations from exact periodicity), however, there is no formal, unequivocal and rigorous definition [10] that allows to develop many jitter variants (Schoentgen and de Guchteneere [11]; Baken and Orlikoff [12]). The computation of Jitter can be done using either the $F0$ contour, or the inversely proportional pitch period $T0 = 1/F0$ contour. Typically, researchers focus on the latter. The possible differences in the quantification of the information in the speech signal using either the $F0$ contour or the $T0$ contour was investigated in Tsanas et al. [13], the authors conclude that neither approach led to improved quantification of vocal severity. Specifically, the jitter variants we used are:

- **The mean absolute difference of F_0 estimates between successive cycles:**

$$Jitter_{F_{0,abs}} = \frac{1}{N}\sum_{i=1}^{N-1}\left|F_{0,i} - F_{0,i+1}\right| \tag{1}$$

Where N is the number of $F0$ computations.

- **$F0$ mean absolute difference of successive cycles divided by the mean $F0$, expressed in percent (%):**

$$Jitter_{F_{0,\%}} = 100 \cdot \frac{\frac{1}{N}\sum_{i=1}^{N-1}\left|F_{0,i} - F_{0,i+1}\right|}{\frac{1}{N}\sum_{i=1}^{N}F_{0,i}} \tag{2}$$

- **Perturbation quotient measures using K cycles (we used K = 5):**

$$Jitter_{F_{0,PQ1,K}} = \frac{\frac{1}{N-K+1}\sum_{i=k_1}^{N-K_2}\left[\frac{1}{K}\sum_{j=i-k_2}^{i+k_2}\left|F_{0,i} - F_{0,i+1}\right|\right]}{\frac{1}{N}\sum_{i=1}^{N}F_{0,i}} \tag{3}$$

- **Perturbation quotient using an autoregressive model**

$$Jitter_{F_0,PQ3,K} = \frac{\frac{1}{N-P}\sum_{i=p+1}^{N}\left[\sum_{j=i-p}^{i} a_j\left(F_{0,j} - \frac{1}{N}\sum_{i=1}^{N} F_{0,i}\right)\right]}{\frac{1}{N}\sum_{i=1}^{N} F_{0,i}} \tag{4}$$

Here, the autoregressive model coefficients are $\{a_j\}_j^P = 1$, their estimation using the Yule-Walker equations is done from the F_0 contour. Following Schoentgen and de Guchteneere's [11] suggestion, we used $p = 5$ coefficients. Instead of quantifying only the average absolute difference between two successive $F0$ estimates, we quantify the absolute (weighted) average difference between the mean $F0$ estimate and the F_0 estimate of the previous p time windows. Thus Eq. (4) is effectively the generalization of Eq. (2).

- **Mean absolute and normalized mean squared perturbations:**

$$Jitter_{F_0,p1} = \frac{1}{N}\sum_{i=1}^{N-1}\left|F_{0,i} - \frac{1}{N}\sum_{j=1}^{N} F_{0,j}\right| \tag{5}$$

$$Jitter_{F_0,p2} = \frac{\frac{1}{N}\sum_{i=1}^{N-1}\left(F_{0,j} - F_{0,i+1}\right)^2}{\left(\frac{1}{N}\sum_{i=1}^{N} F_{0,i}\right)^2} \tag{6}$$

Furthermore, we compute jitter-like measures using the standard deviation of the contour. Additionally, the difference between the mean from the estimation algorithm with the average of age-and gender-matched healthy controls was also calculated. This information was summarized in Fig. 1 [10].

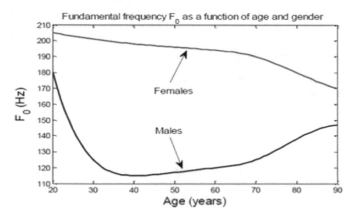

Fig. 1. Life-span changes of the fundamental frequency F_0 as a function of gender for the ages 20–90 years old [5].

In addition, we computed *frequency modulation* (FM) [10]:

$$FM = \frac{\max\left(F_{0,i}\right)_{i=1}^{N} - \min\left(F_{0,i}\right)_{i=1}^{N}}{\max\left(F_{0,i}\right)_{i=1}^{N} + \min\left(F_{0,i}\right)_{i=1}^{N}} \tag{7}$$

Using the nonlinear *Teager-Kaiser energy operator* (TKEO) Ψ [14] the contour was also analysed, and computed the mean, standard deviation and $5th$, $25th$, $75th$ and $95th$ percentile values of $\Psi()$. Where Ψ is defined as:

$$\Psi(X_n) = X_n^2 - X_{n+1} \cdot X_{n-1} \tag{8}$$

The *amplitude modulation* (AM) and the *frequency modulation* (FM) content of an oscillating signal were quantified by TKEO.

2.2   Shimmer Variants

The definition of jitter as the cycle-to-cycle F0 perturbations was done in the preceding section. For the amplitude of the speech signal, Shimmer is the analogue of jitter, rather than F0. The same calculations presented in the preceding section for the jitter variants was used by applying the amplitude A0 contour instead of the F0 contour in Eqs. (1, 2, 3, 4, 5, 6, 7) in order to derive the shimmer variants. In the context of A0 computation, we defined the A0 contour using the maximum amplitude value within each glottal cycle after using DYPSA ([15, 16]) to obtain the glottal cycles. In another ways, the A0 contour can be defined by focusing on signal segments (e.g. 25 ms) instead of within glottal cycles, or using the minimum amplitude values. The difference of the shimmer variants compared to the jitter variants appears by using K = 3, 5, and 11 in Eqs. (3, 5, 4) to conform to traditional amplitude perturbation quotient measures as used by standard reference software programs such as PRAAT. Since this has often been previously used, we computed shimmer in decibels (dB):

$$Shimmer_{dB} = \frac{1}{N}\sum_{i=1}^{N-1} 20 \cdot \left| log_{10}\frac{A_{0,i}}{A_{0,i+1}} \right| \tag{9}$$

3   Proposed Technique

We have calculated and concatenated several variants of JITTER and SHIMMER with two types of speech parameterization which are the MFCC's coefficients and the recently proposed PNCC's coefficients.

We have tested this technique and for the first time on the NEMOURS dysarthric database.

3.1   Results of Speech Recognition with MFCC's and PNCC's Coefficients

We calculated for each 25 ms frame of the dysarthric speech signal (NEMOURS) the MFCC's coefficients and the PNCC's coefficients with their first and secondary derivatives, so in the end we will obtain for each frame a vector of 39 coefficients.

From Table 1 we will notice that word accuracy with MFCC's coefficients are better compared to PNCC's coefficients.

Table 1. Results of speech recognition with coefficients: MFCC's/PNCC's.

Features extraction	Word accuracy (%)	Number of parameters
MFCC_0_D_A	**45.80**	39
PNCC_0_D_A	**44.49**	

3.2   Results of Speech Recognition with Coefficients: MFCC's + JITTER

For each 25 ms frame of the dysarthric speech signal (NEMOURS) we have calculated and concatenated the MFCC's and different JITTER types with their first and secondary derivatives, so in the end we will obtain for each frame a vector of 42 coefficients (see Fig. 2).

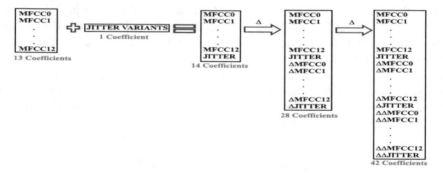

Fig. 2. Concatenation technique of MFCC's coefficients with JITTER.

The results in Table 2 are obtained from a concatenation of the MFCC's coefficients with different types of JITTER, according to Table 2 the best word accuracy is obtained with the MFCC's coefficients + JITTER Relative, and the best word correction, is obtained with MFCC's + JITTER Absolute coefficients, the results are respectively 43.69% and 50.40%.

Table 2. Results of speech recognition with coefficients: MFCC's + JITTER.

Features extraction	Word correction (%)	Word accuracy (%)	Number of parameters
MFCC_0_JITTER Relative (D_A)	48.97	43.69	42
MFCC_0_JITTER Absolute (D_A)	50.40	42.97	
MFCC_0_JITTER Mean absolute difference of successive cycles	47.17	40.72	

3.3 Results of Speech Recognition with Coefficients: MFCC's + SHIMMER

We notice from Table 3 that the best word correction is obtained with the MFCC's + SHIMMER Relative coefficients as well as with the MFCC's + SHIMMER Ampl PQ11 classical Baken coefficients (see Fig. 3), and the best word accuracy is obtained with the coefficients MFCC's + SHIMMER CV, the results are respectively 51.67%, 47.97%.

Table 3. Concatenation technique of MFCC's coefficients with SHIMMER.

Features extraction	Word correction (%)	Word accuracy (%)	Number of parameters
MFCC_0_SHIMMER Relative (D_A)	51.67	45.80	42
MFCC_0_SHIMMER dB (D_A)	49.28	43.77	
MFCC_0_SHIMMER Mean absolute difference of successive cycles (D_A)	50.51	43.38	
MFCC_0_SHIMMER Mean absolute difference of successive cycles expressed in percent (%) (D_A)	50.95	46.14	
MFCC_0_SHIMMER Ampl PQ3 classical Schoentgen (D_A)	51.01	45.51	
MFCC_0_SHIMMER Ampl PQ3 classical Baken (D_A)	50.94	44.64	
MFCC_0_SHIMMER Ampl PQ3 generalised Schoentgen (D_A)	51.02	45.78	
MFCC_0_SHIMMER Ampl PQ5 classical Schoentgen (D_A)	50.36	44.86	
MFCC_0_SHIMMER Ampl PQ5 classical Baken (D_A)	51.09	44.45	
MFCC_0_SHIMMER Ampl PQ5 generalised Schoentgen (D_A)	51.02	45.78	
MFCC_0_SHIMMER Ampl PQ11 classical Schoentgen (D_A)	49.93	45.43	

(*continued*)

Table 3. (*continued*)

Features extraction	Word correction (%)	Word accuracy (%)	Number of parameters
MFCC_0_SHIMMER Ampl PQ11 classical Baken (D_A)	51.67	46.01	42
MFCC_0_SHIMMER Ampl PQ11 generalised Schoentgen (D_A)	51.02	45.78	
MFCC_0_SHIMMER zeroth order perturbation	51.16	45.87	
MFCC_0_SHIMMER CV	51.16	47.97	
MFCC_0_SHIMMER TKEO mean	48.83	43.27	
MFCC_0_SHIMMER TKEO std	49.42	42.91	

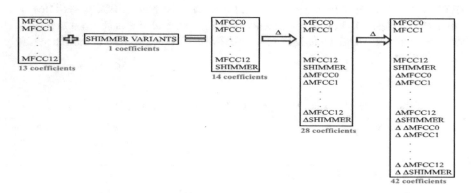

Fig. 3. Concatenation technique of MFCC's coefficients with SHIMMER.

3.4   Results of Speech Recognition with Coefficients: PNCC's + SHIMMER

For each 25 ms frame of the dysarthric speech signal (NEMOURS) we calculated and concatenated the PNCC's coefficients with different variants of SHIMMER then calculated their first and secondary derivatives, we have at the end for each frame a vector of 42 coefficients.

Table 4 shows that the best word accuracy is obtained with the coefficients PNCC's + SHIMMER CV, and the best word correction, is obtained with the coefficients PNCC's + SHIMMER Ampl PQ3 classical Baken, the results are respectively 48.19%, 52.83%.

Table 4. Results of speech recognition with coefficients: PNCC's + SHIMMER.

Features extraction	Word correction (%)	Word accuracy (%)	Number of parameters
PNCC_0_SHIMMER dB (D_A)	51.52	45.43	42
PNCC_0_SHIMMER Relative (D_A)	52.17	46.52	
PNCC_0_SHIMMER Mean absolute difference of successive cycles (D_A)	51.96	46.52	
PNCC_0_SHIMMER Mean absolute difference of successive cycles expressed in percent (%) (D_A)	51.75	45.34	
PNCC_0_SHIMMER Ampl PQ3 classical Schoentgen (D_A)	50.87	44.91	
PNCC_0_SHIMMER Ampl PQ3 classical Baken (D_A)	52.83	46.81	
PNCC_0_SHIMMER Ampl PQ3 generalised Schoentgen (D_A)	50.87	44.64	
PNCC_0_SHIMMER Ampl PQ5 classical Schoentgen (D_A)	51.45	46.23	
PNCC_0_SHIMMER Ampl PQ5 classical Baken (D_A)	52.55	47.45	
PNCC_0_SHIMMER Ampl PQ5 generalised Schoentgen (D_A)	50.87	44.64	
PNCC_0_SHIMMER Ampl PQ11 classical Schoentgen (D_A)	51.74	47.68	
PNCC_0_SHIMMER Ampl PQ11 classical Baken (D_A)	52.40	47.82	
PNCC_0_SHIMMER Ampl PQ11 generalised Schoentgen (D_A)	50.87	44.64	
PNCC_0_SHIMMER zeroth order perturbation	51.24	45.63	
PNCC_0_SHIMMER CV	51.59	48.19	
PNCC_0_SHIMMER TKEO mean	50.29	44.25	
PNCC_0_SHIMMER TKEO std	50.44	44.30	

3.5   Results of Speech Recognition with Coefficients: MFCC/PNCC + JITTER + SHIMMER

For each 25 ms frame of the dysarthric speech signal (NEMOURS) we calculated and concatenated the MFCC's/PNCC's coefficients and different types of JITTER and SHIMMER with their first and secondary derivatives, we have at the end for each frame a vector of 45 coefficients (see Fig. 4).

According to Table 5 that the best word accuracy and word correction is obtained with the PNCC's coefficients + JITTER + SHIMMER RELATIVE the results are respectively 47.59%, 50.37%.

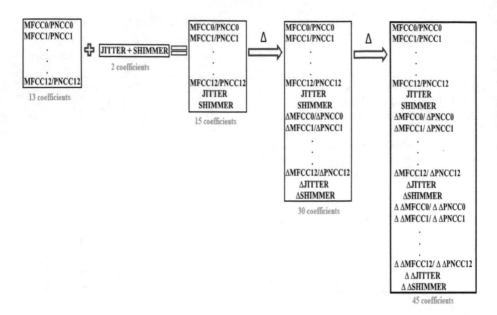

Fig. 4. Results of speech recognition with the coefficients: MFCC's/PNCC's + JITTER + SHIMMER.

Table 5. Results of speech recognition with coefficients: MFCC's / PNCC's + JITTER + SHIMMER.

Features extraction	Word correction (%)	Word accuracy (%)	Number of parameters
MFCC_0_JITTER_SHIMMER Relative (D_A)	50.22	45.22	45
PNCC_0_JITTER_SHIMMER Relative (D_A)	50.37	47.59	

4   Discussions

Table 2 shows the concatenation of the MFCC's with several variants of JITTER, according to this table we will notice that the best parameter of JITTER for automatic recognition of the speech is the jitter Relative.

If we compare Table 1 which represents basic system with Table 2, we will notice that the word accuracy is better with a basic system and more precisely with the MFCC's coefficients.

Table 3 shows the concatenation of the MFCC's coefficients with several SHIMMER variants, according to this table we will notice that the best parameter of SHIMMER in term of word accuracy is obtained with the SHIMMER CV, as well as the best parameter of SHIMMER in term of word correction is obtained with SHIMMER Relative and SHIMMER Ampl PQ11 classical Baken.

Table 4 shows the concatenation of the PNCC's coefficients with several SHIM-MER variants, according to this table we will notice that SHIMMER's best parameter in terms of word accuracy is obtained with the SHIMMER CV, as well as the best parameter of SHIMMER in terms of word correction is obtained with the SHIMMER Ampl PQ3 classical Baken.

If we compare the two Tables 3 and 4, we will conclude that the best results in terms of word accuracy and word correction are obtained with the concatenation of SHIMMER with the PNCC coefficients compared to the MFCC's.

Table 5 shows the concatenation of the MFCC's/PNCC's coefficients with JITTER and SHIMMER, according to this table the best results are obtained with the combination of the PNCC's, JITTER and SHIMMER.

If we compare Tables 2, 3, 4 and 5 with the basic system (Table 1) we will conclude that the best results in terms of word accuracy and word correction are obtained with the concatenation of SHIMMER CV or the SHIMMER Ampl PQ3 classical Baken with the PNCC's.

5 Conclusion

In this paper we have proposed several variants of JITTER and SHIMMER then combined these variants with the MFCC's and PNCC's speech parameterization techniques for the improvement of a system of automatic recognition of the speech of the dysarthria.

According to the results of the recognition and if we compare them with our basic system (Table 1), we will notice that the best results are obtained with the combination of the PNCC's coefficients with the SHIMMER CV or with the SHIMMER Ampl PQ3 classical Baken.

Today our challenge is the improvement of a system of automatic recognition of the dysarthric speech to give a hope to the people having difficulty to speak and to make these people able to communicate with the normal people.

References

1. Kim, C., Stern, R.M.: Power Normalized Cepstral Coefficients (PNCC) for robust speech recognition. IEEE Trans. Audio Speech Lang. Process. **24**, 1315 (2016)
2. Mohammed, A., Mansour, A., Ghulam, M., Mohammed, Z., Mesallam, T.A., Malki, K.H., Mohamed, F., Mekhtiche, M.A., Mohamed, B.: Automatic speech recognition of pathological voice. Indian J. Sci. Technol. **8**, 32 (2015)
3. Tsanas, A.: Accurate telemonitoring of Parkinson's disease symptom severity using nonlinear speech signal processing and statistical machine learning. University of Oxford, June 2012
4. Zaidi, B.F., Selouani, S.A., Boudraa, M., Hamdani, G.: Human/machine interface dialog integrating new information and communication technology for pathological voice. In: IEEE Xplore, Future Technologies Conference (FTC), San Francisco, CA, USA, January 2017

5. Alam, M.J., Kenny, P., Dumouchel, P., O'Shaughnessy, D.: Robust feature extractors for continuous speech recognition. In: IEEE Xplore, European Signal Processing Conference (EUSIPCO), Lisbon, Portugal, November 2014

6. Dua, M., Aggarwal, R.K., Kadyan, V., Dua, S.: Punjabi automatic speech recognition using HTK. Int. J. Comput. Sci. Issues **9**(4), 359 (2012)

7. Young, S., Kershaw, D., Odell, J., Ollason, D., Valtchev, V., Woodland, P.: The HTK Book, version 3.1, pp. 1–277 (2006)

8. Menéndez-Pidal, X., Polikoff, J.B., Peters, S.M., Leonzio, J.E., Bunnell, H.T.: The nemours database of dysarthric speech. J. IEEE (in press)

9. Darley, F.L., Aronson, A.E., Brown, J.R.: Differential diagnostic patterns of dysarthria. J. Speech Lang. Hear. Res. **12**, 246–269 (1969)

10. Titze, I.R.: Principles of Voice Production. National Center for Voice and Speech, Iowa City, USA, 2nd printing (2000)

11. Schoentgen, J., de Guchteneere, R.: Time series analysis of jitter. J. Phon. **23**, 189–201 (1995)

12. Baken, R.J., Orlikoff, R.F.: Clinical Measurement of Speech and Voice, 2nd edn. Singular Thomson Learning, San Diego (2000)

13. Tsanas, A., Little, M.A., McSharry, P.E., Ramig, L.O.: Nonlinear speech analysis algorithms mapped to a standard metric achieve clinically useful quantification of average Parkinson's disease symptom severity. J. R. Soc. Interface **8**, 842–855 (2011)

14. Kaiser, J.: On a simple algorithm to calculate the 'energy' of a signal. In: Proceedings of IEEE International Conference on Acoustics, Speech, and Signal Processing (ICASSP 1990), pp. 381–384, Albuquerque, NM, USA, April 1990

15. Kounoudes, A., Naylor, P.A., Brookes, M.: The DYPSA algorithm for estimation of glottal closure instants in voices speech. In: IEEE International Conference on Acoustics, Speech and Signal Processing, (ICASSP), pp. 349–352, Orlando, FL (2002)

16. Naylor, P.A., Kounoudes, A., Gudnason, J., Brookes, M.: Estimation of glottal closure instants in voices speech using the DYPSA algorithm. IEEE Trans. Audio Speech Lang. Process. **15**, 34–43 (2007)

CapsGAN: Using Dynamic Routing for Generative Adversarial Networks

Raeid Saqur[(⊠)] and Sal Vivona[(⊠)]

Department of Computer Science,
University of Toronto, Toronto, Canada
{raeidsaqur,vivona}@cs.toronto.edu

Abstract. In this paper, we propose a novel technique for generating images in the 3D domain from images with high degree of geometrical transformations. By coalescing two popular concurrent methods that have seen rapid ascension to the machine learning zeitgeist in recent years: GANs (Goodfellow et al.) and Capsule networks (Sabour, Hinton et al.) - we present: **CapsGAN**. We show that CapsGAN performs better than or equal to traditional CNN based GANs in generating images with high geometric transformations using rotated MNIST. In the process, we also show the efficacy of using capsules architecture in the GANs domain. Furthermore, we tackle the Gordian Knot in training GANs - the performance control and training stability by experimenting with using Wasserstein distance (gradient clipping, penalty) and Spectral Normalization. The experimental findings of this paper should propel the application of capsules and GANs in the still exciting and nascent domain of 3D image generation, and plausibly video (frame) generation.

Keywords: Capsule networks · Generative adversarial networks · 3D image generation

1 Introduction

Deep learning has made significant contributions to areas including natural language processing and computer vision. Most accomplishments involving deep learning use supervised discriminative modeling. However, the intractability of modeling probability distributions of data makes deep generative models difficult. Generative adversarial networks (GANs) [4] help alleviate this issue through setting a Nash Equilibrium between a generative neural network model (Generator) and a discriminative neural network (Discriminator). The discriminator is trained to determine whether its input is from a real data distribution or a fake distribution that was generated by the generative network.

Since the advent of GANs, many applications and variants have risen. Most of its applications are inspired by computer vision problems, and involve image generation as well as (source) image to (target) image style transfer. Although GANs have been proven to be successful in the space of computer vision, most use

© Springer Nature Switzerland AG 2020
K. Arai and S. Kapoor (Eds.): CVC 2019, AISC 944, pp. 511–525, 2020.
https://doi.org/10.1007/978-3-030-17798-0_41

Convolutional Neural Networks (CNN) as their discriminative agents. Through convolutions and pooling, CNN are able to abstract high level features (e.g. face, car) from lower level features (e.g. nose, tires). However, CNNs lack the ability to interpret complicated translational or rotational relationships between lower features that make up higher level features [14] - they lack *'translational equivariance'*. CNNs attempt to solve this relational issue through imposing translational invariance while increasing its field of view through max or average pooling. However, these pooling operations are notorious for losing information that can be potentially essential for the discriminator's performance.

Capsules were introduced in [7]. Capsules offers a better model for learning hierarchical relationships of lower features when inferring higher features [14]. These capsules are able to decode aspects such as pose (position, size, orientation), deformation, velocity, albedo, hue, texture etc. Based on the limitations of CNN's, we propose using Capsule Networks as the discriminator network for a simple GAN framework. This will allow the discriminator to abide by inverse-graphics encoding. As a result, this will encourage the generator to generate samples that have more realistic lower to higher feature hierarchies.

2   Related Work

Training GANs architecture has been notoriously difficult [15] - as they suffered from instability due to 'mode collapse'. The original GANs implementation using feed-forward MLP did not perform well generating images in complex datasets like the CIFAR-10, where generated images suffered from being noisy and incomprehensible. An extension to this design using Laplacian pyramid of adversarial networks [3] resulted in higher quality images, but they still suffered from wobbly looking objects due to noise emanating from chaining multiple models. In order to circumvent the issues and make training feasible, a plethora of techniques were proposed in the last few years, most of which extended or altered existing popular architectural designs. Notable and popular ones include 'DCGAN' [13] and 'GRAN' [8] that are based off of CNN and RNN classes of architecture respectively. Following the trend, [12] proposed 'Conditional GANs', which includes an architectural change to GANs where the discriminator also acts as a classifier for class-conditional image generation.

In this paper, we used Radford et al.'s [13] work in the sense that we propose to replace the CNN based architecture of '**DCGAN**' with capsule networks. We perform experiments and validation tests using the MNIST dataset, as it is the dominant test dataset for majority of the papers in this domain, allowing us to juxtapose our results with preceding architectures. As future work, the performance measure can be tested on CIFAR-10,100 datasets (*discussed more in detail in* Sect. 6).

3   CapsGAN

This section details the architecture and algorithm of the baseline DCGAN and proposed CapsGAN used for the rotated MNIST dataset.

3.1   Architecture

For the **baseline**, the DCGAN architecture was used. The following Fig. 1 shows the topological architecture of the baseline GAN architecture used.

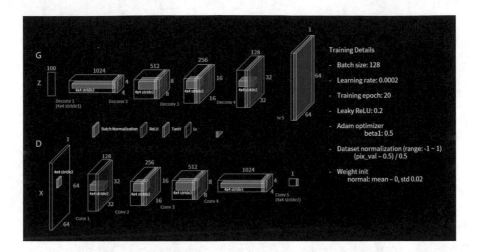

Fig. 1. Topological baseline architecture using DCGAN

For **CapsGAN**, the model follows a vanilla GAN framework, dually training both a generator and discriminator network. The discriminator is represented as a capsule network which uses dynamic routing (please see Fig. 6), with its implementation being a minor variant of the one featured in the original paper [14]. The key difference is the final layer which is one 16 dimensional capsule, while the original architecture featured 10×16 dimensional capsules. This was done, as the discriminator's output should represent the probability of whether the input image was sampled from the real or generated distribution. Thus, this binary information requires one 16-D vector capsule, with its length being correlated to the probability of the input image being sampled from the real data distribution. The dynamic routing process occurs between the primary capsule layer and the output digit capsule. A high-level illustration is shown below in Fig. 2.

The generator involves a 2D transpose convolutional network, which is illustrated in the Fig. 1. The generator network was inspired from the DCGAN framework [13]. It incorporates 5 convolution transpose layers, which are responsible for up-sampling a 128 dimensional vector with each element being independently sampled from a normal distribution of mean 0 and standard deviation 1. Batch Normalization is also applied between each of the deconvolution layers. This is done to regularize the generative network, which would in turn increase training stability [9]. Rectified linear units are used as the non-linearity between all layers with the exception of the final layer. The non-linearity is applied to the output prior to batch normalization. A 'tanh' nonlinearity function is then applied to the final output of the up-sampled image.

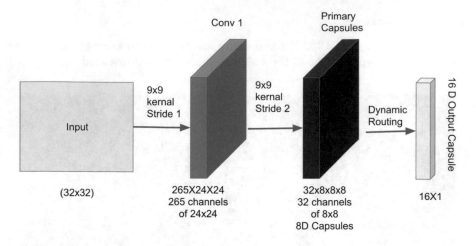

Fig. 2. Capsule discriminator with one output capsule

3.2 Loss Function

While constructing the network's framework, **three types** of loss functions were considered. The first two were binary cross entropy, and mean squared error (MSE). However, during experiments, using these loss functions did not allow the CapsGAN to converge. For the DCGAN baseline, binary cross entropy was used for training. The CapsGAN loss function was chosen to be the margin loss as introduced in the original dynamic routing paper. Using this loss encouraged the model to converge without any mode collapse.

$$v_i = CapsD(x_i) \tag{1}$$

$$Loss = T_i \max(0, m^+ - \|v_i\|^2) + \lambda(1 - T_i) \max(0, \|v_i\| - m^-)^2 + \beta R_{loss} \tag{2}$$

Where T_i is the target value (0 is fake, 1 is real), v_i is the final output capsule vector, and x_i is the input. Through hyper-parameter search, λ, β, m^-, m^+ were set to 0.5, 0.005, 0.1 and 0.9 respectively. The discriminator and generator optimized their weights based on the total loss function over the entire batch of images.

$$D^* = \max_w \{(Loss(CapsD(x_i), T_i^{real} = 1) + \max_w (Loss(CapsD(x_i), T_i^{fake} = 0)\} \tag{3}$$

$$G^* = \max_w \{Loss(CapsD(G(z_i)), T_i^{real} = 1)\} \tag{4}$$

In addition to margin loss, reconstruction loss (R_{loss}) was also added. This was inspired by the original dynamic routing paper which uses fully connected layers to reconstruct image inputs from the output capsule. This was included in the loss function in order to increase the quality of the capsule network classifier.

3.3   Algorithm

The following Algorithm 1 shows the high-level training algorithm for CapsGAN[1].

Algorithm 1. CapsGAN algorithm

1: **arguments**: Generator G_θ, Discriminator $CapsD_\phi$
2: **initialize** The networks and other pertinent hyper-parameters   ▷ number of epochs,
   LR etc.
3: **procedure** PRE-TRAIN DISCRIMINATOR
4:     **for** D-iters steps **do**   ▷ Number of pre-training steps, default $= 1$
5:         Sample minibatch of m noise samples $\mathbf{Z} = \{z^{(1)}, ..., z^{(m)}\}$
6:         Sample minibatch of m samples $\mathbf{X} = \{x^{(1)}, ..., x^{(m)}\}$ from $p_{data}(x)$
7:         Update the discriminator, CapsD, using margin loss.
8: **procedure** ADVERSARIAL TRAINING($G, CapsD$)
9:     **for** number of epochs **do**
10:         **for** number of training steps **do**
11:             Sample minibatch of m noise samples $\mathbf{Z} = \{z^{(1)}, ..., z^{(m)}\}$
12:             Sample minibatch of m samples $\mathbf{X} = \{x^{(1)}, ..., x^{(m)}\}$ from $p_{data}(x)$
13:             Update discriminator by minimizing total loss: ▷ Train the discriminator
14:                 $fake\_loss \leftarrow CapsD(G(Z))$
15:                 $real\_loss \leftarrow CapsD(X)$
16:                 $total\_loss \leftarrow real\_loss - fake\_loss$
17:             Update generator by minimizing fake loss: ▷ Train the generator
18:                 $fake\_loss \leftarrow CapsD(G(Z))$

---

4   Experiments

The model was trained on the MNIST[2] dataset due to its simplicity. This involved tracking the proposed loss function over the number of total trained batches, where each batch contained 32 images with a width and height of 32 pixels. Prior to experimentation, grid hyper parameter search was performed for both the Capsule network as well as the baseline DCGAN model. For more valid comparisons, the number of parameters within the experimental and baseline model was controlled.

4.1   Methods

Both the CapsGAN as well as the baseline were trained until convergence of the loss functions. This typically involved training over 1000 to 2000 batches until model convergence.

   In order to improve training stability, several additions were included. Specifically, these additions included spectral normalization [11], and incorporating Earth-Mover (EM) or Wasserstein distance for loss function optimizations, as

---

[1] https://github.com/raeidsaqur/CapsGAN.
[2] http://yann.lecun.com/exdb/mnist/.

was proposed and outlined by Arjovsky et al. [1]. Experiments were done with both weight-clipping of gradients like the original paper, and also adding a gradient-penalty term as a suggested improvement by Gulrajani [5].

Using grid hyper-parameter search, a single model was selected to represent the baseline and capsule GAN experimental model. After hyper-parameter search, the models were trained until 2000 batch iterations. However, both models appeared to converge at around 1000 iterations. Thus the results in the experiments exhibit the first 1000 iterations of both the baseline and the CapsGAN. Because it is not possible to apply inception scores to the MNIST dataset, the models were compared qualitatively based on their generated images as shown in Tables 1 and 2. Table 1 represents both models being trained on images that have no prior geometric transformation. The models were also trained in two additional scenarios where geometric rotations were applied. This involved rotating the images based on its center axis after sampling an angle from a uniform distribution. The angular bounds of the uniform distribution represented the different training scenarios. The two geometrical rotations were uniformly sampled to have a maximum angle of 15 and 45°. The performance of both the models are then qualitatively compared based on their generative images.

5   Results and Evaluation

5.1   MNIST

The experiments include observations for three different training scenarios. The first scenario is not rotated, while the other two scenarios are randomly rotated up to 15 and 45° from its true orientation. This was done to help measure both of the model's sensitivity to geometric transformations. In the first experiment involves observing the loss function over the number of iteration batches for the CapsGAN. This plot will show the convergence properties of the loss function proposed in Sect. 3.2. The second part of the experiment involves qualitative comparisons between the reconstructions of the CapsGAN and DCGAN baseline. These reconstructions are visualized over numerous stages of training to show the qualitative convergence of the reconstructed images.

Loss Convergence of CapsGAN. The following two figures (Figs. 3 and 4) show the loss of discriminator and generator respectively during the training phase:

As shown in Fig. 3, the Capsule discriminator rapidly converges at around 100 iterations, and continues to fluctuate heavily. Interestingly, the geometric rotations do not appear to affect the capsule networks' convergence. However, the scenario with up to 45° of rotation involves the largest local variations and fluctuation peaks. Nonetheless, despite these local variations, the global trend seems to continue to stabilize while avoiding divergence. For the generator's loss function, a similar trend is exhibited as shown in Fig. 4. The overarching trend is

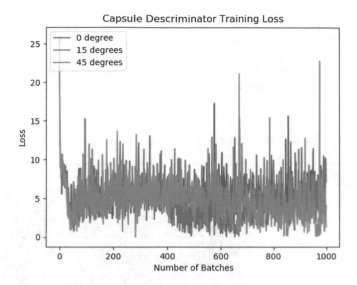

Fig. 3. Loss of capsule based discriminator over the number of trained batches.

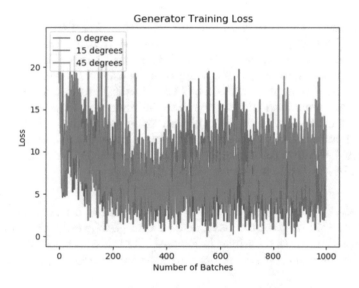

Fig. 4. Loss of capsule based generator over the number of trained batches.

downward and decreasing, indicating improvement in the generator's loss function during training. However, the convergence also exhibits highly variant local fluctuations. In future work, there should be more focus on reducing the variance observed in this study. To help stabilize the convergence, 'Spectral Normalization' [11] and *Wasserstein distance* [1] was implemented separately as well as in combination. Unfortunately, these implementations **did not** appear to improve the results. This is perhaps due to a lack of hyper-parameter tuning with the inclusion of these implementations. Thus for future work, extensive investment must be done for hyper-parameter tuning. This aspect of the experiment was partially hindered from the lack of available local and remote computational resources.

Table 1. CAPS-DCGAN and baseline DCGAN training on non-rotated images.

Batch iterations	500	1000	2000
CAPS-DCGAN (0°)			
DCGAN (0°)			

Reconstruction of Non-transformed MNIST. The first training scenario involves MNIST digits that are not rotated or transformed. Table 1 shows the image reconstructions from the baseline as well as the CapsGAN models. Through a qualitative comparison, it appears the CapsGAN generator converges to higher quality images than the DCGAN. This is shown at batch 2000 where the CapsGAN generated digits appear to be finer and more defined in comparison to the baseline. In order to further prove this claim, more quantitative measures should be done in future experiments. However qualitatively, it appears that CapsGAN converges to higher quality images while training on the same number of batches compare to the DCGAN baseline.

CAPS-DCGAN and Baseline DCGAN Training on Rotated Images. As shown in Table 2, both the baseline and CapsGAN were trained on geometrically rotated data, as described in Sect. 4. The table shows generated images from the baseline and CAPSGAN architecture for images rotated up to 15° as well as for 45°.

For 15° rotations, both models appear to show similar results at 300 and 500 iterations. However, at 1000 iterations, the CAPS-DCGAN exhibits higher quality generated digits compared to the baseline. This is indicated by more inappropriate stroke artifacts in the baseline images than the CapsGAN.

CapsGAN's advantage is more explicit with higher degree of geometric transformations applied to the image - as is illustrated by the experiment results with the images rotated 45°. This is specifically shown at 1000 iterations where the baseline's reconstructions are shown to be unstable, while remaining stable for the CapsGAN. These results were aligned with expectations as Capsule networks

Table 2. CAPS-DCGAN and baseline DCGAN training on rotated images that are rotated at angle that is sampled from a uniform distribution between the positive and negative value of the angle presented. Results show that for small angles, both models preform similar in their reconstructions. Larger angles result in divergence in the baseline at 1000 iterations.

Batch iterations	300	500	1000
CAPS-DCGAN (15°)			
DCGAN (15°)			
CAPS-DCGAN (45°)			
DCGAN (45°)			

are intended to be more robust to geometric transformations. This geometric adversity is due to the capsule's inverse graphics properties as discussed in the original paper [14].

5.2   SmallNORB

The smallNORB [10] is a staple dataset for testing efficacy of generative models in the 3D domain. It is organized in gray-level stereo images of 5 classes of toys (please see Table 3). Also refer to Table 4 for additional generated Small-NORB cars at different training iterations. The multivariate geometric transformations applied to each of the images make it a quintessential dataset to benchmark. Each category is pictured at 9 elevations, 6 lighting conditions (notice the shadows across the samples), and 18 different azimuths. Training and test sets contain 24,300 stereo pairs of 96×96 images.

Baseline DCGAN Training on SmallNORB. The original images were downsampled to 64×64 pixels. The baseline architecture used was similar to the one in Fig. 1. Batch size of 8 and 64 were used with batch normalizations. For the generator, random normal noise vector of size 100 was used as generator's input data. No pre-training was performed for the generator, and was left for future experiments. Lipschitz continuity was enforced to improve training stability by using gradient clipping after each gradient update [1].

Table 3. Small-NORB dataset real dataset samples down-sampled to 64 × 64.

Categories	Batch samples
Animals (id = 0)	
Humans (id = 1)	
Airplanes (id = 2)	
Trucks (id = 3)	
Cars (id = 4)	

Reconstruction Using Baseline DCGAN. The following Table 4 exemplifies the results obtained using our baseline architecture. Here we showcase only the '*cars*' category for brevity, as similar trends in results were observed for all the other categories. Batches of size 8 from the real data and generated data are juxtaposed at random training iteration intervals.

As can be seen, the generic trend shows convergence and the generated images gradually look sharper. However, due to the high degree of geometric transformations (lighting, elevation, azimuth, rotation), the baseline generator struggles to pin-point, i.e. converge on specific features.

This baseline experiment elucidates the weakness of traditional convolutional architecture based GANs to capture geometric transformations in images, and consequently in video frame generations.

Proposed Architecture Using Capsules. Our proposed architectural change for the discriminator is shown in Fig. 5 below. It is closely aligned with the 'capsule routing matrix' structure as outlined in [6].

The model has a initial 5 × 5 convolutional layer with 32 channels (A), with a stride of 2 and ReLU non-linearity. All other layers are capsules. There are 32 primary capsules (B) and 4 × 4 pose of each of these capsules is a learned linear

Table 4. Reconstruction of 'cars' category toy images using baseline DCGAN architecture. The left-most column indicates the type of data (real or generated) and the following number indicates the training iteration.

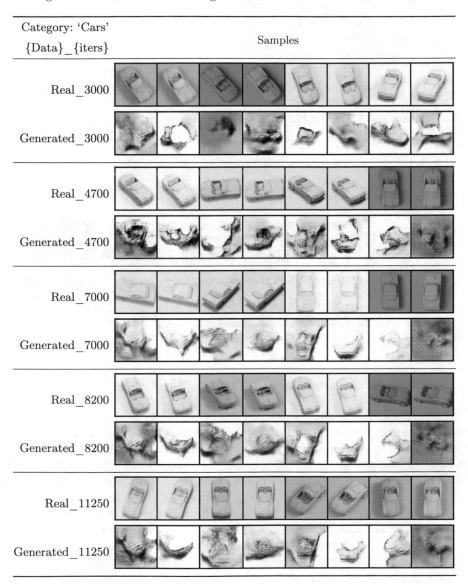

Category: 'Cars' {Data}_{iters}	Samples
Real_3000	
Generated_3000	
Real_4700	
Generated_4700	
Real_7000	
Generated_7000	
Real_8200	
Generated_8200	
Real_11250	
Generated_11250	

transformation from the lower layers. The primary capsules are followed by two 3×3 convolutional capsule layers (K), each with 32 capsule types ($C = D = 32$) with strides of 2 and one, respectively. The last layer of convolutional capsules is connected to the final capsule layer which has one capsule per output class.

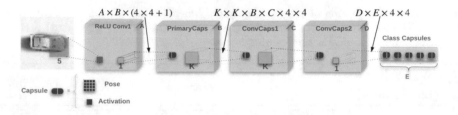

$A \times B \times (4 \times 4 + 1)$ $K \times K \times B \times C \times 4 \times 4$ $D \times E \times 4 \times 4$

Fig. 5. Proposed CapsGAN discriminator architecture for smallNORB dataset.

Results Using CapsGAN. Table 5 shows reconstructed samples using the CapsGAN architecture.

Table 5. Reconstruction of '*cars*' and '*planes*' categorical toy images using CapsGAN architecture.

Category	Reconstructed samples
Cars	
Planes	

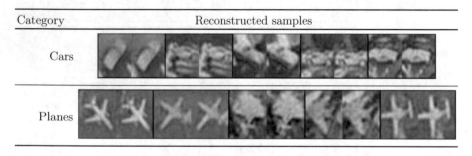

6   Future Work and Limitations

Experiments and results from this paper show CapsGANs to be more robust to geometric transformations than traditional deep CNN based approaches. However, the simplistic image dataset (MNIST) used is a limitation, and more experiments using complex datasets are needed. The main comparative advantage of CapsGAN lies in its superior ability in capture geometric transformations, and thus results using smallNORB (or similar datasets) would certainly reinforce the notion.

Training stability during loss function optimization requires more work as loss convergence exhibited high variance. There is heavy research on GAN training stability, and newer, improved techniques can be explored to alleviate this issue. This may include doing more intensive hyper-parameter searches in order to find a more stable model. This can be encouraged through moving away from grid search methods and more towards Bayesian hyper-parameter search methods [2].

In this paper, we have established groundwork by illustrating the weakness of CNN architecture based GANs in generating images in the 3D domain. As an immediate future work, we will try and replicate the experiment using our proposed capsule based architecture. A more optimistic goal is to apply capsule networks to 3D generative modeling that is similar to the framework of the 3D-GAN [16].

Additionally, it is expected that future work would include more quantitative measures for comparing between the CapsGAN and a diverse set of baselines. It is understandable that although qualitative analysis of the generative 3D construction is intuitive, it does not capture the degree of the difference between the two model's performances.

7 Conclusion

In this paper, we have shown the efficacy and validity of using capsule networks in designing GAN architectures. We have also shown the promise of such GANs' superior capability in generating images from geometrically transformed dataset. Finally, we explored the Gordian Knot in GANs: the training stability - by performing ablation studies and experimentation with some of the more popular techniques proposed in recent machine learning academic literature.

We hope that our work in this paper will ignite interest and further work in the related domains of 3D image generation and the exciting and relatively unexplored domain of video generation.

Appendices

A   Capsule Algorithms and Formulae

A.1   Squashing Function

$$v_j = \frac{||s_j||^2}{(1 + ||s_j||^2)} \frac{s_j}{||s_j||} \tag{5}$$

Here, v_j is the vector output of capsule j and s_j is its total output.

A.2   Margin Loss

The following equation shows the margin-loss used for digit existence:

$$L_k = T_k \ max(0, m^+ - ||v_k||^2 + \lambda(1 - T_k) \ max(0, ||v_k|| - m^-)^2 \tag{6}$$

- Softmax Function

$$c_{ij} = \frac{exp(b_{ij})}{\sum_k exp(b_{ik})} \tag{7}$$

A.3   Routing Algorithm

Procedure 1 Routing algorithm.

1: **procedure** ROUTING($\hat{\mathbf{u}}_{j|i}, r, l$)
2:    for all capsule i in layer l and capsule j in layer $(l+1)$: $b_{ij} \leftarrow 0$.
3:    **for** r iterations **do**
4:        for all capsule i in layer l: $\mathbf{c}_i \leftarrow \texttt{softmax}(\mathbf{b}_i)$ ▷ $\texttt{softmax}$ computes Eq. 3
5:        for all capsule j in layer $(l+1)$: $\mathbf{s}_j \leftarrow \sum_i c_{ij}\hat{\mathbf{u}}_{j|i}$
6:        for all capsule j in layer $(l+1)$: $\mathbf{v}_j \leftarrow \texttt{squash}(\mathbf{s}_j)$ ▷ $\texttt{squash}$ computes Eq. 1
7:        for all capsule i in layer l and capsule j in layer $(l+1)$: $b_{ij} \leftarrow b_{ij} + \hat{\mathbf{u}}_{j|i}.\mathbf{v}_j$
    return $\mathbf{v}_j$

Fig. 6. Routing algorithm used in Capsnet as in [14].

B   Execution

B.1   Public Repository Details

All code pertaining to this research paper has been hosted on Github at author's page: (https://raeidsaqur.github.io/CapsGAN/). All frameworks used and code execution information is available in README.md.

B.2   GPUs Used

For execution, we used one multi-GPU rig (with 2 NVIDIA Titan Xps) and Google Cloud Computing instance with P1000 GPU.

References

1. Arjovsky, M., Chintala, S., Bottou, L.: Wasserstein GAN. arXiv preprint arXiv:1701.07875 (2017)
2. Bergstra, J.S., Bardenet, R., Bengio, Y., Kégl, B.: Algorithms for hyper-parameter optimization. In: Advances in Neural Information Processing Systems, pp. 2546–2554 (2011)
3. Denton, E.L., Chintala, S., Fergus, R., et al.: Deep generative image models using a Laplacian pyramid of adversarial networks. In: Advances in Neural Information Processing Systems, pp. 1486–1494 (2015)
4. Goodfellow, I., Pouget-Abadie, J., Mirza, M., Xu, B., Warde-Farley, D., Ozair, S., Courville, A., Bengio, Y.: Generative adversarial nets. In: Advances in Neural Information Processing Systems, pp. 2672–2680 (2014)
5. Gulrajani, I., Ahmed, F., Arjovsky, M., Dumoulin, V., Courville, A.C.: Improved training of wasserstein GANs. In: Advances in Neural Information Processing Systems, pp. 5769–5779 (2017)
6. Hinton, G., Frosst, N., Sabour, S.: Matrix capsules with EM routing (2018)
7. Hinton, G.E., Krizhevsky, A., Wang, S.D.: Transforming auto-encoders. In: International Conference on Artificial Neural Networks, pp. 44–51. Springer (2011)

8. Im, D.J., Kim, C.D., Jiang, H., Memisevic, R.: Generating images with recurrent adversarial networks. arXiv preprint arXiv:1602.05110 (2016)
9. Ioffe, S., Szegedy, C.: Batch normalization: accelerating deep network training by reducing internal covariate shift. arXiv preprint arXiv:1502.03167 (2015)
10. LeCun, Y., Huang, J.: The small NORB dataset, v1.0
11. Miyato, T., Kataoka, T., Koyama, M., Yoshida, Y.: Spectral normalization for generative adversarial networks. arXiv preprint arXiv:1802.05957 (2018)
12. Odena, A., Olah, C., Shlens, J.: Conditional image synthesis with auxiliary classifier GANs. arXiv preprint arXiv:1610.09585 (2016)
13. Radford, A., Metz, L., Chintala, S.: Unsupervised representation learning with deep convolutional generative adversarial networks. arXiv preprint arXiv:1511.06434 (2015)
14. Sabour, S., Frosst, N., Hinton, G.E.: Dynamic routing between capsules. In: Advances in Neural Information Processing Systems, pp. 3859–3869 (2017)
15. Salimans, T., Goodfellow, I., Zaremba, W., Cheung, V., Radford, A., Chen, X.: Improved techniques for training GANs. In: Advances in Neural Information Processing Systems, pp. 2234–2242 (2016)
16. Wu, J., Zhang, C., Xue, T., Freeman, B., Tenenbaum, J.: Learning a probabilistic latent space of object shapes via 3D generative-adversarial modeling. In: Advances in Neural Information Processing Systems, pp. 82–90 (2016)

Constrainted Loss Function
for Classification Problems

Haozhi Huang and Yanyan Liang[✉]

Faculty of Information Technology,
Macau University of Science and Technology, Taipa, Macau SAR, China
stdio_five@163.com, yyliang@must.edu.mo

Abstract. Producing good representative features is the key to perform high accuracy in conventional methods of computer vision in many
tasks. Our work is to show that whether good features are still critical in
deep learning models. We bring two loss functions so that one works well
in the classification problems and the other achieve good performance
in verification problems, together to see whether they will improve the
performance on classification problems. In literature, loss functions for
classification problems and verification problems are working independently within the deep learning domain, losses of classification emphasis
the discrimininative power in distinguishing data of different classes while
the verification losses focus on establishing invariant mapping within an
embedding via metric learning. In this paper, the major work is to look
for a better balancing for loss functions of different kinds. The verification loss together with the classification loss requires a novel scheme in
achieving better tradeoff while there are conflicts exited between losses to
be involved. Experiments are designed based on the well-known models
of ResNet and VGG-16 evaluating on CIFAR-10 and CIFAR-100.

Keywords: Computer vision · Feature representation · Classification ·
Deep learning

1 Introduction

For the great success of achieving convolutional neural networks (CNNs) on
computer vision community, studies on deep metric learning of formulating new
supervisions as losses to achieve better representative features have been a recent
surge of interest. In deep learning, classification problems or verification problems
are governed by their own corresponding loss sharing the same understructure
of a given CNN model.

The classification loss, usually being considered as the cross entropy that
forms in an multinomial distribution, which focus on separating data of different classes. Losses for verification usually apply metric learning which put their
effort on performing invariant in representing the data of the same class. Before
[1] was proposed, people are using the intermediate bottleneck layer for dimensionality reducing and data visualizing which are considered as the keyinsight in

© Springer Nature Switzerland AG 2020
K. Arai and S. Kapoor (Eds.): CVC 2019, AISC 944, pp. 526–535, 2020.
https://doi.org/10.1007/978-3-030-17798-0_42

addressing the invariance problems [2]. The triplet loss which directly optimize the mapping from input data to the Euclidean space itself that brings a new aspect in solving the verification problems in a deep learning way. Recently a lot of Euclidean based loss functions are proposed to achieve better representative features, one of them known as "center loss" [3] which provides us a fundamental work for our paper—jointly applying both the losses for classification and verification together in training a deep convolutional neural network which allows us to look back of the classification problems by providing compact representative deep features at the same time.

When studying the loss functions individually, we find that the two losses to be joined are actually in conflicts.

1. Without the classification loss (softmax loss) the Euclidean based loss (center loss) focus on invariance and it will lead to a collapsion of all the embedded features fall into a very small region. In that time features of different classes are no longer distinguishable from one class to the other.
2. It cannot provide a compact embedding for features to lie on but the one prone to be separable by applying classification loss only. We find that even through a fine-tuned compact embedding introducing as a pretrained model can not prevent the features from expanding. Finally, the loss will converge to the embedding space of separable only.

And these conflicts can be dated back to 2017 for the tradeoff unknown between invariance and discriminability indicated by [4].

Moreover, there are shortcomings exists in the invariant loss modeling with the cluster centers.

1. Initializing centers of different clusters in random will not always converge to the position just right or closed to the underlying center of the data cluster.
2. Once one of the center is converged to a wrong position, the corresponding cluster may shift towards the incorrect center because of the punishment from the loss itself.

To avoid these issues, on one hand, going deeper in the joint equation formed the two different losses a linear combination. The balancing factor between the loss of classification and verification denoted by a single scalar λ only is far from sufficient. To avoid the conflicts and to look for a better tradeoff, we suggest that in calculating the invariance loss, the performance of the other loss in current iteration should be taken into account along with the existing λ.

On the other hand, we proposes a new Euclidean loss that takes the squared subtraction of every two features belonging to the same class rather than directly penalize the squared distance between features and its center which is waiting to be converged. The idea is motivated by the k-mean clustering of which objective to be modeled by the mean μ from the points within a particular cluster is equivalent to minimizing the pairwise squared deviations of points in the same cluster.

We find that in applying the new loss to well-known architectures, such as VGG-16 [5] and ResNet-50 [6], to solve the classification problems on CIFAR-10 and CIFAR-100, a switching scheme of replacing the original classification loss with the new joint loss proposed in this paper by turns do improve the testing accuracy.

Finally, the main contribution of this paper can be concluded in:

1. We proposed a new mechanism to join different losses of classification as well as verification together. By taking the status of the other loss into account during training will help the entire loss converge to a better solution, especially when conflicts existed between the joint losses.
2. Proposed a new center loss that relies on the data itself rather than the hyperparameter center randomly selected at the very beginning. The new loss has the same underlying interpretation with the idea of center loss.
3. A switching training is proposed for training the model of classification along with a compact embedding space of good representative features.

2   Related Work

The literature on solving classification problems via deep learning models is highly related to the modification of the activation functions. People put their efforts on adding more non-linearity to the network achieving better accuracy. After great succeed has been made by AlexNet [7] in the year of 2012, the study of *Rectified Linear Units* (ReLUs) [8]—the non-linear activation function for deep learning model has seen a recent surge of interest [9–11]. Some are enabling the negative gradient for those units which are not active [9,10]. Some are trying to strengthen the active units by introducing the exponential equations [11].

But few of them think of another objective function with constraints about good representations.

Cases are similar in the community of developing good representative features with deep models. People developing constraints for good representative features addressing the face recognition problems and human re-identification problems [1,3,12,13]. In most of their losses, in order to perform invariance of variety data of the same identity, loss of Euclidean distance are wildly used. But for extra-class distinctions are just modeled by the *minimum margin*, say the two closest data of different clusters should not have the distances smaller than the minimum margin set previously.

3   Problem Formulation

Suppose we are given a vector $x_0 \in \mathbb{R}^d$ by a deep CNN model; the joint loss will be computed in the form

$$\mathcal{L} = -\log \frac{e^{W_{y_0}^T \mathbf{x}_0 + \mathbf{b}_{y_0}}}{\sum_{i=0}^{n} e^{W_{y_i}^T \mathbf{x}_0 + \mathbf{b}_{y_i}}} + \frac{\lambda}{2} \|\mathbf{x}_0 - \mathbf{c}_{y_0}\|_2^2 \qquad (1)$$

according to center loss [3], where y_i, $i = 0, \ldots, n$ is the classification label and $\mathbf{c}_{y_0}$ denotes the hyper-parameter defines the y_0th class center. Considering the two losses represented by f and g, the loss become

$$\mathcal{L} = f(x_0) + \lambda g(x_0)$$

where

$$f(x_0) = -\log \frac{e^{W_{y_0}^T \mathbf{x}_0 + \mathbf{b}_{y_0}}}{\sum_{i=0}^{n} e^{W_{y_i}^T \mathbf{x}_0 + \mathbf{b}_{y_i}}}$$

and

$$g(x_0) = \frac{1}{2} \|\mathbf{x}_0 - \mathbf{c}_{y_0}\|_2^2.$$

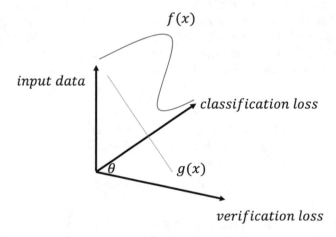

Fig. 1. Relationship between classification loss and representation loss reflecting by the angle θ.

To take the idea of joint supervision further by modeling the relationship between f and g. One intuitive way is to directly take inner-production between the two functions within the term of constraint. Consider the classification loss and the representation loss two different axes within the system, relationship between the loss can be defined by the angle between the axes (Fig. 1). Assumes that the size of mini-batch m is knowing to us, then the constraints of inner-production has the form of

$$f \odot g = \|f\|\|g\|\cos\theta$$

with

$$\|f\| = \mathcal{L}_S = \sum_{i=0}^{m} f(x_i)$$

and

$$\|g\| = \mathcal{L}_C = \sum_{i=0}^{m} g(x_i).$$

Finally we are going to have

$$f \odot g = \Gamma \mathcal{L}_S \mathcal{L}_C$$

where $\{\Gamma | 0 \leq \Gamma \leq 1\}$ reflecting the value of $cos\theta$. The new constraint can be also derived from the AM-GM inequality in an completely different aspect describing in the next subsection.

3.1   Mathematical Lower Bound

In mathematics, there is a property called inequality of arithmetic and geometric means, or more briefly the AM-GM inequality, stipulates that the arithmetic mean of a list of non-negative real numbers is greater than or equal to the geometric mean of the same list; furthermore, the lower bound is defined by the equality that holds if and only if every number in the list is the same. The simplest non-trivial case of AM-GM inequality is

$$\frac{a+b}{2} \geq \sqrt{ab}, \tag{2}$$

when given $a \geq 0$ and $b \geq 0$ with equality holds if and only if $a = b$.

In the fundamental joint loss function (1), given $\mathcal{L}_S \geq 0$, $\mathcal{L}_C \geq 0$ and $\lambda \geq 0$, we will have the formula similar to (2), which

$$\frac{\mathcal{L}_S + \lambda \mathcal{L}_C}{2} \geq \sqrt{\mathcal{L}_S \lambda \mathcal{L}_C}$$

with the lower bound $2\mathcal{L}_S$ that holds if and only if $\lambda = \frac{\mathcal{L}_S}{\mathcal{L}_C}$.

Since the truth values of $\mathcal{L}_S$ and $\mathcal{L}_C$, which requires number of iterations to converge, are unknown, we need to approximate the value of λ by introducing an other hyperparameter Γ with the following equation:

$$\lambda = \lim_{\Gamma \to \frac{1}{\mathcal{L}_C}} \Gamma \mathcal{L}_S,$$

and finally, we have the major loss in the following form that much approaching to its mathematics lower bound when given $\{\Gamma | 0 \leq \Gamma \leq 1, \Gamma \approx \frac{1}{\mathcal{L}_C}\}$,

$$\mathcal{L} = \mathcal{L}_S + \Gamma \mathcal{L}_S \mathcal{L}_C. \tag{3}$$

3.2   Solution Space

Suppose we have $\mathcal{L}_S$ already converged. Then the following relationship is satisfied by introducing ϵ a very small number

$$0 \leq \mathcal{L}_S \leq \epsilon \tag{4}$$

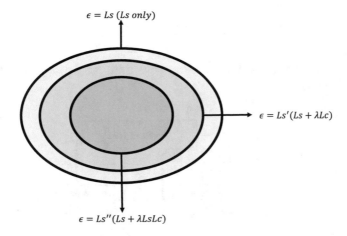

Fig. 2. Different constraints performed in the solution space.

The fundamental loss which can be interpreted as an optimization problem written in

$$minimize \ f(X) = \mathcal{L}_S \tag{5}$$
$$subject \ to \ \min g(X) = \mathcal{L}_C \geq 0$$

where the solutions of (5) is the subset of (4).

By comparing our loss with the fundamental one, we find that they have the relationship formed in

$$2\sqrt{\mathcal{L}_S \lambda \mathcal{L}_C} \leq \mathcal{L}_S + \Gamma \mathcal{L}_S \mathcal{L}_C \leq \mathcal{L}_S + \lambda \mathcal{L}_C$$

providing the solutions much closer to the objective of both representative and discriminative. Solution space that illustrate the relation more intuitive is presented in Fig. 2.

3.3 Hierarchical Loss Layer and Gradients

By rewritten the proposed loss (3) the following form

$$\mathcal{L} = \mathcal{L}_S + \Gamma \mathcal{L}_S \mathcal{L}_C$$
$$= \mathcal{L}_S (1 + \Gamma \mathcal{L}_C).$$

The gradient then can be computed via the chain rule

$$\partial \mathcal{L} = \partial \mathcal{L}_S (1 + \Gamma \mathcal{L}_C) + \partial \mathcal{L}_C \Gamma \mathcal{L}_S$$

Finally, the gradients can be computed individually

$$\frac{\partial \mathcal{L}_S}{\partial x}(1 + \Gamma \mathcal{L}_C) = \frac{\partial \mathcal{L}_S}{\partial W}\frac{\partial W}{\partial x}(1 + \Gamma \mathcal{L}_C)$$

and

$$\frac{\partial \mathcal{L}_C}{\partial x} \mathcal{L}_S = \left(\frac{\partial \mathcal{L}_C}{\partial x_i} + \frac{\partial \mathcal{L}_C}{\partial x_j}\right) \mathcal{L}_S$$

where they yield the hierarchical architecture that can be implemented to the CNN models (Fig. 3).

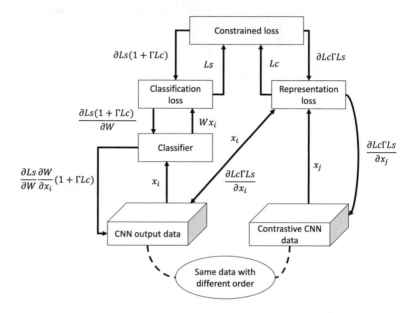

Fig. 3. Hierarchical architecture implementing Eq. (3).

4   Clique Loss vs Center Loss

In k-means clustering, given a set of data $(x_1, \ldots, x_n)$ which is going to partition into k $(k \leq n)$ sets $\{S_1, \ldots, S_k\}$, has the objective in the form of

$$\arg\min_S \sum_{i=1}^{k} \sum_{x \in S_i} \|x - \mu_i\|_2^2 = \arg\min_S \sum_{i=1}^{k} |S_i| \mathbf{Var}(S_i) \qquad (6)$$

where μ denotes the mean of data points in cluster S_i. This Eq. (6) is provided to be equivalent to minimizing the pairwise squared deviations of features in the same order

$$\arg\min_S \sum_{i=1}^{k} \frac{1}{2|S_i|} \sum_{x_i, x_j \in S_i} \|x_i - x_j\|_2^2 \qquad (7)$$

following from the law of total variance. With loss function formed in

$$\mathcal{L}_C = \sum_{i=0}^{m} \frac{1}{2} \|\mathbf{x}_i - \mathbf{c}_{y_i}\|_2^2. \qquad (8)$$

By considering $\mathbf{c}_{y_i}$ equivalent to the mean point μ in Eq. (6), we will have the following form of new loss function which is equivalent to (8).

$$\mathcal{L}_C = \frac{1}{2} \sum_{i,j}^{m} \frac{\|\mathbf{x}_i - \mathbf{x}_j\|_2^2}{2|S_k|}$$

$$= \frac{1}{2} \sum_{i,j}^{m} \frac{\|\mathbf{x}_i - \mathbf{x}_j\|_2^2}{\sum_{i,j}^{m} \delta(y_i = y_j = k)}, \quad \mathbf{x}_i, \mathbf{x}_j \in S_k. \tag{9}$$

where $\delta(\cdot)$ is the condition function defined in

$$\delta(condition) = \begin{cases} 1 & if\ condition\ is\ satisfied \\ 0 & otherwise. \end{cases} \tag{10}$$

Table 1. Classification accuracy of all ResNet models proposed in [6] on the CIFAR-10 test set with red highlight indicates the best of the single loss of classification and the blue highlight represents the best of our loss.

ResNet models on CIFAR-10					
	ResNet-20	ResNet-32	ResNet-44	ResNet-56	ResNet-110
Baseline	91.25%	92.49%	92.83%	93.03%	93.57%
Iteration: 1 ($\mathcal{L}_S$)	91.42%	92.43%	92.59%	92.74%	93.37%
Iteration: 2 ($\mathcal{L}_S + \Gamma\mathcal{L}_S\mathcal{L}_C$)	91.50%	92.57%	93.02%	93.63%	94.14%
Iteration: 3 ($\mathcal{L}_S$)	91.95%	92.77%	93.15%	93.94%	93.98%
Iteration: 4 ($\mathcal{L}_S + \Gamma\mathcal{L}_S\mathcal{L}_C$)	92.16%	92.91%	92.97%	94.27%	93.86%
Iteration: 5 ($\mathcal{L}_S$)	92.25%	92.81%	93.23%	94.18%	94.11%
Iteration: 6 ($\mathcal{L}_S + \Gamma\mathcal{L}_S\mathcal{L}_C$)	92.20%	92.87%	93.22%	94.11%	94.19%
Iteration: 7 ($\mathcal{L}_S$)	92.24%	92.93%	93.40%	93.93%	94.19%
Iteration: 8 ($\mathcal{L}_S + \Gamma\mathcal{L}_S\mathcal{L}_C$)	92.43%	92.75%	93.12%	94.06%	94.14%
Iteration: 9 ($\mathcal{L}_S$)	92.67%	92.81%	93.43%	93.91%	94.14%
Iteration: 10 ($\mathcal{L}_S + \Gamma\mathcal{L}_S\mathcal{L}_C$)	92.36%	92.78%	92.95%	93.78%	93.64%

In contrast to the original center loss (Eq. (8)), the new loss (Eq. (9)) shrinks the cluster formed in clique by minimizing all its edges. This will avoid the inaccuracy from the convergence of hyper-parameter $\mathbf{c}_{y_i}$ which is initialized randomly.

5   Experiment

We studied the impact of the proposed loss on the CIFAR-10 and CIFAR-100 datasets that evaluated the classification performance by comparing the results from both our proposed loss and the conventional softmax loss. Experiments

Table 2. Classification accuracy of ResNet-110 as well as VGG-16 on both the CIFAR-10 and CIFAR-100 test set with the same highlight indicators of Table 1.

Iteration	CIFAR-100		CIFAR-10	
	ResNet-110	VGG-16	ResNet-110	VGG-16
1 ($\mathcal{L}_S$)	70.35%	56.06%	93.37%	87.47%
2 ($\mathcal{L}_S + \Gamma\mathcal{L}_S\mathcal{L}_C$)	70.94%	57.03%	94.14%	88.24%
3 ($\mathcal{L}_S$)	72.09%	56.62%	93.98%	87.98%
4 ($\mathcal{L}_S + \Gamma\mathcal{L}_S\mathcal{L}_C$)	71.48%	57.04%	93.86%	88.29%
5 ($\mathcal{L}_S$)	71.80%	56.23%	94.11%	88.16%
6 ($\mathcal{L}_S + \Gamma\mathcal{L}_S\mathcal{L}_C$)	71.90%	56.67%	94.19%	88.65%
7 ($\mathcal{L}_S$)	71.53%	56.41%	94.19%	88.07%
8 ($\mathcal{L}_S + \Gamma\mathcal{L}_S\mathcal{L}_C$)	72.27%	56.54%	94.14%	88.32%
9 ($\mathcal{L}_S$)	71.35%	55.85%	94.14%	88.09%
10 ($\mathcal{L}_S + \Gamma\mathcal{L}_S\mathcal{L}_C$)	72.03%	55.88%	93.64%	88.28%

are presented on loss switching training scheme. Our main focus is more on the behaviors of the cross training paradigm rather than improvements on the state-of-the-art accuracy.

It turns out that the accuracy does improved comparing with the baseline from [6] in Table 1. Moreover the gaps between the red and blue indicators are getting closer when the network goes deeper. This is because the deeper the network gets, the better descriptors the classifier will receive. And those descriptors will allow the network converging to a very good solution naturally approximating to the lower bound of $\sqrt{\mathcal{L}_S \lambda \mathcal{L}_C}$ without requiring any constraints. Then the gap between the loss with or without constraints getting smaller.

Table 2 reflects the accuracy between ResNet-110 and VGG-16 testing on both CIFAR-10 and CIFAR-100. The results are consistence to the results of Table 1.

6   Conclusion

In the paper, we proposed a new loss function by adding constraints of learning good representative features. It turns out that the performance of the classification will be improved by the proposed loss while applying to the switching loss training scheme. Experiments are in well-known CNN architectures, i.e. ResNet and VGG-16, with the dataset CIFAR-10 and CIFAR-100.

Acknowledgment. The project is funded by Science and Technology of Macau. The authors wish to acknowledge financial support grant, No. 112/1024/A3, 151/2017/A and 152/2017/A.

References

1. Schroff, F., Kalenichenko, D., Philbin, J.: FaceNet: a unified embedding for face recognition and clustering. In: 2015 IEEE Conference on Computer Vision and Pattern Recognition (CVPR), pp. 815–823, June 2015
2. Hinton, G.E., Salakhutdinov, R.R.: Reducing the dimensionality of data with neural networks. Science **313**(5786), 504–507 (2006)
3. Wen, Y., Zhang, K., Li, Z., Qiao, Y.: A discriminative feature learning approach for deep face recognition. In: Leibe, B., Matas, J., Sebe, N., Welling, M. (eds.) Computer Vision - ECCV 2016, pp. 499–515. Springer, Cham (2016)
4. Huang, R., Zhang, S., Li, T., He, R.: Beyond face rotation: global and local perception GAN for photorealistic and identity preserving frontal view synthesis, pp. 2458–2467 (2017)
5. Simonyan, K., Zisserman, A.: Very deep convolutional networks for large-scale image recognition. Comput. Sci. (2014)
6. He, K., Zhang, X., Ren, S., Sun, J.: Deep residual learning for image recognition. In: IEEE Conference on Computer Vision and Pattern Recognition, pp. 770–778 (2016)
7. Krizhevsky, A., Sutskever, I., Hinton, G.E.: ImageNet classification with deep convolutional neural networks. In: International Conference on Neural Information Processing Systems, pp. 1097–1105 (2012)
8. Nair, V., Hinton, G.E.: Rectified linear units improve restricted Boltzmann machines. In: Proceedings of ICML, pp. 807–814 (2010)
9. Maas, A.L., Hannun, A.Y., Ng, A.Y.: Rectifier nonlinearities improve neural network acoustic models (2013)
10. He, K., Zhang, X., Ren, S., Sun, J.: Delving deep into rectifiers: surpassing human-level performance on ImageNet classification. In: IEEE International Conference on Computer Vision, pp. 1026–1034 (2016)
11. Clevert, D.A., Unterthiner, T., Hochreiter, S.: Fast and accurate deep network learning by exponential linear units (ELUs). Comput. Sci. (2015)
12. Baldi, P., Chauvin, Y.: Neural networks for fingerprint recognition. Neural Comput. **5**(3), 402–418 (1993)
13. Chen, W., Chen, X., Zhang, J., Huang, K.: Beyond triplet loss: a deep quadruplet network for person re-identification, pp. 1320–1329 (2017)

Machine Learning and Education in the Human Age: A Review of Emerging Technologies

Catherine A. Bacos[✉]

University of Nevada, Las Vegas, Las Vegas, NV 89154, USA
catherine.bacos@unlv.edu

Abstract. Today's emerging technologies are moving in the direction of the human age. As technologies emerge in the educational setting, humans can evolve in the way they learn because of technology; and machines can evolve in the way they learn because of humans. Technology can be used to effectively observe and assesses human behaviors to better understand and respond to them. Whether the behaviors need to be adjusted or the behaviors are worth modeling, technology can provide support such as tools to track and collect data, assess performance, and provide meaningful feedback to the learner. In the human age of machine learning, the focus is less on technology and more on being human. To adapt to changing educational contexts, more effective applications of emerging technologies are needed. This paper explores the following novel applications of human-centered approaches to using technology in education: the quantified self, affective computing, emotional design, and pedagogical agents.

Keywords: Emerging technologies · Education · e-Learning ·
Artificial intelligence · Quantified self · Affective computing ·
Pedagogical agents

1 Introduction

In *Ex Machina* [1], a film about a humanoid robot with a sophisticated artificial intelligence system, the main character tells the robot a story that describes how machines are different from humans. A reference to Plato's Allegory of the Cave [2], the story explains that while machines are locked in a room and can only see in black and white, humans are on the outside and can see in color. Like the people imprisoned in Plato's Cave who can only see shadows, machines can only see a manufactured reality. To escape this constraint, machines need to become more human.

In recent years, machines have become more human in appearance and behavior by modeling the way humans think, act, and learn. The emergence of technologies designed to look and act in more human-like ways are changing the way people interact with machines and each other. For example, motion capture systems have been used recently to record and model human body movements to create more human-like avatars and ecologically valid social interactions in virtual training environments [3, 4]. Such realistic contexts provide new opportunities for human connection through

K. Arai and S. Kapoor (Eds.): CVC 2019, AISC 944, pp. 536–543, 2020.
https://doi.org/10.1007/978-3-030-17798-0_43

authentic experiences. These experiences can also provide machines new opportunities to understand humans in settings where more realistic interactions are observable. The more human machines become, the more they will be able to adapt to change. An important function of early human development, adaptation is adjusting to the demands of the environment [5]. Humans are equipped with tools to analyze data from their environment, and the results of their analyses inform their behavior. When their environment changes, humans adapt. Like humans, machines can learn to recognize and interpret data to adapt to new contexts.

In the educational context, humans and machines will need to adapt to unforeseen changes created by the integration of emerging technologies. A defining characteristic of emerging technologies is that they are unpredictable and uncertain [6]. This uncertainty associated with emerging technologies makes it difficult to study their effect on human psychology. As technologies become more human-centered and ubiquitous, they will be used with greater frequency and in more creative ways [7]. To adapt to today's unprecedented rate of technological change, people need to use their imagination and creativity to explore more effective applications of emerging technologies. The following novel applications of human-centered approaches to using technology in education are explored: the quantified self, affective computing, emotional design, and pedagogical agents.

2 Characteristics of Emerging Technologies

Before considering the applications of emerging technologies, it is helpful to frame the discussion with a definition of emerging technologies. Whether a technology is considered emerging depends on its place, domain, and application [8]. In the information and communication technology (ICT) domain, a characteristic of an emerging technology is that it does not need to have a limited or fixed life; and even when the technology has been used for a considerable length of time, it can begin to emerge when it is used in novel ways to serve individuals as a means of accomplishing their goals. Like the ICT domain, other fields have also developed definitions to guide research. For example, policy researchers have operationalized emerging technologies with the following attributes: (a) radical novelty, (b) relatively fast growth, (c) coherence, (d) prominent impact, and (e) uncertainty and ambiguity [9].

In the education field, technology has been defined broadly as "tools and resources that are used to improve teaching, learning, and creative inquiry" [11]. Research in this domain has defined emerging technologies with the following characteristics: (a) not necessarily new, (b) still evolving, (c) not yet fully understood or researched, and (d) unfulfilled but promising potential [10]. In addition to exhibiting these characteristics, whether a technology is emergent depends on its context. Defining the characteristics of emerging technologies in educational contexts can help improve technology integration models (e.g., technology acceptance model, the substitution augmentation modification redefinition model, and the technological pedagogical and content knowledge model), which can help understand how technology is adopted. According to 15 years of research on emerging technology trends, several developing technologies are expected to be adopted over the next 5 years, including enabling technologies (i.e., technologies that make tools more useful by expanding their reach) such as affective computing, artificial intelligence, and virtual assistants [11]. These emerging technologies have the potential to change the way people teach and learn as they adapt to changes in technology and education.

3   The Quantified Self: Self-tracking as an Emerging Technology

As humans become more connected through the mediating experiences afforded by technology, they need to enhance their understanding and use of these technologies, and by extension, enhance themselves. One way in which people can enhance their understanding of themselves and potentially improve their behavior is through technologies that support the quantified self, an individual who engages in the self-tracking of any biological, physical, behavioral, or environmental information [12]. Some examples of this emerging technology are smart watches and fitness trackers that can track and analyze a variety of activities such as exercise and sleep patterns. An example of the quantified self for educational purposes is a web application called Stackup. Called by teachers "the Fitbit of reading," Stackup measures students' level of online reading and engagement through big data metrics [13]. An advantage of the quantified self is its focus not just on the quantified data but the qualitative data as well [12]. The tracking and analysis of both quantitative and qualitative data will offer new avenues of research to better understand human behavior.

As educators consider new uses of technology in new contexts, it is important to align these uses with learning theories and empirical evidence showing the impact of these emerging technologies on teaching and learning [14]. For example, the decision to use a technology may be dependent on how the technology can support instructional design strategies that are effective. Regardless of the medium or technology used to teach, the pedagogical strategy should be an important consideration. This will place the focus on using technology to keep the learner motivated and engaged emotionally and cognitively.

4   Emotions and Technology

4.1   Emotion, Cognition, and Technology

To motivate and engage students, technology-enhanced approaches to instruction need to take into consideration both the learner's cognitive and emotional needs. For example, you might hear a student say, "the more I thought about it, the more I felt confused." Such a statement counters the dichotomized view of human psychology that has separated thinking from feeling. Current research indicates cognition and emotion are not mutually exclusive, but rather, they are inherently interconnected [15].

In the same way that emotion and cognition are linked, humans and computers are not necessarily divided by two separate worlds: the real and the virtual. Instead, they are connected. Technologies can mediate a person's experience of reality. This experience serves as an extension of the person's self, which is why it makes sense to think of human experience with computers in terms of human psychology [16]. Technology is more than just a tool; it is inextricably connected to the social and psychological lives of the people who use these tools. Technology affects how people see themselves and the world around them.

4.2   Measuring Emotion

The close connection people have with technology makes it an ideal medium for studying, modeling, and responding to human emotion. To understand the important role of user affects in human-computer interactions, the highly complex and multi-modal phenomenon of emotional expression needs to be investigated. Using technology, there are an increasing number of methods for measuring emotional expressions. Although self-report measures are still considered the gold standard for measuring emotion, facial expression coding and physiological measures such as electrodermal activity (ED) are empirically grounded approaches that are showing some promise in the field of emotion research [17]. Other factors that need to be considered when attempting to understand and measure emotion are the individual differences of the users, from their traits to their self-confidence to the ways in which they use the technology [18]. Being able to track and monitor both quantitative and qualitative user information is another reason technology serves as a useful tool to control for and accurately measure the variables associated with studying human emotion. Through technologies such as affective computing, human emotion can be measured, analyzed, and influenced within educational contexts.

4.3   Affective Computing

Affective computing is the use of technology to purposely influence emotions [19]. Emotion is tied to a cognitive, motivational and behavioral process [20]. Since emotion plays a central role in the control of behavior, environments that influence emotion can greatly control user experiences. This means that instructional design can benefit from considering the effect of emotions on learning.

In the educational setting, affective computing can be considered an emerging technology. For example, a novel application of affective computing in learning management systems would be to collect and analyze emotion data to provide relevant and meaningful feedback to the learner. Not considering a students' emotions may explain why students receive less socio-emotional support in online learning environments, which can lead to a loss of motivation and decreased learning comprehension [21]. Affective computing can emerge as a solution to address these deficits.

4.4   Emotional Design

Although affective computing can be used to influence emotion, what is important is not so much that emotions are induced but how they are induced. For example, designing instruction to induce emotion can support cognition and learning when those emotions are positive [22]. When done effectively, positive emotions can strengthen motivation. Thus, building a learner's positive affect is an essential first step when designing instruction.

While inducing positive emotions can be beneficial for learning, reducing negative feelings can also provide emotional support, and the technology used to provide this level of support does not have to be new [23]. For example, in a study that used pictures of an agent instead of an embodied agent with voice recognition and response capabilities, the facial expressions and textual statements provided to the user were impactful enough to significantly reduce negative feelings [24].

Video is another technology that is not new but can make an impact on learner affect. When used in a novel way, this everyday technology can be considered an emerging technology. A study that used relationship-building strategies in instructional videos found that the affective features in the video (i.e., the instructor used a friendly and warm tone of voice, included colloquialisms, provided anecdotes about own experience, and used encouraging language toward learners) significantly improved the students' attitude toward the instructor and the learning material compared to video presented without these strategies [25].

Another example of emotional design is the use of a framework that integrates both emotional and cognitive design factors. When designing digital media for learning, the use of this integrative model can have a significant impact on learning [26]. This emotional design approach used design features to influence learner's emotions such as the way information is presented and the way interactions in the environment are structured.

More than just recognizing emotions and adjusting the instruction, the learning system needs to provide students with quality feedback. Instructors have the potential to influence a student's emotions, a goal of affective computing [19], but "misleading affect recognition could result in inadequate emotional feedback, ruining learning" [27]. Therefore, designers of learning environments also must be cognizant of the potential negative effects of delivering affective responses to student learning.

5 Pedagogical Agents

Pedagogical agents are another emerging technology in the education field. Although these types of agents have been in existence for some time, the new applications and context in which they are used make them an emerging technology. As the research indicates, when considering the use of affective embodied agents, it is no longer a question of whether they can influence users' behavior in virtual environments, but the question is how and in what context these agents can be effective in engaging and motivating the learner.

Before using pedagogical agents, it is important to first validate the emotional expressions the agent will be using to make sure they will be perceived by the users as intended [28]. This is not unlike considering the user's emotional type, which could be compared to taking a baseline approach prior to calibrating how the effect of an intervention will be measured. If the user is sensitive to emotional expressions, validating them in advance of instruction will aid in understanding the user's behavior. Such an approach is key when studying human-computer interactions as it places emphasis on the human side of the interaction by designing from the user's perspective. When agents play the role of a colleague instead of a tutor, the user's perceptions of the affective agent were enhanced, which resulted in improved learning outcomes [28]. Understanding the user's perceptions is an important first step to determining how to influence them.

The decision to use pedagogical agents should depend on the efficacy of the design and behaviors of the agents [29]. In a meta-analysis of studies comparing the presence of affect in embodied pedagogical agents, the findings indicated that to effectively respond to learner affect, agents' expressions needed to be authentic and student affect had to be detected accurately [30]. Without meeting these two requirements, embodied agents cannot be effective.

One of the reasons affective agents are more effective than non-affective agents in supporting interactions, particularly those that yield desirable learning outcomes, is because embodied agents use abstract symbols better than disembodied agents [31]. Social cognitive theory may provide an explanation for this as the role of embodied agents serve as a model to guide users' behaviors [32]. In addition, embodied agents support cooperative learning, which is known to be more effective at promoting interactive learning than individualistic instruction [33].

In the educational context, pedagogical agents can support learning by catering to the needs of the user who may need empathetic support initially until they are confident enough to self-regulate their learning. Users tend to like and trust pedagogical agents when they express empathetic emotion [24]. People respond to being cared about by computer agents, so caring agents have an advantage in motivating users because agents are perceived as trustworthy. In addition, people are likely to be more compliant toward an agent that reciprocates helpful behavior [34]. This suggests a design approach that incorporates reciprocity with a collaborative task between the agent and human would be helpful. Based on social exchange theory, people may even prefer working with computers to humans, especially when they experience a stable pattern of reciprocity, they may exhibit a higher degree of commitment toward a computer than a human [35].

6   Conclusion

An advantage of an emerging technology is that it presents an opportunity to engage with something novel, an experience that may shed light on the mysteries of human behavior. Technologies that are designed to model human-human interactions based on observations of positive human emotions and behaviors can support instructional objectives such as increasing learner motivation and successfully building student-instructor relationships in online environments. Technology can be used to effectively observe and assesses human behaviors to better understand and respond to them. Whether the behaviors need to be adjusted or the behaviors are worth modeling, technology can provide support such as tools to track and collect data, assess performance, and provide meaningful feedback to the learner. Although technology still exists within the realm of the digital and the artificial, the aim of technology is to move toward the natural and the authentic. From natural user interfaces such as wearable technologies to enabling technologies such as artificial intelligence, today's emerging technologies need to be designed to perform less like a machine and more like a human.

Today's emerging technologies are moving in the direction of the human age. As technologies emerge in the educational context, humans can evolve in the way they learn because of technology; and machines can evolve in the way they learn because of humans. In the human age of machine learning, the focus is less on technology and more on being human.

References

1. Macdonald, A., Reich, A., Garland, A.: Ex machina. Film4 Productions & DNA Films, United Kingdom (2014)
2. Plato, Bloom, A., Kirsch, A.: The Republic of Plato. Basic Books, New York (2016)
3. Bacos, C.A., Carroll, M.: Kinematics for e-learning: examining movement and social interactions in virtual reality. In: Proceedings of E-Learn: World Conference on E-Learning in Corporate, Government, Healthcare, and Higher Education, pp. 413–417. Association for the Advancement of Computing in Education (AACE), Las Vegas, NV (2018)
4. Griffith, T., Dwyer, T., Kinard, C., Flynn, J.R., Kirazian, V.: Research on the use of puppeteering to improve realism in army simulations and training games. In: Lackey, S., Shumaker, R. (eds.) Virtual, Augmented and Mixed Reality, pp. 386–396. Springer, Cham (2016)
5. Piaget, J.: Biology and Knowledge. Edinburgh University Press, Edinburgh (1971)
6. Allenby, B.: Emerging technologies and the future of humanity. Bull. At. Sci. **71**, 29–38 (2015)
7. Ludlow, B.L.: Virtual reality: emerging applications and future directions. Rural Spec. Educ. Q. **34**, 3–10 (2015)
8. Halaweh, M.: Emerging technology: what is it? J. Technol. Manag. Innov. **8**, 108–115 (2013)
9. Rotolo, D., Hicks, D., Martin, B.R.: What is an emerging technology? Res. Policy **44**, 1827–1843 (2015)
10. Veletsianos, G.: Defining characteristics of emerging technologies and emerging practices in digital educaiton. In: Veletsianos, G. (ed.) Emergence and Innovation in Digital Learning: Foundations and Applications. Athabasca University Press, Edmonton (2016)
11. Adams Becker, S., Cummins, M., Davis, A., Freeman, A., Hall Giesinger, C., Ananthanarayanan, V.: NMC Horizon Report, 2017 Higher Education Edition, Austin, Texas (2017)
12. Swan, M.: The quantified self: fundamental disruption in big data science and biological discovery. Big Data **1**, 85–99 (2013)
13. Geisel, N.: Chrome extensions can boost and celebrate literacy among K-12 students (2017)
14. Czerkawski, B.C.: Strategies for integrating emerging technologies: case study of an online educational technology master's program. Contemp. Educ. Technol. **4**, 309–321 (2013)
15. Izard, C.E.: Emotion theory and research: highlights, unanswered questions, and emerging issues. Ann. Rev. Psychol. **60**, 1–25 (2009)
16. Turkle, S.: The Second Self: Computers and the Human Spirit. MIT Press, Cambridge (2005)

17. Harley, J.M.: Measuring emotions: a survey of cutting-edge methodologies used in computer-based learning environment research. In: Tettegah, S.Y., Gartmeier, M. (eds.) Emotions, Technology, Design, and Learning, pp. 89–114. Elsevier Publishers, London (2016)
18. Jokinen, J.: Traits, events, and states. Int. J. Hum. Comput. Stud. **76**, 67–77 (2015)
19. Picard, R.: Affective Computing. MIT Press, Cambridge (1997)
20. Pekrun, R.: Progress and open problems in educational emotion research. Learn. Instr. **15**, 497–506 (2005)
21. Garrison, D.R.: Online community of inquiry review: social, cognitive, and teaching presence issues. J. Asynchronous Learn. Netw. **11**, 61–72 (2007)
22. Um, E., Plass, J.L., Hayward, E.O., Homer, B.D.: Emotional design in multimedia learning. J. Educ. Psychol. **104**, 485–498 (2012)
23. Klein, J., Moon, Y., Picard, R.W.: This computer responds to user frustration. Interact Comput. **14**, 119–140 (2002)
24. Brave, S., Nass, C., Hutchinson, K.: Computers that care: investigating the effects of orientation of emotion exhibited by an embodied computer agent. Int. J. Hum. Comput. Stud. **62**, 161–178 (2005)
25. Kim, Y., Smith, D., Thayne, J.: Designing tools that care: the affective qualities of virtual peers, robots, and videos. In: Tettegah, S., Gartmeier, M. (eds.) Emotions, Technology, Design, and Learning, pp. 114–128. Elsevier Publishers, London (2016)
26. Plass, J.L., Kaplan, U.: Emotional design in digital media for learning. In: Tettegah, S., Gartmeier, M. (eds.) Emotions, Technology, Design, and Learning, pp. 131–162. Elsevier Publishers, London (2016)
27. Moridis, C.N., Economides, A.A.: Toward computer-aided affective learning systems: a literature review. J. Educ. Comput. Res. **39**, 313–337 (2008)
28. Beale, R., Creed, C.: Affective interaction: how emotional agents affect users. Int. J. Hum Comput Stud. **67**, 755–776 (2009)
29. Lane, H.C.: Pedagogical agents and affect: molding positive learning interactions. In: Tettegah, S., Gartmeier, M. (eds.) Emotions, Technology, Design, and Learning, pp. 47–62. Elsevier Publishers, London (2016)
30. Guo, Y.R., Goh, D.H.: Affect in embodied pedagogical agents: meta-analytic review. J. Educ. Comput. Res. **53**, 124–149 (2015)
31. Cassell, J.: Embodied conversational agents. MIT Press, Cambridge (2000)
32. Bandura, A., Bryant, J.: Social cognitive theory of mass communication. Media Eff. Adv. Theory Res. **2**, 121–153 (2002)
33. Atkinson, R.K.: Optimizing learning from examples using animated pedagogical agents. J. Educ. Psychol. **94**, 416–427 (2002)
34. Lee, S.A., Liang, Y.: Reciprocity in computer-human interaction: source-based, norm-based, and affect-based explanations. Cyberpsychol. Behav. Soc. Netw. **18**, 234–240 (2015)
35. Posard, M.N., Rinderknecht, R.G.: Do people like working with computers more than human beings? Comput. Hum. Behav. **51**, 232–238 (2015)

3M2RNet: Multi-Modal Multi-Resolution Refinement Network for Semantic Segmentation

Fahimeh Fooladgar[(✉)] and Shohreh Kasaei

Sharif University of Technology, Tehran, Iran
fahimehfooladgar@ce.sharif.edu, kasaei@sharif.edu

Abstract. One of the most important steps towards 3D scene understanding is the semantic segmentation of images. The 3D scene understanding is considered as the crucial requirement in computer vision and robotic applications. With the availability of RGB-D cameras, it is desired to improve the accuracy of the scene understanding process by exploiting the depth along with appearance features. One of the main problems in RGB-D semantic segmentation is how to fuse or combine these two modalities to achieve more advantages of the common and specific features of each modality. Recently, the methods that encounter deep convolutional neural networks have reached the state-of-the-art results in dense prediction. They are usually used as feature extractors as well as data classifiers with an end-to-end training procedure. In this paper, an efficient multi-modal multi-resolution refinement network is proposed to exploit the advantages of these modalities (RGB and depth) as much as possible. This refinement network is a type of encoder-decoder networks with two separate encoder branches and one decoder stream. The feature abstract representation of deep networks is performed by down-sampling operations in encoder branches leading to some resolution loss in data. Therefore, in the decoder branch, the occurred resolution loss must be compensated. In the modality fusion process, a weighted fusion of "clean" information paths of each resolution level of the two encoders is utilized via the skip connection by the aid of the identity mapping function. The extensive experimental results on the three main challenging datasets of NYU-V2, SUN RGB-D, and Stanford 2D-3D-S show that the proposed network obtains the state-of-the-art results.

Keywords: RGB-D images · Semantic segmentation · Deep learning

1 Introduction

Semantic segmentation of RGB-D images is considered as one of the most central issues for 3D scene understanding in the field of computer vision and robotics. The goal of semantic segmentation is to partition the images into semantically meaningful regions. Without any doubt, this field covers a wide range of applications; such as robot navigation, autonomous driving, inferring support relationships among objects, and content-driven retrieval. Most of researches in this field have been done on outdoor scenes (which are less challenging compared to indoor scenes). Existence of small

K. Arai and S. Kapoor (Eds.): CVC 2019, AISC 944, pp. 544–557, 2020.
https://doi.org/10.1007/978-3-030-17798-0_44

objects, light tailed distribution of objects, occlusions, and poor illumination cause some challenges in indoor scenes, to name a few. By introducing the Microsoft Kinect camera which captures both RGB and Depth images, some indoor semantic segmentation approaches have been concentrated on the RGB-D dataset which alleviate the challenges of the indoor scene. For instance, when RGB images have a poor illumination in some regions, depth images can improve the labeling accuracy. Recently, the convolutional neural networks (CNNs), as both feature extractor and classifier, have formed the most promising approaches in almost all of the computer vision fields; such as image classification [1–3], object detection [4–6], action recognition [7, 8], depth estimation [9, 10], pose estimation [11], and semantic segmentation [12, 13].

The early deep learning methods for semantic segmentation utilize the deep networks to extract features. They then apply a classifier to classify each pixel, superpixel, or region. The fully convolutional network (FCN) [13] changed the CNNs to a fully convolutional one that gets images with any size as the input and gives an output with the same size as the input image. It then prepares a model for the semantic segmentation process. Different types of CNNs have been proposed and they are rapidly going deeper and wider to enjoy more improvements. But, deeper networks emerge two major problems of vanishing gradient and increased number of parameters. Different network architectures such as [14, 15] have been proposed to overcome these two problems by using short path connections from previous layers.

As one of the goals of this paper is the semantic segmentation of RGB-D images, we focus on the main challenges and approaches of RGB-D datasets. The main challenge in RGB-D semantic segmentation is how to represent and fuse the RGB and depth channels so that the strong correlation between the depth and photometric channel are considered. Simple methods of the fusion of RGB and Depth channels are based on the early fusion [16] and late fusion [13].

In this paper, the utilization of the depth channel beside the RGB channel to enable the high resolution prediction via the RefineNet [17] idea with the attention to preserve the specific properties is explored. The more abstract feature representation of CNNs (by pooling operations and stride convolutions) has caused the CNNs to be invariant to most local changes. This is one of the advantages of CNNs in some applications (like image classification and object detection) while it is harmful for dense prediction applications (like semantic segmentation, depth estimation, and surface normal estimation). The proposed method utilizes the feature maps of the down-sampling path extracted from the depth channel alongside the RGB channel to refine the information loss caused by the down-sampling operation. The proposed architecture consists of two encoder branches which simultaneously extract features from RGB and Depth channels via the residual connections, independently. In the decoder path, the multi-level feature maps of these two modalities are jointly exploited to recover the information loss of the encoder parts at that level. These skip connections provide the feature reuse ability for deep networks and solve the vanishing gradients problem. The experimental results on the three challenging RGB-D datasets demonstrate the considerable improvement of the labeling process by adding the RGB and Depth streams to refine the high resolution feature maps, jointly.

Main contributions of this work are listed as:

- Extending the RefineNet for multiple modalities.
- Preserving the long-range residual connection and identity mapping properties.
- Multi-level refining in the multipath multi-modal encoder-decoder network.
- Improving the feature reuse ability of deep networks.
- Efficient end-to-end training by employing the identity mapping in the RefineNet blocks to fuse two modalities.

This paper is organized as follow. In Sect. 2, the pioneer RGB-D semantic segmentation methods are explained. The proposed methods and related definitions are presented in Sect. 3. In Sect. 4, the experimental results of the proposed methods are evaluated on the existing RGB-D datasets. Finally, the concluding remarks are discussed in Sect. 5.

2   Related Work

Recently, the most successful methods in the field of semantic segmentation have been proposed based on CNNs which outperform the traditional methods that are based on engineering features. These traditional methods have been studied in two main categories of having 2D or 3D images. There are many interesting work on traditional approaches [4, 18–26]. Their differences are based on the type of variable extracted features as well as classifiers. To exploit contextual relationships, some post-processing methods might be applied (like conditional random field (CRF) [20, 27–29], Markov random field (MRF) [30], multi-scale CRF [27, 31] or hierarchical CRF [28, 29]). These traditional approaches are not within the scope of this article (as the recent advances are based on deep learning architectures).

The AlexNet [1] started the widespread use of CNNs and actually reintroduced deep neural networks which attained a great popularity in machine learning fields. The ZF-Net [32] was an improvement of the AlexNet in which the hyper parameters of the AlexNet had changed and the more accurate results had achieved. The authors of [3] proposed the GoogLeNet with the idea of using cascaded inception modules. By removing the fully connected layers of the previous network, they significantly reduced the number of network parameters (approximately 12× lower compared with the AlexNet). The GoogLeNet was superior to previous models in case of the lower number of parameters and computational cost as well as the higher accuracy. In 2014, the VGGNet [2] was also proposed in which all of the convolutional layers have 3 × 3 filters with the stride and the padding of 1, proceeding with the 2 × 2 max pooling layer with stride 2. The architecture of the VGGNet is very simple and homogenous. It is formed by a 19 layers model where the early layers of the convolution consume more memory and the late fully connected layer imposes most parameters to the model. The authors of the ResNet [14] have shown that increasing the number of layers with the attention to the vanishing gradients problem improves the accuracy while it has lower time complexity and easy to train, when compared with previous models. They proposed a short cut connection by an element-wise addition that has neither extra parameters nor computational complexities while solving the vanishing gradient and degradation problems. This model has faster convergence than the shallow one and also achieves a higher accuracy. The ResNet model won the ILSVRC 2015 [14] with a significant gain in accuracy and a lower complexity by increasing the depth of the network and solving the degradation problem. The authors of [15] proposed the DenseNet model with the goals of alleviating the vanishing gradient problem, strength features propagation, encouraging features reuse, and reducing the number of parameters. The solution of the DenseNet is similar to the ResNet while all layers are directly connected to each proceeding layer by concatenating the feature maps instead of summation which was proposed in the ResNet. The DenseNet achieved the same accuracy as the ResNet while reducing the number of parameters about 70% by using the deep supervision concept.

All of these network architectures were originally proposed for the image classification task. Unfortunately, the outputs of these models for the semantic segmentation are very coarse. This is because the cascaded down-sampling and stride convolution decrease the resolution of feature maps. To compensate these information loss, Couprie et al. [16] utilized the multiscale CNNs to extract dense hierarchical features for each region (provided by the segmentation tree). They then fed these deep features to the Support Vector Machine (SVM) classifier for super-pixel labeling. The FCN [13] changed the CNNs to a fully convolutional one and prepared the model to be used for dense prediction tasks; such as semantic segmentation, depth estimation [9, 10], and surface normal prediction [9]. In the FCN, the output of the network has the same size as its input. The last fully connected layers are substituted with the convolutional layers. Different approaches are proposed to scale-up the output to reach the localization of the labels; which is the primary goal of semantic image segmentation methods, depth estimation [9, 10], and surface normal prediction [9]. Shelhamer et al. [13] up-sampled the feature maps of the last layers and concatenated them with the corresponding intermediate feature maps in a stage-wise end-to-end training. They extended the AlexNet, GoogleNet, and VGG Net to the fully convolutional neural networks. The authors of [33, 34] utilized the transpose of the convolution in the decoder side (instead of convolution layers in the encoder) and called it the deconvolutional neural network. To preserve the resolution, DeepLab [12] did not use the max-pooling while applied the dilated convolution to enlarge the receptive field of filters. By substituting the down-sampling operation with the dilated convolution, the computational complexity and memory requirements were also exceeded. The SegNet proposed a novel idea to be more efficient in the memory and computational cost. The SegNet designers preserved the indices of the max-pooling layers to incorporate them in the decoder side. The authors of [35] proposed the encoder-decoder types of CNN with the dense block idea of the DenseNet. The RefineNet [17] proposed the refinement blocks which considered the idea of the residual blocks of the ResNet and the fusion of the long-range residual connections in order to restore the resolution loss of the CNN architecture.

The first challenge of the RGB-D semantic segmentation based on CNNs is the lack of large annotated RGB-D datasets for the training process. This shortcoming can be adapted with some proposed tricks like data augmentation, pre-training, transfer learning, and fine tuning methods. Some other methods were also proposed to prevent the overfitting problem (such as the drop-out). There are three known RGB-D datasets for dense per pixel labeling; namely, the NYU-V2 [24], SUN RGB-D [36], and

Stanford 2D-3D-Semantics [37]. The Stanford 2D-3D-Semantic dataset was introduced recently which provides a larger scale data than the two others. The other main challenges are how to represent and fuse the RGB and depth channels so that the strong correlation between the depth and photometric channel is considered. Simple methods for fusion of the RGB and Depth channels are based on the early [16], late [13], and middle fusion [38, 39]. The FuseNet [39] extended the VGG-16 model in two encoder branches with one decoder. One encoder branch for the depth channel and another encoder branch for the RGB and the summation of intermediate depth feature maps with the RGB feature maps. The authors of [34] extended the deconvolution networks idea of [12] for RGB-D images as input. They also proposed a transformation network between convolution and deconvolution parts to learn the common and specific features of the RGB and depth channel. The fusion of the vertical and horizontal long short-term memorized (LSTM) have been applied in the LSTM-CF proposed in [38]. The vertical LSTM exploit the interior 2D global contextual relation for each RGB and depth image, separately. The horizontal LSTM is applied on their fusion. Park et al. in [40] expanded the RefineNet [17] to fuse these RGB and depth modality. They added a fusion stream in which the feature maps of RGB and depth images, produced by the ResNet block, are fused. Then, the cascaded refinement blocks of the RefineNet are utilized. The refinement process is applied on the fusion of the RGB and depth feature maps to emend the resolution loss. Although the fusion blocks learn the fusion of these two modalities, but the long-range residual connection and the identity mapping idea of RefineNet in the refinement blocks have been eliminated via the convolution and nonlinearity layers in the fusion blocks.

3   Proposed Method

The multi-modal multi-resolution refinement network proposes an efficient multi-modal refinement method in the decoder branch of the network to recover the information loss caused by the down-sampling in multi encoder streams. The multi-modal refinement block has been inspired by the RefineNet block in the RefineNet model [17] which is focused on the identity mapping of the short and long range skip connection. In the following, before introducing the proposed 3M2RNet, more details on the RefineNet model are first given.

3.1   RefineNet Overview

The main idea of the RefineNet is to employ the multi-level features to enhance the resolution which is lost by the down-sampling operation performed by the pooling or convolution layers (with stride > 1). The inputs of the RefineNet are the feature maps of the ResNet blocks exploited residual connections following the identity mapping concept. Based on the resolution of the feature maps decreased in the pooling layers, the ResNet model is divided into 4 blocks. Then, the output of these blocks are employed as the skip connections and fed to the 4-cascaded RefineNet blocks in level orders. As such, it actually utilizes the long and short residual connections with the identity mapping idea. The short range ones are the residual connection used in each ResNet block and the long range residual connections are considered between each ResNet and RefineNet block in the same levels. These skip connections, along with the identity mapping, enable an efficient end-to-end training as well as efficient high-resolution predicting. The overall structure of the RefineNet model is illustrated in Fig. 1.

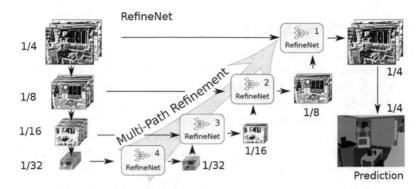

Fig. 1. Architecture of 4-cascaded RefineNet [17].

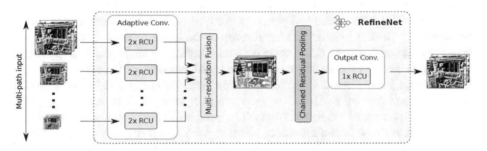

Fig. 2. Architecture of each RefineNet [17].

The RefineNet block consists of three main sub-blocks of residual convolution unit (RCU), multi-resolution fusion (MRF), and chained residual pooling (CRP). The RCU is the same as the residual unit in the ResNet but without the batch normalization layers. The MRF is the fusion block in which the output of the previous RefineNet block fuses with the output of the RCU to perform the refinement and to produce the high resolution feature maps. The idea of CRP sub-block is to capture the context in multiple region sizes with the chained residual pooling. Finally, the output of CRP passes to one RCU.

Lin et al. investigated the multi-variations of the RefineNet architecture and finally the single-scale 4-cascaded version has been specified as the standard model that performs the best in most of the evaluated datasets. The single-scale 4-cascaded RefineNet and its refinement block are illustrated in Figs. 1 and 2, respectively. The RefineNet block takes the feature maps of the corresponding ResNet level with the previous refined feature maps as the inputs. The residual connections with the identity mapping strategy are applied to the most of the individual components of the RefineNet except for the convolution layers in the MRF sub-block before the up-sampling operation and the Relu function in the CRP sub-block. The authors of the RefineNet imply that one single nonlinearity in each RefineNet block cannot affect exponentially to vanish the gradients.

3.2   Proposed 3M2RNet Architecture

The 3M2RNet model is an encoder-decoder CNN architecture with multi-encoder branches to encode multi-modal input streams simultaneously while including one decoder branch. The goal of the decoder is to recover the high resolution prediction from the low-resolution high-level semantic feature maps produced by the cascaded residual blocks with the down-sampling operations come in between of the residual block. All of the encoder branches follow the structure of the ResNet model. Then, the information loss in the multi-modal encoders recovers in the decoder with the fusion of the skip connections achieving from each modality in the same resolution level.

The sequential down-sampling operations (mostly applied by the pooling layers in the encoders) increase the receptive field of the filters to capture more context and also prevent the growth in the number of training weights through the network. Therefore, they preserve the efficient and tractable training. But, the network loses some valuable information. This information loss produces the low-resolution prediction in the dense per pixel classification in which the localization of the semantic labels is more essential than the other applications (such as image classification). This means that the higher level feature maps of the deeper layers in the multi-encoders which encode the high-level semantic information and carry more object level information suffer from the lack of localization information. Here, it is proposed to recover this information loss in the up-sampling process of the decoder branch by weighted summation of the long-range residual connections of multi-encoder streams with the preceding decoder output. Therefore, the decoder part is responsible to recover this resolution loss in cascaded refinement blocks.

The architecture of the 3M2RNet with RGB-D inputs is illustrated in Fig. 3. The ResNet model serves as two separate encoder branches, where 4 CNN blocks (Convi-x) show the 4 ResNet blocks based on the 4 pooling layers of the ResNet. The 3M2R block fuses multi-modal feature maps coming from multi-encoder branches as well as multi-resolution feature maps coming from the preceding 3M2R block. We can simply extend the two modalities and single scale architecture to multi-modal and also multi-scale inputs.

Multi-modal Multi-resolution Refinement Block. The proposed 3M2RNet extends the identity mapping of the short and long-range skip connections in the multi-modal framework. The overall structure of the 3M2R block with three modalities as its inputs is shown in Fig. 4. These three modalities consist of: (i) feature maps extracted from RGB encoder branch, (ii) feature maps extracted from the depth encoder branch at the same resolution level, and (iii) the feature maps of the preceding 3M2R block at the lower resolution. This block consists of 3 modules.

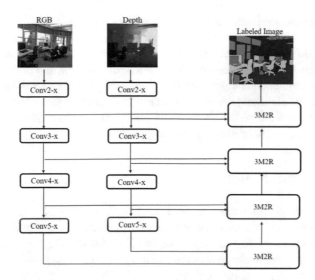

Fig. 3. Proposed multi-modal RefineNet model.

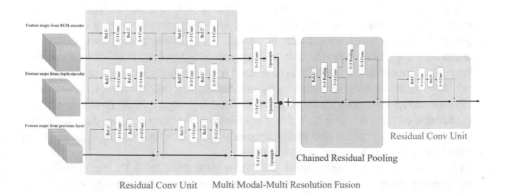

Fig. 4. Structure of proposed multi-modal multi-resolution refinement (3M2R) block.

All of the components of the 3M2R block follow the identity mapping rule. This means that the original information is preserved trough the network and it is not blocked by any nonlinearity function. Exclusively, the convolution layers in the multi-modal multi-resolution fusion (3M2RF) block before the up-sampling operation perform the weighted summation of multi-feature maps. These convolution layers are utilized for dimension adaptation to perform a weighted summation as the multi-modal fusion strategy. In this case, the whole properties of the identity mapping rule and the specific derivations explained in [41] are not satisfied. While these layers are limited to the dimension adaptation (linear transformation), it could not affect the "clean" information stream exponentially to vanish the gradients. The overall structure shows that up to up-sampling process in the 3M2RF, the identity mapping rule has been preserved. Hence, the original information flow in the decoder is not blocked by any nonlinearity via the multi-modal long-range skip connections. Consequently, the performance and stability of the network as well as the accuracy have been improved as the gradients can propagate through the network without suffering from the vanishing problem. In Fig. 4, the red line connections show the flow of this path. As such, none of the residual connections employed the nonlinearity layers in the skip connection. Then, the nonlinearity layers are not considered in the main path of the decoder branch and instead they are utilized in the residual connections. Consequently, the gradients can be propagated easily to early convolution layers.

4   Experimental Results

The proposed method has been evaluated on 3 challenging RGB-D datasets of NYU-V2 [24], SUN-RGBD [36], and Stanford 2D-3D-Semantics (2D-3D-S) [37]. The three import measurements in semantic segmentation [13], namely the global accuracy, mean accuracy, and the intersection-over-union (IoU) score are computed for each dataset. The random cropping, scaling, and flipping are utilized for data augmentation. The weights of the RefineNet are employed as the pre-trained weights of the encoder branches.

The publicly available NYU-V2 dataset contains 1449 RGB-D images with dense per pixel labeling which are categorized into 26 scene types. The dataset contains 894 different classes that are grouped into 40 classes. There is also a void class for regions without a valid ground-truth annotation. The standard split of the NYU-V2 dataset with 795 training images and 654 test images is used to evaluate the accuracy of the method. Also, 20% of the training set is randomly selected as the validation set to perform the evaluation along the training. The proposed method is compared with the most valuable and the state-of-the-art methods. Table 1 summarizes this comparison. The 3M2RNet has achieved the best mean and global accuracy as well as IoU scores. The proposed method has obtained a higher mean accuracy.

One of the most challenging RGB-D datasets is the SUN RGB-D. It contains 10335 synchronized RGB and depth images with dense per pixel labeling in which 5285 RGB-D images are denoted as the training and validation data, 2666 for training and 2619 for validation, and 5050 RGB-D images determined as the test data. The dataset classifies each pixel in one of the 37 valid classes where one class label is assigned to those pixels that do not have a valid ground-truth annotation. The distribution of the labels in the dataset is considerably unbalanced and it is remarkable that approximately 25% of the pixels in the training data have not been assigned to any of 37 valid classes and are set as the void class. To cope with the void class, the random cropping is performed relying upon the class-based sampling. Table 2 shows the comparison of 3M2RNet with previous methods. In particular, 3M2RNet significantly outperforms the best method presented before and achieves the state-of-the-art accuracy in all 3 evaluation metrics.

Table 1. Semantic segmentation accuracy on NYU-V2 dataset.

Methods	Data	Pixel acc.	Mean acc.	IoU
Gupta et al. [42]	RGB-D	–	35.1	–
Eigen et al. [9]	RGB	65.6	45.1	34.1
FCN [13]	RGB-D	65.4	46.1	34.0
Wang et al. [34]	RGB-D	–	47.3	–
Context [43]	RGB	70.0	53.6	40.6
PAG [44]	RGB-D	73.5	–	46.5
D-CNN [45]	RGB-D	–	56.3	43.9
RefineNet [17]	RGB	73.6	58.9	46.5
RDFNet [40]	RGB-D	76.0	62.8	50.1
3M2RNet (Proposed)	RGB-D	76.0	63.0	48.0

Table 2. Semantic segmentation results on SUN RGB-D dataset.

Methods	Data	Pixel acc.	Mean acc.	IoU
Ren et al. [21]	RGB-D	–	36.3	–
B-SegNet [46]	RGB	71.2	45.9	30.7
LSTM-CF [38]	RGB-D	–	48.1	–
FuseNet [39]	RGB-D	76.3	48.3	37.3
Context [43]	RGB	78.4	53.4	42.3
D-CNN [45]	RGB-D	–	53.5	42.0
RefineNet [17]	RGB	80.6	58.5	45.9
RDFNet [40]	RGB-D	81.5	60.1	47.7
3M2RNet (Proposed)	RGB-D	83.1	63.5	49.8

The new and large-scale RGB and depth images along with the corresponding surface normals, 3D point clouds, 3D and 2D annotations, and camera metadata have been collected as the Stanford 2D-3D Semantics dataset. It contains about 102,000 RGB-D data with the 13 categories of objects which are annotated in the instance level in 2D and 3D spaces. To the best of our knowledge, it is the first paper which reports the 2D semantic segmentation results in this dataset. Hence, we created a baseline evaluation on this dataset for the 2D semantic segmentation. The evaluation results of the 3M2RNet are listed in Table 3, for 13 valid classes.

Table 3. Semantic segmentation results on 2D-3D-S dataset.

Method	Data	Pixel acc.	Mean acc.	IoU
3M2RNet (Proposed)	RGB-D	79.8	75.2	63.0

In the experimental results, the two main goals of the 3M2RNet have been evidenced by comparison with the two new and state-of-the-art methods. The results illustrate that using the depth channel in a proper manner improves the accuracy and performance of the network, remarkably. The 3M2RNet explicitly exploits the depth channel along with the down-sampling process as well as in the refinement of the information loss problem in the up-sampling process of the semantic segmentation. The proposed method increased the prediction performance of the RefineNet by exploiting the depth channel in the separate encoder branch. In addition, it is noteworthy that 3M2RNet achieves a higher accuracy than the RDFNet which employs the multi-modal feature fusion (MMF) network to fuse the RGB and depth before the refinement blocks of the RefineNet. In the MMF, the clean information path for long-range skip connections has got weaken by the Relu used after the fusion of two modalities. Hence, the fused features employed for the refinement instead of the clean flows of information from the RGB and depth branches as proposed in our 3M2RNet. On the other hand, the clean flow of information is also diminished in the MMF blocks before and after the fusion. Therefore, the experimental results have shown the importance of using the identity mapping rule which allows an efficient end-to-end training in deep networks by solving the vanishing gradient problem.

5   Conclusion

The fusion of RGB and depth images with the attention on the identity mapping of the long and short-residual connections was investigated. The proposed method is a type of encoder-decoder CNNs with two separate branches for encoding RGB and depth images, simultaneously, and one decoder to fuse and refine the information loss caused by the down-sampling process in encoder branches. The two main challenges of the RGB-D semantic segmentation have been considered and a method has been proposed to overcome these challenges, efficiently, in terms of mean and global accuracy as well as the number of parameters. The proposed method achieves improvements in terms of the IoU score, mean accuracy, and global accuracy in the more challenging SUN RGB-D dataset. It also attains the state-of-the-art results in the NYU-V2, SUN RGB-D, and Stanford 2D-3D-Semantics datasets.

References

1. Krizhevsky, A., Sutskever, I., Hinton, G.E.: Imagenet classification with deep convolutional neural networks. In: Advances in Neural Information Processing Systems, pp. 1097–1105 (2012)
2. Simonyan, K., Zisserman, A.: Very deep convolutional networks for large-scale image recognition (2014). arXiv preprint arXiv:14091556
3. Szegedy, C., Liu, W., Jia, Y., Sermanet, P., Reed, S., Anguelov, D., Erhan, D., Vanhoucke, V., Rabinovich, A.: Going deeper with convolutions. In: IEEE Conference on Computer Vision and Pattern Recognition, pp. 1–9 (2015)
4. Cadena, C., Košecka, J.: Semantic parsing for priming object detection in RGB-D scenes. In: 3rd Workshop on Semantic Perception, Mapping and Exploration (2013)

5. Ren, S., He, K., Girshick, R., Sun, J.: Faster R-CNN: towards real-time object detection with region proposal networks. In: Advances in Neural Information Processing Systems, pp. 91–99 (2015)
6. Szegedy, C., Toshev, A., Erhan, D.: Deep neural networks for object detection. In: Advances in Neural Information Processing Systems, pp. 2553–2561 (2013)
7. Simonyan, K., Zisserman, A.: Two-stream convolutional networks for action recognition in videos, pp. 568–576 (2014)
8. Wang, L., Qiao, Y., Tang, X.: Action recognition with trajectory-pooled deep-convolutional descriptors. In: IEEE Conference on Computer Vision and Pattern Recognition, pp. 4305–4314 (2015)
9. Eigen, D., Fergus, R.: Predicting depth, surface normals and semantic labels with a common multi-scale convolutional architecture. In: IEEE International Conference on Computer Vision, pp. 2650–2658 (2015)
10. Liu, F., Shen, C., Lin, G.: Deep convolutional neural fields for depth estimation from a single image. In: IEEE Conference on Computer Vision and Pattern Recognition, pp. 5162–5170 (2015)
11. Toshev, A., Szegedy, C.: Deeppose: human pose estimation via deep neural networks. In: IEEE Conference on Computer Vision and Pattern Recognition, pp. 1653–1660 (2014)
12. Chen, L.-C., Papandreou, G., Kokkinos, I., Murphy, K., Yuille, A.L.: DeepLab: semantic image segmentation with deep convolutional nets, atrous convolution, and fully connected CRFs. IEEE Trans. Pattern Anal. Mach. Intell. **40**(4), 834–848 (2018). %@ 0162-8828
13. Long, J., Shelhamer, E., Darrell, T.: Fully convolutional networks for semantic segmentation. In: IEEE Conference on Computer Vision and Pattern Recognition, pp. 3431–3440 (2015)
14. He, K., Zhang, X., Ren, S., Sun, J.: Deep residual learning for image recognition. In: IEEE Conference on Computer Vision and Pattern Recognition, pp. 770–778 (2016)
15. Huang, G., Liu, Z., Van Der Maaten, L., Weinberger, K.Q.: Densely connected convolutional networks. In: IEEE Conference on Computer Vision and Pattern Recognition, p. 3 (2017)
16. Couprie, C., Farabet, C., Najman, L., LeCun, Y.: Indoor semantic segmentation using depth information (2013). arXiv preprint arXiv:13013572
17. Lin, G., Milan, A., Shen, C., Reid, I.D.: RefineNet: multi-path refinement networks for high-resolution semantic segmentation. In: IEEE Conference on Computer Vision and Pattern Recognition, p. 5 (2017)
18. Gupta, S., Arbelaez, P., Malik, J.: Perceptual organization and recognition of indoor scenes from RGB-D images. In: IEEE Conference on Computer Vision and Pattern Recognition, pp. 564–571. IEEE (2013). %@ 1063-6919
19. Li, C., Kowdle, A., Saxena, A., Chen, T.: Towards holistic scene understanding: feedback enabled cascaded classification models. In: Advances in Neural Information Processing Systems, pp. 1351–1359 (2010)
20. Muller, A.C., Behnke. S.: Learning depth-sensitive conditional random fields for semantic segmentation of RGB-D images. In: IEEE International Conference on Robotics and Automation, pp. 6232–6237. IEEE (2014)
21. Ren, X., Bo, L., Fox, D.: RGB-(D) scene labeling: features and algorithms. In: IEEE Conference on Computer Vision and Pattern Recognition, pp. 2759–2766. IEEE (2012). %@ 1467312282
22. Shotton, J., Winn, J., Rother, C., Criminisi, A.: Textonboost: joint appearance, shape and context modeling for multi-class object recognition. In: Computer Vision–ECCV 2006, pp. 1–15. Springer (2006). %@ 3540338322

23. Silberman, N., Fergus, R.: Indoor scene segmentation using a structured light sensor. In: IEEE International Conference on Computer Vision Workshops, pp. 601–608. IEEE (2011). %@ 1467300624
24. Silberman, N., Hoiem, D., Kohli, P., Fergus, R.: Indoor segmentation and support inference from RGBD images. In: European Conference on Computer Vision, pp. 746–760. Springer (2012). %@ 3642337147
25. Silberman, N., Sontag, D., Fergus, R.: Instance segmentation of indoor scenes using a coverage loss. In: European Conference on Computer Vision, pp. 616–631. Springer (2014). %@ 3319105892
26. Zheng, S., Cheng, M.-M., Warrell, J., Sturgess, P., Vineet, V., Rother, C., Torr, P.H.S.: Dense semantic image segmentation with objects and attributes. In: IEEE Conference on Computer Vision and Pattern Recognition, pp. 3214–3221. IEEE (2014)
27. Plath, N., Toussaint, M., Nakajima, S.: Multi-class image segmentation using conditional random fields and global classification. In: Proceedings of the International Conference on Machine Learning, pp. 817–824. ACM (2009). %@ 1605585165
28. Reynolds, J., Murphy, K.: Figure-ground segmentation using a hierarchical conditional random field. In: Canadian Conference on Computer and Robot Vision, pp. 175–182. IEEE (2007). %@ 0769527868
29. Yang, M.Y., Förstner, W.: A hierarchical conditional random field model for labeling and classifying images of man-made scenes. In: IEEE International Conference on Computer Vision Workshops, pp. 196–203. IEEE (2011). %@ 1467300624
30. Kindermann, R., Snell, J.L.: Markov Random Fields and Their Applications, vol. 1. American Mathematical Society Providence (1980). %@ 0821850016
31. Lafferty, J., McCallum, A., Pereira, F.C.N.: Conditional random fields: probabilistic models for segmenting and labeling sequence data (2001)
32. Zeiler, M.D., Fergus, R.: Visualizing and understanding convolutional networks. In: European Conference on Computer Vision, pp. 818–833. Springer (2014)
33. Noh, H., Hong, S., Han, B.: Learning deconvolution network for semantic segmentation. In: IEEE International Conference on Computer Vision, pp. 1520–1528 (2015)
34. Wang, J., Wang, Z., Tao, D., See, S., Wang, G.: Learning common and specific features for RGB-D semantic segmentation with deconvolutional networks. In: European Conference on Computer Vision, pp. 664–679. Springer (2016)
35. Jégou, S., Drozdzal, M., Vazquez, D., Romero, A., Bengio, Y.: The one hundred layers Tiramisu: fully convolutional DenseNets for semantic segmentation. In: IEEE Conference on Computer Vision and Pattern Recognition Workshops, pp. 1175–1183. IEEE (2017). %@ 1538607336
36. Song, S., Lichtenberg, S.P., Xiao, J.: Sun RGB-D: A RGB-D scene understanding benchmark suite. In: IEEE Conference on Computer Vision and Pattern Recognition, pp. 567–576 (2015)
37. Armeni, I., Sax, S., Zamir, A.R., Savarese, S.: Joint 2D-3D-semantic data for indoor scene understanding (2017). arXiv preprint arXiv:170201105
38. Li, Z., Gan, Y., Liang, X., Yu, Y., Cheng, H., Lin, L.: LSTM-CF: unifying context modeling and fusion with LSTMS for RGB-D scene labeling. In: European Conference on Computer Vision, pp. 541–557. Springer (2016)
39. Hazirbas, C., Ma, L., Domokos, C., Cremers, D.: FuseNet: incorporating depth into semantic segmentation via fusion-based CNN architecture. In: Asian Conference on Computer Vision, pp. 213–228. Springer (2016)
40. Park, S.-J., Hong, K.-S., Lee, S.: RDFNet: RGB-D multi-level residual feature fusion for indoor semantic segmentation. In: The IEEE International Conference on Computer Vision (2017)

41. He, K., Zhang, X., Ren. S., Sun, J.: Identity mappings in deep residual networks. In: European Conference on Computer Vision, pp. 630–645. Springer (2016)
42. Gupta, S., Girshick, R., Arbeláez, P., Malik, J.: Learning rich features from RGB-D images for object detection and segmentation. In: European Conference on Computer Vision, pp. 345–360. Springer (2014). %@ 3319105833
43. Lin, G., Shen, C., Van Den Hengel, A., Reid, I.: Exploring context with deep structured models for semantic segmentation. IEEE Trans. Pattern Anal. Mach. Intell. **40**(6), 1352–1366 (2018). %@ 0162-8828
44. Kong, S., Fowlkes, C.: Pixel-wise Attentional Gating for Parsimonious Pixel Labeling (2018). arXiv preprint arXiv:180501556
45. Wang, W., Neumann, U.: Depth-aware CNN for RGB-D Segmentation (2018). arXiv preprint arXiv:180306791
46. Kendall, A., Badrinarayanan, V., Cipolla, R.: Bayesian SegNet: model uncertainty in deep convolutional encoder-decoder architectures for scene understanding (2015). arXiv preprint arXiv:151102680

Cognitive Consistency Routing Algorithm of Capsule-Network

Huayu Li$^{(\boxtimes)}$ and Yihan Wang

Northern Arizona University, Flagstaff, AZ 86001, USA
hl459@nau.edu

Abstract. Artificial Neural Networks (ANNs) are computational models inspired by the central nervous system (especially the brain) of animals and are used to estimate or generate unknown approximation functions that rely on a large number of inputs. The Capsule Neural Network [1] is a novel structure of Convolutional Neural Networks (CNN) which simulates the visual processing system of the human brain. In this paper, we introduce a psychological theory which is called Cognitive Consistency to optimize the routing algorithm of Capsnet to make it more close to the working pattern of the human brain. Our experiments show that progress had been made compared with the baseline.

Keywords: Capsnet · Cognitive Consistency · Deep learning

1 Introduction

Convolutional neural networks (CNN) contribute to a series of breakthroughs for image classification task. There are plenty of structures of CNN proven to make an outstanding performance in classification tasks in various domains. Capsnet is a novel variant of CNN proposed by [1]. Capsnet uses the outputs of a group of neurons which is called capsule [2] to represent different properties of the same entity. The mechanism [1] of Capsnet is to ensure that the output of the capsule gets sent to an appropriate parent in the layer above. Initially, the output is routed to all possible parents after scaled down by coupling coefficients that sum to 1. A "prediction vector" which is the product of its output and a weight matrix for each possible parent is computed by the capsule. Top-down feedback increases the coupling coefficient for that parent and decreases coupling coefficient for other parents, determined by the scalar product of this prediction vector and the output of a possible parent. The parent further increases the scalar product of the capsule's prediction with the parent's output as a result of increasing the contribution made by the capsule. In [1], the "agreement" is mentioned as simply the scalar product $a_j = v_j \cdot \hat{u}_{j|i}$ which is treated as if it was a log likelihood and is added to the initial logit, b_{ij} before computing the new values for all the coupling coefficients linking capsule i to higher level capsules.

In this paper, we were inspired by psychological theories and applied them to develop a new method of routing algorithm. According to [3], people have a drive

© Springer Nature Switzerland AG 2020
K. Arai and S. Kapoor (Eds.): CVC 2019, AISC 944, pp. 558–563, 2020.
https://doi.org/10.1007/978-3-030-17798-0_45

to generate consistent cognition and behavior on objects. When the cognition is dissonant, people will feel uncomfortable, and then try to reduce it, reducing a mechanism of dissonance by selectively seeking support information or avoiding inconsistent information.

We simply regard the prediction vectors made by each capsule as the "cognition" it keeps, and our work is to explore an algorithm to unify the cognition between each capsule, which we call the Cognitive Consistency Routing Algorithm. We propose to treat the differences between the input "prediction vectors" from the lower layer and the "prediction vectors" made by the current layer as dissonance coefficients to ensure the coupling between higher capsules and lower layer increases much sooner than if the differences between their predictions are smaller. The motivation behind the Cognitive Consistency Routing Algorithm is to ensure each capsule layer predicts of the target is as consistent as possible. In other words, the capsule layers of the whole network should ultimately tend to a consistent cognition.

2   Related Work

In a conventional convolutional neural network, there are usually multiple aggregation layers. Unfortunately, the operation of these aggregation layers often loses a lot of information, such as the exact location and posture of the target object. If we want to classify the entire image, it's no big deal to lose it, but the missing information is essential for performing more accurate image segmentation or object detection (which requires more precise position and orientation). A key feature called equivariance of the capsule network is the preservation of detailed information about the position and posture of the object in the network. For example, if the image is rotated a small angle, the activation vector will change slightly. The equivariance characteristics of the capsule make it very promising for a variety of tasks.

At the pooling layer, CNNs will actively discard a large amount of information. Pooling layers reduce spatial resolution. As a result, the output will be insensitive to small changes in input. If the network is supposed to save details, such as in a semantic segmentation scenario, CNNs cannot perform as good as Capsnet. Of course, some of the lost information can be recovered by retrieving a complex architecture around CNN. With Capsnet, detailed location information (such as precise object position, rotation, thickness, skew, size, etc.) is preserved throughout the network. The advantage of retaining this information is that small changes in the input directly result in subtle differences in output. As a result, Capsnet can appear in a variety of computer vision tasks simply and consistently.

3   Motivation and Consideration

We consider that since the capsule neural network is imitating the processing of visual information by the human brain, we can introduce the theory of psychology and cognitive science into its routing algorithm. We must realize that artificial intelligence emphasizes a simulation of human behavior, through the existing hardware and software technology to simulate human behavior, including machine learning, image thinking, language understanding, memory, and a series of intelligent behaviors such as reasoning, common sense reasoning, and non-monotonic reasoning. In this paper, we apply the theory of consistency in cognitive science to the routing algorithm of capsule neural networks. In a way, the improvements we made are in line with the concept of agreement in capsule neural network.

4   Cognitive Consistency Routing Algorithm

There are many possible ways to implement the general idea of Cognitive Consistency. The aim of this paper is not to find a method that achieves the state-of-art performance but merely to test and verify the practicability of the Cognitive Consistency Routing Algorithm.

Firstly, we want to get the initial values of the prior probabilities from the lower capsules and let the capsules discriminatively coupled to the lower layers. We ,therefore, use a simple "clip" function for the prior probabilities initially to ensure that the capsules above accept the prediction vectors $\hat{\mathbf{u}}_{j|i}$ from the capsules below and restrict them within a range.

$$b_{ij} = \begin{cases} a_{min}, & \text{if } \hat{\mathbf{u}}_{j|i} < a_{min} \\ \hat{u}_{j|i}, & \text{if } a_{min} < \hat{\mathbf{u}}_{j|i} < a_{max} \\ a_{max}, & \text{if } \hat{\mathbf{u}}_{j|i} > a_{min}, \end{cases} \tag{1}$$

where the $\hat{\mathbf{u}}_{j|i}$ are the prediction vectors from the lower capsules and (a_{min}, a_{max}) is the expected range of b_{ij}. The motivation behind "clip" function is to avoid completely inactivating the prediction vectors. We expect the "bad guys" which are decided by the lower capsule continue to have the right to make predictions in the current layer thus the whole network maintain a consistent cognition without partial dissonance.

The coupling coefficients c_{ij} between capsule i and all the capsules in the layer above sum to 1 are determined by the distribution of the prior probabilities by the "softmax" function.

$$c_{ij} = \frac{exp(b_{ij})}{\sum_k exp(b_{ik})} \tag{2}$$

Our approach to getting the total input to a capsule s_j is as the same as [1].

$$\mathbf{s}_j = \sum_j c_{ij}\hat{\mathbf{u}}_{j|i}, \hat{\mathbf{u}}_{j|i} = \mathbf{W}_{ij}\mathbf{u}_i \tag{3}$$

where $\mathbf{u}_i$ is the output of the capsule in the layer below and $\mathbf{W}_{ij}$ is the weight matrix.

And the vector output of the capsule j is the "squashing" results of $\mathbf{s}_j$.

$$\mathbf{v}_j = \frac{\|\mathbf{s}_j\|^2}{1 + \|\mathbf{s}_j\|^2} \frac{\mathbf{s}_j}{\|\mathbf{s}_j\|} \tag{4}$$

which ensure that the vectors are in the range of zero to one.

The prior probabilities are supposed to be iterated by adding the scalar product $\|\mathbf{v}_j\| \cdot \|\hat{\mathbf{u}}_{j|i}\| \cdot a_{ij}$.

$$b_{ij} = b_{ij} + |\mathbf{v}_j| \cdot |\hat{\mathbf{u}}_{j|i}| \cdot a_{ij},\, a_{ij} = cos((|\mathbf{v}_j| - |\hat{\mathbf{u}}_{j|i}|)^2) \tag{5}$$

where a_{ij} are the consistency ratios that decreases with the increment of difference between the input prediction vectors from the lower layer and the prediction vectors made by the current layer. The curve of a_{ij} is shown as Fig. 1.

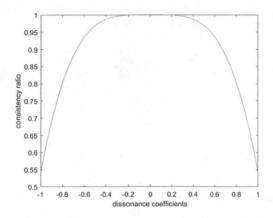

Fig. 1. The curve of consistency ratios shows that the bigger difference between the predictions made by the current layer and the lower layer lead to the smaller consistency ratios which means each layer tends to a consistent cognition to avoid cognitive dissonance.

The overall flow of our algorithm is shown in Algorithm 1.

Algorithm 1. Cognitive Consistency Routing Algorithm of Capsule-network

---

1: For all capsule i in layer l and capsule j in layer $(l+1)$: $b_{ij} \leftarrow clip(\hat{\mathbf{u}}_{j|i}, a_{min}, a_{max})$.
2: **for** each r iterations **do**
3:     for all capsule i in layer l: $c_{ij} \leftarrow \frac{exp(b_{ij})}{\sum_k exp(b_{ik})}$;
4:     for all capsule j in layer $l+1$: $\mathbf{s}_j \leftarrow \sum_j c_{ij}\hat{\mathbf{u}}_{j|i}$;
5:     for all capsule j in layer $l+1$: $\mathbf{v}_j \leftarrow squash(\mathbf{s}_j)$;
6:     for all capsule i in layer l and all capsule j in layer $l+1$: $b_{ij} \leftarrow b_{ij} + \|\mathbf{v}_j\| \cdot \|\hat{\mathbf{u}}_{j|i}\| \cdot cos((|\mathbf{v}_j| - |\hat{\mathbf{u}}_{j|i}|)^2)$;
7: **end for**
8: **return** $\mathbf{v}_j$;

---

5   Experimental Results

Due to the lack of computing resources and slowness of the routing process, we can only test whether our algorithm has improved performance compared to the original algorithm on toy datasets. But from the experimental results on the mnist and fashion-mnist datasets, our algorithm is successful to make an improvement than the original algorithm.

5.1   Our Algorithm on MNIST

We tested our algorithm on MNIST to verify whether our algorithm works or not. Our model has the same architecture as [1] showed in Fig. 2 and was set as 3 times routing, but with batch normalization (BN) [4] after each layer. We tested two algorithms on this model and compared their performance. We observed that a Capsnet with Cognitive Consistency Routing achieves state-of-the-art performance on MNIST which equals to the baseline.

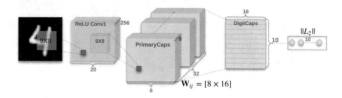

Fig. 2. The structure of the original Capsnet.

5.2   Our Algorithm on Fashion-MNIST

Via experiments on other datasets, we can directly observe that our algorithm get better and higher results than the baseline in more complex data. We use the same model as on MNIST. We evaluated our method and the original algorithm on the Fashion-MNIST dataset [5] which consists of 28×28 grayscale images of 70,000 fashion products from 10 categories, with 7,000 images per category. The models were trained on the 48,000 training images and evaluated on the 12,000 validation images. We also obtained a final result on the 10,000 test images. We tested using 10 models and got the final result by model averaging. We can see from Table 1 that our routing approach achieved better performance than the baseline.

Table 1. Our Algorithm on Fashion-MNIST, the higher the better as the original Algorithm

Iterations	1	3	5	10	15
Sabour et al. [1]	0.7632	0.7993	0.8227	0.8514	0.8644
Our Algorithm	0.7832	0.8163	0.8266	0.8655	0.8803

6   Conclusion

We show that the Cognitive Consistency from psychological and cognitive theories can improve the capsule neural network and demonstrate that introducing the theories from other domains, besides only statistics or computer science is an effective way to perfect the theoretical foundation of deep learning and Artificial Neural Networks. Future work includes improving our algorithm to achieve the state-of-art performance and trying to introduce more theories from psychologic and cognitive science into Artificial Neural Networks. We will also try to improve the overall flow of the routing algorithm to enable the capsule neural network more quickly for practical computer vision projects.

Acknowledgment. At the point of finishing this paper, we would like to express our sincere thanks to the authors of Dynamic routing between capsules [1] who had made significant breakthroughs and outstanding contributions in exploring the new architecture of neural networks. At the same time, we also have to thank Professor. Razi and Professor. Bakke who gave guidance and support to us during the process of completing this paper.

References

1. Sabour, S., Frosst, N., Hinton, G.E.: Dynamic routing between capsules. In: Advances in Neural Information Processing Systems, pp. 3856–3866 (2017)
2. Hinton, G.E., Krizhevsky, A., Wang, S.D.: Transforming auto-encoders. In: International Conference on Artificial Neural Networks, pp. 44–51. Springer, Heidelberg (2011)
3. Festinger, L.: A Theory of Cognitive Dissonance. Stanford University Press, Palo Alto (1962)
4. Ioffe, S., Szegedy, C.: Batch normalization: accelerating deep network training by reducing internal covariate shift. arXiv preprint arXiv:1502.03167 (2015)
5. Xiao, H., Rasul, K., Vollgraf, R.: Fashion-MNIST: a novel image dataset for benchmarking machine learning algorithms. arXiv preprint arXiv:1708.07747 (2017)

Minimizing the Worst Case Execution Time of Diagnostic Fault Queries in Real Time Systems Using Genetic Algorithm

Nadra Tabassam[✉], Sarah Amin, and Roman Obermaisser

University of Siegen, Siegen, Germany
nadra.tabassam@uni-siegen.de

Abstract. The number of embedded systems in safety-critical applications are continuously increasing. These systems requires high level of reliability and have strict timing constraints specially in case of fault occurrence. One method to enhance the reliability and availability of these systems is to introduce the concept of optimization of diagnostic fault queries and real time database management systems. Both of them can be used to trace back failures to faults and trigger suitable recovery actions. Our major concern is the completion of diagnostic query in bounded time in order to satisfy timing constraints for fault recovery (e.g. actuator freezing). For this purpose it is important to provide a solution which can optimize the diagnostic fault queries in a manner that they can complete their execution within the pre-defined deadline of the real time system. Our proposed algorithm optimize the diagnostic fault queries using genetic algorithm, so that the overall Worst Case Execution Time (WCET) of these queries can be minimized. A diagnostic query is represented in the form of (i) Left Deep Tree (LDT) and (ii) Bushy Tree (BT). Each query tree is converted into multiple task graphs by considering different combinations of nodes (in query tree). Our genetic algorithm selects the task graph with minimum make span (scheduling length), so that the goal of fault diagnosis within the defined deadline of the real time system can be achieved. The evaluation based on our results shows that the WCET of the diagnostic queries is better in case of bushy trees and ring topology.

Keywords: Active diagnosis · Real time systems · RTDBMS ·
Diagnostic queries · Pervasive SQL · WCET · Genetic algorithm

1 Introduction

Safety critical systems are different from traditional IT systems. In safety-critical systems, it is important that a fault is determined and the problem is solved within predictable time before the system reaches an unsafe state. Therefor any failure might result in loss of life, damage of environment or damage of property so it is essential that these systems should equipped with a fault-tolerance mechanism. For example in case of autonomous vehicles the performance of all system

© Springer Nature Switzerland AG 2020
K. Arai and S. Kapoor (Eds.): CVC 2019, AISC 944, pp. 564–582, 2020.
https://doi.org/10.1007/978-3-030-17798-0_46

components should be monitored actively along with fault detection mechanism, in order to guarantee the safety of passenger [3].

Another example is a Flight Tolerant Control System (FTCS). An active FTCS system reacts to the malfunctioning of system components by reconfiguring the controller based on the information received from a Fault Detection and Diagnosis (FDD) unit. The major objectives of an active FTCS is to develop an effective FDD scheme to provide information about the fault with minimal uncertainties in a timely manner [4]. Therefore the fault detection based on-line diagnosis improves the overall reliability of these safety critical systems. The active diagnosis in these systems is imperative because these applications have timing constraints for fault recovery (e.g., actuator freezing, limited time until correct service is required again or to prepare for the next fault when combining diagnosis with conventional replication) [12].

There are two solutions in state of art in order to ensure the reliability of these systems. First solution is based on adding the redundant components in the final product. But it is not feasible as this increases the overall cost of the final product [7]. Another solution is the continuous monitoring of the system for temporary or permanent component failures. There are two types of diagnostic techniques in the literature, (i) passive diagnosis that stores the diagnostic information and analyzes it later for fault detection and recovery and (ii) active diagnosis that analyzes the diagnostic information at run time so that an immediate recovery action can be taken [16].

Another fundamental concern of these systems is the effective data processing within time bound of the system. Due to the huge amount of data processing among the different components of the system, it is important to process and store data by considering the timing constraints of the system. In order to meet these timing requirements specifically for data processing in real time systems, numerous real time database management systems like Genesis, Polyhedra, Berkeley DB and TimesTen have designed [6]. These real time database management systems have properties like temporal consistency, timeliness, meeting deadlines, etc.

Therefore the execution of diagnostic fault queries over the database is an effective technique for root cause analysis i.e. to find the fault responsible for a failure and determine an appropriate recovery action. But the processing of fault queries should be completed within the timing constraints of the real time system.

Contributions: Our proposed algorithm optimize the diagnostic fault queries so that their WCET can be minimized and they can complete their execution within the pre-defined deadline of the system. Our major contributions are as follows:

WCET of the diagnostic query is minimized so that it can complete its execution within the defined deadline of the system. This is the major concern of our work. Proposed GA goes through the search space and selects the task graph with minimum make span. Fitness function for GA is based on the calculation of make span of each task graph. For calculating the fitness score of each TG we are invoking scheduler. Our scheduler will map the TGs on end systems considering different topologies.

The rest of the paper is structured as follows. Section 2 describes the related work and gap existing in the state of art. Section 3 describes the steps in proposed algorithm. Section 4 describes the calculation of WCET. Section 5 describes the importance of scheduler in our work. Section 5 describes the illustrative example and Sect. 6 describes the experimental results.

2   Related Work

In case of query optimization, database management system sometimes chooses the worst query execution plan for some queries. This is the case where query optimizer does not considers the amount of time and resources that needs to be efficiently selected [8]. An improved particle swarm optimization technique is used to perform the embedded database query optimization [11]. An experimental embedded system platform using embedded database SQLite3.3.4 has been designed to test the effectiveness of proposed algorithm. Most of the genetic algorithm in case of query optimization only deals with the join ordering problem [9,14]. Selection of best index for query execution has also been implemented by using genetic algorithm [5]. In order to select the best join order for the creation of optimal optimization strategy, the author in [14] has employed the genetic query optimizer (GQO). Author completes the experimentation by associating GQO with DB2, PostgreSQL and MySQL.

In [10] author has presented a framework to optimize the query execution plan while using a flexible communication model. A polynomial time algorithm for finding the optimal plan is proposed. This algorithm minimizes the total communication cost when the communication graph is restricted to be a tree. In [15] authors consider the problem of improving the performance of parallel query tasks based on a tool known as QScheduler which is designed and implemented for controlling the process of the query execution sequence. Proposed QScheduler treats the database system as a black box and decides which query to schedule preferentially according to the different algorithm.

For the query optimization of distributed database, the speed of query depends on the amount of data transfer and order of join. The cost model which can minimize the communication cost is the basic concern of this research. Parallel Genetic Algorithm-Max-Min Ant System is proposed to seek a best query execution plan. This algorithm in [2] combines faster convergence of Genetic Algorithm with the globally search ability of Max-Min Ant System and parallel properties of both.

Although the above describe work is presenting the algorithms for query optimization but in our case our underlying problem is slightly different from them. None of the aforementioned techniques considered the query optimization in stringent timing constraints that are essential for analysis in real-time distributed systems. The fault diagnosis queries are represented in SQL format. These SQL queries are then converted to task graphs. The initial population of TG are then mapped to end systems for processing by our scheduler. Scheduler is invoke in for calculating the fitness function (make span) of these TGs. Therefore this technique has not implemented in state of art and is novel way of optimizing the diagnostic queries so that they can complete their execution within the pre defined deadline of the system.

3   Steps in Proposed Algorithm

Figure 1 shows the steps involved in our proposed algorithm. Next subsections will describe the details about these steps.

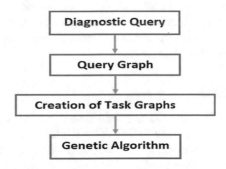

Fig. 1. Steps in proposed algorithm

Algorithm 1. Task Graph Optimization Algorithm (TGO)

    Input:Diagnostic Query (DQ)
    Output:Optimized Task Graph (TG), Makespan of TG
 1: Convert DQ into Relational Algebra (RA) representation.
 2: Convert RA of DQ into Query Tree (QT) representation.
 3: QT can be left depth tree or bushy tree
 4: Generate different Task Graphs (TG) with different combinations of nodes from QT.
 5: Add all these TG into Search Space (SP) of Genetic Algorithm (GA)
 6: Create initial population based on these TG for GA.
 7: **while** the TG with minimum Make Span (MS) is not found **do**
 8:    **for all** TG in initial population **do**
 9:       Calculate fitness function (MS)
10:    **end for**
11:    Select TG with minimum MS
12:    Apply mutation on selected solutions (TG)
13:    Send them to next generation.
14:    Exit when TG with minimum MS is found
15: **end while**

3.1 Diagnostic Query

Diagnostic fault queries are based on the features and symptoms extracted directly from the sensors data. Our diagnostic queries are represented in SQL format. In order to make our scenario simple we are only considering one type of diagnostic query. But considered query is quite complex and have multiple join operations.

3.2 Query Tree

Second step in algorithm is to convert SQL query into query tree. Query tree represents the query in the form of a relational algebra expression. The nodes of the query tree represents the relational algebra operations as well as the tables that are involved in the query. In order to make effective comparison for our results, we have considered two types of trees. These are (i) Left Deep Tree (LDT) (ii) Bushy Tree (BT) (c.f. 2–3).

3.3 Task Graph (TG)

For each query tree different TGs are generated (c.f. 4). Each node of a query trees is considered as the one task in TG. The $WCET(N)$ represents the WCET of task node. And $W(E)$ represents the weight of the edge. $W(E)$ represents the amount of data that is being transferred from parent node to the child node. We assume that each edge has unit weight that is $W(E) = 1$. Whenever the data item of size S is sent across the node, overall cost of the edge becomes $S.W(E)$. Each task graph is considered as a solution for the search space in GA. Each TG contains different number of nodes from one query tree. However it means that each TG has different number of nodes. These TGs are created depending on the type of query tree. They can be LDT based TG or BT based TG. The most simpler TG is the one in which each operation of query tree (within a query node) is consider as a one task.

3.4 Genetic Algorithm (GA)

Last step is to apply the GA in order to find the TG which has the minimum make span. Next section will describe all the necessary steps in this context.

Initial Population. Each TG (individual solution) that is created by considering the different combinations of query tree operations is considered as the part of our solution space (c.f. 5–6). We have used the binary string for encoding the chromosome. Each SQL operation within the query tree is assigned 1 when it is considered as a separate operation in the task graph. And each SQL operation in query tree is assigned 0 if it is considered in combination with the other SQL operation. For example if the parent SQL operation is combined with the child SQL operation (two joins combine together in the form of one SQL query). Then the parent SQL operation is assigned 1 and child SQL operation is assigned 0.

Fitness Function. Calculation of fitness function is the most important part of our work (c.f. 7–10). The fitness score for each solution is calculated by invoking the scheduler. The WCET of each node along with the weights of edges (for each task graph) is given as an input to the scheduler for calculating the make span. The scheduler maps these TGs for processing to the end systems. Scheduler is considering different topologies including Star-Bus, Star-Ring and Star-Star. The calculated make span of a each task graph is considered as our fitness score. The scheduler is invoke whenever the fitness function of the solution (task graph) in search space has to calculate. More information about the type of scheduler is out of context of this paper. For more details please refer to [1].

Selection. Our selection phase selects the fittest individual and passe them to the next generation (c.f. 11–12). For each solution (TG) make span is calculated and then all make spans are compared to each other. The TG with minimum make span are considered as the best one to select.

Mutation. In our case we will not perform the cross over of two solutions because our new individual becomes wrong. We will only perform mutation (c.f. 13). In case of mutation it is important to maintain the order of a task graph. So that when the new TG is generated the considered diagnostic query should not changed. All type of mutation is not allowed because it is possible that the new TG will have different query.

Termination. Our algorithm terminates when our GA keeps on finding the similar make spans or in other words the make span does not decrease more after a certain value.

4   Calculation of Worst Case Execution Time

The WCET is the estimation of the maximum time that a diagnostic query can take to complete its execution [13]. In our case we are concerned about the WCET of whole query which actually means that we have to calculate the WCET of each node within a task graph. Therefore these nodes actually contains the SQL operations with in the diagnostic query. And all of these operations are executed on the cars database, which is created in PSQL. For finding the WCET for each node of a task graph, the SQL operation in this graph node is split from the main query and converted into sub queries. The number of sub queries depends upon the number of operations present in the original query. These sub queries are measured against all the combinations of data. Almost 3,000 test queries are run to find the estimated WCET.

5   Illustrative Example

This sections describes the illustrative implementation of our proposed algorithm with the help of working example.

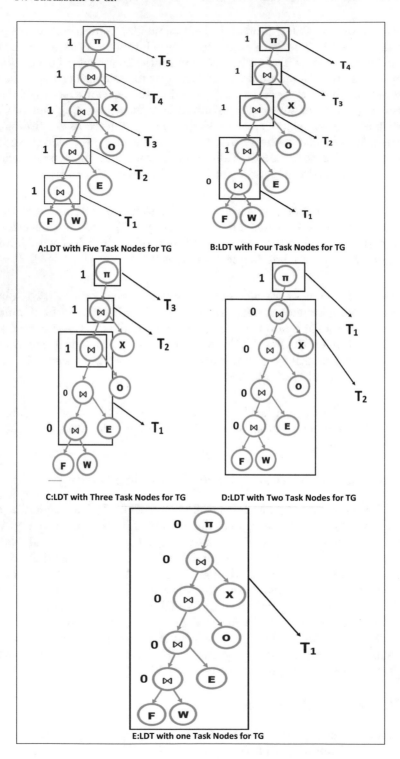

Fig. 2. Different left deep trees

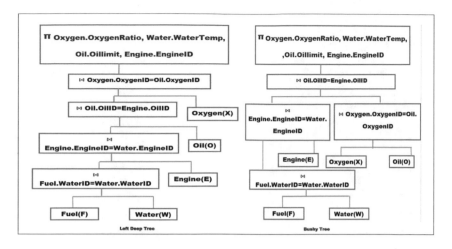

Fig. 3. Query tree for A: Left Deep Tree, B: Bushy Tree

5.1 Diagnostic Query

As described earlier the diagnostic query is represented in SQL format. The example diagnostic query is executed on the database, that we have designed in PSQL. This database comprises of data that is extracted from the cars sensors. The example diagnostic query we are considering is denoted by Q_D. The query Q_D is a join query which is joining five different tables. Defined query is complex because data manipulation is performed on different set of data from different database tables.

Q_D: **select Oxygen.OxygenRatio, Water.WaterTemp, Oil.Oillimit, Engine.EngineID from Oxygen, Oil, Water, Engine, Fuel where Fuel.WaterID = Water.WaterID   and   Water.WaterID =   Engine. EngineID   and   Oil.OilID = Engine.OilID   and   Oxygen.OxygenID = Oil.OxygenID.**

5.2 Query Tree

As described in Sect. 3 that we have considered two types of query trees. These are (i) Left Deep Tree (LDT) and (ii) Bushy Tree (BT). Figure 3 shows the informal representation of query Q_D in the form of LDT and BT. While Fig. 2 shows the different LDT. Figure 2A generates the LDT based TG which will have five nodes. Figure 2B generates the LDT based TG which will have four nodes. Figure 2C generates the LDT based TG which will have three nodes. Figure 2D generates the LDT based TG which will have two nodes. Figure 2E generates the LDT based TG which will have only one node.

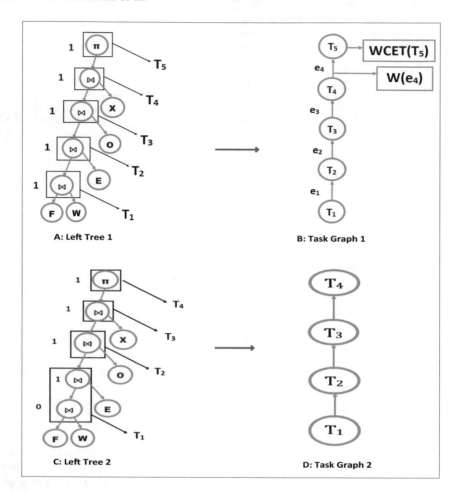

Fig. 4. A: Left Deep Tree 1, B: Task Graph 1, C: Left Deep Tree 2, D: Task Graph 2

5.3   Task Graph

We start creating the most simpler TG which consider all the operations (join, projection) of a query tree as a one separate task (Fig. 4B). For generating other TG we consider more than one operation of a query tree as a one task (Fig. 4D). As mentioned in Sect. 3 that TGs are generated depending on the type of query tree. They can be LDT based TG or BT based TG. Sections 5.4 and 5.5 describes LDT based TG or BT based TG and BT based TG.

5.4   Left Deep Tree Based TG

For example if we consider the Fig. 4A which is actually the LDT for query Q_D. It shows that we have five operations in this LDT including four joins and one projection.

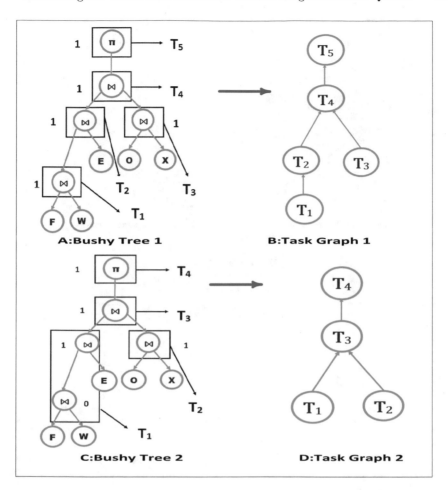

Fig. 5. A: Bushy Tree 1, B: Task Graph 1, C: Bushy Tree 2, D: Task Graph 2

Task Graph 1. If we define the first and most simpler TG for this query tree according to our assumptions, then this TG contains the five nodes.

So if we consider the relational algebra equation for query Q_D then it looks like Eq. 1.

$$\{\Pi_{(oxygen.OxygenRatio, Water.WaterTemp}$$
$$Oil.Oillimit, Engine.EngineID)} \tag{1}$$
$$((((F \bowtie W) \bowtie E) \bowtie O) \bowtie X))\}$$

Now considering Eq. 1 for creating TG, the first TG can be generated by considering each operation of the query tree as a one task. $T_1 = (F \bowtie W)$ $T_2 = \bowtie E$, $T_3 = \bowtie O$, $T_4 = \bowtie X$, $T_5 = \Pi_{(oxygen.OxygenRatio, Water.WaterTemp, Oil.Oillimit, Engine.EngineID)}$. Figure 4B shows the structure of this TG. This task graph comprise of five processing nodes.

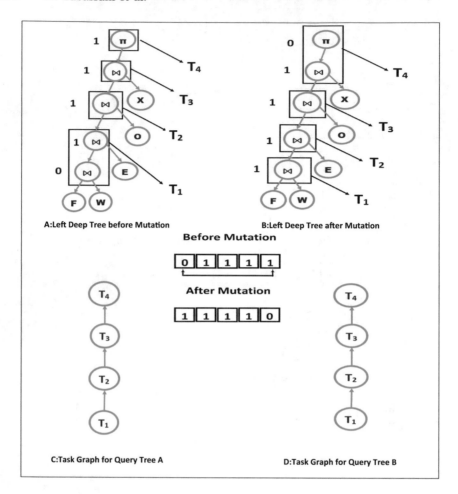

Fig. 6. Left query tree before and after mutation

Task Graph 2. Now we will generate another TG from the query tree shown in Fig. 4C. We consider the same Eq. 1 for creating this TG but this time we have combined the operations $(F \bowtie W)$ and $\bowtie E$ into one task T_1. So the task nodes defined in Fig. 4D becomes $T_1 = (F \bowtie W) \bowtie E$, $T_2 =\bowtie O$, $T_3 =\bowtie X$, $T_4 = \Pi_{(oxygen.OxygenRatio, Water.WaterTemp, Oil.Oillimit, Engine.EngineID)}$.

If we compare LDT in Fig. 4A with LDT in Fig. 4C. We can see that LDT in Fig. 4C contains the two join operations in task node T_1. While LDT in Fig. 4A only contains the one join operation in task node T_1. Therefore the TG in Fig. 4B has five nodes and TG in Fig. 4D has four nodes. And number of these task graphs keeps on increasing when we have more join operations and more relations in diagnostic query. Similarly many other task graphs can be generated in this manner.

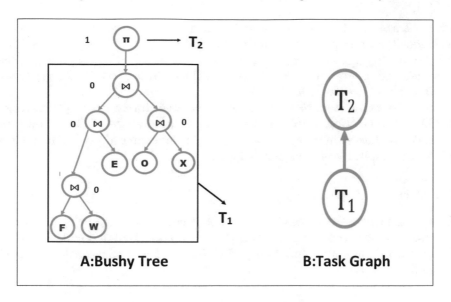

Fig. 7. A: Busy Tree B: TG for A

5.5 Bushy Tree Based TG

Task Graph 1. Now if we consider the Fig. 5A. which is BT then the Eq. 2 shows the relational algebra representation of this query Q_D.

$$\{\Pi_{(oxygen.OxygenRatio,Water.WaterTemp}$$
$$Oil.Oillimit,Engine.EngineID)} \tag{2}$$
$$[((F \bowtie W) \bowtie E) \bowtie (O \bowtie X)]\}$$

On the basis of Eq. 2, we will generate some TGs. The BT for query Q_D is shown in Fig. 5A. Suppose we generate the first TG in which we consider each operation as a one task. Now we consider $T_1 = (F \bowtie W)$, $T_2 = \bowtie E$, $T_3 = \bowtie (O \bowtie X)$, $T_4 = (((F \bowtie W) \bowtie E) \bowtie (O \bowtie X))$, $T_5 = \Pi_{(oxygen.OxygenRatio,Water.WaterTempOil.Oillimit,Engine.EngineID)}$. This TG is shown in Fig. 4B.

Task Graph 2. Now we will create another TG from the query tree shown in Fig. 5C, using the Eq. (2). This TG as shown in Fig. 5D, comprise of less number of tasks as compare to the TG shown in Fig. 5B. These tasks are $T_1 = ((F \bowtie W) \bowtie E)$, $T_2 = \bowtie (O \bowtie X)$, $T_3 = (((F \bowtie W) \bowtie E) \bowtie (O \bowtie X))$, $T_4 = \Pi_{(oxygen.OxygenRatio,Water.WaterTempOil.Oillimit,Engine.EngineID)}$.

Task Graph 3. Similarly another TG can also be created using the same technique. This task graph contains tasks including $T_1 = (F \bowtie W) \bowtie E) \bowtie (O \bowtie X)$, $T_2 = \Pi_{(oxygen.OxygenRatio,Water.WaterTempOil.Oillimit,Engine.EngineID)}$. So this task graph only contains two tasks as shown in Fig. 7B. While the Bushy tree for this TG is shown in Fig. 7A.

So different types of task graphs are generated during the implementation of GA. All these task graphs are added in solution space of GA in different generations. The solution space can be bigger depending upon the number of joins and complexity of a diagnostic query.

5.6   Genetic Algorithm

This section will describe the implementation details of our GA using illustrative example. For our proposed genetic algorithm the size of initial population $= 100$, No of generations $= 100$, mutation probability $= 0.1$ and convergence probability $= 0.2$.

Initial Population. Each TG (individual solution) is created by considering the different combinations of query tree operations. As described in Sect. 5, two types of task graphs are generated depending on the type of query tree. These TG are LDT based TG and BT based TG. The chromosome representation of these TG are shown in Fig. 2. The chromosome representation of TG based on LDT are shown in Fig. 4A and C. The chromosome representation of TG based on the BT are shown in Figs. 5A, C and 7B.

Fitness Function. The fitness score for each TG in the selected solution space is calculated by invoking the scheduler. The WCET of each task node within the TG along with the weight of edges is given as an input to the scheduler for calculation of fitness score (make-span) for TG in solution space. Suppose the initial population in our GA comprised of hundred solutions (task graphs). For each TG its make span is calculated. For elaboration of example we take three graphs (Figs. 4B, D and 5B). The input given to the scheduler for calculating the fitness score (make span) of each TG is shown in Table 1.

Selection. Our selection step selects the fittest individual and passes them to the next generation. For each solution fitness score (make span) is calculated and then all make spans are compared to each other. The solutions with minimum fitness score (make span) are considered as the best one to select. According to Table 1 the Fig. 5B has the smallest make span calculated. And it is considered as the fittest solution as it has the most smallest make span. So this solution (TG) is passed to the next generation.

Mutation. In case of our proposed solution, the order of mutation is really important. After the mutation the query should remain the same and TG should

Table 1. Make span of task graphs in Secs.

Task graph	WCET (ms)	Weight of edge W(e)MB
Fig. 4B	$T_1 = 1.08$	$T_1 \rightarrow T_2 = 58.80$
	$T_2 = 3.12$	$T_2 \rightarrow T_3 = 33.78$
	$T_3 = 4.56$	$T_3 \rightarrow T_4 = 67.24$
	$T_4 = 5.08$	$T_4 \rightarrow T_5 = 45.98$
	$T_5 = 4.23$	
Make Span	7.29 ms	
Fig. 4D	$T_1 = 4.35$	$T_1 \rightarrow T_2 = 74.90$
	$T_2 = 4.56$	$T_2 \rightarrow T_3 = 67.24$
	$T_3 = 5.08$	$T_3 \rightarrow T_4 = 45.98$
	$T_4 = 4.23$	
Make Span	6.09 ms	
Fig. 5B	$T_1 = 1.08$	$T_1 \rightarrow T_2 = 58.80$
	$T_2 = 4.87$	$T_2 \rightarrow T_4 = 56.03$
	$T_3 = 5.48$	$T_3 \rightarrow T_4 = 66.19$
	$T_4 = 4.23$	$T_4 \rightarrow T_5 = 78.90$
	$T_5 = 4.74$	
Make Span	4.18 ms	

remain in the same order. Figure 6A shows the LDT before mutation and Fig. 6B shows the LDT after the mutation. Therefore Fig. 6C shows the TG for LDT in Fig. 6A and D shows the TG for LDT in Fig. 6B. It is clear that both TG shown in Fig. 6C and D are similar in structure but it is also shown that the T_1 in Fig. 6C is different from the T_1 in Fig. 6D. It means that the overall WCET of T_1 in Fig. 6C is different from the WCET of T_1 in Fig. 6D, which changes the make span of both the task graphs.

5.7 Calculation of Worst Case Execution Time Using Example Query

This section shows the calculation of WCET of TG shown in Fig. 4B. For example the query Q_D has four sub-queries and we name them $Q_{D1}, Q_{D2}, Q_{D3}, Q_{D4}$. Q_{D1} represents the task node T_1. Q_{D2} represents the task node T_2. Q_{D3} represents the task node T_3. Q_{D4} represents the task node T_4. We are considering only join queries.

Q_{D1}: select Oxygen.OxygenRatio, Water.WaterTemp, Oil.Oillimit, Engine. EngineID from Oxygen, Oil, Water, Engine, Fuel where Fuel.WaterID = Water.WaterID and Water.WaterID = Engine.EngineID and Oil.OilID = Engine.OilID and Oxygen.OxygenID = Oil.OxygenID

Q_{D2}: select Oxygen.OxygenRatio, Water.WaterTemp, Oil.Oillimit, Engine. EngineID from Oxygen, Oil, Water, Engine, Fuel where Fuel.WaterID = Water.WaterID and Water.WaterID = Engine.EngineID and Oil.OilID = Engine.OilID

Q_{D3}: Select Oxygen.OxygenRatio, Water.WaterTemp, Oil.Oillimit, Engine. EngineID from Oxygen, Oil, Water, Engine, Fuel where Fuel.WaterID = Water.WaterID and Water.WaterID = Engine.EngineID

Q_{D4}: Select Oxygen.OxygenRatio, Water.WaterTemp, Oil.Oillimit, Engine. EngineID from Oxygen, Oil, Water, Engine, Fuel where Fuel.WaterID = Water.WaterID

According to Fig. 4B and Eq. 1 $Q_{D1} = T_5$, $Q_{D2} = T_4$, $Q_{D3} = T_3$, $Q_{D4} = T_2$. For each of these queries (task nodes) the estimated WCET is calculated by running these queries with different amount of data. The WCET calculated for the TG in Fig. 4B is shown in Table 1.

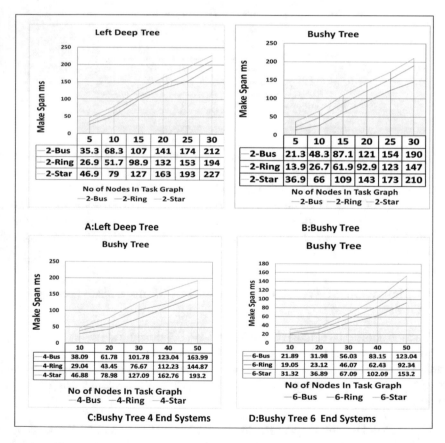

Fig. 8. A: LDT based TG, B: BT based TG, C: BT based TG 4 end systems, D: BT based TG 6 end systems

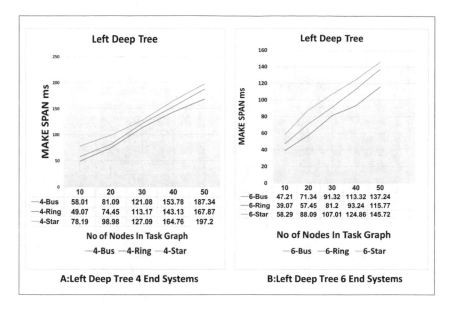

Fig. 9. A: LDT based TG 4 end systems, B: LDT based TG 6 end systems

6   Experimental Results

In our experimental results we have taken into account the three types of network topology including (i) Star-bus (ii) Star-Ring and (iii) Star-Star. All the task graphs present in the solution space are given as an input to the scheduler for calculating the fitness score (make span). Our genetic algorithm selects the task graph with minimum make span in all the generations. The scheduler is invoked when the fitness score (make span) of the TG is calculated. Our scheduler is running over the 8 networked (based on three network topology) distributed system.

Result 1. Figure 8A and B shows the results of make span calculation in case where we have considered two end systems in our topology. If we compare both results, it is clear that the task graphs generated on the basis of bushy trees has better make span as compare to the left deep trees. In case of left deep trees we have a sequential task graphs due to which scheduler can process only one task at one time because each child task has to start its execution after its parent task ended up its execution, so scheduler cannot schedule multiple tasks. On the other hand in case of bushy tree we have more tasks at one level of task graph which are ready so at one time multiple tasks can be assigned to processing nodes due to which overall make span is reduced.

Result 2. If we compare results presented in Fig. 8C and D it is clear that BT gives the better results when we have considered our scheduler with six

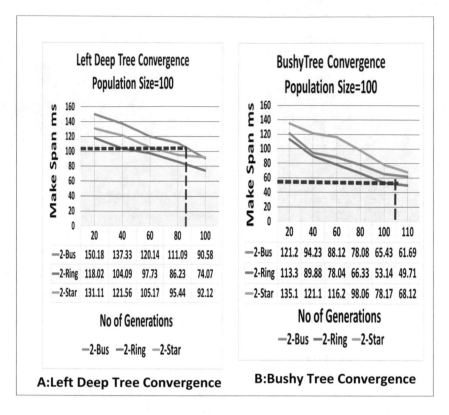

Fig. 10. A: LDT convergence, B: BT convergence

end systems. The make span of task graphs with more than 30 nodes shows remarkable decrease in case when we have more end processing systems. The task graphs withe lesser number of nodes does not show any decrease in make span when more end processing systems are added.

Result 3. If we compare the results in Fig. 9A and B it is clearly shown that even in case of left deep trees the make span is also decreased when the more processing end systems are added in the topology. So the task graphs based on deep left trees have better make spans in case of scheduler based on 6 end systems.

Result 4. If we compare the Figs. 8C and 9A, it is clearly seen that our BT based task graphs have better results in case of 4 end systems as compare to the LDT based TG.

Result 5. Similarly if we compare the Figs. 8D and 9B, it is also clear that BT based TG has lesser make span as compare to the LDT based task graphs in case of scheduler based on 6 end systems.

Result 6. If we compare the Fig. 10A and B, it is clearly seen that our BT converges earlier as compare to the LDT. In case of LDT, genetic algorithm converges after 80th generations while in case of bushy tree it converges after 100th generation but gives us better and minimized make spans.

Result 7. If we consider the results based on different topologies then it is seen that ring topology is giving us better make span as compare to the star and bus topology. As in case of ring topology we have more connections among the processing nodes so overall network load is minimized. Minimized network load increases the chances of schedule ability.

7   Conclusion

This paper presents the genetic algorithm based technique for minimizing the WCET of diagnostic queries, so that the objective of timing constraint and meeting deadlines in real time systems can be achieved. Different task graphs on the basis of diagnostic query trees are generated. These task graphs are considered as a solution for our search space in case of genetic algorithm. The task graph with minimum make span is selected by our proposed GA, so that the deadline constraint of the system can be fulfilled. Different task graphs are tested with different topology. The results shows in context of bushy trees over the ring topology are better as compare to left deep trees. The overall make span of the diagnostic query is reduced in case of bushy trees much more as compare to the left deep trees after the optimization is applied. In our future work we will considered solving the problem of memory hot spots that can cause memory bottle necks, in case of bigger tasks graphs and more complex diagnostic queries.

Acknowledgment. This work has been supported in part by the German Research Foundation (DFG) under project ADISTES (OB384/6-1, 629300).

References

1. Amin, S., Obermaisser, R.: Time-triggered scheduling of query executions for active diagnosis in distributed real-time systems. In: 2017 22nd IEEE International Conference on Emerging Technologies and Factory Automation (ETFA), pp. 1–9. IEEE (2017)
2. Ban, W., Lin, J., Tong, J., Li, S.: Query optimization of distributed database based on parallel genetic algorithm and max-min ant system. In: 2015 8th International Symposium on Computational Intelligence and Design (ISCID), vol. 2, pp. 581–585. IEEE (2015)

3. Bateman, F., Noura, H., Ouladsine, M.: Fault tolerant control strategy based on the DoA: application to UAV. In: 7th IFAC Symposium on Fault Detection Supervision and Safety of Technical Processes (2009)
4. Ducard, G.J.: Fault-Tolerant Flight Control and Guidance Systems: Practical Methods for Small Unmanned Aerial Vehicles. Springer, Heidelberg (2009)
5. Fang, L., Wang, P., Yan, J.: A multi-copy join optimization of information integration systems based on a genetic algorithm. In: The Third International Multi-Conference on Computing in the Global Information Technology, ICCGI 2008, pp. 223–228. IEEE (2008)
6. Idoudi, N., Duvallet, C., Sadeg, B., Bouaziz, R., Gargouri, F.: Structural model of real-time databases: an illustration. In: 2008 11th IEEE International Symposium on Object Oriented Real-Time Distributed Computing (ISORC), pp. 58–65. IEEE (2008)
7. Kandasamy, N., Hayes, J.P., Murray, B.T.: Time-constrained failure diagnosis in distributed embedded systems: application to actuator diagnosis. IEEE Trans. Parallel Distrib. Syst. **16**(3), 258–270 (2005)
8. Kiefer, M., Heimel, M., Breß, S., Markl, V.: Estimating join selectivities using bandwidth-optimized kernel density models. Proc. VLDB Endow. **10**(13), 2085–2096 (2017)
9. Kratica, J., Ljubić, I., Tošić, D.: A genetic algorithm for the index selection problem. In: Workshops on Applications of Evolutionary Computation, pp. 280–290. Springer (2003)
10. Li, J., Deshpande, A., Khuller, S.: Minimizing communication cost in distributed multi-query processing. In: IEEE 25th International Conference on Data Engineering, ICDE 2009, pp. 772–783. IEEE (2009)
11. Mingyao, X., Xiongfei, L.: Embedded database query optimization algorithm based on particle swarm optimization. In: 2015 Seventh International Conference on Measuring Technology and Mechatronics Automation (ICMTMA), pp. 429–432. IEEE (2015)
12. Muenchhof, M., Beck, M., Isermann, R.: Fault-tolerant actuators and drives structures, fault detection principles and applications. Ann. Rev. Control **33**(2), 136–148 (2009)
13. Munnich, A., Farber, G.: Calculating worst-case execution times of transactions in databases for event-driven, hard real-time embedded systems. In: 2000 International Database Engineering and Applications Symposium, pp. 149–157. IEEE (2000)
14. Vellev, S.: An adaptive genetic algorithm with dynamic population size for optimizing join queries (2008)
15. Zhang, Q., Li, S., Xu, J.: Qscheduler: a tool for parallel query processing in database systems. In: 2014 19th International Conference on Engineering of Complex Computer Systems (ICECCS), pp. 73–76. IEEE (2014)
16. Zhang, Y., Jiang, J.: Bibliographical review on reconfigurable fault-tolerant control systems. IFAC Proc. Volumes **36**(5), 257–268 (2003)

MNIST Dataset Classification Utilizing k-NN Classifier with Modified Sliding-Window Metric

Divas Grover$^{(\boxtimes)}$ and Behrad Toghi

University of Central Florida, Orlando, FL, USA
{GroverDivas,Toghi}@knights.ucf.edu

Abstract. The MNIST dataset of the handwritten digits is known as one of the commonly used datasets for machine learning and computer vision research. We aim to study a widely applicable classification problem and apply a simple yet efficient K-nearest neighbor classifier with an enhanced heuristic. We evaluate the performance of the K-nearest neighbor classification algorithm on the MNIST dataset where the $L2$ Euclidean distance metric is compared to a modified distance metric which utilizes the sliding window technique in order to avoid performance degradation due to slight spatial misalignments. The accuracy metric and confusion matrices are used as the performance indicators to compare the performance of the baseline algorithm versus the enhanced sliding window method and results show significant improvement using this proposed method.

Keywords: MNIST dataset · Machine learning ·
Hand-written digits dataset · K-nearest neighbor ·
Sliding window method · Computer vision

1 Introduction

The K-nearest-neighbor (k-NN) classifier is one of the computationally feasible and easy to implement classification methods which sometimes is the very first choice for machine learning projects with an unknown, or not well-known, prior distribution [1]. The k-NN algorithm, in fact, stores all the training data and creates a sample library which can be used to classify unlabeled data. During the 70's, k-NN classifier was studied extensively and some of its formal properties were investigated. As an example, authors in [1], demonstrate that for $k = 1$ case the k-NN classification error is lower bounded by the twice the Bayes error-rate. Such studies regarding mathematical properties of k-NN led to further research and investigation including new rejection approaches in [2], refinements with respect to Bayes error rate in [3], and distance weighted approaches in [4]. Moreover, soft computing [5] methods and fuzzy methods [6] have also been proposed in the literature.

© Springer Nature Switzerland AG 2020
K. Arai and S. Kapoor (Eds.): CVC 2019, AISC 944, pp. 583–591, 2020.
https://doi.org/10.1007/978-3-030-17798-0_47

A vast literature exists for the classification problem. Among which, one can refer to [3] by LeCunn et al. where authors have applied different classification algorithms ranging from k-Nearest Neighbor to SVM and Neural Networks. Authors also have used a different kind of pre-processing to increase the accuracy rate. In this work, our main idea is to skip the pre-processing procedure and finding if without any pre-processing we can increase the accuracy over the normal application of k-NN. Thus, we used 10 fold cross-validation to find optimum value of k and then applied Sliding Window technique which is commonly used in Machine Vision to detect different objects in a frame. Utilizing the sliding window technique, performance degradation, due to a minor spatial displacement between test and training, has been avoided. Accuracy, confidence interval, and confusion matrices are used to evaluate the model's prediction performance.

The rest of the paper is organized as follows: in Sect. 2, we'll see the basic outline of k-NN and how it's implemented. In Sect. 3, we have defined our process of cross-validation and our experiments with simple k-NN which is followed by Sect. 4, explaining the Sliding Window and experiments ran on it. Finally, we compare the above methods in Sect. 5 and conclude the experiments in Sect. 6.

2   Baseline k-NN Implementation

The k-NN algorithm relies on voting among the k nearest neighbors of a data point based on a defined distance metric. The distance metric is chosen considering the application and the problem nature and it can be chosen from any of the well-known metrics, e.g., Euclidean, Manhattan, cosine, and Mahalanobis, or defined specifically for the desired application. In our case, we utilize a Euclidean distance metrics which is defined as follows:

$$
\mathbf{D}^{n \times m}(\mathbf{x^n}, \mathbf{y^m}) = \left[\sum_{i=0}^{28 \times 28 - 1} (\mathbf{x_i^n} - \mathbf{y_i^m})^2 \right]^{\frac{1}{2}}
\tag{1}
$$

where,

$$\mathbf{x^n}; n \in \{0, 59999\} \text{ is the } TrainSet$$

$$\mathbf{y^m}; m \in \{0, 9999\} \text{ is the } TestSet$$

As it is mentioned before, we are using a refined version of the MNIST database of handwritten digits which consists of $60,000$ labeled training images as well as $10,000$ labeled test images. Every data-point, i.e., a test or training example is a square image of a handwritten digit which has been size-normalized and centered in a fixed-size 28×28 pixel image [2]. Figure 1 show an example visualization of the first 50 digits from the above-mentioned test set. Similar digits can have various shapes and orientation in the dataset which means, in the extreme case, distance, i.e., Euclidean distance, between two alike digits can be greater than that of two non-alike digits. This virtue adds more complexity to the classification process and may cause performance degradation of the model. Thus, pre-processing the data can help to mitigate such classification errors and consequently improve the model's performance.

As it is demonstrated in Algorithm 1, we use a 2-dimensional matrix to store all distance pairs between test and training data points, i.e. $[60,000 \times 10,000]$, distances. Once this distance matrix is derived, there is no need to calculate distances on every iteration, for example for different k-values. This significantly increases the time-efficiency and decreases the run times. The ith row of the distance matrix stores the distance between the ith test image and all 60,000 training images. Thus, for every test image, the ith row of the matrix is sorted and k lowest distances are extracted alongside their corresponding indices. These indices can be used for comparison to the original saved labels in order to evaluate the classification accuracy. In order to attain more intuition on the problem, we calculate the average distance between a sample digit and all other digits in the data set; results are shown in Fig. 2.

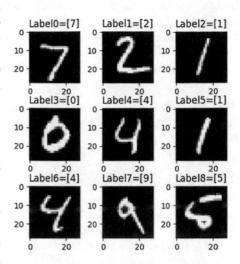

Fig. 1. Sample digits from the testset

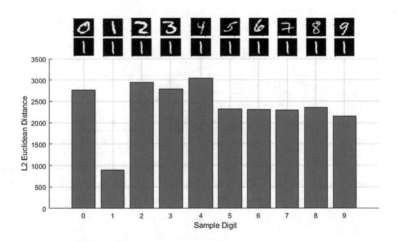

Fig. 2. Average Euclidean distance between a sample "one" digit and other training images

As it is shown in Fig. 2, the distance between two alike digits is obviously less than that of two different digits. However, in some cases where digits are visually similar, e.g., eight and zero, the distance between two not-alike digits can be also small. The next step, after getting the nearest neighbors to a test image, would be reporting the evaluation metrics. We chose accuracy metric, confidence interval, and confusion matrix as our evaluation metrics. Accuracy can be obtained from the following equation:

$$ACC = \frac{TP + TN}{P + N} \tag{2}$$

where TP and TN denote the number of True Positive and True Negative instances, respectively. P and N are total positive and total negative samples respectively.

3   k-fold Cross Validation

As it is mentioned in the previous sections, the choice of the k-value can impact the classifier's performance. Hence, we conduct a k-fold cross-validation procedure to evaluate the obtained accuracies for k-values in the range $[0, 10]$ and consequently choose the optimal k-value [5]. The cross-validation procedure is illustrated in Fig. 3. The cross-validation is conducted over the training set by dividing it into 10 slices, each containing 6,000 images. Then the total accuracy for every k-value is derived by averaging the accuracy values for each fold. Table 2 tabulates the results for cross-validation procedure over the desired k-values also in Fig. 4 we can see how accuracy changes by varying the k value. Results demonstrate that the optimal k-value for the dataset under test should be. We use this value for the rest of our study.

Table 1. Performance analysis k $=3$

Accuracy (%)	97.17
Standard deviation	0.001658
Confidence interval	[0.97004, 0.97335]

Figure 5 illustrates the confusion matrix for the classification model with $k = 3$. Accuracy and confidence interval results are also reported in Table 2. The standard deviation of a Binomial distribution can be estimated as

$$\hat{\sigma} = \sqrt{\frac{\hat{p} \times (1 - \hat{p})}{n}} \tag{3}$$

where $\hat{p}$ is calculated accuracy. Considering 95% Confidence Interval, the z-score will 1.96 which leads to confidence interval as shown below.

$$\text{c.i.:} = \mu \pm 1.96 \times \hat{\sigma} \tag{4}$$

Algorithm 1. Pseudo code of Algorithm

```
1: function <MAIN>
2:     train, test = LOAD_DATASETS()
3:     change train and test to np.array() class
4:     cross_val = 10x10 null array to store accuracies
5:
6:     for i in range (1,10) do
7:         new_train, val_set, ind = CROSS_VALIDATION(train, i)
8:         Extract labels from first column of both new_train and val_set
9:         and store them as lbl_trn and lbl_val
10:
11:        dist_matrix = EUCLIDIAN_DISTANCE(new_train, val_set)
12:        dist_matrix is an array of 6,000 by 54,000, having distance of each validation
    example
13:        to every training example
14:
15:        for k in range (1,11) do
16:            neigh = NEIGHBOR(dist_matrix, k)
17:            neigh is the a matrix of 6,000 by k which contains indices of
18:            k Nearest Neighbors in Training Set.
19:
20:            prd = GET_LABEL(neigh, lbl_trn)
21:            prd is a vector of 6,000 having predicted labels based on the indices stored
    in neigh
22:
23:            crr = no. of correct labels; by comparing prd and lbl_val
24:            accuracy = (crr divided by length of lbl_val)x100
25:            add accuracy of ith validation set and for present value
26:            of k to cross_val(i)(k-1)
27:        end for
28:     end for
29:     save cross_val as csv file
30: end function
```

Table 2. 10-fold cross validation

k-value	1	2	3	4	5
Accuracy (%)	96.53	96.84	97.17	96.64	97.06
k-value	6	7	8	9	10
Accuracy (%)	96.28	97.11	96.84	96.39	96.59

A confusion matrix is often used as a graphic to visualize a classifier's performance. It's a quantitative plot of Actual Classes vs. the Predicted Classes. The actual classes are horizontal and predicted are vertical in Fig. 5. We can clearly see that diagonally we have bigger counts because the classifier maps actual class to correct predicted class most of the times and in other cells, we can see that how many times a certain actual class is misclassified.

4   Sliding Window L2 Metric

The idea in this section is to mitigate the false distance measurements due to the small spatial translations in the image under test or the training images. As mentioned before, every example image is a square 28×28 The idea in this section is to mitigate the false distance measurements due to the small spatial translations in the image under test or the training images. We pad all the training examples by 0s and they become squares of 30×30 then, every extended image is cropped by sliding a square 28×28 window over 9 possible positions. Hence, every image produces 9 versions among which one is the original image itself [4]. For the sake of simplicity, we use a simple black and white diagonal input image to show the sliding window process in Fig. 6. Accuracy and confidence interval are reported in Table 3. After applying the sliding window on the dataset we can also visualize the modified classifier performance through confusion matrix shown in Fig. 7. A modified distance metric $\widetilde{D}^{(n \times m)}$, can be introduced to summarize the sliding window method

$$\widetilde{\mathbf{D}}^{(\mathbf{n} \times \mathbf{m})}(\mathbf{x^n}, \mathbf{y^m}) = min\left[\widetilde{\mathbf{D}}^{(\mathbf{n} \times \mathbf{i} \times \mathbf{m})}(\mathbf{x^{n,i}}, \mathbf{y^m})\right] \tag{5}$$

where $i = 1, ..., 9$.

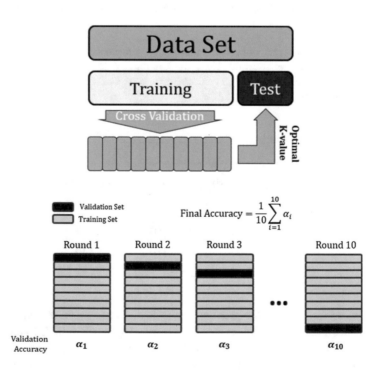

Fig. 3. 10-fold cross validation procedure and deriving optimal k-value

5   Classifier Accuracy

We conduct hypotheses evaluation in order to compare the performance of our two methods, i.e., baseline k-NN and sliding window k-NN. We define the null and alternative hypotheses as follows:

$$\begin{cases} \mathcal{H}_0 : p_{baseline} = p_{sliding} \\ \mathcal{H}_a : p_{baseline} < p_{sliding} \end{cases}$$

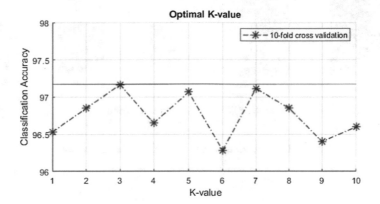

Fig. 4. Obtained accuracy for k values in range $[0, ..., 10]$

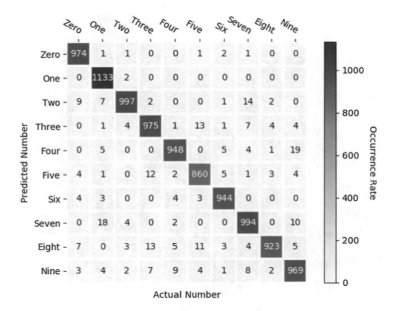

Fig. 5. Confusion matrix k = 3

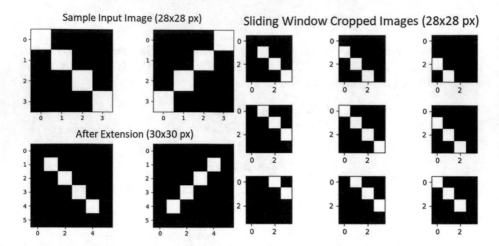

Fig. 6. Sliding window illustration

Table 3. Performance analysis k = 3 sliding window

Accuracy (%)	97.73%
Standard deviation	0.001489
Confidence interval (c.i.)	[0.97581, 0.97879]

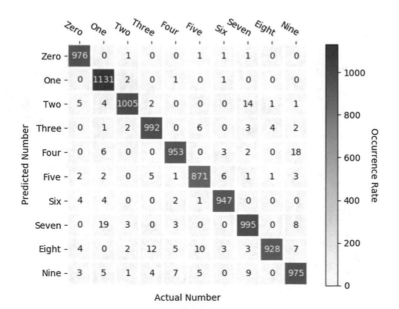

Fig. 7. Confusion matrix k = 3 sliding window

The significance level, α is defined as the test's probability of incorrectly rejecting the null hypothesis:

$$\alpha = P_r(|\hat{d} > z_N \sigma|) \tag{6}$$

or in other words $\alpha = 1 - N$, where N is the confidence interval of 95% in our case. We then reject the hypothesis if the test statistic $z = \frac{|d|}{\sigma_{\hat{d}}}$ is greater than $z_N = 1.96$ (for $n = 95$). Assuming the normal distribution, $\hat{d} = 0.0022$ or 0.2 which can be considered rule of thumb difference criteria. In our case we have $d = 0.9773 - 0.9717 = 0.0056$. Hence $d > \hat{d}$ which means the null hypothesis is rejected and there is evidence for the alternative hypothesis.

6   Concluding Remarks

We implemented a k-NN classifier with Euclidean distance and an enhanced distance metric utilizing the sliding window method. Our results show a significant improvement by employing the enhanced metric. Our algorithm relies on removing the spacial shifts from the test and training set data in a way that avoids misclassification resulting from co-labeled but shifted images. We present our results in terms of performance indicators such as accuracy metric and confusion matrices.

7   Source Code

The source code of our implementation is publicly available on GitHub[1]. The current implementation is compatible with the refined MNIST dataset, also available online [6].

References

1. Peterson, L.E.: K-nearest neighbor. http://scholarpedia.org/article/K-nearest_neighbor
2. LeCun, Y., Cortes, C., Burges, C.J.C.: The MNIST Database of handwritten digits. http://yann.lecun.com/exdb/mnist/
3. LeCunn, Y., et al.: Gradient-based learning applied to document recognition. Proc. IEEE **86**(11), 2278–2324 (1998). http://yann.lecun.com/exdb/publis/pdf/lecun-98.pdf
4. Glumov, N.I., Kolomiyetz, E.I., Sergeyev, V.V.: Detection of objects on the image using a sliding window mode. J. Opt. Laser Technol. **27** (1995). https://www.sciencedirect.com/science/article/pii/003039929593752D
5. Pedregosa, F., et al.: Scikit-learn: machine Learning in Python J. Mach. Learn. Res. (2011). http://www.jmlr.org/papers/v12/pedregosa11a.html
6. LeCunn, Y., et. al.: Refined MNIST dataset of handwritten digits. http://yann.lecun.com/exdb/mnist/

---

[1] https://github.com/BehradToghi/kNN_SWin.

Comparative Feature-Ranking Performance of Machine Learning: Model Classifiers in Skin Segmentation and Classification

Enoch A-iyeh$^{(\boxtimes)}$

University of Manitoba, Winnipeg R3T 2N2, Canada
umaiyeh@myumanitoba.ca

Abstract. Comparative test classification performance of machine learning model classifiers in skin segmentation using entire and sub-feature spaces is studied and presented. The effects of features on classification performance is demonstrated via test performance of classifiers trained on feature spaces and subsets. Model validation is assessed.

Keywords: Classification · Machine learning · Model · Skin · Segmentation · Validation

1 Introduction

Color spaces are generally utilized for segmentation purposes. Some of those frequently explored for skin segmentation include RGB, HSV, YCbCr and CIE [1–3,7,9,10]. Simple thresholding is also used although other methods including fuzzy decision tree models, deep learning [6], random forests [5] etc. supersede simple thresholding [7,11] in terms of classification accuracy. Skin segmentation and classification systems are indispensable components in automated decision systems [6,11] for skin examination and cancer detection as well as for subsequent signal processing, analysis and decision making. As such accurate machine learning implementations and model validation, feature efficacy and efficiency are paramount for the success of subsequent processes that succeed it in application domains. This paper tackles skin segmentation in the framework of machine learning in RGB color spaces employing different model classifiers; both heretofore explored and unexplored in skin segmentation in entire and sub-feature spaces. Thus skin segmentation classifiers are implemented for the problem employing feature ranking. Models are validated and performance traits demonstrated via cross-validation test accuracy and validation plots. The paper is organized as follows; we examine a dataset, obtain diagnostics to help us understand the interplay of variables in the classification task, proceed to identify models, set experimental parameters, finally apply models and present, discuss and validate the results, and end with conclusions.

© Springer Nature Switzerland AG 2020
K. Arai and S. Kapoor (Eds.): CVC 2019, AISC 944, pp. 592–607, 2020.
https://doi.org/10.1007/978-3-030-17798-0_48

2   Dataset and Preparation

A robust skin segmentation dataset must encompass several angles of the problem; sample data points must cover the skin spectrum of all complexions, people of different ages, races, genders in different environments, backgrounds, lighting conditions of images etc. The skin segmentation dataset [4] sufficiently meets these conditions, thus it is used for this study. Each data point is a tuple; $[x1, x2, x3, y]$ where $x1, x2, x3$ are the feature values of the blue, green and red channels respectively, y is the label i.e. whether the data point is a skin pixel or not. Feature values respectively represent the response of probe functions to blue, green and red light. Although color space segmentation has been explored in HSV, YCbCr, CIE spaces, etc. the current framework restricts us to RGB spaces, and subsequently comparison with results on RGB spaces.

2.1   Feature Distribution and Diagnostics

We examine the distributions of the input features via diagnostics; density plots and table plot.

Observe from the density distributions of the feature channels in Fig. 1a–c that the blue $(x1)$ and green $(x2)$ features have somewhat similar functions, both peaking between 0 and 100, 100 and 200. The table plot shows the distribution of the feature vectors alongside their class labels. Notice the shift in the distribution of the feature vectors belonging to different classes. Skin pixels are in the 'yes' class and non-skin pixels are in the 'no' class. Alternatively they are also referred to as the 'true' class and 'false' class, respectively. For examples in the 'yes' class the mean distribution line is shifted and different from the distribution line of the samples in the 'no' class (0–20% of the scale from the top in Fig. 1d). There are 50859 skin samples and 194198 non-skin samples, respectively.

Given the shifts in the distributions of the categorical samples it would be useful to examine the importance of the various variables, investigate how they influence classification and performance. Because of the clear shift, we would expect to be able to train classifiers that efficiently map the decision space and classify most samples correctly. Given the clear shift in distributions and hence the expectation to separate samples efficiently, we explore model and classifier performance in this respect. We explore the problem via ranking of variables by their relative feature importance, feature combinations in vector spaces for segmentation. Sample feature ranking is given in Fig. 2.

We see from the feature importance ranking that $x3$ is the most important feature, followed by $x2$ and then $x1$. This means that the red channel feature is the most important, followed by the green and finally the blue. We will see how this is so in skin segmentation and classification based on test accuracies of one or more feature vector spaces. The effects of an individual feature and combination of features on model performance is demonstrated and model validation assessed.

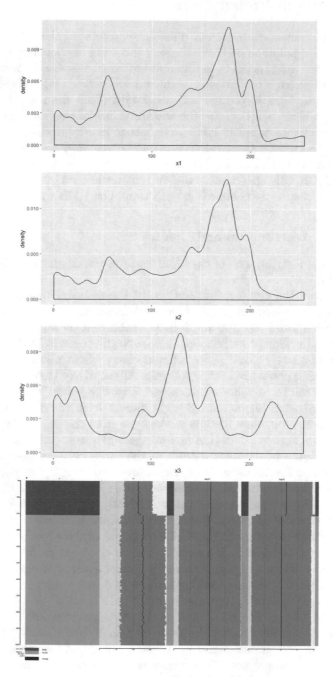

Fig. 1. (a) Blue channel feature distribution, (b) Green channel feature distribution, (c) Red channel feature distribution, (d) Table plot of features and labels

```
Variable Importances:
Variable Relative Importance Scaled Importance Percentage
      x3        96443.632813          1.000000   0.710094
      x2        24907.287109          0.258257   0.183387
      x1        14467.180664          0.150007   0.106519
```

Fig. 2. Feature importance ranking

3   Machine Learning Algorithms in Skin Segmentation and Classification: State of the Art

We explore skin segmentation and classification using state-of-the-art machine learning algorithms such as deep learning (DL) [6], Naive Bayes (NB) and random forests (RF) [5], as well as gradient boosted trees (GBT), and generalized linear models (GLM) trained and tested in both entire feature vector spaces and sub-vector spaces. To our knowledge although machine learning classifiers such as fuzzy decision trees, neural nets and random forests have been investigated for skin segmentation, they have largely been implemented without due consideration given to sub-feature spaces and or the contribution of single or multiple feature spaces on test classification accuracy across models. Moreover machine learning algorithms including GBT, GLM and DL are relatively unexplored for skin segmentation and classification. Hence we provide a somewhat comprehensive study of skin segmentation considering linear as well as non-linear models, ensembles of weak classifiers, tree-based models or otherwise etc. to assess merits of models, understand and inform state-of-the-art.

4   Experiments and Results

Skin segmentation and classification is studied using entire RGB color spaces as well as sub-vector spaces; the two most important features and the most important feature, selected based on feature importance ranking. Skin segmentation and classification models are implemented based on GBT, GLM, RF, NB and DL algorithms and compared. They are compared based on model classification performance (test accuracy), validation plots and the $f - measure$. Models are trained, tested and validated using 10-fold cross-validation, local random seed and shuffled sampling based on 70/30 train-test splits. Although model comparison may be achieved using the $f - measure$, we focus on test classification accuracy for wider comparability between studies.

Besides 10-fold cross-validation, applicable parameters of models are as follows; number of trees (NT) = 90, maximal depth (MD) = 3, minimum number of rows (MR) = 5, number of bins (BN) = 5, learning rate (LR) = 0.1 for GBT, NT = 90, MD = 3 for RF, automatic link and inverse link functions, least squares solver for GLM, softmax activation function, two hidden layers each of size 50, 10 epochs for DL, Laplace correction applied in NB. The models were implemented in RapidMiner Studio Large 7.6.001. The results are presented in the following figures and tables, along with the ensuing discussion. Classification performance reported for RGB spaces in prior works [1,3,4] are comparable to those reported in this study although those here marginally or substantially exceed them and different datasets are used. Because the same dataset is not used in all the works, we directly compare our results with [4]. The test accuracies are 99.50% and 94.10% respectively in this work and [4]. These classification accuracies are based on all three features of the RGB space, and they correspond to decision tree models. Even on a subspace of two features, the GBT model still outperforms the fuzzy decision tree model trained on the entire RGB feature space. Nonetheless, high classification accuracies in the range of 90.00%–99.50% are achieved on skin images in entire color spaces across different classifiers investigated here and elsewhere [1,3,4,8,9].

Table 1. Confusion matrix (all 3 features)

Model	yes class recall (%)	no class recall (%)	yes class precision (%)	no class precision (%)	acc. (%)	f measure (%)
GBT	98.55	99.75	99.62	99.06	**99.50**	**99.81**
GLM	81.94	94.46	79.48	95.23	**91.86**	**80.69**
RF	77.35	96.71	86.05	94.22	**92.70**	**81.47**
NB	73.28	97.40	88.06	93.30	**92.39**	**79.99**
DL	99.66	99.38	97.67	99.91	**99.44**	**98.66**

4.1   Model Validation

We provide Gini curves, lift charts and quantile (confidence) plots to validate models. Gini index is native to economics and income distribution quantification but has been adopted for several tasks including signal processing [12], and now used here for model validation and selection. In this context, it shows the contribution of equal bins of data in class label prediction. The x-axis represents the proportion of data in bins and the y-axis represents the proportion of correct predictions the data is responsible for or generates. If all features are used, we see from validation plots in Figs. 3, 4, 5, 6, 7, 8, 9, 10, 11, 12, 13, 14, 15, 16 and 17 that about 25% of the data is responsible for typically over 80% of correct predictions. The blue lines are the Gini curves of model predictions of the true or yes samples, and the diagonal green lines are the Gini curves of models predicting the same labels randomly (Figs. 3, 6, 9, 12 and 15). Thus we see that the greater the accuracy of the classifiers the greater the percentage of the predictions the first few bins or quantiles of data are responsible for.

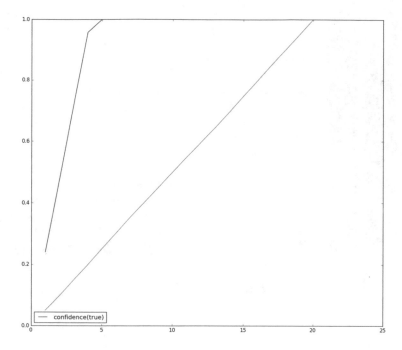

Fig. 3. GBT: Gini curve

Notice that just like receiver operating characteristic curves, Gini curves can be used in model validation and selection as demonstrated herein. The curves rise up sharply and are mostly confined to specific regions for high performing classifiers. The greater the area between the curves corresponding to the Gini index for the model and the random prediction model line, the better the classifier.

Lift charts (Figs. 4, 7, 10, 13 and 16) show the predictive performance of the classifiers based on the model outputs compared to the predictive performance without them (random predictions). With the models, test samples are mostly classified correctly within the first few quantile ranges, thus showing good lift compared to a diagonal lift of a model predicting randomly.

Quantile plots (Figs. 5, 8, 11, 14 and 17) show comparison between the confidence of predictions and the sums of correct predictions. Assuming direct proportionality between number of positive class predictions and quantile intervals, we see that as the quantile range increases, the number of test samples predicted for the true or yes class increases monotonically.

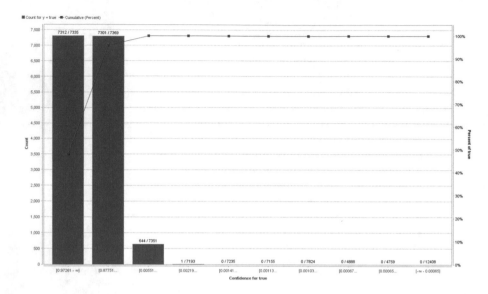

Fig. 4. GBT: Lift chart

Consider the test set performance accuracy of the classifiers based on the entire feature space and its subsets in Tables 1, 2 and 3. For example, with only the red channel feature, the GBT has accuracy of 88.50%. This indicates that feature alone is responsible for 88.50% of the performance accuracy. For the same GBT, feature $x3$ together with $x2$ give a performance accuracy of 98.84%, indicating that the green channel feature is responsible for 10.34% of the classifier performance based on both red and green channel features.

However, for all three features combined, the classifier performance of 99.50% indicates that the blue channel is responsible for 0.66% of the classifier performance. Performance irrespective of classifier is greater than 90.00% in the space of all three features, as well as for the top two features. It is greater than 80.00% across classifiers for the top one feature. The performance of GBT is superior to all others across the entire feature space and its subsets.

The performance of GBT and DL are comparable although the performance of the former is marginally better than the latter and all others across all feature spaces. The GLM is the worst performing model and classifier after GBT, DL, NB and RF in that order in all feature spaces. It however aids our understanding and interpretation of other decision spaces based on linear decision spaces and boundaries since the model is the sum of products of variables and coefficients. Similarly with the $f - measure$ as the performance metric, the comparative performance of the classifiers is apparent and similar ordering pertains. The efficacy of the individual feature spaces is remarkable. The top most important feature in the ranking accounts for a large proportion of the classifier performance. Feature additions contribute substantial or marginal amounts depending on the model by way of classification accuracy.

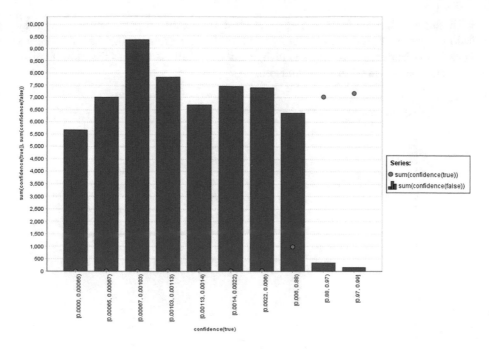

Fig. 5. GBT: Quantile plot

Table 2. Confusion matrix (top 2 features)

Model	yes class recall (%)	no class recall (%)	yes class precision (%)	no class precision (%)	acc. (%)	f measure (%)
GBT	98.35	98.97	96.14	99.57	**98.84**	**97.24**
GLM	72.24	92.92	72.78	92.74	**88.63**	**72.51**
RF	77.41	94.18	77.70	94.09	**90.70**	**77.56**
NB	73.58	96.12	83.23	93.28	**91.44**	**78.11**
DL	92.65	98.72	94.99	98.09	**97.46**	**93.81**

A classifier trained on combined individual red, green and blue channel features has better performance than any two combined feature channels or one channel. The combination of the three primary colours: red, green and blue produces white and may so therefore better approximate different skin complexions in the spectrum. The combination of red and green produces yellow and also better approximates and classifies various skin complexions compared to only red. Generally, the features influence classifier performance by different amounts with the red channel feature being the most impactful in performance accuracy, followed by green and marginally by the blue feature channel. A feature is defined here to be important or impactful if it substantially accounts for classification performance alone or in combination.

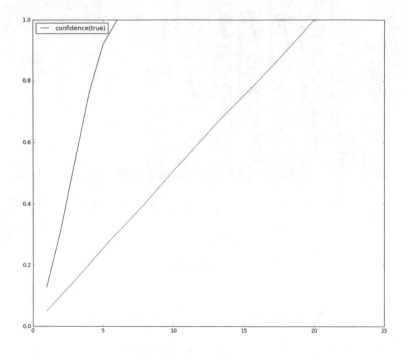

Fig. 6. GLM: Gini curve

Table 3. Confusion matrix (top 1 feature)

Model	yes class recall (%)	no class recall (%)	yes class precision (%)	no class precision (%)	acc. (%)	f measure (%)
GBT	76.43	91.66	70.59	93.69	**88.50**	**73.39**
GLM	63.15	93.31	71.20	90.63	**87.05**	**66.94**
RF	77.42	91.18	69.68	93.91	**88.32**	**73.35**
NB	68.88	92.44	70.47	91.90	**87.55**	**69.66**
DL	77.42	91.18	69.68	93.91	**88.32**	**73.35**

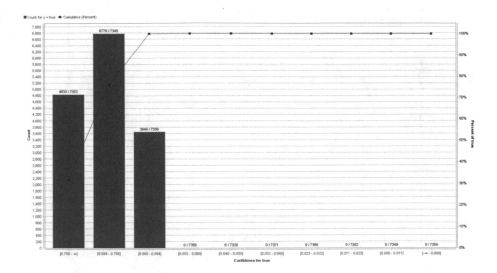

Fig. 7. GLM: Lift chart

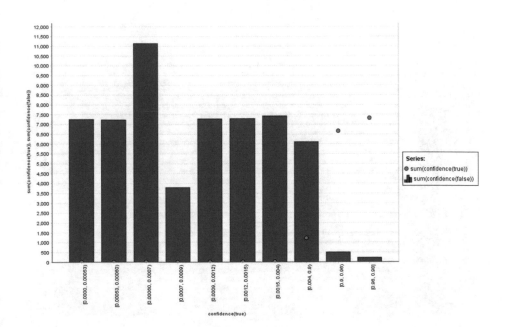

Fig. 8. GLM: Quantile plot

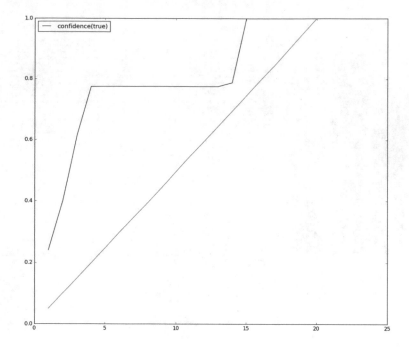

Fig. 9. RF: Gini curve

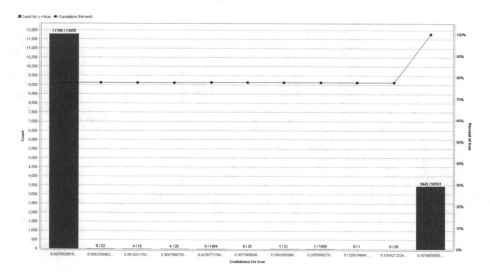

Fig. 10. RF: Lift chart

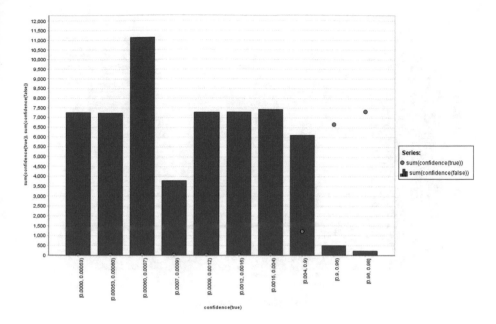

Fig. 11. RF: Quantile plot

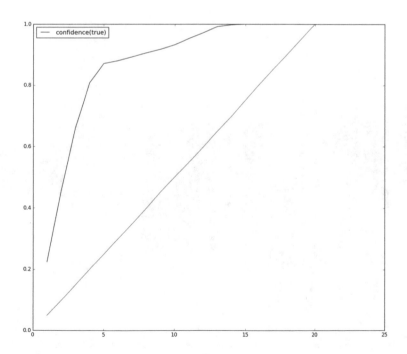

Fig. 12. NB: Gini curve

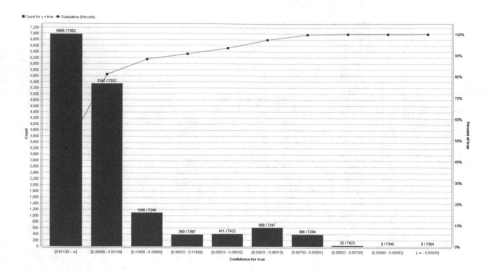

Fig. 13. NB: Lift chart

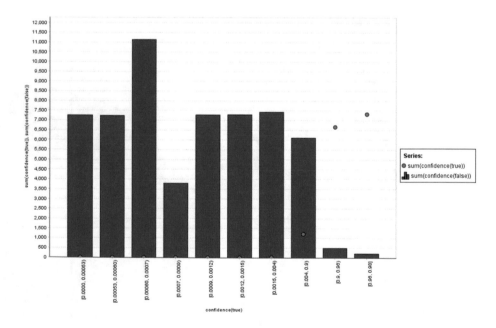

Fig. 14. NB: a. Gini curve b. Lift chart c. Quantile plot

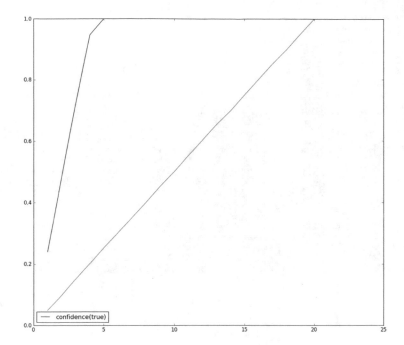

Fig. 15. DL: Gini curve

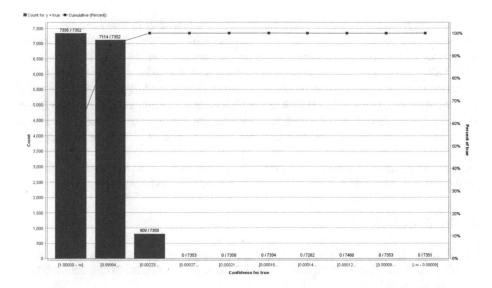

Fig. 16. DL: Lift chart

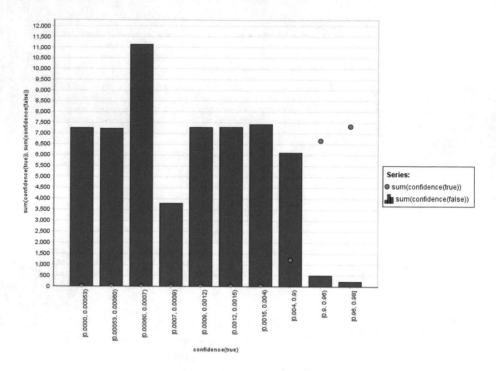

Fig. 17. DL: Quantile plot

5   Conclusion

Machine learning models and classifiers are implemented in the context of skin segmentation and classification. The comparative accuracies of the classifiers based on entire and sub-vector spaces are presented and discussed. Models are also validated using Gini curves, lift charts and quantile plots. The effect of a single feature and feature combinations on classifier design is demonstrated and quantified. Since a GLM is a linear function of input variables, non-linear models, linear functions of non-linear inputs and ensembles of weak classifiers such as GBT and DL better discriminate between skin and non-skin pixels.

Acknowledgment. I thank the anonymous reviewers for their feedback that led to improvements. Last but not least I thank God Almighty and my family.

References

1. Phung, S.L., Bouzerdoum, A., Chai, D.: Skin segmentation using color pixel classification: analysis and comparison. IEEE Trans. Pattern Anal. Mach. Intell. **27**(1), 148–154 (2005)
2. Saini, H.K., Chand, O.: Skin segmentation using RGB color model and implementation of switching conditions. Skin **3**(1), 1781–1787 (2013)

3. Kolkur, S., Kalbande, D., Shimpi, P., Bapat, C., Jatakia, J.: Human skin detection using RGB, HSV and YCbCr color models. arXiv preprint arXiv:1708.02694 (2017)
4. Bhatt, R.B., Dhall, A., Sharma, G., Chaudhury, S.: Efficient skin region segmentation using low complexity fuzzy decision tree model. In: India Conference IEEE-INDICON, pp. 1–4 (2009)
5. Khan, R., Hanbury, A., Stöttinger, J.: Skin detection: a random forest approach. In: 17th IEEE International Conference on Image Processing, pp. 4613–4616 (2010)
6. Esteva, A., Kuprel, B., Novoa, R.A., Ko, J., Swetter, S.M., Blau, H.M., Thrun, S.: Dermatologist-level classification of skin cancer with deep neural networks. Nature **542**(7639), 115 (2017)
7. Shaik, K.B., Ganesan, P., Kalist, V., Sathish, B.S., Jenitha, J.M.M.: Comparative study of skin color detection and segmentation in HSV and YCbCr color space. Procedia Comput. Sci. **57**, 41–48 (2015)
8. Vezhnevets, V., Sazonov, V., Andreeva, A.: A survey on pixel-based skin color detection techniques. Proc. Graph. **3**, 85–92 (2003)
9. Khan, R., Hanbury, A., Stöttinger, J., Bais, A.: Color based skin classification. Pattern Recogn. Lett. **33**(2), 157–163 (2012)
10. Nisar, H., Ch'ng, Y.K., Chew, T.Y. , Yap, V.V., Yeap, K.H., Tang, J.J.: A color space study for skin lesion segmentation. In: IEEE International Conference on Circuits and Systems, pp. 172-176 (2013)
11. Garnavi, R., Aldeen, M., Celebi, M.E., Bhuiyan, A., Dolianitis, C., Varigos, G.: Automatic segmentation of dermoscopy images using histogram thresholding on optimal color channels. Int. J. Med. Med. Sci. **1**(2), 126–134 (2010)
12. A-iyeh, E., Peters, J.F.: Gini Index-based digital image complementing in the study of medical images. J. Intell. Decis. Technol. **9**(2), 209–218 (2015)

Using a 3D Computer Vision System for Inspection of Reinforced Concrete Structures

Sajjad Sayyar-Roudsari[1], Sameer A. Hamoush[2(✉)],
Taylor M. V. Szeto[3], and Sun Yi[3]

[1] Department of Computational Science and Engineering,
North Carolina A&T State University, Greensboro, NC, USA
[2] Department of Civil and Architectural Engineering,
North Carolina A&T State University, Greensboro, NC, USA
`sameer@ncat.edu`
[3] Department of Mechanical Engineering, North Carolina A&T State University,
Greensboro, NC, USA

Abstract. Reinforced concrete (RC) structures need to be frequently inspected. The visual inspection is the practice in the structural engineering field. In the case where the structures are not accessible, using RealSense Camera mounted on a robot plays a vital role to detect external defects of RC members. In this paper, a RealSense Camera is used to inspect two different reinforced concrete beams built with specified surface defects. The First set is control beams by having mostly smooth surface with a small honeycombing area. The second set is RC beams that have excessive honeycombing defects in almost throughout the RC beam's side. Intel RealSense D435 Depth Camera is employed to scan the side of beams and record the data in X, Y and Z (depth) directions while camera is moving on a robot. Also, the MATLAB toolbox is used to convert the matrix data into image processing technique and the Mesh Plot is exploited to capture the images. The results show that the camera's images accurately depict the surface damaged areas and provide accurate representation of depths of the surface indentations.

Keywords: Reinforced concrete structures · Honeycombing defects · RealSense D435 Depth Camera

1 Introduction

Concrete is a common building material that can be used in reinforced concrete structures. In some cases, the reinforced concrete (RC) structures will experience damages due excessive loading or severe environmental exposures. The first step in assessing the conditions of the structures id to perform a visual. There are many types of surface defects can be noted during the inspection. They are honeycombing, spalling, rusting and external cracks. Therefore, evaluating these damages is vital for understanding the safety of structures in their service life [1–4]. In general, the surface inspection is the first approach to evaluate the concrete conditions for possible internal

© Springer Nature Switzerland AG 2020
K. Arai and S. Kapoor (Eds.): CVC 2019, AISC 944, pp. 608–618, 2020.
https://doi.org/10.1007/978-3-030-17798-0_49

or external deficiency [5–7]. But in the inspection standards of concrete structures, there is a very specific - methodology of investigating external defects. In many years, the 3D data acquisition technique has been using in both academic and industry. The techniques used in [8–10] are used for composite materials as three-dimensional computer vision system to assess the surface conditions of the elements. But, using camera and its exclusive features like RGB system and more importantly, this kind of camera has been designed to capture the depth of geometries and damages [11–14]. Carfagni et al. [15] investigated on the performance of depth camera as Intel SR300. Her results indicated that using this particular application can be compatible with most of the cameras. Godycka et al. [16, 17] evaluated the damage reinforced concrete beam by laser scanning method. He carried out some loaded reinforced concrete beams to evaluate the failure modes. Then, the terrestrial camera scanner was employed to record the pictures throughout the beams. The results showed that the mentioned method can be useful to display the cracks of RC beams. Moreover, Carey et al. [18] did study on small-scale 3D camera scanner to calibrate the accuracy at the high speed of moving. He used SR 300 camera scanner which his results illustrated the higher capability of this camera to evaluate the geometry specially depth investigation.

In this paper, reinforced concrete beams have been built in the laboratory of North Carolina Agricultural and Technical State University. These beams were carried out to have external defects. Then, the RealSense D435 Depth Camera is employed to assess the external defects. The scanning results of beam with defect is compared to the scan of RC beam without defects. It shows that Cameras can be mounted on a robot to perform visual inspection with great accuracy.

2 Experimental Test

In this research paper, Full-size reinforced concrete beams have been built in two groups that are control beams (no defects) and beams with external surface damages. The damage beam indicates regarding the beam with rough surface as honeycombing defect. The RC beams have 20×40 as cross-section with the order of width and height and the length of 96-in. Moreover, the concrete compressive strength is considered 4000 Psi for RC beams. Figure 1 shows the constructed reinforced concrete beams. Also, the rough surface and honeycombing is shown in the Fig. 2.

The surfaces of RC beams have been recorded by Intel RealSense D435 Depth Camera. This camera has RGB sensors which contribute researchers to use color map of scanning. The RGB depicts three main color as red, green and blue which are the main color of visualization technique. The depth resolution and FPS (frame per second) are at least 1280×720. It has dual global shutter sensors for up to 90 FPS in which has capability to record an ample number of pictures. In addition, the full HD resolution of RGB can synchronized the data to achieve very good accurate of depth. Moreover, the depth field view (DFV) of this camera is 85.2×58 which is the extend of angular of imaged camera. It has two recording properties as Wide Infrared Projector (WIP) and Wide Stereo Imagers (WSI). It should be noted that the unit of measurement is millimeter for distance and gram for mass [19]. Figure 3 displays the Intel RealSense D435 Depth Camera.

Fig. 1. Reinforced concrete beam specimens

The camera is used to scan the side of RC beams in which the length and width of image are in the term of pixel, also the depth can be measured from a pixel to the next ones to take the depth as millimeter. The distance of camera with surface of RC beam is considered as one meter, in constant. In effect, in order to scan the side of RC beams, the hand-held camera starts to be moved with the constant speed from one side of the beam to reach opposite side. While the camera is scanning, the MATLAB toolbox is exploited to manage data and the outputs are transferred to the MATLAB workspace. This toolbox has capacity to convert the image processing of the data into matrices or vice versa. In this research, the output of camera is used as input data of MATLAB toolbox which the input data is converted to .MAT file format. This format has capability of plotting the data from matrix form into visualization point of view. The Mesh Plot method is employed to make figures and the Colormap format is assigned as JET combination of color.

Fig. 2. Reinforced concrete beam with external defect

Fig. 3. Intel RealSense D435 Depth Camera [19]

3 Result and Discussion

In this section, the output of the scan shown will be the subject of the discussion. Two types beams have been scanned, control beams and beams with external defect. The control beams have almost smooth surface, but there is a small area with an external honeycombing defect. It should be noted that the figures in this section are captured based on X, Y and Z directions which are named as distance measuring by pixel in elongation direction (X), distance (pixel) in the direction of height of the beam (Y) and more importantly the depth of recorded output means the Z direction to evaluate the surface roughness. It must indicate that the resolution of images is 1920 by 1080 pixel by having unit 16 bit. In Fig. 4, the 3D image of control beams is demonstrated. As it is clear in the Fig. 4, the X direction has 2000-pixel, Y and Z directions have 1200 and 2000 pixel (Z direction is depicted as vertical axis). By examining Fig. 4, it can be seen that up to about 1000 pixel in Y axis do not have a high intensity value as the depth. As a point of fact, the values in Z direction indicate the degree of roughness in the surface. It means that the top-left of 3D image has a higher value in pixel than other areas. In fact, the Z-value has a higher intensity, the rougher the surface area, it has more honeycombing. On the other side, in Fig. 5, the X-Z phase is demonstrated the depth of roughness surface of RC beams. As it can be seen, most of the depth (Z) has an intensity value smaller than 800 pixel which indicates a smooth surface. By comparing the smooth values with rough surface values, it can be assent that the rough surface's value (honeycombing area) is twice of smooth area (about 1600 versus 800 pixel). Moreover, the value of Y-Z plane from camera scanner is portrayed in Fig. 6. In this figure, the difference in images of defected and not defected areas are completely clear. By looking at depth value (Z), it is visible that only a part of RC beam has higher pixel meaning the surface is not smooth.

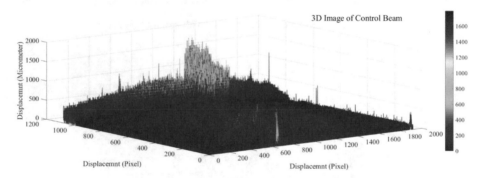

Fig. 4. 3D Image of RealSense D435 Depth Camera – Control RC Beam

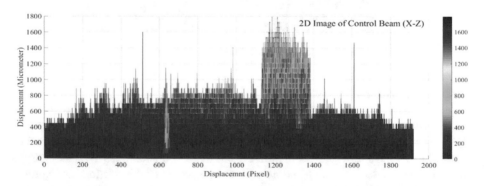

Fig. 5. X-Z Plane - Image of RealSense D435 Depth Camera – Control RC Beam

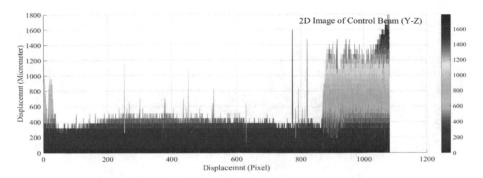

Fig. 6. Y-Z Plane - Image of RealSense D435 Depth Camera – Control RC Beam

On the other hand, the result of damaged beam is shown in Figs. 7, 8 and 9. In Fig. 7, the 3D image of damaged reinforced concrete beam is displayed. The vertical axis which is Z direction is regarding the depth of roughness describing the uniform distribution of damaged area. In fact, this uniform depth has a high pixel value which is presented in Fig. 8. As it is clear in Fig. 8, the value of depth by having more than 500 pixel indicates the honeycombing surface which is almost in all length of the RC beam. This issue is in Fig. 9, too in which the depth value has higher range throughout the RC beam. In these figures which the value in Z direction is much more than other value in this axis demonstrating the surface in more rough than other area.

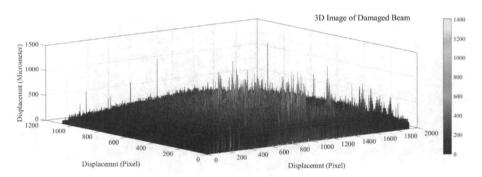

Fig. 7. 3D Image of RealSense D435 Depth Camera – Damaged RC Beam

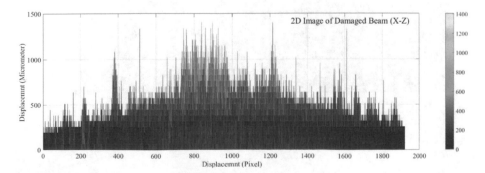

Fig. 8. X-Z Plane - Image of RealSense D435 Depth Camera – Damaged RC Beam

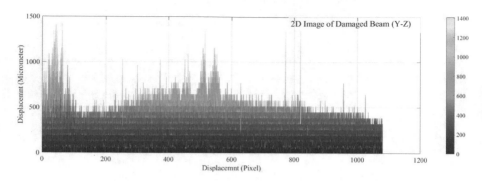

Fig. 9. Y-Z Plane - Image of RealSense D435 Depth Camera – Damaged RC Beam

Also, by comparing Figs. 4, 5, 6 for control RC beam and Figs. 7, 8 and 9 for damaged RC beam, the difference of control beam and damaged beam is completely clear. As a matter of fact, in the control RC beam by having smooth surface in approximately all areas except the left side of the beam which has honeycombing can be seen that the value of undamaged area is lower than 500 pixel while the honeycombing area has more than 500 pixel. Also, there are some summits in Z direction which have a value around 1000 pixel. This phenomenon is happened in the damaged RC beam meaning the honeycombing area should have higher depth value.

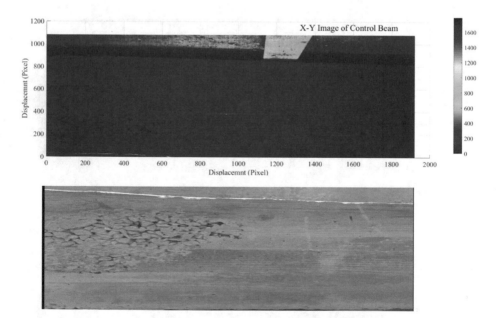

Fig. 10. Visualization of control RC beam

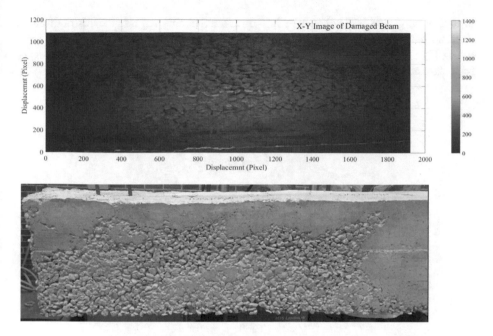

Fig. 11. Visualization of damaged RC beam

After comparing the pixel value in each single direction, the visualization part and real RC beam for the undamaged and damaged reinforced concrete beams are shown in Figs. 10 and 11, respectively. As it is clear in Fig. 10, the experimental and visualization output of undamaged RC beam is depicted so that, the smooth surface and honeycombing area is obvious. Also, in Fig. 11, it can be seen that the camera can capture the defect with high resolution which honeycombing surface is comparable to other area.

4   Indentation of the Honeycombing

One of the goals of this research is to find out the depth of damaged and undamaged surfaces of RC beams. In fact, the relationship of Z direction and the depth of 2D image has been mentioned before but, in this section, the intensity value of depth is computed. In Fig. 12, the intensity value in Z direction can illustrate how much the honeycombing area has rough surface meaning that the depth of roughness is measured by MATLAB toolbox. In this figure, the value in depth direction is between 430 and 490 μm, approximately. In effect, the demonstrated summits are showing the difference of smooth surface and honeycombing areas, in micrometer unit. By considering the value, it can be assent that the honeycombing area has between 0.43 to 0.49 mm more value as depth of smooth surface of concrete. Moreover, in Fig. 13, the highlighted point of honeycombing area is displayed in XY direction. In this regard, the distribution of intensity has very good agreement to the real RC beams. On the other side, by comparing Figs. 13 and 14, the comparison of real RC beams with scanned ones has very good agreements.

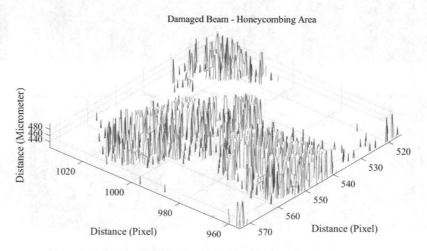

Fig. 12. The honeycombing area of RC beam based on depth value

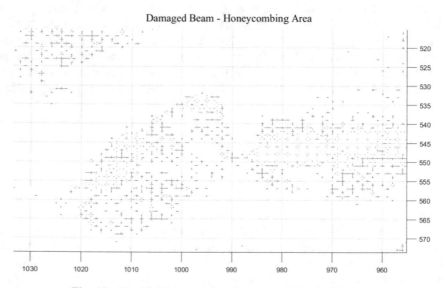

Fig. 13. The highlighted point of honeycombing in X-Y axis

Fig. 14. Honeycombing surface in RC beam

5   Conclusion

The Intel RealSense Depth Camera (D435) is used to perform an inspection tests of surface damage of reinforced concrete beams. The images of the camera are exploited as input of MATLAB file to evaluate by image processing techniques. Eventually, the following results have been taken into consideration:

- The new technique evaluated in this paper can provide the structural engineer with tools to qualify surface damage. The value of depth intensity image can indicate the quality of concrete surface.
- The techniques are shown to provide the inspectors with accurate analysis of the damaged surface and classify them as related to their depths.
- The difference of the depth intensity values can provide a clear indication of having honeycombing or other surface damages such as cracks.
- The D435 Camera Depth Scanner has very good accuracy in detecting concrete roughness and surface conditions.
- Using the RGB color map technique helps to highlight the defect area of concrete surface.

References

1. Galambos, T.V., Ellingwood, B.: Serviceability limit states: deflection. J. Struct. Eng. **112** (1), 67–84 (1986)
2. American National Standard Minimum Design Loads in Buildings and Other Structures. ANSI-A58.1-1982. American National Standards Institute, New York, NY (1982)
3. Ebrahimpour, A., Sack, R.L.: A review of vibration serviceability criteria for floor structures. Comput. Struct. **83**(28–30), 2488–2494 (2005)
4. Building Code Requirements for Structural Concrete (ACI 318-14). American Concrete Institution, September 2014. ISBN 978-0-87031-930-3

5. Beutel, R., Reinhardt, H.-W., Grosse, C.U., Glaubitt, A., Krause, M., Maierhofer, C., Algernon, D., Wiggenhauser, H., Schickert, M.: Comparative performance tests and validation of NDT methods for concrete testing. J. Nondestruct. Eval. **27**(1–3), 59–65 (2008)
6. Popovics, S.: Analysis of the concrete strength versus ultrasonic pulse velocity relationship. The American Society for Nondestructive Testing (2001)
7. Shotaro, K., Takayuki, T., Satoru, D., Shigeru, U., Maria, Q.F.: Detection of voids in concrete by nondestructive testing using microwave, January 2016. https://doi.org/10.1299/jsmermd.2016.2a2-09b7
8. Cristofani, E., Vandewal, M., Matheis, C., Jonuscheit, J.: 3-D radar image processing methodology for non-destructive testing of aeronautics composite materials and structures. In: 2012 IEEE Radar Conference. https://doi.org/10.1109/RADAR.2012.6212248
9. Soldan, S.: Towards 3D thermal imaging: geometric calibration of thermal imaging cameras and registration with 3D surface models using optical motion tracking for camera and object position. Technical report, University of Kassel, Germany and Laval University, Quebec, April 2013, 22 p
10. Teza, G., Galgaro, A., Moro, F.: Contactless recognition of concrete surface damage from laser scanning and curvature computation. NDT & E Int. **42**(4), 240–249 (2009)
11. Janowski, A., Rapinski, J.: M-split estimation in laser scanning data modeling. J. Indian Soc. Remote. Sens. **41**(1), 15–19 (2013)
12. Chen, L., Kondo, K., Nakamura, Y., Damen, D., Mayol-Cuevas, W.W.: Hotspots detection for machine operation in egocentric vision. In 2017 Fifteenth IAPR International Conference on Machine Vision Applications (MVA), pp. 223–226, May 2017
13. Izadi, S., et al.: KinectFusion: real-time 3D reconstruction and interaction using a moving depth camera. In: Proceedings of ACM Symposium on User Interface Software Technology 2011, pp. 559–568 (2011)
14. Beraldin, J.-A., Mackinnon, D., Cournoyer, L.: Metrological characterization of 3D imaging systems: progress report on standards developments. In: Proceedings of 17th International Congress of Metrology, p. 13003 (2015). https://doi.org/10.1051/metrology/20150013003
15. Carfagni, M., Furferi, R., Governi, L., Servi, M., Uccheddu, F., Volpe, Y.: On the performance of the Intel SR300 depth camera: metrological and critical characterization. IEEE Sens. J. **17**(14), 4508–4519 (2017)
16. Nagrodzka-Godycka, K., Szulwic, J., Ziółkowski, P.: The method of analysis of damage reinforced concrete beams using terrestrial laser scanning. In: 14th International Multidisciplinary Scientific GeoConference SGEM 2014 (2014)
17. Janowski, A., Nagrodzka-Godycka, K., Szulwic, J., Ziółkowski, P.: Modes of failure analysis in reinforced concrete beam using laser scanning and synchro-photogrammetry. In: Proceedings of the Second International Conference on Advances in Civil, Structural and Environmental Engineering-ACSEE 2014. Copyright © Institute of Research Engineers and Doctors, USA. All rights reserved. https://doi.org/10.15224/978-1-63248-030-9-04. ISBN 978-1-63248-030-9
18. Carey, N., Werfel, J., Nagpal, R.: Fast, accurate, small-scale 3D scene capture using a low-cost depth sensor. In: 2017 IEEE Winter Conference on Applications of Computer Vision (WACV), pp. 1268–1276, March 2017
19. Intel® RealSense™ Depth Camera D400-Series, (Intel® RealSense™ Depth Camera D415, Intel® RealSense™ Depth Camera D435), September 2017

Specular Photometric Stereo for Surface Normal Estimation of Dark Surfaces

Mengyu Song[✉] and Tomonari Furukawa

Virginia Polytechnic Institute and State University, Blacksburg, VA 24061, USA
{ucemso0,tomonari}@vt.edu

Abstract. This paper presents Specular Photometric Stereo (SPS), which is a Photometric Stereo (PS) technique incorporating specular reflection. The proposed SPS uses multiple images of a surface under different lighting conditions to obtain surface normals similarly to the conventional PS, but uniquely utilizes specular components of a dark surface, which reflects little diffuse light. The proposed framework consists of two sequential numerical steps, which are the conversion of a highly non-linear specular reflection model to a non-linear equation with only one non-linear parameter, and then the iterative removal of the diffuse components. The proposed SPS can estimate normals of dark surfaces, which is not possible by the conventional PS. The proposed SPS was examined using synthesized data and then tested with real-world surfaces. The results of surface normal estimation show that the capability of the proposed SPS over the existing PS in both accuracy and computational cost.

Keywords: Computer vision algorithm · Shape from X · Photometric Stereo

1 Introduction

Photometric Stereo (PS), originally introduced by Woodham [1], uses multiple digital images taken from one viewpoint but under different lighting conditions to determine surface's orientation. Originally, PS has a strong assumption on the reflection property of the concerned surface. By inversing the diffuse reflection model and image formation, surface normal can be solved mathematically. Different from conventional PS, Specular Photometric Stereo uses the specular components from the reflection of a dark surface, which violates the Lambertian assumption of PS. The surface of materials like rubber has a low diffuse albedo but a wide specular reflection region (see Fig. 1), which provides information cue to recover its shape.

Point lighting sources were used in conventional photometric stereo method (Diffuse Photometric Stereo (DPS), [2]). Lambertian reflection model was assumed for DPS so that the image formation procedure could be reversed linearly, easing the procedure of obtaining numerical solutions. However, it also

© Springer Nature Switzerland AG 2020
K. Arai and S. Kapoor (Eds.): CVC 2019, AISC 944, pp. 619–637, 2020.
https://doi.org/10.1007/978-3-030-17798-0_50

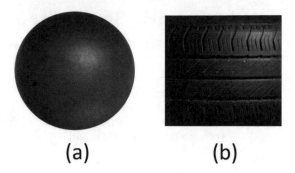

(a)   (b)

Fig. 1. images of reflection from dark surfaces under point illumination: (a) one side of a rubber ball, (b) part of a tire surface

restricts the application of DPS. To apply DPS on complex reflection surface, dichromatic reflectance model (DRM, [3]) was applied by Li and Furukawa [4]. DRM employs the difference of chromaticity between the lighting source and the reflecting surface to separate specular components from diffuse components in the reflection light. It has been proved to be effective for regular surfaces. But for dark surfaces, whose diffuse albedo is small, the separation of reflecting surface's chromaticity using the image from a digital camera contains a large error, resulting in an unsound reflection components decomposition.

When the diffuse reflection is insufficient, specular components can also provide useful visual cue for image-based algorithms [5,6]. Blake and Brelstaff applied two cameras to observe a specular surface under point illumination, and used the position of specular highlight to determine the surface's curvature properties [7]. Ikeuchi employed a Lambertian board to extend a point lighting source to a plane lighting source to illuminate a mirror-type surface. By changing the position of the point lighting source, the lighting condition on the specular surface from the extended lighting source changes accordingly, therefore the surface's orientation was able to be determined [8]. Both cases assumed a narrow specular reflection region, which is valid for surfaces of mirror-like material, like metal. However, this approach cannot be utilized directly on dark surfaces of material like rubber, for its wide specular reflection region.

Although non-parametric PS [9] and [10] extended the application of PS, parametric PS can provide more accurate numerical result. DPS is widely utilized due to its ease of computation. Extending DPS from Lambertian reflectance model to a complex reflectance model introduces a high non-linearity. The main trend to deal with complex reflection surfaces is to fit a parameterized reflection model [11]. Due to the existence of specular components, the parameterized reflection model is highly non-linear. Solving for its parameters, which include the surface orientation, requires a non-linear regression approach and a good initial guess of all the parameters [12]. This approach is classified as Non-linear Iterative Photometric Stereo (NIPS) in this paper, whose performance will be compared with SPS.

To deal with the surface of dark material like rubber, a new pipeline is proposed in this paper to apply PS on dark surface using the specular reflection components. The proposed method, SPS, utilizes the specular components to obtain surface normal, differing from DPS, which utilizes the diffuse components. Blinn-Phong reflectance model is used instead of delta reflectance model (suitable for narrow specular reflection region) in SPS. Unlike NIPS, SPS avoids non-linear regression algorithm due to its heavy computational cost and high dependence on the initial guess.

The paper is organized as follows. Next section presents background knowledge about generic photometric stereo problem formulation, reflectance model, image formation procedure, and numerical solution to conventional PS. The pipeline of SPS on dark surfaces is described with details in Sect. 3. Section 4 discusses experimental results of the proposed SPS method whereas conclusion and future work is summarized in the last section.

2   Photometric Stereo Technique

In this section, the generic photometric stereo problem will be formulated first. Since photometric stereo is a reverse problem of reflection and image formation, the forward procedure of these two will be explained. At last, DPS will be described in details.

2.1   Generic Photometric Stereo Problem Formulation

Figure 2 shows the schematic diagram of generic photometric stereo problem. Photometric stereo uses a digital camera to capture the images of a surface under N different lighting conditions, to obtain pixel-wise surface normal.

To be more specific, given the lighting direction of the kth point illumination at pixel (i,j) as $\bar{l}_k^{(i,j)}$, corrected image intensity of pixel (i,j) under k^{th} point illumination as $\breve{I}_k^{(i,j)}$, the aim of photometric stereo is to derive the unit surface normal of a small patch on the target surface, which is corresponding to the pixel (i,j) at all the images, $\bar{n}_k^{(i,j)}$, by minimizing the following cost function:

$$\text{cost} = \sum_{k=1}^{N} w_k \left[\breve{I}_k^{(i,j)} - f\left(\bar{l}_k^{(i,j)} \cdot \bar{n}^{(i,j)} \right) \right]^2, \tag{1}$$

where $\left[\breve{I}_k^{(i,j)} - f\left(\bar{l}_k^{(i,j)} \cdot \bar{n}^{(i,j)} \right) \right]^2$ is the squared residual for k^{th} measuring data, w_k is the weight of kth data, $f\left(\bar{l}_k^{(i,j)} \cdot \bar{n}^{(i,j)} \right)$ is a function of $\bar{l}_k^{(i,j)}$ and $\bar{n}^{(i,j)}$, which can be referred to as reflectance model.

The discussion in the following parts is regarding to a small patch of surface, which is corresponding to the pixel (i,j) in the images. The problem will be described and solved pixel-wisely. The superscript $\cdot^{(i,j)}$ is omitted in this section for conciseness, if there is no ambiguity.

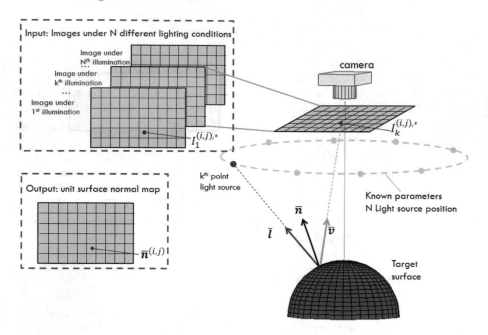

Fig. 2. Problem formulation

2.2 Forward Problem - Reflectance Model

A patch of surface reflects some fraction of incident light. The intensity of reflectance light (L_r) can be separated into two components, diffuse reflectance (L_d) and specular reflectance (L_s):

$$L_r = L_d + L_s. \tag{2}$$

For single point illumination, which is assumed in this paper (the reflectance geometry is shown in Fig. 3), the diffuse reflectance is proportional to the cosine of the angle between the light direction ($\bar{l}$) and surface normal ($\bar{n}$):

$$L_d = E_i p_d \left(\bar{n} \cdot \bar{l} \right), \tag{3}$$

where E_i is the intensity of incident light, p_d is the diffuse reflectance factor, which can be considered as the proportion of diffuse reflectance of the incident light. Notice that the symbol $\bar{\cdot}$ indicates a unit vector.

Specular reflectance, as modeled in Blinn Phong specular reflectance model [13], is formulated as:

$$L_s = E_i p_s \left(\bar{n} \cdot \bar{h} \right)^{\alpha}, \tag{4}$$

where p_s is the specular reflectance factor, which can be considered as the proportion of specular component. α is the shininess constant, which negative correlates to the size of specular reflection region. Halfway vector, $\bar{h}$, points halfway between lighting direction and viewing direction $(\bar{v})$, that is

$$\bar{h} = \frac{\bar{l} + \bar{v}}{\|\bar{l} + \bar{v}\|}. \qquad (5)$$

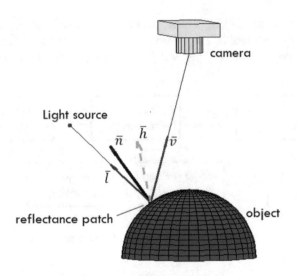

Fig. 3. Reflectance geometry

Notice $\theta_h = \arccos(\bar{n} \cdot \bar{h})$ is the angle between surface normal and halfway vector, which denotes how far the surface normal is away from the maximum specular direction.

Then the reflectance model can be expressed as:

$$L_r = E_i p_d (\bar{n} \cdot \bar{l}) + E_i p_s (\bar{n} \cdot \bar{h})^\alpha. \qquad (6)$$

In Eq. 6, E_i, $\bar{l}$, $\bar{h}$ can be calibrated in advance. L_r is a measuring variable. $\bar{n}$ describes the orientation of this patch of surface, p_d, p_s and α describe the surface's reflectance property. $\bar{n}, p_d, p_s, \alpha$ are the unknown parameters.

The reflectance model is referred as BRDF (bidirectional reflectance distribution function) in most literatures, which describes the relationship between the amount of incident light and reflecting light, and reflectance model parameters (p, including $\bar{n}$ and reflectance-related parameters).

$$BRDF = \frac{L_r}{E_i} = f(\bar{l}, \bar{v}, p). \qquad (7)$$

In this paper, since L_r and E_i will be measured separately, Eq. 6 will be used to explain our reflectance model.

2.3   Forward Problem - Image Formation

All photometric techniques rely on radiance measurement. Digital camera is utilized to measure surface radiance. Since it is difficult to distinguish the difference of chromaticity between the reflecting surface and the lighting source, a monochrome camera model is assumed in this paper instead of a color camera model.

Photometric stereo is the inverse procedure of image formation. To understand photometric stereo, it is necessary to have a basic idea of image formation procedure. Figure 4 shows the general digital image formation procedure.

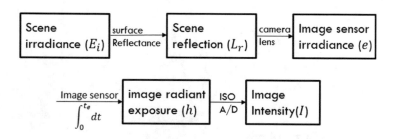

Fig. 4. Digital image formation

As shown in Eq. 6, the scene reflection, L_r, of a certain patch (with surface orientation $\bar{n}$ and reflectance properties p_d, p_s and α) is determined by scene irradiance E_i, lighting direction $\bar{l}$ and viewing direction $\bar{v}$. Since the direction towards the camera is the only viewing direction, $\bar{l}$ is denoted as the direction towards the camera, and $\bar{h}$ as the corresponding halfway vector for the rest of this paper without ambiguity.

The intensity of the reflectance light in the direction towards the camera determines the image irradiance of the camera (e), along with a factor (k_{lens}) depending on lens. To simplify the problem, k_{lens} is assumed to be a constant for a fixed lens.

$$e = k_{lens} L_r. \tag{8}$$

The image irradiance is then integrated over the exposure time (t_e) on an image sensor cell. A static scene is assumed during the exposure, therefore e is time invariant, then

$$h = \int_0^{t_e} e(\tau) d\tau = t_e k_{lens} L_r, \tag{9}$$

where h is image radiant exposure.

Notice each cell of an image sensor is an analog device. It measures the amount of the electrical charge generated from the image radiant exposure h by the photo sensing unit. This analogous data will be boosted by an amplifier and then digitized by an A/D converter. The amplifier is referred as film speed, which describes the sensitivity of the image sensor towards light, and can be measured using ISO standard. For a fixed ISO-number, the image intensity is proportional to the image radiant exposure.

$$I = k_{iso}h = k_{iso}t_ek_{lens}L_r, \tag{10}$$

where k_{iso} is the scaling factor for the fixed ISO-number.

Substitute L_r with the reflectance model (Eq. 6),

$$\begin{aligned} I &= k_{iso}t_ek_{lens}L_r \\ &= k_{iso}t_ek_{lens}E_ip_d\left(\bar{\boldsymbol{n}}\cdot\bar{\boldsymbol{l}}\right) \\ &\quad+k_{iso}t_ek_{lens}E_ip_s\left(\bar{\boldsymbol{n}}\cdot\bar{\boldsymbol{h}}\right)^\alpha. \end{aligned} \tag{11}$$

Define scaled diffuse factor k_d and scaled specular factor k_s as

$$k_d = k_{iso}t_ek_{lens}p_d, \tag{12}$$
$$k_s = k_{iso}t_ek_{lens}p_s. \tag{13}$$

Then the image intensity captured by a digital camera is described with respect to $\bar{\boldsymbol{n}}$, $\bar{\boldsymbol{l}}$, $\bar{\boldsymbol{h}}$, k_d, k_s, α and E_i as:

$$I = E_ik_d\left(\bar{\boldsymbol{n}}\cdot\bar{\boldsymbol{l}}\right) + E_ik_s\left(\bar{\boldsymbol{n}}\cdot\bar{\boldsymbol{h}}\right)^\alpha. \tag{14}$$

I could be measured directly from a digital camera, and E_i can be calibrated in advance. Then define corrected image intensity $\check{I}$ as

$$\check{I} = \frac{I}{E_i}. \tag{15}$$

Notice $\check{I}$ is a known variable, and till now, the full reflectance model for SPS has been derived as:

$$\check{I} = k_d\left(\bar{\boldsymbol{n}}\cdot\bar{\boldsymbol{l}}\right) + k_s\left(\bar{\boldsymbol{n}}\cdot\bar{\boldsymbol{h}}\right)^\alpha. \tag{16}$$

2.4   Conventional Solution - Diffuse Photometric Stereo

According to Lambertian reflectance assumption, k_s is zero, then Eq. 16 becomes

$$\check{I}_k = k_d\left(\bar{\boldsymbol{n}}\cdot\bar{\boldsymbol{l}}_k\right), \tag{17}$$

where subscript k is used to denote different illumination condition, and $k = 1,\ldots,N$.

Then surface normal and scaled diffuse factor can be calculated as:

$$k_d = \left\|\left(\boldsymbol{L}^T\boldsymbol{L}\right)^{-1}\boldsymbol{L}^T\boldsymbol{\mathcal{I}}\right\|, \tag{18}$$

$$\bar{\boldsymbol{n}} = \frac{\left(\boldsymbol{L}^T\boldsymbol{L}\right)^{-1}\boldsymbol{L}^T\boldsymbol{\mathcal{I}}}{k_d}, \tag{19}$$

where $\boldsymbol{L} = \left[\bar{\boldsymbol{l}}_1,\ldots,\bar{\boldsymbol{l}}_N\right]^T$, $\boldsymbol{\mathcal{I}} = \left[\check{I}_1,\ldots,\check{I}_N\right]^T$.

3   Specular Photometric Stereo

A method mainly considering the specular reflectance components to calculate surface orientation is proposed and described in this section.

As described in previous section, the intensity of reflectance light can be separated into diffuse and specular components. Given the linear relationship between reflectance light and corrected image intensity, the corrected image intensity can also be separated into two parts:

$$\breve{I}_k = \breve{I}_{d,k} + \breve{I}_{s,k}, \tag{20}$$

where

$$\breve{I}_{d,k} = k_d \left(\bar{\boldsymbol{n}} \cdot \bar{\boldsymbol{l}} \right), \tag{21}$$

$$\breve{I}_{s,k} = k_s \left(\bar{\boldsymbol{n}} \cdot \bar{\boldsymbol{h}} \right)^{\alpha}. \tag{22}$$

For images of a dark surface, the brightness of bright pixels mainly comes from specular reflectance. The intensity of pixel with little specular reflectance tends to be small. In other words, the diffuse reflectance factor is much smaller than the specular reflectance factor,

$$k_s \gg k_d \approx 0. \tag{23}$$

Due to a relatively large signal error ratio, it is error prone to use Eq. 21 to calculated surface normal, which is adopted by DPS. Comparatively, it is more reasonable to employ specular components.

3.1   Variation of Parameters

Inspired by DPS, Eq. 22 can also be reversed to calculate surface normal. However, the equation itself, along with the unit vector constrain,

$$\|\bar{\boldsymbol{n}}\| = 1, \tag{24}$$

contains a high non-linearity to solve for $\bar{\boldsymbol{n}}$.

As a result, this equation cannot be solved using least square regression (LSR) directly, which is applied by DPS.

To contour this problem, a variation of parameters is proceeded, and the problem becomes a non-linear problem but with only with one non-linear parameter. The procedure is derived as:

First define

$$\boldsymbol{N} = k_s^{1/\alpha} \bar{\boldsymbol{n}}. \tag{25}$$

Notice that

$$k_s = \|\boldsymbol{N}\|^{\alpha}, \tag{26}$$

$$\bar{\boldsymbol{n}} = \frac{\boldsymbol{N}}{\|\boldsymbol{N}\|}. \tag{27}$$

Equation 22 becomes:

$$\breve{I}_{s,k} = \left(\boldsymbol{N} \cdot \bar{\boldsymbol{h}}\right)^{\alpha}. \tag{28}$$

Furthermore, raise both sides to the power of $\frac{1}{\alpha}$, and then move the left hand side of the equation to the right, gives

$$\left(\breve{I}_{s,k}\right)^{1/\alpha} - \boldsymbol{N} \cdot \bar{\boldsymbol{h}} = 0. \tag{29}$$

Notice, the previous high non-linear equations are modified to a 4-unknown (α and 3 in $\boldsymbol{N}$) equations with only one non-linear variable (α), which can be solved using the method describes in next section.

3.2   Numerical Solution for the Non-linear Equation with Only One Non-linear Unknown

A method to solve for non-linear equations with only one non-linear unknown is introduced by Shen and Ypma [14]. The general problem can be described as:

Assume $N+1$ non-linear equations of $N+1$ unknowns, $y \in R$ and $z \in R^N$ of the form

$$\boldsymbol{A}(y)\boldsymbol{z} + \boldsymbol{b}(y) = \boldsymbol{0}, \tag{30}$$

where the $(N+1) \times N$ matrix $\boldsymbol{A}(y)$ and $(N+1) \times 1$ vector $\boldsymbol{b}(y)$ are functions of scalar variable y only.

Shen and Ypma offered a solution to this general problem, which is summarized in Algorithm 1.

Algorithm 1

1: Select a value for y.
2: Solve for the remaining N unknowns $\boldsymbol{z}$ using LSR.
3: Determine whether the selected y^* and corresponding computed $\boldsymbol{z}^*$ satisfy the equations, by compare the cost $\|\boldsymbol{A}(y^*)\boldsymbol{z}^* + \boldsymbol{b}(y^*)^2\|$ and the tolerance ϵ.
4: If $\|\boldsymbol{A}(y^*)\boldsymbol{z}^* + \boldsymbol{b}(y^*)^2\| < \epsilon$, return y^* and $\boldsymbol{z}^*$, otherwise pick a new based on previous solutions, go back to step 1.

The method applies to solve the specific problem in this paper is inspired by Algorithm 1. The difference is instead of calculating the unknowns iteratively, different values are selected for the only one non-linear unknown, and calculate the remaining linear unknowns, as well as the corresponding cost at one time. Then the set of unknowns which gives the minimum cost is picked. The modification is due to the widely usage of graphics processing units (GPU). With applying parallel computing, the proposed method will be much faster than the iterative one.

Algorithm 2

1: Select N_a different values of α, as $\alpha_i \in [1, 1024]$ for $i = 1...N_a$.

2: Calculate $\boldsymbol{N}_i$ as

$$\boldsymbol{N}_i = \left(\boldsymbol{H}^T \boldsymbol{H} \right)^{-1} \boldsymbol{H}^T \boldsymbol{\mathcal{I}}_i, \tag{31}$$

where $\boldsymbol{H} = \left[\bar{\boldsymbol{h}}_1, \ldots, \bar{\boldsymbol{h}}_N \right]^T$, $\boldsymbol{\mathcal{I}}_i = \left[\check{I}_{s,1}^{1/\alpha_i}, \ldots, \check{I}_{s,N}^{1/\alpha_i} \right]^T$.

3: Calculate cost as

$$\mathrm{cost}_i = \sum_{k=1}^{N} \left(\check{I}_{s,k}^{1/\alpha_i} - \bar{\boldsymbol{h}}_k \cdot \boldsymbol{N}_i \right)^2. \tag{32}$$

4: Find the i^* set, which gives the minimum cost:

$$i^* = \operatorname*{argmin}_{i \in [1,...,N_a]} \mathrm{cost}_i. \tag{33}$$

5: Return

$$\alpha = \alpha_{i^*}, \tag{34}$$

$$k_s = \| \boldsymbol{N}_{i^*} \|^{\alpha_{i^*}}, \tag{35}$$

$$\bar{\boldsymbol{n}} = \frac{\boldsymbol{N}_{i^*}}{\| \boldsymbol{N}_{i^*} \|}. \tag{36}$$

To be more specific in the content of the concerning case, the algorithm is described in Algorithm 2.

3.3   Redesign Cost Function

Due to the existence of different sources of error, such as image noise and quantization error, a number of measurement N larger than the number of unknowns is chosen. In other words the system is over-constrained. As a result, the input data will disagree with the fitted model to some extent.

In Eq. 32, each term can be seen as the penalty of the disagreement of the data and fitted model. It is obvious from Eq. 32 that the penalty for each input data have the same power on the fitted model. In other words, bu rewriting Eq. 32 as weighted form:

$$\mathrm{cost} = \sum_{k=1}^{N} \left(w_k \left(\check{I}_{s,k}^{1/\alpha} - \bar{\boldsymbol{h}}_k \cdot \boldsymbol{N} \right) \right)^2, \tag{37}$$

with the same weight for each input data as:

$$w_k = 1, \tag{38}$$

for $k = 1, \ldots N$.

As mentioned by Kay and Caelli [15], the weights have a strong influence on the fitted parameters. If incorrect weights were applied, the result would be away from the ground truth.

For the case of $w_k = 1, k = 1, \ldots, N$, the parameters that minimize cost function 32 is the solution to

$$\check{I}_{s,k}^{1/\alpha} = k_s^{1/\alpha} \left(\bar{h}_k \cdot \bar{n} \right), \tag{39}$$

rather than the original specular reflectance model (Eq. 22), whose solution should be the parameters which minimize the following cost function:

$$\text{cost} = \sum_{k=1}^{N} \left(\check{I}_{s,k} - k_s \left(\bar{h}_k \cdot N \right)^\alpha \right)^2. \tag{40}$$

For small $\check{I}_{s,k}$, the error calculated by $\left(\check{I}_{s,k}^{1/\alpha_i} - \bar{h}_k \cdot N_i \right)^2$ is enlarged compared to the error calculated by $\left(\check{I}_{s,k} - k_s \left(\bar{h}_k \cdot N \right)^\alpha \right)^2$, due to the logarithm. As a conclusion, result using $w_k = 1, k = 1, \ldots, N$ is not a good approximation of result using cost function 40.

However, due to the high non-linearity, cost function 40 can not be solved using the proposed method. Alternatively, $w_k = \check{I}_{s,k}, k = 1, ..., N$ is proposed, then the cost function becomes:

$$\begin{aligned}
\text{cost} &= \sum_{k=1}^{N} \left(\check{I}_{s,k} \left(\check{I}_{s,k}^{1/\alpha} - \bar{h}_k \cdot N \right) \right)^2 \\
&= \sum_{k=1}^{N} \left(\check{I}_{s,k}^{1+1/\alpha} - \check{I}_{s,k} \bar{h}_k \cdot N \right)^2. \tag{41}
\end{aligned}$$

In this function, penalty for each input data is weighted by its own value, which indicates that small $\check{I}_{s,k}$ does not have large penalty, which is similar to the case if cost function 40 is used.

Then previous algorithm is upgraded to Algorithm 3.

3.4　Parameter Refinement

Although for dark surfaces, $k_s \gg k_d$, $\breve{I}_s$ drops faster than $\breve{I}_d$ due to the higher power in its formulation when $\bar{n}$ goes away from $\bar{h}$. So for fixed lighting direction and viewing direction, there are surface whose normal will cause $\breve{I}_s \approx \breve{I}_d$ or even $\breve{I}_s < \breve{I}_d$. For these cases, it is then incorrect to assume that the corrected image intensity from measurement approximately equal to $\breve{I}_s$. Therefore it is reasonable to remove the diffuse part $\breve{I}_d$ from the original measurement $\breve{I}$.

The algorithm to refine the parameter by removing $\breve{I}_d$ is shown in Algorithm 4.

The parameter refinement part is important, especially for a relatively small k_s/k_d case. Since the specular reflectance model is used only to recover the surface normal, the deposition of the diffuse component will make the input data fit the model better compared to the raw data.

Algorithm 3

1: Select N_a different values of α, as $\alpha_i \in [1, 1024]$ for $i = 1 \ldots N_a$.

2: Calculate $\boldsymbol{N}_i$ as

$$\boldsymbol{N}_i = \left(\boldsymbol{H}^T \boldsymbol{H} \right)^{-1} \boldsymbol{H}^T \boldsymbol{\mathcal{I}}_i, \tag{42}$$

where $\boldsymbol{H} = \left[\breve{I}_{s,1} \bar{\boldsymbol{h}}_1, \ldots, \breve{I}_{s,N} \bar{\boldsymbol{h}}_N \right]^T$, $\boldsymbol{\mathcal{I}}_i = \left[\breve{I}_{s,1}^{1+1/\alpha_i}, \ldots, \breve{I}_{s,N}^{1+1/\alpha_i} \right]^T$.

3: Calculate cost for $i = 1, \ldots N_a$ as

$$\text{cost}_i = \sum_{k=1}^{N} \left(\breve{I}_{s,k}^{1+1/\alpha_i} - \breve{I}_{s,k} \bar{\boldsymbol{h}}_k \cdot \boldsymbol{N}_i \right)^2. \tag{43}$$

4: Find the i^* set, which gives the minimum cost:

$$i^* = \operatorname*{argmin}_{i \in [1, \ldots, N_a]} \text{cost}_i. \tag{44}$$

5: Return

$$\alpha = \alpha_{i^*}, \tag{45}$$

$$k_s = \| \boldsymbol{N}_{i^*} \|^{\alpha_{i^*}}, \tag{46}$$

$$\bar{n} = \frac{\boldsymbol{N}_{i^*}}{\| \boldsymbol{N}_{i^*} \|}. \tag{47}$$

The reason why the diffuse component is taken out from the corrected intensity and use specular photometric stereo, instead of taking out specular one and using diffuse ps is specular information provide more powerful image cue to recover the surface normal, especially for the case where specular highlight is obvious. And more error tolerant due to the high power.

Algorithm 4

1: Assume $\breve{I}_{d,k} = 0$, then $\breve{I}_{s,k} = \breve{I}_k$.
2: Use Algorithm 3 to calculate surface normal $\bar{n}$ and specular reflectance parameters k_s and α from $\breve{I}_{s,k}$, $k = 1, ..., N$.
3: Find the lighting direction which will give the smallest specular components among all the N lights:

$$k^* = \operatorname*{argmin}_{k \in [1,...,N]} \left(\bar{n} \cdot \bar{h}_k \right). \tag{48}$$

4: Calculate the diffuse reflectance factor using the k^*-th illumination as:

$$k_d = \frac{\breve{I}_{k^*} - k_s \left(\bar{n} \cdot \bar{h}_{k^*} \right)^\alpha}{\bar{n} \cdot \bar{l}_{k^*}}. \tag{49}$$

5: Update $\breve{I}_{s,k}$ by subtract the diffuse part from the corrected image as:

$$\breve{I}_{s,k} = \breve{I}_k - k_d \left(\bar{n} \cdot \bar{l}_k \right), \tag{50}$$

for all $k = 1, ..., N$.
6: Repeat step 2-5 till $\bar{n}$ converge.
7: Return $\bar{n}, k_d, k_s, \alpha$.

---

4   Experimental Results

The proposed method is verified on a simulated scene first, with comparison of NIPS. Then SPS's performance on real world surface is tested in two real world scene.

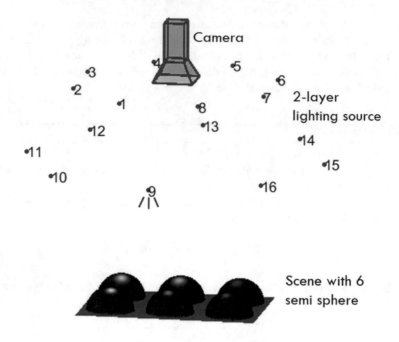

Fig. 5. Simulated scene

4.1   Simulated Scene

A simulated scene (see Fig. 5) with six semi-sphere with different reflectance parameters (see Table 1) was synthesized. 16 point lighting sources are placed in different position above the scene, and only one will be lit at each time. Lambertian reflectance model combined with Blinn-Phong specular reflectance model are used to generate the synthesized images. (See Fig. 6 for the synthesized image under 9th illumination.)

Two methods, SPS and NIPS, were implemented on the synthesized data. The performance were compared in two-ways, angular error and rerendering error.

Angular error can be calculated through the comparison of the ground truth surface normal and calculated surface normal. Since we are using the synthesized data, ground truth can be obtained. To be more specific, the angular error map at pixel (i, j) can be calculated as

$$e_\theta^{(i,j)} = \arccos\left(\bar{\boldsymbol{n}}_{gt}^{(i,j)} \cdot \bar{\boldsymbol{n}}_{cal}^{(i,j)}\right),\tag{51}$$

where $\bar{\boldsymbol{n}}_{gt}^{(i,j)}$ is the surface normal from ground truth, and $\bar{\boldsymbol{n}}_{cal}^{(i,j)}$ is the surface normal from calculation.

Table 1. Simulated scene parameters

Semisphere	1	2	3	4	5	6
k_d	0.05	0.05	0.05	0.05	0.05	0.05
k_s	0.19	0.58	0.69	0.23	0.40	0.82
α	32	32	32	64	64	64

Rerendering error, on other hand, can be used for the case when ground truth is absent. It applies the fitted reflectance model and image formation procedure to rerender the image of the scene, and then compares the rerendered image with the captured image. The formation to calculate the rerendering error at pixel (i, j) can be calculated as

$$
e_I^{(i,j)} = \frac{\left| \breve{I}_{k,gt}^{(i,j)} - f(\overline{l}_k^{(i,j)}, \overline{v}^{(i,j)}, \overline{n}_{cal}^{(i,j)}, p_{cal}^{(i,j)}) \right|}{\breve{I}_{gt}^{(i,j)}}, \tag{52}
$$

where $\breve{I}_{k,gt}^{(i,j)}$ is the corrected intensity of pixel (i, j) under k^{th} illumination, $f(\overline{l}_k^{(i,j)}, \overline{v}^{(i,j)}, \overline{n}_{cal}^{(i,j)}, p_{cal}^{(i,j)})$ is the fitted forward reflectance and image formation model and $p_{cal}^{(i,j)}$ is the fitting parameters for the forward model.

Figure 7 shows the angular error and rerendering error of SPS and NIFS of the synthesized experiment respectively. Table 2 shows the mean error of each semisphere. As it is shown in the results, SPS achieved better performance for this specific application. Also for smaller α, which indicates a wider specular reflection region, SPS achieved smaller error. As for the computational cost, the running time of SPS is about 1/3 the one of NIPS under the same computation environment.

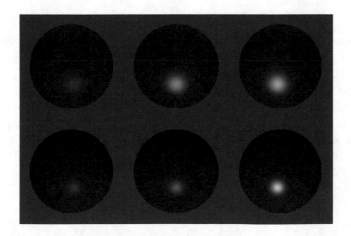

Fig. 6. Synthesized image under 9th illumination

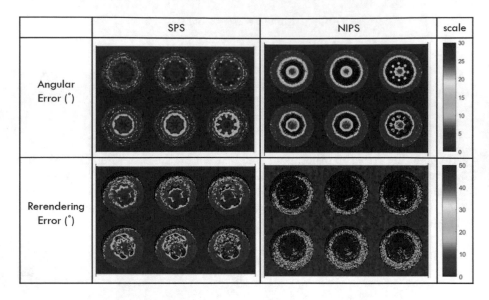

Fig. 7. Experimental results from simulated scene

Table 2. Numerical results from simulated scene

Semisphere		1	2	3	4	5	6
Angular error (°)	SPS	3.38	3.14	3.26	3.72	4.09	4.72
	NIPS	12.7	21.5	21.5	10.8	14.7	16.1
Rerendering error (%)	SPS	9.87	11.8	12.3	13	14.7	16.6
	NIPS	40.7	42.8	43.2	35.6	36	37.9

4.2   Real World Surfaces

To verify SPS, a system to collect data in real world was built (see Fig. 8). Two real world objects, Oreo cookie and auto tire (see Fig. 9), were observed using the constructed system. Due to the absence of ground truth, only rerendering error was calculated. The mean rerendering error for Oreo is 9.3%, while the one for tire is 12.4%. Meanwhile, the mean rerendering errors of NIPS exceed 50 for both cases.% The main reason is the poor quality of initial values for the parameters of the parameterized model. Since NIPS is not designed specifically for the dark surfaces, the performance could be improved if some extra pre-processing steps have been done, which is beyond the scope of this paper.

camera   2 layer lighting sources   scanning surface

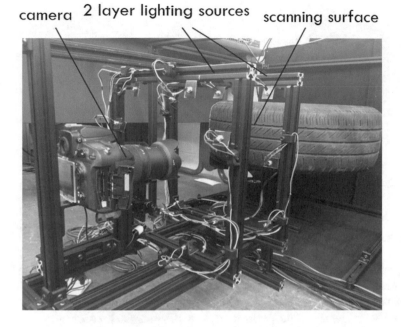

Fig. 8. Experiment setup

(a)   (b)

Fig. 9. Scanning objects: (a) Oreo cookie (b) tire

Furthermore, depth maps were generated using surface normal integration (SNI, [2]) from the calculated surface normal, which is presented in Fig. 10, from which you see that the shapes of both surfaces are recovered with fine detail.

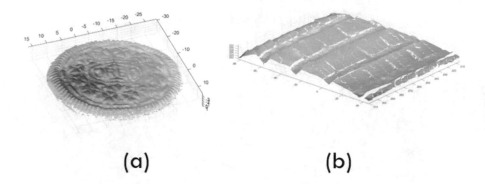

(a)   **(b)**

Fig. 10. Constructed depth map using SNI

5   Conclusion

This paper presented the technical details of SPS, which obtains surface orientation through reversing specular reflection and image formation procedure. The
fundamental of SPS is described first, followed by the numerical solution of SPS
problem with details. The performance of SPS on dark surfaces was compared
with state-of-the-art PS method, NIPS, in both simulated scene and surfaces in
real world. For this specific application, SPS achieved better performance with
shorter processing time.

This paper has mainly focused on the derivative of SPS and much work is
still left open, particularly the exploration of applications of SPS. It is also of
particular interest for parameter studies of how lighting conditions will effect
SPS's performance.

References

1. Woodham, R.J.: Photometric method for determining surface orientation from
   multiple images. Opt. Eng. **19**(1), 191139 (1980)
2. Li, B., Furukawa, T.: Microtexture road profiling system using photometric stereo.
   Tire Sci. Technol. **43**(2), 117–143 (2015)
3. Shafer, S.A.: Using color to separate reflection components. Color Res. Appl. **10**(4),
   210–218 (1985)
4. Li, B., Furukawa, T.: Photometric stereo under dichromatic reflectance framework
   dealing with non-lambertian surfaces. In: 2015 IEEE International Conference on
   Multisensor Fusion and Integration for Intelligent Systems (MFI), pp. 139–144.
   IEEE (2015)
5. Kipman, Y.: Non-contact vision based inspection system for flat specular parts.
   US Patent 6,525,810, 25 February 2003
6. Klijn, S., Reus, N.J., van der Sommen, C.M., Sicam, V.A.D.P.: Accuracy of a novel
   specular reflection technique for measurement of total corneal astigmatism. Invest.
   Ophthalmol. Vis. Sci. **56**(7), 1903–1903 (2015)
7. Blake, A., Brelstaff, G.: Specular stereo. IJCAI **2**, 973–976 (1985)

8. Ikeuchi, K.: Determining surface orientations of specular surfaces by using the photometric stereo method. IEEE Trans. Pattern Anal. Mach. Intell. **6**, 661–669 (1981)

9. Hertzmann, A., Seitz, S.M.: Example-based photometric stereo: shape reconstruction with general, varying BRDFs. IEEE Trans. Pattern Anal. Mach. Intell. **27**(8), 1254–1264 (2005)

10. Alldrin, N., Zickler, T., Kriegman, D.: Photometric stereo with non-parametric and spatially-varying reflectance (2008)

11. Ma, W.-C., Hawkins, T., Peers, P., Chabert, C.-F., Weiss, M., Debevec, P.: Rapid acquisition of specular and diffuse normal maps from polarized spherical gradient illumination. In: Proceedings of the 18th Eurographics Conference on Rendering Techniques, pp. 183–194. Eurographics Association (2007)

12. Goldman, D.B., Curless, B., Hertzmann, A., Seitz, S.M.: Shape and spatially-varying BRDFs from photometric stereo. IEEE Trans. Pattern Anal. Mach. Intell. **32**(6), 1060–1071 (2010)

13. Blinn, J.F.: Models of light reflection for computer synthesized pictures. ACM SIGGRAPH Comput. Graph. **11**, 192–198 (1977)

14. Shen, Y.-Q., Ypma, T.J.: Solving nonlinear systems of equations with only one nonlinear variable. J. Comput. Appl. Math. **30**(2), 235–246 (1990)

15. Kay, G., Caelli, T.: Estimating the parameters of an illumination model using photometric stereo. Graph. Models Image Process. **57**(5), 365–388 (1995)

XMIAR: X-ray Medical Image Annotation and Retrieval

M. M. Abdulrazzaq[1]([✉]), I. F. T. Yaseen[1], S. A. Noah[2],
M. A. Fadhil[3], and M. U. Ashour[4]

[1] KICT, International Islamic University Malaysia, Gombak, Selangor, Malaysia
eng.alobaydee81@gmail.com
[2] FTSM, Universiti Kebangsaan Malaysia, Bangi, Selangor, Malaysia
[3] IT, Philadelphia University Jordan, Jerash, Jordan
[4] FOIT, Majan University College, Muscat, Oman

Abstract. The huge development of the digitized medical image has been steered to the enlargement and research of the Content Based Image Retrieval (CBIR) systems. Those systems retrieve and extract the images by their own low level features, like texture, shape and color. But those visual features did not aloe the users to request images by the semantic meanings. The image annotation or classification systems can be considered as the solution for the limitations of the CBIR, and to reduce the semantic gap, this has been aimed annotating or to make the classification of the image with few controlled keywords. In this paper, we suggest a new hierarchal classification for the X-ray medical image using the machine learning techniques, which are called the Support Vector Machine (SVM) and k-Nearest Neighbour (k-NN). Hierarchy classification design was proposed based on the main body region. Evaluation was conducted based on ImageCLEF2005 database. The obtained results in this research were improved compared to the previous related studies.

Keywords: Machine learning · Support vector machines ·
Medical image analysis

1 Introduction

Massive numbers of medical images are generated daily in hospitals, clinics and medical centers [1]. These medical images can be used in many tasks, such as research and training, surgical planning, diagnosis of diseases, and medical reference. Therefore, there is a necessity to develop techniques that can provide appropriate indexing, efficient search and retrieval results through this large amount of medical images.

The Content Based Image Retrieval (CBIR) has been suggested to lead the users to retrieve relevant images in efficient and/or effective way. The CBIR system has the ability of using the visual contents of images to search images rather than using text to find images. However, CBIR systems are still incapable to interpret the image as how human's eyes can understand it, this issue is called semantic gap, this is the gap in between the low-level features (color, texture and shape) and high-level semantic concept. Recently, researchers moved to use annotation and classification techniques based on hierarchical concepts for X-ray medical image to solve the problem of semantic gap [2–8].

© Springer Nature Switzerland AG 2020
K. Arai and S. Kapoor (Eds.): CVC 2019, AISC 944, pp. 638–651, 2020.
https://doi.org/10.1007/978-3-030-17798-0_51

2   Related Work

Lately, Content Based Medical Image Retrieval (CBMIR) systems were introduced and implemented. However, those systems still have limitations in medical image retrieval. Therefore, various research works focused on feature extraction and the combination of feature extraction based on classification techniques to showed better performance in terms of accuracy rate [9, 10].

Table 1. Summaries of related work

No.	Author	Features	Classifier	Database	Semantic gap	Results
1	RWTH-i6	IDM with X32 thumbnails sizes, Sobel filter	1-NN	ImageCLEF2005	Not addressed	Error rate 12.6%
	RWTH-mi	CCF and IDM	1-NN	ImageCLEF2005	Not addressed	Error rate 13.3%
	Ulg.ac.be	16×16 randomly extracted patches from images	Decision Tree	ImageCLEF2005	Not addressed	Error rate 14.1%
	Geneva-gift	Gabor Texture Filters	5-NN	ImageCLEF2005	Not addressed	Error rate 20.6%
	Infocomm	Texture features	SVM	ImageCLEF2005	Not addressed	Error rate 20.6%
	MIRACLE	weighting function for 20-NN	20-NN	ImageCLEF2005	Not addressed	Error rate 21.4%
	NTU	Gray values	1-NN 2-NN	ImageCLEF2005	Not addressed	Error rate 21.7%
	NCTU-DBLAB	Scaling	SVM	ImageCLEF2005	Not addressed	Error rate 24.7%
	CEA	Sobel filter	3-NN	ImageCLEF2005	Not addressed	Error rate 36.9%
	Mtholyoke	Tamura texture features and Gabor	k-NN	ImageCLEF2005	Not addressed	Error rate 37.8%
	CINDI	Canny Edge detector	SVM	ImageCLEF2005	Not addressed	Error rate 45.3%
	Montreal	Fourier shape and contour descriptors	–	ImageCLEF2005	Not addressed	Error rate 55.7%

(*continued*)

Table 1. (*continued*)

No.	Author	Features	Classifier	Database	Semantic gap	Results
2	[14]	Blob Gray level texture and contrast	SVM	ImageCLEF2005	Not addressed	Accuracy: 89%
3	[2]	–	SVM	LabelMe Corel Images Google Images	Addressed	Very positive
4	[3]	–	SVM	LabelMe Corel Images Google Images	Addressed	Very positive
5	[4]	Texture, Shape, Local & global features	k-NN SVM	ImageCLEF2005	Addressed	Accuracy: 82% for k-NN, 89% for SVM
6	[15]	DWT	SVM	126 computerized lung tomography	Not addressed	Accuracy: 76.74%
7	[16]	Histogram moments and GLCM	k-NN	478 Ultrasound images	Not addressed	Accuracy: 88.12%
8	[5]	Local distribution of edges	Predictive clustering trees	ImageCLEF2008	Addressed	90.64% 82.64
9	[17]	DWT	SVM	60 MRI	Not addressed	Accuracy: 65%
10	[18]	DWT	FP-ANN k-NN	50 MRI	Not addressed	Accuracy: 90% FP-ANN 99% k-NN
11	[19]	Shape features	Bayesian	4937 X-ray	Not addressed	82.87%
12	[20]	Edges and shape	SVM	1169 X-ray	Not addressed	90.88%
13	[21]	Shape and texture	SVM, Euclidean distance, and neural network	4402 X-ray	Not addressed	88.77% 94.2%
14	[22]	Canny edge detection and gabor filter	k-NN	500 CT scan images of brain	Not addressed	52.30%
15	[6]	Shape and texture	k-NN and neural network	2158 X-ray	Addressed	93.6%

(*continued*)

Table 1. (*continued*)

No.	Author	Features	Classifier	Database	Semantic gap	Results
16	[23]	GLCM, Canny Pixel, BoW, LPB	SVM k-NN	ImageCLEF2007	Not addressed	90% for SVM 86% for k-NN
17	[24]	Walsh Hadamard transform and Gabor	k-NN, decision tree, and Neural Networks	lung CT scan	Not addressed	90% by neural network
18	[25]	Gray scale, symmetrical and texture	SVM	50 MRI	Not addressed	84%
19	[7]	–	SVM	PolSAR images	Addressed	–
20	[8]	SIFT	SVM	Pascal VOC Animals with Attribute	Addressed	Hierarchical classification

The "2005 Medical Annotation Tasks", which conducted via twelve teams, as illustrated in Table 1, No. 1 [11–13]. In general, these teams used different feature extraction and classification techniques as shown. Thus, the error rate started as 12.6% by RWTH-i6, and ended as 55.7% by Montreal.

Basically, related works show the use of global or local features are not always a suitable solution for medical image classification as shown in Table 1, they used different techniques to extract features either globally or locally, and applied machine learning techniques to improve the results. But the results still not good enough and there are ways to improve them. Furthermore, the most difficult and important parts are how to find the semantic concepts. Finding the semantic concept among medical images in a dataset can reduce the semantic gap and improve the results of medical image classification and retrieval. Thus, researchers tried to treat a group of image regions with multi-levels categorizations and then annotate images based on these regions.

In conclusion, there is a need to develop hierarchical classification depends on combining local and global features based on using multi classifiers for medical images. To overcome all of the mentioned limitations, the work in this research presented a framework to classify and annotate X-ray medical images automatically depending on multi level of feature extraction and hierarchical relations among medical images based on using two classification techniques.

3 Methodology

In this research X-ray Medical Images Annotation and Retrieval (XMIAR) prototype was proposed based on multi-level of feature extraction, multi classifiers, and hierarchical relations among X-ray medical images. ImageCLEF2005 database was used for evaluation. The database includes 10,000 X-ray images; these images are segmented into two sets; 9000 images as a training predefined set into 57 categories, and 1000 images as testing set [26]. Figure 1 shows samples of the X-ray images from the 57 categories. Figure 2 shows the Architecture of the proposed prototype.

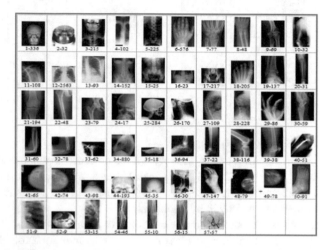

Fig. 1. Samples of X-ray images from ImageCLEF2005

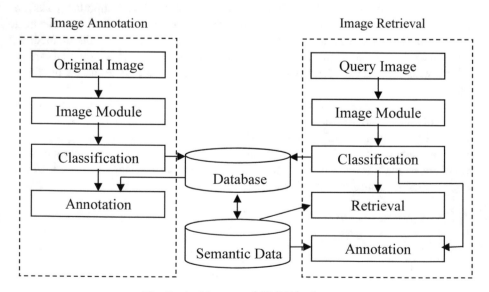

Fig. 2. Architecture of XMIAR prototype

The image module is composed of a collection of processes; image resizing, image enhancement, feature extraction, feature combination and feature reduction, all of these steps are explained in our previous published research work [27–29]. The classification and Annotation part are explained in this paper.

Three levels of image annotation are done; in this paper the first level of the hierarchical is going to be present, which the classification annotation based on the main body region. The basic idea of this level is the annotation of images to construct semantic relations. The training set is categorized into 57 known classes, where each one of those classes was annotated using long description. In practice, the hierarchical design of these classes is based on specific available descriptions (pre-defined).

The training set of ImageCLEF2005 dataset contains nine main body regions, which are "cranium, spine, arm (upper extremity), breast, leg (lower extremity), abdomen, chest, pelvis, and thorax". Thus, the proposed hierarchical concepts based on nine categories (main body region) as the first level of annotation. The hierarchical design of those concepts is shown in Fig. 3. (The figure shows the three hierarchical levels; this paper is presenting the first level only).

Main Body Region	Sub Body Region	Image Orientation		Main Body Region	Sub Body Region	Image Orientation	
1.Cranium	Facial	Coronal	1		Upper Leg	Coronal-FL	30
		Sagittal	2			Sagittal	31
		Other	3			Coronal	32
	Neuro	Sigttal	4		Hip	Coronal	33
		Other	5			Sagittal	34
	Cranium	Coronal	6		Lower Leg	Coronal-FL	35
2.Spine	Cervical	Coronal	7			Sagittal	36
		Sagittal	8			Coronal	37
		Other	9	5.Leg	Ankle	Coronal	38
	Thoracic	Coronal	10			Sagittal	39
		Sagittal	11		Foot	Coronal	40
	Lumbar	Coronal	12			Sagittal	41
		Sagittal	13			Other	42
3.Arm	Upper Arm	Coronal	14		Knee	Coronal-FL	43
	Carpal	Coronal	15			Sagittal	44
		Sagittal	16			Coronal	45
	Elbow	Coronal	17			Axial	46
		Sagittal	18		Abdomen-GA	Coronal	47
	Forearm	Coronal	19	6.Abdomen	Upper Abdomen	Coronal-FL	48
		Sagittal	20			Coronal	49
	Shoulder	Coronal	21		Abdomen-UR	Coronal	50
		Sagittal	22		Bones	Coronal	51
	Hand	Coronal	23	7.Chest	Chest	Sagittal	52
		Sagittal	24			Coronal	53
		Other	25		Pelvis	Coronal	54
4.Breast	Right	Axial	26	8.Pelvis	Pelvis	Coronal-FL	55
		Other	27		Pelvis	Coronal-AN	56
	Left	Axial	28	9.Thorax	Hilum	Coronal-FL	57
		Other	29				

Fig. 3. The hierarchical design

For the proposed hierarchy design, the 57 categories are initially combined into nine categories to carry out the first classification level. The first level of classification includes 9 concepts, which represents the main parts of the human body. Thus, this level can be achieved based on using a training set of 9 classes: $L1 = \{(x1, y1), (x2, y2), ..(x9, y9)\}$; where y_i represents a label of the class that related to the class x_i.

The main benefit of the first-level annotation method is that the annotation way offers more meaningful retrieval process. Recently, the majority of content-based image retrieval systems depend on using the visual similarity only, while some of those systems use the semantic information, but with using low-level semantic similarities, which represents the varied appearances of the same object. As an example, the similarity among a cervical spine image in both views; sagittal and coronal is known as the low-level semantic similarity. On the other hand, the high-level semantic similarity represents the similarity among various objects. As an example, it represents the similarity between the cervical spine and thoracic spine images or among the chest and spine images.

The presented annotation method in this work depends on using both the low and high-level semantic similarities. With the use of this method, a user who has a spine image as a query expects to see the whole spine images kinds, such as "lumbar, thoracic, and cervical spine images" or specific images for the spine, such as the "lumbar spine" image in the sagittal view. One more benefit of the presented method is that it facilitates the mixed retrieval. Generally, the retrieval of images depends on using the texts or contents of those images, while the mixed retrieval is generally rare. On the other hand, the presented method in this work depends on extracting the low-level features and then assigning the annotation. Thus, a match can be found among those features with a possibility to perform text-based retrieval due to the annotation. This in turn permits both types of retrieval; text based and content based.

During the evaluation process two classifiers were performed which are SVM and k-NN for the first-level medical image annotation method. Two classifiers were chosen to perform to demonstrate and assess the ability of the proposed method to solve the semantic gap problem and to obtain the best accuracy result. In addition, the evaluation set assists in developing medical image retrieval processes.

4   Experiment and Results

A series of experiments were conducted during this research. The main goals of these experiments are to evaluate the performance of the proposed XMIAR prototype in order to validate that it plays an important role in X-ray medical image retrieval field, especially in retrieving images based on main body region, such that it is able to reduce the semantic gap problem and to meet the main aim of this research.

Semantic gap represents the gap between the information that one can extract from visual data and the interpretation that the same data have for a user in a given situation [30]. To solve this problem, image annotation was proposed based on the hierarchical image classification as one image may contain different meanings at multiple semantic levels. Moreover, the hierarchical image classification is also strongly expected to achieve better image annotation, which provides better image semantics for an image retrieval system. Normally, the concept hierarchy defines the basic vocabulary of semantic image concepts and their logical relationships. An upper (first) level of a semantic hierarchy covers the main concepts. This experiment involves the first level of hierarchy in which consists of nine main concepts, which represent the nine main regions of the human body.

In this experiment, both of k-NN and SVM classifiers were trained according to the first level of hierarchy, where the model then was generated. These classifiers were trained using images from the ImageCLEF2005 database. Training images were divided into 90% training and 10% test images for each class, where the total number of classes is 9 in the first level.

A more suitable way to illustrate the results of a multi-class classifier is by plotting a precision and recall plot. Precision and Recall are common in information retrieval systems for evaluating the classification performance. In precision and recall space, recall is plotted on the x-axis, while precision is on the y-axis. For evaluation measurements, standard *Recall (R)* and *Precision (P)* are appropriate to rate the performance. They are defined as follows:

$$P = \frac{\psi}{\psi + \xi} \tag{1}$$

$$R = \frac{\psi}{\psi + \zeta} \tag{2}$$

Where ψ is the set of true positive images that are related to the corresponding semantic class and are classified correctly, ξ is the set of true negative samples that are irrelevant to the corresponding semantic class and classified incorrectly and ζ is the set of false positive samples that are related to the corresponding semantic class but they are misclassified. High *Precision* means that few irrelevant images are returned, while high *Recall* means few relevant images are "missed".

In the information retrieval field, the most common used item to evaluate the retrieval results is the PR curve (precision recall curve). Though in the current case this curve is a classification problem, it is used as an evaluation figure. The difference is that there are no curves, but graphs. A 'curve' is used to describe the effect of different parameters on a retrieval method, and each curve corresponds to one class only. On the other hand, a 'plot' can show all the classes' position in the same coordinates simultaneously and describe the effect of the classification method. It is more suitable to illustrate the results of a multi-class classifier.

Tables 2 and 3 show the confusion matrix of level one of classification for the SVM and k-NN, respectively. The confusion matrix explains how the classification of each test image was performed. The confusion matrix was used to calculate the rate of classification accuracy, which was defined as the proportion of the total number of correct predictions. Each column of the matrix represents a specific predicted class, while each row represents the actual class. The overall accuracy rates were 93.05% and 91.94 for SVM and k-NN, respectively.

Table 2. Confusion matrix of level one by using SVM

Classes	1	2	3	4	5	6	7	8	9	Precision
1	80	1	4	0	1	3	0	1	0	88.88
2	3	100	1	0	1	2	1	1	0	91.74
3	1	1	137	0	1	1	0	1	0	96.47
4	2	1	4	336	2	7	0	1	0	95.18
5	1	1	2	1	14	1	0	0	0	70
6	2	3	2	1	1	20	0	1	0	66.66
7	1	0	2	0	0	0	128	0	0	97.71
8	0	0	0	0	1	2	0	27	0	90
9	0	0	0	0	0	0	0	0	1	100
Recall	88.88	93.45	90.13	99.40	66.66	55.55	99.22	84.37	100	

Table 3. Confusion matrix of level one by using k-NN

Classes	1	2	3	4	5	6	7	8	9	Precision
1	80	1	4	0	1	3	0	1	0	88.88
2	3	98	2	0	1	2	2	1	0	89.90
3	1	2	131	0	1	1	5	1	0	92.25
4	2	1	4	340	2	3	0	1	0	96.31
5	1	1	2	1	14	1	0	0	0	70
6	2	3	2	1	1	20	0	1	0	66.66
7	1	2	6	0	0	0	122	0	0	93.13
8	0	0	0	0	1	2	0	27	0	90
9	0	0	0	0	0	0	0	0	1	100
Recall	88.88	90.74	86.75	99.41	66.66	62.5	94.57	84.37	100	

Figures 4 and 5 show the precision and recall plots for the classes of level one of both classifiers. In these figures, G (Good) in the top right area is the best region because the points in this region have high precision and high recall. On the other hand, B (Bad) in the bottom left area is the worst region as its points have low precision and low recall. If most of the classes fall in region G, then it is considered to have high performance.

Figure 4 shows the precision and recall plot of level one classification using the SVM. It shows that all of the nine classes fall in region G with no classes in region B. Whilst, Fig. 5 shows the same plot of precision and recall, but using the k-NN. The plot shows that all of the nine classes fall in region G and no class in region B. Based on these results; all the classes have achieved high performance in the first level of classification.

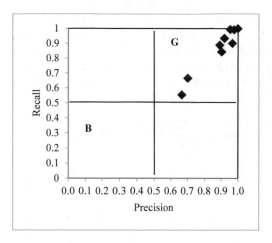

Fig. 4. Precision and recall plot of level one classification by using SVM

In the first level of annotation, one word is assigned to the classified images. It is important to note that once an image is classified, the high level and low-level semantic concepts are assigned to become the semantic label for that image. In case each image has one annotated word from level one of the classification. This automatic image annotation is very useful for the XMIAR prototype to perform semantic retrieval. Based on the obtained results of this experiment, it was found that the SVM classifier performs better than the k-NN one in the first level. Therefore, in the XMIAR proto-type, two classifiers were used for automatic annotation; the SVM is used for the first classification to have better accuracy results.

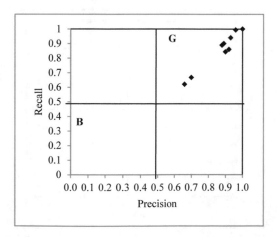

Fig. 5. Precision and recall plot of level one classification by using k-NN

The obtained results in this research show improvement when compared with previous studies for the same dataset. For the same image classification using low-level features and ImageCLEF2005 database, the results in this work show improvements as compared with the results obtained by [14]. Their results show that with using the SVM classifier, 53% of the 57 classes fall in area G as shown in Fig. 6.

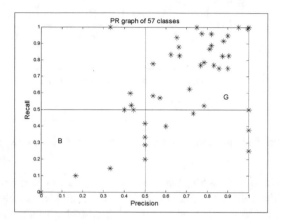

Fig. 6. Precision and recall plot by [14]

Additionally, the results in this work are better than those achieved by [4]. His results demonstrated that 89% of classes fall in region G with no class in region B in the first level of classification as shown in Fig. 7.

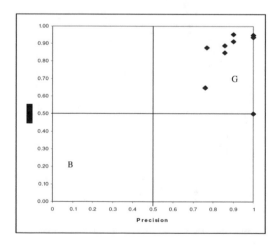

Fig. 7. Precision and recall plot by Mueen [4]

5   Conclusion

The main goal of this research work is to validate that the automatic medical image annotation plays an important role in meaningful image retrieval; specifically, to retrieve images according to body region, sub-body regions, and orientations. This paper presented the fist level of classification, which is based on the main body region. Normally, the classification of an image depends on these low-level features, while the annotation relies on the classification accuracy. Moreover, the multi-level classification yields to multi-level annotation. Therefore, the results show that the hierarchical classification has high accuracy results. Based on the optimal performance of both the SVM and k-NN, these two classifiers were used in the XMIAR prototype for both classification and annotation, where the SVM is used for the first level of classification, because SVM obtained better accuracy results. Moreover. The experiment is conducted to investigate the performance of the XMIAR prototype using ImageCLEF2005 dataset. In the future we going to publish the process of the second and third level of classification, they are based on the sub-body region and image orientation.

References

1. Muller, H., Deserno, T.M.: Content-based medical image retrieval. In: Biomedical Image Processing. Biological and Medical Physics, Biomedical Engineering, pp. 471–494. Springer, Heidelberg (2011)
2. Fan, J., Gao, Y., Luo, H.: Hierarchical classification for automatic image annotation. In: Proceedings of the 30th Annual International ACM SIGIR Conference on Research and Development in Information Retrieval, pp. 111–118. ACM (2007)
3. Fan, J., Gao, Y., Luo, H., Jain, R.: Mining multilevel image semantics via hierarchical classification. IEEE Trans. Multimedia 10(2), 167–187 (2008)
4. Mueen, A.: Multi-level automatic medical image annotation. Doctoral thesis. Faculty of Computer Science and Information Technology, University of Malaya (2009). http://dspace.fsktm.um.edu.my/handle/1812/451. Accessed 30 June 2013
5. Dimitrovski, I., Kocev, D., Loskovska, S., Džeroski, S.: Hierarchical annotation of medical images. Pattern Recogn. 44(10), 2436–2449 (2011)
6. Fesharaki, N.J., Pourghassem, H.: Medical X-ray image hierarchical classification using a merging and splitting scheme in feature space. J. Med. Sig. Sens. 3(3), 150–163 (2013)
7. Liu, F., Shi, J., Jiao, L., Liu, H., Yang, S., Wu, J., Hao, H., Yuan, J.: Hierarchical semantic model and scattering mechanism based PolSAR image classification. Pattern Recogn. 59, 325–342 (2016)
8. Zhao, S., Zou, Q.: Fusing multiple hierarchies for semantic hierarchical classification. Int. J. Mach. Learn. Comput. 6(1), 47–51 (2016)
9. Muller, H., Michoux, N., Bandon, D., Geissbuhler, A.: A review of content-based image retrieval systems in medical applications—clinical benefits and future directions. Int. J. Med. Inform. 73(1), 1–23 (2004)
10. Hersh, W., Kalpathy-Cramer, J., Jensen, J.: Medical image retrieval and automated annotation: OHSU at ImageCLEF 2006. In: Evaluation of Multilingual and Multi-modal Information Retrieval, vol. 4730, pp. 660–669. Springer, Heidelberg (2007)

11. Clough, P., Müller, H., Deselaers, T., Grubinger, M., Lehmann, T.M., Jensen, J., Hersh, W.: The CLEF 2005 cross–language image retrieval track. Accessing Multilingual Information Repositories, vol. 4022, pp. 535–557. Springer, Heidelberg (2006)
12. Deselaers, T., Müller, H., Clough, P., Ney, H., Lehmann, T.M.: The CLEF 2005 automatic medical image annotation task. Int. J. Comput. Vis. **74**(1), 51–58 (2007)
13. Amaral, I.F.A.: Content-Based image retrieval for medical applications. Master thesis. Faculty of Science, University of Porto (2010). http://www.inescporto.pt/~jsc/students/2010IgorAmaral/2010relatorioIgorAmaral.pdf. Accessed 10 Mar 2014
14. Qiu, B., Tian, Q., Xu, C.S.: Report on the annotation task in ImageCLEFmed 2005. In: Cross Language Evaluation Forum 2005 Workshop, Vienna, Austria (2005)
15. Sarikaya, B.: Lung mass classification using wavelets and support vector machines. Master thesis. Institute of Sciences. Computer Engineering Graduate Program (2009). http://libris.bahcesehir.edu.tr/dosyalar/Tez/077821.pdf. Accessed 2 Feb 2014
16. Sohail, A.S.M., Bhattacharya, P., Mudur, S.P., Krishnamurthy, S., Gilbert, L.: Content-based retrieval and classification of ultrasound medical images of ovarian cysts. In: Artificial Neural Networks in Pattern Recognition, vol. 5998, pp. 173–184. Springer, Heidelberg (2010)
17. Othman, M.F.B., Abdullah, N.B., Kamal, N.F.B.: MRI brain classification using support vector machine. In: Proceeding of the 4th International Conference on Modeling, Simulation and Applied Optimization (ICMSAO). IEEE, pp. 1–4 (2011)
18. Rajini, N.H., Bhavani, R.: Classification of MRI brain images using k-nearest neighbor and artificial neural network. In: Proceeding of International Conference on Recent Trends in Information Technology (ICRTIT), pp. 563–568. IEEE (2011)
19. Fesharaki, N.J., Pourghassem, H.: Medical X-ray images classification based on shape features and Bayesian rule. In: 2012 Fourth International Conference on Computational Intelligence and Communication Networks (CICN), pp. 369–373. IEEE (2012)
20. Ghofrani, F., Helfroush, M.S., Danyali, H., Kazemi, K.: Medical X-ray image classification using Gabor-based CS-local binary patterns. In: Proceedings International Conference on Electronics, Biomedical Engineering and its Applications, Dubai, pp. 284–288 (2012)
21. Mohammadi, S.M., Helfroush, M.S., Kazemi, K.: Novel shape-texture feature extraction for medical X-ray image classification. Int. J. Innov. Comput. Inf. Control: IJICIC (Int. J. Innov. Comput. Inf.), **8**(1), 658–676 (2012)
22. Charde, P.A., Lokhande, S.D.: Classification using K nearest neighbor for brain image retrieval. Int. J. Sci. Eng. Res. **4**(8), 760–765 (2013)
23. Zare, M.R., Seng, W.C., Mueen, A.: Automatic classification of medical X-ray images. Malays. J. Comput. Sci. **26**(1), 9–22 (2013)
24. Bhuvaneswari, C., Aruna, P., Loganathan, D.: A new fusion model for classification of the lung diseases using genetic algorithm. Egypt. Inform. J. **15**(2), 69–77 (2014)
25. Nandpuru, H.B., Salankar, S.S., Bora, V.R.: MRI brain cancer classification using Support Vector Machine. In: Proceeding of Students' Conference on Electrical, Electronics and Computer Science (SCEECS), pp. 1–6. IEEE (2014)
26. Lehmann, T.M., Schubert, H., Ott, B., Leisten, M.: ImageCLEF2005 library (2005). http://ganymed.imib.rwth-aachen.de/irma/. Accessed 30 July 2013
27. Abdulrazzaq, M.M., Yaseen, I.F., Noah, S.A., Fadhil, M.A.: Multi-level of feature extraction and classification for X-ray medical image. Indonesian J. Electr. Eng. Comput. Sci. **10**(1), 154–167 (2018)

28. Abdulrazzaq, M.M., Noah, S.A., Fadhil, M.A.: X-ray medical image classification based on multi classifiers. In: 2015 4th International Conference on Advanced Computer Science Applications and Technologies (ACSAT), Kuala Lumpur, pp. 218–223 (2015)

29. Abdulrazzaq, M.M., Mohd, S.A., Fadhil, M.A.: Medical image annotation and retrieval by using classification techniques. In: 2014 3rd International Conference on Advanced Computer Science Applications and Technologies (ACSAT), pp. 32–36. IEEE (2014)

30. Smeulders, A.W., Worring, M., Santini, S., Gupta, A., Jain, R.: Content-based image retrieval at the end of the early years. IEEE Trans. Pattern Anal. Mach. Intell. **22**(12), 1349–1380 (2000)

Visual Percepts Quality Recognition Using Convolutional Neural Networks

Robert Kerwin C. Billones[✉], Argel A. Bandala,
Laurence A. Gan Lim, Edwin Sybingco,
Alexis M. Fillone, and Elmer P. Dadios

De La Salle University, Manila, Philippines
robert.billones@dlsu.edu.ph

Abstract. In visual recognition systems, it is necessary to identify between good or bad quality images. Visual perceptions are discrete representation of observable objects. In typical systems, visual parameters are adjusted for optimal detection of good quality images. However, over a wide range of visual context scenarios, these parameters are usually not optimized. This study focused on the learning and detection of good and bad percepts from a given visual context using a convolutional neural network. The system utilized a perception-action model with memory and learning mechanism which is trained and validated in four different road traffic locations (DS0, DS3-1, DS4-1, DS4-3). The training accuracy for DS0, DS3-1, DS4-1, and DS4-3 are 93.53%, 91.16%, 93.39%, and 95.76%, respectively. The validation accuracy for DS0, DS3-1, DS4-1, and DS4-3 are 88.73%, 77.40%, 95.21%, and 83.56%, respectively. Based from these results, the system can adequately learn to differentiate between good or bad quality percepts.

Keywords: Convolutional neural networks · Image quality ·
Perception-action modeling · Visual perceptions

1 Introduction

Visual perceptions are discrete representation of observable objects [1]. Visual perceptions include an external and internal sensing [2] mechanism to automatically adjust visual parameters whenever bad quality images (percepts) are detected. The external sensing processes raw inputs from the external environment, while the internal sensing is based from the internal processes of the memory, intelligence [3], and control blocks of a cognitive system. Any actions that caused changes in percepts are due to the internal sensing capability of a system. In automatic video analysis of road traffic systems [4], it is essential to capture good quality images for applications such as vehicle detection, and tracking [5, 6], vehicle classification [7], license plate detection, traffic violations detection [8, 9], and pedestrian activity analysis [10–12], etc. The system performance for such applications degrade whenever blurred images [13, 14] or poor-quality videos due to un-calibrated systems are captured by monitoring cameras. Several techniques for image [15–17] and video quality assessment were introduced. An interesting method is the use of a learning mechanism to evaluate image quality [18].

© Springer Nature Switzerland AG 2020
K. Arai and S. Kapoor (Eds.): CVC 2019, AISC 944, pp. 652–665, 2020.
https://doi.org/10.1007/978-3-030-17798-0_52

This study proposed the use of convolutional neural networks (CNN) to learn, recognize, and differentiate between good and bad quality visual percepts. CNN is a machine learning architecture that is used for generic object classification [19–21]. The CNN architecture is remarkably good for image classification. The study also includes a perception-action (PA) model with a memory and learning mechanism. Due to wide range of road context, the system is trained in different road (context) locations.

2   Object Detection Quality Evaluation

Image quality assessment has been a topic of several research works. Some methods include statistical correlation [15, 22] with full reference image, image quality metric using gradient profiles [16] and natural scene statistics [17, 23, 24] with no reference image. Another research work includes a learning mechanism to assess image quality such as adaboosting neural network [18]. Image quality can also be undertaken in the context of object detection and tracking. A good object detector and tracker can consistently identify and track individual objects even with occlusion and illumination changes [25]. Some measures to identify the quality of detected objects are precision and recall. Recall ($\rho_{i,j}$) can be calculated given a ground truth GT_j and an estimate E_i using Eq. 1 below:

$$\rho_{i,j} = \frac{|\varepsilon_i \cap GT_j|}{|GT_j|} \tag{1}$$

Recall measures how much of the ground truth is covered by the estimate. Precision ($v_{i,j}$), on the other hand, measures how much of the estimate covers the ground truth. Precision can be calculated using Eq. 2 below:

$$v_{i,j} = \frac{|\varepsilon_i \cap GT_j|}{|\varepsilon_i|} \tag{2}$$

Both measures can take values between 0 (no overlap) and 1 (fully overlapped). Ideally, we want detected objects with high precision and recall. However, it is also possible to have detected objects with high precision but low recall, and vice-versa. Precision and recall are illustrated in Fig. 1.

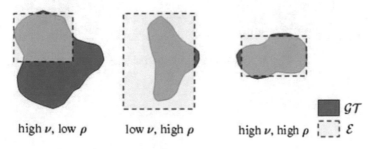

high v, low ρ low v, high ρ high v, high ρ GT ε

Fig. 1. Precision (v) and recall (ρ) [25]

Another type of error measurement for object detection is configuration. Configuration includes false positive (FP), false negative (FN), multiple trackers (MT), and multiple objects (MO). These object detection and tracking errors are illustrated in Fig. 2.

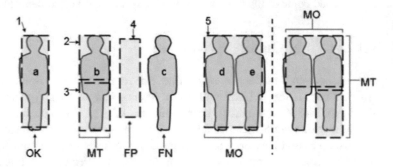

Fig. 2. Configuration errors: FP, FN, MT, and MO [25]

In crowded scenes, one type of object detection and tracking error that degrades the quality of visual perception is occlusion. When two ground truths overlap, estimates which are correctly configured also overlap. To handle multiple occlusions, it is usually defined directionally instead of pairwise, see Eq. 3 below:

$$OCC_j = \begin{cases} 1, & \exists \mathcal{GT}_{k\,s.t.} |\mathcal{GT}_j \cap \mathcal{GT}_k| > t_o \\ 0, & \text{otherwise} \end{cases} \tag{3}$$

Occlusion can be illustrated in Fig. 3.

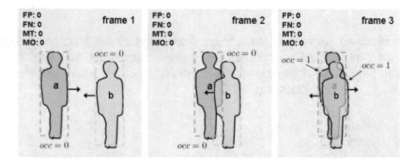

Fig. 3. Occlusion in detected objects [25]

3 Perception-Action Model with Memory and Learning Mechanism

The proposed system used a perception-action (PA) model with a memory and learning mechanism, see Fig. 4. This hybrid model extends the capability of a conventional PA model to learn the quality of visual percepts being fed to the system. The perception module includes an external and internal sensing blocks. A simple object detector can be used in the external sensing block. The output perceptions are regularly calibrated by the internal sensing block, whenever the quality of perceptions fall below a certain threshold. This means that a continuous stream of bad percepts will trigger the perception re-calibration mechanism. Supervised knowledge and learned weights are stored in the memory module. A simple intelligence module consists of a learning mechanism. The learner used in this system is a convolutional neural network. The system is trained to differentiate between a good or bad percept. The control module handles the decision-making based from the inference mechanism.

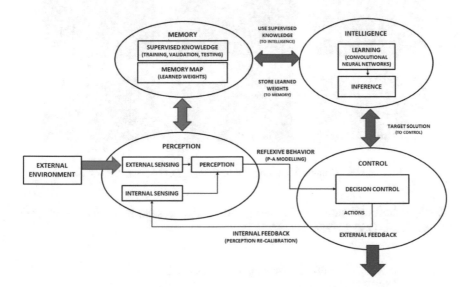

Fig. 4. Simplified diagram of visual perception-action model with memory and learning mechanism

4 Methodology

4.1 Image Data Sets for Training and Validation

The image data set used for this study is composed of four different locations. DS0 is composed of traffic videos captured using low-altitude cameras which are used for traffic violations monitoring (number coding and illegal loading/unloading) of a T-type road intersection. It is further composed of LS1, LS2, LS3, TA1, TA2, and TA3 video data sets. DS3-1 is a traffic video captured using high-altitude camera which is usually used for traffic congestion monitoring of a wide intersection. DS4-1 is a traffic video captured using medium-altitude camera which is usually used for traffic violations (number coding and illegal loading/unloading) and congestion monitoring of a bus stop area. DS4-3 is the same camera used in DS4-1, but the video capture is done during night-time. Figure 5 shows the sample video capture images from the different context locations.

Fig. 5. Context locations: DS0, DS3-1, DS4-1, and DS4-3

Table 1 shows the number of training and validation samples used for each context locations (DS0, DS3-1, DS4-1, DS4-3).

Table 1. Training and validation samples

Location	No. of training samples	No. of validation samples
DS0	900	300
DS3-1	1400	600
DS4-1	1400	600
DS4-3	1400	600

Figures 6, 7, 8 and 9 shows the sample training images for good quality percepts in locations DS0, DS3-1, DS4-1, and DS4-3, while Figs. 10, 11, 12 and 13 shows the sample training images for bad quality percepts. These bad quality percepts are usually the result of occlusion. It also exhibits high precision but low recall problem, and vice versa.

Good (1)   Good (2)   Good (3)   Good (4)   Good (5)

Good (6)   Good (7)   Good (8)   Good (9)   Good (10)

Fig. 6. Sample of good percepts for DS0

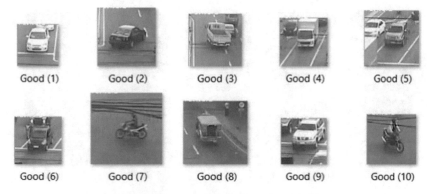

Good (1)   Good (2)   Good (3)   Good (4)   Good (5)

Good (6)   Good (7)   Good (8)   Good (9)   Good (10)

Fig. 7. Sample of good percepts for DS3-1

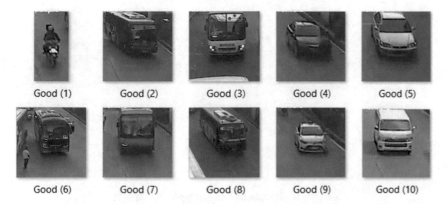

Good (1)   Good (2)   Good (3)   Good (4)   Good (5)

Good (6)   Good (7)   Good (8)   Good (9)   Good (10)

Fig. 8. Sample of good percepts for DS4-1

| Good (1) | Good (2) | Good (3) | Good (4) | Good (5) |

| Good (6) | Good (7) | Good (8) | Good (9) | Good (10) |

Fig. 9. Sample of good percepts for DS4-3

| Bad (1) | Bad (2) | Bad (3) | Bad (4) | Bad (5) |

| Bad (6) | Bad (7) | Bad (8) | Bad (9) | Bad (10) |

Fig. 10. Sample of bad percepts for DS0

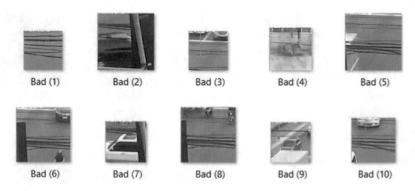

| Bad (1) | Bad (2) | Bad (3) | Bad (4) | Bad (5) |

| Bad (6) | Bad (7) | Bad (8) | Bad (9) | Bad (10) |

Fig. 11. Sample of bad percepts for DS3-1

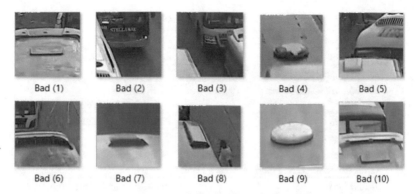

Fig. 12. Sample of bad percepts for DS4-1

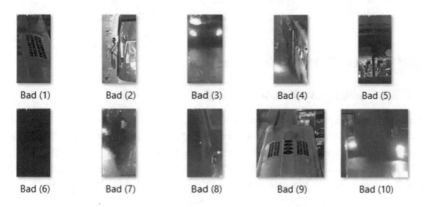

Fig. 13. Sample of bad percepts for DS4-3

4.2 Convolutional Neural Network Parameters

The study used a CNN model with three (3) layers of 2D convolutional (CONV) layers and an output layer of densely-connected neural network. The input to CNN is the images from the training and validation samples, and the binary output is the quality of image (good or bad percepts). The CNN model and its parameters are listed in Table 2.

Table 2. CNN model and parameters

CNN model	Parameters
Input layer (2D convolutional layer)	32 filters 3 × 3 kernel size
Activation	Rectified linear unit (relu)
2D max pooling layer	2 × 2 pool size
2D convolutional layer	32 filters 3 × 3 kernel size

(continued)

Table 2. (*continued*)

CNN model	Parameters
Activation	Rectified linear unit (relu)
2D max pooling layer	2 × 2 pool size
2D convolutional layer	64 filters 3 × 3 kernel size
Activation	Rectified linear unit (relu)
2D max pooling layer	2 × 2 pool size
Core layer (Densely connected neural network layer)	64 filters
Activation	Rectified linear unit (relu)
Dropout	0.5
Densely connected neural network layer	2 filters (output)
Activation	Sigmoid

5   Results and Discussions

5.1   CNN Training and Validation Results

This section provides the simulation results for the training and validation samples shown in IV-A, and the CNN model described in IV-B. Figure 14 shows the perception quality recognition for DS0. Using 900 training samples, and 300 validation samples with 10 epochs, the CNN resulted in 93.53% training accuracy and 88.73% validation accuracy for the binary classification of good and bad quality percepts. Figure 15 shows the perception quality recognition for DS3-1. Using 1400 training samples, and 600 validation samples with 30 epochs, the CNN resulted in 91.16% training accuracy and 77.40% validation accuracy for the binary classification of good and bad quality percepts. In DS3-1, it took around 30 epochs before the training accuracy reach above the 90% mark. The high-altitude camera placement for this type of context makes it harder for smaller detected objects to be classified. Figure 16 shows the perception quality recognition for DS4-1. Using 1400 training samples, and 600 validation samples with 10 epochs, the CNN resulted in 93.39% training accuracy and 95.21% validation accuracy for the binary classification of good and bad quality percepts. Figure 17 shows the perception quality recognition for DS4-3. Using 1400 training samples, and 600 validation samples with 10 epochs, the CNN resulted in 95.76% training accuracy and 83.56% validation accuracy for the binary classification of good and bad quality percepts. The best result for DS4-3 simulation was achieved in the 20 epochs. However, the training accuracy for DS4-3 already reached 93.53% in epoch 10, which is already an acceptable accuracy threshold in this study.

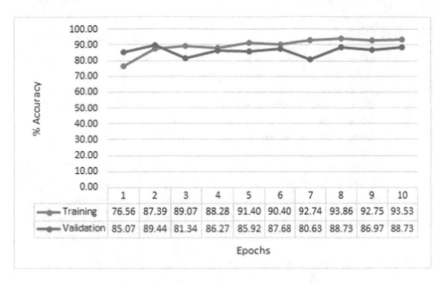

Fig. 14. CNN results for perception quality recognition of DS0

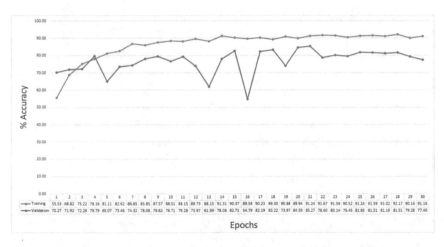

Fig. 15. CNN results for perception quality recognition of DS3-1

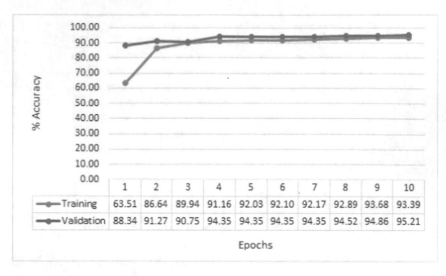

Fig. 16. CNN results for perception quality recognition of DS4-1

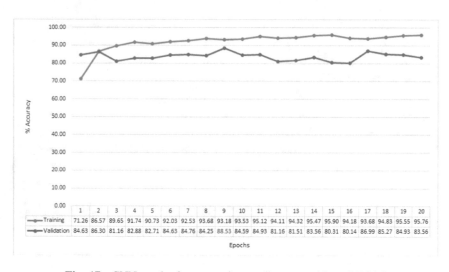

Fig. 17. CNN results for perception quality recognition of DS4-3

5.2 Summary of Perception Quality Recognition Results

Table 3 lists the summary results of perception quality recognition using convolutional neural networks. The table includes the training and validation accuracy, loss function, no. of epochs, and simulation time for each context locations (DS0, DS3-1, DS4-1, DS4-3).

Table 3. Summary of CNN results

Results	DS0	DS3-1	DS4-1	DS4-3
Training accuracy	93.53%	91.16%	93.39%	95.76%
Validation accuracy	88.73%	77.40%	95.21%	83.56%
Training loss function	0.1706	0.2575	0.1975	0.1362
Validation loss function	0.3448	0.6614	0.1532	0.8157
No. of epochs	10	30	10	20
Simulation time (seconds)	394	3418	707	1224

6   Conclusion

The study aims to demonstrate a visual percept (or detected image) quality recognition using a perception-action model with memory and learning mechanism. A CNN model was used to learn, recognize, and differentiate between a good or bad quality percept in four different road context (DS0, DS3-1, DS4-1, DS4-3). Using 900 training samples, and 300 validation samples with 10 epochs for DS0, the CNN resulted in 93.53% training accuracy and 88.73% validation accuracy for the binary classification of good and bad quality percepts. Using 1400 training samples, and 600 validation samples with 30 epochs for DS3-1, the result showed a 91.16% training accuracy and 77.40% validation accuracy. In DS3-1, it took around 30 epochs before the training accuracy reach above the 90% mark. The high-altitude camera placement for this type of context makes it harder for smaller detected objects to be classified. Using 1400 training samples, and 600 validation samples with 10 epochs for DS4-1, the result showed a 93.39% training accuracy and 95.21% validation accuracy. Lastly, using 1400 training samples, and 600 validation samples with 10 epochs for DS4-3, the result showed a 95.76% training accuracy and 83.56% validation. The best result for DS4-3 simulation was achieved in the 20 epochs. However, the training accuracy for DS4-3 already reached 93.53% in epoch 10, which is already an acceptable accuracy threshold in this study. This study was made to help in the video analysis of road traffic activities. Good quality of detected images is essential for analyzing activities such as detection, tracking, and classification of road traffic participants (vehicles and pedestrians). Whenever a stream of bad quality images is captured in the cameras, the system can adjust its parameters based from the output of a visual percept quality recognition system.

Acknowledgment. The authors highly appreciate the Department of Science and Technology - Philippine Council for Industry, Energy, and Emerging Technologies for Research and Development (DOST-PCIEERD) for providing funds for this research study. The authors also acknowledged Dr. Alvin B. Culaba, and Dr. Marlon D. Era for sharing their valuable knowledge for the completion of this research.

References

1. Windridge, D., Shaukat, A., Hollnagel, E.: Characterizing driver intention via hierarchical perception-action modeling. IEEE Trans. Hum.-Mach. Syst. **43**(1), 17–31 (2013)
2. Kim, J.-H., Choi, S.-H., Park, I.-W., Zaheer, S.A.: Intelligence technology for robots that think. IEEE Comput. Intell. Mag. **8**, 70–84 (2013)
3. Bielecki, A.: A model of human activity automatization as a basis of artificial intelligence systems. IEEE Trans. Auton. Ment. Dev. **6**(3), 169–182 (2014)
4. Buch, N., Velastin, S.A., Orwell, J.: A review of computer vision techniques for the analysis of urban traffic. IEEE Trans. Intell. Transp. Syst. **12**(3), 920 (2011)
5. Billones, R.K.C., Bandala, A.A., Sybingco, E., Lim, L.A.G., Fillone, A.D., Dadios, E.P.: Vehicle detection and tracking using corner feature points and artificial neural networks for a visionbased contactless apprehension system. In: Computing Conference 2017, pp. 688–691 (2017)
6. Sivaraman, S., Trivedi, M.M.: A review of recent developments in vision-based vehicle detection. In: 2013 IEEE Intelligent Vehicles Symposium (IV), pp. 310–315 (2013)
7. Billones, R.K.C., Bandala, A.A., Lim, L.A.G., Sybingco, E., Fillone, A.M., Dadios, E.P.: Microscopic road traffic scene analysis using computer vision and traffic flow modelling. J. Adv. Comput. Intell. Intell. Inform. **22**(5), 704–710 (2018)
8. Billones, R.K.C., Bandala, A.A., Sybingco, E., Lim, L.A.G., Dadios, E.P.: Intelligent system architecture for a vision-based contactless apprehension of traffic violations. In: 2016 IEEE Region 10 Conference (TENCON), pp. 1871–1874 (2016)
9. Uy, A.C.P., Bedruz, R.A., Quiros, A.R., Bandala, A., Dadios, E.P.: Machine vision for traffic violation detection system through genetic algorithm. In: 2015 International Conference on Humanoid, Nanotechnology, Information Technology, Communication and Control, Environment and Management (HNICEM), pp. 1–7 (2015)
10. Candamo, J., Shreve, M., Goldgof, D.B., Sapper, D.B., Kasturi, R.: Understanding transit scenes: a survey on human behavior-recognition algorithms. IEEE Trans. Intell. Transp. Syst. **11**(1), 206–224 (2010)
11. Robol, M., Giorgini, P., Busetta, P.: Applying social norms to high-fidelity pedestrian and traffic simulations. In: IEEE Conference Publication, pp. 1–6 (2016)
12. Escolano, C.O., Billones, R.K.C., Sybingco, E., Fillone, A.D., Dadios, E.P.: Passenger demand forecast using optical flow passenger counting system for bus dispatch scheduling. In: 2016 IEEE Region 10 Conference (TENCON), pp. 1–4 (2016)
13. Lin, H.-Y.: Vehicle speed detection and identification from a single motion blurred image. In: Proceedings of the Seventh IEEE Workshop on Applications of Computer Vision (WACV/MOTION 2005), pp. 1–7 (2005)
14. Dailey, F.C.D., Pumrin, S.: An algorithm to estimate mean traffic speed using uncalibrated cameras. IEEE Trans. Intell. Transp. Syst. **1**, 98–107 (2000)
15. Ding, Y., Wang, S., Zhang, D.: Full-reference image quality assessment using statistical local correlation. Electron. Lett. **5**(2), 79–81 (2014)
16. Liang, L., Wang, S., Chen, J., Ma, D.Z.S., Gao, W.: No-reference perceptual image quality metric using gradient profiles for JPEG2000. Sig. Process. Image Commun. **25**(7), 502–516 (2010)
17. Sybingco, E., Dadios, E.P.: Blind image quality assessment based on natural statistics of double-opponency. J. Adv. Comput. Intell. Intell. Inform. **22**(5), 725–730 (2018)

18. Liu, L., Hua, Y., Zhao, Q., Bovik, A.C.: Blind image quality assessment by relative gradient statistics and adaboosting neural network. Sig. Process. Image Commun. **40**(1), 1–15 (2016)
19. Arel, I., Rose, D.C., Karnowski, T.P.: Deep machine learning—a new frontier in artificial intelligence research. IEEE Comput. Intell. Mag. **5**, 13–18 (2010)
20. Huang, F.-J., LeCun, Y.: Large-scale learning with SVM and convolutional nets for generic object categorization. In: Proceedings of Computer Vision and Pattern Recognition Conference (CVPR 2006) (2006)
21. Girshick, R.: Fast R-CNN. In: 2015 IEEE International Conference on Computer Vision, vol. 8, no. 1, pp. 1440–1448 (2015)
22. Sheikh, H.R., Sabir, M.F., Bovik, A.C.: A statistical evaluation of recent full reference image quality assessment algorithms. IEEE Trans. Image Process. **15**(11), 3440–3451 (2006)
23. Sheikh, H.R., Bovik, A.C., Cormack, L.: No-reference quality assessment using natural scene statistics: JPEG2000. IEEE Trans. Image Process. **14**(11), 1918–1927 (2005)
24. Sheikh, H.R., Bovik, A.C., de Veciana, G.: An information fidelity criterion for image quality assessment using natural scene statistics. IEEE Trans. Image Process. **14**(12), 2117–2128 (2005)
25. Smith, K., Gatica-Perez, D., Odobez, J.-M., Ba, S.: Evaluating multi-object tracking. In: IEEE Conference Publication, pp. 1–8 (2005)

Automated Control Parameters Systems of Technological Process Based on Multiprocessor Computing Systems

Gennady Shvachych[1], Boris Moroz[2], Ivan Pobocii[1],
Dmytro Kozenkov[1], and Volodymyr Busygin[1(✉)]

[1] National Metallurgical Academy of Ukraine, Dnipro 49000, Ukraine
busygin2009@gmail.com
[2] National TU Dnipro Polytechnic, Dnipro 49000, Ukraine

Abstract. The paper considers the development features and the use of a multiprocessor computing system with its mathematical focus and software for simulation of heat treatment modes of steel billets. Objective is to develop a simulation for the heat treatment mode of a long steel billet that can be used for recrystallization and spheroidizing annealing of calibrated steel. The proposed use of modern multiprocessor computing technology strives to increase the speed and productivity of computing, which ensures effective control of the process. By using special software, a multiprocessor computing system is able to set and control the required temperature conditions on the entire cross-sectional plane of the billet during heating and holding, and if necessary, it can control the heat treatment of steel in the annealing temperature range. A multiprocessor computing system with special software includes mathematical simulations in the form of the heat equations. Such equations are solved by splitting methods. Thanks to this approach, the solution of a two-dimensional equation reduces to a sequence of integrating one-dimensional equations of a simpler structure. The use of a numerical-analytical method ensures the economic and sustainable algorithms for solving problems of such type. There experiments were conducted with the study of the properties of steel billets. The results of the experiments allow recommending the proposed approach to creating simulations of metal processing speeds for the development of new technological processes.

Keywords: Mathematical simulation · Multiprocessor computing system · Informational bidirectional interface · Metal temperature control

1 Introduction

In the world of today there is a rapid growth of the number of multiprocessor computing systems (MPCS) and their total productivity. This is due to the fact that such systems have become publicly available along with cheap hardware platforms for high-performance computing. At the same time, interest in the problems of computing networks has grown dramatically, and there is widespread awareness that the introduction of such networks will have a huge impact on the development of human society, compared with the impact on it advent of single electric networks at the

K. Arai and S. Kapoor (Eds.): CVC 2019, AISC 944, pp. 666–688, 2020.
https://doi.org/10.1007/978-3-030-17798-0_53

beginning of the century. In this regard, considering the problems of mastering multiprocessor systems, it should be taken into account that they became the first step in creation of such computing networks.

In addition, today practice brings to the application scientists a variety of problems, which full solution in most cases is possible only by applying the multiprocessor computing systems. For example, in the metallurgical industry there is a lot of diverse and interconnected processes. First and foremost, these are the technologies of smelting and casting of iron-carbon alloys, heating, rolling and heat treatment (HT) of metal products, etc. Industrial practice shows that neither the intensification of processes of metallurgical production nor the constructive improvement of various metallurgical equipment is impossible without studying and analyzing the phenomena of heat and mass transfer. At the same time, the solution of these problems by known standard approaches is a complex problem, which overcoming is possible only through the use of modern multiprocessor computing technology. At the same time, one of the main features of the application of such technologies is to increase the speed and performance of computations. High performance computing allows solving multidimensional problems, as well as those that require a large amount of processing time. Speed control enables either effectively manage technological processes, or even create the preconditions for the development of new promising technological processes.

In this regard, the development and use of multiprocessor computing systems with own mathematical and software issues is an urgent problem that can significantly reduce the number of experimental studies and the time it takes to conduct them. All that allows obtaining the necessary information for creation and implementation of various technological innovations.

2 Research Problem Statement

The paper considers the problem of introduction of new technological processes of thermal processing of metal. To do this, it is necessary to create a metal HT simulation, which is used in the manufacture of high-strength fasteners by the cold-die-forging method without the final heat treatment. Such a simulation aims to improve the technological properties of metal rolling through high dispersion and homogeneity of the sample structure throughout its cross-section plane. In addition, the technological process of steel thermal treatment should acquire such advantages as high efficiency, reduced power consumption, improved performance. This can be achieved by a multiprocessor computing system as a separate module. By using the special software, the multiprocessor system is capable of setting and controlling the required temperature regimes on the entire plane of the billet cross section when the metal heated and endured, and if necessary, can control the thermal mode of steel processing in the range of annealing temperatures.

In order to solve the above-mentioned problems, an installation for the HT of long steel billets [1] with the use of MPCS [2] was developed. The multiprocessor computing system with its software by the mathematical simulation of the billet heating process allows controlling the wire heating in the production conditions until its transition to the austenitic state and the temperature onset of recrystallization phase throughout the cross section of the long steel product. Followed by controlling the required mode of isothermal holding in the range of annealing temperatures on the whole plane of the billet's cross-section. The installation application, which ensures the realization of the spheroid annealing regime, causes the uniform distribution of globules of cementite in the ferrite matrix that creates the necessary mechanical properties of the metal required for further cold deformation.

Figure 1 shows the design scheme for the installation of a long steel billet, where (1) is a dismantling device; (2) the correct traction device, equipped with the actuator 3; (4) inductor of the heating device; (5) generator with actuator; (6) pyrometer; (7) Isothermal shutter with actuator 8; (9) pyrometer; (10) is a chamber of regulated welding of the wire with an actuator 11 for adjusting the supply of water-air mixture; (12) pyrometer; (13) Isothermal shutter chamber with actuator 14; (15) pyrometer; (16) a device for intensive spheroidization with an actuator 17; (18) pyrometer; (19) a device for feeding the wire to a further process cycle; (20) Information bi-directional data acquisition interface for devices 3, 7, 9, 12, 15, 18, connected to control unit 21 and to actuators (3, 5, 8, 11, 14, 17) of corresponding devices; (22) information bi-directional communication interface of the control unit MPCS 23.

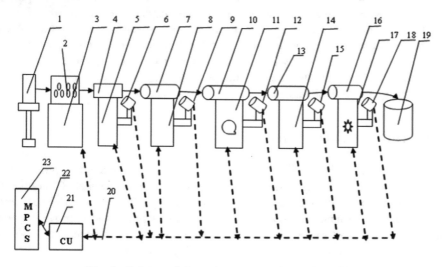

Fig. 1. Scheme of thermal treatment of a steel product.

The installation for HT of a long-steel steel billet works in the sequence described below. From the unwinding device 1 through the correct traction device 2, the wire is fed into the inductor of the heating device 4, which heats up to the transition to the austenitic stage, obtaining the temperature of the recrystallization phase. The heating temperature is controlled by the pyrometer. The temperature control is maintained by the control unit 21 and the MPCS 23. The signal from the pyrometer 6 through the information bi-directional interface 20 enters the control unit 21 and then through the information bi-directional interface 22 at the MPCS 23, where according to the result the solution of the mathematical simulation of the direct heat conduction problem is the generator 5 power regulation. Then the heated wire enters the chamber of isothermal shutter 7, the temperature which regulated by the actuator 8. In this case, the signal from the pyrometer 9 through the information bi-directional interface 20 enters the control unit 21, and then through the information bi-directional interface 22 at the MPCS 23, where according to the results of the mathematical simulation solution of isothermal holding, the temperature regime is controlled by the actuator 8. Next, the wire passes into the regulated chamber 10. Depending on the HT mode, the steel grades and the wire diameter are given the required tempering speed in the range of temperatures from 750 to 700 °C.

In this temperature range, austenite continuously loses carbon, and when it reaches the necessary concentration of the latter, there undergoes the polymorphic transformations, transforming into ferrite, thus austenite disintegration occurs under the abnormal mechanism. The tempering temperature is controlled by the CU 21 and the MPCS 23, wherein the temperature recorded by the pyrometer 12 taking into account, and according to the result of the solution of the mathematical simulation, the actuator 11 supplies the water-air mixture, increasing or decreasing its amount, depending on the given temperature mode of feeding. Then the wire enters the chamber of isothermal shutter 13, which temperature is regulated by the actuator 14. During isothermal holding, the formation of a quasi-ectoteid (perlite) is completed, which includes zones with high concentration of carbon and prepared cementite particles. The signal from the pyrometer 15 through the information bi-directional interface 20 enters the automatic CU 21, and then through the information bi-directional interface 22 in the MPCS 23, where, based on the results of the solution of the mathematical simulation of isothermal holding via the CU 21, the temperature regime is controlled in the chamber 13. After that, the wire is fed into the device of intensive spheroidization 17, where the temperature regime change by 15 to 20 °C/min with the achievement of subcritical temperature Ac1 (at these temperature zone there is an intense spheroidization of cementitious particles). The signal from the pyrometer 18 through the information bi-directional interface 20 enters the automatic CU 21, and then via the information bi-directional interface 22 at MPCS 23, where, the results of the solution of the mathematical simulation of the direct heat conduction problem is taking into account, and the temperature regime is regulated in the intensive spheroidization device 16. Further through the feeder device 19, the wire passes to the next processing cycle.

At the same time, the multiprocessor computing system with special software as a single base should include mathematical simulations in the form of the heat equation.

At the same time, almost simultaneously with the first multiprocessor computing systems advent, there emerged a need to evaluate their efficiency, performance, speed and later the comparison with similar computing systems, considering the listed criteria. It is the speed and performance that draw main focus, including the multiprocessor computing systems design. Such an approach is aimed, for example, to the development of new technological processes (when the time of computations is a critical value), the analysis of environmental pollution, and also owing to its method one can solve a variety of multidimensional non-stationary problems [1–6]. In addition, such problems types often have to be solved in medicine, military equipment, etc.

Consequently, the main purpose of this paper is to develop a simulation of a long-lasting steel product, which can be used for recrystallization and spheroid annealing of calibrated steel on the basis of multiprocessor computing systems. As the simulation base, it was decided to put a HT method of low and medium carbon steels intended for cold heading [3].

One of the main problems of using a multiprocessor computing system to solve such problems considered in this paper can be stated as follows: we have the dimension mesh difference M, the computation time of a problem solved by a single-processor system denoted by the t value. This is a key parameter. It is necessary to significantly reduce the computing time while saving the M value. Consequently, we consider the problem where we aim to reduce the computation time by increasing the nodes number in the multiprocessor computing system. In this case, the computing area is evenly distributed among the multiprocessor computing system nodes.

The application of the automated control system for the technological process parameters of the product heat treatment is aimed at improving technological properties of a billet by ensuring a high dispersion and homogeneity of the sample structure in the entire cross-section area. At the same time, the technological process should acquire such advantages as high efficiency, reduced power consumption, and improved performance. The specified properties of technological process are obtained by a multiprocessor computing system. The multiprocessor computing system is mounted as a separate module and special software allows setting and controlling the necessary temperature regimes in the entire cross-section area of a sample while heating and stretching.

The problems that are solved by the developed automated control system of the technological processes parameters, concentrate in two directions - monitoring (real-time observation of the object's temperature parameters), and retrospective control, analysis and correction of the processing temperature regimes for a long product.

It is also known that the computations parallelization efficiency essentially depends on many factors; one of the most important is the specificity of the data transfer among adjacent nodes of the multiprocessor computing system, since this is the algorithm slowest part can undo the effect of increasing the processors number used. These issues are considered crucial in a simulation of a wide range of problems using modular multiprocessor computing systems, and today those are successfully solved by many researchers [2, 7, 9].

Working methods of efficiency analyzing of multiprocessor computing systems do not allow determining optimal number of its nodes for solving a certain class of problems. At the same time, they did not get proper research development on the network interface impact on the efficiency of modular multiprocessor computing systems. In addition, for computing multiprocessor system evaluation efficiency, the main analytical relations are not provided through parameters of the system being studied.

2.1   Literature Review

Traditional technology of spheroid anisotropic steel annealing involves the saddle furnaces (galericulate or shaft type). The disadvantages of traditional methods of preparing billets for cold heading in detail are described in the well-known I.E. Dolzhenkov papers [2, 3] and are deeply analyzed in the paper [1].

An alternative to the oven heat treatment method is an electrothermal method characterized by a high heating rate due to the effects of electromagnetic induction (induction heating) or electrical resistance (electrocontact heating) [3, 8]. The induction heating in the technological line for the wire HT is already known in industrial practice [6, 7]. For the implementation of such technology, there was developed the installation for the manufacture of high-strength fasteners without a final heat-shrinkage. But in this case, during the heat treatment, the control of temperature regimes of heating, holding and cooling is not carried out due to the lack of means for measuring and controlling the metal temperature.

Another approach to the implementation of the electrothermal method of processing calibrated steel is the installation [9], which provides a thermal chamber and a thermoregulation screen. But the HT process is characterized by a significant duration of the annealing regime, because isothermal holding and making the required cooling regime is carried out in the thermal chamber and takes a long time. According to the authors, the duration of annealing is 30 to 90 min, which does not allow synchronizing the closed cycle of manufacture of three-field products.

3 Presentation of the Main Research Material

3.1 Automated Control Parameters of Technological Processes Based on Multiprocessor Computing System

The HT installation of a long steel product [1] was developed to solve the above-stated problems. Figure 2 shows block diagram of the installation control system of a long steel product heat treatment. Such a control system has the units that allow obtaining the information about current parameters of the managed processes. Its peculiarity is in the fact that at each of the five stages of the sample technological processing, the two-dimensional heat conduction problem is solved. In this case, the MPCS software allows controlling the temperature regimes, as in the entire cross-section area of a sample, and along its length. Control of such temperature regimes is carried out in the center of the cross-section area of a sample.

MPCS with special software, as a single base, includes mathematical simulations as a heat conduction Eq. (1). In this case, the coordinate z boundary conditions may be of the first, second or third kind, depending on the features of the problem to be solved. Solve the problem (1) by splitting methods, allowing to reduce the complex operator (1) to the simplest one. This approach allows integrating this equation as a sequence of integration of one-dimensional equations of a simpler structure. Given the significant complexity of the mathematical simulation (1), the development of economic algorithms for the computation of control functions effects of the proposed installation becomes of great importance. The process of creating these algorithms is covered in the papers [10, 11].

Note that here the problem of control (as well as the problem of synthesis) in its exact formulation refers to the reversed class, since it involves the definition of controlling functional parameters on the basis of a predetermined, desired result (the inverse problem of control). The algorithm for solving inverse problems is the "fork"

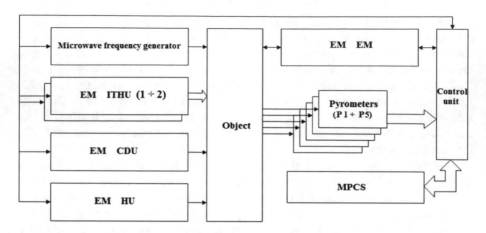

Fig. 2. The block diagram of the installation control system of a long steel product heat treatment.

method with the preliminary definition of some initial segment. The solution of the problem is realized in two stages. At the first, the separation of the non-compliance minimum is realized, while at the second the minimum of the desired control function from the separated intervals determined. This procedure is implemented standard. In other words, if ϑ represents a certain real value of the desired root, that is, when $a \leq \vartheta \leq b$, and $f(\vartheta) = 0$, then one can compute such w number that satisfies the conditions: $a \leq \vartheta \leq b$ and $|\vartheta - w| < \varepsilon$ that is, less than any predefined small number. A similar scheme is included in the structure of the mathematical control unit of the MPCS. The mathematical simulation (1) is used in all metal HT cycles, so the results of the simulation will apply to the cycle of primary metal heating.

The problem of simulation is that in order to provide the required accuracy and stability of computing it is necessary to take a computation grid with a considerable number of nodes and perform a lot of iterations. As a result, the number of arithmetic operations necessary to compute the temperature fields is within $10^7 \div 10^8$ nodes, and when the time increment is 10^{-2} s, then the total nodes number for computations can reach 10^{20} or even more. Single-processor computing systems cannot solve such a problem during simulation in real time, so the most justifiable is the use of multiprocessor systems, which was implemented to improve the technology of a long steel product.

$$\frac{\partial T}{\partial \tau} = \frac{\partial^2 T}{\partial z^2} + \frac{\partial^2 T}{\partial r^2} + \frac{1}{r} \cdot \frac{\partial T}{\partial r} + W, \tag{1}$$

with Fourier criterion $\tau = \frac{at}{R^2}$, if $\tau > 0$, W is specific power as heat sources, W/m^2.

The equation boundary conditions are as follows:

$$T(0, r, z) = f(r, \tau);$$
$$T(\tau, 1, z) = var;$$
$$\frac{\partial T(\tau, 0, z)}{\partial r} = 0;$$
$$T(\tau, 0, z) \neq 0.$$

The last two correlations in the boundary conditions indicate that the temperature value in the area of the cylinder axis during the entire heat transfer process must be finite. According to the z coordinate the boundary conditions, depending on the features of the solvable problem, can be of the first, second or third kind. By solving the problem (1) by splitting methods that reduce the molecular operator (1) to the simplest one. This approach allows integration of this equation as an integration sequence of one-dimensional equations of a simpler structure. With regards to the significant complexity of the mathematical simulation (1), the economic algorithms development for the control functions effects calculation of the proposed installation becomes of great importance. Consider the process of creating such algorithms by constructing the Eq. (1) splitting schemes. At the descriptive level, the idea of constructing splitting schemes along the length and radius of a sample can be described as follows.

The differential problem formulated by the expression (1) is given as:

$$\frac{\partial T}{\partial \tau} = A \cdot f, \tag{2}$$

wherein A is a certain operator of spatial variables, for instance:

$$Af = \frac{\partial^2 T}{\partial z^2} + \frac{\partial^2 T}{\partial r^2}. \tag{3}$$

Valuations $f(z, r, t_p)$ for already known values $f(z, r, t_{p-1})$, if $t_p = p \cdot Dt1$ $(Dt1 = (t_p - t_{p-1}))$, is the distance between grid nodes at a given time interval (p = 1, 2, 3, ... - node numbers), which can be expressed by the formula:

$$f(z, r, t_{p-1} + Dt1) = f(z, r, t_{p-1}) + Dt1 \frac{\partial f}{\partial t} + O(Dt1^2)$$
$$\equiv (E + Dt1 \cdot A)f(z, r, t_{p-1}) + O(Dt1^2),$$

wherein E is a unity operator.

Present the Eq. (3) right part as follows:

$$Af = A_1 f + A_2 f.$$

Then it will be rewritten as follows:

$$\frac{\partial T}{\partial \tau} = A_1 f + A_2 f \qquad (4)$$

This equation allows splitting into the following two equations:

$$\left.\begin{array}{ll} \frac{\partial v}{\partial \tau} = A_1 v, & t_{p-1} \leq t \leq t_p, \\ v(z, r, t_{p-1}) = f(z, r, t_{p-1}), & \end{array}\right\}, \qquad (5)$$

$$\left.\begin{array}{ll} \frac{\partial w}{\partial t} = A_2 w, & t_{p-1} \leq t \leq t_p, \\ w(z, r, t_{p-1}) = v(z, r, t_p). & \end{array}\right\}. \qquad (6)$$

Note that

$$w(z, r, t_p) \equiv f(z, r, t_{p-1}) + O(Dt1^2). \qquad (7)$$

Equality (7) provides a basis for each time interval, when $t_{p-1} \leq t \leq t_p$, instead of Eq. (2), sequentially solve (5) and (6).

For practical solutions of equations (5) and (6) we formally approximate them with any difference schemes. Then there is already some difference scheme splitting that allows in two stages to compute the function value $f(z, r, t_p)$ by already known value $f(z, r, t_{p-1})$.

The first stage is the value $v(z, r, t_p)$ computation for known function $f(z, r, t_{p-1})$ value, and the second solution stage is the expression $f(z, r, t_p) = w(z, r, t_p)$ computation: when from the first stage it is known that $w(z, r, t_{p-1}) = v(z, r, t_p)$.

The stated considerations are heuristic in nature. After the difference scheme of splitting is made for a numerical problem solution, it is necessary by some method to check its approximation and stability. The said splitting of the two-dimensional equation in the expression (2) into two one-dimensional (5), (6) can be interpreted as an approximate replacement of heat transfer process along the area Ozr within the following time limits: $t_{p-1} \leq t \leq t_p (p = 1, 2, 3 \ldots)$, for two processes. In the first of them, which is described by the first equation, there mentally get introduced heat-proof partitions, which prevent heat transfer towards the axis Or. Then, at the onset of time $Dt1$, instead of the stated partitions, the other ones are introduced preventing the heat from transfer towards the axis, and the former partitions get removed. The heat transfer over time is now described by the second equation.

It should be noted that here the control problem (as well as the synthesis problem) in its exact statement refers to the inverse class, since it involves the definition of controlling functional parameters based on a predetermined, desired result (the inverse control problem). The algorithm for solving inverse problems is the "fork" method with preliminary definition of some initial segment. The problem solution is realized by two stages. At the first stage, the separation of the residual minimum is realized, while the second stage is determined by the minimum of the control desired function from the separated interval. This procedure is implemented conventionally. In other words, if ϑ represents a certain real value of the desired radical, that is, when $a \leq \vartheta \leq b$, and

$f(\vartheta) = 0$, then one can compute the number w such that satisfies the conditions: $a \le \vartheta \le b$ and $|\vartheta - w| < \varepsilon$ that is, less than any predetermined small number ε. A similar scheme is included in the structure of the mathematical control unit of the MPCS.

3.2 Major Researches on the Concept of Development and Efficiency Research of the Multiprocessor Computing System of Temperature Mode Control

In order to control the temperature regime of a *billet*, the concept of building a multiprocessor computing complex was developed, which would have the real efficiency and performance peak. In addition, this system should be characterized by increased reliability and high energy efficiency. Blocks of the proposed system were implemented by mass media computing facilities.

The idea of choosing the form factor of computing nodes, for which there are several traditional solutions today, was grounded. The projected cluster system is based on blade technologies. For these reasons, it is a densely packed module with blade type processors installed in the rack. The rack contains nodes, equipment for efficient connection of components, and for controlling the system's internal network, and so on. Each cluster blade operates under the control of its copy of the standard operating system. The composition and the nodes power can be different within a single module, but in this case a homogeneous module was considered. The interaction among nodes of the cluster system is established using the programming interface, that is, specialized libraries of functions. But there was a problem of the number of blades. Some developers of this kind of systems are convinced that the number of computing nodes of a multiprocessor computing system completely depends on the class of problems solved. For example, the chosen mathematical simulations for the study of metallurgical production are based on differential equations in partial derivatives. In view of these circumstances, it was decided to limit the six-node module to the multiprocessor module, and if necessary, a cluster build option would be provided. It was also found that fewer nodes may not produce the desired result. Thus, in developing a multiprocessor computing system, the particular attention was paid to possibility of expansion or modification of cluster in the future.

For this purpose, an analysis of the choice of architecture and parameters of the cluster system computing nodes was carried out, since they largely determine its characteristics. It was at the design stage of the cluster that the main parameters of the motherboard nodes were provided, taking into account the specifics of problems to be solved by such a system. The multiprocessor module contains one master node (*MNode001*) and five computing slave nodes (*NNode001, NNode002, NNode003, NNode004, NNode005*), three control switches (*SW1, SW2, SW3*), intermediate switch memory buffers reconfigured a network for data interchange among computing nodes, virtual local area networks, a mechanism for redundancy of key components, and also provides network boot nodes. The switched network of the multiprocessor computing system operates in two modes. The first models a star type topology, the second - the ring. Such operation modes were focused on the implementation of the boundary data interchange, reflecting features of the problems that are solved by the proposed multiprocessor computing system. Figure 3 presents its block diagram.

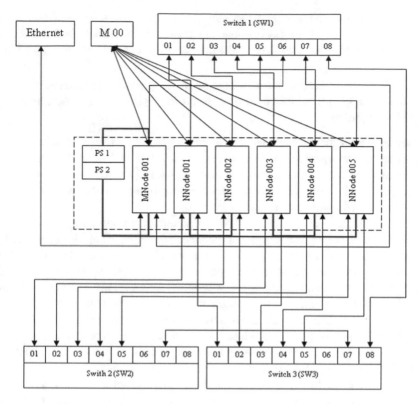

Fig. 3. Block diagram of a multiprocessor computing system.

The operation features of the multiprocessor computing system module. After power supply to the *PS1* unit and the receipt of the external *PUSK* signal from the *M00* panel starts the initialization and operation of the master unit of the multiprocessor system module. The master node (*MNode001*) via the *SW1* switch provides the direction of the data flow associated with diagnosing and receiving/transmitting problem conditions and controlling them to slave nodes. The latter, in accordance with the algorithm of problem solving and the flow of processes, implement the mode of necessary computing. The data interchange among computing nodes is carried out in a separate network, organized by means of the controlled switches *SW2–SW3*. To achieve maximum efficiency of the cluster system, the process of the structure reconfiguration of the second network is carried out in accordance with the specifics of the problems to be solved. The data receiving/transmitting in slave nodes occurs via intermediate memory buffers. The results of computations through the first communication network, via a controlled *SW1* switch, go to the master node (*MNode001*), where the necessary mode of data extraction and processing is carried out.

The technical result achieved with the introduction of the proposed system is that by introducing into the multiprocessor computing system a separate computing network for data interchange, additional managed switchers operating in parallel, intermediate memory buffer switches, as well as by developing a network load processor and the mechanism of redundancy of key components of the module the success was achieved:

- Firstly, to increase the computing speed when solving strongly-connected problems, to provide high-speed access to the cluster node's memory and the data interchange among them, to reduce the download of the channel passing among nodes of the cluster due to the formation of a separate computing network and realization of mechanisms of *channel bonding* and *VLAN*;
- Secondly, at the expense of the intermediate memory buffer to "unload" the *CPU* at the moment of transmission and receiving of data among the cluster nodes, which increased the efficiency of the whole computing system;
- Thirdly, to improve the cluster system efficiency, adapting the structure of its network to the solution of the problems of each particular type;
- Fourthly, due to the modular construction principle, to simplify the development, build-up or replacement of failed clustered nodes, as well as the operation and processing of the whole system;
- Fifthly, due to network boot and *Power on After Power Fail/Former-Sts* mode, the simultaneous launch of the multiprocessor computing module nodes of high-availability from a single power supply unit instead of several, which results in a significant increase in the system's energy efficiency and its cost reduction;
- Sixthly, due to the module network loading of high-availability multiprocessor system, and with redundancy mechanism of its key components significantly reduce the number of its components and thereby significantly increase the reliability of the module operation;
- Finally, to provide the ability to transfer software to other cluster systems to perform such computations.

A series of experimental computations was conducted on the configured computing complex. The effectiveness of the proposed approach is confirmed by the successful resolution of broad-based problems.

Research purpose is to further develop the approach associated with definition of a methodology for evaluating the multiprocessor modular computing system effectiveness. At the same time, the main attention is paid to influence peculiarities on the given indicator of the developed system network interface. It is also necessary to derive analytical relations for determining the optimal nodes number in the application to different modes of its operation. For ease of estimation of the multiprocessor computing system efficiency, it is necessary to derive the main analytical relationships through its parameters.

For a class of problems that are solved in this paper, all computations are performed on the basis of the difference grid. Then, when analyzing the multiprocessor computing system efficiency, the most important parameter was the time to compute a single iteration (T_{it}) in association with the computing field. In terms of multiprocessor computing system, this indicator was determined on the basis of the following ratio:

$$T = T_n + T_{ex}, \tag{8}$$

herein $T_n = \frac{T_{it}}{N}$, this value means the computation time of a single iteration by N computing nodes, in seconds, T_{ex} is the time of boundary data exchange among the system nodes, in seconds. The computation time of the single iteration itself when used in the system N of the computing nodes can be specified by the following formula:

$$T_n = \frac{E_i \cdot E_y \cdot K_R}{N \cdot V_c}. \tag{9}$$

In expression (9) E_i is the array length of the computations boundary field; at the same time, this value determines the difference grid length along the abscissa; E_y is the length of the difference grid along the ordinate; K_R is the one difference cell size of type *Real*8 (64 bits); parameter V_c shows the speed of computing when solving such problems using the proposed processor.

The value T_{ex} was determined by the following formula:

$$T_{ex} = \frac{m \cdot (N - 1) \cdot E_i \cdot K_R}{k \cdot d \cdot V_p}. \tag{10}$$

In expression (10), the m value may be equal to one in the unilateral mode of boundary data exchange, or two when it is two-way; V_p is network bandwidth in the system, Gbit/s; k is the communication channels number of the network operating simultaneously (computing networks number), d - half-duplex ($d = 1$), or duplex ($d = 2$) mode of the computer network in a multiprocessor computing system.

Under these conditions one can compute total computation time of a single iteration that will include, in fact, the computation time of a single iteration when using N nodes of a multiprocessor computing system and the time of boundary data exchange depending on the number of nodes N, that is

$$T = \frac{E_i \cdot E_y \cdot K_R}{N \cdot V_c} + \frac{m \cdot (N - 1) \cdot E_i \cdot K_R}{k \cdot d \cdot V_p}. \tag{11}$$

The relation (11) analysis shows that computation area distribution among the nodes allowed reducing the number of computations performed by each of its blades. Due to the fact that the multiprocessor computing system nodes work in parallel, then the total computing iteration time decreases. At the same time, with the nodes increase in the system, the boundary data volume also increases, and thus, time for the information exchange among the nodes increases.

In order to compute acceleration and efficiency of the system, the commonly accepted concepts in the theory of parallel computations were taken as the basis. An analytical ratio was derived for estimating the efficiency of a multiprocessor computer system through its parameters, i.e.

$$Q = \frac{T_{it} \cdot k \cdot d \cdot V_n \cdot N}{T_{it} \cdot k \cdot d \cdot V_n \cdot N + N^2 \cdot m \cdot (N-1) \cdot E_i \cdot K_R}. \tag{12}$$

The performance indicators of multiprocessor computing system were simultaneously determined by the formulas (10–12) given above and by means of experimental computations. It was observed that the results obtained coincide, which is explained by the computations nature. Figures 4 and 5 depict simulation graphs of dependence of a single iteration computation time on the nodes number of a multiprocessor computing system and computations acceleration dependence on the nodes number of a multiprocessor computing system.

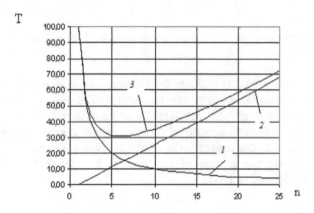

Fig. 4. A single iteration computation time dependence curves on the nodes number of the multiprocessor computing system.

Figure 4 shows a single iteration computation time with increasing number of nodes decreases with hyperbolic dependence (curve 1). However, the boundary data exchange time with the increase in the nodes number of the system increases according to the linear law (line 2). The general picture of the change in a single iteration computation time in a multiprocessor computing system illustrates dependence represented by curve 3. An analysis of such a curve shows that at the first stage, the computations time decreases with increasing nodes number of the system. A similar result was foreseen. However, the reduction of this time occurred to a certain limit. If, for instance, the nodes number exceeds six, then the total computation time begins to increase. This happens along with the increase in the data amount, which get transferred among the nodes. Thus, one can conclude that with the constant grid size the blades number in the system should be no more than six. At the same time, the time spent on solving the problem, decreased from 100 s to 30.81 s.

In the second part of the research, further development methods for analyzing the multiprocessor modular computing system efficiency have been developed. Moreover, the main attention was paid to the peculiarities of the network interface interference of the developed system on its efficiency values. Initially, the network interface architecture peculiarities of the multiprocessor computing system and the basic modes of its

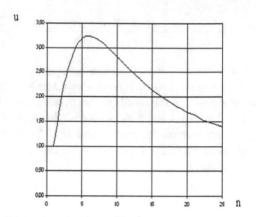

Fig. 5. Computations acceleration dependence curves on the nodes number of a multiprocessor computing system.

operation were considered. To evaluate the processes occurring in the system during the information flows transmission, the network system bandwidth and the switch throughput were compared. This procedure was necessary for optimal selection of components of the network interface of a multiprocessor computing system. In this regard, for the research convenience, the total bandwidth parameter of the multiprocessor network system was introduced according to the specification of the manufacturer (V_S). It was defined as follows:

$$V_s = V_p \cdot N \cdot d. \tag{13}$$

Herein N is the nodes number in the system, and V_p is its network protocol throughput, Gbit/s. With this approach it is already possible to compare the values of the total network interface bandwidth of the system (V_S) and the switch (V_b) bandwidth. For further analysis of the network interface of multiprocessor computing system, the throughput coefficient of the network system (k_s) was introduced. Its value was determined by the following ratio:

$$k_s = \frac{V_s}{V_b}. \tag{14}$$

By taking into account formula (14)

$$k_s = \frac{V_p \cdot N \cdot d}{V_b}. \tag{15}$$

To further use this approach, the concept of switching capacity coefficient (k_b) was proposed and a formula was derived for its definition, i.e.

$$k_b = \frac{V_b}{V_p \cdot N \cdot d}. \tag{16}$$

To illustrate broad picture of the processes under research, some definitions were introduced, and then, with their account, a more detailed analysis of the main network characteristics of multiprocessor computing system was performed.

Definition 1. *The network interface deficiency mode of a multiprocessor computing system* is such an option of its network functioning, where there is an inequality: $k_s < k_b$.

Definition 2. *The network interface surplus mode of the multiprocessor computing system* is such an option of its network functioning, where there is an inequality: $k_s > k_b$.

Obtaining the Optimal Nodes Number in the Cluster System. It should be noted that the numerical-analytical schemes of the higher accuracy order, considered in the papers [12, 13], serve as the computational methods for solving the heat conduction problem. As a basis for determining the boundary data exchange time in a cluster system while working in its network interface deficiency mode, the relation (10) becomes as follows:

$$T_{ex1} = \frac{m \cdot (N - 1) \cdot E_i \cdot K_R \cdot K_{pr}}{k \cdot d \cdot V_p}, \tag{17}$$

To compute the boundary data exchange time in the cluster system in its network interface surplus mode, we apply the relation of the following form:

$$T_{ex2} = \frac{m \cdot (N - 1) \cdot E_i \cdot K_R \cdot N \cdot K_{pr}}{k_m \cdot d \cdot V_b} \tag{18}$$

Further, considering the multiprocessor computing system in the performed experiment conditions, we establish in it the nodes number that provides the most effective solution to the problem. At the same time, [9, 10] show that the computational speed will increase by about the moment when

$$T_{calc} \approx T_{ex}. \tag{19}$$

Thus, on the basis of the relation (19), it is possible to compute the nodes number in the cluster computing system needed to effectively solve the problem. Note that this research phase aims to reduce the total computation time by parallelizing the program. Obviously, at the same time, the overall grid size does not depend on the computing nodes number in the cluster system. Taking into account the relation (16), we obtain analytical expressions for determining the optimal nodes number of a cluster system, when it operates in a network interface deficiency mode, that is,

$$\frac{T_{calc}}{N} \approx \frac{m \cdot (N - 1) \cdot E_i \cdot K_R \cdot K_{pr}}{k \cdot d \cdot V_p}. \tag{20}$$

We also have the following expression for the network interface surplus:

$$\frac{T_{calc}}{N} \approx \frac{m \cdot (N-1) \cdot E_i \cdot K_R \cdot N \cdot K_{pr}}{k_m \cdot d \cdot V_b}. \tag{21}$$

Using expressions (20) and (21), we can obtain two equations in relation to N to determine the optimal nodes number in a cluster system, where the total computational required time for solving the problem will be minimal. Thus, the Eq. (20) reduces to a quadratic form, i.e.

$$N^2 - N - \frac{E_y \cdot k \cdot d \cdot V_p}{m \cdot V_c} = 0. \tag{22}$$

For analysis convenience, we will write the Eq. (22) as follows:

$$N^2 - N - \lambda = 0. \tag{23}$$

In Eq. (23) $\lambda = f(E_y, V_p, 1/V_c)$, this value can be interpreted as the capabilities coherency coefficient of the selected processors, the network interface and the computing area value when the system operates in a network interface deficiency mode. In addition, it should be emphasized that the correspondence of the cluster system capabilities to the nature of the problems to be solved requires the coordination of all parts that is included in the value λ. Let's analyze this coefficient. At first glance, the result turned out to be somewhat paradoxical. It shows that the coefficient of consistency λ, as well as the optimal blades number in the cluster system do not depend on the data area size. Such an assertion can be explained by the fact that the computation domain distribution among the cluster system nodes was carried out at its constant size. This means that the ratio of the time spent on processing the data in this area and the transfer time also remained unchanged and independent of its size. The second very important conclusion is that the optimal blades number in the cluster system, which provides its highest speed, decreases with increasing computing power of the processors included in it. Such a statement becomes quite clear when one considers that the network data exchange among the cluster system nodes is more likely to impede the overall computation process, the less time will be spent directly on solving a specific problem.

Thus, the Eq. (22) solution will be two radicals, one of them is negative, and the other one is positive. Proceeding from the set physical conditions of the problem, a positive radical is adopted, which value is equal to eight, hence $N = 8$. Note that this result satisfies the inequality that establishes the conditions for the cluster system to function in the network interface deficiency mode [13].

Equation (18) is reduced to a cubic form, i.e.

$$N^3 - N^2 - \frac{E_y \cdot k_m \cdot d \cdot V_b}{m \cdot V_c} = 0. \tag{24}$$

For the convenience of analysis, we will write it in this way

$$N^3 - N^2 - \mu = 0. \tag{25}$$

In Eq. (25) $\mu = f(E_y, V_b, 1/V_c)$, and this value can be considered as the capabilities coherence coefficient of selected processors, the network interface and the value of the computing area when the system operates in the network interface surplus mode. Let's analyze the value of this coefficient. One can conclude that the optimal number of blades in the cluster system is capable of providing its highest performance, will depend on the size of the computing area, the switching capabilities and computing power of the processors which the cluster system is composed from. These parameters variation allows selecting the appropriate blades number when operating the system in the network interface surplus mode.

As a result of the Eq. (24) solution three radicals are obtained, in particular, two imaginary, and one valid. The actual radical corresponds to the nodes number: $N = 33$. However, this result analysis indicates that it does not satisfy the condition of the cluster system functioning in the network interface surplus mode [9]. Having analyzed the simulation results, we can conclude that under the conditions of the problem being researched the optimal blades number of the cluster system would correspond to $N = 8$.

4   Experimental Research

Several experiments were carried out to test the functions of the proposed installation, when a 20 mm diameter wire from 20G2P steel was put for HT. Let us consider one of the characteristic experiments.

The ferrite-perlite structure of the billet was taken as the initial one. The process of treatment was carried out by heating the billet within the intercritical temperature zone limit. For the given material the following values of critical points are set: $Ac_1 = 725\ °C$; $Ac_3 = 795\ °C$.

The heating occurred up to the following value: $Ac_1 + (10 - 30\ °C)$. During the next stage of processing, the isothermal holding process was implemented for 45 s. Further, the product was cooled at a temperature of 20–30 °C/s to a temperature of 620 °C, followed by an isothermal holding for 45 s. Finally, at the last processing stage, the billet was heated at a rate of 15–25 °C/s to subcritical temperatures. Graphic interpretation of metal HT mode is shown in Fig. 6.

The mechanical characteristics were determined by the results of measurements of the billet hardness. The stretching test was carried out on the *FU10000ez* machine. The metal microstructure study was carried out on a *Neophot*-2 light metallographic microscope using an *Epiquant* structural analyzer, which was additionally equipped with the Anasonic device for digital image registration.

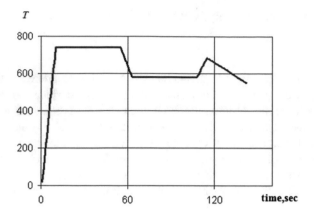

Fig. 6. Schedule of the 20G2P steel thermal treatment.

4.1   Results

According to the experiments results, the temperature distribution curves of the billet along the plane of its cross section were obtained (see Fig. 7), where T_H is the billet heating temperature of the billet surface, T_K is the temperature of the phase metal transformation (Ac_1) controlled by MPCS along the cross section plane of the billet. Simulation of such temperature fields is carried out taking into account the change in the thermophysical properties of the material while being heated.

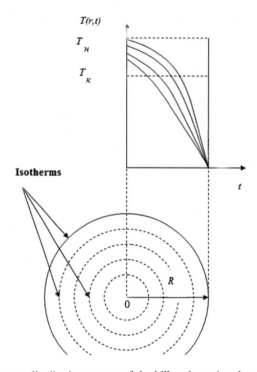

Fig. 7. The temperature distribution curves of the billet along the plane of its cross section.

The temperature distribution curves of the billet along its length while heating are depicted in Fig. 8, where 1 indicates the surface temperature (T_H), and 2 is the temperature in the center of the cross-section plane. Here, zone I reflects the billet heating process to a given temperature on its surface, and zone II shows the output of the given temperature mode in the center of the cross section plane of the billet.

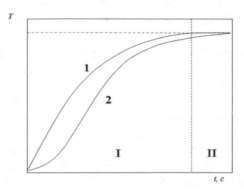

Fig. 8. The temperature distribution graphs of the billet along its length during the heating process.

Figure 9a, b demonstrates the billet microstructure before and after the spheroidization, while the hardness of the billet after the HT has reached the values of 150–169 HB.

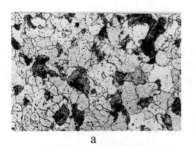
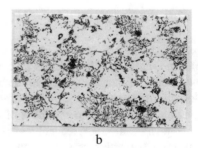

a   b

Fig. 9. 20G2P steel microstructure: a - initial ferrite-pearlitic structure, x 500; b - structure after annealing - granite perlite (grade 5), x500.

4.2 Discussion

The carbide phase spheroidization of the metal in the corresponding HT modes provides material with the structure of granular perlite. Moreover, high-speed spheroidization causes a more even distribution of cementite globules in a ferrite matrix (Fig. 9b). Samples from steel of almost identical hardness after the HT have a fine-grained structure, which provides a higher level of metal ductility. Due to the rapid heating of the sample and incomplete steel austenitization, certain changes occur in the morphology of the carbide phase from the lamellar to the fine-grained globular.

The achieved technical result proves that with the proposed system ensures high dispersion and homogeneity of the billet structure all over the plane of its cross-section, while the HT technological process of the steel is characterized by high productivity, low power consumption, and improved operational characteristics. The device application for spheroid annealing regime causes uniform distribution of cementite globules in the ferrite matrix, and thus provides the necessary mechanical properties of the metal for its subsequent cold deformation.

Note that solving problems defined by this paper, as a rule, is based on use of the apparatus of difference equations, which involves the obligatory replacement of derivatives by difference equations. The studies carried out in this paper show that methods for solving problems of this class should be not only diverse, but also combine the evaluation of quantitative indicators with the possibilities of qualitative analysis. Nowadays, there have been some trends in the development of numerical-analytical methods with a complex logical structure, but they have a higher order of accuracy and the possibility of constructing algorithms adapted to the order of approximation in comparison with the piecewise differences methods. From a computational point of view, such an approach is somewhat cumbersome, but it provides a peculiar benchmark for comparison with other practical methods. But, given the benefits of a computational experiment with a multiprocessor computing system, it can be claimed that the fact that hindered the development of a numerical-analytical approach is currently losing its relevance. With this regard, in order to solve the simulation (1), this paper further develops the idea of progressing schemes of higher order of accuracy on the basis of a numerical-analytical approach to solving many of the problems under study.

5 Conclusions

Improvement of the existing and creation of new technological processes of metal-working require significant costs associated with conducting a large number of field experiments in laboratory, experimental and industrial equipment, as well as in industrial conditions. The reduction of the number of experimental studies and the time to conduct them with the obtaining of the necessary information for the construction and implementation of technological developments can be achieved through the application of multiprocessor computing systems. The paper covers actual problem of controlling the temperature regimes of recrystallization process and spheroid annealing of calibrated steel by using a multiprocessor system; which allows coordinating the time intervals of the annealing technological process.

The scientific novelty of the conducted researches is that for the first time on the basis of the multiprocessor computing system a high-speed HT simulation a long-length steel billet in real time was created for the purpose of recrystallization and spheroidized annealing of calibrated steel and manufacture of high-strength fasteners by the cold-die-forging method without the final HT. The proposed approach provides an opportunity to control the technological parameters in the modes of metal HT, in particular the temperature at the section center of the metal billet, which provides the material with the necessary properties, and along the entire plane of cross-section and length of a billet. This was achieved by a multiprocessor computing system that has the form of a separate module, and by special software, which is able to set and control the necessary temperature regimes on the entire plane of the cross section of a billet when heated and endured. Compared to traditional approaches, there was realized the opportunity of improving the technological properties of metal rolling via dispersion and homogeneity of the structure of a billet along its cross-section plane.

The practical value of the results obtained is that it has been possible to improve the HT technological process by appropriate mathematical simulation and a set of programs. The application of mathematical simulations that are processed on a multiprocessor computing system allows controlling the temperature field of a metal in the process of its heating, holding and cooling, and thus provides a rapid adaptation of the production to the consumer's requirements.

Implementation of the developed approach for the HT of the metal on the basis of the multiprocessor computing system creates a problem of reconciling the capabilities of processors and the network interface of a multiprocessor system. Consequently, further promising researches are the ways of solving this problem by multiprocessor computing systems with processors of different types. Under these conditions it is necessary to derive analytical relations for establishing the optimal number of nodes of a multiprocessor computing system, taking into account the computational capabilities of certain processors.

References

1. Ivashchenko, V., et al.: Patent of Ukraine, Plant for the heat treatment of long-length steel products/; Owners: National Metallurgical Academy of Ukraine, Donetsk National Technical University. - No. u 201014225; Bulletin No. 15, 29 November 2010. Published 10.08.2011
2. Bukatov, A., et al.: Programming of multiprocessor computing systems, vol. 208. Publishing house LLC TsVVR, Rostov-on-Don (2003)
3. Caruso, M., et al.: Microstructural evolution during spheroidization annealing of eutectoid steel: effect of interlamellar spacing and cold working. In: Advanced Materials Research, vol. 89, pp. 79–84 (2010)
4. Dolzhenkov, I.: Thermal and deformation-heat treatment of metal. Theor. Pract. Metall. **3**, 30–36 (2002)
5. Dolzhenkov, I.: The effect of plastic deformation and other preprocessing on spheroidization of carbides in steels. Theor. Pract. Metall. **1**, 66–68 (2007)
6. Bobylev, M., Greenberg, V., Zakirov, D., Lavrinenko, Yu.: Preparation of the structure during electrothermal processing of steels used for the planting of high-strength fasteners. Steel **11**, 54–60 (1996)
7. Bobylev, M., Zakirov, D., Lavrinenko, Yu.: Optimization of annealing modes with induction heating of 20G2R and 38HGNM steels. Steel **4**, 67–70 (1999)
8. Shvachych, G., Ivaschenko, O., Busygin, V., Fedorov, Y.: Parallel computational algorithms in thermal processes in metallurgy and mining. Naukovyi Visnyk Natsionalnoho Hirnychoho Universytetu **4**, 129–137 (2018)

688   G. Shvachych et al.

9. Zakirov, D., Bobylev, M., Lavrinenko, Yu., Lebedev, L., Syuldin, V.: Patent of RF 2137847, cl. C 21 D 1/32, C 21 D 9/60, C 21 D 11/00. Installation for heat treatment of calibrated steel/; Patentee: Avtonormal Open Joint-Stock Company. - № 98117255/02, 16 September 1998. Published 09/20/1999
10. Khokhlyuk, V.: Parallel Algorithms for Integer Optimization, vol. 224. Radio and Communications, Moscow (1987)
11. Shvachych, G., Busygin, V.: Effective algorithms for solving coefficient problems of high accuracy order. In: Proceedings of the System Technologies, Ukraine, vol. 4, no. 117, pp. 86–94 (2018)
12. Tikhonov, A., et al.: Equations of Mathematical Physics, vol. 724. Science, Moscow (1966)
13. Ivaschenko, V., Shvachych, G., Kholod, Ye.: Extreme algorithms for solving problems with higher order accuracy. In: Applied and Fundamental Research, pp. 157–170. Publishing House Science and Innovation Center, Ltd., St. Louis (2014)

Image Retrieval Method
Based on Back-Projection

Kohei Arai[⌧]

Department of Information Science, Saga University, Saga, Japan
arai@cc.saga-u.ac.jp

Abstract. Image retrieval method based on back-projection with histogram matching is proposed. The distance measures are compared among correlation, Chi-square, intersection, and Bhattacharrya. Through preliminary experiment, it is found that the most appropriate distance measures are Chi-Square or Bhattacharrya, which depends on the nature of the images in concern though. Also it is found that image retrievals with back-projection does works almost 100% in terms of success ratio.

Keywords: Image retrieval · Image matching · Histogram matching

1 Introduction

A histogram generally means a bar graph, and it is often used to represent the distribution of population and grades for each age. However, it is also used as a method to express data related to images such as color and brightness. In OpenCV, in order to handle the distribution of densities such as color and lightness in the image, it has a function to process the histogram. Below, density histograms of color image and grayscale image are shown using OpenCV.

As an index showing similarity of different images, there is a method of observing each histogram and measuring the distance. There are several methods for distance measurement, OpenCV can handle distance by correlation, distance by chi square, distance by intersection, Bhattacharyya (Hellinger) distance. For each function that calculates the distance between histograms, it is possible to calculate the distance by each method by passing the method to be used as an argument. Although each method was used at the time of execution, the source code to be posted later specifies the method of Bhattacharrya distance. Since other methods only change the argument of the function part, omit it.

Indexing Method for Image Database Retrievals by Means of Spatial Features is proposed [1]. Image portion retrievals in large scale imagery data by using online clustering taking into account pursuit algorithm based Reinforcement learning and competitive learning is also proposed [2]. Visualization of 3D object shape complexity with wavelet descriptor and its application to image retrievals is proposed [3] together with Wavelet based image retrieval method [4].

© Springer Nature Switzerland AG 2020
K. Arai and S. Kapoor (Eds.): CVC 2019, AISC 944, pp. 689–698, 2020.
https://doi.org/10.1007/978-3-030-17798-0_54

DP matching based image retrieval method with wavelet Multi Resolution Analysis: MRA which is robust against magnification of image size is proposed [5]. Method for image portion retrieval and display for comparatively large scale of imagery data onto relatively small size of screen which is suitable to block coding of image data compression is also proposed [6]. Content based image retrieval by using multi-layer centroid contour distance is also proposed [7].

Remote sensing satellite image database system allowing image portion retrievals utilizing principal component which consists spectral and spatial features extracted from imagery data is proposed [8]. Image retrieval and classification method based on Euclidian distance between normalized features including wavelet descriptor is also proposed [9].

Image retrieval based on color, shape and texture for ornamental leaf with medicinal functionality is proposed [10]. Comparison contour extraction based on layered structure and Fourier descriptor on image retrieval is also proposed [11]. On the other hand, Pursuit reinforcement competitive learning (PRCL) based online clustering with tracking algorithm and its application to image retrieval is proposed [12]. Image retrieval method utilizing texture information derived from discrete wavelet transformation together with color information is also proposed [13].

Image retrieval method based on back-projection with histogram matching is proposed. The distance measures are compared among correlation, Chi-square, intersection, and Bhattacharrya.

In the next section, the proposed method for image retrievals using histogram matching of back projection is described. Then, the distance measure for the histogram matching is discussed with a preliminary experiment. After that, conclusion is described together with some discussions. Finally, future works are followed.

2 Proposed Method

2.1 Back Projection

Back projection is the inverse projection. Generally, projection refers to an operation to reduce information by dropping information. As an example, there is a shade of trees made by sunlight. It can be said that trees, which are three-dimensional objects, are projected on a two-dimensional plane with dimensions reduced. For inverse projection, the inverse operation can be performed, and the image can be reconstructed from the data.

OpenCV's back-projection function calculates the back-projection of the histogram. For each tuple of pixels in the same position of all input single-channel images, the value of the bin of the histogram corresponding to that tuple is set as the pixel value of the output image.

(a)Original

(b)Back Projection

Fig. 1. Back projection image

Statistically, it can be said that each pixel value of the output image is the occurrence probability of the observation tuple in the given distribution (histogram). Hereinafter, a histogram of the original image is calculated, and a back projection image obtained by converting the density with higher frequency of appearance is brighter and the lower density is darker. Figure 1 shows the original image of "Lena" in the SIDBA database and the back projection image of "Lena".

2.2 Back Projection Patch

In the back projection patch, a part having the same histogram as the designated template image is extracted from the input original image. In this method, even if the template image is rotated or reversed, the histogram itself does not change so that the same portion can be searched for. The search method is shown in Fig. 2.

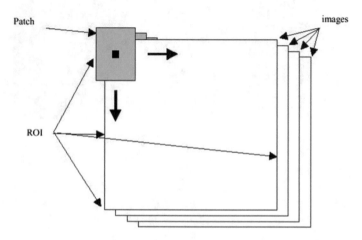

Fig. 2. Back projection patch

3   Preliminary Experiment

3.1   Distance Between Histograms

As an index showing similarity of different images, there is a method of observing each histogram and measuring the distance. There are several methods for distance measurement, OpenCV can handle distance by correlation, distance by chi square, distance by intersection, Bhattacharyya (Hellinger) distance.

(a)Reference

(b)Current

Fig. 3. Reference and current images for histogram distance calculations

Table 1. Normalized histogram distances

Model	Correlation	Chi-Square	Intersection	Bhattacharrya
Complete_Match	1.0	0.0	1.0	0.0
Half_Match	0.7	0.67	0.5	0.55
Complete_Unmatch	−1.0	2.0	0.0	1.0

For each function that calculates the distance between histograms, it is possible to calculate the distance by each method by passing the method to be used as an argument. Although each method was used at the time of execution, the source code to be posted later specifies the method of Bhattacharrya distance. Since other methods only change the argument of the function part, omit it. First, two original images used for comparison are shown in Fig. 3. The histogram distances between two images are calculated as examples.

For OpenCV, you can use four distances as correlation distance, chi-squared distance, distance by intersection and Bhattacharyya distance. When normalizing the histogram to 1.0, the criterion of the comparison result is shown in Table 1.

(1) Correlation Distance

The distance by correlation extracts the histogram of the two images and uses the two correlation coefficients as the distance. The formula for obtaining the correlation coefficient is as follows:

$$d(H_1, H_2) = \frac{\sum_I H'_1(I) * H'_2(I)}{\sqrt{\sum_I H'_1(I)^2 * \sum_I H'_2(I)^2}} \tag{1}$$

where

$$H'_k(I) = H_k(I) - \frac{1}{N} * \sum_J H_k(J) \tag{2}$$

Below is the result of computing the distance by the correlation between the two images.

$d(H_1, H_2) = 0.9754$

(2) Chi-Square Distance

In this method, the i-th element of each histogram is extracted, and the distance between the histograms is expressed as the total sum of the Chi square values. Calculations on OpenCV are as follows:

$$d(H_1, H_2) = \sum_I \frac{(H_1(I) - H_2(I))^2}{H_1(I) + H_2(I)} \tag{3}$$

The calculated Chi-Square distance is as follows:

$d(H_1, H_2) = 34561$

(3) Intersection Distance

Comparing the values of the i-th histogram of the two images, selecting the smaller one and adding the added sum is called the distance by the intersection. Calculation formulas are as follows.

$$d(H_1, H_2) = \sum_I \min(H_1(I), H_2(I)) \tag{4}$$

The calculated intersection distance is as follows,
$d(H_1, H_2) = 472213$

(4) Bhattacharrya Distance

This distance measurement method between histograms is also called Hellinger distance. This method can be used only with the normalized histogram, and the calculation formula on OpenCV is as follows:

$$d(H_1, H_2) = \sqrt{1 - \sum_I \frac{\sqrt{H_1(I) * H_2(I)}}{\sqrt{\sum_I H_1(I) * \sum_I H_2(I)}}} \tag{5}$$

The calculated Bhattacharrya distance is as follows:
$d(H_1, H_2) = 0.168$
The normalized distance between two comparative images was computed and outputted in all methods as follows:

Correlation: 0.9755
Chi-Square: 0.1000
Intersection: 0.8454
Bhattacharrya Distance: 0.1679

In OpenCV, it is said that it is better to use the distance by intersection when requesting speed of calculation, Bhachattarrya distance or Chi square distance when obtaining accuracy.

3.2 Back Projection Patch Calculation

At this time, when setting the corresponding pixel (the central pixel of the patch image) of the target image, a patch that slides on the plane of the input image is used. In the model using the normalized histogram, the result image can also be viewed as a probability distribution map representing the possibility of existence of the object. The input original image, template image, and search result are shown in Fig. 4.

(a)Original

(b)Image for retrieve

(c)Retrieved result

Fig. 4. Back projection patch calculation result

4 Conclusion

Image retrieval method based on back-projection with histogram matching is proposed. The distance measures are compared among correlation, Chi-square, intersection, and Bhattacharrya. Through preliminary experiment, it is found that the most appropriate distance measures are Chi-Square or Bhattacharrya, which depends on the nature of the images in concern though. Also it is found that image retrievals with back-projection does works almost 100% in terms of success ratio.

As for future works, currently, MRI (magnetic resonance imaging) used for imaging inside the body at a medical site is an examination method using a nuclear magnetic resonance phenomenon, utilizing each magnetic resonance phenomenon of hydrogen atoms in the body. When imaging is performed, first, a static magnetic field is generated by a superconducting magnet so that hydrogen nuclei in the body are oriented in a certain direction. At this time, by generating a gradient magnetic field, it is possible to identify which part of the body was imaged by considering gradient magnetic field.

In consideration of this, radio waves of different frequencies are irradiated for each part, and radio waves after passing through the human body are measured to acquire data necessary for imaging. At this time, by generating a gradient magnetic field different from the above-mentioned while varying the timing and the duration, the frequency and phase of the hydrogen nuclear spin change in each part of the cross section, and the position of each part in the same slice Information is obtained.

Thereafter, imaging is performed based on this data, but back projection, that is, imaging by inverse projection is performed at this time. An in-vivo image is reconstructed by performing two-dimensional or three-dimensional Fourier transformation on the obtained data. In the two-dimensional conversion, it is necessary to select a slice, but in the three dimensions it is possible to perform the whole reconstruction without selecting the slice. The imaging result becomes a binary image or a gray scale image, and abnormality can be observed from shading and the like. Also, unlike conventionally used CT scans, imaging of soft tissues became possible, so that image diagnosis can be performed more accurately.

Acknowledgment. The author would like to thank Mr. Mitsuhiro Suenaga of former student in the Information Science Department of Saga University for his experimental efforts.

References

1. Etoh, H., Yamamoto, T., Arai, K.: Indexing method for image database retrievals by means of spatial features. J. Jpn. Soc. Photogramm. Remote Sens. **39**(3), 14–20 (2000)
2. Arai, K., Kenkyo, B.: Image portion retrievals in large scale imagery data by using online clustering taking into account pursuit algorithm based reinforcement learning and competitive learning. J. Image Electron. Soc. Jpn. **39**(3), 301–309 (2010)
3. Arai, K.: Visualization of 3D object shape complexity with wavelet descriptor and its application to image retrievals. J. Vis. **15**, 155–166 (2011). https://doi.org/10.1007/s12650-011-0118-6

4. Arai, K., Rahmad, C.: Wavelet based image retrieval method. Int. J. Adv. Comput. Sci. Appl. **3**(4), 6–11 (2012)
5. Arai, K.: DP matching based image retrieval method with wavelet Multi Resolution Analysis: MRA which is robust against magnification of image size. Int. J. Res. Rev. Comput. Sci. **3**(4), 1738–1743 (2012)
6. Arai, K.: Method for image portion retrieval and display for comparatively large scale of imagery data onto relatively small size of screen which is suitable to block coding of image data compression. Int. J. Adv. Comput. Sci. Appl. **4**(2), 218–222 (2013)
7. Arai, K., Rahmad, C.: Content based image retrieval by using multi-layer centroid contour distance. Int. J. Adv. Res. Artif. Intell. **2**(3), 16–20 (2013)
8. Arai, K.: Remote sensing satellite image database system allowing image portion retrievals utilizing principal component which consists spectral and spatial features extracted from imagery data. Int. J. Adv. Res. Artif. Intell. **2**(5), 32–38 (2013)
9. Arai, K.: Image retrieval and classification method based on Euclidian distance between normalized features including wavelet descriptor. Int. J. Adv. Res. Artif. Intell. **2**(10), 19–25 (2013)
10. Arai, K., Abdullah, I.N., Okumura, H.: Image retrieval based on color, shape and texture for ornamental leaf with medicinal functionality. Int. J. Image Graph. Signal Process. **6**(7), 10 (2014)
11. Rahmad, C., Arai, K.: Comparison contour extraction based on layered structure and Fourier descriptor on image retrieval. Int. J. Adv. Comput. Sci. Appl. **6**(12), 71–74 (2015)
12. Arai, K.: Pursuit Reinforcement Competitive Learning: PRCL based online clustering with tracking algorithm and its application to image retrieval. Int. J. Adv. Res. Artif. Intell. **5**(9), 9–16 (2016)
13. Arai, K., Rahmad, C.: Image retrieval method utilizing texture information derived from discrete wavelet transformation together with color information. Int. J. Adv. Res. Artif. Intell. **5**(10), 1–6 (2016)

Investigating the Effect of Noise on Facial Expression Recognition

Muhannad Alkaddour and Usman Tariq$^{(\boxtimes)}$

Department of Electrical Engineering, American University of Sharjah,
PO Box 26666, Sharjah, UAE
utariq@aus.edu

Abstract. In this paper we investigate the effect of noise on automated recognition of facial expressions. We take images from a publicly available data set; corrupt them with different noise levels and use them for training and testing algorithms for expression recognition. We do recognition using a variant of AlexNet. We do training and testing on same noise levels and also, train on clean images and test on images with added noise. We show that the recognition performance is fairly robust for reasonable levels of noise, however it degrades considerably after that.

Keywords: Noise · Facial expression recognition · Deep learning

1 Introduction

Face expression recognition is a challenging and well-established problem in computer vision literature. Face expressions convey non-verbal information which is key in understanding context and in effective communication [11]. Automated recognition of facial expression recognition has a broad range of applications. These include human behavior interpretation [40], electronic customer relationship management (ECRM) [6], smart automobile systems, socially aware robots and entertainment industry [41].

The research in the community has begun to move in the direction of emotion recognition in the wild. Some of the recent works in this regard include [19] and [42]. Here, we no longer restrict ourselves to laboratory controlled imaging conditions. The facial images of people photographed without any regard to lighting, head-pose orientation are taken into consideration. The images are more often than not downloaded from the internet. However, if we were to recognize expression of a person photographed "in-the-wild", perhaps we also need to understand how the image quality effects the recognition performance. In this work, we strive to achieve just that. We take images from a publicly available corpus, Multi-PIE [16]. It has hundreds of thousands of images with a lot of variation in pose, lighting and demographics. We corrupt the images with controlled noise at different levels. We then compare performance of recognition algorithm in two distinct cases; when the training and testing have similar noise levels and,

© Springer Nature Switzerland AG 2020
K. Arai and S. Kapoor (Eds.): CVC 2019, AISC 944, pp. 699–709, 2020.
https://doi.org/10.1007/978-3-030-17798-0_55

when the recognition model is trained on clean images and followed by testing on noisy images. We show that the recognition performance is fairly robust for reasonable noise levels, however it degrades significantly at higher noise levels.

We use deep learning for recognizing expressions. The motivation behind using deep models is that they have achieved state-of-art performance on facial expression recognition, among other problems, in recent past. Such an extensive study in exploring the effect of noise on facial expression recognition, and more specifically using deep models, is first of its kind in the literature. This problem has been explored in face recognition, though; e.g. [15] is one such good example. There also, AlexNet was one of the top models, as regards the performance in noisy situations is concerned. Hence, justifying our use of AlexNet in this work.

The rest of the paper is organized as follows. We first discuss briefly the relevant literature and explore the sources of noise in imaging. This is then followed by the description of the dataset and experiments, which is then followed by results and discussion.

2 Background

There may be different sources of noise during image acquisition through digital cameras [38]. We are particularly interested in the noise at the camera sensor. These may be of different types. For instance, same amount of light falling on the camera sensor may lead to different sensed intensities, primarily due to manufacturing inconsistencies [38]. This is known as *fixed pattern noise*. This can be corrected by scaling perceived intensities in relation to the *fixed pattern noise*. Then there is *shot noise*, which is due to tiny fluctuations of packets of light arriving at different times at different locations on the camera sensor. This can be approximated by a Poisson distribution. Note that Poisson distributed approaches a Gaussian distribution when the packets of light arriving at the sensor are in fairly a large number. And last but not the least, there is *thermal noise* which is modeled by an additive zero mean Gaussian distribution. Apart from these, there are further sources of noise while quantization of signal amplitudes and image compression.

We are primarily interested in thermal noise here, as this is a dominant contributor of image noise when we increase image sensor's sensitivity (ISO level) in low lighting conditions to have a better lit image. As pointed out earlier, we can model it with a zero mean Gaussian distribution (Additive White Gaussian Noise (AWGN)) with a standard deviation that is proportional to the camera sensor's sensitivity. And this is what we do in the subsequent experiments and examine the effect of noise on expression recognition performance. This is particularly useful to study, for instance if we want to do facial expression recognition in dimly lit faces.

Effect of noise on face recognition has been studied in earlier works. Some of the examples include [8, 12, 15]. There they examine the effect of various kinds of noise on face recognition performance. [15] explores it more specifically with reference to deep models. Noise also has an effect on human visual perception of faces as well, as discussed in [7], where the authors found the results to be similar among healthy individuals and schizophrenic patients.

The works that deal with noise on expression recognition include [2,21,25,37]. [2] deals the noise effects with regards to audio-visual emotion recognition and shows classifier fusion helps in performance. For visual channel, they use local binary patterns as features. [21] describes de-noising Electroencephalogram (EEG) signals for emotion recognition. And [37] describes a *Cany-HOG* descriptor, which performs better than the traditional HOG descriptor. Our work is different in regards that we deal specifically with deep models. It is important to understand how deep models degrade in performance for expression recognition with noise, as these are now stat-of-art in face expression classification. We do an in-depth analysis. We not only deal with the case where training set is clean and test set has images corrupted at various noise levels; we also study cases where both training and testing sets are corrupting with similar noise level. We show that when noise level is similar in both training and testing cases, the performance for expression recognition remains fairly robust. From this we conjecture that some transfer learning based approach [24] can help expression classification where original model is trained on clean images and then model parameters may be updated through noisy images to help in recognition in those situations.

In the following we will briefly review techniques used for facial expression recognition. Facial expression recognition has come a long way since the seminal work by Ekman and Friesen [14]. The standard algorithmic pipeline for automatic facial expression recognition can be divided into pre-processing (face-alignment etc), feature extraction and classification. However, in more recent works many of these steps are embedded in one framework: deep learning. This algorithmic pipeline learns features and classifiers jointly from the given examples.

In the following we first discuss the *shallow* approaches. i.e. those that do not employ deep learning. There may be two major categories of features employed for automated facial expression recognition using such methods [43]; geometric features and appearance features. The geometric features are extracted from the shape or salient point locations of important facial components. Some of the works in this category include [5,20,26,39]. And, some example works for expression recognition using appearance features representing texture and other facial miniatures include [1,3,4,10,17].

Note that, much of the literature focuses on expression recognition from frontal or near-frontal face images [43,44]. Expression recognition from near-frontal faces is particularly important for expression recognition "in-the-wild". Some works in this category include [17,29,30,45].

Some works extract dense appearance features on detected faces and stay aside of key-point recognition, as this may be tricky, particularly in noisy cases like the ones we deal with in this work. Some example works in this regard are [22,23], Zheng et al. [45], Tang et al. [33] and Tariq et al. [35,36].

Recently, much of the research in expression recognition has been moved towards Deep learning (DL). It is a data intensive method to learn visual hierarchies and features. The interest in DL techniques sky-rocketed when a deep learning algorithm by Krizhevsky et al. [18] achieved state-of-the-art accuracy on the ImageNet dataset. Some of the notable works on deep networks in facial expression recognition include [13, 19, 24, 27, 28, 34].

For a comprehensive survey on facial expression recognition the readers are referred to [9, 31, 42].

As pointed out earlier, this is the first attempt to study noise in deep models for facial expression recognition. Earlier approaches, either discussed shallow models or dealt with different signal sources such as EEG signals. In the following we will first discuss the database used in this work. It is then followed by experiments, discussion and concluding remarks.

3   Database

We used a publicly available Multi-PIE [16] database in this work. We take 299 subjects from the first three sessions. The images used in our experiments come from 5 camera angles and 19 lighting conditions. We divide the faces into four expression categories. These include; neutral, positive (smile), negative (squint, disgust) and surprise. In this manner we get around 152,000 images in total. It is perhaps, one of the largest datasets in the facial expression recognition community if we consider the number of images as well as image variations. Some example faces from CMU Multi-PIE are presented in Fig. 1.

One may argue to use an image database from "the-wild" such as the EmotionNet dataset [42]. However, the problem with that is we have no control on the image quality. As we want to assess the effect of noise on expression recognition, we should ensure that all the images used in training and testing have similar quality to begin with. Hence, when we degrade them, different images may be degraded by the same amount for a similar noise level.

In the following, we describe our experiments.

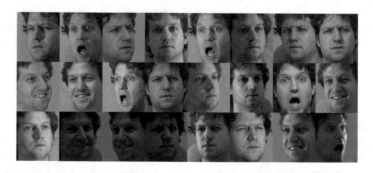

Fig. 1. Example images from CMU Multi-PIE database [36].

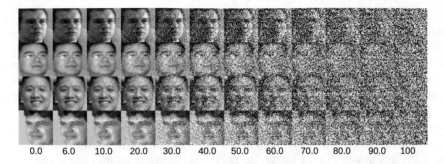

0.0   6.0   10.0   20.0   30.0   40.0   50.0   60.0   70.0   80.0   90.0   100

Fig. 2. Example images from CMU Multi-PIE database corrupted with varying noise levels. The numbers below each column of images show the standard deviation of AWGN they are corrupted with.

4   Experiments

As pointed out earlier, the purpose of this work is to assess the effect of noise on facial expression recognition. Hence, we corrupt the images with varying additive white Gaussian noise levels to simulate the camera sensor noise, as discussed in Sect. 2. We add zero mean Gaussian noise with standard deviation (σ) of 1.0, 2.0, 4.0; then in the increments of 2.0 until 32.0, then with σ value 40.0, 50.0 and then in increments of 10.0 until 100.0. Some images corrupted with varying noise levels are shown in Fig. 2. Since pixel values range from 0 till 255, we clip any value below 0 or above 255 after noise addition. One can hardly see any discernible patterns for expression recognition in images corrupted with noise with standard deviation more than 50.0. It is surprising to see that the deep network in this work can still perform well above chance at such high noise levels. Such high noise levels may not be expected in practice, they have only been included for experimental purposes. However, one may expect images with corruptions that look like the noise with standard deviation of 20.0 in Fig. 2 in practice (in consumer grade cameras set at a higher ISO level).

In this work, we do two sets of experiments. In the first set, we train our recognition model using clean images and then subsequently test it on images which have added noise to a varying degree. In the second set of experiments, we train on noisy images and test on noisy images, where similar amount of noise is added in both training and testing. To simulate various noisy conditions, we add noise by varying its standard deviation. Thus, for the first case of experiments, where we train on clean images and test on noisy images, there are 25 testing cases (including testing on clean images), as described earlier. Also for the second set of experiments, there are 25 training cases and 25 testing cases. We use half of the 152,000 images as training and the other half as testing. We then repeat each experiment twice with re-randomized training and testing partitions and report average results. We ensure that the dataset partitions are subject-independent; i.e. no subject who has appeared in training set appears in validation/testing set and vice-versa.

Since we are interested in deep networks, we used a deep convolutional networks which is a variant of Alexnet [18]. For selecting the network architecture, we took a subset of training set as validation and experimented with various options. Our final network has three sets of convolution; max-pooling; normalization layers. It is then followed by two fully connected layers. Our three convolution layers learn 32, 64 and 128 filters respectively. The filters are learnt on a local 3×3 neighbourhood. In between the convolution layers, we down-sample the filtered results on a 2×2 grid. This down-sampling is done using max-pooling. It is then followed by a normalization layer. The last two fully connected layers have 384 and 192 nodes respectively. Once we find the best network architecture and set of parameters from various possibilities via cross-validation, we fix most of the parameters across our experiments as changing them does not make any sizeable difference in performance. We only change learning rate across our experiments while cross-validating on a subset of training set in each experiment on a log scale. We use a batch size of 256 and do 40,000 iterations. More iterations do not improve the performance as training loss converges. We detect faces from our images and down-sample them to 50×50 pixels.

5   Results and Discussion

The results for the two sets of experiments are plotted respectively in Figs. 3 and 4. The figures are best viewed in color. The black curves show the average performance. One can see that noise significantly effects the expression recognition performance in both cases. However, for lower levels of noise the expression recognition performance is fairly robust in both cases. But for higher levels there is a significant drop. One interesting observation is the case where training and testing is done on images with similar level of noise, i.e. Figure 4. One can see the expression recognition performance stays fairly robust for a wide range of noise levels. It is as high as 80% when the standard deviation of added noise is 40.0!. To have a sense of the kind of images the network is dealing with, please refer to Fig. 2. This shows that if the training and testing set have similar noise levels, the network can learn discriminating features that are reasonable enough for good classification.

Apart from the average performance, there are interesting patterns in performance degradation of various expressions. For example, in Fig. 3, while the recognition performance drops significantly beyond the added noise of $\sigma = 20$ for all the expressions, the surprise expression drops significantly more compared to others. The neutral expression drops initially but then plateaus. And negative expression seem to perform better amongst the others for more noisy situations.

Figure 4 shows the average performance when training was done on noisy images and testing images also had a similar level of noise. The trends here are very different from the trends in Fig. 3. For instance here, negative expression continues to be the worst performer while the other three expressions perform significantly better. The average performance drop seems to be linear rather than exponential and is at a much slower rate compared to Fig. 3. This shows that if you have images from similar noise levels in the training set, your recognition algorithm will tend to perform better.

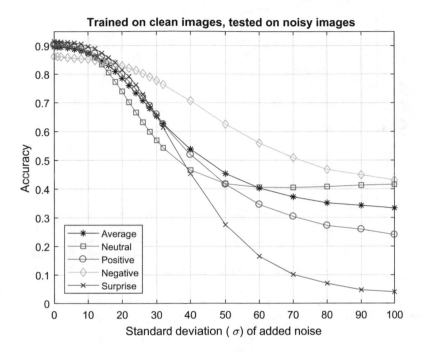

Fig. 3. Average results of training on clean images while testing on noisy images for different expressions

We also compare our results on clean images on Multi-PIE with an earlier work [36] who did experiments in a similar setting. They achieved a recognition rate of 81.8% using supervised super vector encoding. Our model (which is not significantly deep) in comparison, achieves a recognition rate of ~90%. We do not compare to other works in the literature as they follow very different experimental protocol in training and testing, apart from including a vigorous face registration step which may increase the recognition performance anywhere from 4–10% [32]. Also, the purpose of this work is not to push the state-of-art, rather to assess the effect of noise on one of the top performing deep learning architectures. We present the class confusion matrix for the results where training and testing was done on clean data in Table 1 for one of the randomizations. One can see the expression with the highest recognition rate is surprise, followed by positive, neutral and then negative expressions.

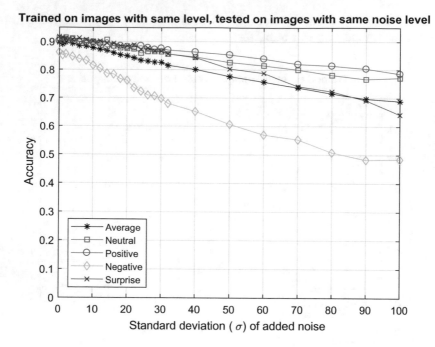

Fig. 4. Average results of training and testing on noisy images with similar level of added noise for different expressions

Table 1. Confusion matrix for recognition performance for training/testing on clean images

Both		Predicted			
		NEU	POS	NEG	SUR
Ground truth	NEU	**89.9%**	4.6	5.0	0.5
	POS	6.2	**91.1%**	2.0	0.8
	NEG	6.4	6.3	**87.0%**	0.2
	SUR	3.7	2.8	0.3	**93.3%**

6   Concluding Remarks

In short, we have presented an extensive study of the effect of additive white Gaussian noise on facial expression recognition. Such a study is first of its kind involving deep networks for facial expression recognition. The results are signifi- cant in understanding the behaviour of the selected deep network in presentence of noise. This noise was chosen to simulate the camera sensor noise. The sensor noise increases when we increase the sensor's sensitivity (ISO level) to have a well-lit image in low-light conditions. We simulated this scenario by having added noise with various standard deviations. This scenario is particularly relevant in facial expression recognition in dimly lit environment. Our results show that the expression recognition using a variant of AlexNet is fairly robust, particularly when training and testing images have similar noise levels. This strongly suggests that we can improve recognition performance by a transfer learning based approach where the beginning model is trained on clean and lower noise images and then adapted with a limited set of higher noise images.

Acknowledgments. This work was supported in part by two Faculty Research Grants; FRG15-R-42 and FRG-17-R-44 from the American University of Sharjah, UAE.

References

1. Anderson, K., McOwan, P.W.: A real-time automated system for the recognition of human facial expressions. IEEE Trans. Syst. Man Cybern. Part B (Cybern.) **36**(1), 96–105 (2006)
2. Banda, N., Robinson, P.: Noise analysis in audio-visual emotion recognition. In: Proceedings of the International Conference on Multimodal Interaction, pp. 1–4 (2011)
3. Bartlett, M.S., Littlewort, G., Braathen, B., Sejnowski, T.J., Movellan, J.R.: A prototype for automatic recognition of spontaneous facial actions. In: Advances in Neural Information Processing Systems, pp. 1295–1302. MIT (1998, 2003)
4. Berretti, S., Del Bimbo, A., Pala, P., Amor, B.B., Daoudi, M.: A set of selected sift features for 3D facial expression recognition. In: 2010 20th International Conference on Pattern Recognition (ICPR), pp. 4125–4128. IEEE (2010)
5. Chang, Y., Hu, C., Feris, R., Turk, M.: Manifold based analysis of facial expression. Image Vis. Comput. **24**(6), 605–614 (2006)
6. Chen, R.-H., Lin, R.-J., Yang, P.-C.: The relationships between eCRM, innovation, and customer value - an empirical study. In: 2011 IEEE International Summer Conference of Asia Pacific Business Innovation and Technology Management, pp. 299–302, July 2011
7. Chen, Y., McBain, R., Norton, D.: Specific vulnerability of face perception to noise: a similar effect in schizophrenia patients and healthy individuals. Psychiatry Res. **225**(3), 619–624 (2015)
8. Cherifi, D., Radji, N., Nait-Ali, A.: Effect of noise, blur and motion on global appearance face recognition based methods performance. Int. J. Comput. Their. Appl. **16**(6), 4–13 (2011)
9. Corneanu, C.A., Simon, M.O., Cohn, J.F., Guerrero, S.E.: Survey on RGB, 3D, thermal, and multimodal approaches for facial expression recognition: history, trends, and affect-related applications. IEEE Trans. Pattern Anal. Mach. Intell. **38**(8), 1548–1568 (2016)
10. Dhall, A., Asthana, A., Goecke, R., Gedeon, T.: Emotion recognition using PHOG and LPQ features. In: 2011 IEEE International Conference on Automatic Face and Gesture Recognition and Workshops (FG 2011), pp. 878–883. IEEE (2011)
11. Donato, G., Bartlett, M.S., Hager, J.C., Ekman, P., Sejnowski, T.J.: Classifying facial actions. IEEE Trans. Pattern Anal. Mach. Intell. **21**(10), 974–989 (1999)

12. Dutta, A., Veldhuis, R.N.J., Spreeuwers, L.J.: The impact of image quality on the performance of face recognition. In: 33rd WIC Symposium on Information Theory in the Benelux. Centre for Telematics and Information Technology (CTIT) (2012)
13. Ebrahimi Kahou, S., Michalski, V., Konda, K., Memisevic, R., Pal, C.: Recurrent neural networks for emotion recognition in video. In: Proceedings of the 2015 ACM on International Conference on Multimodal Interaction, pp. 467–474. ACM (2015)
14. Ekman, P., Friesen, W.V.: Constants across cultures in the face and emotion. J. Pers. Soc. Psychol. **17**(2), 124 (1971)
15. Grm, K., Štruc, V., Artiges, A., Caron, M., Ekenel, H.K.: Strengths and weaknesses of deep learning models for face recognition against image degradations. IET Biometrics **7**(1), 81–89 (2018)
16. Gross, R., Matthews, I., Cohn, J., Kanade, T., Baker, S.: Multi-pie. Image Vis. Comput. **28**(5), 807–813 (2010)
17. Hu, Y., Zeng, Z., Yin, L., Wei, X., Tu, J., Huang, T.S.: A study of non-frontal-view facial expressions recognition. In: 19th International Conference on Pattern Recognition, ICPR 2008, pp. 1–4. IEEE (2008)
18. Krizhevsky, A., Sutskever, I., Hinton, G. E.: ImageNet classification with deep convolutional neural networks. In: Advances in Neural Information Processing Systems, pp. 1097–1105 (2012)
19. Levi, G., Hassner, T.: Emotion recognition in the wild via convolutional neural networks and mapped binary patterns. In: Proceedings of the 2015 ACM on International Conference on Multimodal Interaction, pp. 503–510. ACM (2015)
20. Lucey, S., Ashraf, A.B., Cohn, J.F.: Investigating spontaneous facial action recognition through AAM representations of the face. INTECH Open Access Publisher (2007)
21. Molavi, M., Yunus, J.: The effect of noise removing on emotional classification. In: 2012 International Conference on Computer Information Science (ICCIS), vol. 1, pp. 485–489, June 2012
22. Moore, S., Bowden, R.: Local binary patterns for multi-view facial expression recognition. Comput. Vis. Image Underst. **115**(4), 541–558 (2011)
23. Moore, S., Bowden, R.: The effects of pose on facial expression recognition. In: Proceedings of the British machine vision conference, pp. 1–11 (2009)
24. Ng, H.-W., Nguyen, V.D., Vonikakis, V., Winkler, S.: Deep learning for emotion recognition on small datasets using transfer learning. In: Proceedings of the 2015 ACM on international conference on multimodal interaction, pp. 443–449. ACM (2015)
25. Oh, D., Osherson, D.N., Todorov, A.: Robustness of emotional expression recognition under low visibility
26. Pantic, M., Bartlett, M.S.: Machine analysis of facial expressions. I-Tech Education and Publishing (2007)
27. Reed, S., Sohn, K., Zhang, Y., Lee, H.: Learning to disentangle factors of variation with manifold interaction. In: Proceedings of the 31st International Conference on Machine Learning (ICML-14), pp. 1431–1439 (2014)
28. Rifai, S., Bengio, Y., Courville, A., Vincent, P., Mirza, M.: Disentangling factors of variation for facial expression recognition. In: European Conference on Computer Vision, pp. 808–822. Springer (2012)
29. Rudovic, O., Pantic, M., Patras, I.: Coupled gaussian processes for pose-invariant facial expression recognition. IEEE Trans. Pattern Anal. Mach. Intell. **35**(6), 1357–1369 (2013)

30. Rudovic, O., Patras, I., Pantic, M.: Regression-based multi-view facial expression recognition. In: 2010 20th International Conference on Pattern Recognition (ICPR), pp. 4121–4124. IEEE (2010)
31. Sariyanidi, E., Gunes, H., Cavallaro, A.: Automatic analysis of facial affect: a survey of registration, representation, and recognition. IEEE Trans. Pattern Anal. Mach. Intell. **37**(6), 1113–1133 (2015)
32. Szegedy, C., Liu, W., Jia, Y., Sermanet, P., Reed, S., Anguelov, D., Erhan, D., Vanhoucke, V., Rabinovich, A.: Going deeper with convolutions. In: Proceedings of the IEEE Conference on Computer Vision and Pattern Recognition, pp. 1–9 (2015)
33. Tang, H., Hasegawa-Johnson, M., Huang, T.: Non-frontal view facial expression recognition based on ergodic hidden markov model supervectors. In: 2010 IEEE International Conference on Multimedia and Expo (ICME), pp. 1202–1207. IEEE (2010)
34. Tang, Y.: Deep learning using linear support vector machines. arXiv preprint arXiv:1306.0239 (2013)
35. Tariq, U., Yang, J., Huang, T.S.: Maximum margin GMM learning for facial expression recognition. In: 2013 10th IEEE International Conference and Workshops on Automatic Face and Gesture Recognition (FG), pp. 1–6. IEEE (2013)
36. Tariq, U., Yang, J., Huang, T.S.: Supervised super-vector encoding for facial expression recognition. Pattern Recogn. Lett. **46**, 89–95 (2014)
37. Tong, Y., Shen, Y., Gao, B., Sun, F., Chen, R., Xu, Y.: A noisy-robust approach for facial expression recognition. KSII Trans. Internet Inf. Syst. (TIIS) **11**(4), 2124–2148 (2017)
38. Tsin, Y., Ramesh, V., Kanade, T.: Statistical calibration of ccd imaging process. In: Proceedings Eighth IEEE International Conference on Computer Vision, ICCV 2001, vol. 1, pp. 480–487 (2001)
39. Valstar, M.F., Gunes, H., Pantic, M.: How to distinguish posed from spontaneous smiles using geometric features. In: Proceedings of the 9th International Conference on Multimodal Interfaces, pp. 38–45. ACM (2007)
40. Weerachai, S., Mizukawa, M.: Human behavior recognition via top-view vision for intelligent space. In: ICCAS 2010, pp. 1687–1690 (2010)
41. Wimmer, M., MacDonald, B.A., Jayamuni, D., Yadav, A.: Facial expression recognition for human-robot interaction–a prototype. In: International Workshop on Robot Vision, pp. 139–152. Springer (2008)
42. Zafeiriou, S., Papaioannou, A., Kotsia, I., Nicolaou, M., Zhao, G.: Facial affect "in-the-wild": a survey and a new database. In: The IEEE Conference on Computer Vision and Pattern Recognition (CVPR) Workshops, June 2016
43. Zeng, Z., Pantic, M., Roisman, G.I., Huang, T.S.: A survey of affect recognition methods: audio, visual, and spontaneous expressions. IEEE Trans. Pattern Anal. Mach. Intell. **31**(1), 39–58 (2009)
44. Zheng, W., Tang, H., Huang, T.S.: Emotion recognition from non-frontal facial images. In: Konar, A., Charkraborty, A. (eds.) Emotion Recognition: A Pattern Analysis Approach, 1 edn., pp. 183–213 (2014)
45. Zheng, W., Tang, H., Lin, Z., Huang, T.S.: A novel approach to expression recognition from non-frontal face images. In: 2009 IEEE 12th International Conference on Computer Vision, pp. 1901–1908. IEEE (2009)

The Language of Motion MoCap Ontology

Marietta Sionti[1(✉)], Thomas Schack[1,2], and Yiannis Aloimonos[3,4]

[1] Neurocognition and Action-Biomechanics Group,
Excellence Cluster-Cognitive Interaction Technology, Bielefeld University,
33501 Bielefeld, Germany
{marietta.sionti,thomas.schack}@uni-bielefeld.de
[2] Research Institute for Cognition and Robotics CorLab, Bielefeld University,
33501 Bielefeld, Germany
[3] Computer Vision Laboratory, Center for Automation Research,
Department of Computer Science, University of Maryland,
College Park, MD 20742, USA
yiannis@cs.umd.edu
[4] Perception and Robotics Group, Institute for Advanced Computer Studies,
University of Maryland, College Park, MD 20742, USA

Abstract. We present a systematically organized MoCap Collection, especially designed to serve grounding language to action and provide linguistics with objectively measured method and parameters (among others path, manner, trajectory, direction), in order to solve long lasting theoretical questions, such as the minimum conceptual representation of verbal events, the binary nature of argument-adjunct, cross-lingual typologies, etc. We enriched the ontology with video data and avatars in various formats for behavioral studies. In the remaining of the paper, we present applications related to the present ontology.

Keywords: Language · Ontology · MoCap · Videos · Avatars · English · German · Greek · Events

1 Goal

The *Language of Motion* MoCap Ontology contains systematically recorded kinematic data, which accurately correspond to verbs of locomotion in three languages; American English (AE), German (HochDeutsch-D) and Modern Greek (MG). Its distinctiveness lays on the initial design to combine experimentally obtained kinematic data with linguistic knowledge, in order to specify the minimum conceptual representation of a motion event that distinguishes it from all other events. Therefore, the motor data are collected by measuring the performance of native speakers of the above mentioned languages when they are asked to performed the action described by 25 motion verbs.

The main reason for collecting the current MoCap Collection is to support linguistics with a concrete and objective quantitative method for depicting lexical semantic relations, since several linguistic phenomena are largely presented in a vague way depending on different theories. One example is the distinction between *argument* and *adjunct* whose binary nature has been questioned across theories [1], namely, whether the *argument* or the *adjunct* is more crucial to cover the verb's

© Springer Nature Switzerland AG 2020
K. Arai and S. Kapoor (Eds.): CVC 2019, AISC 944, pp. 710–723, 2020.
https://doi.org/10.1007/978-3-030-17798-0_56

semasiosyntactical *slots* - can probably be better understood by delineating conceptual representations of motion events in this way (both linguistic and motor). Similarly, cross-language analysis' terms, such as lexical aspect, classifications, typological differences between path and manner languages, etc. can equally benefit if they are linked to specific kinematic parameters, e.g. velocity, iteration, vectors, joint-ankles - which universally exhibit stability. These notions are firm constituents of the verbal meaning and have not been fully grounded, leaving so the grounding research without a holistic solution concerning the linguistic structure.

The grounding of language in motion is a philosophical question since the metaphysics of Aristotle and Kant [2]. Centuries later, the boom of artificial intelligence and the robotic needs of better human computer interaction/communication reintroduce the necessity to determine the causal relationship between the signifier and the external reference point [3]. Gibson [4] attributes this relationship to the "affordances", the specific physical entities that are anchored to perceptual representations and through them to the actions that they *afford*. But today literature understands that it's a problem beyond affordances. Jackendoff [5] among other cognitive linguists support that the best way to match the entities of motion and their linguistic representations is the use of sensorimotor data. A human being begins to recruit sensorimotor data as a baby, during the sensory kinetic period [6], which offers an objective representation both of the outside world and the image schemas, i.e. the symbolic representations of movements in brain [7]. Language, contrary to sensorimotor depictions, cannot fully describe images schemas, since they are using different symbolic systems and it fails to provide full metacognitive description. Conclusively, cognitive linguistics recognize the need for combining motion with linguistic data.

2   Structure

The Language of Motion Ontology is divided in three subdirectories. The first one includes the current paper as documentation. The second subfolder is further divided in six directories:

- 700 MoCap data both in BVH and C3D format. The motion data correspond to 25 verbs (Table 1) and were performed by 4 American English, 4 Modern Greek and 20 German native speakers, who decided the minimum length and prototypical representation of each verb. The data are grouped as
  - MoCap German (D)
  - MoCap English (AE)
  - MoCap Greek (MG)
  - MoCap per verb/action (all languages)
  - Videos (AVI) per verb/action (all languages together, N = 25)
  - Avatars per verb/action (all languages) are the motion data saved in MPEG4, AVI, MPG, GIF formats for behavioral studies that only need the 'point light' and the 'skeleton-like' avatars (N = 200; 50 in each format)

Table 1. Verbs/actions collected in three languages

Modern Greek [8]	American English [9, 10]	German [11]	
Perpato	Walk	Gehen	1
Vimatizo	March	Marchieren	2
Pisopato	Step back	Zurucktreten	3
Triklizo	Stagger	Torkeln	4
Dhraskelizo	Stride/step over	Hinübersteigen	5
Busulo	Crawl	Kriechen	6
Trexo	Run	Rennen	7
Aneveno	Go up (step)	Hochsteigen	8
Aniforizo	Go up (ramp)		9
Kateveno	Go down (step)	Herundersteigen	10
Katiforizo	Go down (ramp)		11
Hamilono (body)	Crouch	Hocken	12
Phido (epi topou)	Jump/hop	Springen	13
Phido (pano apo)	Jump over	Überspringer	14
Phido (apo kapou)	Jump down	Runderspringen	15
Katevazo	Pick up and put on (lower)/ lower onto	Aufheben/unterlegen	16
Anevazo	Pick up and put on (higher)/lift onto	Hochheben/ablegen	17
Sikono	Lift/raise	Aufnehmen	18
Ipsono	Lift high	Hochhalten	19
Jirizo1 (antitheti katefthinsi)	Turn around	Umdrehen	20
Jirizo1 (jiro apo kati)	Circle (e.g. chair)	Umkreisen	21
Jirizo 2 (e.g. page)	Turn (e.g. page)	Blättern	22
Peristrefo	Rotate	Umkreisen	23
Anapodojirizo	Turn over	Umdrehen (+obj)	24
Kilo	Roll	Rollen	25

- 36 Mocap actions that correspond to path vs. manner scenarios (Table 2) and were performed by one actor, since the data were collected to be shown to participants for behavioral experiment. Apart from the Mocap files, we include the video files (N = 36) and the avatars in the same formats as above (N = 36 each format), for every verb/action.

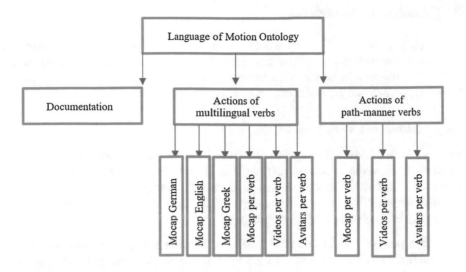

Table 2. Verbs/actions that differentiate in path or manner information

Climb up	Hop between chair	Jump up
Jump back	Jump over two steps	Step back
Jump over	March between	Step over step
Put down	Walk between	Put up
Put out	Put in	Run up to
Raise up two hands	Roll out	Run between back
Roll in	Run out door	Stagger in
Run between	Jump around	Stagger to
Run between forward	Walk out door	Stagger between back
Run in	Stagger	Step up sideway forward
Step up forward	Climb down	Turn around chair 1
Turn around	Walk between forward	Turn a chair around 2
Turn chair around	Throw down	Walk to chair
Walk between	Jump chair around	Walk up
Walk down	Jump down	Walk out
Walk in	Raise down two hands	March in door
Walk in door	Jump over	Walk in door
Walk out door	Step up the other way	Run to

3   Collection Methodology

This section presents the kinematic data collection process, concerning the correlation between locomotion verbs of three typologically different languages. It consists of the verb selection criteria which were uttered to the actors: methodology, equipment, sampling, phases and further comments on the action performance.

3.1   Multilingual Verb/Action Criteria

Twenty five predicates (Table 1) were selected from Antonopoulou's [8] long list of Modern Greek verbs of motion and position (1987). Both intransitive and transitive verbs were included because we assumed zero knowledge of verbal valence. The selected verbs fulfilled requirements imposed by the nature of sensorimotor experiments and lab limitations. More specifically:

- Such a venture would be doomed to failure, if it did not begin with the literal verbal meaning, by isolating the metaphoric uses, which would make the implementation of movements complex or impossible (e.g. running the code).
- Each action is performed by one human; for instance, *kalpazo* (*gallop*) is excluded because it applies only to horses, while *akolutho* (*follow*) was excluded since it violated the one participant prerequisite.
- In addition to the above limitation, there is difficulty in selecting a reference object or receiver of the movement result, where it was necessary to fill the verb's meaning (*slot*), because many objects would make measurement and analysis chaotic.
- Passive verbs were also avoided, because their logical representation is the same as in the active voice, e.g. *it is hurled, persecuted, abandoned*, etc.
- Finally, in the case of verbs with different morphological representation but similar kinetic performance, e.g. *hurl* and *fly* (as transitive) the most representative one was adopted.

The most challenging part of the verb choice was the identification of the equivalent English and German verbs that would be performed in the same manner as the Greek verbs. However, we realized that the dictionary-based translation would not be the preferred one, in terms of everyday communication. English and German translational equivalents of Greek verbs traditionally provided by Greek-English-German dictionaries were either obsolete or did not align one to one with their suggested Greek equivalents. To be more specific, the verb *aneveno* is traditionally translated as *ascend*. However, the AE participants for our collection consider the romance (deriving from Latin) and obsolete *ascend* as distant, while they prefer *step up*, which encodes their popular manner (as a manner or satellite-framed language).

Therefore, at an early stage, we adopted the following method for identifying the English and German equivalents of selected Greek motion verbs: First, we videotaped the Greek participants who performed the selected set of actions. Then, we split the video into segments, each of which corresponded to a single action according to Greek semantics. Next, we presented the segments to ten American English native speakers each one of which was asked to name the performed action according to his/her intuition. Interestingly enough, the responses to seventeen videos were not homogeneous. In order to overcome this impasse, we designed a multiple choice questionnaire. Each question had three choices: two of them were the most frequent answers from the previous phase and the third one was selected from WordNet's lemma (Figs. 1 and 2). A new group of ten native speakers were shown each video segment before filling in the new questionnaire.

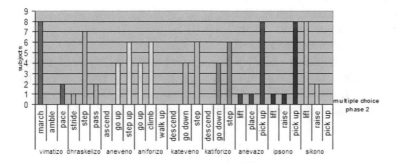

Fig. 1. Histogram of answers in the verb multiple choice task (a)

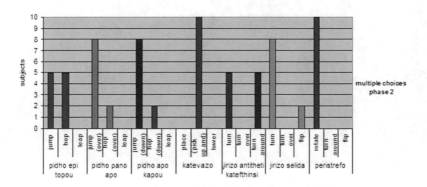

Fig. 2. Histogram of answers in the verb multiple choice task (b)

Finally, we collected the frequent verbs for each action from both phases and we formed our English verb group (Table 1). These verbs were performed by four American English native speakers-according to the following described procedure- but still in case of ambiguity the participants of the sensorimotor experiment were asked to act according to their intuition.

Equipment

For the English and Greek collection we used a full body Moven system containing 16 inertial motion trackers. Each sensor module comprises 3D gyroscopes, 3D accelerometers and 3D magnetometers. With advanced sensor fusion algorithms (Moven Fusion Engine) the inertial motion trackers give absolute orientation values which are used to transform the 3D linear accelerations to global coordinates which in turn give the translation of the body segments. The advanced articulated

Fig. 3. X-Movens Suit (English and Greek corpus)

body model (23 segments and 22 joints biomechanical model) implements joint constraints to eliminate any integration drift or foot-sliding (Fig. 3). For the German collection we employed 12 Vicon MX-F20 cameras (max. 480 images per second, 2.0 megapixels) (Fig. 4) and analyzed it with Nexus 1.7.

Fig. 4. Vicon MX F20 (German corpus)

Phases

The sensorimotor experiment is divided in two phases; capturing of the main dataset for Greek (N = 4), English (N = 4) and German (N = 20) participants.

Subjects

The age range is 25–30 years old, 15 men and 13 women. All subjects were native Greek, American English and German speakers, so they were encouraged to implement each meaning according to their intuition.

Action Performance

Each verb was uttered by the experimenter and then the subject performed the corresponding action. When the verb was performed only with the body of the subject, the action was limited to a floor area restricted by a quadrangle. In order to normalize the distance, subjects were encouraged to start acting at a specific corner of the quadrangle (Fig. 3). Effort was made to involve as few objects as possible. Still, the subjects asked for items that could be found in the lab:

- a step (verbs 8, 10, 15, Table 1),
- a ramp (9, 11),
- one or several balls (5, 14)
- table, book, cylinder, chair (22–25)
- chair (20, 21)

In order to standardize the procedure, the same objects were used throughout the capturing, whenever an object was required.

Methodological Considerations

The correct use of necessary objects received much consideration. We avoided the usage of random objects due to the promiscuous posture of the arm, the leg and the whole body would significantly change the outcome, e.g. when grasping. Therefore, we used the most *unmarked* objects and the same were used by all subjects. In this way, we standardized the procedure without focusing to the objects used. Ideally, we would like to capture every verb with a relatively large set of the objects that are denoted by the verb's (syntactic) dependents; this is left to future research.

Description and Remarks on the Execution of Movements

All participants were asked to be careful during the movement performance and avoid distraction. According to the initial instructions, the execution of each action is limited to a flat space enclosed by a square (Figs. 3, 4), starting from the upper left corner to the bottom right end, for better normalization of the process. This suggestion was not absolutely binding due to several actions that could not be limited in this space. The description of actions per verb is as follows:

Perpato ~ Walk ~ Gehen

Every participant regardless of the language group limited their action at the two corners of the square. When they were completing the action that corresponded to the dictated verb, the experimenter uttered additional sentences such as walk slowly/quickly while participants fluctuated their speed based on their perception of normal walking speed. They seemed to be surprised by the sentence walk from A to B, since they had already executed the initial action between the points with names A (upper left corner) and B (lower right corner). However, if we pay closer attention to the data, we notice that they stomped the last step on point B, while in the initial execution of walking they stopped softer at the same point that they had considered as end of the predetermined space.

Vimatizo ~ Pace/March ~ Marchieren

The prototypical/dominant meaning of *vimatizo* was not directly transparent to the participants. Most of the participants asked for clarification, whether they should do *walk like in a parade/band.* They were encouraged to perform it according to their naïve speaker's intuition and they ended up to pace as in parades. Eventually, *vimatizo* was characterized by a slower ceremonial gait with emphatic arm movement while in *March* there was a long stride and height of the knees. Nonetheless, *vimatizo* is prototypically referred to dictionaries as *walk back and forth* but all our participants selected the secondary meaning of marching as the main one. Due to this change of the Greek verb's dominant meaning and action performance, the American and German speakers –who watched the video- completely rejected *pace* as 1:1 equivalent of *vimatizo* in the verb *pace.*

Dhraskelizo ~ Stride/Step over ~ Hinübersteigen

Greek participants asked for an object/obstacle to make a big step above it, in order to avoid it. For normalization purposes, four tennis balls were placed equidistantly. On the contrary, American and German speakers did not ask any object for *stride*, they just emphasized the long steps, while for *step over*, they asked for objects to overpass them neglecting the manner, e.g. long or short step. After the completion of the *stride* and *step over* verbs' performance, the American and German participants were questioned if they could realize both verbs in exactly the same manner. Their negative answer categorically distinguished the two verbs, rejecting WordNet's opposite association as synonyms.

Trexo ~ Run ~ Rennen

In order to perform running, the participants asked to start before the square of picture 1, in order to enter it after the acceleration phase. Of course, they stopped after the square so that the final deceleration was not measured and thus the action could be normalized as much as possible. All participants waved their hands to boost the torso.

Pidho ~ Jump/Hop ~ Springen

Because *jump* is the case with more than one literal meanings, the experimenter uttered the verb *jump* and asked the participants to perform the first move that came to their mind, in order to detect the prototypical one. All participants bounced on the spot. Afterwards, they were asked to perform other literal meanings that correspond to *jump.* All except from one took a tennis ball to serve as an obstacle/point of reference and jumped. Finally, the experimenter pointed the stairs, so they jumped from there.

Another variation concerning *jump* is the use of the legs: (a) both legs for jumping on the spot (prototypical meaning), (b) 70% of the participants used both legs and 30% used only one leg for jumping over an object, (c) all participants both legs for jumping from a place to another. Concerning cross-language diversities, in English language, the prototypical use of the *jump* is the same in all three languages (jumping on the spot), while *hop* is perceived by all participants as recurring small jumps, which head mainly forward and occasionally on the spot. *Jump* and *hop* may occupy both legs simultaneously. Only in the case of *jump over,* the participants perform the movement with one leg. Instead, *hop over* and *jump down* are performed with both feet.

Jirizo ~ Turn ~ Umdrehen

Prototypically, the verb *jirizo* indicates the rotation in the opposite direction 180° from the starting point. The second literal meaning is the rotation around a fixed reference point, e.g. a chair. In *turn* and *umdrehen*, the concept of *turnaround* was considered to be prototypical, displaying some diversity in the degrees: 60% of the Americans turned 180°, 30% prefer a rotation of 90° and 10% 360° rotation. For close synonyms, such as *rotate*, a participant was verbally outlining his thought before realizing the action because he heard *move around* while another participant recalled *circle around*. Therefore the emphasis is on the preposition *around,* which ultimately constitutes the only fixed element while the verb alternates.

Peristrefo ~ Rotate ~ Umkreisen

The verbs belong to the linguistic class (Antonopoulou [8], Levin [9], FrameNet [10], im Walde [11]) with head the verb *turn*. Participants either used one arm or both for five gyrations. The movement remains circular but the choice of one or two arms fluctuates.

Sikono/Ipsono ~ Lift/Raise/Lift over head ~ Aufnehmen/Hochhalten

For *lift*, the participants raised the object up to the middle of the sternum while for *raise* they lifted it over their heads. In addition, 70% of the participants used both arms, 20% used one arm and finally 10% used both hands in the *lift* action and one hand in the *raise* one. In the justification, the participants stated that for the first action they weighted the object, whereas they considered that they could execute the second one with only one arm. After all it is expected that prior adjustment and action planning affects the performance. *Lift* and *raise* appear as synonyms in WordNet, with no substantial differences. A participant has the feeling that *lift* requires pressure, while another believed that he should examine the surface below the object (book). Respectively for *raise*, the first participant moves it to a reasonable reading distance, while the second one he lifts it so high that he cannot see the backcover. Only, *lift overhead* indicates elevation in relation to the rest of the body and is therefore matched to the *raise*.

Pisopato ~ Step back ~ Zurucktreten

Although *pisopato* does not appear to dictionaries, mainly due to statistical insignificance in everyday speech, it can easily reveal its meaning based on its compounds. At the beginning the Greek participants hesitated to perform it, but eventually they took a step back, except for a one participant, who confused it with walking backwards (*pisoperpato*).

Anevazo, Katevazo ~ Pick up, put on/down ~ Hochheben

In order to carry out upward and downward movement the participants used both arms at the same time, but it is remarkable that they used only one hand when the experimenter reuttered the verb adding the adverbs quickly/slowly. Most likely, unconsciously they weighted the object, when they first handled it. English participants asked for more clarification concerning the end point of the motion, because the plain verbs could not indicate the final goal of the path trajectory, e.g. the table to place the book on. When the object (table) was absent or undetermined, participants replaced the book in the same place where they had raised it, without declaring upward or downward motion but the goal of each move. Instead, no one preferred to change the verbs with *lift on to*, which at least lexicographically would have the same meaning as putting it up and would emphasize the direction of the movement.

Aneveno/Aniforizo ~ Go up ~ Hochsteigen

The mental representation of *aneveno* was very clear for all participants, who asked for a step, while for *aniforizo* they headed towards the ramp. For the A.E. equivalent verb the process appeared more complicated, because the experimenter uttered the plain verb without adding any particular object. Therefore, they were wondering whether they should use the stair or the ramp, because the plain verb could not imply the goal of the action in a similar way as the Greek one. The A.E. participants would prefer the *manner* indicating verb *step up*, instead of *go up*. However, the latter one is more general and still entirely acceptable in terms of understanding. Similarly for the ramp ascending action, two of them would prefer the verb *walk up*, without expressing any further confusion for *go up*, too.

Kateveno/katiforizo ~ Go down ~ Herundersteigen

Kateveno has the same semasiosyntactic and kinematic behavior as its antonym *aneveno*. However, in English, the same people who had accepted *go up* were puzzled with *go down*. On the other hand, in the correspondence process to find the equivalent verbs in these three languages (Sect. 3.1), go down was selected by a large number of participants.

3.2   Path-Manner Verb/Action Criteria

It is clear that our languages enable us to communicate a lot of things, when we have the intention to do that, but (i) they do not enable us to communicate everything and (ii) they somehow filter what we communicate. In fact, some linguists have gone so far as to claim that our languages eventually filter (or express the filter that affects) the way we perceive reality. There is an ongoing debate on this issue: recent studies in cognitive linguistics suggest that *sensorimotor input is the same for all humans* and that *language-independent categorization of events converges among native speakers of typologically different languages* (despite the fact that the corresponding linguistic description does not). Rather than filtering perception, it is suggested that languages favour some available perspectives of viewing events. On the other hand, different lexicalization patterns of motion events in different languages may predict how their speakers perform in non-linguistic tasks [12]. So, typological studies seem to agree that linguistic divergences follow a certain pattern. At this point, an example drawn from the domain of motion will illuminate important aspects of the multi-linguality problem. Languages notoriously diverge in the representation of motion events (in the wider sense of motion). So, if native speakers, one of English one for German and one for Greek for example, see the event *a child walking to the nearby school*, they are likely to communicate their experience verbally as follows:

(1) Greek: *το παιδί πήγε στο   σχολείο*
        the child went to the school
(2) English: *the child walked to the school*
        The difference is that Greek focuses on the path and not on the manner of motion while English and German tends to describe manner of motion. Of course, Greek has a verb for "walk" but (3) or (4), that employs some form of *perpato* (*walk*) would be used with an emphasis on manner, in a context that requires it.

(3) Greek: *το παιδί πήγε στο   σχολείο περπατώντας*
     the child went to the school by walking

(4) Greek: *το παιδί περπάτησε ως το σχολείο*
     The child walked up to the school

The phenomenon affects the vocabulary of the languages: English seems to offer more verbs that describe how one moves as compared to Greek that seems to have a richer vocabulary to describe where one moves. It also affects translational correspondences. Consider (2) and (3) above. It is not obvious how to establish a structural match between these two structures that have been considered translational equivalents. In fact, this is one of the notorious problems (the *head switching problem*) in machine translation. There are events that can be described in one language but not in another, exactly because of preferences concerning manner and path in motion. So, there is no Greek equivalent for (5), although it is attested that Greeks often show it with an appropriate gesture:

(5) *He passed his hand through his hair.*

The typological difference between Greek and English exemplified with (1) and (2) above could be modelled as the selective linguistic tagging (naming) of the sensorimotor experience (common to all humans), or as tagging at different levels of generalization (because the Greek verb *πηγαίνω σε* (*to go to*) (1) is the representative verb for motion along a path, short of general term for *walk to, run to, jump to*, etc.). Actually, the question asked is how the semantic relation between the 'most general' verb *to go* and the 'more specific' verb *to walk* can be modeled in terms of conceptual grounding and/or embodiment.

We collected the MoCap data that correspond to the following verbs (Table 2), in order to show them in triads. One would be the anchor verb/action and the other two would differentiate in path of manner information. Based on the selection of native speakers from path or manner languages, we could explain their view of the world.

4   Applications

The above mentioned kinematic data have served several studies, such as:

- Cognitive: Processing of motion verbs, based on the finding of Hauk, Johnsrude and Pulvermüller [13] brain imaging studies concerning the. Since the distribution of the semantic imaging of the movement verbs in the brain during processing is localized in the parts used in the movement (i.e. in the hands and feet movement), we extended the experiment by showing avatars and videos with the same verbs as Hauk and colleagues did in their study.
- Linguistics: Classifications of the motion verbs (Table 1) from different languages (Modern Greek [8]; American English [9, 10]) based on different theories and criteria (prototype theory, semantic grouping and valence alternations, conceptual

structure) exhibit strong similarities. After analyzing the MoCap data with Principal Components, we noticed that the visualization of the first two components resembled the grouping of the three different linguistic classifications, which could imply that the representation of the actions in the natural world influences linguistic perception, even if linguists consider other.

- Grounding: The creation of a Motion-Verb ID such as the following example of verb *anapodojirizo-turn over*

Characteristics		Verbal description	Instrumentality	Plot of representative action
Path		straight	Hips x,z,y Knees y	
Direction	Forward-backward	forward		path
	Right-Left			
	Upward-downward			
Action duration (lexical aspect)		repetition		direction
Speed		acceleration every 1 and ½ steps		velocity
Gaze		precedes 0.2 sec		gaze

- E-learning: Combination of MoCap data with linguistic notions in a game-like intelligent tutoring system, in order to help elementary school students to better differentiate literal from metaphorical uses of motion verbs, based on embodied information. In addition to the thematic goal, the young students' attention and spatiotemporal memory was improved, by presenting the sensorimotor data discussed in this paper [14].

Acknowledgements. The authors would like to thank all of the participants of the sensorimotor collection. We would particularly like to thank Dr Leonardo Claudino, Dr Stella Markantonatou and Maaike Esselaar. Moreover, MS is funded by the Cluster of Excellence Cognitive Interaction Technology 'CITEC' (EXC 277) at Bielefeld University, which is funded by the German Research Foundation (DFG).

References

1. Andrew, G., Grenager, T., Manning, C.: Verb sense and subcategorization: using joint inference to improve performance on complementary tasks. In: Proceedings of the 2004 Conference on Empirical Methods in Natural Language Processing (EMNLP 2004), pp. 150–157 (2004)
2. Kant, I.: Prolegomena to Any Future Metaphysics That Will Be Able to Come Forward as Science, Revised edition edn. Cambridge University Press, Cambridge (2004). Translated and edited by Gary Hatfield
3. De Saussure, F.: Course in General Linguistics. Columbia University Press, New York (2011)
4. Gibson, J.J.: The Ecological Approach to Visual Perception Classic. Psychology Press, UK (2014)
5. Jackendoff, R.: Foundations of Language: Brain, Meaning, Grammar, Evolution. Oxford University Press, Oxford (2002)
6. Piaget, J.: The Origins of Intelligence in Children. International Universities Press, New York (1952). M. Cook Trans
7. Lakoff, G., Johnson, M.: Metaphors We Live By. University of Chicago Press, Chicago (2003)
8. Antonopoulou, E.: Prototype theory and the meaning of verbs, with special reference to modern greek verbs of motion. Doctoral Dissertation, School of Oriental and African Studies, University of London (1987)
9. Levin, B.: English Verb Classes and Alternations: A Preliminary Investigation. University of Chicago press, Chicago (1993)
10. Ruppenhofer, J., Ellsworth, M., Petruck, M.R., Johnson, C.R., Scheffczyk, J.: FrameNet II: Extended theory and practice. Institut für Deutsche Sprache, Bibliothek (2016)
11. im Walde, S.S.: Human associations and the choice of features for semantic verb classification. Res. Lang. Comput. 6(1), 79–111 (2008)
12. Gennari, S.P., Sloman, S.A., Malt, B.C., Fitch, W.T.: Motion events in language and cognition. Cognition 83(1), 49–79 (2002)
13. Hauk, O., Johnsrude, I., Pulvermüller, F.: Somatotopic representation of action words in human motor and premotor cortex. Neuron 41(2), 301–307 (2004)
14. Sionti, M., Schack, T., Aloimonos, Y.: An embodied tutoring system for literal vs. metaphorical concepts. Front. Psychol. 9. (2018). https://doi.org/10.3389/fpsyg.2018.02254

An Improved Compressive Tracking Approach Using Multiple Random Feature Extraction Algorithm

Lanrong Dung[✉] and Shih-Chi Wang

National Chiao Tung University, Hsinchu 30010, Taiwan
lennon@nctu.edu.tw

Abstract. This paper presents an object-tracking algorithm with multiple randomly-generated features. We intent to improve the compressive tracking method whose results are fluctuated between good and bad. Because the compressive tracking method generates the image features randomly, the resulting image features varies from time to time. The object tracker might fail for missing some significant features. Therefore, the results of traditional compressive tracking are unstable. To solve the problem, the proposed approach generates multiple features randomly and chooses the best tracking results by measuring the similarity for each candidate. In this paper, we use the Bhattacharyya coefficient as the similarity measurement. The experimental results show that the proposed tracking algorithm can greatly reduce the tracking errors. The best performance improvements in terms of center location error, bounding box overlap ratio, and success rate are from 63.62 pixels to 15.45 pixels, from 31.75% to 64.48%, and from 38.51% to 82.58%, respectively.

Keywords: Object tracking · Feature extraction · Compressive tracking

1 Introduction

This paper presents an object-tracking algorithm with multiple randomly-generated features. We intent to improve the compressive tracking method whose results are fluctuated between good and bad. Because the compressive tracking method generates the image features randomly, the resulting image features varies from time to time. The object tracker might fail for missing some significant features. Therefore, the results of traditional compressive tracking are unstable. To solve the problem, the proposed approach generates multiple features randomly and chooses the best tracking results by measuring the similarity for each candidate, using the Bhattacharyya coefficient as the similarity measurement.

2 Background/Significance

Object tracking is an essential application in the field of computer vision. It is commonly used in monitoring systems or human-computer interaction. A variety of tracking algorithms have been proposed. In the IVT method [1], the incremental principal component analysis is used to reduce the image space. The IVT method learns

© Springer Nature Switzerland AG 2020
K. Arai and S. Kapoor (Eds.): CVC 2019, AISC 944, pp. 724–733, 2020.
https://doi.org/10.1007/978-3-030-17798-0_57

a target model that can be continually updated to adapt to track the changing target. In [2], the target model is decomposed into several basic target models constructed by sparse principal component analysis. The tracker tracks the target with a set of additional basic motion models. It can deal with appearance transformation or movement. However, these methods have a huge amount of computation and make it difficult to run in real time.

The compressive tracking method [3] is a fast tracking algorithm using compressed sensing theory. It projects the high-dimensional image features into the lower-dimensional image space by a very sparse projection matrix, and track the target with the low-dimensional image features generated by random projection. It reduces a lot of image features needed to compare and greatly reduces the computational complexity of the algorithm. However, the image features generated by the projection matrix are completely random. Even in the same testing video, each time the image features have a considerable change. It makes the results of each execution sometimes good and sometimes bad and difficult to use effectively.

To solve the problem, an object-tracking algorithm with multiple randomly-generated features is proposed in the present paper. Tracking with the additional and different image features can produce a number of different tracking results. If we choose the most ideal tracking result as the final target position, there will be more opportunities to produce a better result than the original algorithm. The experimental results show that the proposed tracking algorithm can greatly reduce the tracking errors. The best performance improvements in terms of center location error, bounding box overlap ratio, and success rate are from 63.62 pixels to 15.45 pixels, from 31.75% to 64.48%, and from 38.51% to 82.58%, respectively.

3   Method

In this paper, the proposed tracking algorithm is showed in Fig. 1. In order to solve the drift problem caused by occlusion, we refer to [4] and apply the sub-region classifiers to our algorithm. We reduce the number of sub-region classifiers to speed up the algorithm. Only nine sub-region classifiers are used for tracking. In addition to this, the locations of the sub-region classifiers are evenly distributed in order to avoid excessive concentration or excessive dispersion of the sub-region classifiers. In the process of tracking, each sub-region classifier independently tracks the specified part of the target. If the target is partially occluded, only the occluded part of the tracking will be affected. As a result, the drift problem due to occlusion can be avoided. In the classifier update phase, each sub-region classifier is decided whether to update based on the respective classifier scores in order to prevent the target model from being updated by occlusions. If the score of the classifier is less than zero, it indicates that the probability of the region being judged as belonging to a non-target object is relatively large. Therefore, the target model of the sub-region is not updated to retain the object information.

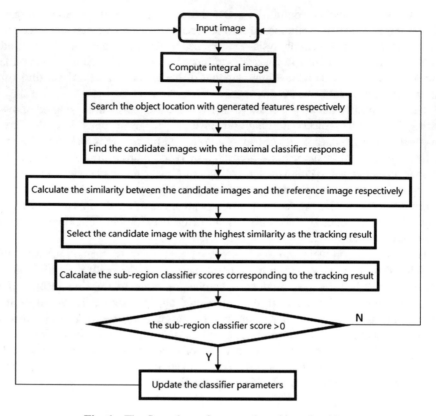

Fig. 1. The flow chart of proposed tracking algorithm.

Figure 2 illustrates examples of the distribution of sub-regions and Fig. 2(a) shows all of the sub-regions. There are some overlapping regions between sub-regions. Figure 2(b) shows the lower right corner sub-region. The position of each sub-region is shown in Eq. (1), where x and y are the coordinates of the upper left corner of the region of interest, w_s and h_s are the width and height of the sub-regions and T_{ij} is the coordinates of the upper left corner of the sub-regions.

$$T_{ij} = \left[x + i \times \frac{w_s}{2}, y + j \times \frac{h_s}{2} \right] \tag{1}$$

During the establishment of the target model stage, we use the method proposed in [5] to assign different weightings according to the importance of the positive samples. The target and background model is established as Eqs. (2)–(4), where $p(y = 1|V^+)$ and $p(y = 0|V^-)$ are the target and background model. $p(y = 1|v_{1j})$ is the posterior probability for sample v_{1j}. N is the number of positive samples and L is the number of negative samples. l is the location function and c is a normalization constant. w is a constant.

$$p(y = 1|V^+) = \sum_{j=0}^{N-1} w_{j0} p(y_1 = 1|v_{1j}) \qquad (2)$$

$$w_{j0} = \frac{1}{c} e^{-|l(v_{1j})-l(v_{10})|} \qquad (3)$$

$$p(y = 0|V^-) = \sum_{j=N}^{N+L-1} w p(y_0 = 0|v_{0j}) \qquad (4)$$

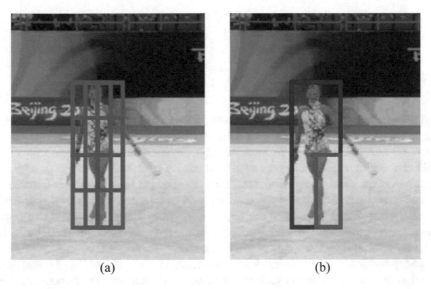

(a)   (b)

Fig. 2. Examples of the distribution of sub-regions. (a) All sub-regions. (b) A single sub-region.

In the proposed tracking algorithm, we use multiple sets of randomly-generated and different image features to track respectively. After each time the highest classifier scores for candidate positions are calculated, we select the optimal tracking result as the final target location. Because of the multiple sets of image features, there are additional opportunities to produce the better results than the original. Therefore, if the best candidate can be selected from the candidate images, it is possible to obtain a better tracking performance than the conventional one. In the proposed tracking algorithm, the optimal tracking result is determined by calculating the Bhattacharyya coefficient between the candidate image and the reference image. The Bhattacharyya coefficient is defined as Eq. (5), where N is the total number of indices of the histogram. The target image and the candidate image model are proposed in [6] and shown in Eqs. (6)–(9), where δ is the Kronecker delta function. C and C_h are the normalization constants.

The large value of the Bhattacharyya coefficient indicates that the candidate image has high similarity with the target image. Therefore, after the end of each tracking we select the largest Bhattacharyya coefficient corresponding to the candidate image position as the tracking result.

$$\rho[p, q] = \sum_{u=1}^{N} \sqrt{p^{(u)} q^{(u)}} \tag{5}$$

$$q^{(u)} = C \sum_{i=1}^{n} k(\|x_i^*\|^2) \delta(b(x_i^*) - u) \tag{6}$$

$$C = \frac{1}{\sum_{i=1}^{n} k\left(\|x_i^*\|^2\right)} \tag{7}$$

$$p^{(u)} = C_h \sum_{i=1}^{n_h} k(\left\|\frac{y - x_i}{h}\right\|^2) \delta(b(x_i) - u) \tag{8}$$

$$C_h = \frac{1}{\sum_{i=1}^{n_h} k\left(\left\|\frac{y - x_i}{h}\right\|^2\right)} \tag{9}$$

The low-dimensional image features used in [3] have scale invariance. This paper also integrates the multi-scale tracking function into the proposed algorithm. In this paper, we also use the image features of large, invariant and small scale to track. To avoid changes in target size is very small and cannot be detected, we use an additional target model for scale detection and tracking. The second target model is updated less frequently and will not be updated until the end of every fifth frame. Slower update frequency is intended to preserve the target image information before the five frames. The detected image will be more different from the second target model and it is relatively easy to detect changes in the target size.

The proposed multi-scale tracking algorithm is showed in Fig. 3. Multi-scale detection is performed at the end of the trace phase and executes once every five frames. If scale detection is required, the image features of different scales and the second target model are used to track again. If the highest classifier score is derived from a larger or smaller scale image feature, it represents a change in size of the target. Therefore, the target position is determined by the highest classifier score obtained in the final tracking. If the highest classifier score is derived from the invariant scale image features, the tracking result obtained before the scale detection is taken as the target position. Because the previous result is tracked with the first target model updated each frame, it is more accurate than the result tracked with the second target model.

In addition, the dramatic change in the size of the object is usually the case in the distance between the target and the observer has a huge change. The color of the target will change due to the influence of the medium between the target and the observer.

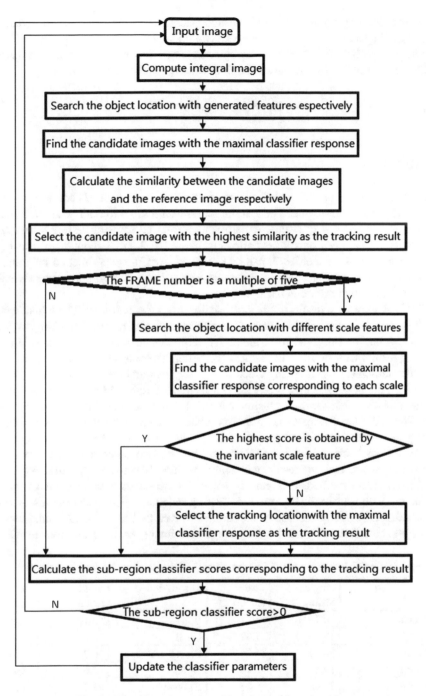

Fig. 3. The flow chart of multi-scale tracking algorithm.

Therefore, the target model used in the multi-scale detection needs to be updated to reduce the impact of target color changes. The updating method is proposed in [7] and shown in Eq. (10). Where $q_t^{(u)}$ is the updated target model in frame t, λ is learning parameter and $p^{(u)}$ is the model of the final tracing results in frame t.

$$q_t^{(u)} = (1 - \lambda)q_{t-1}^{(u)} + \lambda p^{(u)} \tag{10}$$

4 Results

The experiment parameters used in this paper are consistent. There are 10 weak classifiers in each sub-region. The learning parameter λ is set to 0.85. The scale change parameter δ is set to 0.1. Twenty experiments are tested for each testing video. Measurements are averaged over 20 experiments. Tables 1, 2 and 3 show the results of the experiment with three metrics. It can be observed from the experimental data that the proposed two sets of feature tracking algorithm have a significant improvement, regardless of which metric is used.

Table 4 shows the results of the multi-scale tracking experiment. The left side is the result of tracking with single feature. The right side is the result of tracking with two features. It can be observed from the above table that in most of the test videos, the tracking results with two sets of features are better. There is significant progress in the difficult examples, such as testing video *Bus*, *Car_silver* and *Car_scooter*. In the testing video *Bus*, the target has undergone the changes in the light and shadow caused by the bridge and the short-term occlusion caused by the scooter. In the testing video *Car_silver*, the small target and the large shadow area make the tracking difficulty greatly increased. The difficulty of the testing video *Car_silver* is from the long-term partial occlusion caused by the scooter. In the last two cases, the results with two features are worse than the results with single feature. Because we use the image color as a discriminant method, it is easy to select an erroneous candidate image when a similar color appears in the background. Testing video *Freeway* is an obvious example. The tracker drifts away because of the interference caused by the background color. It causes the classifier to update by the background image, so there are poor results.

Table 1. Center location error (CLE) (in pixels) of the single-scale tracking experiments.

Testing video	Single feature	Two feature	Testing video	Single feature	Two feature
Campus	6.73	5.21	Ball	62.12	28.25
T-junction	14.69	7.37	Hand	16.05	8.92
Station	24.14	14.26	Person_crossing	46.76	26.31
Fighting	10.53	6.15	Gym	63.62	15.45
Fighting-disappearance	35.96	9.05	Basketball	62.26	22.54
Police	14.76	11.40	Person_floor	63.80	20.02
Worker	18.57	9.15	Skating	60.14	32.12

Table 2. Bounding box overlap ratio (BBOR) (%) of the single-scale tracking experiments.

Testing video	Single feature	Two feature	Testing video	Single feature	Two feature
Campus	73.16	80.31	Ball	29.07	52.94
T-junction	55.77	57.86	Hand	53.88	63.95
Station	50.02	59.88	Person_crossing	48.19	68.21
Fighting	67.47	76.38	Gym	31.75	64.48
Fighting-disappearance	60.01	76.00	Basketball	43.83	67.78
Police	71.21	73.82	Person_floor	44.75	67.93
Worker	69.19	69.49	Skating	42.41	59.60

Table 3. Bounding box overlap ratio (BBOR) (%) of the single-scale tracking experiments.

Testing video	Single feature	Two feature	Testing video	Single feature	Two feature
Campus	92.57	99.37	Ball	24.26	59.13
T-junction	55.23	63.53	Hand	62.76	78.81
Station	51.34	73.25	Person_crossing	55.06	74.81
Fighting	84.98	97.37	Gym	38.51	82.58
Fighting-disappearance	69.19	93.25	Basketball	50.04	87.37
Police	91.54	89.07	Person_floor	52.75	81.25
Worker	75.40	89.80	Skating	47.30	71.36

Table 4. The multi-scale tracking experiments.

Testing video	Single feature			Two feature		
	CLE	BBOR	SR	CLE	BBOR	SR
Bus	28.67	56.37	68.66	23.44	64.59	76.64
Car_silver	27.76	59.60	71.87	10.20	63.03	81.65
Car_boulevard	5.78	69.58	90.78	5.38	72.89	93.88
Van	10.59	79.18	98.15	7.65	81.80	99.99
Car_blue	18.67	66.80	84.61	6.12	68.33	86.76
Car_scooter	4.61	74.79	90.82	3.72	78.00	95.31
Car_white	7.10	81.57	99.97	4.02	90.01	97.99
Car_red	4.80	81.51	98.81	5.62	80.11	99.58
Freeway	6.60	78.05	99.68	27.23	46.78	30.15

Table 5. Average frame per second (FPS) of the single-scale tracking experiments.

Testing video	Single feature	Two feature	Image size
Campus	18.4077	15.4067	1280 × 480
T-junction	24.3139	19.5400	768 × 576
Station	24.7422	19.4261	720 × 576
Fighting	28.5837	21.4797	640 × 480
Fighting-disappearance	28.5216	21.5955	640 × 480
Police	36.2956	25.3854	640 × 360
Worker	40.5832	27.9352	528 × 360
Bus	34.7153	20.5242	640 × 360
Car_silver	35.3681	21.3057	
Car_boulevard	35.2754	21.7842	
Van	35.6460	19.9841	
Car_blue	34.9502	22.1828	
Car_scooter	35.0141	20.4013	
Car_white	34.0907	20.9430	

Table 5 shows the speed of the proposed tracking algorithm. The experimental results show that the speed of the algorithm is related to the image size, and the larger the image size, the slower the speed. The proposed single scale tracking algorithm can achieve near real-time speed when the image size is not too large. In addition, the speed of the multi-scale tracking algorithm is not reduced to one third of the speed of the single scale algorithm.

5   Conclusions

The tracking algorithm proposed in this paper is mainly aimed at improving the tracking results in compressive tracking. We use a number of different sets of image features to track and produce better tracking performance by selecting the best tracking results. For the choice of tracking results, we experiment with the Bhattacharyya coefficient. According to the experimental results, using the Bhattacharyya coefficient to judge the results can produce very good experimental results. The color information of the object is usually no dramatic change. Therefore, in most tracking such as occlusion, deformation, or similar background can be overcome by selecting best tracking result. Significant improvements can be seen in the three metrics used to measure performance. The experimental results show that the proposed tracking algorithm can greatly reduce the tracking errors. The best performance improvements in terms of center location error, bounding box overlap ratio and success rate are from 63.62 pixels to 15.45 pixels, from 31.75% to 64.48% and from 38.51% to 82.58%, respectively. Moreover, when the image size is not large, it will be able to achieve real-time computing.

References

1. Ross, D.A., Lim, J., Lin, R.S., Yang, M.H.: Incremental learning for robust visual tracking. Int. J. Comput. Vision **77**(1–3), 125–141 (2008)
2. Kwon, J., Lee, K.M.: Visual tracking decomposition. In: IEEE Conference on Computer Vision and Pattern Recognition, pp. 1269–1276 (2010)
3. Zhang, K., Zhang, L., Yang, M.H.: Fast compressive tracking. IEEE Trans. Pattern Anal. Mach. Intell. **36**(10), 2002–2015 (2014)
4. Zhu, Q., Yan, J., Deng, D.: Compressive tracking via oversaturated sub-region classifiers. IET Comput. Vis. **7**(6), 448–455 (2013)
5. Wang, W., Xu, Y., Wang, Y., Zhang, B., Cao, Z.: Effective weighted compressive tracking. In: The 17th IEEE International Conference on Image and Graphics, pp. 353–357 (2013)
6. Comaniciu, D., Ramesh, V., Meer, P.: Real-time tracking of non-rigid objects using mean shift. In: IEEE Conference on Computer Vision and Pattern Recognition, vol. 2, pp.142–149 (2000)
7. Babu, R.V., Perez, P., Bouthemy, P.: Robust tracking with motion estimation and local kernel-based color modeling. Image Vis. Comput. **25**(8), 1205–1216 (2007)

Classical Algorithm vs. Machine Learning in Objects Recognition

Jakub Czygier$^{(\boxtimes)}$, Piotr Tomaszuk$^{(\boxtimes)}$, Aneta Łukowska$^{(\boxtimes)}$,
Paweł Straszyński$^{(\boxtimes)}$, and Kazimierz Dzierżek$^{(\boxtimes)}$

Department of Automatic Control and Robotics,
Faculty of Mechanical Engineering, Bialystok University of Technology,
Wiejska 45C, 15-351 Białystok, Poland
czyrzwa@gmail.com, ptom9515@gmail.com,
aalukowska@gmail.com, pawell93a@gmail.com,
k.dzierzek@pb.edu.pl

Abstract. This article focuses on two most popular methods of detecting regular shapes in the pictures. Over the past years, image processing and object recognition are entering our lives more and more. The difference between a classical algorithm and machine learning was analyzed, in case of a tennis ball recognition on photos. It has been created on own dataset to avoid a uniform background and bright colors. Images were taken with a low-cost camera in different conditions. Creating the classical algorithm and machine learning and comparing the accuracy, false flags and performance of the two methods are described.

Keywords: Object recognition · Machine learning · Image processing · Classification · Neural network

1 Introduction

Image processing and detection of the objects is used more often in every area of our lives, from everyday activities to typically industrial ones and security-related solutions [1, 2]. In the presented topic, was decided to confront two paths leading to the detection of a popular, but not so evident for the algorithms or neural networks, an object that is a tennis ball, which served as an example in this research. On the one hand, we have a classical solution - an algorithm that processes image, mathematical transformations that allow obtaining satisfactory results under controlled conditions. How will it behave in the absence of the programmer's intervention during much more dynamic tests than those in which they achieved satisfactory results? A uniform background and bright colors have been abandoned, in favor of the most natural conditions, a low-end camera and low contrast. The second method was carried out by the neural network. A solution known for a long time recently began to be used in practical cases was tried not to favor any solutions, that is why dataset was limited to 2000 photos. It is definitely not enough for the operation to be close to what is known from the best solutions from world leaders. However, was wanted a solution that each of us will be able to reproduce and will not devote weeks, or need a computing cluster, but only a better laptop.

© Springer Nature Switzerland AG 2020
K. Arai and S. Kapoor (Eds.): CVC 2019, AISC 944, pp. 734–745, 2020.
https://doi.org/10.1007/978-3-030-17798-0_58

2 Methods

2.1 Classical Algorithm

The classical algorithm was based on OpenCV image processing library. The first step, in preparing the frame from the camera, is the Gaussian blur of the image to compensate the noise from the camera. A 5 × 5 pixel brush creates the blur. After blurring, the image is converted to HSV model. It increases the resistance for changing the scene lightning much better, thereby less error susceptibility resulting from the wrong white balance setting by the camera software. Next step, is creating a mask – filtering pixels with a specific value, which was set in advance before starting the object search-received the black and white image. Using the erode function, one has gotten rid of individual pixels coming from the noise or tiny fragments of objects with the color spectrum. The masked image is blurred again to avoid pixel reconciliation and smooth out the whole picture. The prepared image is processed using Hugh transform [3, 10]. Transformation parameters were selected experimentally during the creation of the algorithm. The Hugh function returns a list of objects along with their probability of existence. On this basis, was marked the most probable objects as interesting shapes in the image. This way of processing the image is burdened with errors. The main drawback is the white balance changing during the lighting and scene changes, and the spectrum of color that is associated with our object. The camera may interpret the green object as a yellow or cyan object depending on the environment. The lack of white and black points intensifies this chaos and in case of lack of lighting or mechanisms compensating for the erroneous auto-calibration of the image recording device, has been forced to deepen the complexity of the algorithm and the time loss for its implementation. It is a questionable practice due to the complexity of the subject matter of recognizing a given object [9].

During the implementation and testing of the classical algorithm, the same images were used as during the neuron network training, as well as photos taken by outsiders found on the internet. This was to ensure that the anticipation of the algorithm was as accurate as possible. In the image, the algorithm placed a maximum of 2 balls detected by it, sorting them according to the probability of occurrence, from the number 0.

2.2 Machine Learning

Machine learning was carried out using a neural network (convolutional neural network). *TensorFlow* [7] and *scikit learn* [8] libraries, which are the most popular tools nowadays providing solutions in the field of artificial intelligence, were used to build the neural network model. Have been decided to use a classical division of a training and testing set [4]. 80% of data set images were used for training and 20% of images from the whole dataset were used for validation (testing set). The images were taken out of the training set to calculate accuracy during the training process.

The CNN network had a few layers. The first layer is an input layer, next one consists of a 4-D tensor, in the form of [n-images width height num_channels]. The next layer is the flattened layer, a one-dimensional layer with the size of the multiplied dimensions of the previous one (Fig. 1).

The next step is a fully connected layer. This layer assumes coefficients: weight and biases, and its result is the value of the function:

$$y = wx + b \tag{1}$$

where
y — function result
x — input
w — weight
b — bias - coefficient

The fully connected layer is characterized by the connection of each neuron to each one in the previous layer. The output of this layer is calculated by the matrix multiplication followed by bias offset. The optimization of the calculation and weight optimization gradient was performed using *AdamOptimizer* (Fig. 2).

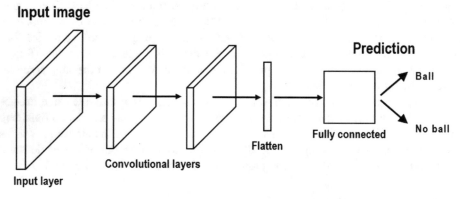

Fig. 1. The diagram of the used machine learning structure.

A database of over 1000 photos, evenly divided in the context of the inclusion of the ball or its absence, was used for training our neural network. The photographs were taken in a changing environment with different lighting. The pictures were taken while the camera was moving and in a stable position. The ball was placed at different distances, in different places of the frame, and at different angles to the lens. The pictures were taken in similar proportions, not favoring any of the camera's position configurations.

During the training, different settings of network layers were tested, from smaller sizes to larger ones. Finally, the maximum size of the matrix for the computer used was assumed.

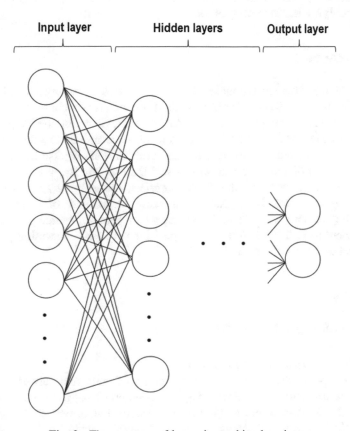

Fig. 2. The structure of layers in machine learning.

3 Testing Data

Testing data was a collection of 120 images, 60 of which contained a tennis ball. The images containing the search object were selected from among several hundred pictures taken in similar conditions for the pictures used as data in the training set of the neural network. The selected photos have a different composition, i.e., the location of the ball, the distance between the ball and the camera. The lighting conditions were changed from daylight to fluorescent lighting. The angle of light incidence concerning the object and the camera also changed during the changes of the ball and the camera position. The images have not been processed. They were a raw image from the camera after compression made by the camera software. The aim was to consider non-standardized photographs, thereby to study the use of the algorithm when working in real time on data provided directly by the camera. The photographs reflect the natural working conditions of the camera and what may be encountered during a non-isolated environment, i.e., during the practical use of the detection methods described in the article. It should also be mentioned that the distance from the camera to the ball was different and the diameter of the ball in the camera image has changed, which was another factor aimed at verifying the flexibility of the developed solutions.

4 Performance

The classical algorithm can recognize objects in real time assuming about 10 fps, even on a very inefficient unit. Only writing an algorithm and its validation can absorb more time. Machine learning requires about 1 s for the model to determine the probability of occurrence or non-object, taking into account the same unit as the algorithm. This is unsatisfactory if is needed a live recognition solution. It takes less time to prepare the model and layers of the neural network itself than to test and correct errors in the algorithm, but after preparing the network project, the learning process takes place, which in our case took 14 h. One can increase the number of iterations (epoch). However, it does not always affect the accuracy of accuracy; at some point the changes are insufficient to justify the further training of the network. Concerning performance, the classical algorithm is definitely a winner.

5 Results

5.1 Classical Algorithm Results

Research Using the Classical Algorithm
The same sample of data as used in the testing set of the neural network was used in testing the classical algorithm. Each image has been processed in the same way. At the beginning of the detection of an object using a classical algorithm, it was necessary to determine the Hough transformation parameters experimentally [4]. Parameters cannot bear any traces of machine learning, although, the subject of the research is the superiority of human or machine solutions over the competitive side. After determining the Hough transformation parameters, Gaussian blur or eroding function, in such a way that the test results are as satisfactory as possible. It was left to determine the color spectrum that will be filtered using a simple mask. The spectrum was determined on the basis of the highest and lowest values collected among 20 samples of the search object from 5 photographs selected randomly from the testing set of the neural network [5]. After determining the mentioned assumptions, the research has been started. It was also assumed that the search object should not be lower than the 2nd place in the probability distribution of the regular shape, which is the result of the Hough transformation. Thus, the remaining findings were skipped in the considerations (Figs. 3 and 4).

Fig. 3. Selected pictures without tennis balls from the testing set of the classical algorithm.

Fig. 4. Selected pictures with tennis balls from the testing set of the classical algorithm.

The graph in Fig. 5 is comparing the accuracy of a ball recognition in different distances on the context of the area occupied by the object.

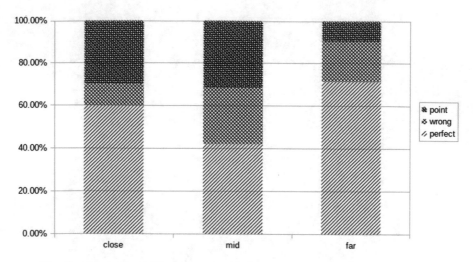

Fig. 5. Comparison of the accuracy of ball recognition in the different distance.

Where:

- *close* – means occupying over 30% of pixels by the ball
- *mid* – from 10 to 30% of pixels
- *far* – under 10% of pixels
- *perfect* – the ball was correctly recognized by determining the radius and position.
- *point* – the position of the ball was well recognized, but the radius of the ball was wrong
- *wrong* – the recognized area where the ball did not occur.

In 81.67% cases, the ball was detected in the image, but it is not a full success, because only 66.7% cases of the actual positioning of the ball were indicated as the most probable, and only 51.6% of the recognized balls had a specific radius approximating the actual state [6]. However, in the absence of a ball in the picture and unchanged parameters, in 71.6% cases a ball was detected, and only in 20.9% of them, only one ball was detected. So, the algorithm works well when detecting objects when they are actually in Fig. 6.

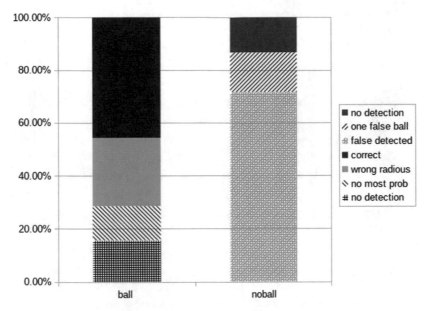

Fig. 6. Results of recognizing the classical algorithms.

Where:

- Ball:
  - *correct* – the ball has been recognized correctly in all cases.
  - *wrong radius* – the detected ball had a poorly defined radius
  - *no most prob* – the detected ball was not considered in the most likely area
  - *no detection* – the ball was not detected
- No ball:
  - *no detection* – the ball was not detected
  - *one false ball* – only one false ball was detected
  - *false detected* – all detections are false balls

5.2 Machine Learning Results

In histograms, the machine learning method has a relatively high level of accuracy, but it did not avoid errors in the form of false positives. In the absence of a ball on the machine learning composition, it most often predicted the lack of an object in the image. This is highly satisfactory considering the size of datasets and the lack of optimization, testing and improvement (Figs. 7 and 8).

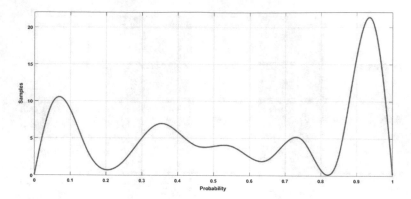

Fig. 7. The probability of the distribution for the dataset with the object.

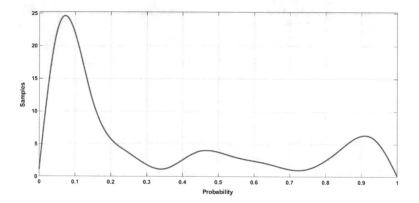

Fig. 8. The probability of the distribution for dataset without the object.

5.3   Results Summary

The heatmaps showed the fields in which the detection of the ball was the most common in both methods. The fields were added together and transformed with a Gauss filter. The correct recognition of an object by a classical algorithm concentrates around one point, whereas machine learning was able to recognize an object on a large span covering the entire area of interest. The incorrect detection for the classical algorithm occurred in many different places, especially on the broader areas of the recognized area, when most of the erroneous results of the machine learning were presented with the classical algorithm recognizing the object (Figs. 9, 10, 11, and 12).

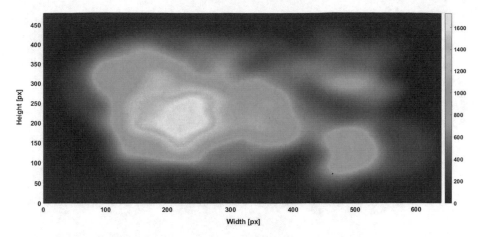

Fig. 9. The correctly recognized objects by the algorithm.

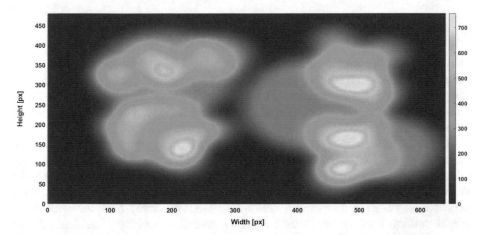

Fig. 10. The correctly recognized objects by machine learning.

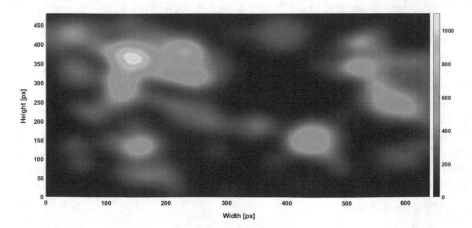

Fig. 11. The incorrectly recognized objects by the algorithm.

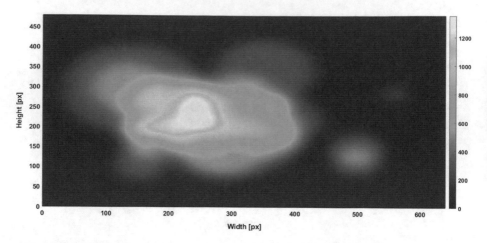

Fig. 12. The incorrectly recognized objects by machine learning.

6 Conclusions

The trivial task for humans turns out to be quite a challenge for the robot – the computer. The research which was carried out illustrates how big the gap between the computer and the human remains. None of the ways of recognizing the ball had satisfactory effectiveness. In the case of a human, this efficiency is 100% and is as fast as the computer methods.

The classical algorithm proved to be good at recognizing the object, assuming that the object actually appeared in the image. In cases when the object was not an algorithm because of its construction - filter with color, and then searching for regular shapes, it detected many false balls even though there were not any in the picture. However, this algorithm dealt well with the balls located away from the camera and with a uniform background. In case of many objects, especially of similar color or shape, was received many false detections. Of course, one could adjust the algorithm to our needs, change the color spectrum in which was searched for the object or Hough transformation parameters. Many factors have influenced the results and with a sufficient input of work and tests, could probably get a more reliable diagnosis. However, each time the environment changes this process should be repeated.

When analyzing the data obtained with the use of the machine learning method, have been paid particular attention to the effectiveness of the method when detecting balls regardless of their size and location.

As one can see when analyzing heatmaps, both methods could successfully complement each other. Where the classical algorithm most often detected the presence of the machine learning the object did not perform so well, but where the classical algorithm skipped the balls, the machine learning noticed their presence. The total machine learning is also more accurate, that is, it has fewer errors in predicting the presence of the ball and its lack of the composition. In the case of the classical algorithm, one can notice a big problem in the form of a high percentage of missed hits in the absence of an object in the image.

Looking at the results obtained during the research could propose a combination of both methods when solving problems similar to those described in the article.

References

1. Hassanein, A.S., Mohammad, S., Sameer, M., Ragab, M.E.: A survey on Hough transform, theory, techniques and applications. CoRR abs/1502.02160 (2015)
2. De Vries, J., et al.: Object recognition: a shape-based approach using artificial neural networks. Department of Computer Science, University of Utrecht (2006)
3. Jaroslav, B.: Circle detection using Hough Transform. Technical report, University of Bristol, U.K. (2003). https://www.borovicka.org/files/research/bristol/hough-report.pdf
4. Bennamoun, M., Mamic, G.J.: Object Recognition: Fundamentals and Case Studies, pp. 13–14. Springer, London (2002)
5. Grauman, K., Leibe, B.: Visual Object Recognition, pp. 1–5. Morgan & Claypool Publishers, San Rafael (2011)
6. Freitas, A.A.: Data Mining and Knowledge Discovery with Evolutionary Algorithms, pp. 19–20. Springer, Heidelberg (2002)
7. Abadi, M., Barham, P., Chen, J., Chen, Z., Davis, A., Dean, J., Devin, M., Ghemawat, S., Irving, G., Isard, M., Kudlur, M., Levenberg, J., Monga, R., Moore, S., Murray, D.G., Steiner, B., Tucker, P., Vasudevan, V., Warden, P., Wicke, M., Yu, Y., Zheng, X.: TensorFlow: a system for large-scale machine learning. In: 12th USENIX Symposium on Operating Systems Design and Implementation (OSDI 2016), pp. 265–283 (2016)
8. Pedregosa, F., Varoquaux, G., Gramfort, A., Michel, V., Thirion, B., Grisel, O., et al.: Scikit-learn: machine learning in Python. J. Mach. Learn. Res. **12**, 2825–2830 (2011)
9. Treiber, M.A.: An introduction to object recognition: selected algorithms for a wide variety of applications (2010)
10. Shin, M.C., Goldgof, D.B., Bowyer, K.W.: Comparison of edge detector performance through use in an object recognition task (2001)

Segmentation of Color-Coded Underground Utility Markings

Peining Che and John Femiani[✉]

Miami University, Oxford, OH 45056, USA
femianjc@miamioh.edu

Abstract. We present results of using Unet to segment underground utility markings. These markings are painted on the ground to indicate the locations of buried gas, sewage, and power lines before construction. Segmenting the markings is a first important step in updating or creating digital maps of underground assets, which will help manage legal liability, future planning and construction needs. The proposed approach demonstrates quality results even with markings on complex materials, in semitransparent ink, and on surface such as grass and gravel. Even with a limited size dataset we are able to achieve a mean recall of F_1-measure of 72% with visually plausible results.

Keywords: Convolution Neural Networks (CNNs) ·
Image processing · Computer vision · U-Net · Image segmentation

1 Introduction

Underground utility markings are used to indicate the location of underground water and sewage pipes, electric lines, and communication cables before construction that involves digging in the ground. While Geographic Information Systems (GISs) record approximate locations, the exact placement of underground utilities relative to features on the surface is often considered unreliable [4]. It is common practice to use sensors to detect these underground utilities and mark them with a color-coded spray-paint prior to construction work that may involve digging.

We are interested in inexpensive and automatic ways to detect these markings, and therefore identify the underground assets, using simple cameras that include Geographic Positioning System (GPS) capabilities (such as GoPro cameras or cell phones) that can be mounted on the dash of vehicle and used to detect utility markings on the ground.

Specifically, we want to take color images captured in an uncontrolled manner, not directly pointed at utility markings. Then we aim to create a color coded mask that highlights each category of underground utility. The scope of this work is to determine how well we can segment the images and assign a label to each pixel. An example is shown in Fig. 1, and additional results are shown in Fig. 3.

© Springer Nature Switzerland AG 2020
K. Arai and S. Kapoor (Eds.): CVC 2019, AISC 944, pp. 746–751, 2020.
https://doi.org/10.1007/978-3-030-17798-0_59

Fig. 1. An example of the input (left) and the actual output (right) of the proposed approach to underground utility marking segmentation. Different colors are used to indicate the type and location of buried objects.

2   Background/Significance

Some hand-written text recognition CNN models have been developed [1,3,8] however, these methods aim to recognize marks rather than to segment images. Our task is to segment markings on the ground (not perpendicular to the camera) and from complex natural backgrounds by classifying each pixel in the image. Each label stands for a specific underground utility type. Thus, the semantic segmentation approach can help to automatically locate underground utilities via markings in images.

The *Interval HSV* approach [7] used saturation and value as features to extract handwritten annotations from complex backgrounds. Unlike the proposed approach, Interval HSV did not consider the structure of handwritten marks and it focused on annotated and re-scanned photographs. The proposed approach works on marks placed in the scene itself and is sensitive to the structure as well as the color of markings.

The general task of semantic segmentation using CNNs has receive significant attention in computer vision research. SegNet [2] is an efficient CNN architecture for image segmentation. SegNet works by pooling to downsample images while keeping track of the indices at each max-pooling step. Then a matched set of unpooling and convolution, using the saved indices, allows SegNet to label every pixel in an image. SegNet has been used for a variety of urban datasets [6], however a more recent approach, called Unet [11] often produces better results by upsampling and then concatenating (or adding) the matched feature-maps themselves in order to reconstruct labeled images with precise details. The Unet architecture has been used to segment a wide variety of features, but it is unclear whether thin, distorted, and partially transparent paint used to for utility marking can be distinguished.

We make the following contributions: We use Unet to achieve high quality segmentation of underground utility markings, the first step in automatically digitizing utility markings. We identify painted strokes on the ground in a variety of colors that indicate the type of underground utility. Finally, we are able to identify faded marks and marks on complex surfaces such vegetation and gravel.

3   Method

Our approach is to use a CNN to learn the structure and color of underground utility markings. We follow a process of data collection and annotation, choosing a model architecture, and training the model using early stopping [10].

The architecture we chose is Unet [11], a fully convolutional network developed for semantic segmentation tasks. The Unet architecture takes several convolution an pooling layers downsample the input image, and then uses upsampling followed by convolution in order to predict labels for every pixel. What distinguishes Unet from similar architectures such as Segnet [2] is that after each upsampling step, the corresponding feature map from downsampling is concatenated in order to allow the network to learn how to transfer precise details.

We use the standard cross-entropy loss with weights. Due to a large imbalance within the data, especially for the background class, the weights of each sample are multiplied by $w_c = \lg median\_freq / \lg freq(c)$ where $freq(c)$ the the number of pixels of class c and $median\_freq$ is the median of these frequencies. This is similar to the weights used in other segmentation approaches [2,5] except that we choose weights proportional to $1/\lg freq(c)$ rather than $1/freq(c)$. We find that an extremely low weight for the background class favors recall, but hurts precision.

We manually acquired 42 photos and one video from Pheonix, AZ. Each photograph was manually annotated using a photo-editor (the GiMP) to isolate each type of mark. The images are separated into two groups: 24 photos were selected for training dataset and 18 in validation dataset. Although the precise meaning for colors may change depend on city or state ordinances, we identified 7 classes of mark used in AZ. We used a distinct color for each mark and black to indicate the background. One type of mark (indicating reclaimed water) was not present in our data-set and so we removed the label. In addition, the paint material used to mark utilities loses saturation and fades over time but remains visible. We introduced a new class for "faded" marks. In addition we observed other marks that did not fit into one of the existing categories, which we assigned a special label so they would not be included in evaluation calculations.

Due to the very small size of our dataset, we increase variety using data augmentation. We use random cropping [12] of an original image to produce 256×256 pixel sized inputs to our model, with a minibach size of 5, the largest size that fits on a Quadro K2200 GPU. We optimize the weights using Adam, with weight decay of 1e−4 and a learning rate of 1e−3. Each input image is re sized so that the shortest dimension is 600 pixels before cropping, and still many crops will be completely of the background class. Especially for some rare types of annotation using a single crop from each image would never encounter certain labels at all. Therefor we consider one epoch to be 1000 crops (each image is used approximately than 42 times per epoch on average). We also use random crops on the evaluation images and enlarge the evaluation set to include 100 crops (around 5 crops per images on average). We train the network using early stopping [10], which stops at around 108 epochs. In order to fine tune the solution, we reduce the learning rate by a factor of 10 and then continue to train with early stopping starting from the iteration with the best validation score.

Table 1. Classification metrics

	Precision	Recall	Fmeasure	Support
k:None	1.00	1.00	1.00	32455395
r:Power	0.92	0.78	0.84	118245
y:Gas	0.29	0.83	0.43	10633
w:Excavation	0.52	0.85	0.64	33792
b:Water	0.65	0.91	0.76	21264
g:Sewer	0.78	0.59	0.67	16085
l:Faded	0.65	0.73	0.69	112586
Avg/total	0.69	0.81	0.72	32768000

4 Results

We present a qualitative example of the results in Fig. 3, which shows several complete images from our validation set. Even after resizing the image so to a height of 600 pixels, we were limited by GPU ram so each image is processed as a set of tiles that each overlap by 50% and then adding the model outputs before softmax. We observe that errors of omission happen for certain classes of mark that may have been rare in the training set. Errors of commission seem happen in distant regions near the top of the images and also for some distant road markings. These types of error are promising given that we trained on a small dataset, so it seems likely that we will obtain better results if more data is collected.

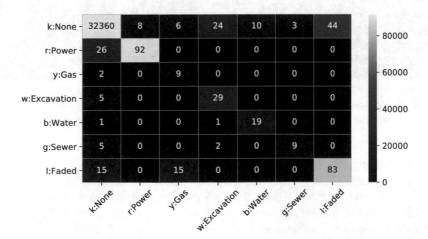

Fig. 2. The confusion matrix (in thousands). The labels on y-axis are the target labels and labels on x-axis are predicted labels. Each entry (i, j) is the number of times class i is predicted as class j.

Fig. 3. Some example images from the evaluation set: original images are shown on top and the color-coded markings are shown on the bottom with some transparency.

We use precision, recall, and the F_1-measure [9] to evaluate our results. Our overall F_1-measure is 71%. Quantitative evaluation is shared in Table 1 and also presented as a confusion matrix in Fig. 2. Each label-name is presented as a color (k = black, r = red, y = yellow, w = white, b = blue, g = green, ℓ = lightgray/olive) and also based on the type of underground utility indicated by the mark. The rarest class (*Gas*) has the lowest F_1-measure, which is largely due to confusion between the *Gas* and much more prevalent *Faded* classes. In addition we note that the white *Excavation* and gray *Faded* classes are often confused with background, which is understandable because white gravel, lane lines, and sidewalks make this color difficult. The blue *Water* label is confused with the background as well, and this seems to be at least in part due to confusion with the sky.

5 Conclusion

Recognition of underground utility markings is an important problem that has the potential to help with city planning and construction. GIS information is much easier to manage than it is to acquire, and the approach presented here is a first step in automatic digitization of utility locations based on computer vision. While many existing methods exist for processing ink marks in document images, few of them target markings painted onto a scene itself. The proposed approach based on Unet is able to achieve high quality and useful results even with a very small training set, with an average recall 81% of precision just shy of 70%. Our results indicate that automatic utility marking recognition is an achievable goal.

Acknowledgment. This research was supported by the Arizona Salt River Project (SRP).

References

1. Obaid, A.M., El Bakry, H.M., Eldosuky, M.A., Shehab, A.I.: Handwritten text recognition system based on neural network. Int. J. Adv. Res. Comput. Sci. Technol. (IJARCST) **4**(1), 72–77 (2016)
2. Badrinarayanan, V., Kendall, A., Cipolla, R.: SegNet: a deep convolutional encoder-decoder architecture for image segmentation. IEEE Trans. Pattern Anal. Mach. Intell. **39**(12), 2481–2495 (2017)
3. Bartz, C., Yang, H., Meinel, C.: STN-OCR: a single neural network for text detection and text recognition (2017)
4. Costello, S.B., Chapman, D.N., Rogers, C.D.F., Metje, N.: Underground asset location and condition assessment technologies. Tunn. Undergr. Space Technol. **22**(5), 524–542 (2007)
5. Eigen, D., Fergus, R.: Predicting depth, surface normals and semantic labels with a common multi-scale convolutional architecture, November 2014
6. Femiani, J., Para, W.R., Mitra, N., Wonka, P.: Facade segmentation in the wild, May 2018
7. Femiani, J.C., Razdan, A.: Interval HSV: extracting ink annotations. In: IEEE Conference on Computer Vision and Pattern Recognition, CVPR 2009, pp. 2520–2527 (2009)
8. Perwej, Y., Chaturvedi, A.: Neural networks for handwritten English alphabet recognition, May 2012
9. Powers, D.M.: Evaluation: from precision, recall and f-measure to ROC, informedness, markedness and correlation, December 2011
10. Prechelt, L.: Early stopping - but when? In: Orr, G.B., Müller, K.-R. (eds.) Neural Networks: Tricks of the Trade, pp. 55–69. Springer, Heidelberg (1998)
11. Ronneberger, O., Fischer, P., Thomas, B.: U-Net: convolutional networks for biomedical image segmentation (2015)
12. Wang, L., Xiong, Y., Wang, Z., Qiao, Y.: Towards good practices for very deep two-stream ConvNets, July 2015

Character Localization Based on Support Vector Machine

Jin-Hui Li[1], Ming-Ming Shi[1], Xiang-Bo Lin[1(✉)], Jia-Jun Zeng[1],
Zhong-Liang Wang[1], and Zuo-Jun Dai[2]

[1] Faculty of Electronic and Electrical Engineering,
Dalian University of Technology, Dalian, China
linxbo@dlut.edu.cn
[2] Beijing Chuangyuanweizhi Software Co. Ltd., Beijing, China

Abstract. The purpose of character localization is to detect and locate a certain character from the picture, which can be used in automatic testing task. In this paper, we propose a new virtual keyboard character localization pipeline, mainly including image preprocessing, feature extraction, training classification and identification and positioning. It combines HOG feature extraction algorithm and SVM classifier to complete the design and implementation of character positioning, including image sample acquisition and processing, image feature extraction, SVM training, target character recognition and positioning. The experiments are done on characters in different virtual keyboards, and get more accurate results than CNN based text detection method.

Keywords: Small isolated characters · SVM · HOG · CNN

1 Introduction

The development of Internet and smart mobile phone expedites diversified application software. Ensuring basic functions is the basic requirement for these applications. Function testing should be done before a software goes online. Here our work focuses on such testing task, and provides an automatic character localization pipeline based on image object localization technology for virtual keyboard module.

Template based matching is a common method for character localization and recognition, which extracts effective points and then computes the correspondence between their descriptors. Extracting character features is usually the important step in template matching. Statistical features are usually the common and stable features among the same classes, and structural features consist of a sequence of character structural elements. Recently, CNN has been widely used for feature extraction and object recognition. However, the requirements of large amount of training data and high computational complexity hinder its practicality. Moreover, our experiments got poor results on character localization in the virtual keyboard using the state of the art CNN based text detection model, the optimized EAST detector [1].

Simple lines and dots form the character in the virtual keyboard, which is lack of easily distinguished features. Some popular feature descriptors, such as SURF, are unsuitable to solve our problem. Inspired by pedestrian recognition using HOG&SVM

© Springer Nature Switzerland AG 2020
K. Arai and S. Kapoor (Eds.): CVC 2019, AISC 944, pp. 752–763, 2020.
https://doi.org/10.1007/978-3-030-17798-0_60

method [2], we propose a new character localization pipeline. The general process of the proposed pipeline can be divided into image preprocessing, feature extraction and selection, SVM classification and localization.

In short, the highlights of our work lies as discussed below.

Apply the combination of the HOG feature descriptor and the SVM classifier to localize the specified character using a given template, where this strategy hasn't been discussed on this application. The proposed method is easy to understand, and it is quite effective. It can provide a good reference for practical applications.

Localization characters in a virtual keyboard seem to be a simple task, because it has a simple background. However, simple character shape with monochrome background makes it lack of contexts, thus in fact it is not an easy task for an automatic model. The published deep neural network models are trained for natural text detection. When we use the trained model to detect the mentioned isolated character in virtual keyboards, they show poor performance. When the deep neural networks are re-trained for this specific application, the performance might be improved. But they need large amount of data, while SVM can get good performance using small samples.

2   Related Works

2.1   Feature Extraction and Selection

Feature extraction is an important step in template based matching. Numerous operators have been proposed, such as Gabor transformation for texture feature, gradient feature, SIFT, SURF, Histograms of Oriented Gradients (HOG), and so on.

This article focuses on the local features of the object. Abdulmunim et al. [3] used the region-based features and the SURF descriptors to identify the Logo. In the first stage, according to the external shape of the Logo, the global features were extracted gradually. In the second stage, the SURF descriptors were used for Logo matching, and the local invariant feature was used to identify the scaling transformation. Mistry [4] compared SIFT and SURF descriptors on rotation invariant, blurring and curve transformation. Hanif et al. [5] using Mean Difference Feature (MDF), Standard Deviation (SD) and HOG features for text detection and localization in complex scene images. These published results show that different features are suitable for different applications.

2.2   Classification and Identification

The classifier uses a known class of training samples to learn classification rules, and then classifies the unknown feature vectors. The quality of the classification model is crucial to the results of the experiment. So the design of the classifier is very important. According to whether there is a class label in the training sample or not, the classifier can be classified as supervised classification, unsupervised classification and semi-supervised classification. Supervised classification means that the training samples using the learning classification model contain category labels, such as SVM [6], neural networks, and so on. Unsupervised classification refers to the use of self-attributes directly to perform classification in this process, such as K-means algorithm and mean C fuzzy algorithm [7] and other clustering algorithms. The semi-supervised classification combines the two forms of supervised and unsupervised classification. In the training process, only a part of the training samples has class labels, such as semi-supervised SVM [8] and manifold learning based LapSVM [9]. The SVM proposed by Vapnik and others based on the study of statistical learning theory is widely used. Csurka and others [10] used the Bag-of-words model (BoW model) to represent the images. They compared the Bayesian classifier [11] and SVM respectively, which confirmed the superiority of the SVM classifier.

HOG feature combined with SVM classifier had been successfully applied in pedestrian detection [2], but was not often discussed in character localization. In pedestrian detection, as long as the pose is in normal upright walking, tiny movements have no effects on detection results. This method inspires many researchers to detect and locate targets in other application scenarios, such as bird region detection [12], wood identification [13], IRIS image segmentation [14], and so on.

3   HOG Implementation

HOG descriptor is a feature descriptor used for object detection in computer vision and image processing. It constitutes characteristics by calculating statistical histogram of gradient direction in local area of image.

The main idea is that in an image the appearance and shape of a local target can be well described by the directional density distribution of the gradient or edge. The concrete implementation steps are showed in Fig. 1.

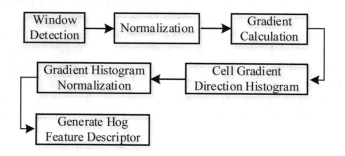

Fig. 1. The implementation steps of HOG operator.

3.1   Window Detection

The image sample pixels in the data set are usually too large to be trained directly. In our applications, only the character regions are the candidate targets, which are relatively small regions. There is no need to consider the other regions. Therefore, the window detection should be done to find the area contain characteristics first.

3.2 Normalization

The normalization step transforms the original RGB images into gray images, and then normalizes the gray images into a specified range. Here we use the Eqs. (1) and (2) to get the normalized gray image.

$$G = 0.39 \times R + 0.5 \times G + 0.11 \times B \tag{1}$$

$$g(i,j) = a + \frac{\ln(G(i,j)+1)}{b \ln c} \tag{2}$$

Where a, b, and c are adjustable parameters.

3.3 Gradient Calculation

Image gradient can be divided into horizontal gradient and vertical gradient, which is usually handled by using the first order differential derivative. The horizontal edge operator $\begin{bmatrix} -1 & 0 & 1 \end{bmatrix}$ and the vertical edge operator $\begin{bmatrix} -1 & 0 & 1 \end{bmatrix}^{\mathrm{T}}$ are used to calculate the horizontal gradient direction value $G_x(x,y)$ and the vertical of gradient direction value $G_y(x,y)$ for each pixel position. The pixel gradient amplitude and direction are:

$$|G(x,y)| = \sqrt{G_x(x,y)^2 + G_y(x,y)^2} \tag{3}$$

$$\alpha(x,y) = \tan^{-1}\left(\frac{G_y(x,y)}{G_x(x,y)}\right) \tag{4}$$

3.4 Histogram of Cell Gradient Direction

The image is decomposed into multiple adjacent, non-overlapping cell units, and the cell gradient is divided into 9 bins. For the nine directional bins, all pixels in the cells will be weighted using linear voting. The [0, 180] is divided into nine regions, namely every 20° for a region. Take the center point as the center value of the histogram of each region. The resulting cell histogram is a vector with nine nonnegative entries.

3.5 Gradient Histogram Normalization Within the Block

Image is sensitive to the influence of local illumination changes, which will make the range of the gradient histogram expanded. It is necessary to compress the range of the gradient histogram. Normalization within the block will further compress the edges, intensities and shadows. The blocks are made up of cell grouping. Two horizontally or vertically consecutive blocks overlap by two cells. That means each internal cell is covered by four blocks. Concatenate the four cell histograms in each block into a single block feature v and normalize the block feature using Eq. (5):

$$f = \frac{v}{\sqrt{||v||_2^2 + e^2}} \tag{5}$$

Where, e is a small positive constant.

3.6   Hog Feature Descriptor Generation

Since each cell is covered by up to four blocks, each histogram is represented up to four times with up to four different normalizations. The normalized block features are concatenated into a single HOG feature vector h, which is further normalized using Eq. (6):

$$h \quad \leftarrow \quad \frac{h}{\sqrt{||h||^2 + e}} \tag{6}$$

The final normalization makes the HOG feature independent of overall image contrast.

The resulting HOG feature is the concatenation of four times as many cell histograms as there are blocks. It conveys information that is somewhat like that of an edge map, except that some of the gradient magnitude information is retained and the location of the edges is only recorded to cell resolution.

4   SVM Classification and Localization

The Support Vector Machine (SVM) is a two-class model. Its learning strategy maximizes the interval and minimizes the error. It transforms the problem into a convex quadratic programming problem. The problems encountered in general are divided into two categories: the linear separable problem and the linear inseparable problem. Linear separable problems are generally easy to solve, but we usually encounter linear inseparable problems. For nonlinear problems, it is hard to find a straight line to distinguish the two categories. For this nonlinear classification problem, it usually introduce a kernel function to map the data from the low-dimensional feature space to the high-dimensional feature space. In the high-dimensional feature space, the data becomes linear, separable, and it is possible to find a suitable classifier (hyper-plane) to divide the data into two categories. The training data usually does not appear independently. The data appears in the form of inner product. The advantage of using the SVM to train the classifier is that the adjustable parameters in the representation are inconsistent with the number of the data attributes. The inner product are calculated by using appropriate kernel functions between the data points. Hiding the map itself while directly calculating the inner product of the high-dimensional feature space vector in the low-dimensional feature space will significantly reduce the computational complexity. The kernel function performs the inner product calculation in the low space in advance, and expresses the substantial classification effect to a high level. For nonlinear problems, the basic solution in SVM is to map the input data to a high-dimensional inner product space through a nonlinear mapping, and perform linear classification in the high-dimensional space. By using the kernel function, it makes all necessary calculations in the input space.

The kernel function plays important roles in the non-linear data classification. Many kernel functions have been proved to be effective, such as Gaussian kernel, Laplace kernel, ANOVA kernel, and so on. Here we use the Gaussian kernel, shown in Eq. (7):

$$k(x,y) = \exp\left(-\frac{||x-y||^2}{2\sigma}\right) \tag{7}$$

It should be noted that the parameter σ plays an important role. If σ is too large, the weights of the high-dimensional features will actually decay very quickly, which is equivalent to the low-dimensional subspaces. Conversely, if σ is too small, any data can be mapped as linearly separable, but severe over fitting will occur.

SVM is a typical classifier with excellent classification performance. Our purpose is to determine the character's position in an image, so we need to do a conversion from the localization problem to the classification problem. Let the specified character be the target, and the other regions be the background. Set up the positive samples related to the target and the negative samples related to the background. Train the SVM classifier, and then the class model will be set up. For an inquiry character, use the trained SVM model to determine its label on the basis of the extracted features. Slide the window in the target image, extract the features in the corresponding region, and then get the label from the trained SVM. Finally, the inquiry character will be located in the target image by label comparison. Figure 2 shows the detailed character localization procedure.

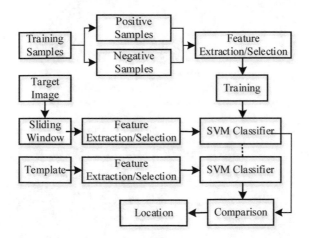

Fig. 2. The character localization procedure.

5   Experiments and Results

This experiment aims at the realization of the detection and positioning functions of letters and special characters in different mobile phone virtual keyboard. Our experimental samples include 600 images containing letters and special symbols as the positive samples, and 1200 images without targets letters and special characters as negative samples. In addition, we have data augmentation in the training sets by 100 samples which are hard to be recognized. These hard samples are used to improve the recognition performance of the classifier. All of the sample pictures are from the app account input interface of different brands and different models of mobile phones.

In the HOG descriptor setting, considering that the target character picture is small and square, the window size is set to 32×32, the block size is set to 16×16, the number of bins in the cell unit is set to 9, and the cell size is set to 8×8.

In the SVM setting, we use the Gaussian kernel function. Parameter C is used to punish training errors. If the value of C is larger, it means that the penalty for training errors is greater. Although the classification error for training data can be controlled well, it also results in a too small classification interval. If the interval between classifications is too small, the generalization ability will be affected when the classifier obtained from the training data is generalized to the test samples. A suitable C value will be helpful to balance the accuracy and the generalization. We allow some training samples to be erroneous in classification to obtain a larger categorization interval, so that the classifier will be more effective when it is extended to the test samples. Figure 3 shows the impacts on localization letter 'a' with different C value, where C is set to 1 and 0.01 respectively. In the following experiments, the C value is set to 1.

(a) C = 0.01   (b) C = 1

Fig. 3. Comparison chart of changing parameter on character 'a'

The trained SVM classifier locates the characters in the virtual keyboard of 38 different mobile phone login screens. The test results are evaluated using the accuracy, which are shown in Tables 1 and 2.

Table 1. Special character localization results.

Character	Success times	Number of test samples	Accuracy (%)
/	38	38	100
&	38	38	100
%	38	38	100
#	38	38	100
.	36	38	94.7
(	38	38	100
!	38	38	100
;	38	38	100
@	38	38	100
~	38	38	100
=	38	38	100
*	38	38	100
+	37	38	97.4
[	38	38	100
]	38	38	100
{	37	38	97.4
}	38	38	100

Table 2. English letter localization results.

Character	Success times	Number of test samples	Accuracy (%)
a	38	38	100
b	38	38	100
c	38	38	100
d	38	38	100
e	38	38	100
f	38	38	100
g	38	38	100
h	38	38	100
i	38	38	100

(*continued*)

Table 2. (*continued*)

Character	Success times	Number of test samples	Accuracy (%)
j	37	38	97.4
k	38	38	100
l	38	38	100
m	37	38	97.4
n	36	38	94.7
o	38	38	100
p	38	38	100
q	37	38	97.4
r	38	38	100
s	38	38	100
t	38	38	100
u	38	38	100
v	38	38	100
w	38	38	100
x	38	38	100
y	38	38	100
z	38	38	100

It can be seen from Table 1 that the proposed method has high accuracy for locating special characters. For some smaller special characters such as ".", the locating accuracy rate is relatively low. This may be due to the fact that the target is too small to have enough features. For too small and too simple special characters, the localization accuracy still needs to be improved.

The results in Table 2 indicate that the proposed method has high accuracy for locating 26 English letters. Most of the letters can be located accurately. Only a few characters are missed for individual virtual keyboards. This may be due to the presence of strong interference in the entire image. The extracted features result in an incorrect recognition when they are sent to the SVM classifier. Increasing the training data might solve the problem and improve the localization accuracy.

CNN based text detection models become popular in recent years, and exhibit good performance in difficult scene text detection [15–17]. However, these models didn't work well in our application, although the task looks very simple. Figure 4 shows some typical character detection results using the well trained EAST model [1]. The EAST model has been proved to be able to detect both large and small text of arbitrary orientations successfully, but the character detection results for all experimental data in our application are poor.

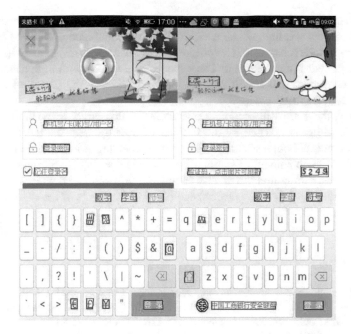

Fig. 4. Typical character localization results using the well trained EAST model.

6 Discussion

Text detection and recognition is an active research field in computer vision, and can be used in many real-world applications. Most works focus on text detection in complicated scenes, such as clutter background, arbitrary text orientations. Because CNNs have achieved great success in general object detection, more and more researches adopted CNNs to detect more and more challenging scene text, using large amount of labeled training data. However, in our opinion, texts are often printed on a simple background in some real-world applications, especially in industrial applications. The main challenge lies in the small sizes, the simple shape and the lack of the context. To our knowledge, there is no published CNN model to solve this problem specially. Through our research, we find that traditional methods can exhibit good performance in such applications. Particularly, SVM is a small sample based learning algorithm, and there is no need for large amount of training data. We think it is valuable to reconsider the traditional classical methods in specific applications.

7 Conclusion

The purpose of this paper is to realize the localization of various characters in virtual keyboard of various types of mobile phones. These characters are made up of simple lines and dots, which is lack of easily distinguished features. Thus regular template matching methods can't give accurate localization results, because those feature descriptors, such as SURF, SIFT, ORB, and so on, produce poor feature representation. We propose a new pipeline, which uses HOG feature combined with SVM classifier to localize the specified character. The proposed new strategy is inspired by its successful application in pedestrian detection. The experimental results with high accuracies both on special characters and English letters have proved that the proposed method has very good performance.

However, there are still obvious deficiencies in this target localization algorithm. First, the algorithm is affected by the number of training samples, the distribution of samples. Second, although SVM has its unique advantages in small sample training, the demand for large amount of samples by machine learning is still critical. To obtain better classifiers, setting up high quality sample set will still be necessary. It is believed that as the in-depth study, the performance of various aspects of the algorithm will be gradually increased.

References

1. Zhou, X., et al.: EAST: an efficient and accurate scene text detector. In: IEEE Conference on Computer Vision & Pattern Recognition (CVPR 2017), Honolulu, Hawaii, pp. 2642–2651 (2017). https://doi.ieeecomputersociety.org/10.1109/CVPR.2017.283
2. Dalal, N., Triggs, B.: Histograms of oriented gradients for human detection. In: IEEE Computer Society Conference on Computer Vision & Pattern Recognition (CVPR 2005), pp. 886–893. IEEE Computer Society, San Diego (2005). https://doi.ieeecomputersociety.org/10.1109/CVPR.2005.177
3. Abdulmunim, M.E., et al.: Logo matching in Arabic documents using region based features and Surf descriptor. In: 2017 Annual Conference on New Trends in Information & Communications Technology Applications (NTICT), Baghdad, Iraq, pp. 75–79 (2017). https://doi.org/10.1109/ntict.2017.7976128
4. Mistry, D., Banerjee, A.: Comparison of feature detection and matching approaches: SIFT and SURF. Glob. Res. Dev. J. Eng. 2(4), 7–13 (2017)
5. Hanif, S.M., Prevost, L.: Text detection and localization in complex scene images using constrained AdaBoost algorithm. In: 10th International Conference on Document Analysis and Recognition (ICDAR 2009), pp. 1–5, Catalonia, Spain (2009). https://doi.ieeecomputersociety.org/10.1109/ICDAR.2009.172
6. Freund, Y., Schapire, R.E.: A short introduction to boosting. J. Jpn. Soc. Artif. Intell. 14(5), 771–780 (1999)
7. Xu, R., Wunsch, D.: Survey of clustering algorithms. IEEE Trans. Neural Netw. 16(3), 645–678 (2005). https://doi.org/10.1109/TNN.2005.845141
8. Joachims, T.: Transductive inference for text classification using support vector machines. In: Proceedings of the Sixteenth International Conference on Machine Learning (ICML 1999), Bled, Slovenia, pp. 200–209 (1999)
9. Belkin, M., et al.: Manifold regularization: a geometric framework for learning from labeled and unlabeled examples. J. Mach. Learn. Res. 7(1), 2399–2434 (2006)
10. Csurka, G., et al.: Visual categorization with bags of keypoints. In: Workshop on Statistical Learning in Computer Vision, ECCV, 1(1–22), Prague 1, Czech Republic, pp. 1–2 (2004)

11. Yasin, H., Khan, S.A.: Trained table based recognition & classification (TTRC) approach in human motion recontruction & analysis. In: 4th International Conference on Emerging Technologies (ICET2008), Islamabad, Pakistan, pp. 217–222 (2008). https://doi.org/10. 1109/icet.2008.4777503

12. Kumar, R., et al.: Bird region detection in images with multi-scale HOG features and SVM scoring. In: Proceedings of 3rd International Conference on Computer Vision & Image Processing, Advances in Intelligent Systems and Computing, Jabalpur, Madhya Pradesh, pp. 353–364 (2018). https://doi.org/10.1007/978-981-10-7898-9_29

13. Sugiarto, B., et al.: Wood identification based on Histogram of Oriented Gradient (HOG) feature and Support Vector Machine (SVM) classifier. In: International Conferences on Information Technology, Information Systems and Electrical Engineering, 2018, Yogyakarta, Indonesia, pp. 337–341 (2018). https://doi.org/10.1109/icitisee.2017.8285523

14. Radman, A., et al.: Automated segmentation of IRIS images acquired in an unconstrained environment using HOG-SVM and GrowCut. Digit. Signal Proc. **64**(5), 60–70 (2017). https://doi.org/10.1016/j.dsp.2017.02.003

15. Lyu, M., Liao et al.: Mask TextSpotter: an end-to-end trainable neural network for spotting text with arbitrary shapes. In: Europe Conference on Computer Vision (ECCV 2018), Munich, Germany (2018). arXiv:1807.02242

16. Liao, M., et al.: TextBoxes++: a single-shot oriented scene text detector. IEEE Trans. Image Proc. **27**(8), 3676–3690 (2018). https://doi.org/10.1109/TIP.2018.2825107

17. Gomez-Bigorda, L., Karatzas, D.: TextProposals: a textspecific selective search algorithm for word spotting in the wild. Pattern Recogn. **70**, 60–74 (2017). arXiv ID: arXiv:1604. 02619

Author Index

© Springer Nature Switzerland AG 2020
K. Arai and S. Kapoor (Eds.): CVC 2019, AISC 944, pp. 765–767, 2020.
https://doi.org/10.1007/978-3-030-17798-0

Printed in the United States
By Bookmasters